Unconventional Resources

The oil and gas sector is a vital player in the energy transition. With their vast resource potential, unconventional shale plays will be an essential part in enabling this change. *Unconventional Resources* serves as a comprehensive reference covering the latest technologies, methodologies, and applications of unconventional shale resources in the oil and gas industry and their role in the evolution of the sector's energy transition.

This book:

- Offers an overview of geophysics, geology, and reservoir characterization in unconventional resources
- Discusses drilling, well stimulation and completion, production engineering, and artificial lift
- Covers reservoir management and surveillance, recovery enhancement, production forecasting, and surface facilities and testing
- Details technical and technological advances, including machine learning, AI, data analytics, and Industry 4.0
- Explores the latest methods/workflows in performance analysis in unconventional plays
- Employs integrated and hybrid approaches to the energy transition

The book provides surface and subsurface technical professionals in the oil and gas industry a thorough overview of unconventionals along with the integrated/hybrid applications that will enable them to stay current with the industry's transition.

Unconventional Resources

Advances in Energy Transition

Edited by

Cenk Temizel, Cengiz Yegin,
Jihoon Kim, Luigi Saputelli, and Ole Torsæter

CRC Press
Taylor & Francis Group
Boca Raton London New York

CRC Press is an imprint of the
Taylor & Francis Group, an **informa** business

Designed cover image: Shutterstock.

First edition published 2025
by CRC Press
2385 NW Executive Center Drive, Suite 320, Boca Raton FL 33431

and by CRC Press
4 Park Square, Milton Park, Abingdon, Oxon, OX14 4RN

CRC Press is an imprint of Taylor & Francis Group, LLC

Library of Congress Cataloging-in-Publication Data
Names: Temizel, Cenk, editor. | Yegin, Cengiz, editor. | Kim, Jihoon (Petroleum engineer), editor. | Saputelli, Luigi A., editor. | Torsæter, Ole, editor.
Title: Unconventional shale resources : advances in energy transition / edited Cenk Temizel, Cengiz Yegin, Jihoon Kim, Luigi Saputelli and Ole Torsæter.
Description: First edition. | Boca Raton : CRC Press, 2025. | Includes bibliographical references and index. | Summary: "The oil and gas sector is a vital player in the energy transition. With their vast resource potential, unconventional shale plays will be an essential part in enabling this change. Unconventional Shale Resources serves as a comprehensive reference covering the latest technologies, methodologies, and applications of unconventional shale resources in the oil and gas industry, and their role in the evolution of the sector's energy transition. Offers an overview of geophysics, geology, and reservoir characterization in unconventional resources. Discusses drilling, well stimulation and completion, production engineering, and artificial lift. Covers reservoir management and surveillance, recovery enhancement, production forecasting, and surface facilities and testing. Details technical and technological advances, including machine learning, AI, data analytics, and Industry 4.0. Explores the latest methods/workflows in performance analysis in unconventional plays. Employs integrated and hybrid approaches to the energy transition The book provides surface and subsurface technical professionals in the oil and gas industry a thorough overview of unconventionals along with the integrated/hybrid applications that will enable them to stay current with the industry's transition."— Provided by publisher.
Identifiers: LCCN 2024035770 (print) | LCCN 2024035771 (ebook) | ISBN 9781032322933 (hardback) | ISBN 9781032335155 (paperback) | ISBN 9781003319955 (ebook)
Subjects: LCSH: Oil shale reserves. | Energy transition. | Shale oils. | Shale oils—Refining. | Oil-shale industry.
Classification: LCC TP699 .U53 2025 (print) | LCC TP699 (ebook) | DDC 333.8/2321—dc23/eng/20241031
LC record available at https://lccn.loc.gov/2024035770
LC ebook record available at https://lccn.loc.gov/2024035771

ISBN: 978-1-032-32293-3 (hbk)
ISBN: 978-1-032-33515-5 (pbk)
ISBN: 978-1-003-31995-5 (ebk)

DOI: 10.1201/9781003319955

For Product Safety Concerns and Information please contact our
EU representative: GPSR@taylorandfrancis.com
Taylor & Francis Verlag GmbH,
Kaufingerstraße 24,
80331 München, Germany.

Dedicated to my family.

Cenk Temizel

This book is dedicated to the ones who make life a beautiful adventure: my wife, Johimar; my incredible children, Andres, Gabriella, and Emilio; my ever-supportive sisters, Lucia and Adriana; and my parents, Emilio and Juia.

Luigi Saputelli

Contents

Foreword

THE UNCONVENTIONAL FRONTIER

As someone who has spent more than a decade immersed in the world of unconventional resources, I've had the privilege of witnessing firsthand the remarkable evolution of this industry. What once seemed like an ambitious pursuit, driven by the need to explore beyond conventional reservoirs, has now become a critical pillar of global energy supply. Unconventional resources, with their unique challenges and opportunities, have transformed not only the technical landscape but also the strategic direction of the energy industry.

This book is a testament to the progress we've made and the knowledge we've gained over the years. It is a comprehensive guide that captures the intricate science, engineering innovations, and strategic thinking that are necessary to navigate the complexities of unconventional resource development. From the initial stages of exploration to the final phases of production optimization, this work offers a detailed and nuanced understanding of the processes involved.

Unconventional resources require more than just technical expertise; they demand a deep appreciation for the variability and complexity of the subsurface, as well as an ability to adapt and innovate in the face of new challenges. Over the years, we've learned that success in this field hinges on a collaborative approach, combining insights from geology, engineering, data science, and economics to unlock the full potential of these resources.

This book is not just for those new to the field; it is an essential resource for seasoned professionals as well. It encapsulates the lessons learned, the strategies developed, and the technologies that have been refined through years of trial and triumph. Whether you are an engineer, a geologist, a data scientist, or a decision-maker, you will find this book to be an invaluable tool in your work.

As we look to the future, the importance of unconventional resources in meeting global energy demands cannot be overstated. The insights contained within these pages will undoubtedly play a crucial role in shaping the next chapter of our industry. I am confident that this book will serve as a key reference for anyone involved in the exploration, development, and optimization of unconventional resources, and it will inspire the next generation of professionals to continue pushing the boundaries of what is possible in our field.

Srimoyee Bhattacharya
Team Lead for Data Science, Americas Portfolio, Shell

Preface

The energy transition is poised to significantly impact global oil demand. Depending on the modeled scenarios, oil demand could decrease by up to 50% by 2050. This decline is attributed to factors like improved energy efficiency and the widespread adoption of electric vehicles (EVs). However, even in the most accelerated scenario, investment in a diverse energy mix, including oil and gas, will persist to ensure supply security and meet demand in sectors such as chemicals, aviation, and heavy trucking.

The emergence of unconventional hydrocarbon sources, including US shale/tight oils, Canadian tar sands, Venezuela's extra-heavy oils, and Brazil's pre-salt oils, is significantly influencing the energy landscape. These resources contribute to meeting energy demand while navigating the transition to cleaner alternatives. Achieving a sustainable energy future involves striking a balance between traditional oil and innovative solutions, with unconventional sources as a critical component.

In the early 1990s, I embarked on my petroleum engineering career in Venezuela's national oil industry. During that era , extra-heavy oil and shale oil existed, but they remained largely off the radar—unexplored frontiers awaiting discovery. My professional focus centered on integrated production modeling, spanning from reservoir dynamics to the final point of sales for conventional oil and gas reservoirs.

It was during my tenure at Halliburton (2003–2009) and later at Hess (2009–2012) that my fascination with unconventional oil production truly ignited. I delved into modeling productivity enhancement techniques, specifically multistage hydraulic fracturing, for various shale plays, including the Eagle Ford Shale, Bakken Shale, and the Wolfcamp and Bone Spring shales within the Permian Basin. The Utica and Marcellus shale formations also captured my attention.

Each new shale play presented a formidable challenge—an opportunity to decipher unexpected outcomes. We soon realized that these unconventional reservoirs defied conventional rules, demanding innovative approaches and a fresh perspective. It was a thrilling journey—one in which the earth's hidden treasures revealed themselves, reshaping the energy landscape.

Certainly, unconventional oil has profoundly reshaped North America's energy sector. The US unconventional energy boom successfully reversed the decline in domestic oil production. The extraction of tight oil (shale oil) and shale gas witnessed remarkable growth, bolstering energy security, lowering natural gas consumer prices, reducing import dependence, and fostering economic growth and local employment.

Hydrocarbon liquids, primarily petroleum, transformed the 20th-century energy landscape, revolutionizing global production and transportation. Geopolitics, orchestrated by dominant oil companies and OPEC nations, shaped global oil production, prices, and strategies. Unconventional oil sources now mark a pivotal moment as we seek sustainable energy solutions, with new players like China, Argentina, Algeria, Canada, Mexico, Russia, and Libya potentially disrupting energy markets.

The debate surrounding unconventional oil is intense, with varying perspectives on its implications, advantages, and risks. Critics contend that unconventional extraction methods—such as fracking—pose environmental threats, including altered groundwater due to faulty well design and chemical leaks. Proponents argue for coexistence with cleaner energy sources. Transitioning to cleaner energy alternatives while responsibly managing unconventional oil resources requires informed dialogue, evidence-based policies, and responsible resource management. Balancing environmental concerns with economic imperatives is crucial. Policymakers, industry stakeholders, and communities must weigh the benefits against the ecological costs.

This book outlines the scientific roadmap and transformative breakthroughs in unconventional oil and gas exploration. Key advancements in exploration area include enhanced measurement techniques allowing better understanding of geologic factors controlling hydrocarbon distribution and deliverability and predictive modeling to identify resource "sweet spots." Extraction topics around drilling, completion, and production include hydraulic fracking with extended reach horizontal wells able to access larger reservoir areas. Cutting-edge imaging technologies including seismic surveys, 3D modeling, and microseismic monitoring help locate resource-rich zones and optimize well placement. Sophisticated techniques analyze rock properties, porosity, and permeability, providing understanding reservoir heterogeneity guides, well design, and production strategies. Noteworthy data analytics, machine learning, and artificial intelligence processes aid reservoir modeling and production optimization while enabling decision-making, reducing exploration risks. Mitigation technologies addressing environmental concerns related to unconventional extraction, including water recycling, reduced emissions, and sustainable practices promote responsible development.

This book serves as the definitive reference for professionals in the unconventional oil and gas industry. It shall aid the reader in answering these key questions:

- What distinguishes unconventional oils from conventional crudes?
- How do we classify oils along the continuum from conventional to unconventional?
- What geologic and geophysical factors govern the development of unconventional reservoirs?
- How do we evaluate and optimize exploration, drilling, completion, and production techniques for unconventional resources?
- What technological and market transformations are occurring?

Dr. Luigi Saputelli
Houston, Texas, USA
May 20, 2024

Editors

Cenk Temizel is a seasoned reservoir engineer boasting nearly two decades of extensive experience in reservoir simulation, smart fields, optimization, unconventional resources, field development, and enhanced oil recovery. His career has spanned roles at industry giants such as Saudi Aramco, Schlumberger and Halliburton across the Middle East, the United States, and the UK, and a notable tenure as a reservoir engineer at Aera Energy LLC (a Shell–ExxonMobil affiliate) in Bakersfield, California, USA. Prior to his industry roles, he contributed as a teaching and research assistant at the University of Southern California and Stanford University. Throughout his career, Cenk has garnered numerous accolades, including the SPE 2022 Regional Reservoir Description and Dynamics Award. He secured second place at the SPE Global R&D Competition at SPE ATCE 2014 in Amsterdam and the Technical Excellence Awards at the 22nd World Petroleum Congress in 2017 in Istanbul. In 2012, he received the Halliburton Applause Award for Innovation. He is also a holder of several US patents and has a portfolio of publications in areas of reservoir engineering, enhanced oil recovery processes, machine learning, and smart fields. Cenk holds a BS degree (honors) from the Middle East Technical University, Ankara (2003), and an MS degree (2005) from the University of Southern California (USC), both in petroleum engineering. A committed member of SPE since 2000, he has actively contributed to several program committees.

Cengiz Yegin is Cofounder/CTO, Incendium Technologies LLC, Round Rock, Texas. He received his PhD in materials science and engineering from Texas A&M University.

Jihoon Kim is Associate Professor, Harold Vance Department of Petroleum Engineering, Texas A&M University. He received his PhD in energy resources engineering from Stanford University.

Luigi Saputelli is Upstream Analytics and Artificial Intelligence Team Leader at Abu Dhabi National Oil Company (ADNOC), UAE. He is also President/Reservoir and Production Engineering Advisor, Frontender Corporation, Houston, Texas. He earned his PhD in chemical engineering from the University of Houston.

Ole Torsæter is Professor of Reservoir Engineering in the Department of Geoscience and Petroleum, Norwegian University of Science and Technology, and Adjunct Professor in the Department of Physics, University of Oslo. He is also Research Associate at PoreLab, Norwegian Center of Excellence. He earned his PhD from the Norwegian University of Science and Technology.

Contributors

Cilia Abdelhamid
University of Wyoming
Laramie, Wyoming, USA

Ekrem Alagoz
R&D Department of Turkish Petroleum Corporation (TPAO)
Ankara, Turkey

Yaser Alghawi
Department of Petroleum Engineering
Texas A&M University
College Station, Texas, USA

Fatimah AlNasser
Saudi Aramco
Khobar, Saudi Arabia

Mubarek Alpkiray
Petroleum and Natural Gas Engineering
New Mexico Institute of Mining and Technology
Socorro, New Mexico, USA

Yohana Andrawus
Department of Geology
Federal University of Lafia
Nasarawa, Nigeria

Muhammad Arif
Department of Petroleum Engineering
Khalifa University
Abu Dhabi, United Arab Emirates

Tao Bai
School of Energy Resources
University of Wyoming
Laramie, Wyoming, USA

Karthik Balaji
Aethoro
Grand Forks, North Dakota, USA

Artur Davletshin
Hildebrand Department of Petroleum and Geosystems Engineering
The University of Texas at Austin
Austin, Texas, USA

Emre Can Dundar
R&D Department of Turkish Petroleum Corporation (TPAO)
Ankara, Turkey

Anzaku Y. Ibrahim
Department of Geology
Federal University of Lafia
Nasarawa, Nigeria

Shams Kalam
Department of Petroleum Engineering
Khalifa University
Abu Dhabi, United Arab Emirates

Aliyu Mohammed Kaura
Department of Geology
Federal University of Lafia
Nasarawa, Nigeria

Mohammadreza Keshavarz
School of Petroleum Engineering
State University of Campinas
Campinas, Brazil

Abdeldjalil Latrach
University of Wyoming
Laramie, Wyoming, USA

Mohammed A Mohammed
Department of Chemistry
Federal University of Lafia
Lafia, Nigeria

Minou Rabiei
University of Wyoming
Laramie, Wyoming, USA

Amirsaman Rezaeyan
Department of Earth, Energy, and Environment
University of Calgary
Calgary, Alberta, Canada

Hosein Rezvani
Department of Chemistry
University of Hull
Hull, United Kingdom

Baris Tali
Technische Universität Clausthal
Clausthal-Zellerfeld, Germany

Son Tran
Department of Earth, Energy, and Environment
University of Calgary
Calgary, Alberta, Canada

Kalyan Venugopal
University of Wyoming, Laramie
Wyoming, USA

Ying Yu
Center for Economic Geology Research
University of Wyoming
Laramie, Wyoming, USA

Mesut Yurukcu
The University of Texas Permian Basin
Midland, Texas, USA

Leilei Zhang
Department of Chemical and Biomolecular Engineering
Rice University
Houston, Texas, USA

Yihuai Zhang
James Watt School of Engineering
University of Glasgow
Glasgow, United Kingdom

1 Introduction

Cenk Temizel, Cengiz Yegin, Jihoon Kim, Luigi Saputelli, and Ole Torsæter

1.1 SHALE

Shale is a delicate sedimentary rock made up of small clastic grains that are less than 0.0625 millimeters in size, combined with clay and organic matter. It has a shaley or thinly laminar bedding structure that gives it a distinctive appearance (Jiang et al., 2016; King, 2020). It is sometimes categorized as mudstone (Boggs Jr., 2006), with the only distinguishing feature being that mudstones break into blocky pieces while shales break into thin chips with parallel tops and bottoms. Shale comprises the largest sedimentary rocks, making up approximately 70% of this type of rock in the Earth's crust (Speight, 2013). Shale is more technically defined as a sedimentary rock composed mostly of silt and clay particles (Boggs Jr., 2006). A "fissile" describes its ability to split into thin sheets along bedding. At the same time, "terrigenous" refers to its origins as a product of rock weathering.

Shales are sedimentary rocks composed of fine-grained particles, like silt and clay. They are created through transportation, deposition, and compaction of these detrital materials. Typically, shales are formed in environments with very slow-moving water, such as lakes and lagoons, river deltas, floodplains, and offshore beach sands. Due to their fine-grained nature, shales often exhibit a layered or laminated appearance and can contain fossilized remains of plants and animals (Ahmed et al., 2016). They can also be deposited in sedimentary basins and on the continental shelf in relatively deep, quiet water. Shale is the most abundant sedimentary rock found in sedimentary basins worldwide. Mineral compositions of shale are complex, comprising mainly brittle minerals such as quartz, feldspar, calcite, and other clay minerals like montmorillonite, illite/smectite, and illite, kaolinite, as well as a minor amount of pyrite (Jiang et al., 2016).

Shale is a prevalent type of sedimentary rock, accounting for roughly 55% of all such rocks. There are various kinds of shale, including black, carbonaceous, siliceous, ferruginous, and calcareous. Of these, carbonaceous and siliceous shale are the most gas bearing and prone to fracturing (Zou, 2013). The mineral composition of shale is complex and consists of clastic grains, clay minerals, and organic matter. Clastic grains like quartz, feldspar, carbonate rock, and pyrite are present, as are clay minerals like illite, chlorite, kaolinite, smectite, and hydromica. The differing amounts of these elements give rise to a wide range of shale properties (Zhang et al., 2018). Organic-rich shale, such as black and carbonaceous shale, is the primary rock type that produces shale gas.

DOI: 10.1201/9781003319955-1

Shale occurs in various colors, including red, brown, green, gray, and black. The color of shale is mainly determined by its composition. Typically, shales with higher organic content tend to be darker in color. Reddish and purple hues are created by hematite and limonite while ferrous iron–rich mineral components give rise to blue, green, and black shades (Speight, 2013). Calcareous shales, on the other hand, are usually light gray or yellowish. Black shale formations are considered valuable sources of oil and gas as they contain organic debris that, when preserved and heated, can produce natural gas and oil. Shale formations such as the Barnett, Marcellus, Haynesville, and Fayetteville shales are well known for their natural gas production and are characterized by their dark gray or black appearance (Zou, 2012).

Black shale contains substantial amounts of organic matter as well as fine, dispersed pyrite and siderite, with a total organic carbon content typically ranging from 1% to 15% or higher. Black shale is a type of shale that contains more than 1% organic carbon. It is a good hydrocarbon source rock, containing up to 20% organic carbon. Algae is usually responsible for the influx of carbon in black shale. When it decays, an ooze known as sapropel is formed (NEB, 2009). When subjected to precise conditions of pressure and temperature, the ooze transforms, resulting in the formation of a substance known as kerogen. This process occurs at depths ranging from 3 to 6 kilometers (1.8–3.7 miles) below the Earth's surface and at temperatures typically between 90 and 120°C.

1.2 CHARACTERISTICS OF SHALE

Shale is an unconventional resource that presents unique challenges for extraction due to its distinct characteristics. One of the primary characteristics differentiating shale from mudstone is that it has a laminated structure and is fissile: i.e., it tends to split into thin layers that are usually parallel to the bedding-plane surface (Figure 1.1). Fissility in shales appears directly related to the parallel orientation of mineral grains in the rock fabric. The property is determined by the presence of clay minerals in shale. Clay minerals are phyllosilicates with a platy habit and tend to align parallel during deposition. In mudrocks, however, clay minerals are only poorly or partially oriented.

Shales that are resource plays are typically deposited in a low-energy environment, which allows for the accumulation of very fine-grained particles. However, high-energy deposits such as channel sands and turbidites can also lead to shale deposition (Ahmed et al., 2016). The relatively slow change in geological and petrophysical properties over large distances in marine shales contrasts with the higher levels of heterogeneities often seen in sandstones and carbonates. Petrophysicists can take advantage of this by normalizing log responses in marine shales when analyzing logs of varying vintages and from varying suppliers.

1.2.1 Total Organic Carbon (TOC)

The formation of shales through sedimentation involves depositing fine-grained organic material, often accompanied by silt, mud, and clay particles. These rocks typically boast a high TOC percentage by weight. Due to their rich organic carbon content, shales can serve as valuable reservoir rocks for oil and gas deposits under specific temperature and pressure conditions.

TOC values of hydrocarbon reservoirs generally range from approximately 2% to 10%. They can vary widely and at several scales throughout a reservoir. Reservoirs with TOC < 2% do not have sufficient organics to form attractive hydrocarbon reservoirs while those with TOC > 10% are not sufficiently mature to be produced (Alexander et al., 2011). Shales contain organic matter from living organisms like plants, animals, and bacteria such as algae, plankton, and decomposed plant tissue. The transformation of organic matter into kerogen is facilitated by diagenesis, primarily influenced by mineralogy and redox conditions. Intricate thermal changes mark kerogen conversion (Passey et al., 2010).

1.2.2 Thermal Maturity

The thermal maturity of the organics will control the types of hydrocarbons present in a shale reservoir. Maturity is governed mainly by the burial history of the sediments in a basin. It is a function of the burial depth, geothermal gradient during basin evolution, and the time scales over which the sediments experienced different temperatures.

As a sedimentary layer is buried deeper, higher temperatures affect the organic matter it contains. When temperatures are below 50°C, the organic matter converts into kerogen, and some methane is produced by microbial activity. As temperatures range between 50°C and 150°C, the kerogen undergoes catagenesis and transforms into oil and more high-liquids-content gas. At higher temperatures (> 150°C), metagenesis produces dry gas (Alexander et al., 2011).

1.2.3 Permeability

The permeability of shale is extremely low due to its extremely small sediment particles. Shale permeabilities can be lower than 0.01 to 0.0001mD and range over several orders of magnitude (Yao, 2019). This matrix permeability is generally too low to permit commercial oil or natural gas production rates unless unconventional extraction techniques (horizontal drilling and hydraulic fracturing) are used. Such physical properties as permeability and plasticity largely depend on the constituent minerals' grain sizes. However, the presence of sets of continuous, laterally extensive natural fractures can provide additional permeability that can substantially impact the ability of a shale reservoir to be produced via hydraulic stimulation, which can intersect these natural fracture systems (Zou, 2013).

1.2.4 Porosity

Pores found in rocks play a crucial role in the accumulation and storage of oil and gas. The amount of free gas present solely depends on the rock's porosity. Matrix pores of shale are responsible for storing 50% of shale gas (Zou, 2013). Shale comprises various micro- to nanoscale pores, such as intergranular pores, dissolved matrix internal pores, and organic pores. The sizes of these pores range from 1 to 3 nm to 400 to 750 nm, with an average size of 100 nm (Hackley & Cardott, 2016). This unique structure results in a large specific surface area that can absorb significant amounts of gas. Despite its low porosity and permeability, shale can still serve as a high-quality reservoir (Zou, 2013). Shale is known for its extremely low permeability, meaning it does not readily allow fluid flow. This low permeability poses challenges for extracting oil or gas from shale formations.

1.2.5 Mineralogy

The composition of shale is a crucial factor in determining the characteristics of its matrix pores, microfracture development, and fracture mode. The mineralogy of shales varies widely, particularly in the proportions of clays, carbonates, and quartz. High proportions of clay minerals result in relatively high ductility, which limits the ability of hydraulic fractures to propagate continuously through the rock (Zou, 2013). In contrast, shales that contain more than 50% quartz and/or carbonate minerals are more brittle and generally reflect better production rates due to the ability of hydraulic fractures to propagate for longer distances (Passey et al., 2010; Zou, 2013). Understanding these characteristics is crucial for effectively testing and extracting resources from shale formations. Advanced techniques like hydraulic fracturing and horizontal drilling are necessary to efficiently capture the oil and gas in these unconventional resources.

1.2.6 Thickness

The thickness of the shale unit controls the gross reservoir thickness available to contain the hydrocarbons. Because shales are deposited in parasequences (which have variable grain sizes, TOC contents, and other characteristics), not all the shale unit is considered productive; a fraction of the total thickness of the shale unit will have unfavorable reservoir properties. Thus, the net thickness, rather than the gross thickness, will affect the productivity and resource capability of the reservoir (Zou, 2012).

1.2.7 Color

Shale is a rock that can come in various colors, such as gray, black, red, green, or brown. These colors are largely determined by the composition and environment in which the shale was formed. Shales with higher organic content tend to be darker, while those with mineral components rich in ferric iron can have blue, green, or black tones. Shales containing hydrated ferric oxides like hematite and limonite often have reddish or purple hues. On the other hand, calcareous shales with a significant amount of calcite usually appear light gray or yellowish.

Shale formations may exhibit significant heterogeneity, meaning that different layers within the formation can vary in their lithology, porosity, and fluid content. This complexity makes the optimization and characterization of shale play a challenging task. Shale formations with potential commercial exploitation as oil and gas reservoirs share certain distinguishing characteristics. They possess a high TOC content, typically exceeding 2.0%. Furthermore, they exhibit a high degree of thermal evolution characterized by a R^o value of 1.0% to 2.0% and high brittleness, as indicated by over 40% of quartz and other brittle minerals, which is conducive to fracturing. Additionally, they contain high quantities of oil and gas, which favor accumulation and extraction in significant volumes (Zou, 2013).

1.2.8 Shales as Resource Plays

Shale formations have become increasingly important in the energy industry due to their potential as a source of hydrocarbons. In the realm of "resource play," shale is identified as a precious natural resource with immense possibilities for commercial utilization. Shale formations are unique and necessitate specific techniques like hydraulic fracturing and horizontal drilling to access the hydrocarbons confined within the rock.

The energy landscape in the United States has been significantly transformed by the shale boom, particularly in regions such as the Permian Basin, Bakken Formation, and Eagle Ford shale. The development of shale resources has yielded numerous benefits, including increased domestic oil and gas production, reduced import reliance, and economic growth for regions with significant shale reserves. Shale's resource potential presents companies with exciting exploration, development, and production opportunities for unconventional reservoirs. However, shale extraction raises environmental and social concerns, such as potential impacts on water resources, air quality, and seismic activity.

1.3 UNCONVENTIONAL RESOURCE TRIANGLE

Unconventional petroleum refers to natural resources in continuous or subcontinuous accumulations, but their natural industrial production is not achievable through traditional techniques. The economic extraction of these resources is impossible without using new techniques that can improve the reservoir permeability or fluid viscosity (Zou, 2012). Unconventional petroleum presents unique characteristics that require innovative approaches to its extraction. Unlike conventional petroleum, it is typically found in extensive, continuous formations without distinct trap boundaries. Additionally, its natural industrial production is unstable and often features indistinct Darcy flow. These factors challenge the industry to develop new and creative methods to extract and process unconventional petroleum resources.

Unconventional petroleum can be identified through geological perspectives, such as the paragenesis of the source reservoir, its wide distribution in basin slopes and centers, blurred trap boundaries, and hydrodynamics. These features provide useful insights for studying unconventional petroleum and its potential for development (Zou, 2012). As energy demand continues to rise, we have developed innovative methods to extract petroleum. These include large-scale horizontal well fracturing, platform-based factory production, and advanced nano and gas displacement techniques. With these methods, we are able to increase the amount of petroleum we can recover, ensuring a stable energy supply for the future.

Global petroleum exploration is currently shifting from conventional to unconventional hydrocarbon resources. Unconventional hydrocarbon resources (including tight oil/gas, shale oil/gas, and coalbed gas) are becoming significant components of world energy consumption (Geri et al., 2019). Generally, unconventional resources (e.g., tight oil, shale oil, oil sands, shale gas, tight gas, coalbed methane, and gas hydrates) are locked in geological structures where extraction is not economically or technologically feasible; however, with technological advancements, these can turn into viable fossil fuel reserves (Hobson, 2008). Unconventional reservoirs are defined as oil and gas formations with several specific inherent characteristics, such as ultra-low reservoir permeability in the range of micro- to nano-Darcy, low porosity, poor connectivity, non-Darcy flow, high TOC, and desorption behavior of the rock surface, as well as requirement of special production operations outside the conventional methods (Geri et al., 2019).

The concept of the unconventional resource triangle was introduced by Masters and Gray in 1979 to illustrate the huge potential of unconventional gas resources, such as coalbed methane, shale gas, and tight gas (Ahmed & Meehan, 2016). Since then, the concept has been expanded to include unconventional oil resources, such as heavy oil, oil sands, shale oil, and tight oil (Ahmed & Meehan, 2016). The resource triangle describes the relationship between the abundance, quality, and difficulty of extracting hydrocarbon resources. The triangle has three sides: conventional, tight, and unconventional resources. In the resource triangle, conventional gases are placed at the top, given their superior reservoir characteristics and quality (Holditch, 2006). The development of this gas is made easier by using conventional technology, leading to a relatively smooth development process. However, the availability of conventional gas is limited due to its existence in small volumes. As one delves deeper into the triangle, below tight gas and coalbed methane (CBM), shale gas (and gas hydrates) can be found at the base. This progression is marked by a decline in permeability and reservoir quality while the complexity of development and associated technological requirements increase (Holditch, 2006; Ahmed & Meehan, 2016). Nevertheless, significant quantities of these valuable resources can be found. The resource triangle is a universal principle for all hydrocarbon-producing basins worldwide. In Figure 1.1, a scale on the right side provides an idea of the formation permeability values commonly associated with tight gas sands. However, it is important to remember that other low-quality resources, like gas shales, coalbed methane, and gas hydrates, may exhibit different permeability scales.

Martin and Makrodimopoulos (2008) utilized a computer program to confirm the validity of the resource triangle concept. They also expanded the resource triangle to include liquid and solid hydrocarbons, adding heavy oil and oil shale and referencing work from Gray (1977). Figure 1.1 shows that shale resource development is located in the bottom left corner, indicating the vast resource. However, the reservoir quality is declining, so more advanced and complex technologies are required to extract the resource.

1.4 WORLD SHALE GAS, OIL PLAYS, AND BASINS

Shale is a sedimentary rock rich in organic matter and can be a source of oil and natural gas. Shale basins are areas where shale deposits are found. Shale plays and basins are important sources of natural gas and oil, and they are found in various countries

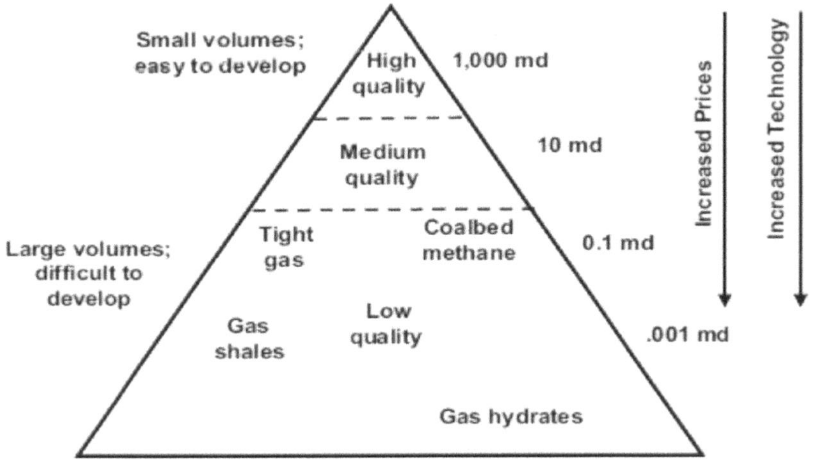

FIGURE 1.1 The resource triangle unconventional resources by source.

Source: Ahmed & Meehan, 2016; Holditch, 2006

around the globe. The United States has been a leader in shale gas production, but shale gas resources are also found in other parts of the world. The Energy Information Administration (EIA) has reported that China, Argentina, Algeria, Canada, and Mexico have the highest technically recoverable shale gas resources. These nations possess extensive shale gas reserves that can be extracted using advanced drilling techniques, including hydraulic fracturing, making them significant players in the global energy market (U.S. EIA, 2012). The EIA also reports that Russia, China, Argentina, and Libya are the countries with the largest technically recoverable shale oil resources. Some notable shale plays and basins include the Permian Basin in the United States, which is one of the most prolific and actively developed shale regions. The Eagle Ford shale in Texas and the Bakken shale in North Dakota contribute significantly to US shale production.

In Canada, the Montney and Duvernay formations in Alberta are important shale plays, while the Vaca Muerta formation in Argentina has attracted attention for its vast shale oil and gas resources (EIA, 2010). The Santos Basin in Brazil and the Karoo Basin in South Africa are also emerging as potential shale gas areas. The world is experiencing a shale gas revolution; over 20 countries outside North America have performed pre-evaluation and pilot tests for the exploration and development of shale gas resources (Figure 1.2), and shale gas has been discovered in Australia, China, Russia, and various countries in Europe. These resources have the potential to significantly impact global energy markets and contribute to the long-term energy security of many countries. With horizontal drilling and hydraulic fracturing techniques, natural gas producers have tapped into low-permeability geological formations, such as shale formations, with greater efficiency and profitability. With this development, they can now tap into new sources of natural gas and extract it more efficiently, which is a promising step towards meeting the world's increasing energy demands.

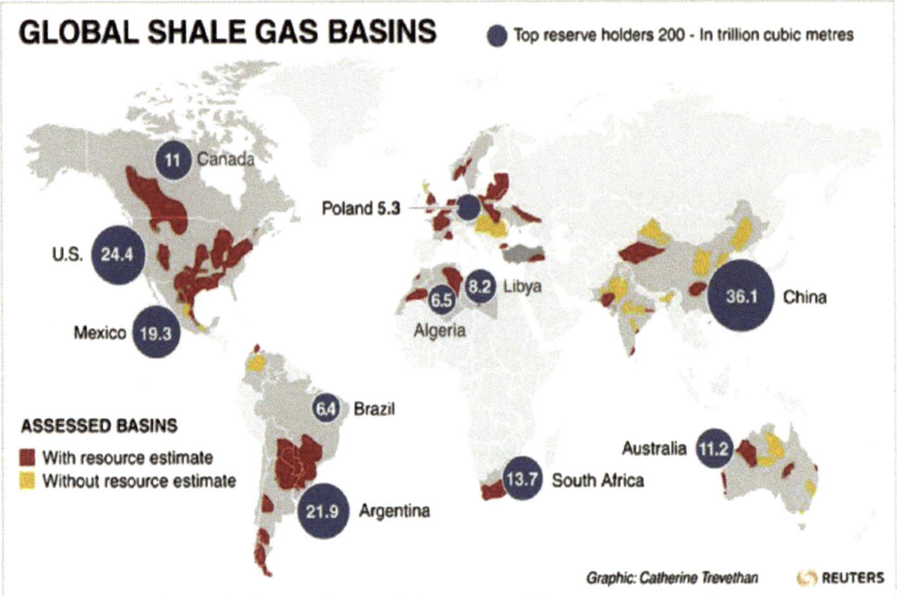

FIGURE 1.2 Worldwide distribution of shale gas resources by source.

Source: EIA, 2010

1.5 US SHALE PLAYS

Over the past few years, the US has experienced significant growth in shale gas production. This is due to advancements in drilling and extraction techniques and increased exploration efforts in the natural gas industry (Mauger et al., 2011). In 2008, US shale production reached 2.02 trillion cubic feet, a 71% increase from the previous year (EIA, 2011). By 2009, shale gas production had grown by 54% to 3.11 trillion cubic feet while the proven US shale reserves also increased by 76% to 60.6 trillion cubic feet by the end of that year (EIA, 2010). The US Energy Information Administration (EIA) more than doubled its estimate of technically recoverable shale gas reserves in 2011, projecting that the number would reach 827 trillion cubic feet, up from 353 trillion cubic feet. This was due to the incorporation of drilling results from new shale fields, such as the Marcellus, Haynesville, and Eagle Ford shales, in its Annual Energy Outlook for that year. The EIA also predicts that shale production will make up 49% of total US gas production by 2035, up from 23% in 2010 (EIA, 2012).

Shale energy extraction has the potential to impact US energy security by reducing the need to purchase large quantities of crude oil and refined petroleum products from global markets. In addition, shale gas contributes to the country's natural gas independence. As a fuel source, natural gas, primarily composed of methane, is considered to have lower air pollution emissions than coal or oil. Some view it as a transitional energy source towards a less greenhouse-gas-intensive energy future.

TABLE 1.1
Some Notable Shale Plays Worldwide by Source

Country	Shale Play	Basin
Argentina	Vaca Muerta	Neuquén
	L Inoceramus-Magnas Verdes	Astral-Magallanes
	Aguuada Bandera	San Jorge
Mexico	Eagle Ford	Burgos
	Pimienta	Tampico
	Maltrata	Veracruz
Columbia	Gacheta	Llanos
	La Luna/Tablazo	Middle Magdalena Valley
Brazil	Ponta Grossa	Parana
	Jandiatuba	Solimoes
Australia	Canning	Goldwyer
	Roseneath-Epsilon	Cooper
	Murteree	
UK	Carboniferous	North UK
	Lias Shale	South UK
Poland	Llandovery	Baltic/Warsaw Trough
Tunisia	Frasnian	Ghadames
Turkey	Hamitabat	SE Anatolia
China	Longmaxi	Sichuan
	Qiongzhusi	Sichuan
	L. Cambrian	Tarim
UAE	Shilaif	Rub' Al-Khali
India	Cambay	Cambay
Pakistan	Sembar	Lower Indus
Indonesia	Aifam	
Algeria	L. Silurian	Tindouf
Libya	Sirte/Rachmat	Sirte
South Africa	Whitehill	Karoo
Egypt	Khatatba	Abu Gharadig
Chad	L. Cretaceous	Doseo
Algeria	Frasnian	Timimoun
Venezuela	La Luna/Capacho	Maracaibo/Catatumbo
France	Permian-Caboniferous	Paris
Germany	Posidonia	Lower Saxony

Source: EIA, 2010

However, it is worth noting that the development of shale energy directly impacts water resources.

Shale gas and oil plays have revolutionized the US natural gas market, significantly increasing shale gas production. From just 0.39 trillion cubic feet in 2000, the production rose to 4.87 trillion cubic feet in 2010, accounting for 23% of the total dry gas production in the United States. This remarkable development has positively impacted the country's energy industry, providing abundant and affordable natural gas to power homes, businesses, and industries (U.S. EIA, 2012). The overall effort has been to identify shale gas "sweet spots" and avoid "dead zones." Accordingly, multilaterals and fracturing techniques are designed to maximize access to "sweet spots." The following describes several shale plays in the United States and Canada and their characteristic features.

The US Energy Information Administration (EIA) monitors the country's availability and usage of energy resources. According to EIA records (2001), there were roughly 22 active shale plays in the contiguous United States (Soeder & Borglum, 2019). Although no new shale plays have been identified since then, the focus has shifted towards different targets and production methods. These shale gas and oil plays are in various parts of the country. Shale energy deposits come in different sizes, depths, and quality across the United States. They are found in large-scale basins or accumulations of sedimentary rocks that may also contain other oil and gas resources. Currently, shale gas production is taking place in 16 states. Notable key shale formations in the lower 48 states are illustrated in Figure 1.3, and the features of these shale plays are shown in Table 1.1 and briefly discussed.

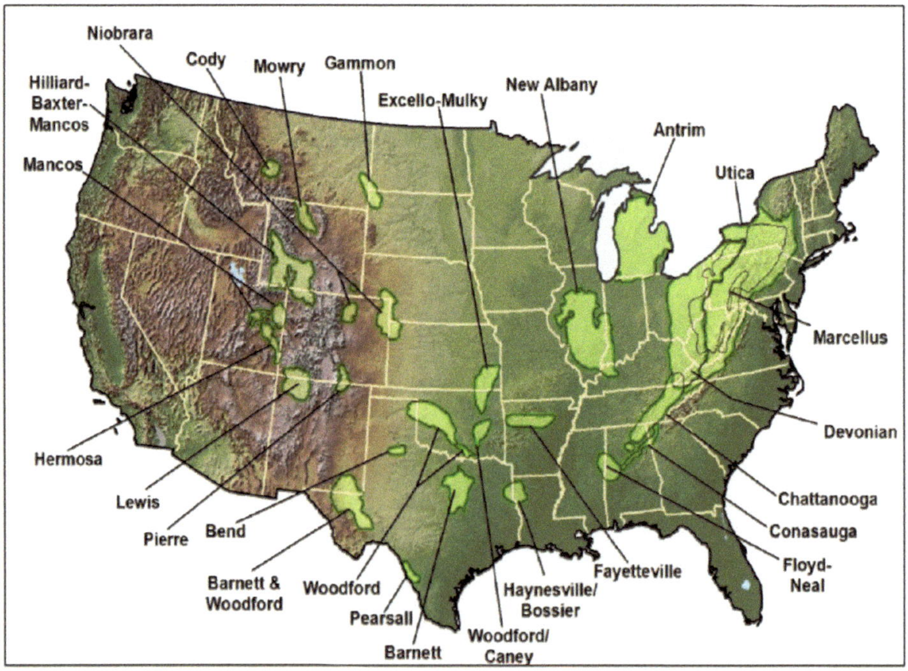

FIGURE 1.3 Map of shale gas plays in the United States by source.

Source: EIA, 2011

1.5.1 Marcellus Shale Gas Play

The Marcellus shale is the largest shale gas play in the northeastern United States, covering an expansive area of 95,000 square miles across West Virginia, Pennsylvania, and New York. The Marcellus shale formation is geographically divided into two primary units: the Active Area and the Undeveloped Area. The Active Area spans West Virginia and Pennsylvania, covering an extensive area of 10,622 square miles (EIA, 2011). The Undeveloped Area encompasses the remaining territory of 84,271 square miles, which has yet to be explored for natural gas and oil reserves.

The Marcellus shale formation is estimated to be between 4,000 and 8,500 feet deep and has a thickness ranging from 50 to 200 feet. It is a Middle Devonian–age shale formation, bounded above by Hamilton Group shale and below by Tristates Group limestone. According to available data, the formation's technically recoverable resources are estimated to be 262 trillion cubic feet (tcf). However, like Haynesville, the play's potential estimates are frequently revised upward due to its early stage of development. Overall, the Marcellus shale formation is a significant resource for the United States, with the potential for long-term production and economic growth.

1.5.2 Big Sandy Shale Gas Play

The Big Sandy shale gas play in the Devonian period consists of the Huron, Cleveland, and Rhinestreet formations. The main producing area of the Devonian Big Sandy shale gas play straddles the state lines of Virginia, West Virginia, and Kentucky but extends through central Ohio and along Lake Erie into the panhandle of Pennsylvania. The Big Sandy was divided into two main units for modeling purposes: the Developed Area and the Undeveloped Area.

The Big Sandy shale play covers an area of 10,669 square miles (6,828,000 acres). Each well in the shale play is estimated to produce an average of 0.325 billion cubic feet of gas, and the total technically recoverable gas is estimated to be around 7.4 trillion cubic feet (Speight, 2019). According to Speight (2019) and EIA (2011), the active area of Big Sandy covers approximately 8,675 square miles, with an undeveloped area of 1994 square miles. The well spacing for this area is 80 acres per well. The Devonian Big Sandy has an average estimated ultimate recovery (EUR) of 0.325 billion cubic feet (bcf) per well and an area of 8,675 square miles, with a well spacing of 8 wells per square mile. The TRR for this area is 6.47 tcf. The Devonian Big Sandy shale play has an average depth of 3,800 feet and a thickness of 175 feet, with a porosity of 10% and a total organic content of 3.75% by weight.

The Big Sandy shale play encompasses a vast area of 10,669 square miles (6,828,000 acres), with an EUR of 0.325 bcf per well with a well spacing of 8 wells per square mile and technically recoverable gas of approximately 6.47 tcf. The active area covers around 8,675 square miles, while the remaining 1994 square miles remain undeveloped, with an 80-acre well spacing (Speight, 2019). However, the undeveloped area has a TRR of 0.92 tcf. The Devonian Big Sandy shale play has an average depth of 3,800 feet, thickness of 175 feet, porosity of 10%, and total organic content of 3.75 %, as per general properties (EIA, 2011).

1.5.3 New Albany Shale Gas Play

The New Albany shale is a noteworthy geological formation that spans the southeastern parts of Illinois, southwestern Indiana, and northwestern Kentucky. It is a shale formation similar to the Antrim shale and can be found at depths ranging from 500 to 2,000 feet. Due to its relatively shallow depth and water-filled composition, its unique CBNG-like character distinguishes it from other gas shales (Speight, 2019). The New Albany formation is a Devonian-to-Mississippian-age shale that is bounded above by Rockford Limestone and below by North Vernon Limestone, adding to its geological importance. The shale formation comprises a diverse range of sedimentary rocks, including brown, black, and green shale, as well as minor beds of dolomite and sandstone. This formation was deposited under anoxic marine conditions and contains a common accessory mineral: pyrite. Interestingly, certain portions of the shale boast an impressive 4% or more organic carbon by weight. Additionally, the black shale layers are known for their unusually high levels of radioactivity, phosphorus, and heavy metals (Ettensohn & Barron, 1981).

The New Albany shale is a significant shale gas play that spans approximately 43,500 square miles. The wells in this region are spaced about 80 acres apart. Although similar to the Antrim shale, the average net pay thickness is thinner, ranging from 50 to 100 feet. The wells found in the New Albany play typically produce an average of 5 to 500 bbls of water daily. The measured gas content in the New Albany shale is roughly 40 scf/ton to 80 scf/ton. The original gas for this formation is estimated to be around 160 tcf, with technically recoverable resources estimated to be less than 20 tcf. The New Albany shale boasts total organic content porosity of 12% and 13% by weight, respectively.

1.5.4 Antrim Shale Gas Play

The Antrim shale is a component of a vast, organic-rich shale deposition system that spread across significant portions of the North American continent during the middle-to-late Devonian period. The Michigan basin, situated along the Eastern Interior Seaway, comprises several depocenters that make up this shale play. Over time, sediment has filled the basin to a depth of over 17,000 feet. Within this depth, 900 feet is made up of the Antrim shale and other rocks from the Devonian-Mississippian period (Speight, 2017). The Antrim shale, located at the center of the modern structural basin, begins approximately 2,400 feet below sea level (EIA, 2011; Speight, 2019).

Antrim shale play has a total area of approximately 12,000 square miles. The play's total area encompasses both developed and undeveloped sections. The developed portion measures roughly 527 square miles, leaving the remaining 11,743 square miles undeveloped. Due to the significant disparity in size between the two areas, the Antrim shale is treated as a single unit. This shale gas play yields an average EUR of 0.28 bcf per well and boasts a technically recoverable gas supply of around 19.9 tcf. The Antrim formation ranges in depth from 600 to 2,200 feet and is 70 to 120 feet thick. The area spans 12,000 square miles, with a well spacing of 7 TRR wells per square mile and a total estimated gas supply of 19.93 tcf (EIA, 2011; Speight, 2019).

The Antrim shale play is widely recognized as one of the most actively developed shale gas plays, following the Barnett. This unique formation spans approximately 12,000 square miles and boasts several distinct features, such as its shallow depth, small stratigraphic thickness, and high volumes of produced water. Additionally, the Antrim shale play is characterized by its 9.5% porosity and 9% by weight of the total organic content, making it an exciting exploration and production area.

1.5.5 Haynesville Shale Gas Play

The Haynesville shale gas play, also called the Haynesville/Bossier shale play, is a shale formation that covers a large area alongside the border of eastern Texas and northwestern Louisiana, close to the Arkansas line. This region is commonly known as the ArkLaTex. The Haynesville shale, which spans an expansive area of approximately 9,000 square miles, caught the attention of energy experts in 2007 due to its high shale gas well production rates. It was believed that the shale formation held substantial gas reserves, which was confirmed by estimates of a technically recoverable resource of 74.7 tcf (EIA, 2011). The Haynesville shale is a significant natural gas reservoir, with an estimated average of 3.6 billion cubic feet per well. The shale formation lies at depths ranging between 10,500 and 13,500 feet, with a thickness of 200 to 300 feet, and exhibits a porosity of 8.5%. Additionally, the shale contains a total organic content of 2.25%. The reservoir is underlain by the Cotton Valley Group sandstone and overlain by the Smackover Formation limestone, making it a highly valuable and strategically important geological formation (GWPC, 2009).

The Haynesville region is divided into two distinct zones: active and undeveloped. The active zone includes the acreage currently held by companies and is potentially under development. The undeveloped zone comprises the acreage not currently held by companies. The active zone spans over 3,574 square miles and has a total resource potential of 53.3 tcf, translating into an estimated 6.5 bcf per well. Reported EURs at the well level range between 4 and 10 bcf. However, the undeveloped zone has a total resource potential of 19.41 tcf or 1.5 bcf per well.

1.5.6 Eagle Ford Shale Gas and Oil Play

The Eagle Ford shale, Late Cretaceous age, is a calcareous shale extending along the western Gulf Coast from East Texas to Mexico. The shale is stratigraphically located below the Austin Chalk and above the Buda Limestone and Woodbine Formation (Soeder & Borglum, 2019). The composition of the play is predominantly liquid, resulting in the identification of three distinct zones: namely, an oil zone, a condensate zone, and a dry gas zone, as reported by EIA (2011) and Speight (2019). The dry gas zone covers approximately 200 square miles, while the condensate zone covers 890 square miles. The oil zone covers a much larger area of roughly 2,233 square miles.

The shale play has an average estimated ultimate recovery (EUR) for each well reported to be between 4 and 6 bcf, with the oil zone having an average EUR of 750 MBO bcf per well. The condensate zone has an average EUR of 4.5 bcf per well while the dry gas zone has an estimated EUR of 5.5 bcf per well. The well spacing in the play varies from 4 to 8 wells per square mile, with the oil zone having 5 wells per square mile. The total recoverable resources (TRR) for the play is estimated at 20.8 tcf of gas and 3.35 billion barrels of oil (BBO). The oil zone alone has a TRR of 3.35 BBO, the condensate zone has an impressive TRR of 16.4 tcf, and the dry gas zone has a TRR of 4.4 tcf. The shale play has a depth of 7,000 feet and a thickness of 240 feet, with a porosity of 9% and TOC of 4.25% by weight (EIA, 2011) and permeability of > 1 mD (Islam, 2014), making it a good candidate for gas recovery.

1.5.7 Caney Shale

The Caney shale is a significant stratigraphic formation in the Arkoma Basin of Oklahoma. It is comparable to the Barnett shale located in the Fort Worth Basin and has gained popularity as a gas-producing formation following the success of the Barnett shale formation. The Caney shale was formed during the Chesterian period and is part of a series of basins that formed progressively westward along the Ouachita Fold Belt. These basins extend from Mississippi's Black Warrior Basin to basins in southwest Texas (Speight, 2019). The Caney shale formation dips southward from a depth of 3,000 feet in northern McIntosh County, Oklahoma, to 12,000 feet north of the Choctaw thrust. It thickens as it moves toward the southeast along the Choctaw fault. It is identified by six intervals that can be divided based on the characteristics of the density and resistivity logs.

The Caney formation boasts significant potential for organic carbon extraction, as evidenced by average values that range from 5% to 8%. Density exhibits a strong linear correlation with these values, while mud log gas shows a robust correlation with desorbed gas values that fall in the range of 129 to 150 cubic feet per ton of shale. Additionally, gas estimates in place reveal great potential, ranging from 30 to 40 bcf. These positive findings suggest a bright future for the Caney formation.

1.5.8 Woodford Shale Gas Play

The Woodford shale is a highly significant geological formation that spans the south-central region of Oklahoma. It is divided into two parts: the Woodford Central, situated in the Ardmore Basin, and the Woodford Western, located in the Arkoma Basin. The shale play in the Arkoma Basin covers an extensive area of 2,900 square miles while the Ardmore Basin spans 1,800 square miles, with a well spacing of 160 acres per square mile. It is renowned for its impressive average EUR of 2.5 bcf per well and boasts an incredible 22.2 tcf of technically recoverable gas in Western and Central Woodford. The Woodford Western, located in the Arkoma Basin, has a depth range of 6,000 to 13,000 feet and a thickness of 150 feet, with an impressive EUR of 4.0 bcf per well. Meanwhile, the Woodford Central has a depth of 5,000 feet and an EUR of 1.0 bcf per well. Overall, the Woodford shale is a highly productive formation with a vast amount of technically recoverable gas.

The Woodford shale has an average thickness ranging from 150 to 250 feet throughout the play, making it a highly promising location for natural gas production. Additionally, the porosity and TOC content of the Woodford shale are impressively high, with porosity reaching 7% and TOC at 6.5% by weight. These characteristics further enhance the appeal of the Woodford shale as a prime location for natural gas exploration and production (EIA, 2011; GWPC, 2009; Speight, 2017; Soeder & Borglum, 2019). The Woodford formation exhibits a significant variation in thickness, ranging from approximately 125 feet in the northern regions to as much as 900 feet in the deeper sections of the Anadarko Basin. This diverse thickness of the formation may have significant implications for resource exploration and production in the region (Soeder & Borglum, 2019).

1.5.9 Fayetteville Shale

The Fayetteville shale, an unconventional natural gas reservoir located on the Arkansas side of the Arkoma Basin, is a massive formation divided into two main units—Central and Western—based on the location of the shale. The Central unit stretches from the Arkansas-Oklahoma border to the east of Johnson, Logan, and Yell counties, spanning 4,000 square miles. The remaining shale, Fayetteville Western, which covers approximately 5,000 square miles, completes the shale gas play's overall area, estimated to be around 9,000 square miles.

The Fayetteville shale is a fissile black shale containing thin interbeds of bluish-black lithographic limestone with septarian concretions (Soeder & Borglum, 2019). The shale gas play is known for its impressive average EUR of 1.7 bcf per well and has approximately 31.96 tcf of technically recoverable gas. The shale is 6,500 feet deep from surface outcrops and ranges in thickness from 50 to 550 feet (Soeder & Borglum, 2019), making it an ideal natural gas reservoir (EIA, 2011). The Fayetteville shale play has a phase porosity of 5% and TOC of 6.9% by weight (EIA, 2011). These unique characteristics make the Fayetteville shale play an attractive source of natural gas for many energy companies.

1.5.10 Floyd Shale

The Floyd shale, of Upper Mississippian origin, is comparable to the Barnett shale in the Fort Worth Basin. Within the Floyd shale's lower section is an organic-rich layer referred to as the Neal shale, which is a starved-basin deposit of great significance. This layer is the principal source rock for conventional hydrocarbon derivatives in the Black Warrior Basin.

The Floyd shale, a marine shale, is situated below the Mississippian Carter sandstone and above the Mississippian Lewis sandstone (EIA, 2011). The Carter and Lewis sandstones have historically been the most productive gas-producing zones in Alabama's Black Warrior Basin region. However, the Floyd shale has not been previously explored for production. The Chattanooga shale lies below the Floyd, typically separated from it in most areas by the Tuscumbia Limestone and the Fort Payne Chert. The Floyd shale, mostly composed of shale and limestone, extends from the Appalachian Thrust Belt of Georgia to the Black Warrior Basin of Mississippi and shows similarities to the prolific Barnett shale of the Fort Worth Basin and the Fayetteville shale of the Arkoma Basin, making it an area of significant interest for exploration. Floyd shale in Georgia is composed of strata analogous to the Tuscumbia Limestone. In Alabama and Mississippi, the Floyd is situated above the Tuscumbia Limestone, Pride Mountain formation, or

Hartselle Sandstone and below the initial sandstone layer in the Parkwood formation, with complex facies relationships. It should be noted that not all facies of the Floyd formation are viable gas reservoirs. The TOC content of the Floyd formation ranges from 0.5% to 10.2% by weight (Pawlewicz & Hatch, 2007).

1.5.11 Conasauga Shale

The Conasauga shale is a significant geologic formation dating back to the Cambrian period, and it is located in the Appalachian thrust and fault region that spans Alabama, Georgia, and Tennessee. The rock formation comprises shales and carbonates, which have undergone multiple orogenic events, resulting in a structural thickness of over 15,000 feet (4,600 m) in certain areas throughout the southern Appalachians. The shale is particularly thick, and as a result, the estimated natural gas reserves suggest that this field has the potential to be one of the largest in the world. This shale play was first discovered in the 1980s, and production began in 2005. The rock formation comprises shales and carbonates that have structurally thickened over time, providing an excellent opportunity for the exploration and production of natural gas.

The Conasauga shale overlies the Rome formation and, in turn, is overlain by the Knox formation. The Conasauga shale is present as far south as Talladega County. It extends north to Calhoun County, Cleburne County, St. Clair County, Etowah County, Cherokee County, and DeKalb County, Alabama, and Northwest Georgia. The shale has TOC ranging from 0.5% to 1.5%. The Schlumberger Montage shale log filed with the Alabama Oil and Gas Board calculates 330.2 bcf per square mile of gas in place for every 4270 feet thickness of shale. This area has a unique subsidence and thrust zone under it and is classified as a "mushwad."

1.5.12 Devonian Low Thermal Maturity Play and Greater Siltstone Shale Gas Play

The shale gas reserve known as the Devonian Low Thermal Maturity or Northwestern Ohio shale is in the Appalachian Basin, which spans Kentucky, New York, Ohio, Pennsylvania, Tennessee, and West Virginia. The Greater Siltstone formation is also located with the Appalachian Basin, which also spans New York, Ohio, Pennsylvania, Virginia, and West Virginia (Hackley et al., 2013). The USGS report of 2002 reveals that Devonian Low Thermal Maturity shale covers a total area of 45,844 square miles while the Siltstone shale play spans a total area of 22,914 square miles (EIA, 2011). In the realm of natural gas exploration, the Low Thermal Maturity formation boasts an average estimated ultimate recovery of 0.3 billion cubic feet per well, with a technically recoverable gas reserve of around 13.5 tcf. Meanwhile, the Devonian Greater Siltstone formation has an estimated ultimate recovery of 0.19 billion cubic feet per well and a technically recoverable reserve of approximately 8.5 tcf. These two formations, the Devonian Low Thermal Maturity, and the Greater Siltstone, hold immense potential for natural gas production.

The Low Thermal Maturity and Greater Siltstone shale plays have undergone thorough evaluation regarding key properties like depth, thickness, and porosity. It has been discovered that the Low Thermal Maturity shale can be found at a depth of 3,000 feet, while the Greater Siltstone shale is estimated to be at a depth of 2,911 feet. The Low Thermal Maturity shale has a thickness of 371 feet, whereas the Greater Siltstone shale is thicker at 623 feet. Regarding porosity, the Low Thermal Maturity shale has a 7.5% rating while the Greater Siltstone shale has a porosity of 5.8% (EIA, 2011).

1.5.13 Barnett Shale Gas Play

The Barnett shale, situated in the Fort Worth and Permian Basins in Texas, is the largest active onshore natural gas field in the United States. The field contains over 2 tcf of natural gas reserves and an estimated 30 tcf of natural gas resources. Despite being known for many years, the field's significance has risen sharply in recent years due to advancements in recovery techniques and high gas prices (Perryman, 2007). The production area covers a considerable portion of north and central Texas, with over 99.9% of the production occurring in a 14-county area until 2006. However, development is now expanding beyond this core region.

This vast natural gas reserve is sandwiched between two limestone formations—Marble Falls Limestone above and Chappel Limestone below. The shale is of Mississippian age and has been the site of over 10,000 wells drilled, making it the most significant shale gas play in the US. The development of the Barnett shale has been a showcase for modern tight-reservoir development techniques typical of gas shales in the US.

The total area of the Barnett is estimated as 5,000 square miles. This area is subdivided into two sections: the Greater Newark East Frac-Barrier Continuous Barnett Shale Gas (1,555 square miles) and the Extended Continuous Barnett Shale Gas (4,903 square miles) (Speight, 2013). The depth of Barnett shale ranges from 6,500 to 8,500 feet, while the thickness varies from 100 to 600 feet (Brathwaite, 2010; Islam, 2014). The productive region also increased as drilling and exploration expanded in the Barnett shale formation. Nonetheless, there are still vast unexplored areas of the Barnett shale. The Barnett shale gas play, comprising both the productive and unexplored zones, is estimated to have an average estimated ultimate recovery (EUR) of 1.4 bcf of natural gas per well and a total of approximately 43.37 tcf of technically recoverable natural gas (EIA, 2011). Barnett shale has porosity of 6% and low permeabilities of 250nD (Islam, 2014).

1.5.14 Hermosa Shale

The Hermosa shale play is a potential source of natural gas in the eastern Paradox Basin of Colorado and Utah. It consists of several shale formations, such as the Chimney Rock, Gothic, and Hovenweap shales, deposited in a marine environment during the Pennsylvanian and Permian periods. The Hermosa shale play has not yet been extensively explored or developed, but some outcrop analogs have been identified in the Hermosa Mountain area north of Durango, Colorado.

The Hermosa Group, located in Utah, is composed of black shale, which is made up of a combination of clay-size quartz, dolomite, and other carbonate minerals, as well as various clay minerals. The dominant type of clay in this shale is illite, with small amounts of chlorite and mixed-layer chlorite-smectite also present, as reported by Hite et al. (1984).

The black shale of the Hermosa Group is mainly concentrated in the northeastern section of the Paradox Basin, specifically within the fold-and-fault belt region. This area is characterized by a complex system of geological structures formed by tectonic activity over millions of years. This area is characterized by thick halite deposits in the Paradox formation, forming narrow salt walls and broad interdome depressions. As we move towards the southwest of this strategically controlled structural zone, the black shale intervals become thinner and less frequent, and they lack the exceptional seals provided by the halite. This region encompasses the eastern parts of Wayne and Emery Counties, the southern part of Grand County, and the northeastern third of San Juan County (Schamel, 2006). The kerogen in the shale is predominantly gas-prone humic type III and mixed type II-III (Nuccio & Condon, 1996).

The Paradox formation has been a sought-after exploration target for nearly a century, with a thickness ranging from 500 to 5,000 feet. The Cane Creek has produced more than 9.0 million barrels of oil (MMBO). In December 2018, only 22,000 barrels of oil (BO) were produced from 29 wells. Despite this, the US Geological Survey reports that Cane Creek has a mean undiscovered resource of 215 MMBO and 4,723 bcf of gas, making it a highly valuable target for future exploration efforts (Speight, 2019).

1.5.15 Avalon and Bone Springs Shale Oil Play

The Avalon shale is above the First Bone Spring Sand and underneath the Bone Spring Lime. The Leonard-Avalon play ranges in depth from 6,000 to 13,000 feet in the Permian Basin area. The Bone Spring formation comprises different types of rocks, including siliciclastic, carbonate, and shale rocks, with a total height of up to 4,000 feet. The formation is divided into several sandstone and carbonate intervals, with the sandstone members at the base of each interval being labeled as the First, Second, and Third Bone Spring. Additionally, the First Bone Spring Carbonate contains the Avalon shale (EIA, 2019).

The Avalon and Bone Springs area has an active surface of approximately 1,313 square miles. The shale in this area has good TOC, which ranges from 5% to 6%, and low permeability, generally less than 0.1 millidarcy. The shale has a thickness ranging from 900 to 1,700 feet and a reported depth of 6,000 to 13,000 feet. The porosity is approximately 16% to 18%, and each well has an estimated ultimate recovery (EUR) of 300 MBO and an estimated 1.58 billion barrels of technically recoverable oil (EIA, 2011).

The play produces oil and gas from layers of rocks that contain a mixture of organic-rich and organic-lean materials. These layers are found throughout a 3,500-foot section, including carbonate debris flows and turbidite reservoirs. The layers are shale or carbonate, intermixed with sandstone or carbonate. According to the 2018 US Geological Survey (USGS) report, the Bone Spring and Avalon assessment units in the Delaware Basin are estimated to contain undiscovered continuous hydrocarbon resources. The report indicates that these resources include 14 billion barrels of oil, 32 trillion cubic feet of natural gas, and 2.3 billion barrels of natural gas liquids in one unit and 2.7 billion barrels of oil, 27.5 trillion cubic feet of natural gas, and 2.8 billion barrels of natural gas liquids in the other (Gaswirth et al., 2018; EIA, 2019).

1.5.16 Hilliard-Baxter-Mancos Shale Gas Play

The Hilliard-Baxter-Mancos shale gas play is located in Wyoming and Colorado, covering an area of 16,416 square miles. The shale gas play has an average expected ultimate recovery of 0.18 billion cubic feet per well and approximately 3.77 trillion cubic feet of technically recoverable gas. The shale ranges from 10,000 to 19,500 square miles in depth and is 2,850 to 3,300 feet thick.

In the Hilliard-Baxter-Mancos shale gas play, the target formation is the organic-rich Mancos shale, which lies beneath the Hilliard and Baxter shale. This play is characterized by complex geology, high pressure, and low permeability (EIA, 2011). The play has been in development since the late 1990s, but it has faced challenges such as low gas prices, environmental regulations, and infrastructure limitations. However, the play has the potential for horizontal drilling and hydraulic fracturing, which could improve natural gas recovery.

1.5.17 Neal Shale

The Neal shale is an organic-rich facies of the Upper Mississippian–age Floyd shale formation located in the Black Warrior Basin of Mississippi and Alabama and has long been recognized as the principal source rock that charged conventional sandstone reservoirs (EIA, 2011; Speight, 2017). The play lacks a primary operator that can fully reveal its potential. In the Black Warrior Basin, the Neal shale is predominantly found in the southwest region and shares a facies relationship with the Pride Mountain formation, Hartselle Sandstone, the Bangor Limestone, and the lower Parkwood formation. These strata are in close proximity to the Neal shale and contribute to the overall geological makeup of the area.

The shale play represents a promising source of unconventional petroleum, boasting thickness ranging from 25 to 350 feet, TOC levels of 2.27% by weight, and impressive porosity. Of particular interest is the Neal shale, which holds vast resource potential, with an estimated 460 tcf of adsorbed gas and recoverable free gas estimates ranging from 227 to 4,943 bcf (Legg et al., 2016).

1.5.18 Huron Shale

Upper Devonian in age, the shale is an unconventional natural gas play in the Appalachian Basin. It includes the Huron, Cleveland, and Rhinestreet formations. The shale is a stacked interval roughly three times as thick as the Marcellus shale, which sits just below it. Intervals within the Upper Devonian include the Cleveland, Dunkirk, Geneseo/Burket, Middlesex, Pipe Creek, and Rhinestreet shales. Most early unconventional drilling in the formation has thus far targeted the Geneseo/Burket shales.

The Huron shale formation is a rock layer that lies underground, between 1,000 and 7,000 feet deep. It can be found in West Virginia, Ohio, and Northeast Kentucky. Most of the development and production of this resource is happening in West Virginia. The thickness of the rock layer varies between 200 and 2,000 feet.

1.5.19 Lewis Shale Gas Play

The Lewis shale gas play is an area for natural gas exploration located in the San Juan Basin, spanning approximately 7,506 square miles across Colorado and New Mexico. The shale is between 1,640 and 8,202 feet deep and is about 200 to 300 feet thick. The shale gas play has an average EUR of 1.3 bcf per well and approximately 11.6 tcf of technically recoverable gas. The well spacing for Lewis is estimated at 200 acres per well. The shale play has a 3.5% porosity and a 1.4% weight of total organic carbon.

1.5.20 Bakken Shale

The Bakken shale oil play (Devonian to Early Mississippian) is located in the Williston Basin in Montana and North Dakota. It extends into Canada and hosts several rock layers that produce significant hydrocarbons. The shale play covers approximately 6,522 square miles in the United States. The Bakken formation has three members: lower shale, middle sandstone, and upper shale. Each member is bigger than the one below it. The lower and upper shale are consistent in their geology and are the source of crude oil. The middle sandstone, on the other hand, has varying thickness, geological, and petrophysical properties. The Bakken shale is unique because it is an oil reservoir of dolomite sandwiched between two shale layers. It is located at depths ranging from about 8,000 to 10,000 ft and produces oil, gas, and natural gas liquids (Speight, 2019).

The Bakken formation is a highly productive shale formation located in North America. It has an average TOC content of 8% by weight and can contain up to 30% by weight. This formation has been a source of hydrocarbons for many rock units in the basin, including the Middle Bakken and underlying Three Forks. The Bakken shale is a significant contributor to hydrocarbon production in the region, estimated to contain 3.65 billion barrels of oil, 1.85 trillion cubic feet of natural gas, and 148 million barrels of natural gas liquids. It is a valuable source of energy resources (EIA, 2011; Speight, 2019; USGS, 2008, 2012). The shale ranges in thickness from 4,500 to 7,500 feet, with a mean of 6,000 feet and an average thickness of 22 feet. The well spacing ranges from 320 to 1,280 acres per well with a mean of 640 acres per well and an average depth of 6,000 feet, thickness of 22 feet, and porosity of 8% (EIA, 2011).

1.5.21 Niobrara Shale and Pierre Shale

The Niobrara formation (late Cretaceous) is in the Denver-Julesburg Basin, spanning Northeast Colorado, Northwest Kansas, Southwest Nebraska, and Southeast Wyoming. It is overlain by the organic-rich Pierre shale (Soeder & Borglum, 2019).

These formations can be found at significant depths within a number of structural basins in the Western Interior Seaway (WIS), where thermal maturity levels are sufficient for the creation of thermogenic hydrocarbons. However, on the eastern platform of the WIS, the Niobrara is relatively shallow, measuring less than 3,000 feet, and the Pierre Shale above it is even shallower. Despite its shallowness, the Niobrara still contains significant biogenic gas resources.

The Niobrara formation is generally less than 300 feet thick across the WIS. In contrast, the Pierre shale can be as thick as 2,000 feet (Soeder & Borglum, 2019). The Niobrara has better reservoir properties than the Pierre due to its chalky nature. The Pierre, on the other hand, is a clay-rich, fissile, dense, muddy shale. The Pierre shale acts as a caprock and seals on top of the Niobrara, creating a stratigraphic trap for hydrocarbons generated and contained within the Niobrara. In this sense, the Pierre and Niobrara formations are unconventional- and conventional type reservoirs, conventionally trapping gas but requiring horizontal drilling and fracking. This exciting oil shale play is being compared to the Bakken shale resource in North Dakota (Speight, 2013). Rock-eval analyses of Niobrara cores revealed TOC contents as high as 6% by weight (Soeder et al., 2015).

1.5.22 Monterey Oil Play

The Monterey/Santos shale oil play, middle to late Miocene in age, is the largest shale oil formation in the United States, estimated to hold 15.4 billion barrels or 64% of the total shale oil resources in the country (EIA, 2011). The Monterey shale play is the main source rock for the conventional oil reservoirs in the Santa Maria and San Joaquin Basins of Southern California. The shale play remains active in an estimated 1,752 square miles in the San Joaquin and Los Angeles Basins. The Monterey formation is present in the San Joaquin Basin, reaching impressive thicknesses of over 2,000 feet and depths of up to 14,000 feet (Soeder & Borglum, 2019). The Monterey/Santos shale oil play has an average estimated ultimate recovery (EUR) of 550 MBO per well. It is believed that there are approximately 15.42 billion barrels of technically recoverable oil in this shale play. The shale play is located at a depth of 11,250 feet and has a thickness of 1,875 feet. The shale has a porosity of 11% and a total organic content of 6.5% by weight (EIA, 2011).

1.5.23 Pearsall Shale

The Pearsall shale underlies much of the prolific Eagle Ford shale and is a gas play at depths of 7,000 to 12,000 feet, with a thickness of 600 to 900 feet (Brathwaite, 2010). The western sections of the play are much deeper and gassier than the northeastern portions (Conant & Swanson, 1961). The Pearsall formation is in the dry gas window of hydrocarbon generation and contains moderate levels of total organic carbon. Operators have completed approximately 50 wells in the Pearsall shale in the Maverick Basin of south Texas (Speight, 2013).

The porosity of Pearsall shale ranges from 4% to 8%, while TOC content ranges from 0.2% to 2.97%. Pearsall shale has relatively low porosity and TOC content compared to other shale plays. However, it is still considered a potential shale gas resource as the organic matter in the shale can generate gas when heated.

1.5.24 Mancos Shale Gas Play

The Mancos shale gas play spans the Uinta Basin in Colorado and Wyoming, encompassing 4,217,000 acres (6,589 square miles). According to reports, the estimated ultimate recovery (EUR) for the Mancos formation, in comparison to other formations in the Uinta Basin, such as the Mesaverde and Wasatch, is approximately 1.0 bcf per well (EIA, 2011).

The Mancos shale play has a depth of 15,250 feet and a thickness of 3,000 feet, indicating a significant potential for oil and gas reserves. The porosity of the shale play has been measured at 3.5%, which is considered relatively high and favorable for the extraction of hydrocarbons. Additionally, the shale play's TOC has been measured at 14%, indicating a high concentration of organic materials that can be converted into oil and gas. Overall, the available data suggests that the Mancos shale play has great potential for future profitable oil and gas extraction.

1.5.25 Utica Shale

The Utica play spans about 60,000 square miles across Ohio, West Virginia, Pennsylvania, and New York. The Utica shale is a stratigraphical unit of Middle Ordovician age in the Appalachian Basin. It underlies much of the northeastern United States and adjacent parts of Canada.

According to a report by the US Energy Information Administration in 2012, the Utica shale reserves in the United States are believed to hold an estimated 15.7 trillion cubic feet of unproved, technically recoverable gas. On average, each well is estimated to produce 1.13 billion cubic feet of gas. The same year, the US Geological Survey conducted a study revealing that the Utica shale also contains an estimated 38.2 trillion cubic feet of undiscovered technically recoverable gas, 940 million barrels of oil, and 208 million barrels of natural gas. The Utica shale in Quebec potentially holds 4×10^{12} cft at 1×106 cft per day production rates. The Utica has been recognized as a petroleum source rock for associated conventional reservoirs for decades. It is about 500 feet thick, with organic content of 13%.

1.5.26 Chattanooga Shale

The Chattanooga shale of Devonian to Mississippian age spreads over an area of approximately 86,000 square miles across Alabama, Arkansas, Georgia, Kentucky, Missouri, and Tennessee (Swanson & Conant, 2010). The Chattanooga shale is a massive flat-lying, siliceous, pyritic marine black shale, on average about 100 feet thick and located at a depth from 300 to 6,400 feet. The thickness of the Chattanooga formation exhibits notable variation across the Black Warrior Basin. The shale layer measures less than 10 feet thick and is sporadically absent in many Lamar, Fayette, and Pickens Counties regions.

The Chattanooga shale rests unconformably on Leipers Limestone and is overlain by the Maury formation. It includes three members. At the base, the thin Hardin Sandstone member, which is only locally developed, is of particular significance as it pertains to the conventional extraction of oil and gas from the basin (Speight, 2013). The Chattanooga shale has a belt that extends northwestward from Blount County into Franklin and Colbert Counties. In Tuscaloosa and Greene Counties, the shale is thicker than 30 feet and can reach up to 90 feet. The shale's TOC content ranges from 2.4% to 12.7 % by weight.

1.6 CANADIAN SHALE PLAYS

The production of hydrocarbons from Canadian shales began in 2005, but it was not until recent years that it has significantly increased (Rivard et al., 2014). Currently, most natural gas is being produced from Devonian shales in the Horn River Basin and from the Triassic Montney shales and siltstones in northeastern British Columbia. In addition, the Devonian Duvernay formation in Alberta is also being explored for natural gas. Other shales that show potential for natural gas production are being evaluated, including the Upper Ordovician Utica shale in southern Quebec and the Mississippian Frederick Brook shale in New Brunswick, located in eastern Canada. Evaluating these shales is an important step towards meeting the increasing demand for natural gas in Canada and reducing the country's dependence on imports.

Canada has several prominent shale gas plays, including the Horn River Basin and Montney shales in the northeastern British Columbia. Their characteristics are in Table 1.4. Additionally, the Colorado Group spanning Alberta and Saskatchewan, the Utica shale in Quebec, and the Horton Bluff shale located in New Brunswick and Nova Scotia are among the major shale gas plays in the country (NEB, 2009). Other shales with significant potential include the Lower Besa River shale and the Duvernay shale. Some Canadian shales are discussed next.

1.6.1 Horn River Basin

The Horn River Basin covers 7,100 square miles in northern British Columbia and the Northwest Territories. The basin is bordered to the west by the Bovie Fault, which separates it from the Liard Basin. The shale section thins out to the north in the Northwest Territories while the shale deposit's pinch-out defines the southern border. The Slave Point/Keg River Uplift and the thinning of the shale deposit determine the eastern border.

According to Speight (2019), the Horn River Basin shale formations are Devonian shale that was deposited in deep waters at the foot of the Slave Point carbonate platform in northeast British Columbia. This region has been producing conventional natural gas for many decades. These shale formations are approximately 450 feet thick and are rich in silica (approximately 55% v/v silica), with a total organic content of 1% to 6 %. The rocks are mature and have been heated far into the thermogenic gas window. The most notable shale formation in the Horn River Basin contains a series of organic-rich shales, including Upper and Middle Devonian Muskwa/Otter Park and Evie/Klua (Dong et al., 2017). These two shale units were mapped in the Horn

TABLE 1.2
Characteristics of Some Selected Shale Gas Plays in the US

	Shale Gas Play	Age	Location	Area (mi²)	EUR (Bcfe/well)	TRR (Tcf)	Depth (Ft)	Thickness (m)	Porosity %	TOC (% wt)
1	Marcellus	Devonian	New York, Ohio, Pennsylvania	10,622	3.5	177.90	6,750	125	8	12
2	Devonian Big Sandy	Devonian	Virginia, Kentucky, and central Ohio	8,675	0.325	6.47	3,800	175	10	3.75
3	Devonian Low Thermal Maturity		New York, Ohio, Pennsylvania, Virginia	45,844	0.2942	13.53	3,000	371	7	-
4	Greater Siltstone			22,914	0.193	8.46	2,911	623	5.8	-
5	New Albany	Devonian-Mississippian	Illinois Basin, Illinois, Indiana	1,600	1.1	10.95	2,750	200	12	13
6	Antrim	Late-Devonian	Michigan Basin, Michigan	12,000	0.28	19.93	1,400	95	9	11
7	Haynesville	Jurassic	Louisiana, east Texas	3,574	6.5	53.30	12,000	250	8.5	2.25
9	Floyd-Neal/Conasauga	Late Mississippian	Black Warrior Basin, Alabama, Mississippi	2,429	0.9	4.37	8,000	130	1.6	1.8
10	Fayetteville (Western)	Mississippian	Arkoma Basin,	5,000	1.15	4.64	4,000	110	5	6.9
	(Central)		Arkansas	4,000	2.25	27.32				
11	Woodford (Western)	Devonian-	Oklahoma, Texas	2,900	4.0	19.26	9,500	150	7	6.5
	(Central)	Mississippian		1,800	1.0	2.95	5,500	250	6	4
12	Cana Woodford			688	5.2	5.77	13,500	200	7	6
13	Barnett	Mississippian	Fort Worth and Permian basins, Texas	4,075	1.6	23.81	7,500	300	5	
15	Baxter	Late Cretaceous	Vermillion Basin, Colorado	16,416	0.18	3.77	14,750	3,073	4.25	1.75
16	Lewis	Late Cretaceous	Colorado, New Mexico	7,506	1.3	11.63	4,500	250	3.5	–
	Mancos	Cretaceous	San Juan Basin, New Mexico, Uinta Basin, Utah	6,589	1.0	21.02	15,250	3,000	3.5	14
17	Niobrara shale	Late Cretaceous	Denver Basin, Colorado							

Source: EIA, 2011; GWPC, 2009; Islam, 2014; Speight, 2019

TABLE 1.3
Characteristics of Some of the Shale Oil Plays in the US

Shale Gas Play	Age	Location	Area (mi^2)	EUR (MBO/well)	TRR (BBO)	Depth (Ft)	Thickness (m)	Porosity %	TOC (% wt)
Eagle Ford	Late Cretaceous	Maverick Basin, Texas	2,233	300	3.35	7,000	200	9	4.25
Avalon and Bone Springs			1,313	300	1.58	8,750	1,300	–	–
Bakken Shale			6,522	550	3.59	6,000	22	8	–
Monterey/Santos			22,914	0.193	8.46	2,911	623	5.8	–

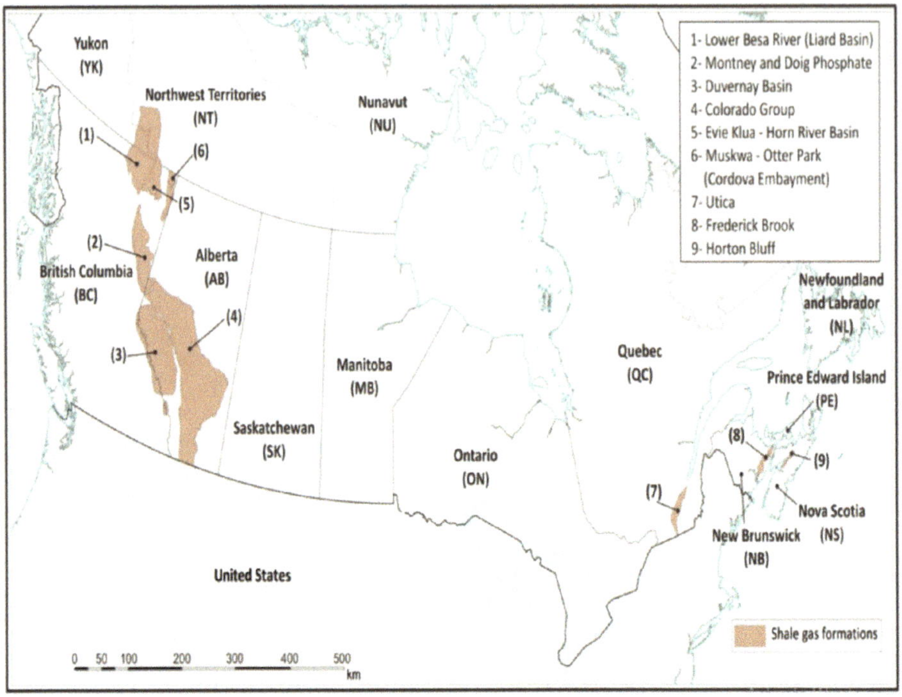

FIGURE 1.4 Shale gas formations in Canada containing dry gas. Note that, the most promising are found in five provinces: British Columbia (BC), Alberta (AB), Quebec (QC), New Brunswick (NB), and Nova Scotia (NS).

Source: Rivard et al., 2014

River Basin to establish a prospective area with sufficient thickness and resource concentration favorable for shale gas development. The British Colombia Ministry of Energy and Mines reported that the medium case estimates for gas in place and marketable natural gas in the Horn River Group shale are 448 and 78 tcf, respectively.

1.6.1.1 Muskwa/Otter Park Shale

Muskwa/Otter Park shale, Devonian in age, is located in the Horn River in northeast British Columbia. The shale is the main shale gas target in the Horn River Basin (EIA, 2013). The Muskwa/Otter Park Shale can be found at varying depths in the prospective area, ranging from 6,300 to 10,200 feet and averaging at around 8,000 feet (EIA, 2013). The Muskwa/Otter Park shale boasts moderate overpressure in the basin's center. Muskwa/Otter Park shale is rich in organic matter, with a gross thickness of 420 feet and a net pay of 380 feet. The net shale thickness investigated reveals an average TOC of 3.5% in the prospective area. With a high thermal maturity averaging about 3.5%, this shale gas falls within the dry gas window. However, the in-place shale gas contains 11% CO_2 due to the high thermal maturity of the prospective

TABLE 1.4
Characteristics of Some Prominent Canadian Gas Shale Sources

Parameters	Horton Bluff	Montney	Colorado	Utica	Horn River
Depth (m)	1,120 to 2, 000+	1,700 to 4, 000	300	500 to 3, 300	2,500 to 3, 000
Thickness (m)	150+	up to 300	17 to 350	90 to 300	150
Total organic content (%)	10	1 to 7	0.5 to 12	0.3 to 2.25	0.5 to 6.0
Gas-filled porosity (%)	2	1.0 to 6.0	less than 10	2.2 to 3.7	3.2 to 6.2
Silica (%)	38	20 to 60	sand and silt	5 to 25	45 to 65
Maturity (R_0)	1.53 to 2.03	0.8 to 2.5	biogenic	1.1 to 4	2.2 to 2.8
Calcite or dolomite (%)	significant	up to 20	-	30 to 70	0 to 14
Clay (%)	42	less than 30	high	8 to 40	20 to 40
Free gas (%)	-	64 to 80	-	50 to 65	66
Adsorbed gas (%)	-	20 to 36	-	35 to 50	34
GIP/section (Bcf)	72.4 to 600+	8 to 160	22 to 62	25 to 210	60 to 318+
Play area GIP (Tcf)	> 130	80 to 700	> 100	> 120	144 to 600+

Source: NEB, 2009

region. Additionally, the Muskwa/Otter Park shale is characterized by high quartz and low clay content (EIA, 2013).

The Muskwa/Otter Park and shale region is known for its significant concentration of valuable resources, estimated to be around 151 billion cubic feet per square mile. When considering the favorable mineralogy and other properties of the reservoir, EIA (2013) estimated that there may be as much as 94 trillion cubic feet of technically recoverable shale gas in the Muskwa/Otter Park Shale region. Additionally, a risked gas in place of 376 trillion cubic feet has been identified, not including CO_2. These impressive figures highlight the potential for substantial energy production in this area.

1.6.1.2 Evie/Klua Shale

The Evie/Klua shale found in the Horn River Basin is a promising shale gas source. It is located about 500 feet below the top of the Muskwa/Otter Park shale, separated by a non-organic rock layer. The Evie/Klua shale is around 160 feet thick and has a high organic matter content, with an average TOC of 4.5%. Its thermal maturity, or R_0, is high at about 3.8%, placing this shale gas in the dry gas window. The CO_2 content is estimated at 13%. The low clay content of the Evie/Klua shale makes it a good candidate for hydraulic stimulation. An EIA report (2013) revealed that Evie/Klua shale has a resource concentration of 62 bcf/mi^2 and 154 tcf of risk gas in place, excluding CO_2, with technically recoverable shale gas resource of 39 tcf.

1.6.2 Lower Besa River Shale

Lower Besa River shale, Devonian in age, is the dominant shale gas formation in the Liard Basin and is equivalent to the Muskwa/Otter Park and Evie/Klua shales in the Horn River Basin (EIA, 2013). Additionally, the basin's Upper Devonian– and Mississippian-age shales, including the Middle Besa River shale, which is equivalent to Fort Simpson, and the Upper Besa River shale, which is equivalent to Exshaw/Banff, contain less organic matter and are less promising. Despite the limited data available on this shale play, experts have mapped a prospective area of 3,300 square miles for the Lower Besa River shale in the central region of the basin (EIA, 2013).

The shale in this region can be found at depths ranging from 6,600 to 13,000 feet, with an average depth of 10,000 feet. The Lower Besa River section is rich in organic matter; the gross thickness is 750 feet, and the net thickness is 600 feet. The prospective area boasts a TOC of up to 5% locally, with an average of 3.5% for the investigated net shale interval. With an average R_0 of 3.8%, the thermal maturity of the area is high, resulting in an estimated CO_2 content of 13% in the in-place shale gas (EIA, 2013). The geology of the Besa River shale is intricate, with numerous faults and thrusts. The Lower Besa River shale is rich in quartz, with intermittent dolomite and more extensive clay intervals.

1.6.3 Utica Shale

The Utica shale in the Upper Ordovician between Montreal and Quebec City was formed in deep waters near the Trenton carbonate platform. During the early Appalachian Mountain growth, the shale became faulted and folded on its southeastern side. With a thickness of around 150 meters and a 1% to 3% total organic content, the Utica has been a known petroleum source rock for conventional reservoirs for many years. However, unlike other Canadian gas shales, the Utica contains higher calcite levels, reducing its silica content. While calcite is still brittle, hydraulic fractures do not transmit as easily through it.

1.6.4 Colorado Shale

The Colorado Group comprises several shale-containing formations that were deposited across southern Alberta and Saskatchewan during the middle Cretaceous period when sea levels were high worldwide (Beaton et al., 2008). Notable among these formations are the Medicine Hat and Milk River sandstones, which have been sources of natural gas for more than a century, and the Second White Speckled shale, which has been yielding natural gas for several decades.

The Colorado Group is a formation that stretches from southwestern Alberta to the Manitoba Escarpment. It thins out as it goes east, with a height of about 700 meters in the west and 200 meters in the east. In northwest Alberta, the Colorado Group is over 1500 meters thick because it sits on top of the Peace River Arch, which sank during much of the Cretaceous period (Halbertsma, 1994). Compared to shale formations in the Horn River Basin and the Utica Group of Quebec, the Colorado Group shale produces through thin sand beds and lamina, making it a hybrid gas shale similar to the Montney shale. Moreover, the gas produced in the Colorado shale has biogenic origins instead of thermogenic, which suggests low potential for natural gas liquids and an underpressured reservoir. This makes hydraulic fracturing more challenging. Additionally, the shale formations in the Colorado Group are sensitive to water, which makes them responsive to fluids used during hydraulic fracturing. Estimating the total volume of gas in the Colorado Group is challenging due to the shale's vast lateral extent and reservoir variability and the lack of independent and publicly available analyses. Nevertheless, it is believed that there may be a minimum of 100 tcf of gas in place (EIA, 2013).

Within the Colorado Group, several shales possess significant hydrocarbon-generating potential. These include the First and Second White Speckled shales, the Fish Scales zone, and the shale located at the base of the Shaftesbury formation (Halbertsma, 1994). These shales are characterized by high total organic carbon contents, making them valuable resources for energy production. The depth of the Second White Speckled (2WS) and Fish Scale shales varies between 5,000 and 10,000 feet in the prospective area, with the Fish Scale shale being around 200 feet deeper than the 2WS. The interval between the top of the 2WS and the bottom of the Fish Scales Shale ranges from 300 to over 1,000 feet, with an average of 523 feet. The shale's total organic carbon content ranges from 2% to 3%. In this area, the shale's thermal maturity is low (R_0 of 0.5% to 0.6%). However, the presence of biogenic gas seems to have produced adequate volumes of gas generation. The shale's rock mineralogy is low to moderate in clay (31%), making it favorable for hydraulic fracturing.

1.6.5 Duvernay Shale

The Duvernay shale play (Devonian-Frasnian) is a geological formation in Alberta, Canada, containing natural gas and oil. It is part of the Western Canadian Sedimentary Basin, which covers most of Alberta and Saskatchewan and parts of British Columbia, Manitoba, and the Northwest Territories (NEB, 2009).

The Duvernay shale, rich in organic materials, can be found in the East and West Shale Basin. The basin is located in central Alberta and covers an area of over 50,000 square miles (EIA, 2013; NEB, 2009). The Deformed Belt marks the western boundary of the shale deposit, while the Peace River Arch defines the northern boundary, and the Leduc Shelf defines the southern boundary. The Grosmont Carbonate Platform serves as the eastern boundary. In this extensive area of shale deposits, the Duvernay shale is most promising within a 23,450-square-mile region, primarily situated in the central and western parts of the basin.

The Duvernay shale, which existed during the Upper and Middle Devonian period and is located just north of the Montney shale, is widely distributed throughout central Alberta. However, it is absent in areas where Leduc reef growth occurred, except for beneath the Duhamel reef, where it may be present as a thin layer of calcilutite. This type of rock comprises non-siliceous calcareous flour and is typically dolomite or limestone. Stratigraphically, the Duvernay shale is comparable to the Muskwa shale, which is situated in northwest Alberta and northeast British Columbia (EIA, 2013; NEB, 2009). The Duvernay shale is predominantly composed of organic-rich limestone within the East Shale Basin. Conversely, the West Shale Basin exhibits a gradient in composition, ranging from a carbonate-rich mudstone in the eastern region to a progressively porous and organic-rich shale in the western region. The Duvernay shale–type section in the East Shale Basin is 174 feet thick. It thickens to 246 feet east and southeastward towards the Southern Alberta Shelf. In the subsurface, towards the northeast, the formation reaches 394 feet at its truncation at the pre-Cretaceous unconformity. In the West Shale Basin, it is an average of 197 feet thick and thickens northward, reaching over 820 feet east of Lesser Slave Lake.

The Duvernay shale play is divided into two geological sub-plays: the Fox Creek sub-play in the west and the Innisfail sub-play in the east. The Fox Creek sub-play is more developed and highly productive than the Innisfail sub-play. The large shale potential resource base located in the Liard Basin, Horn River Basin, Cordova Embayment, and deeper portions of the Montney formation is almost all dry gas (Slattery & Atkinson, 2012).

The Duvernay shales are dominated by marine-derived type-II organic matter with TOC content of 0.1% and 11.1%; the highest values are generally found in the East Shale Basin (Slattery & Atkinson, 2012). However, when the organically lean rock is excluded using the net-to-gross ratio, the average TOC drops to 3.4%. The thermal maturity (R_0) of the shale varies based on depth, with the east exhibiting immature shale (below 0.8% R_0) while the west is characterized by dry gas (1.3% to 2% R_0) (EIA, 2013). This suggests that the Duvernay shale has significant oil potential in the east, a promising wet gas/condensate area in the center, and a smaller but notable dry gas area in the west.

Based on petrophysics calibrated to core and cuttings samples, the Duvernay is characterized by porosity of 6.0 to 7.5% and permeability of 236 to 805 nD, with low clay content (26% w/w) (Fowler et al., 2001) on the southeastern side. According to the NEB (2017) report, the Duvernay shale's economic oil resource is 156 million m^3 or 1.0 billion barrels while the Duvernay shale's economic natural gas resource is 339 billion m^3 (12.0 tcf). The Duvernay shale's economic natural gas liquid (NGL) resource is 216 million m^3 (1.4 billion barrels).

1.6.6 Horton Bluff Group Shale

The Horton Bluff shale is an Early Mississippian geological formation in Nova Scotia, Canada. It is composed of shale, siltstone, and sandstone and is part of the Maritimes Basin (Ferguson & Colwell, 1992; NEB, 2009). The Middle Shale Unit of the Horton Bluff formation is considered a lake deposit. It consists of dark-colored laminated and unlaminated silty shale and siltstone with interbeds of dolostone and sandstone (Ferguson & Colwell, 1992). The Horton Bluff shale is a budding play in the Maritimes Basin and features the Upper Devonian and Lower Mississippian Horton Bluff shale. The shale has high-quality properties, with total organic carbon contents of around 10% and well above 50% brittle minerals, mostly silica, which means the rock can be fractured (Ferguson & Colwell, 1992).

The Frederick Brook shale found in the Horton Bluff Group in New Brunswick has a high silica content of 38% v/v but also a high clay content of 42% v/v. In Nova Scotia, the Frederick Brook member has an organic content of 10% v/v, higher than other Canadian gas shale formations. The pay zone is over 500 feet thick and can reach up to 2,500 feet in New Brunswick (Speight, 2019). However, most of the gas is attached to the clay and organic matter, which makes it difficult to extract. Reservoir stimulation is necessary to achieve significant production from Nova Scotia shale formations. It is unclear how much of the gas is attached to the clay and organic matter in the New Brunswick shale formations (Speight, 2019).

The analysis of Horton Bluff shale indicates that 67 tcf of free gas in place is present in the Frederick Brook shale of the Sussex/Elgin sub-basins of southern New Brunswick, and 69 tcf of gas is present on the Windsor land block in Nova Scotia (Speight, 2019).

1.6.7 Montney Shale

The Montney shale formation is a deposit of shale rock located deep underneath east-central British Columbia, Canada. It can be found in the Dawson Creek area, just south of the Horn River and Duvernay shale formations. This tight shale formation contains large amounts of natural gas (Williams & Kramer, 2011). The Montney Formation in northeastern British Columbia offers diverse depositional environments, including shallow-water sands to the east and offshore muds to the west. Natural gas production is currently taking place from conventional shallow-water shoreface sandstones at the eastern edge of the Montney and deep-water tight sands at the foot of the ramp. Overall, the Montney formation presents a promising opportunity for natural gas development in the region.

The Montney shale formation is a valuable gas resource found in Canada. It is a mixture of tight and shale gas and can be over 1,000 feet thick in some areas. The formation comprises sandy mudstone and dates back to the Triassic period. It can be found beneath the Doig formation at depths ranging from 5,500 to 13,500 feet. The Montney formation consists of both upper and lower zones, and due to their differing mineralogy, the formation evaluation data is impacted differently for each zone. The lower Montney is particularly challenging to assess, as conventional logging techniques have historically led people to believe that the lower Montney is very tight. However, core data has revealed higher porosity than expected, despite its having less porosity than the upper Montney formation (Speight, 2019; Williams & Kramer, 2011).

The Montney formation is estimated to hold an immense amount of unconventional petroleum resources, with a projected total of 449 trillion cubic feet of marketable natural gas, 14,521 million barrels of marketable natural gas liquids, and 1,125 million barrels of marketable oil. This vast reserve presents a significant opportunity for energy production and could play a crucial role in meeting global energy demands in the future (NEB, 2009, 2013; Speight, 2019). According to Speight (2019), the Montney Formation boasts abundant silt and sand, which share similarities with tight gas. Interestingly, the natural gas found within the formation originates from its organic matter, much like shale plays. However, because of the prevalence of siltstone and sand, the Montney formation's permeability is severely limited, necessitating greater levels of fracture stimulation.

1.6.8 Utica Group

The Utica shale, found between Montreal and Quebec City, was formed in deep waters at the base of the Trenton carbonate platform during the Upper Ordovician period. It is about 150 meters thick and has a 1% to 3% organic content. The Utica shale is known for its oil and gas reserves, but it has a higher concentration of calcite than other Canadian gas shales (Thériault, 2008). This means that hydraulic fractures do not transmit as well through calcite, even though it is still brittle. During early Appalachian Mountain growth, the shale was folded and faulted on its southeastern side (NEB, 2009).

In Utica, biogenic gas is commonly found in relatively shallow areas, whereas medium-deep and structured shales are known to contain thermogenic methane. This means that depending on the location and characteristics of the shale formation, different types of gas can be found, each with its unique properties and potential uses (NEB, 2009). The reservoir has an advantage over others in that it is folded and faulted, which increases the potential for natural fractures.

1.7 SHALE PLAYS AND THEIR CHARACTERISTICS THROUGHOUT THE WORLD

Shale plays across the globe exhibit diverse geological attributes, which are determined by many factors, such as the region's depositional environment, sediment source, and tectonic history. A shale play comprises identified (proved), unidentified (probable), and possible oil and gas reservoirs with similar geological characteristics. These reservoirs are located in geological basins and often contain a mix of sedimentary accumulations, including shale, carbonate, and sand. It is worth noting that there are several other shale plays worldwide, each with its own unique set of characteristics.

Shale formations across the globe may contain an estimated 23,000 tcf of natural gas, according to research (Rajput & Thakur, 2016). A range of geochemical techniques, including TOC, gas volume, and capacity measurements, are extensively used to evaluate the characteristics of these shale reservoirs. The shale formations' TOC values vary significantly from region to region, with values between 2% and 10%. A TOC value greater than 4% is considered to be a promising prospect (Brooks et al., 2011). In addition, petrophysical properties, such as gamma ray, resistivity, porosity, and acoustic measurements, are also used for shale gas exploratory studies to determine porosity and gas concentration in the pore spaces (increase in resistivity), fracture distribution, and fluid migration.

Table 1.5 illustrates a comprehensive overview of the key characteristics of various shale plays worldwide. It provides a detailed analysis of each shale play's geological, geographical, and environmental features, along with the estimated reserves, production rates, and major players operating in the area. The table also highlights the challenges and opportunities of shale exploration and production in each region.

TABLE 1.5
Characteristics of Shale Plays around the World

Country	Basin	Shale Play	Age	Prospective Area (mi²)	Thickness (ft)	Depth (ft)	Cay Content	Environment of Deposition	TOC (%wt)	Thermal Maturity[b] (R_o)	Gass in Place (tcf)	TRR (Tcf)
Argentina	Neuquén	Vaca Muerta	L. Jurassic	3,550	500	8,000	Low/ Medium	Marine	8	1.25	645.1	193.5
	Atral-Magallanes	L Inoceramus-Magnas Verdes	E. Cretaceous	4,620	800	13,500	Low/ Medium	Marine	1.6	1.3	303	75.6
Venezuela	Maracaibo/ Catatumbo	La Luna/Capacho	L. Cretaceous	5,840	1000	12,000	Low	Marine	5.6	1.25	522.5	130.7
Colombia	Llanos	Gacheta	U. Cretaceous	1,820	600	10,000	Low	Marine	2	0.85	18.2	1.8
Brazil	Parana	Ponta Grossa		22,840	1,000	14,000	Low/ Medium	Marine	2.5	1.6	260.4	50.1
Mexico	Burgos	Eagle Ford	L. Cretaceous	6,700	300	10,500	Low	Marine	5	1.3	767.5	230.2
	Tampico	Pimienta	Jurassic	9.000	500	6,200	Low	Marine	2	1.5	58.5	4.7
China	Sichuan	Longmaxi	Silurian	56,875	300–1600	10,700	Low	Marine	3	2.4	1,373	343
	Tarim	L. Cambrian	L. Cambrian	6,520	380	14,620	Low	Marine	2	2	175.9	44.0
UAE	Rub' Al-Khali	Shilaif	M. Cretaceous	9,670	660	6,500	Medium		4	0.7	367.5	22.04
India/ Pakistan	Cambay	Cambay	U. Creta.-Tertiary	580	1,500	14,500	Low/ Medium	Marine	3	1.1	78.8	19.8
	Lower Indus	Sembar	E. Cretaceous	31,320	1,000	13,000	Low	Marine	2	1.25	310	62.2
Indonesia	Bintuni	Aifam	U. Miocene	3,340	1,000	114.3	Low	Marine	1.5	1.5	9,500	28.6
Jordan	Hamad	Batra	Silurian	3,300	160	8,500	Medium	Marine	2	1.3	33.4	6.7
Algeria Murtania	Tindouf	L. Silurian	L. Silurian	12,380	68	11,000	Medium	Marine	4	1.5	55.5	13.6

(continued)

TABLE 1.5 (continued)
Characteristics of Shale Plays around the World

Country	Basin	Shale Play	Age	Prospective Area (mi^2)	Thickness (ft)	Depth (ft)	Cay Content	Environment of Deposition	TOC (%wt)	Thermal Maturity[b] (R_o)	Gass in Place (tcf)	TRR (Tcf)
Libya	Sirte	Sirte/Rachmat	U. Cretaceous	35,240	2,000	11,000	Medium		5	1.15	349.8	28.0
South Africa	Karoo	Whitehill	L. Permian	60,180	200	8,000	Low	Marine	2.5	2.25	845.8	211.3
Egypt	Abu Gharadig	Khatatba	M. Jurassic	6,840	1,500	12,000	Low/ Medium	Marine	4	1.15	352.7	66.1
Chad	Doseo	L. Cretaceous		5,530	660	12,000	Medium	Marine			165.4	13.2
Tunisia	Ghadames	Frasnian	M. Devonian	850	197	9,500	Medum	Marine	4.2	1.15	44.1	8.8
Algeria	Timimoun	Frasnian	Silurian	32,040	200	6,000	Medium	Marine	3	1.15	467.1	93.4

Country	Basin	Shale Play	Age	Area (mi²)	Thickness (ft)	Depth (ft)	Clay content	Environment of Deposition	TOC (%wt)	Thermal maturity (R_o)	Gass in Place (Tcf)	TRR (Tcf)
Poland	Baltic	Llandovery	L. Ord. -Silu.-Cambrian	3,030	575	8,000	Medium	Marine	3.9	2.85	24.5	2.4
France	Paris	Permian-Caboniferous		17,940	500	14,000	Medium	Lacustrine	9.0	1.60	352	70.4
Russia	West Siberian	Bazhenov Central	U. Jurassic—L. Cretaceous	116,200	100	8,200	Low	Marine	10	0.85	1,196	143.5
Bulgarria, Romaania, Ukraine	Dniepr-Donets	L. Carboniferous	L. Carboniferous	1,460	700	11.000	Low	Marine	4.5	0.9	13.3	0.66
Spain	Basque-Cantabrian	Jurassic	L-M. Jurassic	2.100	600	11,000	Medium	Marine	3	1.15	41.8	8.4
Germany	Lower Saxony	Posidonia	L. Jurassic	1,390	100	14.500	Low/ Medium	Marine	8	0.85	56,5	11.8
Australia	Goldwyer	Canning	M. Ordovician	22,880	1,300	13,500	Low	Marine	3	0.8	748.7	148.7
	Cooper	Roseneath-Epsilon-Murteree (Nappamerri)	Permian	625	8,500	6.000	Low	Lacustrine	2.6	0.85	10.6	0.63
UK	North UK	Carboniferous	Permian	5,100	820	8,500	Medium	Marine	3	1.3	125.6	25.1
	South UK	Lias	L. Jurassic	1,735	165	5,000	Medium	Marine	3	0.85	8	0.6
Turkey	SE Anatolia	Hamitabat	M. Eocene	2,950	328–1,300	8,200	medium	Marine	3.5	1.75	43	9

REFERENCES

Ahmed, U., & Meehan, D. N. (2016). Unconventional oil and gas resources: Exploitation and development. *Unconventional Oil and Gas Resources: Exploitation and Development*, 1–862. https://doi.org/10.1201/b20059

Ahmed, U., Meehan, D. N., & Hughes, B. (2016). Characteristics of unconventional oil and gas resources. *Unconventional Oil and Gas Resources: Exploitation and Development*. https://doi.org/10.1201/b20059-5

Alexander, T., Baihly, J., Boyer, C., Clark, B., Waters, G., Jocken, V., Le Calvez, J., Lewis, R., Miller, C., Thaeler, J., Toelle, B. (2011). Shale gas revolution. *Oilfield Review*, *23*(3), 40–55.

Beaton, A. P., Pawlowicz, J. B., Anderson, S. D. A., & Rokosh, C. D. (2008). Rock Eval™, Total organic carbon, adsorption isotherms and organic petrography of the Colorado group: Shale gas data release. *Energy Resources Conservation Board, ERCB/AGS Open File Report*, *11*(2009), 88.

Boggs, Jr. S. (2006). *Principles of sedimentology and stratigraphy* (Vol. 662). Pearson Prentice Hall.

Brathwaite, L. D. (2010, February). *Shale-deposited natural gas: A review of potential*, pp. 1–42. www.energy.ca.gov/2009publications/CEC-200-2009-005/CEC-200-2009-005-SF.PDF

Brooks, B. D., Alexander, T., Baihly, J., Boyer, C., Clark, B., Waters, G., Jochen, V., Le Calvez, J., Lewis, R., Miller, C. K., Thaeler, J., Toelle, B. E., & Brooks, B. D. (2011). Shale gas revolution. *Oilfield Review*, *23*(3), 10–12. www.slb.com/~/media/Files/resources/oilfield_review/ors11/aut11/04_shale_gas_revolution.pdf

Conant, L. C., & Swanson, V. E. (1961). Chattanooga shale and related rocks of central Tennessee and nearby areas. *USGS Professional Paper*, *357*, 91.

Dong, T., Harris, N. B., Ayranci, K., & Yang, S. (2017). The impact of rock composition on geomechanical properties of a shale formation: Middle and Upper Devonian Horn River Group shale, Northeast British Columbia, Canada. *AAPG Bulletin*, *101*(2), 177–204. https://doi.org/10.1306/07251615199

EIA. (2010). *US energy information administration, "summary: US crude oil, natural gas, and natural gas liquids proved reserves.* www.eia.gov/energyexplained/natural-gas/where-our-natural-gas-comes-from.php

EIA. (2011, July). Review of emerging resources: U.S. Shale gas and shale oil plays. *U.S. Energy Information Administration*, 105 pp. ftp://ftp.eia.doe.gov/natgas/usshaleplays.pdf

EIA. (2012). AEO2012 early release overview. *Energy*, *2012*, 1–13.

EIA. (2013, September 29). *Technically recoverable shale oil and shale gas resources: Canada.* www.eia.gov/analysis/studies/worldshalegas/

EIA. (2019). Permian basin Wolfcamp and bone spring shale plays geology review. *Energy.* www.eia.gov/maps/pdf/Wolfcamp_BoneSpring_EIA_Report_July2019.pdf

Ettensohn, F. R., & Barron, L. S. (1981). *Depositional model for the Devonian-Mississippian black-shale sequence of North America: A tectono-climatic approach* (No. DOE/ET/12040-2). Department of Geology, Kentucky University, Lexington, USA.

Ferguson, J. A., & Colwell, S. A. (1992). Geology of the Horton Bluff Formation at Blue Beach. *Prepared of the geological association of Canada and mineralogical association of Canada joint annual meeting in Wolfville, Nova Scotia.* https://ees.acadiau.ca/ft1.html

Fowler, M. G., Stasiuk, L. D., Hearn, M., & Obermajer, M. (2001). Devonian hydrocarbon source rocks and their derived oils in the Western Canada Sedimentary Basin. *Bulletin of Canadian Petroleum Geology*, *49*(1), 117–148. https://doi.org/10.2113/49.1.117

Gaswirth, S. B., French, K. L., Pitman, J. K., Marra, K. R., Mercier, T. J., Leathers-Miller, H. M., Schenk, C. J., Tennyson, M. E., Woodall, C. A., Brownfield, M. E., Finn, T. M., & Le, P. A. (2018). Assessment of undiscovered continuous oil and gas resources in the

Wolfcamp Shale and Bone Spring Formation of the Delaware Basin, Permian Basin Province, New Mexico and Texas, 2018. *U.S. Geological survey fact sheet 2018–3073*, 4. https://pubs.er.usgs.gov/publication/fs20183073

Geri, M. B., Ellafi, A., Flori, R., Belhaij, A., & Alkamil, E. H. K. (2019). New opportunities and challenges to discover and develop unconventional plays in the Middle East and North Africa: Critical review. *Society of petroleum engineers—Abu Dhabi international petroleum exhibition and conference 2019, ADIP 2019*. https://doi.org/10.2118/197271-MS

Gray, J. K. (1977). Future gas reserve potential western Canadian sedimentary basin. *3rd national technical conference*. Canadian Gas Association.

GWPC. (2009, April). Modern shale gas development in the United States: A primer. *Modern shale gas development in the United States: A primer*, p. 116. www.all-llc.com/public-downloads/ShaleGasPrimer2009.pdf

Hackley, P. C., & Cardott, B. J. (2016). Application of organic petrography in North American shale petroleum systems: A review. *International Journal of Coal Geology, 163*, 8–51. https://doi.org/10.1016/j.coal.2016.06.010

Hackley, P. C., Ryder, R. T., Trippi, M. H., & Alimi, H. (2013). Thermal maturity of northern Appalachian Basin Devonian shales: Insights from sterane and terpane biomarkers. *Fuel, 106*, 455–462. https://doi.org/10.1016/j.fuel.2012.12.032

Halbertsma, H. L. (1994). Devonian Wabamun group of the Western Canada sedimentary basin. *Atlas of the Western Canada Sedimentary Basin, 4*, 203–220.

Hite, R. J., Anders, D. E., & Ging, T. G. (1984). Organic-rich source rocks of Pennsylvanian Age in the Paradox Basin of Utah and Colorado. *Hydrocarbon source rocks of the greater Rocky Mountain region, Figure 1.1*, pp. 255–279. http://search.datapages.com/data/rmag/SourceRocks84/hite.pdf

Hobson, A. (2008). Energy, environment, and climate. *American Journal of Physics, 76*(12), 1174–1175. https://doi.org/10.1119/1.2982629

Holditch, S. A. (2006). Tight gas sands. *Journal of Petroleum Technology, 58*(6), 86–93. https://doi.org/10.2118/103356-jpt

Islam, M. R. (2014). *Unconventional gas reservoirs: Evaluation, appraisal, and development* (pp. 1–624). https://doi.org/10.1016/C2013-0-13422-8

Jiang, Z., Zhang, W., Liang, C., Wang, Y., Liu, H., & Chen, X. (2016). Basic characteristics and evaluation of shale oil reservoirs. *Petroleum Research, 1*(2), 149–163. https://doi.org/10.1016/S2096-2495(17)30039-X

King, H. M. (2020). Shale is the most abundant sedimentary rock and is in sedimentary basins worldwide. *Geoscience News and Information, 1*. https://geology.com/rocks/shale.shtml%0Ahttp://geology.com/rocks/shale.shtml

Legg, J. A., Donahoe, R. J., & Martens, R. (2016). Assessment of the petroleum generation potential of the neal shale in the black warrior basin, Alabama. *Society of petroleum engineers—SPE/AAPG/SEG unconventional resources technology conference*. https://doi.org/10.15530/urtec-2014-1922181

Martin, C. M., & Makrodimopoulos, A. (2008). Finite-element limit analysis of Mohr–Coulomb materials in 3D using semidefinite programming. *Journal of Engineering Mechanics, 134*(4), 339–347.

Mauger, S., Geol, P., & Bozbiciu, D. (2011). How changing gas supply costs leads to surging production. *Energy, 403*.

NEB. (2009, November). *A primer for understanding Canadian shale gas—energy briefing note*. National Energy Board, 27.

NEB. (2013). *The ultimate potential for unconventional petroleum from the montney formation of British Columbia and Alberta—energy briefing note*. National Energy Board, 17.

NEB. (2017, November). *Duvernay shale economic resources—energy briefing note*. National Energy Board.

Nuccio, V. F., & Condon, S. M. (1996). *Burial and thermal history of the Paradox Basin, Utah and Colorado, and petroleum potential of the middle Pennsylvanian paradox formation. US Geological Survey Bulletin, 2000.* US Government Printing Office.

Passey, Q. R., Bohacs, K. M., Esch, W. L., Klimentidis, R., & Sinha, S. (2010). From oil-prone source rock to gas-producing shale reservoir—geologic and petrophysical characterization of unconventional shale gas reservoirs. *Society of Petroleum Engineers—International Oil and Gas Conference and Exhibition in China 2010, IOGCEC, 3*, 1707–1735. https://doi.org/10.2118/131350-MS

Pawlewicz, M. J., & Hatch, J. R. (2007). Petroleum assessment of the Chattanooga Shale/Floyd shale-paleozoic total petroleum system, Black Warrior Basin, Alabama and Mississippi. In J. R. Hatch & J. J. Pawlewicz (Eds.), *Geologic assessment of undiscovered oil and gas resources of the Black Warrior Basin province, Alabama and Mississippi.* US Geological Survey Digital Data Series DDS-69-I, 23.

Perryman. (2007). *Bounty from below: The impact of developing natural gas resources associated with the barnett shale on business activity in fort worth and the surrounding 14-county area*, 138. http://web.archive.org/web/20110707212648/www.barnettshale-expo.com/docs/Barnett_Shale_Impact_Study.pdf

Rajput, S., & Thakur, N. K. (2016). *Geological controls for gas hydrates and unconventionals* (pp. 1–366). Elsevier.

Rivard, C., Lavoie, D., Lefebvre, R., Séjourné, S., Lamontagne, C., & Duchesne, M. (2014). An overview of Canadian shale gas production and environmental concerns. *International Journal of Coal Geology, 126*, 64–76. https://doi.org/10.1016/j.coal.2013.12.004

Schamel, S. (2006). Shale gas resources of Utah: Assessment of previously undeveloped gas discoveries. *Annual AAPG convention.* Long Beach, CA. https://doi.org/10.34191/ofr-499

Slattery, S. R., & Atkinson, N. (2012). Energy resources conservation board/Alberta geological survey threedimensional geological modelling workshop. *ERCB/AGS Information Series Report, 141.*

Soeder, D. J., & Borglum, S. J. (2019). *The fossil fuel revolution: Shale gas and tight oil* (pp. 1–336). https://doi.org/10.1016/C2017-0-03254-8

Soeder, D. J., Sawyer, J. F., Freye, A., & Singh, S. (2015). Assessment of hydrocarbon potential in the Niobrara Formation, Rosebud Sioux Reservation, South Dakota. *Society of Petroleum Engineers—Unconventional Resources Technology Conference, URTeC 2015*, 152. https://doi.org/10.2118/178546-ms

Speight, J. G. (2013). Origin of shale gas. In *Shale gas production processes* (pp. 1–23). https://doi.org/10.1016/b978-0-12-404571-2.00001-7

Speight, J. G. (2017). Gas and oil in tight formations. In *Deep shale oil and gas* (pp. 1–61). https://doi.org/10.1016/b978-0-12-803097-4.00001-2

Su, K., Torres, J., Perl, Y. S., Barlet, P., Onaisi, A., & Vidal-Gilbert, S. (2017). Tests of fracture water and gas permeability on vaca muerta gas shale. *SPE/AAPG/SEG unconventional resources technology conference 2017.* https://doi.org/10.15530/urtec-2017-2671318

Swanson & Conant. (2010). Chattanooga black shale region, Kentucky–Tennessee–Alabama. *Uranium deposits of the world*, 379–380. https://doi.org/10.1007/978-3-540-78943-7_14

Thériault, R. (2008). Caractérisation géochimique et minéralogique et évaluation du potentiel gazéifère des shales de l'Utica et du Lorraine, Basses-Terres du Saint-Laurent. *Résumé des conférences et des photoprésentations, Québec exploration, 2008-06.*

USEIA. (2012, September). *Where is shale gas found in the United States?* U.S. Energy Information Administration (EIA). www.eia.gov/analysis/studies/worldshalegas/pdf/fullreport.pdf.

USGS. (2008). Assessment of undiscovered oil resources in the devonian-mississippian bakken formation, williston basin province, Montana and North Dakota, 2008. In *The Bakken formation as an oil resource* (pp. 39–44). USGS

USGS. (2012). Assessment of undiscovered oil resources in the devonian-mississippian bakken formation, williston basin province, Montana and North Dakota, 2008. In *The Bakken formation as an oil resource* (pp. 39–44). USGS

Williams, J., & Kramer, H. (2011). *Montney shale formation evaluation and reservoir characterization case study well comparing 300m of core and log data in the upper and lower montney*. Gulf Professional Publishing.

Yao, S. (2019). *Modeling the fluid flow in low: Permeability unconventional reservoirs across scales* (Issue May) [Doctoral dissertation, Faculty of Graduate Studies and Research, University of Regina].

Zhang, Y. Y., He, Z. L., Jiang, S., Lu, S. F., Xiao, D. S., Chen, G. H., & Zhao, J. H. (2018). Controls on the organic carbon content of the lower Cambrian black shale in the southeastern margin of Upper Yangtze. *Petroleum Science*, *15*, 709–721.

Zou, C. (2012). *Unconventional petroleum geology* (pp. 1–373). https://doi.org/10.1016/C2011-0-06250-6

Zou, C. (2013). *Unconventional petroleum geology*. https://doi.org/10.1016/C2011-0-06250-6

2 Geophysics, Geology, Formation Evaluation, and Reservoir Characterization in Unconventional Resources

Aliyu Mohammed Kaura, Yohana Andrawus, and Anzaku Y. Ibrahim

2.1 SURFACE SEISMIC APPLICATIONS IN SHALES

Surface seismic is a seismic acquisition technique in which the seismic source and sensors or geophones are located at the Earth's surface. Elastic energy is generated and propagated through the ground. Small ground vibrations generated by changes in the elastic characteristics of underlying materials are captured by the receiver or geophone. Some of the wave's energy is reflected in the receivers. The remainder of the wave continues until it reaches another layer of rock. The time required for the waves to travel from the source to the receivers is used to determine the distance between them and, hence, the thickness of the rock layers. The strength of the reflected wave gives information about the density of the reflecting rock. Each time the seismic pulse meets a change in rock properties, such as shale to a sand layer, part of the pulse will be reflected to the surface.

Surface seismic data have proven invaluable for organizations producing hydrocarbons from unconventional resource plays (e.g., shale play). Initially, a primary benefit of surface seismic was the ability to locate and avoid drilling into zones with faults, fractures, and karsting, which adversely affected the ability to complete the well successfully (Heiland-Hansen et al., 1997). Recent advancements in interpreting pre-stack seismic data reveal features that appear to be related to stress fields, rock strength, and formation lithology. Effective use of these features and awareness of them could aid in maximizing drilling and completion operations.

DOI: 10.1201/9781003319955-2

2.2 SEISMIC SURVEY DESIGNS

Two-dimensional (2D) seismic is a linear arrangement controlled by terrain. Typically, the source and receiver are aligned directly. Arrays may be multi-dimensional but most often are also in the survey line. In other words, 2D seismic data are acquired individually instead of the multiple closely spaced lines that constitute 3D seismic data.

Two-dimensional seismic surveys are as linear as the terrain permits. The source and receiver are typically in direct line with one another (Figure 2.1). Arrays may be multi-dimensional, arranged along the survey line. The 2D seismic data are collected along parallel lines, often spaced several kilometers away. The distance between individual lines varies, ranging from 2-mile or 5-mile spacing to even 15-mile spacing. However, in a regional study, it might be a 50-mile spacing (Kiran, 2005; Onajite, 2013). However, a 3D seismic layout (Figure 2.1b) is where the seismic

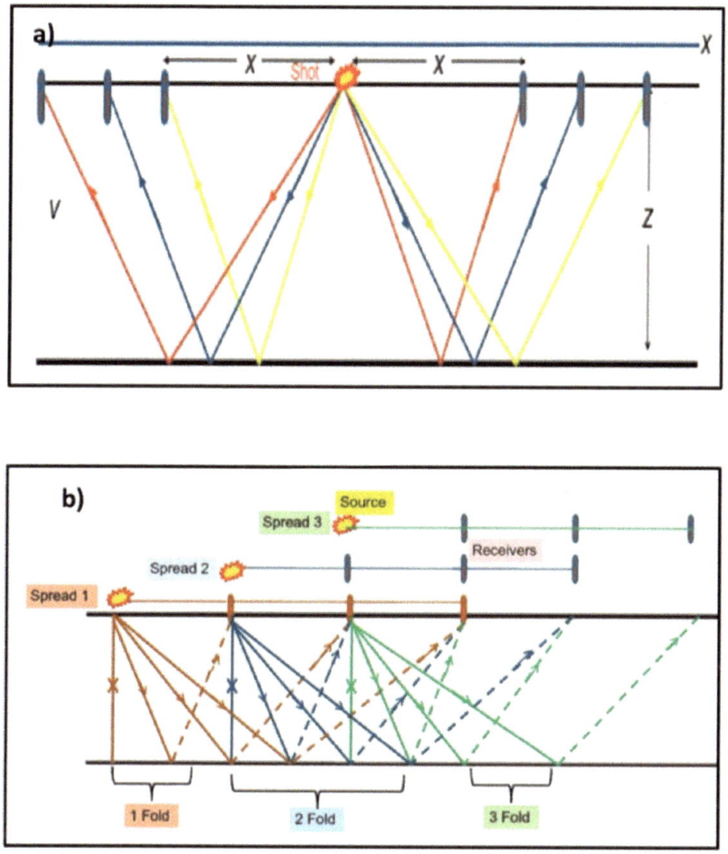

FIGURE 2.1 Survey layout for a) 2D seismic b) 3D seismic.

Source: Onajite, 2013

data are acquired so densely that the data are processed and interpreted as a volume or cube of data. Modern standard 3D surveys have spacing between traces in each direction of about 12.5 meters or less (Onajite, 2013). During 3D surveys, the source line and receiver lines are aligned orthogonally. The shot is within the grid of many lines of receivers, with each line separated by tens of meters. In other words, a 3D seismic volume is produced by shooting a tightly spaced grid of 2D lines and interpolating between them to produce a three-dimensional volume of data known as a cube (Onajite, 2013). Land and marine 3D data are obtained using several sources and receiver arrays to ease the acquisition of huge volumes of data.

2.3 SEISMIC PROCESSING

Seismic processing is modifying the data acquired to remove noise, improve the reported seismic trace, and migrate the seismic trace to its correct space and time position. The main objectives are to eliminate all noise and irregularities introduced by the seismic acquisition method and to develop a seismic section as close as possible to the Earth's subsurface model. Seismic data processing steps include static field correction, noise attenuation, amplitude compensation, deconvolution to extract the seismic reflectivity series for layer boundaries, velocity field establishment, residual static correction, and migration of common midpoint (CMP) and common depth point (CDP), which can be done before or after stacking and high-resolution processing, etc. However, stacking, deconvolution, and migration are the three fundamental operations that process seismic reflection data. All other processing operations are designed to improve the effectiveness of these three operations, either by attenuating noise or by making the data better conform to the assumptions inherent in the operations (Aminzadeh & Dasgupta, 2013a).

Deconvolution is a means of manipulating the characteristics of the signal wavelet, in particular, reversing some of the detrimental consequences of its passage through the subsurface (Dentith & Mudge, 2014). For example, deconvolution often improves temporal resolution by collapsing the seismic wavelet and removing a particular type of distortion, approximately a spike, and undesirable characteristics of the seismic wavelet resulting from its passage through the subsurface effects of the Earth (Aminzadeh & Dasgupta, 2013a). These may involve changing its shape into one that makes interpretation of the data easier: for example, changing it to zero phases or shortening it to improve resolution.

Stacking, also known as normal move-out and statics correction, converts the data to an easier-to-interpret zero offset form and suppresses non-reflected arrivals, random and transient noise, and most kinds of coherent noise (Dentith & Mudge, 2014). The process utilizes the survey location of data collected and velocity to correct the travel delay in the distance for the source-receiver geometry. The process of stacking significantly improves the signal-to-noise ratio of the data. This implies that post-stack processing operations are applied to the drastically reduced number of traces, resulting in significant computer resource savings compared to pre-stack processing operations. However, stacking makes significant assumptions about the nature of the subsurface, and if these assumptions are violated significantly, the data may be degraded. In these instances, pre-stack processing is preferable. Stacking also provides useful information about seismic velocities in the subsurface required for migration and depth conversion.

Migration enables seismic events to be shifted in either space or time to the place where the event occurred in the subsurface rather than the location where it was recorded at the surface, resulting in a more accurate image of the subsurface (Dentith & Mudge, 2014). Migration collapses diffractions and moves dipping events to their supposedly true subsurface locations. In other words, migration is an imaging process.

2.4 ATTRIBUTE INTERPRETATION

Seismic attributes are intrinsic properties of a seismic wave signal retrieved or produced from seismic data that can be analyzed to improve the information that may be more subtle in a typical seismic image, resulting in a better geological or geophysical interpretation of the data. Seismic reflection waveforms carry concealed valuable subsurface geologic information embedded in them, and extraction of attributes and their analysis provide means to retrieve this information (Chopra & Marfurt, 2007; Nanda, 2016). The attributes allow the geoscientist to interpret structures (faults and fractures) and stratigraphic features (channels), recognize the depositional environment, and explain the structural deformation promptly (Chopra & Marfurt, 2018; Maleki, 2014). Attributes are computed by mathematical manipulation of the original seismic data. They can be derived from a single trace by comparison with multiple traces or from an interpreted seismic horizon.

Several authors classify attributes in many ways. Figure 2.2 (Brown, 1996; Sheriff, 2002a) classifies seismic attributes as time, amplitude, frequency, and/or attenuation measurements. Generally, time-derived attributes provide structural information, and amplitude-derived attributes provide stratigraphic and reservoir information. Frequency-derived, frequency-based attributes are not clearly understood in stratigraphy and reservoir characterization. However, attenuation attributes are usually very uncertain. Nanda, 2016 categorized them as primary attributes (amplitude, frequency, phase, polarity) and geometric attributes (dip, azimuth, curvature, and coherency). Li and Zhao (2014a) grouped attributes as horizon, interval, and surface attributes. Subrahmanyam grouped attributes based on domain characteristics as pre-stack and post-stack derived attributes computed before or after time migration, respectively.

Attributes are computed by smoothing and averaging over windows of various sizes, finding residuals; peak values; measuring the distribution within a window mean, median, kurtosis; percent greater/smaller than a threshold; sums; residuals; scatter, etc.; continuity; edges; smoothness; linearity or curvature; gradients or other derivatives; absolute values; polarity changes; zero-crossings; and peak-trough differences (Sheriff, 2002b). In addition, relations may be measured over windows spectra, correlation, semblance, covariance, etc. Attributes can be measured along a single trace, within a volume, or in other ways. Seismic attributes are used for qualitative analysis (e.g., data quality, seismic facies mapping) and quantitative analysis (e.g., net sand, porosity prediction).

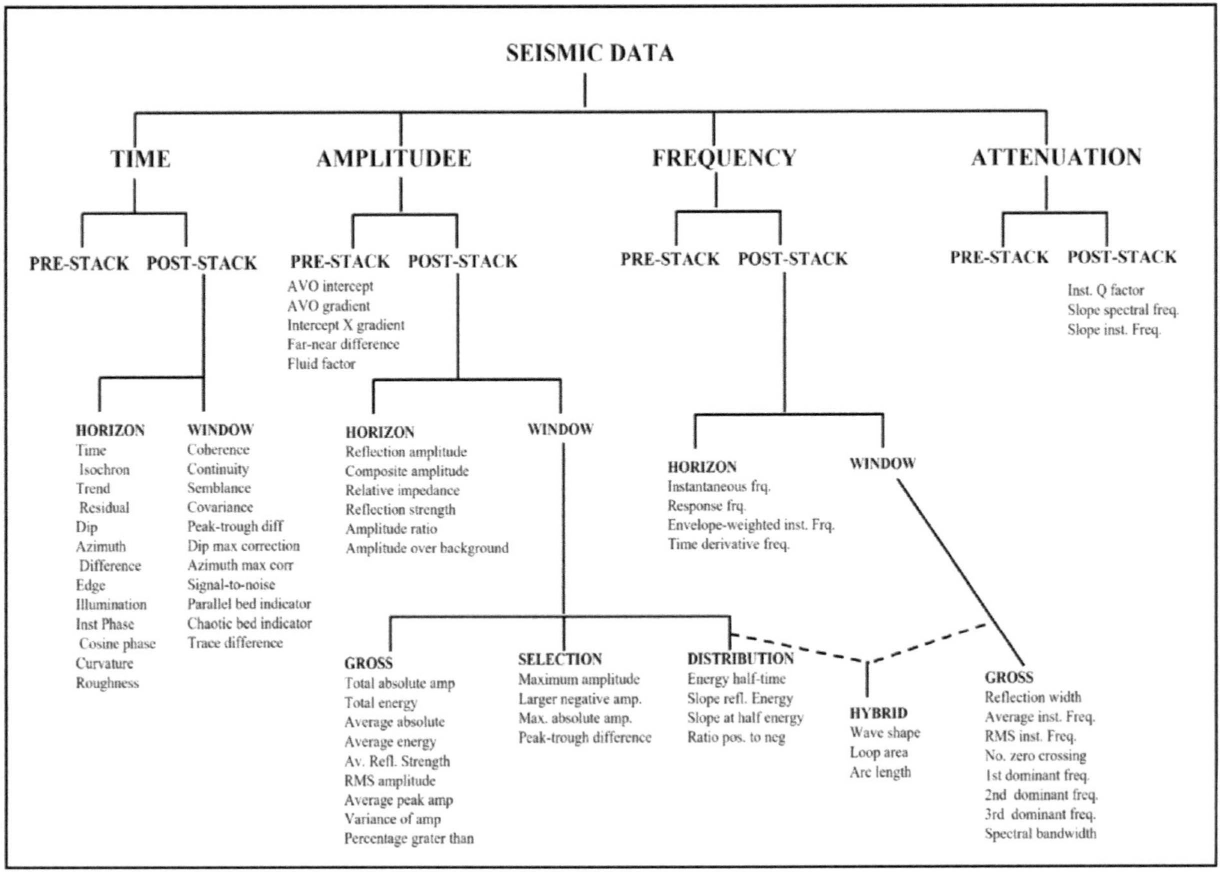

FIGURE 2.2 Seismic attributes. "Windows" can be constant time intervals, intervals hung from one horizon, or intervals between horizons.

Source: Brown, 1996; Sheriff, 2002b

2.5 VALIDATION TECHNIQUES

The geomechanical characteristics of the subsurface are systematically captured by the mechanical Earth model (MEM), but for practical decision-making, they need to be calibrated and validated. Comparison of computed and measured attributes constitutes calibration. Validation entails applying the estimated MEM parameters in a simulator or other technical computations to compare the modeled results to the observed outcomes. The calibration data for conventional and unconventional reservoirs frequently diverges (Ma & Holditch, 2015). Rock strength and mechanical characteristics are typically calibrated using measurements made in the lab using core data. Since it might be difficult to determine pore pressure directly, calibration is frequently derived from drilling events. Due to the necessity of hydraulic fracturing in producing hydrocarbons from unconventional reservoirs, the minimum horizontal stress is often calibrated to a closure measurement using a small fracture injection test.

Typically, wellbore stability assessments or hydraulic fracture models are employed during validation to compute geomechanical characteristics and parameters for unconventional reservoirs (Ma & Holditch, 2015). Wellbore stability assessments use geomechanical parameters and downhole pressures caused by mud and drilling procedures to forecast wellbore instabilities such as breakouts or drilling-induced tensile fractures. Afterward, the estimated near-well stresses are compared to shear and tensile failure criteria (Ma & Holditch, 2015). The modeled instability is then immediately compared to any observed instability in the image logs, caliper logs, or drilling report. A hydraulic fracture simulator employs the geomechanical properties, pumped fluid, materials, and rates to estimate the hydraulic fracture pressure and geometry, then compare to the measured pressure during treatment. When the calculated instability or pressure matches the observed instability or pressure, the model is verified and can guide decisions on subsequent drilling and completion (Ma & Holditch, 2015).

Validation is also applied to seismic attributes (Schuelke & Quirein, 1998). The method is very simple. First, the seismic properties are calibrated against the well-log response using control data. Existing well controls and seismic responses at well locations provide the control data. Next, a portion of the control data is withheld from the calibration procedure and used to validate the prediction. The validation data is concealed well data. Assume that information from ten different wells is available. Assume that the seismic-to-well calibration has been changed so that, utilizing all control data, the best "n" attributes will be selected (Schuelke & Quirein, 1998). For calibration and ranking, this procedure employs a multivariate linear regression method. The optimization technique will identify the appropriate seismic property for predicting the control log data. This seismic attribute correlates most strongly with the predicted log property. The method then finds the appropriate seismic attribute to add to this single log control attribute to maximize the error reduction between the predicted and actual log data. Then, the three attributes that minimize prediction error the most when combined are identified. This ranking continues until it is no longer possible to lower the prediction error on the control data. This strategy ensures that each additional seismic attribute is accompanied by statistically independent information. Therefore, redundant information would not be utilized to reduce estimation error.

The subsequent stage is recalibrating, holding out one well at a time. This validation stage assists in determining the number of rated qualities to be utilized in the prediction. If adding a seismic characteristic does not improve the ability to forecast the buried data, it will not be chosen. Using hidden data to test the prediction removes the overfitting issue sometimes associated with using excessive seismic properties, particularly with nonlinear techniques such as neural networks. One technique to automate this validation method is to remove one well from the calibration at a time in a round-robin fashion and then predict that well. Perform eleven unique calibrations: one for all wells and one for each well removed from the population, making ten calibrations for each of the nine well datasets (one well held out per trial). Compare, for each well, the accuracy gained by using all wells against the accuracy obtained by excluding that well. Therefore, there would be ten correlation coefficient comparisons or the mean squared error between the actual log value and the anticipated seismic value for this case. If, for a significant number of wells, the difference between "all wells accuracy" and "leaving one out accuracy" is too big, reset the calibration parameters to minimize the number of characteristics and repeat the experiment.

2.6 FORMATION EVALUATION

Formation evaluation (FE) is the process used to identify and estimate oil and gas reserves in the rock adjacent to the well by interpreting a series of measurements obtained inside a wellbore. FE involves the analysis of subsurface formation characteristics, such as lithology, porosity, permeability, and saturation, by methods such as wireline well logging and core analysis. It also provides reservoir engineers with the geological and physical data of the formation required for building a reservoir fluid-flow model. FE is used to determine the presence of hydrocarbon reservoirs, appraise reservoirs for potential hydrocarbons, and estimate the volume of those reserves. It is also used to build reservoir mechanical characteristics models in drilling and production. Laboratory fluid property measurements, laboratory (rock) core physical property measurements, drill cuttings descriptions (strip log), mud logs, wireline log measurements, measurements while drilling (MWD), logging while drilling (LWD), and formation flow tests are all necessary for FE (Aminzadeh & Dasgupta, 2013b).

2.6.1 Rock Composition Quantification

In petroleum geosciences and engineering, unconventional reservoir rocks are described as "shale" (Zoback & Kohli, 2019). This convention is derived from studies of conventional oil and gas reservoirs, in which the term "shale" is attributed to fine-grained, ultra-low-permeability units or lithofacies acting as the source or cap-rocks (Ulmer-Scholle et al., 2015). These tiny grains are usually clay minerals such as illite, kaolinite, and smectite. Shale usually contains other clay-size mineral particles, such as quartz, chert, and feldspar. Other constituents might include organic particles, carbonate minerals, iron oxide minerals, sulfide minerals, and heavy mineral grains (Hobart, 2022).

Quantifying the mineral composition and lithology of the reservoir is the first step in understanding the characterization of unconventional reservoirs. At the same time, mineral composition is the most important factor influencing reservoir productivity (Cameron, 2009). Shales can be described mineralogically using a ternary diagram with silica, clay, and carbonates as the three dominant mineral components. In conjunction with core and well-log data, geochemical analysis can help determine the location of a particular shale sample or reservoir on a ternary plot lot (Ma & Holditch, 2015; Zoback & Kohli, 2019). Figure 2.3 shows how some well-known unconventional reservoirs are related to this mineralogical composition framework.

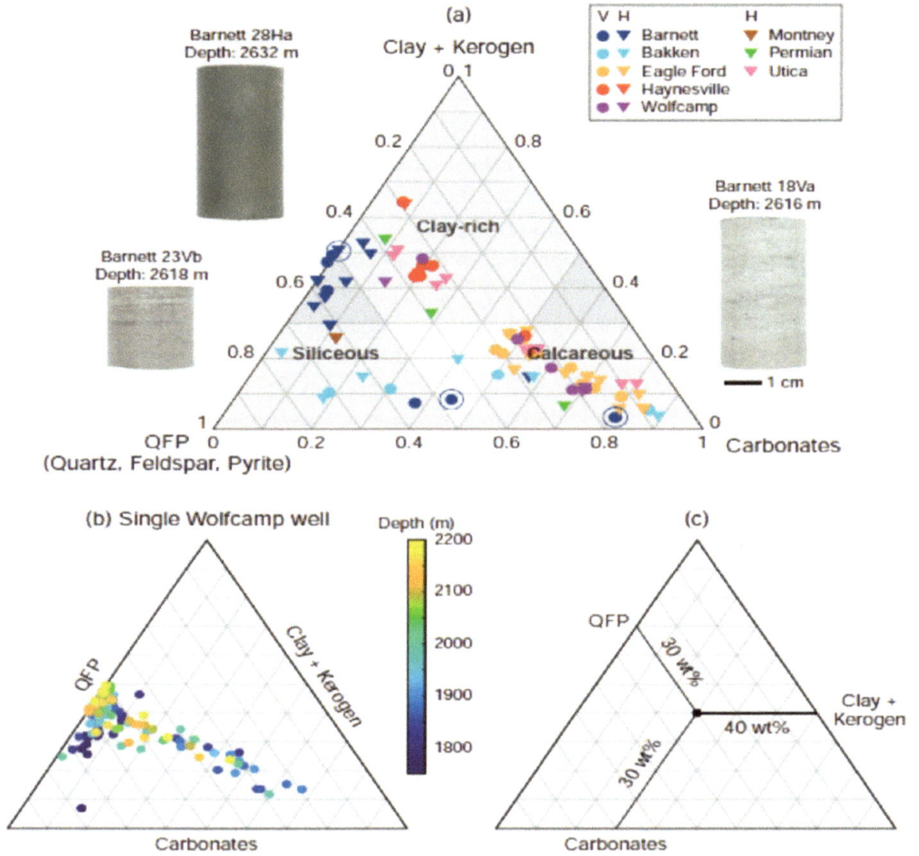

FIGURE 2.3 Ternary classification for the composition of unconventional reservoir rocks. All values in weight fraction. (a) Select shale samples from major US unconventional basins from the Stanford Stress and Crustal Mechanics Group collection. Photographs of circled core samples from a single Barnett shale well highlight meter-scale variations in composition; (b) Samples from a single Wolfcamp shale well show systematic variations in composition with depth; (c) Guide for reading values from ternary source.

Sources: Sone (unpublished); Zoback & Kohli, 2019

These known reservoirs generally contain less clay than the average shale, calculated from all shales that contain organic and inorganic rocks. Notably, these reservoirs have varying mineral and lithological compositions, but they have all been used in successful hydrocarbon projects, indicating that mineralogy is not the only factor determining reservoir quality.

2.6.2 Geological Characterization

The occurrence of petroleum requires several geologic factors, including the presence of source rock, reservoir rock, caprock, a trap, appropriate pressure and temperature (Fanchi, 2010), porosity and permeability of the hydrocarbon-bearing rocks, their vertical and lateral variations, and, finally, the amount of fluids to be extracted in the times (Casnedi, 2005). The ability to manage subsurface resources depends in part on knowledge of the geologic history of the Earth and theories of the Earth's formation and maturation. One can go wrong for lack of geological knowledge or, more specifically, if the geological model developed does not reflect the controls on subsurface geology. Characterization of geologic systems and demarcation of their borders are carried out in accordance with the goal indicated for each specific scenario and study objective (Houlding, 1994). Various factors, including genetics, location, engineering-technological activities, economics, etc., can determine geologic formations' boundaries. The first step in adequately characterizing the geology controls is having adequate, high-quality fundamental data. Data interpretation and robust modeling by professional and certified geologists are critical to bringing to light crucial economic controls on the project, and only then can risk be assessed.

The main goal of geological characterization is to forecast the spatial variation of one or more geological characteristics, which are the properties of the geological subsurface that exhibit spatial variability and can be measured in terms of real numeric values (Houlding, 1994). Typically, the subsurface is anisotropic and especially complex, and sedimentary bodies are internally heterogeneous, representing the spatial variation. During geological characterization, data can be derived from various sources, including borehole core logs, well logs, geological mapping of exposed surfaces, topographical surveys, geophysical borehole logs, and remote geophysical surveys (Figure 2.4).

2.6.3 Total Organic Carbon (TOC) Quantification

TOC quantification is an important step in finding possible organic shale reservoirs, and its amount can be related to the reservoir quality since the producible pore fluid is contained inside the organic matter (Ambrose et al., 2010). There are a variety of different techniques available for quantifying carbon dioxide during TOC analysis. However, before the detection and quantification, the sample must first be analyzed using the analyzer. The process involves sample acidification and oxidation down to CO_2. Then acidification is used to remove inorganic carbon; oxidation occurs either through direct, high-temperature (1,300°C) or catalytic (680°C) combustion or through a technique involving either ultraviolet radiation or thermal oxidation in the presence of persulfate (Lab Manager, 2011).

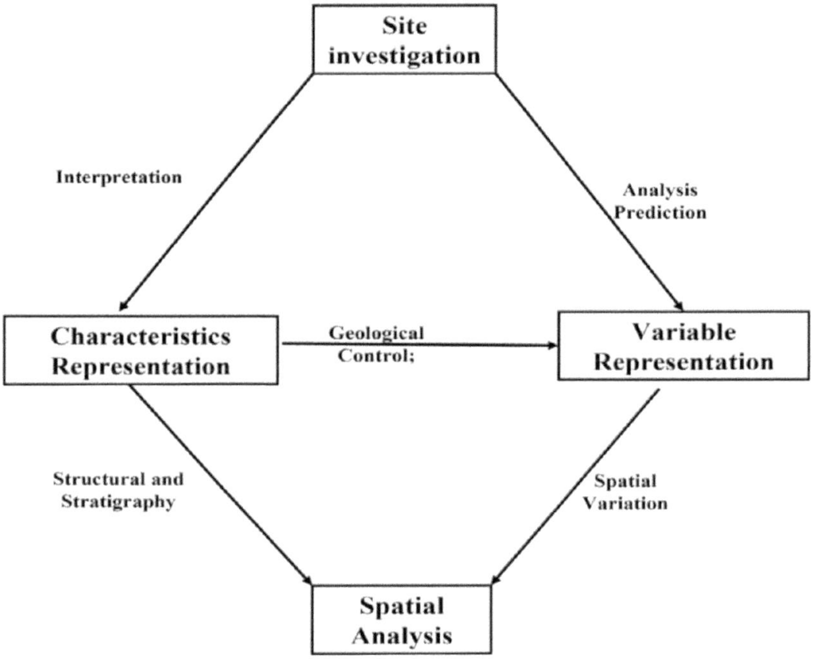

FIGURE 2.4 Summary of the geometric process of geological characterization.

Source: Houlding, 1994

After the sample combustion phase is completed, different quantification techniques can be applied to detect the quantity of organic carbon in the soil or sediment. These techniques include titration, gravimetric, manometric, spectrophotometric, gas chromatographic, thermal conductivity, and coulometry (EPA, 2002; Bisutti et al., 2004). However, the most common techniques are conductivity and non-dispersive infrared (NDIR) (Attribute, 2006; Lab Manager, 2011). Over the years, several well-log-derived TOC indicators based on measurements such as gamma ray, spectral gamma ray, resistivity, density, sonic, neutron, and nuclear magnetic resonance (NMR) have been developed with varying degrees of success, depending on the specific ORR and where it occurs within a basin.

2.6.4 Porosity and Permeability Quantification

Permeability and porosity are two of the primarily related properties that control the movement and storage of fluids in rocks (Šperl & Trčková, 2008). Characterization of porosity and permeability is of fundamental importance for properly evaluating hydrocarbon reservoirs. Both porosity and permeability are related to the number, size, and connections of openings in the rock. Their knowledge is critical to accurately predicting fluid transport and estimating the total volume of oil and gas in place in a reservoir. As a result, scientists and engineers are greatly interested in quantifying permeability characteristics. The porosity of natural and artificial materials has many practical applications related to oil extraction, groundwater flow, and contaminant transport (Šperl & Trčková, 2008).

Unconventional reservoirs, such as shale, contain complex pore throat structures of very tiny interparticle and intraparticle particles. Also, vertically and horizontally scattered natural fractures exist in the reservoir (Lock et al., 2012). Typical laboratory techniques for determining porosity and permeability are inapplicable for unconventional reservoirs, such as shale reservoirs. Typically, the overall porosity of shale ranges from 5% to 12% (Ganat, 2019). The reservoir's varying amounts of total organic carbon and inorganic mineral components will make it difficult to determine shale porosity using conventional log technology. Utilizing elemental spectroscopy and nuclear magnetic resonance can enhance the accuracy of logging devices. The porosity must be measured from the core and compared to the porosity determined from the log. Mercury injection capillary pressure (MICP), gas research institute technique (GRI), standard crushed porosity, scanning electron microscopy (SEM), and focused ion beam (FIB) tomography are some of the laboratory methods used for measuring porosity. However, shale permeability is tested on the nano-Darcy scale and is determined in the laboratory using core analysis data calibrated with core responses.

2.6.5 Fluid Saturation Quantification

In conventional reservoirs, hydrocarbons are only stored within the matrix's pores. The hydrocarbon saturations are computed on laboratory tests or resistivity and porosity data from wireline or LWD logs. Typically, hydrocarbons are trapped as free oil and gas in unconventional reservoirs' cracks and shale matrix pores. In addition, the sorbed gas and oil are either adsorbed or absorbed to the kerogen and mineral surfaces within the fractures and matrix rock, respectively (Ganat, 2019). The dissolved gas can also be stored in the hydrocarbon liquids existing in the bitumen.

In unconventional reservoirs, a combination of laboratory studies is the principal method for evaluating gas or oil quantities and saturation. Fluid saturations are determined using Retort or Dean-Stark analysis on crushed core samples. In gas shale reservoirs, adsorption and desorption isotherm investigations determine adsorbed gas quantities and total gas volumes (Bustin et al., 2009). Typically, water saturation is determined using the same methods and procedures as conventional reservoirs, such as a resistivity and porosity log, utilizing the Shaly sand or Archie equation. Comparing log-derived fluid saturations with adequate core saturation measurements is necessary.

2.6.6 Geomechanical Characterization

Geomechanical characterization plays a significant role in decisions regarding all phases of exploration and production of oil and gas, particularly in geomechanics; influences prospect appraisal; field development; and primary, secondary, and tertiary production activities and hydraulic fracturing of unconventional resources (Aadnøy & Looyeh, 2011a; Teklu et al., 2012). A mechanical Earth model (MEM) can be used to characterize the geomechanical behavior of the subsurface numerically

(Higgins-Borchardt et al., 2016), which can estimate the mechanical characteristics of reservoir rock, such as strength, pore pressure, and stresses.

Hydrocarbon recovery changes reservoir pore pressure. Pore pressure reduction due to pumping and production leads to stress transfer from the pore fluid to the rock-solid skeleton. Similarly, high increases of stress on the rock due to wave propagation, temperature, and injection of enhanced oil recovery (EOR) fluids such as polymer, steam, and gas/CO_2 affect reservoir stress redistribution and re-orientation in the field (Teklu et al., 2012). This increase in pressure may lead to reservoir compaction, reduce reservoir permeability, and disturb the mechanical equilibrium of overlaying strata and faults. Increases in the magnitude of stresses and stress anisotropy may also cause rock failure near the wellbore and produce fine solid particles that may plug and damage pumps and surface facilities (Teklu et al., 2012).

The geomechanical assessment compares rock mechanical characteristics and strengths to in situ stresses (Addis, 2017). When the magnitudes of the stresses exerted on a rock are less than its yield strength, the rock behaves elastically, and tiny deformations occur. However, suppose the changes in stress caused by excavation during drilling, pore pressure, or thermal variations are greater than the compressive yield, peak, or pore collapse strengths. In such a circumstance, the rock will fail via shearing or compaction, resulting in nonlinear irreversible deformation (Addis, 2017). If, on the other hand, the tensions reach tensile, the formation could fracture. A brief timeline of the significant advancements in geomechanical engineering applied to oil and gas extraction (Figure 2.5) demonstrates that these concerns have been explored since the earliest days of extensive hydrocarbon extraction, garnering

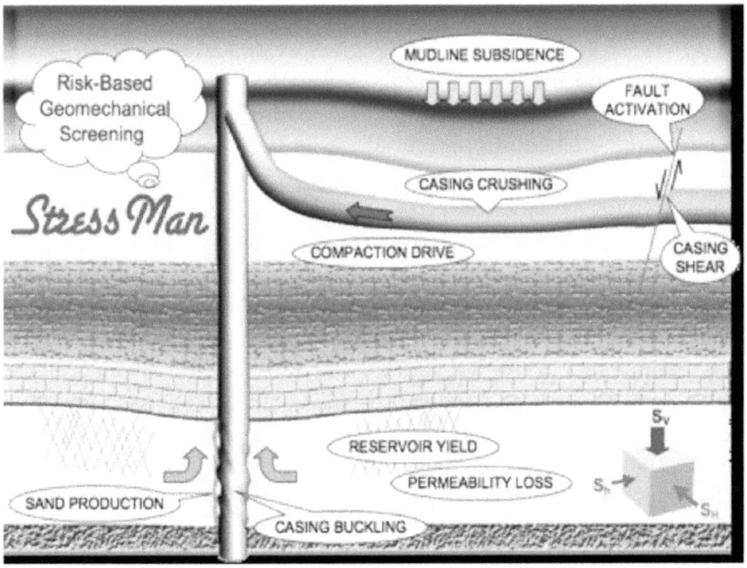

FIGURE 2.5 Reservoir depletion and pore pressure change.

Source: Addis, 2017

the attention of the industry's most prominent figures. The tests are most frequently employed to assess fundamental rock mechanics parameters, including strength, elastic properties, hardness, ultrasonic behavior, index features, and in situ stresses. Folding, faulting, fracturing, and diapirism are geological manifestations of the geomechanical or structural processes that form many oilfields. In numerous reservoirs, the present-day starting stresses and textures are governed by the stress regimes and strains accompanying these deformations.

However, oilfield development causes geomechanical changes:

1. Rock excavation during drilling causes wellbore stability and sand production issues.
2. Removing pore fluid and pressures from the rock causes compaction and subsidence.
3. Injection into a reservoir increases the pressure and induces thermal changes, resulting in rock fracturing or shearing to increase the natural geologically controlled permeability (Addis, 2017).

2.7 TECHNOLOGIES

2.7.1 Elemental Spectroscopy Logging

Elemental spectroscopy logging is geochemical logging (Colson et al., 2007; Grau et al., 1987) that measures the elemental concentrations from which the geochemistry of the formation may be derived. Geochemical logging methods determine the chemical content of subsurface formations using nuclear spectroscopy physics principles. In turn, the formation chemistry calculates matrix mineralogy and total organic carbon (Hertzog et al., 1989; Pemper et al., 2018). The first such logging string was introduced in 1987 (Hertzog et al., 1989; Pemper et al., 2018), followed by the elemental capture spectroscopy sonde in 1995 (Herron & Herron, 1990). Understanding the elemental concentrations and mineralogy of the formation can also provide improved quantitative knowledge of properties such as porosity, grain density, and even permeability (Herron, 1986).

Several logs, such as natural gamma ray spectroscopy, pulsed neutron spectroscopy, elemental capture spectroscopy, and aluminum activation, provide information on elemental weight concentrations (Schlumberger, 2023a). A geochemical log combines some or all of its outputs since it contains data on most of the major elements present in sedimentary rocks. Quantitative results require absolute concentrations; relative elemental yields are provided by pulsed neutron spectroscopy. Absolute concentrations can be calculated using a model like the oxide-closure model or by calibrating to the core (Peters et al., 2016). Using a model that identifies the minerals present, the absolute elemental concentrations can subsequently be transformed into mineral concentrations (Schlumberger, 2023b).

The focus of elemental spectroscopy logging changed over the next two decades to encompass open-hole formation evaluation, where technological advancements produced new instruments for determining element concentrations and interpreting formation mineralogy and nuclear properties. Aluminum, calcium, carbon, chlorine, hydrogen, iron, magnesium, oxygen, potassium, silicon, sulfur, thorium, titanium, and uranium are some of the main elements the system has identified. These elements are utilized to identify the minerals present in both conventional and unconventional formations (Colson et al., 2007; Hertzog et al., 1989).

The recent development of a new high-performance neutron-induced gamma-ray spectroscopy geochemical logging tool (Radtke et al., 2012) is one of three critical developments that finally enable the prediction of detailed and quantitative mineralogy from well logs. However, Pemper et al. (2018) and Radtke et al. (2012) use pulsed neutron generator sources that allow simultaneous acquisition of the capture X-ray spectrum and a spectrum of X-rays produced by inelastic scattering reactions, which allows measurement of formation carbon and computation of TOC.

2.7.2 Core/Fluid Lab Core Analysis

Core analysis describes the lab examination of the core sample taken from the core barrel. Understanding a reservoir rock's characteristics, such as its porosity, permeability, and wettability, requires analysis of the rock and its formation. Geologists and engineers can learn crucial details about the state of wellbore conditions and their potential productivity through core analysis.

2.7.3 Routine Core Analysis (RCAL)

Routine core analysis (RCAL) is often called basic or conventional core analysis. RCAL attempts to give only the very basic properties of the unpreserved core. It involves fluid saturation measurements and petrophysical measurements on dry samples, normally at ambient or nominal stress and temperature, conditioned and tested to reflect reservoir-appropriate saturations and often reservoir conditions (McPhee et al., 2015a).

RCAL plugs are usually collected once per foot or three to four times per meter. Variations in pore system development or lithology require more frequent sampling. Sample density should adequately define net pay, hydrocarbon-water transition zones, contact levels, and formation boundaries. Sampling can be done statistically at the midpoint of each foot, or the most representative sample of each foot can be selected. Routine data can be a helpful guide to well and reservoir performance if its limitations are recognized. These limitations arise because routine porosity and permeability measurements are always made with gases on a cleaned, dried core at room conditions (Glover, 2001).

2.7.4 Hybrid Workflow Approaches

The digital transformation in the oil and gas industry mostly focuses on emerging technologies such as artificial intelligence (AI), robotic process automation (RPA), and the Industrial Internet of Things (IIoT) (Kellton, 2020). Workflow automation in oil and gas driven by technology is rapidly gaining traction. It helps transform the energy industry, supporting optimization, cost reduction, and easy data management. Automated workflow offers a better customer experience and has various advantages that every firm may utilize. Oil and gas firms may be able to manage many locations through the use of a central dashboard, distributing tasks for routine and recurring maintenance, such as the automatic issuance of work orders to reduce safety hazards. Other workflow automation applications in oil and gas include back-office and supply-chain automation to manage procurement and invoicing. Field personnel inspecting wells must take notes on leakage or other issues before returning to the trailer and entering the data into a system. However, with workflow automation, the worker can tell their smart device to automatically recognize the equipment's kind and location and issue a work order using RPA software. Paperless operations assist in limiting the risk of inaccurate data transfer while also saving time (Kellton, 2020).

2.8 ORGANIC CONTENT QUANTIFICATION

Sedimentary rocks commonly contain minerals and organic matter, with the pore space occupied by water, bitumen, oil, and/or gas (Peters, 1986). Organic matter in sedimentary systems, including its carbon content as well as secondary elements such as hydrogen, oxygen, nitrogen, and sulfur, originates in living systems native to the water column (autochthonous OM) and surrounding land surfaces (allochthonous OM) (Curiale, 2017). The amount of this organic matter present in sedimentary rocks is almost always measured as the TOC content (Ronov, 1958; Waples, 1985a; Curiale, 2017), which has been widely used to assess the potential of hydrocarbon resources in the sedimentary basins (Srikant & Shanmugam, 2015). Hence, it is an indicator of source richness in relation to the amount of hydrocarbon the sediment may generate.

Quantification of TOC is an important step in identifying possible organic shale reservoirs (Herron, 2021). Its quantity can be correlated with the reservoir quality of gas shale reservoirs as the producible pore fluid is contained within the organic matter (Ambrose et al., 2010; Herron, 2021). The methods commonly used for the estimation of TOC are geochemical and wireline methods. The geochemical methods are based on combustion, followed by the detection of the generated carbon dioxide using thermal conductivity, coulometry, or spectroscopy (Bisutti et al., 2004; Peters et al., 2016). The wireline method is the alternative developed by earlier petrophysicists. There are four popular and widely used techniques for this method. These are regression of core TOC with core bulk density, ΔlogR technique developed by Passey et al. (1990). The geochemical methods are fully discussed by Bisutti et al. (2004), while wireline and other methods are discussed by Wimberley (1969), Peters and Simoneit (1982), Engleman et al. (1985), Lafargue et al. (1998), Gonzalez et al. (2013), Elias et al. (2013), Herron et al. (2014), and Peters et al. (2016).

2.8.1 KEROGEN CHARACTERISTICS

Kerogen is defined as hydrocarbons that are insoluble in common solvents like carbon tetrachloride but, when heated, produce liquid or gasses (Tissot et al., 1978; Selley, 2005; Vandenbroucke & Largeau, 2007). Chemically, kerogen consists of various complex hydrocarbons with traces of numerous additional elements, such as sulfur, nitrogen, and different radioactive and heavy metals. Kerogen is typically deposited in anoxic, stagnant, oxygen-depleted environments, such as those found in marshes, swamps, meres, salt marshes, and lagoons, and is especially prevalent in deltas (Selley, 2005). In these settings, vegetation can accumulate as horizontally wide strata of peat that are several meters thick. Peat undergoes substantial compaction and diagenesis during subsequent burial, transforming into brown coal (lignite), bituminous coal, anthracite, and finally graphite as it enters the realm of metamorphism. The form of kerogen, known as coal, is a significant energy source. Mudstones contain different levels of kerogen, which is less obvious but significant. This dispersed kerogen originated from plant and animal remains and is the mother of petroleum. Subject to sufficient thermal maturity, shale becomes a possible petroleum source rock when its kerogen content exceeds 1.5%. Kerogen produces petroleum gas and oil, depending on its chemistry and maturity level (Selley, 2005).

The amount of kerogen is measured as the weight percentage of the rock. The amount of organic matter can be described as fair, good, or very good. The chemical and physical properties of kerogen are heavily influenced by the sort of biogenic molecules. Plotting the hydrogen and oxygen indices on a simplified van Krevelen diagram (Peters et al., 2016; Speight, 2019) can allow you to identify the kerogen type in a source rock (Figure 2.6).

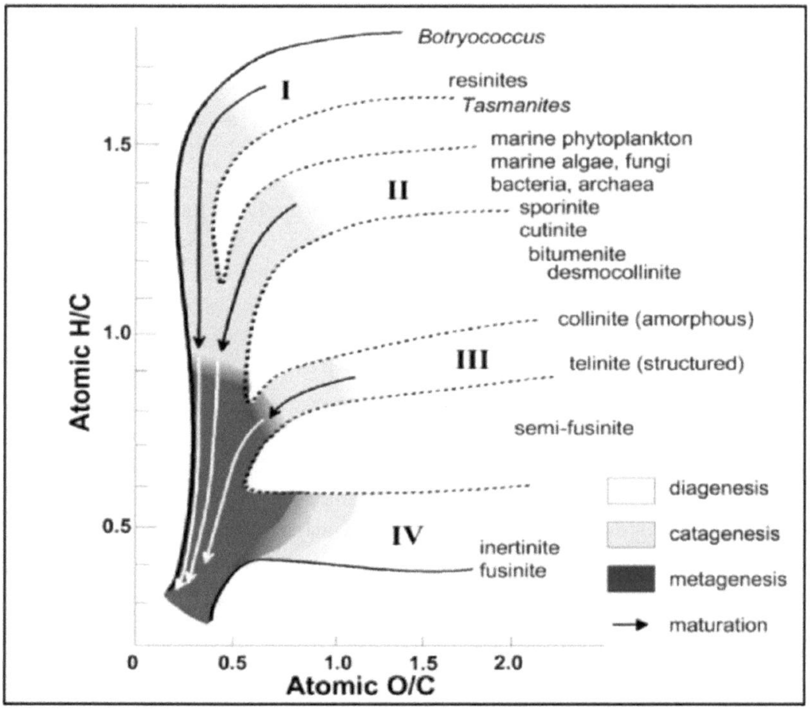

FIGURE 2.6 The van Krevelen diagram depicts the distribution of kerogen types and some of their precursor macerals in relation to their H/C and O/C atomic ratios. During diagenesis, a significant amount of their oxygen content is lost. During the process of catagenesis, the H/C ratio of the remaining kerogens decreases, resulting in oil production. When the H/C ratio <0.5, the remaining kerogens can produce methane.

Source: Walters, 2015

2.8.2 Kerogen Types

Kerogen is a collective name for organic material insoluble in organic solvents, water, or oxidizing acids (Bjørlykke, 2010). The portion of the organic material soluble in organic solvents is called bitumen, which is oil in a solid state. Kerogen quality is mainly a function of hydrogen content, oil-prone exinite, and alginite, as well as hydrogen-poor inertinite (Rezaee, 2015); kerogens with high H/C ratios (> 1.2) are oil prone while those with lower H/C ratios (0.5 to 0.8) tend to generate mostly gas (Walters, 2015). Kerogen consists of very large molecules and is a kind of polymer. When exposed to sufficient time and temperature, these large molecules will crack into smaller molecules, mostly petroleum. When the temperature is about 100°C, a long geological time is required. In rapidly subsiding basins, the exposure time is shorter, and oil generation may only start at about 140 to 150°C. In the North Sea basin, the "oil window" may typically be between 130 and 140°C (Bjørlykke, 2010).

Many approaches try to categorize kerogen types based on their morphology, maceral and palynological assemblages, and/or bulk chemical content (Walters, 2015). The most widely used method is a modified method established by Espitalie et al. (1977), which is utilized by coal petrologists to plot the H/C and O/C ratios. Kerogens are designated as type I, type II, type III, or type IV, based on their position on the plot. Types I and II are oil prone, type III is susceptible to gas, and type IV is inert carbon. The H/C ratios of these organoclays fell within certain ranges, as determined by elemental analysis (Selley & Sonnenberg, 2014). This classification was initially designed for coal but is currently applied to study other kerogen-rich sedimentary deposits.

Type I kerogen is essentially algal in origin/Sapropelic. It has high initial hydrogen-to-carbon (H/C) and low initial oxygen-to-carbon (O/C) ratios. The H/C ratio is about 1.65 (Table 2.1). This kerogen is rich in lipid-derived material and is usually generated from algal organic matter in lacustrine (freshwater) settings but not always. On a mass basis, rocks containing type I kerogen release the most hydrocarbons when pyrolyzed. As a result, from a theoretical standpoint, shales containing type I kerogen are the most promising deposits for conventional oil retorting (Tissot & Welte, 1984).

Type II or liptinitic kerogens are of intermediate composition (Selley & Sonnenberg, 2014). Intermediate initial H/C and O/C ratios distinguish this kerogen type. Type II kerogen primarily comprises marine organic elements deposited in reducing sedimentary settings. The chemical structure is shown in Figure 2.7. The original organic matter of type II kerogen consisted of algal detritus and contained material derived from zooplankton and phytoplankton. The sulfur level of type II kerogen is typically higher than that of other kerogen types, and significant amounts of sulfur are present in the associated bitumen (Walters, 2015). Although pyrolysis of type II kerogen releases less oil than type I kerogen, the amount still suffices for type II–bearing sedimentary formations to qualify as petroleum source rocks.

2.8.2.1 Type III: Humic

Type III kerogens, called humic kerogens, are characterized by low initial H/C ratios and high initial O/C ratios. Chemically, it is low in aliphatic compounds but rich in aromatic ones (Selley & Sonnenberg, 2014). Type III kerogens are derived from terrestrial plant matter, specifically the lignin of the higher woody plants, which grow on land. It is this humic material that, if buried as peat, undergoes diagenesis to coal. Coal is an organic-rich sedimentary rock that is largely made of this form of kerogen. On a mass basis, type III kerogens generate the lowest oil yield of principal kerogen types. The ratio of H/C and O/C is < 1. All kerogen types except type IV can yield commercial amounts of petroleum that will crack to gas. Type IV organic matter will only yield minor amounts of dry gas but could act as a carbon catalyst.

The van Krevelen diagram depicts the distribution of kerogen types and some of their precursor macerals regarding their H/C and O/C atomic ratios (Figure 2.6), whereas Figure 2.7 shows the molecular structure of (A) type I, or algal, kerogen; (B) type II, or liptinitic, kerogen; and (C) type III, or humic, kerogen.

2.8.3 Maturation Process

The chemistry of organic materials in sedimentary rocks changes with time, reflecting the temperature and burial history of the rock. This change, or maturation, can be quantified and paired with data on quality and richness to estimate the quantity of hydrocarbon produced by organic matter. The maturity level results from several factors, including tectonic setting, burial history, and thermal history (Law, 2019). Maturity can be related to the stages of hydrocarbon generation and can be used to constrain geological models (Cornford, 2005). It is defined in terms of temperature and time, using measured indicators such as rock-eval T_{max}, vitrinite reflectance, and spore color indices. The maturation of petroleum occurs in three stages— diagenesis, catagenesis, and metagenesis—which result in the separation of small molecules and the formation of kerogen, a more resistant residue (Tissot & Welte, 1984).

Diagenesis is considered a relatively low-pressure, low-temperature alteration process than the one responsible for metamorphism. Catagenesis is next stage after diagenesis. It usually refers to the stages of kerogen decomposition during which oil and wet gas are produced. Metagenesis, which occurs after catagenesis, represents

TABLE 2.1
Chemistry of Kerogen

	Weight Percent					Ratios		
	C	H	O	N	S	H-C	O-C	Petroleum Type
Type I Algal	75.9 (7 samples)	8.84	7.66	1.97	2.7	1.65	0.06	Oil
Type II Liptinic	77.8 (6 samples)	6.8	10.5	2.15	2.7	1.28	0.1	Oil and gas
Type III Humic	82.6 (3 samples)	4.6		2.1	0.1	0.84	0.13	Gas

Source: Adapted from Selley & Sonnenberg, 2014

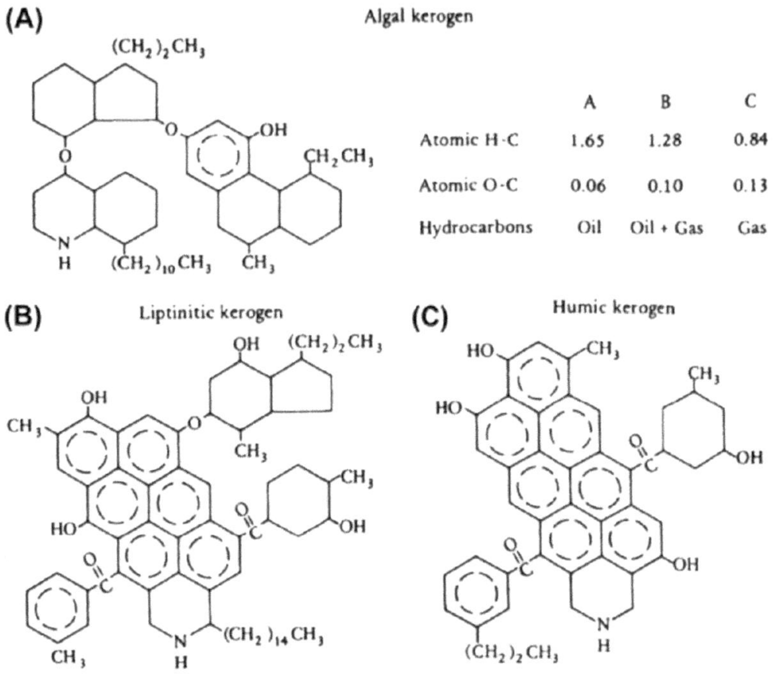

FIGURE 2.7 The molecular structure of (A) type I, or algal, kerogen; (B) type II, or liptinitic, kerogen; and (C) type III, or humic, kerogen.

Source: Selley & Sonnenberg, 2014)

a dry-gas generation. Despite its name, metagenesis is not equivalent to "metamorphism." Metagenesis begins long before true rock metamorphism but continues through the metamorphic stage (Tissot & Welte, 1984; Waples, 1985b). Although catagenesis and oil generation are often used synonymously, they are not precisely equivalent. Catagenesis and hydrocarbon generation occur concurrently, representing different aspects of the same process. Catagenesis refers to transformations of kerogen molecules, whereas hydrocarbon generation focuses on the smaller product molecules. In principle, however, they represent fundamentally different perspectives.

The thermal evolution of the source rock throughout diagenesis, catagenesis, and metagenesis alters several physical and chemical characteristics of organic matter. These characteristics may be used to predict maturation. The parameters most typically utilized in petroleum exploration include optical inspection of kerogen, a physicochemical study of kerogen, and chemical analysis of extractable bitumen (Tissot & Welte, 1984).

2.8.4 Thermal Maturity

Determining the degree of kerogen maturity in the source rocks of an area exposed to petroleum exploration is crucial. Thermal maturity refers to the extent to which heat-driven reactions change the composition of organic matter (e.g., conversion of sedimentary organic matter to petroleum or cracking of oil to gas) (Peters et al., 2012). During the thermal maturation, kerogen breaks in high temperature pyrolysis operations to create lower-molecular-weight products such as bitumen, oil, and gas. The thermal maturation level determines the product type, with lower thermal maturities yielding predominantly bitumen or oil and higher thermal maturities generating gas. Different geochemical indexes, such as vitrinite reflectance, pyrolysis T_{max}, and biomarker maturity ratios, can be used to indicate the level of thermal maturity of organic matter. Many geochemical scales, including vitrinite reflectance, pyrolysis T_{max}, and biomarker maturity ratios, can be used (Herron et al., 2014; Vandenbroucke & Largeau, 2007; Waples, 1985a).

Thermal maturation data used to model these parameters are usually derived from fossils. Thermal maturity describes the organic matter decompositions in the oil or gas window. While the oil window is often considered exclusive of gas, some gas is generated in all source rocks. However, the oil window predominantly contains liquid petroleum products with varying amounts of gas relative to oil, depending on the kerogen type. In contrast, the gas window has only remnants of liquid petroleum and much more gas, depending on how mature the system is.

Crude oil and natural gas are generated in source rocks that contain the necessary amount of organic matter. The best example of source rock is shale rock. To generate crude oil or biogenic or thermogenic gas, organic matter found in shale rock must first be processed by specialized groups of bacteria or buried deep enough and heated to achieve the thermal maturity required for thermogenic gas generation (Grotek & Janas, 2014).

First, thermal maturity begins with diagenetic changes in organic matter trapped in source rocks that form hydrocarbons, progressing from the immature phase to the primary phase of oil and gas generation and finally to the post-maturity phase (Waples, 1985a; Grotek & Janas, 2014). The level of thermal maturity is mostly determined by the highest paleotemperature applied to the rocks during their geological history, as well as the period spent exposed to that temperature. Maturity increases with burial depth because of the greater temperatures found in the Earth's crust. The rise is strongly related to the geothermal gradient, often represented as degrees Celsius per kilometer. However, pressure has little effect on the thermal maturity of organic matter since pressure slows the chemical reactions involved in maturation processes. Nevertheless, it affects physical qualities such as porosity and particle structural layout. With increasing concentrations of stable organic components, the activity leads the biomass to emit mobile products (gases and liquid hydrocarbons). In later maturation (diagenesis) phases, these components lose their functional groups with oxygen, sulfur, and nitrogen while becoming more carbon rich. These processes modify the optical properties of organic materials. Figure 2.8 shows a thermal maturity indicator with a hydrocarbon generation window (Abeed, 2020).

2.9 CORE AND GEOCHEMISTRY ANALYSIS

Geochemical and petrographic techniques determine organic carbon's quantity, quality, and maturity.

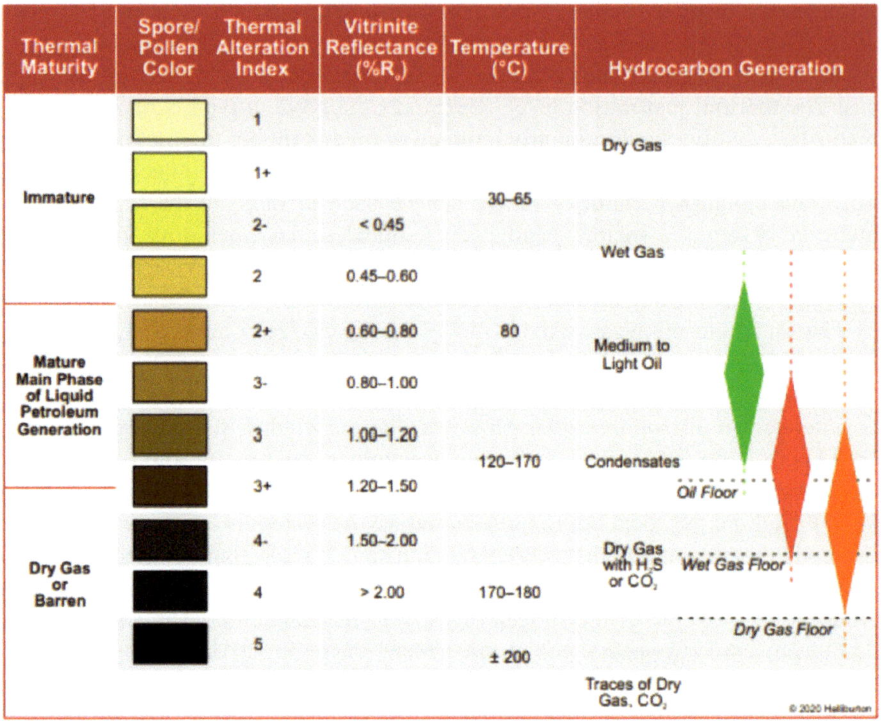

FIGURE 2.8 Thermal maturity indicators and hydrocarbon generation window.

Source: Abeed, 2020

2.9.1 Pyrolysis Analysis

The decomposition occurs due to limited thermal stability in chemical bondings and can break due to heat. This decomposition often results in the synthesis of smaller molecules while the resultant fragments may occasionally interact and produce larger compounds than the initial molecule. Some molecules decompose due to heating at temperatures lower than 250 to 300°C; these processes are often called thermal decompositions rather than pyrolysis (Moldoveanu, 2019). Organic geochemists use pyrolysis to assess the richness and maturity of potential source rocks. Figure 2.9 is a schematic diagram of the rock-eval pyrolysis apparatus.

Pyrolytic processes may be performed intentionally at the laboratory or industrial scale or may take place unintentionally. A broad classification of pyrolysis types can be based on the conditions in which the process takes place, as follows: (1) industrial scale preparative pyrolysis, (2) analytical pyrolysis, (3) preparative pyrolysis at laboratory scale, (4) pyrolysis associated with burning, and (5) pyrolysis performed to simulate a specific process.

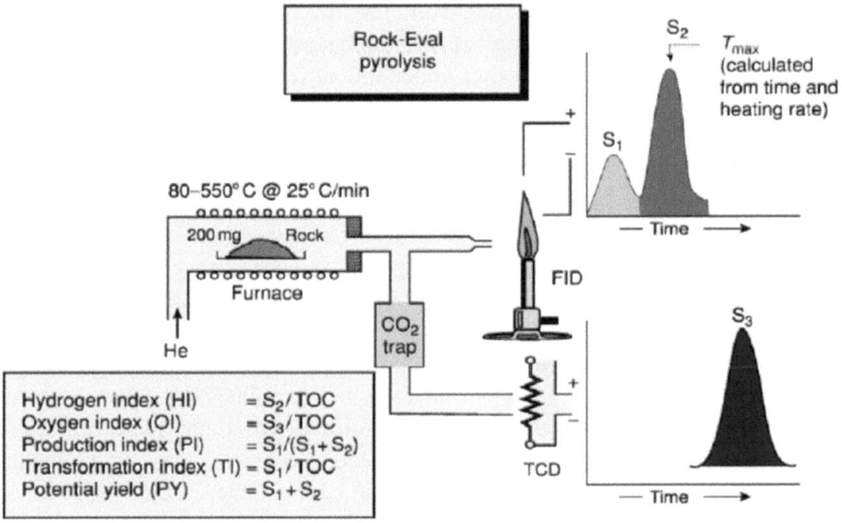

FIGURE 2.9 Schematic of the rock-eval pyrolysis apparatus, output, and (with TOC) derived parameters.

Source: Cornford, 2005

2.9.1.1 LECO and Rock-Eval

Rock-eval pyrolysis is used in petroleum explorations to measure the quantity, quality, and thermal maturity of organic matter in rock samples (Espitalie et al., 1977; Peters, 1986), monitoring the presence of migrated hydrocarbons in suspected reservoir intervals (Espitalie et al., 1977). It can help determine the phase of generated hydrocarbon by interpreting kerogen type and maturity (T_{max}). In unconventional systems, pyrolysis can answer many other questions, such as calculating the potential for hydrocarbons to remain trapped in the source rock (Abeed, 2020). Programs every 30 to 60 feet (9–18 meters) are necessary for accurate results, along with knowledge of interpretive errors and supporting information such as visual kerogen, vitrinite reflectance, and elemental analysis (Peters, 1986). When combined with TOC measurements, this is the most efficient and cost-effective screening method for large samples. Pyrolysis/TOC data are frequently complemented with vitrinite reflectance and spore color (TAI) data to generate geochemical logs of wells.

In rock-eval pyrolysis, for example, rock-eval6, a sample is placed in a vessel and gradually heated to 550°C in an inert environment. During the analysis, the hydrocarbons already present in the sample are volatilized at a low temperature. The volume of hydrocarbons is observed and recorded as a peak known as S1. Kerogen in the sample is then pyrolyzed, producing hydrocarbons and hydrocarbon-like substances (recorded as the S2 peak), CO_2, and water (Tissot & Welte, 1984). The CO_2 generated is recorded as the S3 peak. Residual carbon is also observed and is recorded as S_4. The parameters most frequently reported are TOC, S_1, S_2, S_3, and T_{max}, together with several derived parameters (HI, OI, PP, and PI).

Table 2.2 is the result of the rock-eval. The revealed organic matter in effective source rocks for petroleum satisfies three geochemical requirements. The rock-eval parameters are compared with other geochemical indicators, including TOC, bitumen and hydrocarbon yields, vitrinite reflectance, and thermal alteration index. It was never intended for pyrolysis to be utilized independently from geochemistry (Ma & Holditch, 2015; Peters, 1986). To assess conventional targets, rock-eval pyrolysis and TOC measurements are recommended approximately every 10 meters (32.8 ft) in each well. However (Ma & Holditch, 2015), closer sample spacing (e.g., 1 meter or less) may be required to identify sweet spots in unconventional rock units properly (Ma & Holditch, 2015). More on rock-eval analysis, including pitfalls, is discussed in Peters (1986 and Peters and Mary (1994).

The Laboratory Equipment Corporation (LECO) is a method of estimating TOC using an instrument, a LECO carbon analyzer, to determine TOC values by combusting the organic carbon and measuring the resulting carbon dioxide produced (Law, 2019). Rock-eval pyrolysis has nearly completely replaced the LECO technique. However, data from previous analyses may still be available. Before LECO analysis, samples are pulverized, weighed, and chemically treated to remove the rock's inorganic carbon (carbonate). The sample is then combusted in the presence of extra oxygen, forming carbon dioxide from the free (organic) carbon in the rock. The amount of carbon dioxide produced is proportional to the amount of organic carbon in the rock, also known as the TOC. However, if sulfur compounds, water, and carbonate are present and not removed prior to analysis, the TOC value may be exaggerated. The TOC measured by the LECO method does not account for free hydrocarbons in the sample. The free hydrocarbons are volatilized when samples are dried after acid methods for removing inorganic carbonate minerals. As a result, if a sample contains a significant concentration of free hydrocarbons, the LECO TOC value will

TABLE 2.2
Rock-Eval Pyrolysis Peaks and What They Represent

Parameter	Definition	Measurement	Application
TOC	Total organic carbon	Wt % of Corg/rock	Richness
S_1	Free bitumen	Wt % of HC/rock	Oil or gas shows
S_2	Hydrocarbon potential	Wt % of HC/rock	Richness/quality
S_3	CO_2 potential	Wt % CO_2/rock	Quality
T_{max}	Temp of maximum generation rate (i.e., S_2)	°C	Maturity
HI	Hydrogen index	100*S2/TOC	Kerogen quality/typing
OI	Oxygen index	100*S3/TOC	Kerogen typing
PP	Petroleum potential	S1 + S2	Oil and gas potential
PI	Production index	S1/(S1 + S2)	Maturity/transformation ratio

Source: English, 2020

be lower than a rock-eval TOC value, which includes free hydrocarbons (S1) in the TOC calculation (Law, 2019). LECO techniques were applied to determine source rock richness in different basins (Aladesanmi et al., 2018; Musa Adamu et al., 2020; Abubakar et al., 2023).

2.9.1.2 Source Rock Analyzer (SRA)

The source rock analyzer (SRA) is a lab-based geochemical pyrolysis instrument that is stable and rugged enough for wellsite applications. Geochemical logging of active or previously drilled wells can aid in the determination of pay zones, identify the source rock and its maturity, and estimate the dead oil content from the cuttings for reservoir identification and thickness. Based on the hydrocarbon type and physical parameters, it is possible to evaluate the pay zones in conventional or unconventional reservoirs, especially those with long horizontal legs. As a result, the instrument may offer information that is close to real time on reservoir parameters and details about the best intervals to stimulate and complete, all of which can and will have a significant impact on drilling operations and output on a per-well basis.

The SRA was designed to help identify and characterize source rock and reservoir rock by heating geologic samples (i.e., outcrops, cuttings, conventional cores, and sidewall cores) to a programmed temperature in an inert atmosphere. The SRA quantitatively determines the amount of free hydrocarbons (S1) and the amount of hydrocarbons generated through thermal cracking of nonvolatile organic matter (S2) using a flame ionization detector (FID). The SRA also determines the temperature at which the greatest release of hydrocarbons from kerogen cracking occurs during pyrolysis (T_{max}) and TOC. Jin et al. (2012) and Reservoir (1998) performed anhydrous pyrolysis analysis with the SRA.

2.9.1.3 Hydrocarbon Analyzer with Kinetics (HAWK)

The hydrocarbon analyzer with kinetics (HAWK) analyzes rock samples (core, cuttings, and outcrops) to determine TOC, oil (S1), hydrocarbons yield from kerogen pyrolysis (S2), pyrolysis measurements of both CO and CO_2 (S_3CO and $S3CO_2$), drilled depth maturity (T_{max}), and non-generating organic carbon content as determined by the oxidation cycle's CO and CO_2 (S_4CO and S_4CO_2) (Maende & David Weldon, 2013). The instrument is built on rock-eval and source rock analyzer fundamentals. HAWK produces the traditional source rock pyrolysis/oxidation values with much more reproducibility and reliability. It can determine all the classical pyrolysis parameters using only about 100 milligrams of 60 mesh-size ground rock samples (Reservoirgroup, 2019). In addition, HAWK pyrolysis measures inorganic carbon, giving you the carbonate carbon content of your rock samples (Chiadikobi & Chiaghanam, 2018). After pyrolysis measurements have been completed, a few rock samples are examined using the HAWK to determine kinetic parameters. These kinetic parameters include a distribution of activation energies (Ea) and a single Arrhenius factor (A). These are then used to quantify kerogen decomposition rates utilizing rate equations and an assumed first-order reaction (Maende & David Weldon, 2013).

2.9.1.4 X-Ray Fluorescence (XRF) Spectrometry

An X-ray fluorescence spectrometry is a physical technique that involves using an XRF spectrometer for routine, relatively nondestructive chemical analyses of rocks, minerals, sediments, and fluids. The elemental analysis in geological materials using X-ray fluorescence is possible by the behavior of atoms when they interact with radiation (*X-Ray Fluorescence (XRF)*, 2020). The X-rays penetrate the sample and encounter and excite elements. As each excited element returns to its ground state, it emits X-rays with a unique energy distribution, and X-rays are detected by the instrument. Because the energy of the produced photon essentially indicates a transition between specific electron orbitals in a specific element, the resulting fluorescence X-rays can be utilized to detect the abundance of elements present in the environment (Spencer & Weedmark, 2015; *X-Ray Fluorescence (XRF)*, 2020). Hence, the obtained data are proportional to the number of atoms of a particular element in the stimulated volume. However, because the equipment is inaccurate, the data obtained must be calibrated to standards. Once the data is collected and calibrated, the mineralogy is determined using normative mineral algorithms. Reservoir properties are determined using a "particular mineral interaction" model (Spencer & Weedmark, 2015). The general formula is:

$$PR = \alpha\left(RP\right) + \Sigma\beta_i^0 DV_i + \Sigma\Sigma\beta_{i.j}^1 DV_i DV_j + \Sigma\gamma_i CV_i + \Sigma\theta_i RV_i \tag{2.1}$$

where PR is a reservoir property that includes permeability, porosity, Young's modulus, Poisson's ratio, etc.; $\alpha(\mathrm{RP})$ is a function of the reservoir property while β^0, β^1, γ and θ are fabric-dependent phenomenological coefficients vary as a function of T, P (depth), and rock composition. DV_i is a solid phase I detrital volume fraction, CV_i is the cement volume fraction of solid phase I, and RV_i is the recrystallized volume fraction of solid phase i.

X-ray fluorescence (XRF) analyses provide a large amount of data from oil sand cores quickly, with minimal sample preparation and reasonable cost. XRF analyses provide highly precise and, if properly calibrated, reliable data on the bulk chemistry of a material. In order to provide mineral compositions and reservoir parameters, the chemical data can be transformed into formats more useful to geoscientists and engineers (Spencer & Weedmark, 2015).

The bulk chemical composition of rocks may be determined inexpensively using XRF analysis. This approach can directly analyze rock powders, cuts, slabs, or core faces. Slabbed oil sand core surfaces can be immediately analyzed with portable XRF devices, avoiding the requirement to remove bitumen from the sample before examination. Elemental data derived from XRF analysis can be used to determine a chemical stratigraphy. Geoscientists and engineers, on the other hand, are more conversant with and work more with mineralogy and reservoir qualities than elemental composition. Spencer and Weedmark (2015) developed normative mineral and reservoir property algorithms to transform the elemental data into these more familiar forms (Figure 2.10).

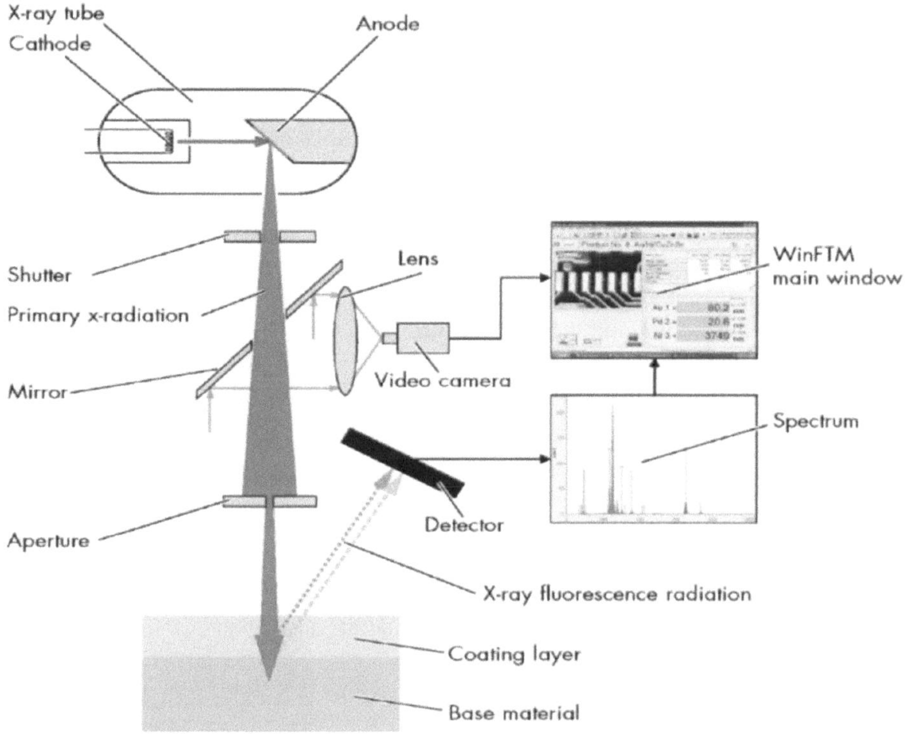

FIGURE 2.10 Functional principle of X-Ray fluorescence spectroscopy (XRF) instrument.

Source: Spencer & Weedmark, 2015

2.10 TOC WEIGHT PERCENTAGE DETERMINATION

TOC measurement and rock-eval pyrolysis analysis are the most time- and cost-effective methods in petroleum geochemistry to answer whether or not a source rock is present in a basin by measuring the quantity, quality, and maturity of its organic matter content. TOC analysis (wt%) measures the combined quantity of soluble (bitumen) and insoluble (kerogen) organic matter present in a rock; it does not represent the amount of kerogen alone. The kerogen is the part of organic matter that can produce hydrocarbons upon heating. This measurement is performed on finely ground rock samples in a combustion apparatus utilizing an induction furnace and a thermal conductivity cell to measure evolved CO_2 (Tissot & Welte, 1984; Abeed, 2020).

TOC analysis is a good starting point for source rock evaluation, and all good source rocks must have (or have had) a high TOC. However, exploration geoscientists should be careful not to draw a quick and unrealistic source rock evaluation based on the initial results of this analysis, ignoring other factors like lithology, kerogen type, organic matter preservation, and maturity. Not all organic matter is created equal;

some organic matters are capable of generating liquid hydrocarbons (oil prone), some can only generate gas (gas prone), some can generate a mixture of hydrocarbons, and some are inert and cannot generate anything (Tissot et al., 1978; Dembicki, 2009; Peters et al., 2016).

To find the total amount of organic matter, multiply TOC by 1.2 and assume that organic matter contains 83 wt% carbon. The minimum TOC for carbonate versus noncarbonate source rocks does not significantly differ, in contrast to certain past findings (Brooks & Welte, 1987). Excellent quantities of TOC rock for thermally immature source rock are typically > 4 wt%. To compensate for this, appropriate TOC levels for very mature, unconventional source rock are > 2 wt%. TOC, however, diminishes with thermal maturity. Moreover, TOC analysis is maturity sensitive, so TOC (wt%) decreases as hydrocarbons are generated and expelled, making it a misleading indicator in mature source rock (Abeed, 2020).

$$TC(\%) = \frac{Carbon(g)}{W(g)} \tag{2.2}$$

$$TOC(\%) = \frac{Organic\,Carbon(g)}{W(g)} \tag{2.3}$$

$$TIC(\%) = TC(\%) - TOC(\%) \tag{2.4}$$

2.10.1 Passey Delta Log R (ΔLogR) Method

This method is an empirical relationship introduced by Passey et al. (1990) to quantify TOC that is driven by the elevated formation resistivity and elevated apparent porosity typical in organic shales, as described earlier. The technique involves using well logs to obtain the organic richness. The $\Delta LogR$ method relies on the response of well logs to organic matter using an overlay of porosity logs and resistivity logs (Passey et al., 1990). A porosity curve, typically sonic porosity, is overplotted with resistivity and usually scaled at $(-50\mu s / f\,or - 164\mu s / m)$(Kenomore et al., 2017). The difference between these curves, called $\Delta LogR$, is equated to TOC content. This method showed that a combination of the sonic curve and resistivity curve provided the best results, as the sonic log is not severely affected by in-hole conditions compared to the density or neutron readings (Kenomore et al., 2017). An overlay of sonic and resistivity curves over mature organic-rich intervals causes a separation: i.e., a $\Delta LogR$ separation occurs (Eq. 2.2). The $\Delta LogR$ separation is directly proportional to TOC and is a function of maturity (Passey et al., 1990) (eqs. 2.3 and 2.4) that is an estimate of thermal maturity, and kerogen type is necessary to convert $\Delta LogR$ to TOC (Gonzalez et al., 2013). This means that the separation can be converted to TOC, providing the level of maturity (LOM) is known—the latter is dependent on estimating thermal maturity through T_{max} or R_o (Eq. 2.3) (Gonzalez et al., 2013).

$$\Delta LogR_{sonic} = log_{10}(R / R_{baseline}) + 0.2 \times (\Delta t - \varnothing t_{baseline}) \tag{2.5}$$

$$Level\ of\ maturity(LOM) = 0.09890.0989^{**5} - 2.1587^{*} R_o^{**} 4 + 12.392^{*} R_o^{**} 3 - 29.032^{*} R_o^2 + 32.53^{*} R_o - 3.0338 \quad (2.6)$$

$$TOC = (\Delta logR) \times 10^{(2.297 - 0.1688^{*} LOM)} \quad (2.7)$$

The methodology is based on the assumption that the matrix properties of the baseline shale sections and the organic shale are comparable in terms of resistivity and apparent porosity (Gonzalez et al., 2013). This resulting TOC estimate is sensitive to the thermal maturity estimate (Issler et al., 2002); however, if the prior value of TOC content is measured, the thermal maturity can usually be changed to create a good TOC log representation (Gonzalez et al., 2013). However, the presence of expandable clays, with low resistivities due to increased clay-bound water, can lead to a significant underestimation of TOC in zones where they are encountered. $\Delta LogR$ was later modified (Passey et al., 2011) to estimate TOC for shales within the gas window (Gonzalez et al., 2013). However, the existence of expandable clays with poor resistivities due to increased clay-bound water might result in a considerable underestimation of TOC in zones where they occur. LogR was then adjusted to estimate TOC for shales within the gas window (Gonzalez et al., 2013; Passey et al., 2011).

2.11 PULSED NEUTRON ELEMENTAL SPECTROSCOPY

Pulsed neutron elemental spectroscopy (PNES) is an elemental spectroscopy tool that utilizes a pulsed neutron source to permit the measure of gamma rays emitted by the interactions with the neutrons in both the inelastic and capture energy spectra. Pulsed neutron devices are electronic devices that create high-energy neutron pulses. These high-energy neutrons bombard the formation, losing energy when interacting with rock atoms. Then, the neutrons lose enough energy to be trapped (generally by chlorine that exists as part of the salt dissolved in the formation waters). Whenever a neutron is captured, a gamma ray is emitted. The detectors in the pulsed neutron tool are designed to measure these "capture" gamma rays. Thus, a "capture cross-section" of the formation through casing can be determined. This formation property allows one to estimate the water saturation when porosity and formation water salinity are known. In certain types of reservoirs, the amount of elemental carbon measured exceeds the amount necessary for the mineralogy of the rock itself, which is presented as "excess carbon." Carbon and excess carbon measurement, along with mineralogy determination, have applications for heavy oil and shale gas reservoir evaluation.

PNES technology is used in wireline tools to emit neutrons into the near-wellbore zone and measure the neutron count decay resulting from scattering and capturing. The primary function of this logging equipment is to estimate current oil saturation and identify channeling in perforated and nonperforated intervals behind the casing. Engineers can forecast future perforation intervals in depleted reservoirs if the findings of PNN logging are correctly interpreted (Zaker et al., 2020).

2.12 LITHOLOGY AND MINERALOGY DETERMINATION

2.12.1 Conventional Log Measurements

Conventional well logs are often acquired during and after drilling a well. They provide valuable information about the formation within a few feet of the wellbore. The properties of the formation are inferred from the well-log response. The well-log records are shown as a graph, with the rock's physical property on one axis and depth on the other. Various properties may be displayed on the same graph. By examining well logs, a geologist can identify and infer the depths and characteristics of formations of interest and create a contour map of the formation, as well as the thickness of the formation. The most important geophysical logs are gamma, neutron, density, acoustic, caliper log, and resistivity. The conventional well logs are discussed in detail by the following authors: Segesman (1980), Tittman (1986), Science (1988), Asquith et al. (2004), Baker et al. (2015b), Evenick (2008), Joshua (2012), Kobr (2011), Luthi (2001), Ofwona (2010), Aminzadeh and Dasgupta (2013a) and Ahmed (2018).

2.12.2 Caliper Logging

Caliper logging is used to measure the diameter of a borehole and how it changes with depth. The caliper tool, which has two or more articulated arms that push against the borehole wall, measures the change in borehole diameter as it is pulled from the bottom of the hole. Each arm is often attached to a potentiometer, which varies the resistance as the borehole's diameter changes, resulting in a varying electrical signal representing the borehole's changing shape.

2.12.3 Gamma Ray Log

The gamma ray log is radioactive logging that measures the naturally occurring gamma radiation from the rock around the borehole or drill. Shales emit more gamma radiation than other forms of rock. Because of the difference in radioactivity between shales and other rocks like sandstones/carbonate rocks, the gamma-ray instrument can differentiate between shales and non-shales, but it cannot distinguish carbonates and sandstone as they both have similar deflections on the gamma ray log (Assaad, 2009). The log is usually recorded in track 1 along with the caliper. Scales are local, but 0–100 or 0–150 are common. The American Petroleum Institute (API) is the accepted unit for radioactivity logging.

2.12.4 SP Log

The SP log records the natural potential differences or self-potentials that evolved due to salinity variations between a drilling mud containing salt water and the formation fluid (Asquith et al., 2004). SP log response is created by electric voltages resulting mostly from electrochemical variables within the borehole and nearby rock. Salinity variations between mud filtrate and the formation of water within permeable beds cause these electrochemical variables (Asquith et al., 2004). In practice, salinity is expressed by mud filtrate and formation water resistivity, which are inversely proportional to fluid salinity. Because high voltages require a conductive fluid in the borehole to be generated, the SP cannot be utilized in nonconductive drilling muds or air-filled holes.

2.12.5 Resistivity Log

Resistivity logging is a type of well logging that involves determining the electrical resistivity of the rock or sediment within a borehole in ohm. Resistivity logging is occasionally employed in mineral groundwater exploration. However, oil exploration is used for formation evaluation to differentiate hydrocarbon- and non-hydrocarbon-bearing formations. A high-porosity or hydrocarbon-bearing formation has high resistivity; a low-porosity or saltwater formation has low resistivity (Evenick, 2008).

Numerous resistivity logs exist, most of which correspond to the depth of the measurement: for example, shallow, medium, and deep resistivity. Shallow readings record the interface of the borehole and the drilling fluid while deep readings correspond to true or uninvaded formation resistivity (Assaad, 2009). Hydrocarbons may be present in a formation where deep resistivity is greater than shallow resistivity. Shales typically have low resistivity, and sandstones and carbonates have high resistivity (Asquith et al., 2004). The resistivity of a well is typically measured in ohm meters (Ωm) and is indicated on the fourth track of the well log.

2.12.6 Neutron Log

A neutron log measures a formation's porosity based on the quantity of hydrogen present in the formation. Neutron logs are primarily used to delineate porous rocks and to determine their porosity (Kennedy, 2015). The neutron porosity log is largely sensitive to the number of hydrogen atoms present in a given formation, which is generally equivalent to rock porosity. Similarly, fluids within a formation can be distinguished from gases due to the similar hydrogen content of water and oil, but gases have less hydrogen per unit volume (Assaad, 2009; Kennedy, 2015). Consequently, if a formation contains gas, the compensated neutron log (CNL) will underestimate porosity. Similarly, salt units' neutron porosity and bulk density measurements are typically low.

2.12.7 Sonic Log

The speed at which the rock transmits sound energy is related to the formation's porosity. Sonic logging provides the interval transit time for the speed of the acoustic wave through a formation and is denoted as ΔT. This interval transit time measures how quickly elastic seismic compressional and shear waves pass through the formation (Evenick, 2008). The acoustic travel time is the reciprocal of the sound velocity across the formation. Δt is measured in microseconds per foot (sec/ft) or millionths of a second per foot.

The sonic log measurements can be interpreted in a variety of ways. The Wyllie time-average equation is a common interpretation formula:

$$\varnothing_{sl} = \frac{\Delta t_{sl} - \Delta t_{ma}}{\Delta t_f - \Delta t_{ma}} \tag{2.8}$$

$\varnothing_{sl}$ is the porosity measured from the sonic log, fraction; Δt_{sl} is the value of the acoustic transit time measurement by the sonic log, µsec/ft; Δt_{ma} is the value of the rock matrix acoustic transit time (µsec/ft) measured in the laboratory; and Δt_f is the value of the saturating fluid acoustic transit time (µsec/ft) measured in the laboratory.

For a given formation, the interval transit time varies with lithology and rock textures and, most notably, decreases with increasing effective porosity and increases with increasing effective confining stress. This means that if the seismic velocity of the rock matrix and pore fluid are known, a sonic log can be used to compute the porosity, confining stress, or pore pressure of a deposit, which is extremely valuable for oil exploration (Kennedy, 2015).

A sonic log is a good indicator of density and the presence of gas (Asquith & Gibson, 1983). It also determines porosity in consolidated sandstones and carbonates with intergranular porosity (grainstones) or intercrystalline porosity (sucrosic dolomites) (Assaad, 2009). Sonic measurements will be lower and have locally irregular signatures where gas is present because the gas has a slower transit time. Hydrocarbons in the reservoir rock cause the sonic log to overestimate porosity, and some modifications may be required. These modifications take the form of

$$Gas : \varnothing_{sonic} = 0.7\varnothing_{sl} \tag{2.9}$$

or

$$Oil : \varnothing_{sonic} = 0.9\varnothing_{sl} \tag{2.10}$$

2.12.8 Elemental Spectroscopy Measurements

As the petroleum industry expands due to the increase in the global need for energy and the search for hydrocarbon is an increasingly complex reservoir, an accurate understanding of a formation's elemental composition and mineralogy is an essential factor in the completion decision. Determining matrix mineral composition and TOC can be invaluable to geoscientists and engineers. Elemental spectroscopy is a quick and nondestructive analytical technique that can identify various elements in a sample. There are various techniques of elemental spectroscopy. However, the choice of the method depends on the type of analysis to be carried out.

In gamma-ray spectroscopy (natural and induced gamma rays), a detector measures the energy of incident gamma rays. By comparing the measured energy to the known energy of gamma rays produced by radioisotopes, the identity of the emitter can be determined. This technique has many applications, particularly when rapid nondestructive analysis is required (Raja & Barron, 2021).

2.12.9 Lithology and Mineralogy Characteristics

Lithology logs are used to determine rock type (Glover, 2001; Xu et al., 2013). The SP and gamma ray logs are common logs used in lithology identification. Natural gamma rays are measured by gamma ray logs and are used to distinguish between reservoir-forming and non-reservoir rocks. Shale volume is determined using gamma ray logs, which are also used to correlate zones and define lithologies. Low radioactive material concentrations in sandstones and carbonates, which result in low gamma ray measurements, are found in shale-free rocks (Asquith & Krygowski, 2014). The gamma ray log response in Table 2.3 increases as shale content increases due to the concentration of radioactive material in shale. However, potassium-feldspar, micas, and glauconite produce a high–gamma ray response in clean or shale-free sandstones. In this case, a "spectral" can be run in addition to the gamma ray log because it breaks the natural radioactivity of a formation into the different types of radioactive material such as potassium, thorium, and uranium. Gamma ray, diameter, and induction/dual induction logs are good for saltwater and oil-mud systems. The gamma log is usually presented in track 1 along with the caliper. Scales are local, but 0–100 or 0–150 are common. The American Petroleum Institute (API) is the accepted unit for radioactivity logging.

A simple gamma ray log is used for in-depth correlation in different operations in a well and is also used for lithology identification and calculation of the gamma ray index and the volume of the shale (Assaad, 2009; Mandeep, 2020). Calculation of the volume of shale:

$$Gamma\,ray\,index = \frac{GR(log) - GR(min)}{GR(max - GR(min)} \tag{2.11}$$

Like the gamma log, the SP log determines gross lithology through its ability to distinguish permeable zones (such as sandstones) from impermeable zones (such as shales). It also correlates zones between wells (Asquith et al., 2004). A shale baseline is a straight line that describes how shales respond, which is generally constant. From the shale baseline, SP curve deflections are calculated. The SP curve can also detect hydrocarbons by suppressing the SP response. A permeable formation's SP deflection is decreased by shale. "Hydrocarbon suppression" refers to situations in which

TABLE 2.3
The Principal Gamma Ray Responses

Lithology	Log Shapes Gamma Ray Values (in API Units)
Sandstone (quartz)	15–30 (rarely to 200)
Limestone	10–40
Dolomite	15–40 (rarely to 200)
Shale	60–150
Organic-rich shale	100–250
Anhydrite, halite	8–15
Sylvite (KCI)	350–500
Coal	15–150 (any value possible)

Source: Morton-Thompson & Woods, 1993

SP reduction in an area is larger than the volume of shale (Assaad, 2009; Morton-Thompson & Woods, 1993). Thin beds, shaliness, and the presence of gas can suppress the SP curve. The spontaneous potential can be estimated from the empirical relationship (Eq. 2.12) (James, 1999; Selley & Sonnenberg, 2014).

$$E_{SP} = klog\frac{R_{mf}}{R_w} = \left(a_1 + a_2T\right)log\frac{R_{mf}}{R_w} \tag{2.12}$$

where

k is a linear function of temperature T with parameters a_1 anda_2.

$R_{mf}$is the resistivity of the mud filtrate.

$R_w$is the resistivity of formation water.

The temperature dependence of resistivity must be considered when using Eq. 2.12.

2.13 POROSITY DETERMINATION

Rock porosity is the ratio of the volume of space between the solid particles of the rock to the total rock volume. The space includes all pores, cracks, vugs, and inter- and intra-crystalline spaces (Glover, 2000). Conventionally, porosity is given the symbol ∅ and expressed as either a fraction between 0 and 1 or a percentage between 0% and 100%. The porosity of a rock is mathematically defined by equation 2.13:

$$\text{The porosity of rock, \%} = \frac{\text{Volume of porous spaces in rock}}{\text{Bulk volume of rock}} \times 100 \tag{2.13}$$

The development of void spaces, some of which are nonconnected or have isolated pores (vugs, moldic pores, etc.), as a result of the formation of the rocks and the deposition of sediments, is what is referred to as "noneffective" fluid movement inside the rock. Therefore, the proportion between the connected pore volume and the overall rock volume is effective or interconnected porosity. In describing reservoirs, it is crucial to make the following distinctions: Hence, absolute and effective porosity are mathematically defined by equations 2.14 and 2.15, respectively.

$$\text{Absolute porosity, \%} = \frac{\text{Volume of all poresand voids in rock}}{\text{Bulk volume of rock}} \times 100 \tag{2.14}$$

$$\text{Effective porosity, \%} = \frac{\text{Volume of interconnected pores in rock}}{\text{Bulk volume of rock}} \times 100 \tag{2.15}$$

It should be noted that effective porosity is lower than absolute porosity because it only considers connected pores. Understanding effective porosity is crucial because only an interconnected porous network can allow the flow of oil and gas (Satter & Iqbal, 2015). Therefore, any fluid in dead-end pores does not contribute to the production.

Porosity is calculated using traditional core and well-log data. Because porosity is determined directly, core data are typically more accurate (Hook, 2003). LWD or NMR logs can be used to calculate porosity using well-log data. In LWD, neutron, density, and acoustic logging techniques are frequently utilized to measure porosity. Typically, the subsurface instruments used in boreholes to evaluate different rock qualities consist of a source and one or more receptors. The source releases energy into the formation, such as acoustic, electric, or radioactive energy, which bounces back to the receptors with a signature, depending on the properties of the rocks and the reservoir's fluid saturations. Similarly, the NMR approach is based on the observation that the NMR signal is inversely proportional to the quantity of hydrogen atoms in the fluid in the rock pores. The NMR tool's answer shows how often the formation fluid's hydrogen nuclei are present (Hook, 2003).

In the laboratory, porosity is calculated using the following equation (Ahmed, 2019):

$$porosity = \varnothing = \frac{V_p}{V_b} = \frac{V_b - V_m}{V_b} \tag{2.16}$$

where V_p= pore space volume, V_m = matrix (solid rock) volume, and V_b= bulk volume = $(V_p + V_p)$. We need two out of these three values to determine porosity. Bulk volume (V_b) can be determined directly from core dimensions if we have a fluid-saturated regularly shaped core (normally cylindrical), or by fluid displacement methods by weight where the density of the solid matrix and the displacing fluid are known, or directly by volume displacement.

However, the core porosities may be more accurate unless every well is cored. For each well that has been cored and logged, it is advised to compare the porosities of the core and log at the same depth. By modifying the inputs into the log calibration (such as matrix density or matrix transit time) or by employing the proper averaging, the log porosities are changed to match the core data (Baker et al., 2015b).

2.14 PERMEABILITY DETERMINATION

The permeability of rock describes how easily reservoir fluids can travel across the porous network. Permeability is one of the most important elements influencing reservoir performance, and it is sometimes the most difficult reservoir metric to assess precisely due to its great variability (Baker et al., 2015a). The permeability of a rock is determined by its effective porosity, which is influenced by rock grain size, grain shape, grain size distribution (sorting), grain packing, and the level of consolidation and cementation (Tiab & Donaldson, 2016a). The type of clay or cementing material between sand grains also affects permeability, especially where fresh water is present. Some clays, particularly smectites (bentonites) and montmorillonites, swell in fresh water and have a tendency to partially or completely block the pore spaces.

Permeability can be measured by letting a liquid or gas flow through a cylindrical rock sample under pressure. (See Figure 2.11.)

Conventional reservoir permeability is higher (usually between D and nD) than unconventional reservoirs (μD to nD). Rock matrix permeability in unconventional reservoirs is found to be extremely low, rendering traditional production methods ineffective. Technological advancements enable profitable oil and gas recovery from unconventional resources, where natural or artificial conduits for fluid flow play an important role (Tiab & Donaldson, 2016a).

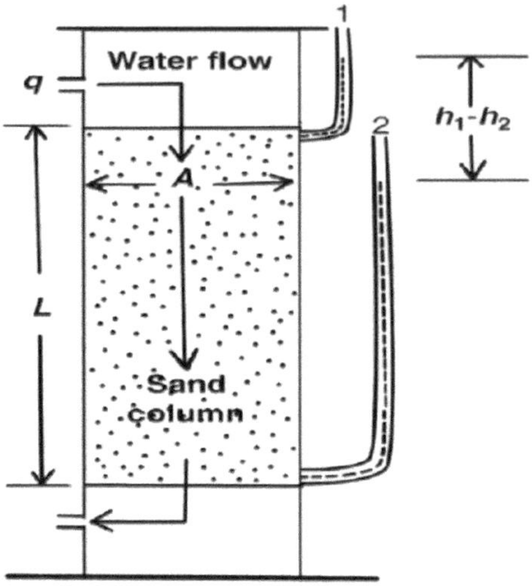

FIGURE 2.11 Schematic of hydraulic conductivity measurement apparatus.

Source: Ghulam & Iqbal, 1992

Permeability is estimated using traditional core data, permeability-porosity correlations, pressure transient analysis, and production rates (Baker et al., 2015a). None of these methods, however, allow a direct assessment of in situ permeability and permeability distribution. The most difficult component of reservoir characterization is undoubtedly the reconciliation and extrapolation of permeability data. The best results are obtained when all four permeability-determining methods are applied together. The procedures are summarized next. Relatively high values of rock permeability lead to good well productivity and better recovery efficiency from the reservoir as a general rule. It connects the fluid flow rate with the applied pressure gradient and viscosity (Kantzas et al., 2005).

2.15 FLUID SATURATION DETERMINATION

In most oil formations, the geological reservoir rocks are fully saturated with water before the oil migration and trapping in the formation. Less dense hydrocarbons are believed to migrate to hydrostatic and dynamic balance positions by displacing the original water. The oil will not displace all the water that was previously present in these pores. As a result, reservoir rocks typically contain both hydrocarbons and water (Kantzas et al., 2005).

The porosity of a reservoir rock is critical because it measures the rock's ability to store fluids (oil, gas, and water). Equally important is the extent to which the pores of the rock are filled with specific fluids. This property is called fluid saturation and is expressed as the fraction, or percent, of the total pore volume occupied by the oil, gas, or water. Thus, saturation Si fluid i in porous rock:

$$S_i = \frac{volume\ of\ fluid}{volume\ of\ pores} \tag{2.17}$$

For a gas-oil-water system, the saturation ratio for every phase in the porous medium is expressed in equations 2.20 and 2.21:

$$S_w = {}^{V_w}\!/_{V_p}, S_o = {}^{V_o}\!/_{V_p}, and\ S_g = {}^{V_g}\!/_{V_p} \tag{2.18}$$

$$S_w + S_o + S_g = 1 \tag{2.19}$$

where S_w is water saturation, S_o is oil saturation, and S_g is gas saturation.

Two procedures are available to determine rock saturation. The first method is evaporation of the fluids in the rock (retort method), and the second is solvent extraction of the fluids from the rock (ASTM method). Both techniques are described in the next section.

2.16 GEOLOGICAL AND GEOMECHANICAL CHARACTERIZATION

2.16.1 Rock Mechanical Properties

Rock mechanical properties are essential for reservoir development and management. Such properties are directly measured from lab tests on formation plugs. The elastic nature of rocks permits wave propagation; in this manner, acoustic-wave velocity can be related to rock elastic properties. This serves as the basis for mechanical-property evaluation by acoustic logs. The log-derived, dynamic rock properties should be calibrated to core-derived static properties (from lab tests) because the static measurements more accurately represent the in situ reservoir mechanical properties (Montmayeur & Graves, 2007).

2.16.2 Dynamic and Static Elastic Properties and Rock Hardness

Determination of a reservoir rock's mechanical properties is central to drilling programs, well placement, wellbore stability, and well-completion design. Furthermore, the accuracy of geomechanical modeling is highly dependent on the rock-mechanical properties, particularly static properties, employed in the model. Typically, two laboratory procedures for determining rock mechanical properties are static and dynamic.

Static properties are mechanical properties of rocks that can be measured in the laboratory using triaxial tests, whereas dynamic properties can be estimated using wireline measurements of compressional and shear acoustic velocities and rock density (Yale & Jamieson, 1994; Hui et al., 2021; Jiang et al., 2021). The static and dynamic moduli of the same rock may differ significantly due to the deformation (strain) amplitude difference between the dynamic and static experiments.

Furthermore, rocks are heterogeneous, so there are considerable differences between static elastic constants and the corresponding dynamic ones.

Usually, static test results are more representative than dynamic methods, but static tests are time consuming and costly. Dynamic methods are nondestructive and less expensive, which is practical in the laboratory and field (Jiang et al., 2021). Similarly, in terms of stiffness, dynamic stiffness is greater than static stiffness (Cheng & Johnston, 1981; Eissa & Kazi, 1988; Fjær, 2009; Lozovyi & Bauer, 2019b; Sone & Zoback, 2013; Tutuncu et al., 1998).

Static moduli are often used in wellbore stability and in situ stress applications to evaluate the possibility of breakouts, elevated pore pressure, and tectonic stress distribution. For example, a common method of calculating the horizontal stress in the Earth is by assuming that the earth is elastic and does not deform in the horizontal direction (Hui et al., 2021).

2.16.2.1 Rock Hardness

The physical and dynamic properties of a rock are governed by the nature of the solid mineral grains, the nature and extent of voids within the mineral aggregate, and the nature of the bond, if any, between the mineral grains (Gerrard, 1988). Rock hardness is an important parameter that characterizes the strength properties of the rock mass (Boutrid et al., 2015). Hardness, as it is usually known today, is resistance to permanent indentation. The hardness of a mineral is intimately tied to its chemistry and atomic structure, and it reflects the material's physical and mechanical qualities to some extent.

Various hardness tests have been established, measuring the material's resistance to scratching or indentation. Mohs proposed a hardness scale based on a standard set of ten minerals (Mohs scale of hardness) to which relative hardness numbers were assigned (Gandhi & Sarkar, 2016). The hardness of a test object is determined by examining whether or not it is scratched by one of the mineral standards. Scratch hardness is a handy tool for a quick but rough assessment and has the advantage of not requiring any instrumentation; nonetheless, it leaves much to be desired as a basis for quantitative measurement and is rarely used in engineering (Boutrid et al., 2015).

2.16.3 Rock Brittleness and Strength

2.16.3.1 Rock Brittleness

Brittleness is one of the important properties of rock, which is of great significance in rock engineering. Correct identification of the brittleness characteristics of the rock mass is important in many field applications, including shale gas recovery through hydraulic fracturing (Zhang et al., 2016). It is a characteristic of most geomaterials in which the pre-existing heterogeneities among the mechanical and geometrical properties of the constituent materials (e.g., grains cementing materials and voids) and loading conditions promote the non-homogeneous distribution of the stresses inside the failing mass and eventually along the potential failure plane (Hajiabdolmajid & Kaiser, 2002). Zhang et al. (2016) define brittle rock as rock subjected to sudden failure with tensile/shear fractures, creating a large strength drop with small inelastic strain.

Brittleness is often employed in rock engineering applications to describe the potential failure characteristics of a rock mass (Zhang et al., 2016). However, there is still much uncertainty about the relation between rock brittleness and rock failures such as hydraulic fracturing and rock bursts in coal mines and tunnels (Xia et al., 2022). The brittleness of rocks is usually quantified by applying the brittleness index (B). It is defined as the ratio of uniaxial compressive strength to tensile strength (Altindag, 2002; Goktan & Yilmaz, 2005).

$$\frac{compressive\,strength}{tensile\,strength} = \frac{\sigma_c}{\sigma_T} \tag{2.20}$$

The brittleness index is the most widely used parameter for quantifying rock brittleness. The higher the magnitude of BI, the more brittle the rock. A general rock classification based on brittleness index values is given in Table 2.4.

2.16.3.2 Rock Strength

The ability of a rock to withstand force without succumbing or fracturing is referred to as its strength. It is affected by the mineralogy of the rock particles and the nature of the particle interactions. These characteristics result from the numerous deposition, diagenesis, and catagenesis processes that created the rock, further modified by folding, faulting, fracturing, jointing, and weathering.

The two most common laboratory methods used to measure the strength of rocks are triaxial or confined compressive strength tests and uniaxial compressive strength tests (Darvishpour et al., 2019). The ultimate strength of a rock, or the highest value of stress reached before failure, is assessed using uniaxial compressive strength tests. One of the easiest metrics of strength to obtain is the uniaxial strength. Its use is restricted, although it typically occurs only when comparing different rocks is necessary. Size and shape of the test sample, rate of loading, amounts and types of fluid contained in the rock sample, mineralogy, grain size, grain shape, grain sorting, and loading rate are some elements that affect uniaxial compression testing.

When test data from core samples are unavailable, a number of relationships involving rock strength have been proposed to assess the wellbore's geomechanical properties. These correlations typically involve measurements that directly impact

TABLE 2.4
Classification of Rock Brittleness

Class	Brittleness Index (BI)	Description
1	>25	Very brittle
2	$15< BI <25$	Brittle
3	$10< BI <15$	Moderately brittle
4	BI<10	Low brittleness

Source: Goktan & Yilmaz, 2005

rock strength, including the elastic modulus and porosity (Bradley, 1979). Of them, the relation between uniaxial compressive strength (UCS) and elastic modulus (E) can be expressed as

$$USC = 2.28 + 4.1E \tag{2.21}$$

2.16.3.3 Pore Pressure

Pore pressure is the pressure of the fluid in the pore spaces within the subsurface. In unconventional reservoirs, pore pressure is important but challenging to quantify. It has a direct impact on the state of stress; it impacts flow rates and production; during drilling, it can cause kicks from permeable formations; and it influences borehole instability in all formations (Higgins-Borchardt et al., 2016).

Pore pressure is normally pressured if it equals the hydrostatic pressure of a water column from the surface to a particular depth. Similarly, it is said to be underpressured if the pressure is less than hydrostatic and overpressured if the pressure is larger than hydrostatic. Underpressuring is frequently caused by depletion. Overpressuring can be generated by rapid sediment burial with retained fluid, fluid formation driven by kerogen maturation, uplift and exhumation, temperature changes, or diagenetic processes (Higgins-Borchardt et al., 2016).

2.16.4 Poroelastic Properties

Poroelasticity refers to the study of porous materials whose mechanical behavior is significantly influenced by the pore fluid. It describes the behavior of porous materials and pore fluid with respect to applied displacements and stresses (Bachrach, 2020). Poroelasticity jointly treats the solid frame and pore fluid to calculate changes in pore pressure, solid displacements, and fluid displacements due to stresses and displacements associated with external or internal processes. Poroelasticity is often called biot theory, and the work of M. A. Biot is considered to have established the complete foundation for the theory (Bachrach, 2020). Biot (1941) and Dormieux et al. (2006) proposed that fluid flow is influenced by the medium's deformation and vice versa. A theoretical extension of soil consolidation models generated to compute the settlement of structures placed on fluid-saturated porous soils proposed the idea.

There are two major parts to poroelasticity: static and dynamic theories. Static poroelasticity is a phenomenon in which water flow and solid skeleton deformation interact concurrently. As a result, it is known as hydromechanics. This static poroelasticity hypothesis is a generalization of the soil mechanics consolidation theory. As a result, poroelasticity is frequently used interchangeably with the consolidation theory or to refer to static poroelasticity in a narrow sense. Dynamic poroelasticity is developed for wave propagation in both the liquid and solid phases of saturated porous materials (Zhen, 2023).

2.16.5 Shale Anisotropy

Anisotropy is defined as the variation in the physical properties of a rock fluid system with respect to direction (Sheriff, 2002b). Two types of anisotropy are present in rocks: intrinsic (inherent) and induced. The preferential orientation of clay platelets, sediment grains, microcracks, fractures, and pores causes intrinsic anisotropy. Intrinsic anisotropy in sedimentary rocks is normally in the form of transverse isotropy. Therefore, understanding anisotropy and its causes is crucial because they strongly influence analyses/interpretations of seismic surveys, sonic logs, and microseismic monitoring (Xu et al., 2005), which can improve reservoir characterization, especially in high-angle and horizontal wells.

Shales frequently exhibit seismic anisotropy resulting from active geological processes before, during, and after deposition. However, the relationship between shale elastic behavior and petrophysical characteristics is more complex (Pavlo & Douglas, 2003).

A very strong anisotropic response characterizes the kerogen-rich shale (Vernik, 1993), which may be related to a very fine, bedding-parallel lamination of organic matter and the preferred orientation of clay particles in the rock matrix. The elastic behavior of shales is commonly connected with the features of a transversely isotropic (TI) medium, which may be simplified to very simple formulae.

2.16.5.1 Stress-Strain Modeling

In rock mechanics, the strength of the material is determined by its capacity to sustain an applied load without failing or deforming plastically. This strength of materials analysis considers the relationship between the external loads applied on a material and the resulting deformation or alteration in the material's dimensions. Regardless of how strong the body is or how light the load is, it is warped or deformed when it is subjected to a load (force). If the load is light, the distortion will most probably fade away when the load is removed. Strain is defined as the overall deformation of a material caused by some applied stress and is evaluated as the proportional dimensional change (intensity or degree of distortion). A deformation is considered elastic if the stress is a linear function of strain. In other words, stress and strain follow Hooke's law. Beyond the linear region, stress and strain show nonlinear behavior. This inelastic behavior is called plastic deformation (Bataee & Irawan, 2014).

2.16.5.2 Seismic Anisotropy

Seismic anisotropy has gained significant attention from academia and industry over the last two decades due to breakthroughs in anisotropy parameter estimation, the change from post-stack imaging to pre-stack depth migration, and the greater offset and azimuthal coverage of 3D surveys. Many seismic processing and inversion approaches now employ anisotropic models, considerably improving seismic imaging quality and resolution. In addition, integrating the anisotropic velocity model with seismic imaging has reduced uncertainty in internal and bounding-fault placements, lowering the danger of investment decisions mainly relying on seismic interpretation. Furthermore, discovering a relationship between anisotropy parameters, fracture orientation, and density resulted in effective reservoir characterization approaches. Suppose fractures are evaluated during the drilling choice process. In that case, the gathering of such information, fracture spatial distribution and density, and the drainage area of each producing well can be greatly boosted (Hemsing & Schmitt, 2006).

2.16.5.3 Log Anisotropy

A rock physics mapping method was developed in order to determine anisotropic parameters from conventional well logs (Li, 2002). Electrical anisotropy suggests a dependence of rock resistivity on current flow direction. When the wellbore is not perpendicular to the plane of anisotropy, anisotropy causes induction logs to read too large. The anisotropy coefficient and the dip angle determine the effects. The effects are minor for angles less than 30°, moderate for angles between 30° and 45°, and significant for angles more than 45° (Klein, 1991).

2.16.5.4 Core Anisotropy

Full-diameter core analysis measurements made on rocks from almost any reservoir suggest that permeability is an anisotropic property. Vertical permeability is almost always lower than permeability measured in any horizontal direction. There are usually sound petrographic reasons why this should be so. However, the practice of making measurements on plugs that are of a length greater than their diameter magnifies true anisotropy (Lishman, 1970).

2.17 IN SITU STRESS

In situ stress is the stress within a rock formation before it is disturbed by excavation or other external factors. An important consideration in the design of underground buildings is in situ stress. In undisturbed rock formations, in situ stresses typically consist of three mutually orthogonal stresses: two components that act horizontally and one vertical component (Aadnøy & Looyeh, 2011b).

Stress, as the force acting on a given area, can have normal and shear components. The in situ stress state is a critical component of any geomechanical model and substantially impacts both unconventional hydrocarbon reservoir drilling and completion. Three perpendicular primary stresses can be used to illustrate the stress condition. If the overburdened rock strata are uniform in the vertical depth direction (z), the vertical stress (v) can be calculated:

$$\int_{o}^{z} \rho(z).dz \tag{2.22}$$

Where g is the acceleration due to gravity, z is the depth, and ρz is the density at depth, the other two are horizontal. Accurate modeling of the in situ stress regime depends on estimating horizontal stresses (Gholilou et al., 2017).

The horizontal stresses in all directions are roughly equivalent in isotropic conditions. Typically, this instance occurs in rock formations not subject to significant regional tectonic movements that result in localized folds and faults in the Earth's crust (Immerstein, 2013). However, in major geologic structures like anticlines, salt domes, or other tectonically active zones, the two main horizontal stresses are distinct, and maximum horizontal and lowest horizontal stress directions can be defined (Das & Chatterjee, 2017). Assessment of horizontal stresses is essential for correct modeling of the in situ stress regime. Under isotropic conditions, horizontal stresses are roughly identical in all directions, which typically occurs in rock. Also, the maximum and minimum horizontal stresses can be calculated using a poroelastic horizontal strain model (PHSM) (Aadnøy & Looyeh, 2011a; Maleki et al., 2014). When the in situ stress is close to zero, joints form, making excavation easier. In circumstances where the in situ stress is severe, any disturbance to the rock might release stored energy and cause the rock to break (Trenchless Pedia, 2018).

2.18 PORE PRESSURE CHARACTERIZATION

In oilfield geomechanics, the most common methods for determining pore pressure use empirical methods that relate velocity to pressure by calculating the effective stress acting on the rock (Ben Eaton & Spe-aime, 1975). For some reservoirs, a rock physics technique can connect velocity and effective stress (Ciz et al., 2005). Without basin modeling or quality velocity versus effective stress equations, a continuous pore pressure profile can be constructed by fitting trends to calibration data.

Drill stem tests (DST) are the most reliable and accurate methods of measuring pore pressure (Darvishpour et al., 2019). Repeat formation tests (RFT) and modular formation dynamics tests (MDT) can also quickly and precisely evaluate formation pressure. Pore pressure can be measured at precise spots on the wellbore wall using this equipment. RFT and MDT equipment uses wireline to deliver reliable pore pressure data, especially when several tests are performed regularly in close proximity in specific formations. Extrapolation can be used to obtain the formation pore pressure of a certain formation in a specific field by charting the relationship curve of pore pressure and depth of many zones in many wells.

Figure 2.12 depicts a typical continuous pore pressure profile for an unconventional reservoir. Each reservoir has its own level of overpressure and transitions from normal pressure to overpressure and back.

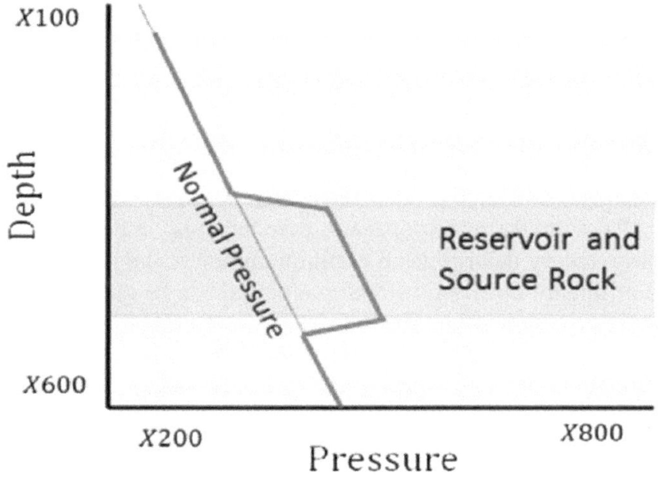

FIGURE 2.12 Typical pore pressure profile for an unconventional reservoir.

Source: Higgins-Borchardt et al., 2016

2.18.1 Pore Pressure Estimation

Pore pressure is significant in unconventional reservoirs but difficult to quantify. It directly impacts the condition of stress, flow rates, and production, and it can generate kicks from permeable formations during drilling. It also influences borehole instability in all formations. If the pore pressure is equal to the hydrostatic pressure of a column of water from the surface to a specific depth, it is normal pressure. Similarly, if the pressure is less than hydrostatic, it is said to be underpressured, and if the pressure is larger than hydrostatic, it is said to be overpressured. Depletion is a typical cause of underpressuring. Overpressuring can occur due to rapid sediment burial with trapped fluid, fluid formation due to kerogen maturation, uplift and exhumation, temperature fluctuations, or diagenetic processes. In a passive margin setting, such as along the Gulf of Mexico coast, the mechanisms for overpressuring are strongly associated with undercompaction and diagenetic processes (Sayers et al., 2005).

2.18.2 Overburden Stress

Overburden stress is the vertical principal stress, caused by the weight the overburdened rock exerts on the formation (Higgins-Borchardt et al., 2016). It can be calculated by integrating the density of the materials overlying the depth of interest as expressed in equation 2.23. The overburden stress increases with depth as the weight of the overburdened rock increases.

$$\int_{o}^{z} \rho(z).dz \tag{2.23}$$

where g is the gravitational coefficient, and z corresponds to depth. The best measurement of overburden is derived from density logs, integrating their results in a plot of overburden versus depth (Mavko et al., 2009). This stress goes vertically downwards from the surface to the depth of interest.

2.18.3 Minimum Horizontal Stress σ_h

Minimum horizontal stress (σ_h) is a principal stress generated by confined subsurface formations. Hydraulic fracture propagates perpendicular to the least principal stress. The poroelastic theory determines the minimum horizontal stress (Trenchlesspedia, 2018a). The minimum horizontal stress magnitude can be measured from a leak-off test or minifracs or a small injection test (Higgins-Borchardt et al., 2016), in which fluid is pumped into a formation, a small fracture is created, the well is shut in, and the pressure at which the fracture closes is measured. In general, fluid is pushed into the wellbore to pressurize a short length of exposed rock until the rock fractures; at this point, the fracture is propagated a short distance away from the wellbore by continuing pumping (Mavko et al., 2009). When an image log is run in a vertical well, the minimum horizontal stress orientation is mainly obtained from wellbore breakouts.

It is possible to construct several transverse fractures that are hydraulically induced and increase output by drilling numerous horizontal wells parallel to the direction of the formation's minimal horizontal stress. However, the fracture closes when the pressure within it is lower than the pressure required for fracture closure (Trenchlesspedia, 2018a). Hydraulic fractures propagate perpendicular to the least principal stress, which in some formations is the overburden stress, resulting in a hydraulic fracture in the horizontal direction. In very deep reservoirs, the least principal stress will mostly be horizontal, creating a vertical hydraulic fracture. Minimum horizontal stress changes with depth due to the pore fluid pressure in the vertical direction and rock lithology. This stress can be measured by acoustic log data analysis or diagnostic fracture testing (Higgins-Borchardt et al., 2016).

2.18.4 Maximum Horizontal Stress σ_H

Maximum horizontal stress is critical in wellbore stability modeling and drilling optimization. The extended leak-off test (XLOT) with a fracture reopening test, developed from the Kirsh solution for a circular hole subjected to internal pressure, can be used to estimate maximum horizontal stress (Trenchlesspedia, 2018a). Underground formations are subjected to stress because of their confined nature. The Kirsch equation calculates the magnitude of in situ maximum horizontal stress utilizing parameters such as minimum horizontal stress, unconfined compressive strength, borehole breakouts, friction angle, and cohesion. It can also be measured using the breakdown pressure of the formation XLOT test. The XLOT with fracture reopening test is based on the Kirsch solution for a circular hole under internal pressure in an isotropic, homogeneous, and linearly elastic media (Higgins-Borchardt et al., 2016).

2.18.5 Stress Direction

Hydraulic fractures usually propagate perpendicular to the minimum in situ stress (Higgins-Borchardt et al., 2016). In a typically stressed environment, a planar fracture will tend to propagate in the maximum horizontal stress direction if the major stresses are vertical and horizontal. There are several methods for determining the maximum horizontal stress direction. With an image log in a vertical well, a breakout will be detected in the direction of minimum horizontal stress. However, drilling-caused tensile fractures will be detected in the direction of maximum horizontal stress. Oriented caliper logs can also be used to indicate the direction of a breakout. In addition, where shear wave splitting is induced by stress, acoustic logs can be utilized to determine the greatest horizontal stress direction (Higgins-Borchardt et al., 2016).

2.19 WELLBORE STABILITY

Wellbore stability is the structural integrity of a wellbore that prevents it from failure. Mud loss is a common occurrence during drilling and occurs when the mud weight required to maintain wellbore stability and well control exceeds the fracture resistance of the formation, particularly in depleted reservoirs (Trenchlesspedia, 2018b). Due to their confined character, underground formations are vulnerable to vertical, minimum, and maximum horizontal stress. Maximum horizontal stress (σ_H) is an important parameter for modeling wellbore stability and drilling optimization.

Maintaining wellbore stability is critical during drilling operations. To avoid an unstable wellbore, the shape and orientation of the wellbore must be controlled while drilling (Trenchlesspedia, 2018b). A proper balance must exist between controllable and uncontrollable agents in a drilling operation. Unpredictable factors include pore pressure, earth stresses, and rock strength; controllable factors include wellbore fluid pressure and the chemical composition of the mud.

In unconventional reservoirs, maintaining wellbore stability while drilling well is one of the common challenges (Higgins-Borchardt et al., 2016). The MEM and wellbore stability assessments can also calculate the downhole pressure necessary to drill a stable wellbore without taking a kick or inflow, having the hole collapse, or losing drilling fluid. The term "safe and stable mud weight windows" is frequently used to describe the ranges of downhole pressures that avoid issues of this nature. The range between the pressure at which a kick could occur and the pressure at which losses will start is known as the safe mud weight window. The stable mud weight window is between the pressure required to prevent breakout and wellbore damage from shear failure and the pressure at which formation breakdown will occur. See, for instance, Figure 2.13.

2.19.1 Required Mud Weight Determination

The mud weight required to regulate a well reflects the pore pressure of any permeable formations that have been drilled (Beaumont & Fiedler, 2019). To manage a well, operators typically employ a mud weight with a pressure close to the estimated pore pressure. Increasing the mud weight can cause the typically compressed formations in the hole to break down. High pressure necessitates heavier-than-usual mud weight. However, the actual weight utilized may have been greater than necessary (i.e., the formation pressure might not be as high as the mud weight implies) (Beaumont & Fiedler, 2019).

If substantial underpressure is discovered, the formation will break down (crack and take mud), and the situation will be documented in the drilling record. Normal

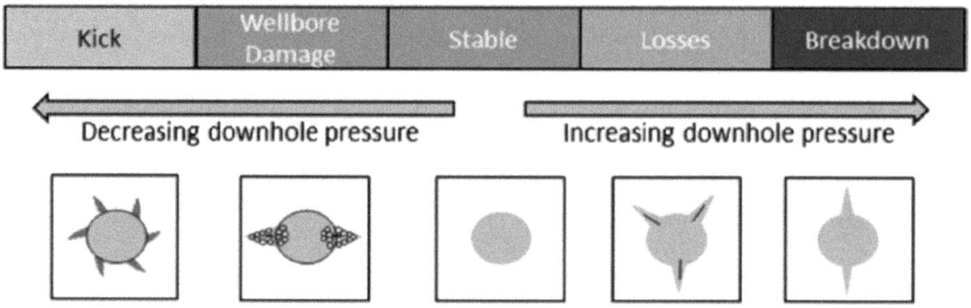

FIGURE 2.13 Mud weight window schematic.

Source: Higgins-Borchardt et al., 2016

mud weight can fracture the formation in underpressured formations. However, if the weight is reduced, fluid from uphole formations can flow into the well bore.

2.19.2 Influence of Multiple Weak Planes

Hydraulic fracturing experiments revealed that the strength of the interface is essentially important to hydraulic fracture height containment (Al-Tammar & Sharma, 2017; Daneshy, 1976). With a weak bonding, fracture containment is possibly associated with shear slip at the interface. Small-scale laboratory experiments by Hanson et al. (1981) indicated that decreasing the friction at the interface reduced the tendency of fractures to propagate through the interface. Laboratory testing by Teufel and Clark (1984) suggests that hydraulic fracture propagation can be hindered at an interface with weak shear strength.

At weak interfaces, fractures will be arrested or diverted into the interface due to shear slip. Hence, fracture propagation is arrested, and height is contained. Barree and Winterfeld (1998) developed an elastically decoupled shear slip model to account for shear slip and failure behavior at the interface. In that scenario, shear slip at the interface causes discontinuous fracture geometry, which minimizes fracture breadth and stress concentration at the fracture tip. When compared to conventional models, the predicted fracture shape is better matched with field observations, and the model gives better height containment and higher treatment pressure. A two-dimensional boundary element model was also used by Zhang et al. (2007) to analyze the propagation and deflection of a fluid-driven fracture at frictional bedding contacts.

Shales at the borehole wall, when intersected with fractures (Figure 2.14), can form multiple planes of weakness; in other words, abundant planes of weakness

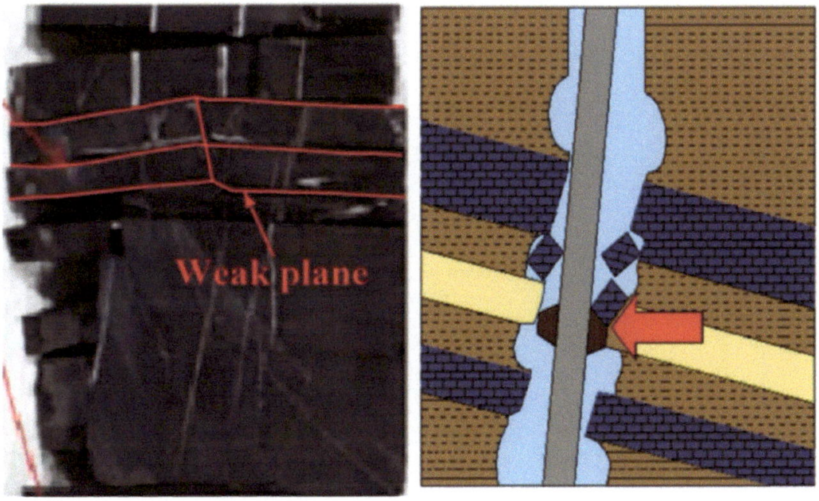

FIGURE 2.14 a) Multiple weak plane shale formation; b) damaged wellbore

Source: Zhao et al., 2021

should be considered in the wellbore stability analysis (Zhang et al., 2021b). As the number of weak planes in a wellbore increases, the collapse pressure does as well (Zhao et al., 2021). The safe drilling window shrinks as there are more weak planes, which raises the drilling operation's difficulty. With the change in the drilling azimuth, the weak plane dip region where the maximum collapse pressure is located for different weak plane groups transforms.

2.19.3 Fracture Containment

Hydraulic fracturing treatments are designed to contain the resulting fracture within a specified geologic interval (Figure 2.15). The parameters that are already in place, those that can be set up during installation, and those that can be managed throughout execution are combined in fracture containment (API, 2015). Fracture containment occurs when fracture propagation terminates at layer interfaces weaker than the surrounding rock.

Different types of interaction between hydraulic fractures and bedding planes are frequently investigated. When hydraulic fractures encounter bedding planes (Thiercelin et al., 1987), fractures can either pass through bedding planes or halt at the interface due to tip blunting. Aside from the severe cases stated earlier, hydraulic fractures can be deflected toward the interface, or a new fracture can be started at the opposite layer to form an offset pattern.

Variations in the stress state, caused by the complex interplay of layer thickness and mechanical properties, make it difficult to anticipate whether fractures propagate through many layers or are contained in one layer. As the fracture tip approaches the interface of two distinct elastic materials, the stress intensity factor begins to vary

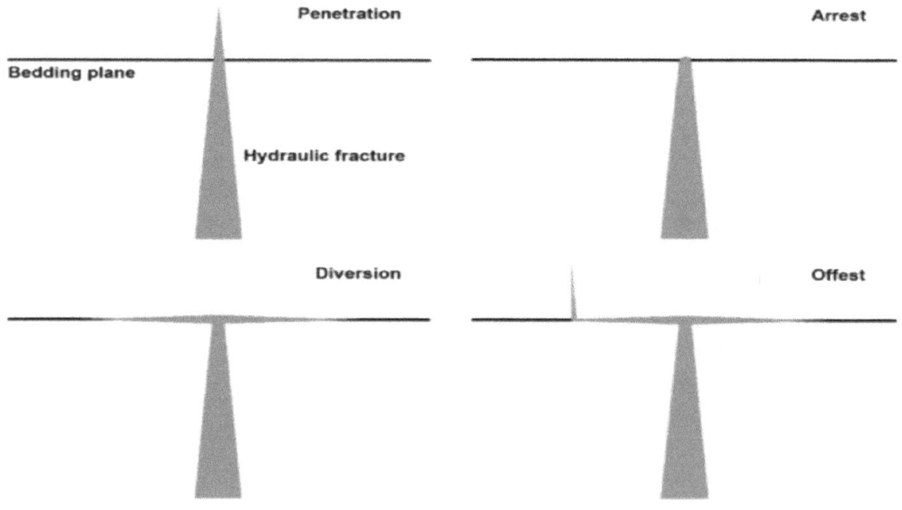

FIGURE 2.15 Four types of HF-BP intersection modes.

Source: Li et al., 2016

(Thiercelin & Plumb, 1994). If the opposing material is stiffer, the stress intensity factor approaches zero as the tip approaches the interface. As a result, the opposing material works as a perfect barrier, preventing fracture penetration. However, this severe fracture mechanics solution necessitates the presence of a geologically rare "sharp" boundary at the interface (Smith et al., 2007).

2.19.3.1 Shale Stiffness

Rocks do not behave as purely elastic media, and their stiffness is influenced by factors such as stress state, the magnitude of stress change, loading rate, drainage conditions, fluid saturation, and scale. In shales, stiffness is related to their elastic properties and the contrast in fracture toughness between adjacent strata, which reflects the material's resistance to crack propagation (Lozovyi & Bauer, 2019a). Fractures are more difficult to propagate in materials with higher fracture toughness or stiffness, such as shale. Numerical analyses by Garcia et al. (2013) and Thiercelin et al. (1989) showed that fracture toughness significantly affects fracture height growth. Two key mechanisms influence shale stiffness: (1) a reduction in undrained static stiffness as stress amplitude increases and (2) the frequency dependence (or dispersion) of dynamic stiffness.

2.19.3.2 Proppant Embedment

Proppant embedment is another way of arresting fracture. The effect of mechanical properties on fracture height containment is related to variations in fracture width, which can be important for fluid flow, proppant conveyance, and net pressure. It was once thought that tougher outer layers impeded hydraulic fracture penetration into the outer layer (van Eekelen, 1982). The investigation found that fracture breadth will narrow in stiffer layers where fractures propagate. This, in turn, lowers fluid flow into the fracture tip in the stiff layers, increasing the rate of fracture height.

Various research (Dewprashad et al., 1993; Michael et al., 2020; Zoveidavianpoor et al., 2018) reported an extensive use of polymers in the petroleum industry to optimize the drilling of wells. They have recently been employed for proppant coating to increase their strength during hydraulic fracturing. Proppants are classified as conventional or advanced proppants (Katende et al., 2021). Sand, ceramics, nutshells, and glass beads are conventional proppants, whereas polymer-coated proppants are more advanced. Modern proppants are substantially more expensive than conventional proppants, which also perform reasonably well. Resin-coated proppants are not frequently employed in shale wells due to issues with limited permeability, although ceramics are the most modern proppants (Figure 2.16).

2.19.4 Coring Methods

The procedure of acquiring representative samples of the productive formation in order to undertake a number of laboratory tests is referred to as well coring. Core samples are obtained using various techniques following the drilling operation, including standard diamond bit coring, rubber sleeve coring, pressure coring, sidewall coring, and cuttings recovery (Lyons, 2010). Coring provides reservoir properties on a much smaller scale than other geoscience or exploration techniques such as logging or

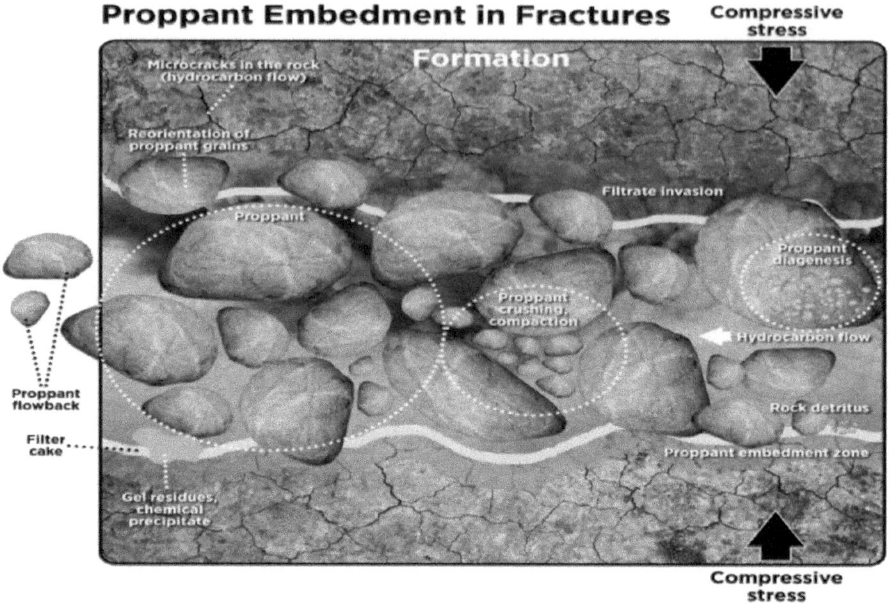

FIGURE 2.16 Illustration of physical phenomena that affect an effectively packed fracture due to proppant embedment after the hydraulic fracturing of a shale reservoir.

Source: Katende et al., 2021

seismic surveys. In the petroleum industry, the main aim of coring is to practically identify formations with a commercial scale of oil and gas content. Coring is vital in exploration (either in petroleum, mining, or any geo-related industry) since the data obtained from the core analysis are considered the ground truth and are thus used for calibrating the data from other exploration methods. The main objective of coring is to provide the qualitative and quantitative geological and mechanical data required for reservoir characterization and decision-making. There are many types of coring; however, the most common and basic methods are sidewall coring, wireline continuous coring, and conventional coring (Ashena & Thonhauser, 2018).

2.19.4.1 Conventional Coring

Conventional coring is usually done in competent formations on diameter cores (Lyons, 2010). It is a rotary coring method in which the inner tube/barrel carrying the core sample is returned to the surface along with the outer tube assembly through a conventional trip (Ashena & Thonhauser, 2018). Because the core bit is coupled to the outer tube assembly and drill pipes, typical core drilling uses a core bit/head that works on the same principle as a drilling bit. The core bit cuts a hollow cylindrical rock and hence has smaller bearings and cutters than a drilling bit. The inner tube of the core barrel is where the sample is placed while the outer tube, along with the drill collar and jar that sit on top, serves as the bottom-hole assembly (BHA) during coring.

2.19.4.2 Sidewall Coring

A sidewall is a core taken from the side of the wellbore. In sidewall coring, a formation sample is acquired with a wireline tool from which a hollow cylindrical bullet is launched into the formation and retrieved by a cable attached to the bullet (Sheriff, 2002a). Sidewall cores may be retrieved by utilizing percussion or mechanical drilling. During percussion drilling, a hollow bullet is fired into the formation to extract the cores. The fasteners that hold the bullets to the tool allow it to be raised, releasing both the fasteners and the bullets with their cores inside. In contrast to the conventional capacity of 20 to 30 bullets per percussion coring tool, two or three tools can be used simultaneously in a hole. Different bullet types and charge sizes maximize recovery in various formations. However, hollow rotary drills are used in mechanical drilling to cut and subsequently remove core plugs. On a single run, up to 75 plugs can be recovered. Typical percussion tool cores measure 1 inch (2.5 cm) in diameter by 1 3/4 inches (4.4 cm) in length with full recovery, whereas mechanical tool cores measure 0.91 inches (2.3 cm) in diameter by 2 inches (5 cm) (Sidewall Core I Energy Glossary, 2023).

2.19.4.3 Wireline Continuous Coring

Wireline continuous coring is another fundamental coring method that uses wireline for tripping. This technique is used when it is more cost effective to retrieve the core to the surface by pulling the inner barrel out through the inside of the drill string using a wireline rather than tripping the entire drilling string out of the hole (Patrick Murphy et al., 1996). The wireline coring method consists of two drilling and coring modes that can be quickly switched between by employing a slick line or wireline. The inner tube component is lowered into the outer tube assembly by a slick line when coring is active (Javadova, 2018). The inner tube component is then hydraulically locked to the outer tube component, and mud circulation is initiated so that core cutting can start. When core drilling is finished or if coring has been stopped, the rock sample is delivered to the surface through the outer tube assembly via a slick line. The inner drilling assembly with the drill bit insert/plug at its bottom end is then run in the hole again using a slick line to continue drilling.

2.19.5 Core Preservation (Wellsite, Lab Preservation)

Core preservation is one of the most crucial steps in obtaining reliable core analysis. The objective is to obtain rock that is representative of the formation while minimizing physical and chemical alteration during coring and handling as well as to maintain the proper sequence and core orientation (Resources, 2023). The goal of core preservation is to keep a core in the same condition it was when it was retrieved from the core barrel during shipment and storage. There are multiple methods to use depending on the formation and fluid types encountered (Core Laboratories). Wellsite core preservation ensures the rock's wettability and initial fluid saturation are preserved for routine and special core analysis testing.

Unconsolidated rocks that require particular coring and wellsite preservation processes include those containing both heavy and light oil, vuggy carbonates, evaporites, fractured rock, clay mineral–rich rocks, shale, low-permeability rock (tight gas sand), coal, and diatomite (Skopec, 1994). In the laboratory, cores are stored in the correct sequence to avoid breakage during shipment and storage, prevent core alteration, and maintain the volume and distribution of the core fluids. This is especially crucial for soft or weakly consolidated cores.

2.19.6 Sedimentology

Sedimentology studies sedimentary rocks' physical and chemical properties and the processes that lead to their formation, such as sediment movement, deposition, and lithification (conversion to rock). The aim of sedimentology, studying sediments, is to derive information on the depositional conditions that acted to deposit the rock unit and the relation of the individual rock units in a basin into a coherent understanding of the evolution of the sedimentary sequences and basins and, thus, the Earth's geological history as a whole.

2.19.6.1 Depositional Fabric

Depositional fabric is a rock fabric or a fabric element or structure formed during sediment deposition. This rock's fabric describes the spatial and geometric arrangement of the elements (minerals, textures, fossils, layers) that make up the rock. In sedimentary rocks, the fabric developed depends on the depositional environment. The fabric of a sedimentary rock controls the rock's porosity and permeability and, therefore, its ability to hold and/or transmit fluids such as oil and water.

2.19.6.2 Optical Microscopy

Optical microscopy studies how light passes through thin sections of rock cut and polished to about 0.3 mm thickness. It involves the diffraction, reflection, or refraction of electromagnetic radiation or electron beams interacting with a specimen, followed by the collection of this scattered radiation or another signal to generate an image (Di Gianfrancesco, 2017). This process can be carried out by scanning a narrow beam over the sample or irradiating the sample in a wide field (as in transmission electron microscopy and regular light microscopy) (e.g., confocal laser scanning and scanning electron microscopy). In scanning probe microscopy, a scanning probe interacts with an object's surface.

The optical microscope (Figure 2.17), often known as the "light optical microscope," employs visible light and a lens system to magnify images of small materials. Basic optical microscopes can be extremely simple, but numerous complex designs strive to improve resolution and sample contrast (Di Gianfrancesco, 2017). Common light-sensitive cameras can capture the image from an optical microscope to create a micrograph. Images were captured on photographic film, but contemporary advances in complementary metal-oxide-semiconductor and charge-coupled device (CCD) cameras enable the recording of digital images.

2.19.6.3 CT Scanning

Computed tomography refers to a computerized x-ray imaging procedure. X-ray-computed tomography (CT) is a nondestructive technique that allows visualization of the internal structure of objects, determined mainly by variations in density

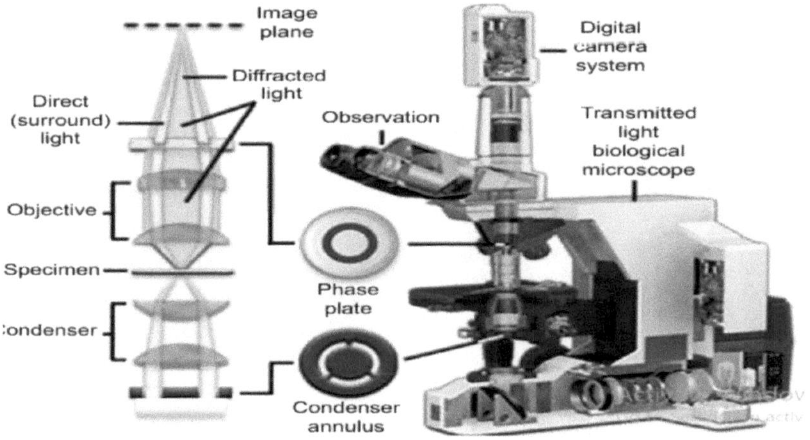

FIGURE 2.17 Optical microscope.

Source: Di Gianfrancesco, 2017

and atomic composition. It is a useful technology with many applications in geoscience REF. It is a nondestructive, high-resolution 3D imaging and analysis method that provides useful information on the porosity, internal structure, and composition of geological samples by density and atomic composition variations. CT scanning can help identify any anomalous material, sample damage, and/or significant heterogeneity in the samples of plugs (Josh et al., 2012; McPhee et al., 2015b).

CT scan technique relies on the contrast in X-ray signal attenuation coefficients for different lithologies/compositions as the X-ray passes through the sample, and thus attenuation is a function of the density, atomic number, and thickness of the sample being analyzed (Josh et al., 2012). The interactions responsible for this attenuation are mainly Compton scattering and photoelectric absorption. The contribution of the photoelectric effect depends on the effective atomic number and is especially important at low energies. At high energies, the Compton effect predominates, and density determines attenuation. Figure 2.18 is an example of CT scan image.

2.19.6.4 Natural Fracture

Fractures are displacement discontinuities in rocks, which appear as local breaks in the natural sequence of the rock's properties. Natural fractures are caused by stress in the formation, usually the result of tectonic forces such as folds and faults. They are apparent at most rock ridges, and most reservoirs likely contain some natural fractures. Most geological formations in the top section of the earth's crust have some degree of fracture. As a result of natural geological stresses like tectonic movement, drilling activity, thermal stresses, high fluid pressure, lithostatic pressure changes, and even fluid withdrawal—since fluid also helps support some of the weight of the overburdened rock—the rock has fractured. This mechanical failure of the rock is indicative of natural geological stresses.

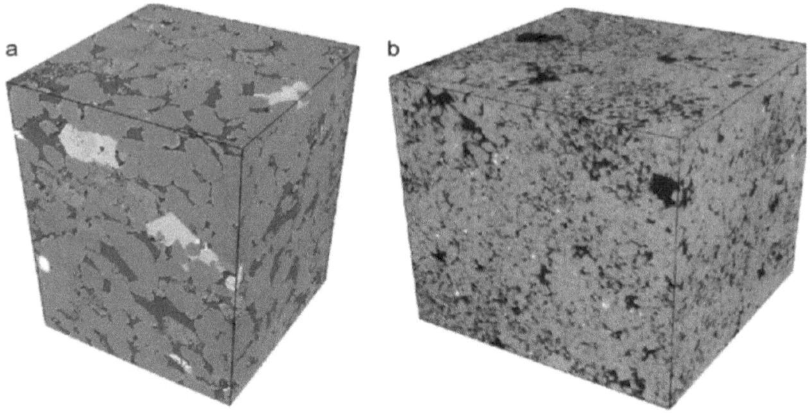

FIGURE 2.18 CT scan volumes of (a) Berea sandstone and (b) carbonate sample. Sizes are 724 × 724 × 1,024 and 1,024 × 1024 ×1024, respectively.

Source: Andrä et al., 2013

Conventional (unfractured) reservoirs are distinct from naturally fractured reservoirs. They are heterogeneous, made up of matrix blocks separated from one another by a fracture system (Chaudhry, 2004), as shown in Figure 2.19. The original rock that was there prior to fracturing is used to make the matrix blocks. The permeability and porosity of the matrix define it. The permeability and porosity of the fracture system are distinguishing features. They denote that a naturally fractured reservoir has double permeability and double porosity.

Natural fractures are more common in carbonate rocks. Fractures occur in preferential directions, determined by the direction of regional stress. This is usually parallel to the direction of nearby faults or folds, but in the case of faults, they may be perpendicular to the fault, or there may be two orthogonal directions. Tiab and Donaldson (2016b) categorized naturally fractured rocks into three main types on the basis of porosity: intercrystalline-intergranular, fracture-matrix, and vugular-solution.

2.19.6.5 Fracture-Healing Minerals

Fractures are often a high-permeability path in a low-permeability rock. However, these fractures are close or heal when material precipitates within them or when they are filled with a cementing material, such as calcite, which decreases the rock's permeability and limits fluid flow. It is important to distinguish between open and healed fractures. The total volume of fractures is often small compared to the reservoir's total pore volume (Bons et al., 2012).

Ju et al. (2020) investigated the self-healing behavior, referring to the decrease in naturally occurring water permeability of fractured rocks exposed to water-CO_2-rock interaction (WCRI). The study revealed that sandrock, rich in feldspars, dissolved and involved in ionic exchange reactions with CO_2 and groundwater, generating secondary minerals (such as kaolinite, quartz, and sericite) or $CaSO_4$ sediments. These

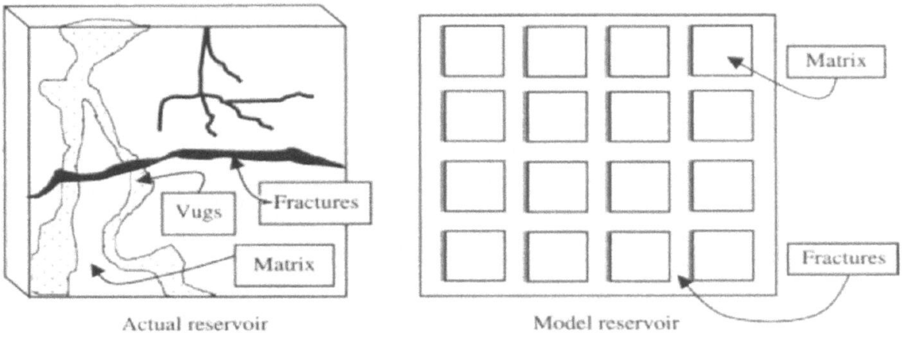

FIGURE 2.19 Idealization of a naturally fractured heterogeneous porous media

Source: Chaudhry, 2004

derivative matters adhered to the fracture surface, thereby gradually repairing fractures and decreasing the water permeability of the fractured rocks.

2.19.7 Helium Porosity and Core Plugs

The porosity of rock samples is traditionally calculated from helium porosimeter measurements, from which an accurate value of grain density is obtained using Boyle's law expansion as the main principle (Figure 2.20). The instrument uses helium drawn into a reference volume and stabilized at around 100 psi. Core samples are held in a matrix cup and isolated by valves, and then helium is expanded into the sample chamber. When the pressure stabilizes, the pressure is recorded. The resulting increase in the volume of gas, which represents the connected pore volume of the sample, is calculated by noting the increase in pressure in the chamber and then applying Boyle's law. Porosimetry can be useful for measuring the total porosity of a material using this equation (Ghulam & Iqbal, 1992; Nowroozi et al., 1999).

2.19.8 Helium Porosity and Permeability on Crushed Samples

Helium porosity and permeability on crushed samples is an approach to measure matrix porosity and permeability through pressure decay. This technique uses a narrow sieve cut to obtain a relatively homogeneous particle size after crushing shale core samples. A schematic (diagram) of the laboratory tools utilized in this procedure may be found in Figure 2.21. Pressure abruptly decreases as helium is expanded from the reference cell to the sample cell because the sample cell has dead space. After that, as helium enters the crushed samples' matrix pores, the pressure progressively drops to a lower level (Luffel et al., 1993). The measured pressure decay can be used to calculate the gas permeability. The permeability values for both methods (pressure decay and pulse decay) can be calculated using simulation-based history matching or an analytical approach.

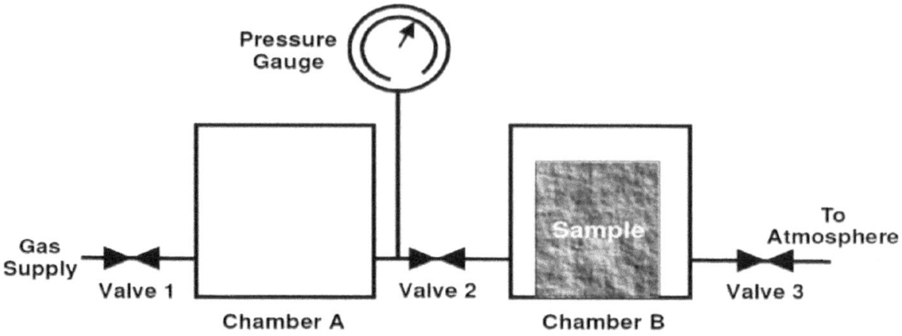

FIGURE 2.20 A schematic diagram of the helium porosimeter apparatus.

Source: Hosseini et al., 2019

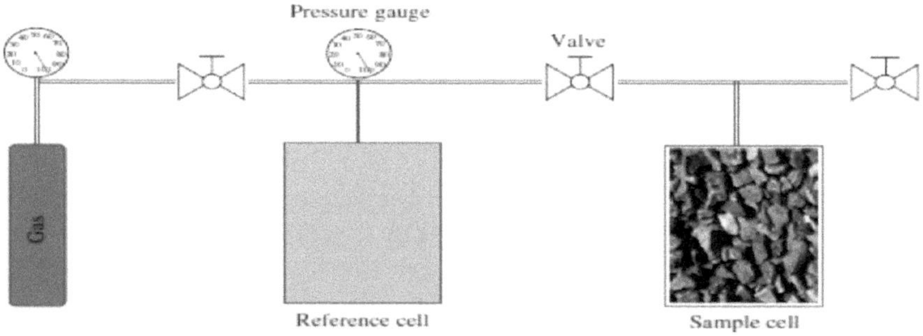

FIGURE 2.21 Schematic sketch of a pycnometer apparatus used for permeability measurement of the crushed shale samples.

Source: Rezaee, 2015

2.19.9 High-Pressure Mercury Porosimetry

Mercury porosimetry uses the non-wetting properties of mercury to gain information on the porous characteristics of solid materials: porosity, pore volume, pore-size distribution, and apparent density. Mercury porosimetry measures pore size distribution by applying controlled pressure to a sample immersed in mercury to force liquid mercury into the pore network and tracking the quantity needed for penetration. Mercury is used because of its high interfacial tension; consequently, it wets few materials (Diamond, 2000). The contact angle quantifies the measure of the resistance to wetting. The size of the pores has an inverse relationship with the amount of pressure needed to penetrate them.

2.19.10 Neutron Scattering

Neutron scattering techniques effectively analyze magnetic materials and magnetic exchange interactions on the nanoscale. The method involves creating a beam of neutrons, passing that beam through a sample, and noting where (and possibly when) neutrons scattered from the sample hit a detector. There are four principal neutron-nucleus interactions: nuclear fission, neutron capture, inelastic neutron scattering (neutron spectroscopy), and elastic neutron scattering (neutron diffraction) (Hosseini et al., 2021). In the first two interactions, known as absorption interactions, the neutron is absorbed into the nucleus, and something different emerges. Nuclear fission is when an excited compound nucleus divides into two main fragments plus several neutrons. The most important cases for nuclear fission are isotopes of uranium and plutonium. A nucleus may capture a neutron and emit a proton, an alpha particle, or a gamma photon (Hosseini et al., 2021). In the remaining interactions, known as neutron-scattering interactions, a neutron emerges from the reaction due to the scattering contrast between different components within the medium.

Neutron logging has been used in the petroleum industry for many years (Neuman et al., 1988). Using acoustic and nuclear methods, one can determine the formation porosity, defined as the volume percentage of the rock matrix occupied by the pores. The measurement depends on the scattering and absorption of gamma rays, which are influenced by the formation's electron density. The density can be determined by measuring the scattered gamma ray flux. The formation's rock matrix and fluid type must be known to transform its density into its porosity (Murugesu, 2017).

2.19.11 NMR Porosity and Pore Distribution

The evolution and application of nuclear magnetic resonance (NMR) techniques in formation evaluation provide fresh insights into pore space distribution and behavior (Schön, 2015). They are applied by reservoir engineers to define critical pore and fluid properties of the reservoir core samples (Allen et al., 1997). The method is based on the observation that the NMR signal corresponds with the number of hydrogen atoms in the fluid in rock pores. The NMR tool's response reveals how frequently the formation fluid's hydrogen nuclei collide with the grain surface. NMR measures the magnetization (M) and relaxation of hydrogen nuclei contained in the pore fluids. The magnetization is proportional to the number of hydrogen nuclei in the sensitive region of the sensor and can be scaled to provide an NMR porosity (Allen et al., 1997; Kenyon, 1992). In principle, the NMR porosity responds directly to the hydrogen content of the formation and thus is unaffected by the lithology effects in density and neutron logging (Kenyon, 1992). The size distribution of fluid-filled pores is another piece of essential information on rock samples that can be obtained from low-field NMR (Chen et al., 2017). The following shows how this distribution and the NMR signal are related.

The NMR technique identifies a wide range of properties, including permeability, water saturation, wettability, porosity, and pore-size distribution. The composite NMR signal is subsequently processed by inverting the NMR data to divide it into constituent elements. This T_2 distribution serves as a model for the pore-size distribution in the core since long T_2s result from large pores and short T_2s from small pores. NMR measures the fluid mobility, permeability, capillary pressure, and gas/water or oil/water contents. Currently, NMR technology helps measure pore structure and wettability properties and optimize logging parameters. The developed technology can obtain NMR petrophysical parameters at the well and laboratory. NMR logs measure the magnetic response of fluids present in the pore spaces of the reservoir rocks. In so doing, these logs measure porosity, permeability, and the types of fluids in the pore spaces.

2.19.12 Field Emission Scanning Electron Microscopes and Porosity

The field emission scanning electron microscope (FE-SEM) is an electron microscope that scans the sample surface with a high-energy electron beam in a raster scan pattern to image it (Alyamani et al., 2012). FE-SEM is applied to observe small structures (as small as one nanometer) on the surface of materials (Havrdova et al., 2014), measurement of sample microscopic features, surface morphology, characterization of coatings, fine structure analysis, study of microstructures, fracture and structural defects analysis (Akhtar et al., 2018), and micropores in unconventional resources.

2.19.13 Water Immersion Porosity Measurement Technique

The saturation and immersion technique determines porosity by saturating a sample with a liquid of known density and calculating the pore volume from the weight difference between the fully saturated and dry states (Kuila et al., 2014). The sample volume is determined by immersing it in the fluid and applying Archimedes' principle. As a result, the same experimental sample is used to estimate bulk density, grain density, and porosity. The American Petroleum Institute has approved this procedure as a standard method of measuring porosity (Kuila et al., 2014). The technique involves determining the water mass that fills the pores after immersion of the sample for two hours in boiling water (Andreola et al., 2000).

2.19.14 Dean Stark Solvent Extraction

Dean stark solvent extraction is a method for measuring fluid saturations in a core sample by distillation extraction. The technique involves vaporization and distilling water from the sample using toluene or xylene extraction (McPhee et al., 2015a). The boiling solvent vaporizes the water in the sample. The solvent condenses and flows back over the sample, extracting the oil. Extraction should last at least two days or until the extracted solvent is clean or the sample no longer fluoresces. The sample's weight is measured both before and after extraction. The volume of oil is then estimated by subtracting the weight loss of the sample from the weight of the water extracted from it. Volumes are used to compute saturation (Schlumberger, 2022a).

2.20 MINERALOGY

2.20.1 XRD

Powder X-ray diffraction is one of the most powerful techniques for mineral identification. Most minerals are crystalline and scatter X-rays in regular, characteristic ways depending on their structure. Each mineral produces a unique diffraction pattern and can be recognized from it like a fingerprint (Moore & Reynolds, 1997).

The structure of minerals is defined by regular, repeating planes of atoms that form crystals. When a focused X-ray beam interacts with these planes, part of the beam is transmitted, the mineral sample absorbs part, part is refracted and scattered, and another part is diffracted. X-rays are diffracted by each mineral differently, depending on what atoms make up the crystal lattice and how the atoms are arranged (Herein, 2008).

The XRD technique has also proven to be a very effective tool in the quantification of mineralogical data (Ruan & Ward, 2002). Quantification is established on the fact that the peak intensities from an individual mineral are proportional to the mineral's content in the sample (Al-Jaroudi et al., 2007). Peak intensities should thus provide information on the relative amount of the corresponding mineral phase. However, other parameters such as the varied absorbance coefficients of distinct mineral phases in a combination, particle size, degree of crystallinity, and preferred orientation of the sample all influence peak intensities (Ouhadi & Yong, 2003).

2.20.2 SEM-EDS

Scanning electron microscopy (SEM) scans a solid sample from the millimeter to the nanoscale range using a concentrated electron beam. Secondary electrons are produced in the sample and collected to create a secondary emission map (Backer, 1975). An SEM consists of a column, a specimen chamber, a display, and an operating device. The column's interior is kept at a high vacuum, and the electron beam produced by the electron gun converges into a fine beam using electromagnetic lenses (condenser and objective lenses) (García Veigas et al., 2012). By applying scan signals to the deflection coils, the electron beam is scanned along the sample surface. Some signals are produced due to electron-matter interaction, including secondary electrons, backscattered electrons, characteristic X-rays, cathodoluminescent light, and others (Chen et al., 2015).

Most SEM systems can generate an electron beam with a spot size of less than 10 nm that contains enough probe current to construct an adequate specimen image. The beam emerges from the last lens and enters the specimen chamber, interacting with the specimen to a depth of about one meter and forming the picture signals. These signals can be collected and amplified by different detectors (Chen et al., 2015). The detection system's electronics transform the signals into spot-by-spot intensity changes on the viewing screen, which result in an image. The quality and yield of the signals that the detector collects determine the contrast of an image (Worden & Utley, 2022).

2.20.3 Elemental Composition by X-Ray Fluorescence

X-ray fluorescence (XRF) is a rapid, relatively nondestructive analytical technique used to determine chemical or elemental analysis of rocks, minerals, sediments, fluids, and soils. XRF analyzers determine a sample's chemistry by measuring the fluorescence (or secondary) X-ray generated by the sample when excited by a main

X-ray source. Each element present in a sample emits a unique set of characteristic fluorescence X-rays ("a fingerprint"), which is why XRF spectroscopy is an ideal tool for qualitative and quantitative material composition analysis (*FIS-B and How Does It Work*, 2020).

2.21 WETTABILITY MEASUREMENT

Wettability is a fundamental property of the interactions between the reservoir rock components and the pore-filling fluids. A fluid tends to spread on or adhere to a solid surface in the presence of other immiscible fluids (Ahmed, 2018; Ghulam & Iqbal, 1992). Wettability is a controlling factor on the location of the reservoir fluids within its porous structure and upon the flow and subsequent distribution of fluids during the dynamic production process. It is, therefore, an essential factor during laboratory experimentation since it will have a controlling influence on all laboratory tests attempting to imitate reservoir characteristics and processes, especially capillary pressure and relative permeability. It is necessary to attempt to ensure that core material has reservoir-representative wettability during laboratory analyses; otherwise, the resultant data could be meaningless (Ahmed, 2018). Physically, wettability signifies a balance of forces that occur at the interface between three phases, one of which is a solid (Wheaton, 2016; Ahmed, 2018; Ghulam & Iqbal, 1992; McPhee et al., 2015c; Tiab & Donaldson, 1999; Wheaton, 2016; Wolfman, 2005).

2.22 THIN SECTION MICROSCOPY

A thin section (also known as a petrographic thin section) is a thin slice of a rock or mineral sample that has been prepared in a laboratory for use with an electron microscope, polarizing petrographic microscope, and electron microprobe (Kemet, 2016). Thin-section microscopy is a type of polarized optical microscopy applied to study thin (or ultra-thin) sections of rock, ore, mineral, or other geological material (Critchley, 2019b). It is a nondestructive method used mostly with solid materials that can evaluate crystalline and amorphous samples. It is a high-resolution microscopy technique that can be used in conjunction with other high-resolution microscopy techniques to investigate the phases in the texture of the sample, such as the structure, phase assemblage, fabric, phase relationships, and reaction textures of the material's surface, by identifying specific optical and morphological characteristics on the material's surface. These various phases on the substance's surface can also be used to identify an unknown sample (Critchley, 2019a). Thin section microscopy can also be used to evaluate the chemical makeup of a sample (Raith et al., 2011; Critchley, 2019b).

2.23 ELECTRON MICROSCOPY

Electron microscopy is one of the available techniques that allows a high-resolution image of the surface of a specimen to be obtained. This technique utilizes the electron interaction with the specimen to generate topological images. With higher magnification than the light microscope, electron microscopy allows the observation of matter in finer detail, past the capabilities of the naked human eye (Abd Mutalib et al., 2017). An electron microscope uses a beam of accelerated electrons as an illumination source. It provides a higher resolving power than light microscopes. Since the wavelength of an electron can be up to 100,000 times shorter than that of visible light photons, it may expose the structure of smaller objects. It is used in petroleum science research to investigate the detailed structure of rocks and minerals and is also applied in the characterization of micropores.

There are two main types of electron microscopy: transmission electron microscopy (TEM) and electron microscopy (SEM). SEM is done with an electron microscope that uses a focused beam of electrons that react with the sample to produce a topological image and relative composition. Upon contact with the sample, the focused beam of electrons will produce secondary electrons (SEs), backscattered electrons, and characteristic X-rays, which are then detected with respective detectors and displayed on the monitor. The main components in a typical SEM include the electron source, a column that contains electromagnetic lenses, the electron detector, the sample chamber, and the computer display (Abd Mutalib et al., 2017). A newly developed SEM backscattering micropore image quantification, with magnification continuously adjustable between dozens and tens of thousands of folds, allows direct observation of micropore distributions that occur in the intergranular clay minerals, intergranular fragments, leached and dissolved feldspar or rock fragments, and intercrystals (Zou, 2013).

2.24 FLUID COMPATIBILITY

Incompatibility can occur between introduced fluids and reservoir pore fluid (for example, emulsion blocks formed by combining mud acid and some crude oils, scale) either between fluids introduced into the well (e.g., between various additives in a chemical stimulation package) (Ezenweichu & Laditan, 2015). When oil or water-based mud filtrates invade the near wellbore region during overbalanced drilling techniques, they might react negatively with hydrocarbons or waters in the matrix to generate compounds that limit permeability. The formation of insoluble precipitates or scales between incompatible waters, the de-asphalting of the in- situ crude or hydrocarbon-based drilling fluid caused by the blending of incompatible oils, or the formation of highly viscous stable water in oil emulsions due to turbulent blending of invaded filtrates with either in situ water or oil would all be problems.

However, fluid compatibility is the ability of fluids to mix without degradation of properties or performances. Incompatibility of injected fluids can be avoided through proper treatment design and quality control. The incompatibility of reservoir fluid and introduced fluid can be reduced by using preflush/afterflush procedures and limiting the residence duration in the reservoir (but this may be rendered only partially effective by irregular distribution of fluids or reactive rock in the near-wellbore region (Ezenweichu & Laditan, 2015). Also, a test or series of tests is performed to check that no undesirable reactions occur with a specific fluid. The testing process may include checks for compatibility with other treating fluids, wellbore fluids, reservoir fluids, and reservoir formation. In extreme cases, mixing seemingly benign fluids can create significant reactions that may permanently damage the reservoir permeability. Fluid-shale compatibility testing is as old as the drilling fluid industry itself, and it continues to be a hot topic as drilling applications go into new, more complex areas. Fluid incompatibilities with clay-rich shale deposits can cause various operational challenges, from small dispersion and accretion concerns to significant stuck pipe and production impairment events (Oort, 2018).

2.25 MECHANICAL ROCK PROPERTIES

Rock mechanical properties such as strength (compressive, tensile, and shear), Young's modulus, and Poisson's ratio are crucial in wellbore stability, fracture forecasting, and other engineering procedures. The mechanical characteristics of rocks are often determined using static and dynamic methods (Al-Shayea, 2004; Baoping & Hongzhi, 2005). Static methods are generally conducted in the laboratory with specific test equipment that contains core specimens (Shan & Liu, 2000). The specimens are constantly compressed until they eventually fail. Stress-strain curves are recorded simultaneously on a computer, and mechanical characteristics can be derived from the curves. Dynamic methods are typically used to estimate compressional wave velocities (VP) and shear wave velocities (VS) from logs or in the laboratory (Baoping & Economides, 2002; Guo & Liu, 2014; Ranjbar-Karami et al., 2014). Various studies regarding the differences between static and dynamic methods have demonstrated that static methods are more direct and realistic while dynamic methods are easier and more continuous (Chang et al., 2006; Fjaer et al., 1992; Mavko, 2003). This study used static tests because of their ability to control the loading pace on test specimens, providing useful information about specimen response on the stress curve. It is important to know and understand natural materials such as rock's properties and behavior during loading (Hudson & Harrison, 1997).

2.26 CANISTER GAS AND ADSORPTION ISOTHERM

Adsorption occurs when a liquid or gas particle attaches to the surface of an adsorbent, causing an atomic layer to develop on the adsorbate (Cui et al., 2022).

Several studies have used limited canister desorption data and petrophysical well log data to estimate the volume of desorbed gas during production in the Montney formation (Cui et al., 2022) and investigate the properties of canister core desorption gas in order to improve the assessment of hydrocarbons in place in unconventional reservoirs (Kang et al., 2021), used canister desorption data from two samples of Longmaxi gas shale to run a numerical simulation to determine the lost gas volume and total gas content. Han and Zhang (2021) described an efficient technique for evaluating shale canister desorption. Dang et al. (2018) compared different methods for "lost" gas estimation based on shale gas.

2.27 DIGITAL ROCK ANALYSIS

Digital rock physics (DRP) has become a complementary part of the characterization of reservoir rocks due to the nondestructive testing character of this technique (Balcewicz et al., 2021). This emerging technique has rapidly become an indispensable tool for estimating elastic properties. The success of DRP mainly depends on three factors: acquiring a 3D rock structure image, accurately identifying 3D minerals, and using a proper numerical simulation method. The concept of DRP, also called digital rock analysis in petroleum engineering, permits the effective properties to be computed to allow pore structures to be linked to the fluid flow characteristics (Andrä et al., 2013).

The digital rock analysis technique involves gathering high-resolution image data of rock samples, which gives an alternate basis for examining rock properties compared to laboratory testing or theoretical models that mimic an actual geometry (GeoExpro, 2023). A sample object is scanned in a standard digital rock physics workflow, with micro-CT typically producing the highest-quality images. Scanners reconstruct image projections to create a stack of images that can be exported in different formats, such as BMP, TIF, or PNG, which then somehow need to be transformed so that the properties of the core can be modeled and analyzed (GeoExpro, 2023).

This kind of analysis can potentially reduce the amount of time spent waiting for results in the lab in various situations, including reservoir characterization and hydrocarbon production. However, non-experts may find it challenging to understand software solutions for digital rock physics, necessitating the use of integrated procedures (GeoExpro, 2023). Shales also present a substantial challenge for DRP, owing to their heterogeneous structure, composition, and properties from micron to centimeter scale. To obtain a sufficiently large field-of-view (FOV) image of a sample that reflects the detailed and representative internal structure and composition, we have developed a new DRP workflow to obtain large-FOV high-resolution digital rocks with 3D mineralogical information.

2.28 RESERVOIR CHARACTERIZATION DURING EXPLOITATION AND APPRAISAL

Reservoir characterization is preparing a quantitative representation of a reservoir using data from various sources and disciplines. The reservoir properties include reservoir thickness, number of reservoir units, porosity, permeability, pressure distribution, fracture distribution in the case of unconventional reservoirs, and fluid saturation (oil, gas, water) (Robertson, 1989). Technically, characterizing the reservoir's geomechanical, petrophysical, geological, and geophysical characteristics can increase its output and lifespan (Elnaggar et al., 2018). These characteristics can be determined using a mix of core and wireline logging analysis data (Elnaggar et al., 2018). All information collected at various scales in the reservoir characterization process must be integrated into a single, comprehensive, and consistent representation of the reservoir. Exploration is the initial stage in petroleum operation that includes generating prospects or play or both and drilling an exploration well (Schlumberger, 2023a).

The appraisal phase follows successful exploratory drilling and links the discovery of petroleum during exploration to the decision of whether or not to develop the field. The data gathered during appraisal helps determine if field development is commercially viable and, if so, how it should be approached. This phase typically involves drilling additional wells (Figure 2.22), each with a comprehensive data collection plan. Additional seismic data is also acquired at this stage. The objective is to estimate the amount of oil (stock-tank oil initially in place, STOIIP) or gas (gas initially in place, GIIP) present and determine how much can be economically recovered, i.e., the petroleum reserves of the field (Gluyas, 2018). Reservoir characterization during appraisal aims to understand reservoir architecture by combining seismic, well log, and core data, often supplemented with analog information from outcrops and modern depositional environments. This includes analyzing logs and cores to evaluate reservoir porosity, permeability, and mechanical properties.

2.29 DETERMINING SHALE MINERALOGY AND ROCK QUALITY

The fine grain size of the constituent minerals largely determines the properties of shales. Shale is a broad category of rocks made of extremely fine-grained particles with very low porosity and permeability values in the few porosity units and nano-Darcy range. Shale formations are highly complicated at the core size, with considerable vertical variations in lithology and total organic carbon (TOC) at a small scale, making core characterization and sweet spot detection challenging. Shale formations are also quite complex at the nanoscale pore level, where the pores have various porosity types that are found within the kerogen volume.

The mineralogy of shales varies greatly. Aside from clay minerals (60%), the average shale comprises quartz and other types of silica, most notably amorphous silica and cristobalite (30%), feldspars (5%), and the carbonate minerals calcite and dolomite (5%). Iron oxides and organic matter (approximately 0.5% and 1%, respectively) play a crucial role. Older estimations severely underestimated clay minerals due to improper potassium assignment to feldspar minerals (Britannica, 2016). Illite

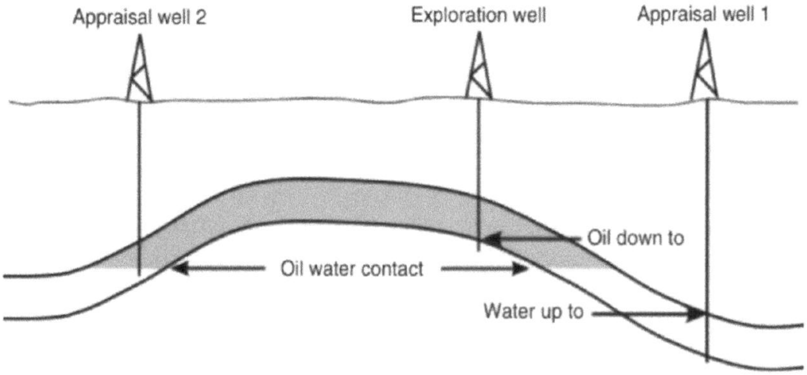

FIGURE 2.22 "Oil down to" established by an exploration well, "water up to" by a first appraisal well, oil-water contact penetrated by the second appraisal well.

Source: Gluyas, 2018

is the most prevalent clay mineral, followed by montmorillonite and mixed-layer illite-montmorillonite, kaolinite, chlorite, chlorite-montmorillonite, and vermiculite. The quartz-to-feldspar ratio is similar to that of adjacent sands. However, feldspar may be formed from local volcanic sources in pelagic (deep-sea) sediments, but quartz may be brought from the continents by wind, disturbing simple patterns. A large number of accessory minerals occur in shales. Some of these are detrital, but diagenetic or in situ varieties (e.g., pyrite, siderite, and various phosphates) and volcanically derived varieties (e.g., zeolites, zircon, and biotite) have been noted (Pettijohn, 1957; Krumben & Sloss, 1963; Passey et al., 2011; Britannica, 2016; Dembicki-Jr, 2017).

2.30 RESOURCE EVALUATION AND GENERATION OF FIELD DEVELOPMENT PLAN

Resource evaluation occurs when the viability of the discovery has been confirmed. A reservoir simulation model is built to calculate the reservoir's initial hydrocarbon volume, simulate reservoir fluid flow behavior, and optimize the field development scenario (type of wells, number of producer and injector wells, location of wells, capacity of field production, etc.). Appraisal wells are drilled to enhance the reservoir description by acquiring more data. Field development is one of the most critical stages after a field discovery, defining the field. This stage of petroleum operation takes place after the appraisal period and before the beginning of the field production. Field development plans (FDPs) provide the necessary support for field optimization and include all activities and processes required to develop a field optimally (Opus Kinetic, 2019). They provide an excellent technical answer and a strategy for enhancing a field's development and output. A large number of criteria related to the geological and physical properties of the reservoir, to operational planning, and to the economic scenario are included in the development plan. The development of approaches that can result in better hydrocarbon recovery strategies and increase the project's profitability is of utmost importance. To choose the finest ideal development model, the studies of the development plan provide important guidance and information to assess whether the project is economically feasible under all conceivable development project scenarios, uncertainties, and dangers. The primary factor in the decision-making process is net present value (Ganat, 2020).

2.31 OPTIMIZATION OF HORIZONTAL WELL PLACEMENT IN A POTENTIAL ZONE

Horizontal drilling is a directional drilling procedure that intersects an oil or gas reservoir at the "entry point" with a near-horizontal inclination and continues to the intended bottom-hole site. Horizontal drilling makes more contact with the reservoir formation than vertical drilling, allowing for increased hydrocarbon production from a given wellbore. However, well optimization is a critical component of field development techniques aimed at increasing hydrocarbon recovery and the economic viability of new field development projects. Given the scarcity of new oilfield discoveries, optimizing oil production and net present value (NPV) have emerged as key factors (hence referred to as "critical factors") in reservoir engineering. As a result, well-optimization research has evolved into a distinct topic in its own right (AlQahtani et al., 2012). Recent efforts by academic and industrial researchers have centered on constructing good, efficient optimization models capable of predicting strategies for managing existing oil and gas fields and establishing new fields with the potential to maximize important parameters. Horizontal wells provide a highly efficient way to maximize contact with the reservoir target and increase overall recovery by allowing a larger drainage pattern. Identifying optimal horizontal well locations traditionally involves domain expertise across multiple disciplines and takes a long time to complete.

Horizontal wells are frequently used in a variety of reservoirs and have produced positive development outcomes as a result of the advancement and sophistication of horizontal well drilling and completion technology (Zhang et al., 2021a). Compared with vertical wells, horizontal wells have increased the wellbore and reservoir of contact zone and improved drainage efficiency, and bottom water can complement the energy of reservoir damage during the development process. This can make the horizontal well smaller to earn a relatively high yield production pressure difference, significantly improve the single well production and recovery factor, and, at the same time, reduce the risk of a horizontal section of the store being sand (Daiyu & Tongwen, 2004; Garrouch et al., 2004). Therefore, horizontal well technology is widely used in bottomwater sandstone reservoirs.

2.32 MICROSEISMIC FRACTURE MONITORING SERVICES

Microseismic or passive seismic methods are feeble earth tremors due to natural causes or human activities. Microseismic events are small-scale earthquakes (Maxwell & Urbancic, 2001) with a magnitude usually less than 0 and frequencies in the 0.1–10 kHz band. Microseismic events can be triggered or induced (McGarr, 1994; McGarr & Simpson, 1997). Triggered microseismicity is caused by a small stress change (i.e., tectonic loading plays a primary role). In contrast, induced seismicity occurs when the causative activity generates most of the stress change or energy required to produce the seismicity. Virtually all subterranean engineering creates artificially induced (stimulated) fractures.

2.33 SWEET SPOT DETERMINATION

Sweet spots are intervals within organic shales that possess the highest relative hydrocarbon yield for drilling purposes (Glaser et al., 2013). It has become difficult to determine the sweet spots necessary in unconventional reservoirs where permeability and viscosity are low. Potential sweet spots are discovered by observing shale characteristics such as reservoir quality and completion quality, which predict relative reservoir producibility. Sweet spot identification is a method of selecting the best drilling location for unconventional wells by using surface seismic surveys and borehole geophysics to determine reservoir attributes such as reservoir quality and completion quality. It is an economically attractive and easy-to-access part of the reservoir (Giles & Tennant, 2014).

The determination of sweet spots requires the integrated analysis of petrophysics, geomechanics, geophysics, and structural data (Alshakhs & Rezaee, 2019). Upon integration and generation of the 3D spatial framework using geostatistical methods, sweet spots can be delineated in space (Liang et al., 2016). Reserve estimation can then be done using the discriminant analysis of the geological, geomechanical, and petrophysical characteristics (Aliouane & Ouadfeul, 2014). Various researchers have used several methods to determine the sweet spot. Changzi et al. (2014), Glaser et al. (2014), Qingcai et al. (2018), and Sena et al. (2011) applied seismic to identify the sweet spots' petrophysical (Kormaksson et al., 2015), and geomechanical characterization (Fernandez Rojas et al., 2016; Rahman & Gui, 2016). Other authors (Aliouane & Ouadfeul, 2014; Chen et al., 2017; He et al., 2020; Ismail et al., 2020; Yang et al., 2015) use seismic and petrophysical characteristics to evaluate sweet spots.

2.34 LATEST ADVANCES, TECHNIQUES, AND METHODS

Over the last five years, technological advancements have spurred industrial progress, resulting in a major change in the sector. Technological advancements will be crucial in addressing global energy demand because they enable the discovery of new resources, access to harsh or isolated regions, and the development of difficult reservoirs that were previously uneconomic to produce. However, software has been at the forefront when it comes to developing innovative technology in the oil and gas industry. Many organizations in the industry have been able to keep good track of their available inventory and supplies thanks to modern oil and gas production software. Furthermore, the most recent oil and gas production software has enabled these organizations to cut costs and save money. The most recent breakthroughs in oil and gas technology have resulted in a few key developments that have benefited the sector in recent years. Artificial intelligence, big data analytics, electric monitoring, and drone technology are among the oil and gas software solutions available (Ali, 2019).

Artificial intelligence refers to machines that can learn how to do specific activities and solve problems. It surveys facilities, monitors machines, forecasts weather, and improves safety. Geophysical and petrophysical research, as well as data processing, are examples of operations in exploration. Geophysical and petrophysical studies often include (1) reservoir-scale seismic surveying, (2) well logging, (3) lab core analysis, and, in certain situations, (4) digital core analysis. Geokinetics developed an AquaVib marine vibrator, a competitor to the seismic air gun with a considerably lower decibel footprint, using AI to save the lives of aquatic species.

Big data analytics is another key advancement in the oil and gas business. Oil and gas firms can use big data to expand the variety and volume of specific data. They may also produce massive amounts of data from 3D seismic surveys, oil drilling, and production. The Internet of Things, (IoT) has become a key trend in recent years. It is the extension of internet connectivity through various hardware devices. Companies will now utilize specific gadgets and hardware connected to the internet as part of the Internet of Things. As a result, they can use internet connections and networks to operate cutting-edge technical devices.

Drone technology is another development in the oil and gas industry. These are objects that fly over a surface to complete certain tasks. Since oil and gas companies look to complete evaluations and inspections quickly, drones will become more commonly used. With drones, companies can get more comprehensive scans of facilities in a shorter period of time, resulting in quicker evaluations daily.

REFERENCES

Aadnøy, B., & Looyeh, R. (2011a). In-situ stress. In *Petroleum rock mechanics* (pp. 107–129). https://doi.org/10.1016/b978-0-12-385546-6.00008-5

Aadnøy, B., & Looyeh, R. (2011b). *Petroleum rock mechanics*. https://doi.org/10.1016/C2009-0-64677-8

Aadnøy, B. S., & Looyeh, R. (2011c). *Petroleum rock mechanics: Drilling operations and well design* (1st ed.). Gulf Professional Publishing. www.scribd.com/book/413520471/Petroleum-Rock-Mechanics-Drilling-Operations-and-Well-Design

Abd Mutalib, M., Rahman, M. A., Othman, M. H. D., Ismail, A. F., & Jaafar, J. (2017). Scanning electron microscopy (SEM) and energy-dispersive X-ray (EDX) spectroscopy. In *Membrane characterization* (pp. 161–179). https://doi.org/10.1016/B978-0-444-63776-5.00009-7

Abeed, Q. (2020, July). The application of organic geochemistry in petroleum exploration. *Halliburton*, 20–29.

Abubakar, R., Adomako-Ansah, K., Marfo, S. A., Fenyi, C., & Owusu, J. A. (2023). Integrated geochemical and statistical evaluation of the source rock potential in the deep-water, Western Basin of Ghana. *Journal of Petroleum Science and Engineering*, *220*, 111164. https://doi.org/10.1016/j.petrol.2022.111164

Addis, M. A. (2017). The geology of geomechanics: Petroleum geomechanical engineering in field development planning. *Geological Society Special Publication*, *458*(1), 7–29. https://doi.org/10.1144/SP458.7/ASSET/87720A35-9EC0-4778-BC41-336BA331D6D7/ASSETS/GRAPHIC/SP458-1863F12.JPEG

Ahmed, T. (2018). Reservoir engineering handbook. In *Reservoir engineering handbook*. Elsevier. https://doi.org/10.1016/C2016-0-04718-6

Ahmed, T. (2019). Reservoir-fluid properties. In *Reservoir engineering handbook* (pp. 29–121). https://doi.org/10.1016/B978-0-12-813649-2.00002-5

Akhtar, K., Khan, S. A., Khan, S. B., & Asiri, A. M. (2018). Scanning electron microscopy: Principle and applications in nanomaterials characterization. In *Handbook of materials characterization* (pp. 113–145). https://doi.org/10.1007/978-3-319-92955-2_4

Aladesanmi, A. O., Olowookere, A., Ogundana, A., & Ochai, J. (2018). LECO analytical study of subsurface tertiary shale offshore Niger delta. *IOSR Journal of Applied Geology and Geophysics*, *6*(5). https://doi.org/10.9790/0990-0605017581

Ali, U. (2019). *Five exploration technologies for the offshore industry*, 1–7. www.offshore-technology.com/features/offshore-exploration-technologies/

Aliouane, L., & Ouadfeul, S. A. (2014). Sweet spots discrimination in shale gas reservoirs using seismic and well-logs data. A case study from the Worth basin in the Barnett shale. *Energy Procedia*, *59*, 22–27. https://doi.org/10.1016/j.egypro.2014.10.344

Al-Jaroudi, S. S., Ul-Hamid, A., Mohammed, A. R. I., & Saner, S. (2007). Use of X-ray powder diffraction for quantitative analysis of carbonate rock reservoir samples. *Powder Technology*, *175*(3), 115–121. https://doi.org/10.1016/j.powtec.2007.01.013

Allen, S. G., Stephenson, P. C. L., & Strange, J. H. (1997). Morphology of porous media studied by nuclear magnetic resonance. *Journal of Chemical Physics*, *106*(18), 7802–7809. https://doi.org/10.1063/1.473780

AlQahtani, G., Vadapalli, R., Siddiqui, S., & Bhattacharya, S. (2012). Well optimization strategies in conventional reservoirs. *Society of petroleum engineers—SPE Saudi Arabia section technical symposium and exhibition 2012*, 833–845. https://doi.org/10.2118/160861-MS

Alshakhs, M., & Rezaee, R. (2019). Sweet-spot mapping through formation evaluation and property modelling using data from the Goldwyer Formation of the Barbwire Terrace, Canning Basin. *Petroleum*, *5*(1), 13–29. https://doi.org/10.1016/j.petlm.2018.06.003

Al-Shayea, N. A. (2004). Effects of testing methods and conditions on the elastic properties of limestone rock. *Engineering Geology*, *74*(1–2), 139–156. https://doi.org/10.1016/j.enggeo.2004.03.007

Al-Tammar, M. J., & Sharma, M. M. (2017). Effect of geological layer properties on hydraulic fracture initiation and propagation: An experimental study. *Society of petroleum engineers—SPE hydraulic fracturing technology conference and exhibition 2017*, 813–839. https://doi.org/10.2118/184871-ms

Altindag, R. (2002). The evaluation of rock brittleness concept on rotary blast hole drills. *Journal of the South African Institute of Mining and Metallurgy*, *102*(1), 61–66.

Alyamani, A., Lemine, O. M., Alyamani, A., & Lemine, O. M. (2012). FE-SEM characterization of some nanomaterial. *Scanning electron microscopy*. https://doi.org/10.5772/34361

Ambrose, R. J., Hartman, R. C., Diaz-Campos, M., Akkutlu, I. Y., & Sondergeld, C. H. (2010). New pore-scale considerations for shale gas in place calculations. *SPE unconventional gas conference 2010*, 167–183. https://doi.org/10.2118/131772-ms

Aminzadeh, F., & Dasgupta, S. N. (2013a). *Fundamentals of petroleum geophysics* (Vol. 60). https://doi.org/10.1016/B978-0-444-50662-7.00003-2

Aminzadeh, F., & Dasgupta, S. N. (2013b). *Geophysics for petroleum engineers* (Vol. 60). Elsevier. https://doi.org/10.1016/B978-0-444-50662-7.00007-X

Andrä, H., Combaret, N., Dvorkin, J., Glatt, E., Han, J., Kabel, M., Keehm, Y., Krzikalla, F., Lee, M., Madonna, C., Marsh, M., Mukerji, T., Saenger, E. H., Sain, R., Saxena, N., Ricker, S., Wiegmann, A., & Zhan, X. (2013). Digital rock physics benchmarks-part I: Imaging and segmentation. *Computers and Geosciences*, *50*, 25–32. https://doi.org/10.1016/j.cageo.2012.09.005

Andreola, F., Leonelli, C., Romagnoli, M., & Miselli, P. (2000). Techniques used to determine porosity. *American Ceramic Society Bulletin*, *79*(7), 49–52.

API. (2015). Hydraulic fracturing—Well integrity and fracture containment. *American Petroleum Institute, ANSI/API RECOMMENDED PRACTICE 100–1*. www.api.org/~/media/Files/Policy/Exploration/100-1_e1.pdf

Ashena, R., & Thonhauser, G. (2018). Fundamental coring methods. In *Coring methods and systems* (pp. 17–22). https://doi.org/10.1007/978-3-319-77733-7_3

Asquith, G., & Gibson, R. C. (1983). *Basic well log analysis for geologists* [George Squith, Charles Gibson] (M. K. Horn, Ed.; 2nd ed.). The American Association of Petroleum Geologist.

Asquith, G., & Krygowski, D. (2014). Basic well log analysis, 2nd edition. *AAPG Methods in Exploration Series*, *28*, 244.

Asquith, G., Krygowski, D., Henderson, S., & Hurley, N. (2004). *Basic well log analysis*. https://doi.org/10.1306/mth16823

Assaad, F. A. (2009). Geophysical well logging methods of oil and gas reservoirs. *Field Methods for Petroleum Geologists*, 41–54. https://doi.org/10.1007/978-3-540-78837-9_5

Attribute, C. (2006). Item 5 standard practice for real-time release of pharmaceutical water for the total organic. *Www.Astm.Org*. www.astm.org/Standards/E2656.htm

Bachrach, R. (2020). Poroelasticity. *Encyclopedia of earth sciences series*, 1–4. https://doi.org/10.1007/978-3-030-10475-7_26-1/COVER

Backer, H. C. (1975). *Scanning electron microscopy.* Lubrication. https://testing.mri.psu.edu/materials-characterization-lab/characterization-techniques/scanning-electron-microscopy-sem-0

Baker, R. O., Yarranton, H. W., & Jensen, J. L. (2015a). *Practical reservoir engineering and characterization* (pp. 349–434). https://doi.org/10.1016/b978-0-12-801811-8.00010-9

Baker, R. O., Yarranton, H. W., & Jensen, J. L. (2015b). Openhole well logs—log interpretation basics. In *Practical reservoir engineering and characterization* (pp. 297–343). https://doi.org/10.1016/B978-0-12-801811-8.00009-2

Balcewicz, M., Siegert, M., Gurris, M., Ruf, M., Krach, D., Steeb, H., & Saenger, E. H. (2021). Digital rock physics: A geological driven workflow for the segmentation of anisotropic Ruhr sandstone. *Frontiers in Earth Science, 9.* https://doi.org/10.3389/feart.2021.673753

Baoping, L., & Hongzhi, B. (2005). Advances in calculation methods for rock mechanics parameters. *Petroleum Drilling Techniques, 33*(5), 44–47.

Baoping, Z., & Economides, M. J. (2002). *Reservoir stimulation* (3rd ed.). Petroleum Industry Press.

Barree, R. D., & Winterfeld, P. H. (1998). *Effects of shear planes and interfacial slippage on fracture growth and treating pressures.* https://doi.org/10.2523/48926-ms

Bataee, M., & Irawan, S. (2014). Review of geomechanical application in reservoir modeling. *Journal of Applied Sciences, 14*(10), 981–990. https://doi.org/10.3923/jas.2014.981.990

Beaumont, E. A., & Fiedler, F. (2019). Formation fluid pressure and its application. In *Exploring for oil and gas traps.* https://doi.org/10.1306/trhbk624c5

Ben Eaton, B. A., & Spe-aime, M. (1975). *The equation for geopressure prediction from well logs.* https://doi.org/10.2118/5544-MS

Biot, M. A. (1941). General theory of three-dimensional consolidation. *Journal of Applied Physics, 12*(2), 155–164. https://doi.org/10.1063/1.1712886

Bisutti, I., Hilke, I., & Raessler, M. (2004). Determination of total organic carbon—An overview of current methods. *TrAC—Trends in Analytical Chemistry, 23*(10–11), 716–726. https://doi.org/10.1016/j.trac.2004.09.003

Bjørlykke, K. (2010, September). *Petroleum geoscience: From sedimentary environments to rock physics* (pp. 1–508). https://doi.org/10.1007/978-3-642-02332-3

Bons, P. D., Elburg, M. A., & Gomez-Rivas, E. (2012). A review of the formation of tectonic veins and their microstructures. *Journal of Structural Geology, 43,* 33–62.

Boutrid, A., Bensihamdi, S., Chettibi, M., & Talhi, K. (2015). Strength hardness rock testing. *Journal of Mining Science, 51*(1), 95–110. https://doi.org/10.1134/S1062739115010135

Bradley, W. B. (1979). Mathematical concept—stress cloud—can predict borehole failure. *Oil & Gas Journal, 77*(8). www.osti.gov/biblio/6016808

Britannica, T. E. (2016). *Shale | rock | Britannica.* Encyclopedia Britannica. www.britannica.com/science/shale

Brooks, J. (James), & Welte, D. (1987). Advances in petroleum geochemistry. *Journal of Petroleum Geology, 10*(4). https://doi.org/10.1111/j.1747-5457.1987.tb00586.x

Brown, A. R. (1996). Seismic attributes and their classification. *Leading Edge (Tulsa, OK), 15*(10), 1090. https://doi.org/10.1190/1.1437208

Bustin, R., Bustin, A., Ross, G., Chalmers, V., Murthy, C., Laxmi, & Cui, X. (2008). Shale gas opportunities and challenges. AAPG Annual Convention, 40382. https://www.searchanddiscovery.com/documents/2009/40382bustin/ndx_bustin.pdf

Cameron, M. S. I. W. (2009). Evaluation of unconventional gas shale reservoirs. *AAPG.* www.searchanddiscovery.com/abstracts/html/2009/annual/abstracts/cameron.htm

Casnedi, R. (2005). *Geological characteristics of hydrocarbon reservoirs, I,* 85–116. http://webcache.googleusercontent.com/search?q=cache:JPWx16JrjM8J:www.treccani.it/export/sites/default/Portale/sito/altre_aree/Tecnologia_e_Scienze_applicate/enciclopedia/inglese/inglese_vol_1/pag085-116ING3.pdf+&cd=1&hl=en&ct=clnk&gl=ir

Chang, C., Zoback, M. D., & Khaksar, A. (2006). Empirical relations between rock strength and physical properties in sedimentary rocks. *Journal of Petroleum Science and Engineering*, *51*(3–4), 223–237. https://doi.org/10.1016/j.petrol.2006.01.003

Changzi, P., Jun, P., Yanhui, C., & Hanrong, Z. (2014). Seismic prediction of sweet spots in the Da'anzhai shale play, Yuanba area, the Sichuan Basin. *Natural Gas Industry B*, *1*(2), 185–191. https://doi.org/10.1016/j.ngib.2014.11.010

Chaudhry, A. U. (2004). Well testing methods for naturally fractured reservoirs. In *Oil well testing handbook* (pp. 254–286). https://doi.org/10.1016/B978-075067706-6/50098-9

Chen, L., Xu, J., & Chen, J. (2015). Applications of scanning electron microscopy in earth sciences. *Science China Earth Sciences*, *58*(10), 1768–1778. https://doi.org/10.1007/s11430-015-5172-9

Chen, S., Zhao, W., Ouyang, Y., Zeng, Q., Yang, Q., Hou, H., Gai, S., Bao, S., & Li, X. (2017). Prediction of sweet spots in shale reservoir based on geophysical well logging and 3D seismic data: A case study of Lower Silurian Longmaxi Formation in W4 block, Sichuan Basin, China. *Energy Exploration and Exploitation*, *35*(2), 147–171. https://doi.org/10.1177/0144598716679961

Cheng, C. H., & Johnston, D. H. (1981). Dynamic and static moduli. *Geophysical Research Letters*, *8*(1), 39–42. https://doi.org/10.1029/gl008i001p00039

Chiadikobi, K. C., & Chiaghanam, O. I. (2018). Biomarker distributions of the Campano-Maastrichtian Nkporo group of Anambra Basin, Southeastern Nigeria. *Petroleum and Coal*, *60*(6), 1161–1175.

Chopra, S., & Marfurt, K. J. (2007). *Seismic attributes for prospect identification and reservoir characterization*. https://doi.org/10.1190/1.9781560801900

Chopra, S., & Marfurt, K. J. (2018). Seismic attributes for stratigraphic feature characterization. 78th society of exploration geophysicists international exposition and annual meeting. In *SEG technical program expanded abstracts 2008* (pp. 1590–1594). Society of Exploration Geophysicists.

Ciz, R., Urosevic, M., & Dodds, K. (2005). Pore pressure prediction based on seismic attributes response to overpressure. *The APPEA Journal*, *45*(1), 449–458. https://doi.org/10.1071/AJ04035

Colson, J. L., Grau, J. A., Herron, M. M., Herron, S. L., Schweitzer, J. S., Hertzog, R. C., Preeg, W. E., Nurmi, R. D., & Johnston, J. (2007). *Applications using geochemical logs*. https://doi.org/10.2523/17963-ms

Core Laboratories. (2023). *Wellsite core characterization & sampling services*. www.corelab.com/ps/wellsite-core-characterization-sampling-services

Cornford, C. (2005). The petroleum system. *Petroleum Geology*, 269–285. https://doi.org/10.1306/m60585c1

Critchley, L. (2019a). *What is in-section microscopy*? www.azomining.com/Article.aspx?ArticleID=1502

Critchley, L. (2019b). *What is thin-section microscopy?* LinkedIn Facebook TwitterTRUSTe AZoMining.

Cui, X., Jiang, C., Nassichuk, B., & Wilson, J. (2022). Characteristics of canister core desorption gas from unconventional reservoirs and applications to improve assessment of hydrocarbons-in-place. *Minerals*, *12*(10), 1226. https://doi.org/10.3390/MIN1210 1226

Curiale, J. A. (2017). Total Organic Carbon (TOC). In *Encyclopedia of petroleum geoscience* (pp. 1–5). Springer International Publishing. https://doi.org/10.1007/978-3-319-02330-4_3-1

Daiyu, Z., & Tongwen, J. (2004). Study on watered-out performance and mode of horizontal well in Bottom Water Reservoir. *Journal of Petroleum Sciences*, *25*(6), 73–77.

Daneshy, A. A. (1976). Rock properties controlling hydraulic fracture propagation. *Society of petroleum engineers—SPE European Spring meeting 1976*. https://doi.org/10.2118/5752-MS

Dang, W., Zhang, J. C., Tang, X., Wei, X. L., Li, Z. M., Wang, C. H., Chen, Q., & Liu, C. (2018). Investigation of gas content of organic-rich shale: A case study from Lower Permian shale in southern North China Basin, central China. *Geoscience Frontiers*, *9*(2), 559–575. https://doi.org/10.1016/j.gsf.2017.05.009

Darvishpour, A., Cheraghi Seifaabad, M., Wood, D. A., & Ghorbani, H. (2019). Wellbore stability analysis to determine the safe mud weight window for sandstone layers. *Petroleum Exploration and Development*, *46*(5), 1031–1038. https://doi.org/10.1016/S1876-3804(19)60260-0

Darvishpour, A., Masoud, C. S., Wood, D. A., & Ghorbani, H. (2019). Wellbore stability analysis to determine the safe mud weight window for sandstone layers. 石油勘探与开发, *46*(5), 1031–1038. https://doi.org/10.1016/S1876-3804(19)60260-0

Das, B., & Chatterjee, R. (2017). Wellbore stability analysis and prediction of minimum mud weight for few wells in Krishna-Godavari Basin, India. *International Journal of Rock Mechanics and Mining Sciences*, *93*, 30–37. https://doi.org/10.1016/J.IJRMMS.2016.12.018

Dembicki-Jr., H. (2009). Three common source rock evaluation errors made by geologists during prospect or play appraisals. *AAPG Bulletin*, *93*(3), 341–356. https://doi.org/10.1306/10230808076

Dembicki-Jr., H. (2017). Petroleum—Geochemistry for exploration and production. In *Practical petroleum geochemistry for exploration and production* (p. 342). Elsevier.

Dentith, M., & Mudge, S. (2014). Geophysics for the mineral exploration geoscientist. In *AusIMM bulletin* (Issue 6). https://doi.org/10.1017/cbo9781139024358

Dewprashad, B., Abass, H. H., Meadows, D. L., Weaver, J. D., & Bennett, B. J. (1993). A method to select resin-coated proppants. *All Days*, 3–6. https://doi.org/10.2118/26523-MS

Di Gianfrancesco, A. (2017). Technologies for chemical analyses, microstructural and inspection investigations. In *Materials for ultra-supercritical and advanced ultra-supercritical power plants* (pp. 197–245). https://doi.org/10.1016/B978-0-08-100552-1.00008-7

Diamond, S. (2000). Mercury porosimetry. *Cement and Concrete Research*, *30*(10), 1517–1525. https://doi.org/10.1016/s0008-8846(00)00370-7

Dormieux, L., Kondo, D., & Ulm, F. J. (2006). Microporomechanics. *Microporomechanics*, 1–328. https://doi.org/10.1002/0470032006

Eissa, E. A., & Kazi, A. (1988). Relation between static and dynamic Young's moduli of rocks. *International Journal of Rock Mechanics and Mining Sciences & Geomechanics Abstracts Elsevier*, *25*(6), 479–482. https://doi.org/10.1016/0148-9062(88)90987-4

Elias, R., Duclerc, D., Le-Van-Loi, R., Gelin, F., & Dessort, D. (2013). A new geochemical tool for the assessment of organic-rich shales. *Unconventional resources technology conference 2013, URTC 2013*. https://doi.org/10.1190/urtec2013-206

Elnaggar, A. A., Kassab, M. A., & Abass, A. E. (2018). Reservoir characterization utilizing core and wire line logging data for Kareem sandstone, Ashrafi oil Field, Gulf of Suez, Egypt. *Egyptian Journal of Petroleum*, *27*(4), 1013–1027.

Engleman, E. E., Jackson, L. L., & Norton, D. R. (1985). Determination of carbonate carbon in geological materials by coulometric titration. *Chemical Geology*, *53*(1–2), 125–128. https://doi.org/10.1016/0009-2541(85)90025-7

EPA. (2002). Methods for the determination of total organic carbon (TOC) in soils and sediments. *Carbon*, *32*(April), 25.

Espitalie, J., Madec, M., Tissot, B., Mennig, J. J., & Leplat, P. (1977). Source rock characterization method for petroleum exploration. *Proceedings of the annual offshore technology conference*, *1977-May*, 439–444. https://doi.org/10.4043/2935-MS

Evenick, J. (2008). *Introduction to well logs and subsurface maps*. PennWell Corporation. https://doi.org/978-1-59370-138-3

Ezenweichu, C. L., & Laditan, O. D. (2015). The causes, effects and minimization of formation damage in horizontal wells. *Petroleum & Coal, 57*(2), 169–184. http://search.ebscohost.com/login.aspx?direct=true&db=a9h&AN=108333548&lang=es&site=ehost-live

Fanchi, J. R. (2010). *Integrated reservoir asset management: Principles and best practices* (pp. 1–361). https://doi.org/10.1016/C2009-0-62240-6

Fernandez Rojas, L., Quintero Peña, Y., & Calderon Carrillo, Z. H. (2016). Brittleness analysis: A methodology to identify sweet spots in shale gas reservoirs. In *Society of petroleum engineers—SPE argentina exploration and production of unconventional resources symposium*. https://doi.org/10.2118/180955-ms

FIS-B and how does it work. (2020). www.thermofisher.com/blog/ask-a-scientist/what-is-xrf-x-ray-fluorescence-and-how-does-it-work/

Fjær, E. (2009). Static and dynamic moduli of a weak sandstone. *Geophysics, 74*(2), WA103–WA112. https://doi.org/10.1190/1.3052113

Fjær, E., Holt, R. M., Horsrud, P., Raaen, A. M., & Risnes, R. (1992). *Petroleum related rock mechanics*. https://doi.org/10.1016/0920-4105(93)90066-n

Ganat, T. A. A. O. (2019). *Fundamentals of reservoir rock properties* (pp. 1–156). https://doi.org/10.1007/978-3-030-28140-3/COVER

Ganat, T. A. A. O. (2020). Technical guidance for petroleum exploration and production plans. *SpringerBriefs in Applied Sciences and Technology*, 1–93.

Gandhi, S. M., & Sarkar, B. C. (2016). *Essentials of mineral exploration and evaluation* (pp. 1–406). https://doi.org/10.1016/C2015-0-04648-2

Garcia, X., Nagel, N., Zhang, F., Sanchez-Nagel, M., & Lee, B. (2013). Revisiting vertical hydraulic fracture propagation through layered formations—A numerical evaluation. *47th US Rock Mechanics/Geomechanics Symposium 2013, 4*, 2501–2511.

García Veigas, F. J., Prats Miralles, E., Domínguez Ximénez, A., & Villuendas Latorre, A. (2012). Advanced applications of scanning electron microscopy in geology. In *Capítol Del Llibre: Handbook of instrumental techniques for materials*. Chemical and Biosciences Research, Centres Científics i Tecnològics, Universitat de Barcelona.

Garrouch, A. A., Lababidi, H. M. S., & Ebrahim, A. S. (2004). An integrated approach for the planning and completion of horizontal and multilateral wells. *Journal of Petroleum Science and Engineering, 44*(3–4), 283–301. https://doi.org/10.1016/j.petrol.2004.03.007

GeoExpro. (2023). *Simplifying digital rock physics workflows*. https://geoexpro.com/simplifying-digital-rock-physics-workflows/

Gerrard, A. J. (1988). Rock strength and resistance. *Rocks and Landforms*, 85–106. https://doi.org/10.1007/978-94-011-5983-8_4

Gholilou, A., Behnoud far, P., Vialle, S., & Madadi, M. (2017). Determination of safe mud window considering time-dependent variations of temperature and pore pressure: Analytical and numerical approaches. *Journal of Rock Mechanics and Geotechnical Engineering, 9*(5), 900–911. https://doi.org/10.1016/j.jrmge.2017.02.002

Ghulam, A. S., & Iqbal, G. M. (2015). Reservoir engineering: *The fundamentals, simulation and management of conventional and unconventional recoveries*. 1st Ed. Gulf Professional Publishing. https://books.google.com/books/about/Reservoir_Engineering.html?id=84hCCQAAQBAJ

Giles, M. R., & Tennant, S. H. (2014). Sweet spots: What are they, where are they, how are they created and are they important anyway? *Society of Petroleum Engineers—European Unconventional Resources Conference and Exhibition 2014: Unlocking European Potential, 2*, 732–737. https://doi.org/10.2118/167760-ms

Glaser, K. S., Miller, C. K., Johnson, G. M., Kleinberg, R. L., & Pennington, W. D. (2014). Seeking the sweet spot: Reservoir and completion quality in organic shales. *Oilfield Review, 25*(4), 16–29. www.slb.com/resources/publications/industry_articles/oilfield_review/2013/or2013win02_sweet_spot.aspx

Glaser, K. S., Miller, C. K., Johnson, G. M., Toelle, B., Kleinberg, R. L., Miller, P., & Pennington, W. D. (2013). Seeking the sweet spot: Reservoir and completion quality in organic shales. *Oilfield Review*, *25*(4), 16–29.

Glover, P. (2000). Petrophysics. *Petrophysics MSc course notes*, 32–54. www.passeidireto.com/arquivo/93087126/petrophysics-glover-paul-w-j

Glover, P. (2001). Formation evaluation MSc course notes. *Formation Evaluation M.Sc. Course Notes*, *9*(February), 84–94.

Gluyas, J. G. (2018). *Petroleum field appraisal*, 1–10. https://doi.org/10.1007/978-3-319-02330-4_22-1

Goktan, R. M., & Yilmaz, N. G. (2005). A new methodology for the analysis of the relationship between rock brittleness index and drag pick cutting efficiency. *Journal of the South African Institute of Mining and Metallurgy*, *105*(10), 727–732.

Gonzalez, J., Lewis, R., Hemingway, J., Grau, J., Rylander, E., & Pirie, I. (2013). Determination of formation organic carbon content using a new neutron-induced gamma ray spectroscopy service that directly measures carbon. *Unconventional resources technology conference 2013, URTC 2013*. OnePetro. https://doi.org/10.1190/urtec2013-112

Grau, J., Ellis, D., Schweitzer, J., Herron, M., Hertzog, R., Colson, L., Seeman, B., O'Brien, M., Scott, H., McKeon, D., & Wright, P. (1987). Geochemical logging with spectrometry tools. *SPE Formation Evaluation*, *4*(02), 153–162.

Grotek, I., & Janas, M. (2014). Thermal maturity of organic matter. *Polish Geological Institute*. https://infolupki.pgi.gov.pl/en/gas/thermal-maturity-organic-matter

Guo, J., & Liu, Y. (2014). A comprehensive model for simulating fracturing fluid leakoff in natural fractures. *Journal of Natural Gas Science and Engineering*, *21*, 977–985. https://doi.org/10.1016/j.jngse.2014.10.020

Hajiabdolmajid, V., & Kaiser, P. (2002). Brittleness of rock and stability assessment in hard rock tunneling. *Tunnelling and Underground Space Technology*, *18*(1), 35–48. https://doi.org/10.1016/S0886-7798(02)00100-1

Han, J., & Zhang, H. (2021). Canister desorption method for shale gas content measurement. *Advanced Resources Research*, *1*, 20–27.

Hanson, M. E., Shaffer, R. J., & Anderson, G. D. (1981). Effects of various parameters on hydraulic fracturing geometry. *Society of Petroleum Engineers Journal*, *21*(04), 435–443.

Havrdova, M., Polakova, K., Skopalik, J., Vujtek, M., Mokdad, A., Homolkova, M., Tucek, J., Nebesarova, J., & Zboril, R. (2014). Field emission scanning electron microscopy (FE-SEM) as an approach for nanoparticle detection inside cells. *Micron*, *67*, 149–154. https://doi.org/10.1016/J.MICRON.2014.08.001

He, X., Zhang, P., He, G., Gao, Y., Liu, M., Zhang, Y., Fang, D., & Li, Y. (2020). Evaluation of sweet spots and horizontal-well-design technology for shale gas in the basin-margin transition zone of southeastern Chongqing, SW China. *Energy Geoscience*, *1*(3–4), 134–146. https://doi.org/10.1016/j.engeos.2020.06.004

Heiland-Hansen, D., Magnus, I., Edvardsen, A., & Hansen, E. (1997). Seismic inversion for reservoir characterization and well planning in the Snorre Field. *The Leading Edge*, *16*(3), 269. https://doi.org/10.1190/1.1437616

Hemsing, D., & Schmitt, D. (2006). Experimental anisotropy results in Alberta Shales. *CSPG/CSEG/CWLS Convention 2006, #90211. https://www.searchanddiscovery.com/abstracts/pdf/2015/90211cspg/abstracts/ndx_Hemsing.pdf*

Herein, D. (2008). X-ray powder diffraction. In *Handbook of heterogeneous catalysis*. https://doi.org/10.1002/9783527610044.hetcat0039

Herron, M. M. (1986). Mineralogy from geochemical well logging. *Clays & Clay Minerals*, *34*(2), 204–213. https://doi.org/10.1346/CCMN.1986.0340211

Herron, M. M., & Herron, S. L. (1990). Geological applications of geochemical well logging. *Geological Society Special Publication*, *48*, 165–175. https://doi.org/10.1144/GSL.SP.1990.048.01.14

Herron, M. M., Loan, M. E., Charsky, A. M., Herron, S. L., Pomerantz, A. E., & Polyakov, M. (2014). Kerogen content and maturity, mineralogy and clay typing from drifts analysis of cuttings or core. *Petrophysics*, *55*(05), 435–446.

Herron, S. L. (2021). In situ evaluation of potential source rocks by wireline logs. *Source and migration processes and evaluation techniques*. https://doi.org/10.1306/trhbk543c13

Hertzog, R., Seeman, K., O'Brien, M., Scott, H., McKeon, D., Wraight, P., Grau, J., Ellis, D., Schweitzer, J., & Herron, M. (1989). Geochemical logging with spectrometry tools. *SPE Formation Evaluation*, 153–162.

Higgins-Borchardt, S., Sitchler, J., & Bratton, T. (2016). Geomechanics for unconventional reservoirs. In *Unconventional oil and gas resources handbook: Evaluation and development* (pp. 199–213). https://doi.org/10.1016/B978-0-12-802238-2.00007-9

Hobart, M. K. (2022). Shale. *Web*. https://geology.com/rocks/shale.shtml

Hook, J. R. (2003). An introduction to porosity (TUTORIAL). *Petrophysics*, *44*(3), 205–212.

Hosseini, E., Hajivand, F., & Tahmasebi, R. (2019). The effect of alkaline–surfactant on the WETTABILITY, relative permeability and oil recovery of carbonate reservoir rock: Experimental investigation. *Journal of Petroleum Exploration and Production Technology*, *9*(4), 2877–2891. https://doi.org/10.1007/S13202-019-0671-4

Hosseini, M., Arif, M., Keshavarz, A., & Iglauer, S. (2021). Neutron scattering: A subsurface application review. *Earth-Science Reviews*, *221*. https://doi.org/10.1016/J.EARSCIREV.2021.103755

Houlding, S. W. (1994). The geological characterization process. *3D Geoscience Modeling*, 7–26. https://doi.org/10.1007/978-3-642-79012-6_2

Hudson, J. A., & Harrison, J. P. (1997). *Rock engineering mechanics: An introduction to the principles*. Elsevier.

Hui, L., Dong, C., Yu, H., Zhao, X., Li, Y., Cao, L., & Qu, M. (2021). Static and dynamic mechanical properties of organic-rich gas shale. *Geofluids*. https://doi.org/10.1155/2021/6695975

Immerstein, T. (2013). *Wellbore stability and rock mechanics study on the Ula field*. University of Stavanger. http://brage.bibsys.no/xmlui/bitstream/handle/11250/183589/Immerstein_Tonje.pdf?sequence=1

Ismail, A., Raza, A., Gholami, R., & Rezaee, R. (2020). Reservoir characterization for sweet spot detection using color transformation overlay scheme. *Journal of Petroleum Exploration and Production Technology*, *10*(6), 2313–2334. https://doi.org/10.1007/s13202-020-00913-5

Issler, D. R., Hu, K., Bloch, J. D., & Katsube, T. J. (2002). Organic carbon content determined from well logs: Examples from Cretaceous sediments of western Canada. *Geological Survey of Canada Open File*, *4362*, 19.

James, D. (1999). SELLEY, R. C. 1998. Elements of Petroleum Geology, 2nd ed. xvi + 470 pp. San Diego, London, Boston, New York, Sydney, Tokyo, Toronto: Academic Press. Price £39.95 (hard covers). ISBN 0 12 636370 6. *Geological Magazine*, *136*(3), 331–340. https://doi.org/10.1017/s0016756899362646

Javadova, A. (2018). *Coring and core analysis in reservoir geology*. Azerbaijan Society of Petroleum Geologists.

Jiang, Y., Killough, J. E., Wu, X., & Cui, Y. (2021). Multiscale flow and optimal production control techniques in smart unconventional reservoirs. *Geofluids*, *2021*. https://doi.org/10.1155/2021/6653442

Jin, H., Sonnenberg, S. A., & Sarg, J. F. (2012). Source rock evaluation for the Bakken Petroleum System in the Williston Basin, North Dakota and Montana. *Search and Discovery, 20156*, 1–4.

Josh, M., Esteban, L., Delle Piane, C., Sarout, J., Dewhurst, D. N., & Clennell, M. B. (2012). Laboratory characterisation of shale properties. *Journal of Petroleum Science and Engineering, 88–89*, 107–124. https://doi.org/10.1016/J.PETROL.2012.01.023

Joshua, C. P. E. (2012). *Basic well log interpretation*, PDH Online Course K132, PDH Center. https://pdhonline.com/courses/k132/k132content.pdf

Ju, J., Li, Q., & Xu, J. (2020). Experimental study on the self-healing behavior of fractured rocks induced by water-CO2-rock interactions in the shendong coalfield. *Geofluids, 2020*. https://doi.org/10.1155/2020/8863898

Kang, S., Lu, L., Tian, H., Yang, Y., Jiang, C., & Ma, Q. (2021). Numerical simulation based on the canister test for shale gas content calculation. *Energies, 14*(20). https://doi.org/10.3390/en14206518

Kantzas, A., Bryan, J., & Taheri, S. (2005). *Fundamentals of fluid flow in porous media*. https://perminc.com/resources/fundamentals-of-fluid-flow-in-porous-media/

Katende, A., O'Connell, L., Rich, A., Rutqvist, J., & Radonjic, M. (2021). A comprehensive review of proppant embedment in shale reservoirs: Experimentation, modeling and future prospects. *Journal of Natural Gas Science and Engineering, 95*. https://doi.org/10.1016/j.jngse.2021.104143

Kellton. (2020). *Top benefits of workflow automation in oil & gas*. www.kellton.com/kellton-tech-blog/benefits-of-workflow-automation-in-oil-and-gas-industry

Kemet. (2016). *Rock thin sections (petrographic thin section preparation)—Kemet*. www.kemet.co.uk/blog/petrography/rock-thin-sections

Kennedy, M. (2015). Logs part II: Porosity, resistivity and other tools. *Developments in Petroleum Science, 62*, 107–149. https://doi.org/10.1016/B978-0-444-63270-8.00005-0

Kenomore, M., Hassan, M., Dhakal, H., & Shah, A. (2017). Total organic carbon evaluation of the Bowland Shale Formation in the Upper Bowland of the Widmerpool Gulf. *Journal of Petroleum Science and Engineering, 150*, 137–145. https://doi.org/10.1016/J.PETROL.2016.11.040

Kenyon, W. E. (1992). Nuclear magnetic resonance as a petrophysical measurement. *Nuclear Geophysics, 6*(2), 153–171.

Kiran, K. T. (2005). *2D and 3D land seismic data acquisition | PDF | reflection seismology | amplifier*. Andhra University. www.scribd.com/doc/7256110/2D-and-3D-Land-Seismic-Data-Acquisition

Klein, J. D. (1991). Induction log anisotropy corrections. *SPWLA 32nd annual logging symposium 1991*. The Log Analyst.

Kobr, M. (2011). Geophysical Well Logging. In: Gupta, H.K. (eds) *Encyclopedia of Solid Earth Geophysics. Encyclopedia of Earth Sciences Series*. Springer, Dordrecht. https://doi.org/10.1007/978-90-481-8702-7_173.

Kormaksson, M., Vieira, M. R., & Zadrozny, B. (2015). A data driven method for sweet spot identification in shale plays using well log data. *Society of petroleum engineers—SPE digital energy conference and exhibition 2015*, 622–630. https://doi.org/10.2118/173455-ms

Krumben, W. C., & Sloss, L. L. (1963). *Stratigraphy and sedimentation*. Freeman. https://doi.org/10.2973/dsdp.proc.9.122.1972

Kuila, U., McCarty, D. K., Derkowski, A., Fischer, T. B., & Prasad, M. (2014). Total porosity measurement in gas shales by the water immersion porosimetry (WIP) method. *Fuel, 117*(PARTB), 1115–1129. https://doi.org/10.1016/J.FUEL.2013.09.073

Lab Manager. (2011). TOC analyzers—"detecting and quantifying carbon content". *Lab Manager*. www.labmanager.com/lab-product/toc-analyzers-detecting-and-quantifying-carbon-content-17940

Lafargue, E., Marquis, F., & Pillot, D. (1998). Applications of rock-eval 6 in the exploration and production of hydrocarbons, and in the studies of soil contamination. *Oil & Gas Science and Technology, Review of the French Petroleum Institute. Scientific Research Publishing*, *53*, 421–437. www.scirp.org/(S(351jmbntvnsjt1aadkozje))/reference/referencespapers.aspx?referenceid=1317696

Law, C. A. (2019). Evaluating source rocks. *Exploring for Oil and Gas Traps*. https://doi.org/10.1306/trhbk624c6

Li, H., Zou, Y., Valko, P. P., & Ehlig-Economides, C. (2016). Hydraulic fracture height predictions in laminated shale formations using finite element discrete element method. *Society of petroleum engineers—SPE hydraulic fracturing technology conference, HFTC 2016*. https://doi.org/10.2118/179129-ms

Li, M., & Zhao, Y. (2014a). Geophysical exploration technology: Applications in lithological and stratigraphic reservoirs. In *Geophysical exploration technology: Applications in lithological and stratigraphic reservoirs*. Elsevier Inc. https://doi.org/10.1016/C2012-0-03328-5

Li, Y. (2002). Anisotropic well logs and their applications in seismic analysis. *2002 SEG annual meeting*. https://doi.org/10.1190/1.1817216

Liang, X., Wang, G., Xu, Z., Zhang, J., Chen, Z., Xian, C., Lu, H., Liu, C., Zhao, C., & Xiong, S. (2016). Comprehensive evaluation technology for shale gas sweet spots in the complex marine mountains, South China: A case study from Zhaotong national shale gas demonstration zone. *Natural Gas Industry*, *36*(1), 33–42. https://doi.org/10.3787/j.issn.1000-0976.2016.01.004

Lishman, J. R. (1970). Core permeability anisotropy. *Journal of Canadian Petroleum Technology*, *9*(2). https://doi.org/10.2118/70-02-01

Lock, E., Ghasemi, M., Mostofi, M., & Rasouli, V. (2012). An experimental study of permeability determination in the lab. *WIT Transactions on Engineering Sciences*, *80*, 221–230. https://doi.org/10.2495/PMR120201

Lozovyi, S., & Bauer, A. (2019a). From static to dynamic stiffness of shales: Frequency and stress dependence. *Rock Mechanics and Rock Engineering*, *52*(12), 5085–5098. https://doi.org/10.1007/s00603-019-01934-1

Lozovyi, S., & Bauer, A. (2019b). From static to dynamic stiffness of shales: Frequency and stress dependence. *Rock Mechanics and Rock Engineering*, *52*(12), 5085–5098. https://doi.org/10.1007/S00603-019-01934-1/TABLES/4

Luffel, D. L., Hopkins, C. W., & Schettler Jr, P. D. (1993, October). Matrix permeability measurement of gas productive shales. *SPE annual technical conference and exhibition* (p. SPE-26633). SPE.

Luthi, S. M. (2001). *Geological well logs*. https://doi.org/10.1007/978-3-662-04627-2

Lyons, W. C. (2010). Formation evaluation. In *Working guide to reservoir engineering* (pp. 97–232). https://doi.org/10.1016/B978-1-85617-824-2.00002-2

Ma, Y. Z., & Holditch, S. A. (2015). *Unconventional oil and gas resources handbook: Evaluation and development* (pp. 1–536). https://doi.org/10.1016/C2014-0-01377-9

Maende, A., & David Weldon, W. (2013). Pyrolysis and TOC identification of tight oil sweet spots. *Unconventional resources technology conference 2013, URTC 2013*. https://doi.org/10.1190/urtec2013-268

Maleki, S., Gholami, R., Rasouli, V., Moradzadeh, A., Riabi, R. G., & Sadaghzadeh, F. (2014). Comparison of different failure criteria in prediction of safe mud weigh window in drilling practice. *Earth-Science Reviews*, *136*, 36–58. https://doi.org/10.1016/J.EARSCIREV.2014.05.010

Mandeep, K. (2020). *Basic well logging*. Lecture Note. www.coursehero.com/file/57355449/Basic-Well-Logging-By-Mandeep-Kumarpdf/

Mavko, G. (2003). *The Rock Physics Handbook: Tools for Seismic Analysis in Porous Media*. 329.

Mavko, G., Mukerji, T., & Dovorkin, J. (2009). *The rock physics handbook—Tools for seismic analysis of porous media* (2nd ed., p. 522). Cambridge University.

Maxwell, S. C., & Urbancic, T. I. (2001). The role of passive microseismic monitoring in the instrumented oil field. *Leading Edge (Tulsa, OK)*, *20*(6), 636–639. https://doi.org/10.1190/1.1439012

Mcgarr, A. (1994). Case histories of induced and triggered seismicity. In *International handbook of earthquake and engineering seismology* (pp. 0–12). Academic Press.

McGarr, A., & Simpson, D. (1997). A broad look at induced and triggered seismicity. *Rockbursts and Seismicity in Mines*, 385–396. http://scholar.google.com/scholar?hl=en&btnG=Search&q=intitle:A+broad+look+at+induced+and+triggered+seismicity.#0

McPhee, C., Reed, J., & Zubizarreta, I. (2015a). Routine core analysis. *Developments in Petroleum Science*, *64*, 181–268. https://doi.org/10.1016/B978-0-444-63533-4.00005-6

McPhee, C., Reed, J., & Zubizarreta, I. (2015b). Preparation for special core analysis. *Developments in Petroleum Science*, *64*, 269–312. https://doi.org/10.1016/B978-0-444-63533-4.00006-8

McPhee, C., Reed, J., & Zubizarreta, I. (2015c). Wettability and wettability tests. *Developments in Petroleum Science*, *64*, 313–345. https://doi.org/10.1016/B978-0-444-63533-4.00007-X

Michael, F. M., Krishnan, M. R., Li, W., & Alsharaeh, E. H. (2020). A review on polymer-nanofiller composites in developing coated sand proppants for hydraulic fracturing. *Journal of Natural Gas Science and Engineering*, *83*. https://doi.org/10.1016/j.jngse.2020.103553

Moldoveanu, S. C. (2019). General information about pyrolysis. *Pyrolysis of Organic Molecules*, 1–33. https://doi.org/10.1016/B978-0-444-64000-0.00001-9

Montmayeur, H., & Graves, R. M. (2007). *Prediction of static elastic/Mechanical properties of consolidated and unconsolidated sands from acoustic measurements: Basic measurements.* https://doi.org/10.2523/14159-ms

Moore, D. M., & Reynolds, R. (1997). *X-ray diffraction and identification and analysis of clay minerals* (2nd ed.). Oxford University Press.

Morton-Thompson, D., & Woods, A. M. (1993). *Development geology reference manual.* https://doi.org/10.1306/mth10573

Murugesu, M. P. (2017). Pore structure analysis using subcritical gas adsorption method. *Proceedings—SPE annual technical conference and exhibition*. https://doi.org/10.2118/189292-stu

Musa Adamu, L., George Obaje, N., Pudentiana Ngozi, O., & Mohammed Umar, U. (2020). Integrated geochemical and multivariate statistical examination of source rocks in the Sokoto and Anambra Basins, Nigeria: Implications for hydrocarbon prospectivity. *Petroleum Science and Engineering*, *4*(2), 39. https://doi.org/10.11648/j.pse.20200402.11

Nanda, N. C. (2016). *Seismic data interpretation and evaluation for hydrocarbon exploration and production.* https://doi.org/10.1007/978-3-319-26491-2

Neuman, C. H., Salaita, G. N., & Mahdavi, M. (1988). Neutron scattering for porosity determination. *IEEE Transactions on Nuclear Science*, *35*(1), 812–816. https://doi.org/10.1109/23.12837

Nowroozi, A. A., Horrocks, S. B., & Henderson, P. (1999). Saltwater intrusion into the freshwater aquifer in the eastern shore of Virginia: A reconnaissance electrical resistivity survey. *Journal of Applied Geophysics*, *42*(1), 1–22. https://doi.org/10.1016/S0926-9851(99)00004-X

Ofwona, C. (2010). Introduction to geophysical well logging and flow testing, Geothermal Development Company Ltd., Short Course V on Exploration for Geothermal Resources, http://large.stanford.edu/courses/2018/ph240/owusu-agyeman2/docs/ofwona-2010.pdf

Onajite, E. (2013). *Seismic data analysis techniques in hydrocarbon exploration*. Elsevier B.V. https://doi.org/10.1016/C2013-0-09969-0

Oort, E. van. (2018). *How to test for compatibility between fluids and shales*. https://doi.org/10.2118/189633-ms

Opus Kinetic. (2019). *What is field development planning (FDP)*. www.opuskinetic.com/2019/06/what-is-field-development-planning-fdp/

Ouhadi, V. R., & Yong, R. N. (2003). Impact of clay microstructure and mass absorption coefficient on the quantitative mineral identification by XRD analysis. *Applied Clay Science*, *23*(1–4), 141–148. https://doi.org/10.1016/S0169-1317(03)00096-6

Passey, Q. R., Bohacs, K. M., Esch, W. L., Klimentidis, R., & Sinha, S. (2010). From oil-prone source rock to gas-producing shale reservoir—Geologic and petrophysical characterization of unconventional shale-gas reservoirs. International Oil and Gas Conference and Exhibition, SPE-131350-MS, https://doi.org/10.2118/131350-MS

Passey, Q. R., Creaney, S., Kulla, J. B., Moretti, F. J., & Stroud, J. D. (1990). A practical model for organic richness from porosity and resistivity logs. *American Association of Petroleum Geologists Bulletin*, *74*(12), 1777–1794. https://doi.org/10.1306/0c9b25c9-1710-11d7-8645000102c1865d

Patrick Murphy, D., Chilingarian, G. V., & Jalal Torabzadeh, S. (1996). Chapter 3 Core analysis and its application in reservoir characterization. *Developments in Petroleum Science*, *44*(Part 2), 105–153. https://doi.org/10.1016/S0376-7361(96)80025-2

Pavlo, Y. C., & Douglas, R. S. (2003). Seismic anisotropy of shales. *CSEG Recorder*, *28*(7). https://csegrecorder.com/articles/view/seismic-anisotropy-of-shales

Pemper, R., Pereira, A., Hou, G., Dolliver, D., Tudge, J., Kharrazi, J., Chok, H., Schmid, G., Mekic, N., Blankinship, T., Epstein, R., Cave, T., & MacPherson, A. (2018). A new geochemical logging tool for determination of formation chemistry and mineralogy in both conventional and unconventional reservoirs. *Proceedings—SPE annual technical conference and exhibition, 2018-September*. https://doi.org/10.2118/191411-MS

Peters, K. E. (1986). Guidelines for evaluating petroleum source rock using programmed pyrolysis. *AAPG Bulletin*, *70*(3), 318–329. https://doi.org/10.1306/94885688-1704-11D7-8645000102C1865D

Peters, K. E., Kacewicz, M., & Curry, D. J. (2012). An overview of basin and petroleum system modeling: Definitions and concepts. *Basin Modeling: New Horizons in Research and Applications*, *4*, 1–16.

Peters, K. E., & Mary, R. C. (1994). Applied source rock geochemistry: Chapter 5: Part II. *Essential Elements*, *77*, 93–120. http://archives.datapages.com/data/specpubs/methodo2/data/a077/a077/0001/0050/0093.htm

Peters, K. E., & Simoneit, B. R. T. (1982). Rock-eval pyrolysis of quaternary sediments from Leg 64, Sites 479 and 480, Gulf of California. *Initial Reports of the Deep Sea Drilling Project*, *64*. https://doi.org/10.2973/dsdp.proc.64.138.1982

Peters, K. E., Xia, X., Pomerantz, A. E., & Mullins, O. C. (2016). Geochemistry applied to evaluation of unconventional resources. In *Unconventional oil and gas resources handbook: Evaluation and development* (pp. 71–126). https://doi.org/10.1016/B978-0-12-802238-2.00003-1

Pettijohn, F. J. (1957). Sedimentary rocks. *Geological Magazine*, 2nd ed., *94*(6). Cambridge University Press. https://doi.org/10.1017/s0016756800070254

Qingcai, Z., Chen, S., He, P., Yang, Q., Guo, X., Chen, P., Dai, C., Li, X., Gai, S., Deng, Y., & Hou, H. (2018). Quantitative prediction of shale gas sweet spots based on seismic data in Lower Silurian Longmaxi Formation, Weiyuan area, Sichuan Basin, SW China. *Petroleum Exploration and Development*, *45*(3), 422–430. https://doi.org/10.1016/S1876-3804(18)30047-8

Radtke, R. J., Lorente, M., Adolph, B., Berheide, M., Fricke, S., Grau, J., Herron, S., Horkowitz, J., Jorion, B., Madio, D., May, D., Miles, J., Perkins, L., Philip, O., Roscoe, B., Rose, D., & Stoller, C. (2012). A new capture and inelastic spectroscopy tool takes geochemical logging to the next level. *SPWLA 53rd annual logging symposium 2012*. PWLA-2012-103

Rahman, K., & Gui, F. (2016). Geomechanical sweet spot identification in unconventional resources development. *Society of petroleum engineers—SPE Asia Pacific oil and gas conference and exhibition 2016*. https://doi.org/10.2118/182247-ms

Raith, M. M., Raase, P., & Reinhardt, J. (2011). *Guide to thin section microscopy*. Citeseer.

Raja, P. M. V., & Barron, A. R. (2021). Principles of gamma-ray spectroscopy and applications in nuclear forensics radioactive decay kinetics and equilibria. In *Physical methods in chemistry and nano science*. https://chem.libretexts.org/Bookshelves/Analytical_Chemistry/Book%3A_Physical_Methods_in_Chemistry_and_Nano_Science_(Barron)/01%3A_Elemental_Analysis/1.17%3A_Principles_of_Gamma-ray_Spectroscopy_and_Applications_in_Nuclear_Forensics

Ranjbar-Karami, R., kadkhodaie-Ilkhchi, A., & Shiri, M. (2014). A modified fuzzy inference system for estimation of the static rock elastic properties: A case study from the Kangan and Dalan gas reservoirs, South Pars gas field, the Persian Gulf. *Journal of Natural Gas Science and Engineering*, *21*, 962–976. https://doi.org/10.1016/j.jngse.2014.10.034

Reservoir, S. (1998). *Source rock analyzer (SRA)*. www.stratumreservoir.com/wp-content/uploads/2019/11/SRNO-PN-1064-Stratum-SRA-Source-Rock-Analyzer.pdf

Reservoirgroup. (2019). *HAWK programmed pyrolysis*. www.reservoirgroup.com/wp-content/uploads/2019/04/HAWK_DS.pdf

Resources, O. (2023). *Core handling and preservation*. https://orogenic.com.my/our-services/core-storage-core-analysis/core-handling-and-preservation/

Rezaee, R. (2015). *Fundamentals of gas shale reservoirs*. https://doi.org/10.1002/9781119039228

Robertson, J. D. (1989). Reservoir management using 3-D seismic data. *Leading Edge*, *8*(2), 25–31. https://doi.org/10.1190/1.1439607

Ronov, A. B. (1958). Organic carbon in sedimentary rocks (in relation to the presence of petroleum). *Geokhimiya (Translation)*, *5*, 510–536.

Ruan, C. De, & Ward, C. R. (2002). Quantitative X-ray powder diffraction analysis of clay minerals in Australian coals using Rietveld methods. *Applied Clay Science*, *21*(5–6), 227–240. https://doi.org/10.1016/S0169-1317(01)00103-X

Satter, A., & Iqbal, G. M. (2015). *Reservoir engineering: The fundamentals, simulation, and management of conventional and unconventional recoveries* (pp. 1–472). https://doi.org/10.1016/B978-0-12-800219-3.00025-5

Sayers, C., Den Boer, L., Nagy, Z., Hooyman, P., & Ward, V. (2005). Pore pressure in the Gulf of Mexico: Seeing ahead of the bit. *World Oil*, *226*(12), 55–58.

Schlumberger. (2022a). Dean-stark extraction. *Energy Glossary*. https://glossary.slb.com/en/terms/d/dean-stark_extraction

Schlumberger. (2023a). Exploration. *Energy Glossary*. https://glossary.slb.com/en/terms/e/exploration

Schlumberger. (2023b). Geochemical log. *Energy Glossary*. https://glossary.slb.com/en/Terms/g/geochemical_log.aspx

Schön, J. H. (2015). Nuclear Magnetic Resonance (NMR): Petrophysical properties. *Developments in Petroleum Science*, *65*, 85–107. https://doi.org/10.1016/B978-0-08-100404-3.00003-2

Schuelke, J. S., & Quirein, J. A. (1998). Validation: A technique for selecting seismic attributes and verifying results. *Seg Technical Program Expanded Abstracts*. https://doi.org/10.1190/1.1820645

Science, P. (1988). *Fundamentals of well-log interpretation*. Elsevier B.V.

Segesman, F. F. (1980). Well-logging method. *Geophysics*, *45*(11), 1667–1684. https://doi.org/10.1190/1.1441057
Selley, R. C. (2005). Sedimentary rocks | mineralogy and classification. In *Encyclopedia of geology* (pp. 25–37). https://doi.org/10.1016/B0-12-369396-9/00304-X
Selley, R. C., & Sonnenberg, S. A. (2014). *Elements of petroleum geology* (3rd ed., pp. 1–507). https://doi.org/10.1016/C2010-0-67090-8
Sena, A., Castillo, G., Chesser, K., Voisey, S., Estrada, J., Carcuz, J., Carmona, E., & Hodgkins, P. (2011). Seismic reservoir characterization in resource shale plays: Stress analysis and sweet spot discrimination. *Leading Edge (Tulsa, OK)*, *30*(7), 758–764. https://doi.org/10.1190/1.3609090
Shan, Y. M., & Liu, W. G. (2000). Experimental study on dynamic and static mechanics parameters of rocks under formation conditions. *Journal of the Chengdu Institute of Technology*, *27*(3), 249–254.
Sheriff, R. E. (2002a). *Encyclopedic dictionary of applied geophysics* (E. F. Scherrer, Ed., 4th ed., p. 318). Society of Exploration Geophysicists.
Sheriff, R. E. (2002b). Encyclopedic dictionary of applied geophysics. In *Encyclopedic dictionary of applied geophysics*. Society of Exploration Geophysicists. https://doi.org/10.1190/1.9781560802969
Sidewall core | Energy Glossary. (2023). https://glossary.slb.com/en/terms/s/sidewall_core
Skopec, R. A. (1994). Proper coring and wellsite core handling procedures: The first step toward reliable core analysis. *Journal of Petroleum Technology*, *46*(4), 280–280. https://doi.org/10.2118/28153-PA
Smith, M. B., Bale, A. B., Britt, L. K., Klein, H. H., Siebrits, E., & Dang, X. (2007). *Layered modulus effects on fracture propagation, proppant placement, and fracture modeling*. https://doi.org/10.2523/71654-ms
Sone, H., & Zoback, M. D. (2013). Mechanical properties of shale-gas reservoir rocks—Part 1: Static and dynamic elastic properties and anisotropy. *Geophysics*, *78*(5), D381–D392. https://doi.org/10.1190/geo2013-0050.1
Speight, J. (2019). *Shale oil and gas production processes* (pp. 1–1031). https://doi.org/10.1016/C2015-0-02189-X
Spencer, R., & Weedmark, T. (2015). Application of X-ray flourescence (XRF) analyses to the characterization of oil sand reservoirs. *Gas Mexico congress and exhibition held in Viahermosa*, Mexico (pp. 1–3).
Šperl, J., & Trčková, J. (2008). Permeability and porosity of rocks and their relationship based on laboratory testing. *Acta Geodynamica et Geomaterialia*, *5*(1), 41–47.
Srikant, D., & Shanmugam, P. (2015). Evaluation of cretaceous sediments for organic richness and level of maturity using petrophysical data. *JIAPUR 11th biennial international conference & exposition*. https://spgindia.org/11_biennial_form/evaluation-of-cretaceous-sediments-for-organic-richness-and-level-of-maturity-using-petrophysical-data-jaipur-2015.pdf
Teklu, T. W., Graves, R. M., Tutuncu, A. N., Kazemi, H., Alameri, W., & AlSumaiti, A. M. (2012). Geomechanics considerations in enhanced oil recovery. *Society of Petroleum Engineers—SPE Canadian Unconventional Resources Conference 2012, CURC 2012*, *2*, 931–942. https://doi.org/10.2118/162701-ms
Teufel, L. W., & Clark, J. A. (1984). Hydraulic fracture propagation in layered rock: Experimental studies of fracture containment. *Society of Petroleum Engineers of AIME, (Paper) SPE*, 449–456. https://doi.org/10.1016/0148-9062(84)91172-0
Thiercelin, M., Jeffrey, R. G., & Naceur, K. Ben. (1989). Influence of fracture toughness on the geometry of hydraulic fractures. *SPE Production Engineering*, *4*(4). https://doi.org/10.2118/16431-pa

Thiercelin, M., Roegiers, J., Boone, T. J., & Ingraffea, A. R. (1987). An investigation of the material parameters that govern the behavior of fractures approaching rock interfaces. *6th ISRM Congress 1987*, 263–269. https://doi.org/10.1016/0148-9062(88)92709-x
Thiercelin, M. J., & Plumb, R. A. (1994). Core-based prediction of lithologic stress contrasts in east Texas formations. *SPE Formation Evaluation*, *9*(4), 251–258. https://doi.org/10.2118/21847-pa
Tiab, D., & Donaldson, E. C. (1999). *Petrophysics: Theory and practice of measuring reservoir rock and fluid transport properties* (4th ed.). Elsevier B.V. https://doi.org/http://dx.doi.org/10.1016/B978-0-12-803188-9.00001-2
Tiab, D., & Donaldson, E. C. (2016a). Porosity and permeability. *Petrophysics*, 67–186. https://doi.org/10.1016/b978-0-12-803188-9.00003-6
Tiab, D., & Donaldson, E. C. (2016b). Naturally fractured reservoirs. *Petrophysics*, 415–481. https://doi.org/10.1016/B978-0-12-803188-9.00008-5
Tissot, B. P., & Welte, D. H. (1984). Petroleum formation and occurrence. *Petroleum Formation and Occurrence*. https://doi.org/10.1007/978-3-642-87813-8
Tissot, B. P., Welte, D. H., Tissot, B. P., & Welte, D. H. (1978). *Sedimentary processes and the accumulation of organic matter* (2nd ed.). Springer-Verlag. https://doi.org/10.1007/978-3-642-96446-6_5
Tittman, J. (1986). *Geophysical well logging* (Vol. Part 5). Academic Press Inc.
TOC and rock-eval—English. (2020). www.apt-int.com/glossary/toc-and-rock-eval
Trenchless Pedia. (2018). *What is tuberculation?—Definition from trenchlesspedia*. www.trenchlesspedia.com/definition/3419/tuberculation
Trenchlesspedia. (2018a). *What is minimum horizontal stress (SH)?—Definition from trenchlesspedia*. www.trenchlesspedia.com/definition/4328/minimum-horizontal-stress-sh-in-situ-stress
Trenchlesspedia. (2018b). *What is wellbore stability?—Definition from trenchlesspedia*. www.trenchlesspedia.com/definition/4257/wellbore-stability
Tutuncu, A. N., Podio, A. L., Gregory, A. R., & Sharma, M. M. (1998). Nonlinear viscoelastic behavior of sedimentary rocks, Part I: Effect of frequency and strain amplitude. *Geophysics*, *63*(1), 184–194. https://doi.org/10.1190/1.1444311
Ulmer-Scholle, D. S., Scholle, P. A., Schieber, J., & Raine, R. J. (2015). *AAPG memoir 109: A color guide to the petrography of sandstones*. American Association of Petroleum Geologists.
van Eekelen, H. A. M. (1982). Hydraulic fracture geometry: Fracture containment in layered formations. *Society of Petroleum Engineers Journal*, *22*(3), 341–349. https://doi.org/10.2118/9261-PA
Vandenbroucke, M., & Largeau, C. (2007). Kerogen origin, evolution and structure. *Organic Geochemistry*, *38*(5), 719–833. https://doi.org/10.1016/j.orggeochem.2007.01.001
Vernik, L. (1993). Microcrack-induced versus intrinsic elastic anisotropy in mature HC-source shales. *Geophysics*, *58*(11), 1703–1706. https://doi.org/10.1190/1.1443385
Walters, C. C. (2015). *The origin of petroleum*. ExxonMobil Research & Engineering Co. https://doi.org/10.1007/978-0-387-25789-1_2
Waples, D. W. (1985a). Geochemistry in petroleum exploration. *Geochemistry in Petroleum Exploration*. https://doi.org/10.1016/0375-6742(86)90089-0
Waples, D. W. (1985b). Geochemistry in petroleum exploration. *Geochemistry in Petroleum Exploration*. https://doi.org/10.1016/0375-6742(86)90089-0
Wheaton, R. (2016). Fundamentals of applied reservoir engineering: Appraisal, economics and optimization. In *Fundamentals of applied reservoir engineering: Appraisal, economics and optimization* (pp. 1–233). https://doi.org/10.1016/C2015-0-04617-2

Wimberley, J. W. (1969). A rapid method for the analysis of total and organic carbon in shale with a high frequency combustion furnace. *Analytica Chimica Acta*, *48*(2), 419–423. https://doi.org/10.1016/S0003-2670(01)85509-4

Wolfman, L. S. B. A. (2005). Geology and geochemistry of oil and gas. *Journal of Chemical Information and Modeling*, *53*(9), 1689–1699.

Worden, R. H., & Utley, J. E. P. (2022). Automated mineralogy (SEM-EDS) approach to sandstone reservoir quality and diagenesis. *Frontiers in Earth Science*, *10*. https://doi.org/10.3389/feart.2022.794266

Xia, Y., Zhou, H., Zhang, C., He, S., Gao, Y., & Wang, P. (2022). The evaluation of rock brittleness and its application: A review study. *European Journal of Environmental and Civil Engineering*, *26*(1), 239–279. https://doi.org/10.1080/19648189.2019.1655485

X-Ray Fluorescence (XRF). (2020). https://serc.carleton.edu/research_education/geochemsheets/techniques/XRF.html

Xu, C., Torres-Verdín, C., & Steel, R. J. (2013). Geological attributes from conventional well logs: Relating rock types to depositional facies in deepwater turbidite reservoirs. *Proceedings—SPE Annual Technical Conference and Exhibition*, *2*, 1298–1312. https://doi.org/10.2118/166178-ms

Xu, S., Wu, X., Huang, X., & Yin, H. (2005). Evaluation of anisotropic rock properties in sedimentary rocks from well logs. *Proceedings of the annual offshore technology conference*, *2005-May*, 624–638. https://doi.org/10.4043/17251-ms

Yale, D. P., & Jamieson, W. H. (1994). Static and dynamic rock mechanical properties in the Hugoton and Panoma fields, Kansas. *Proceedings of the mid-continent gas symposium*, 209–219. https://doi.org/10.2523/27939-ms

Yang, Z., Hou, L., Tao, S., Cui, J., Wu, S., Lin, S., & Pan, S. (2015). Formation conditions and "sweet spot" evaluation of tight oil and shale oil. *Shiyou Kantan Yu Kaifa/Petroleum Exploration and Development*, *42*(5), 555–565.

Zaker, S., Mohamadi Nafchi, S., Rastegarnia, M., Bagheri, S., Sanati, A., & Naghibi, A. (2020). Prediction of new perforation intervals in a depleted reservoir to achieve the maximum productivity: A case study of PNN logging in a cased-well of an Iranian oil reservoir. *Petroleum*, *6*(2), 170–176. https://doi.org/10.1016/J.PETLM.2019.06.003

Zhang, D., Ranjith, P. G., & Perera, M. S. A. (2016). The brittleness indices used in rock mechanics and their application in shale hydraulic fracturing: A review. *Journal of Petroleum Science and Engineering*, *143*, 158–170. https://doi.org/10.1016/j.petrol.2016.02.011

Zhang, G., Wang, P., Chen, R., Li, L., Jiewen Sun, G., & Zhang, X. (2021a). An optimization method of horizontal well completion control water and sand. *IOP Conference Series: Earth and Environmental Science*, *621*(1). https://doi.org/10.1088/1755-1315/621/1/012072

Zhang, M., Fan, X., Zhang, Q., Yang, B., Zhao, P., Yao, B., & He, L. (2021b). Influence of multi-planes of weakness on unstable zones near wellbore wall in a fractured formation. *Journal of Natural Gas Science and Engineering*, *93*. https://doi.org/10.1016/j.jngse.2021.104026

Zhang, X., Jeffrey, R. G., & Thiercelin, M. (2007). Effects of frictional geological discontinuities on hydraulic fracture propagation. *SPE—Hydraulic Fracturing Technology Conference 2007*, *2007*, 268–278. https://doi.org/10.2118/106111-ms

Zhao, W., Liu, Y., Wang, T., Ranjith, P. G., & Zhang, Y. (2021). Stability analysis of wellbore for multiple weakness planes in shale formations. *Geomechanics and Geophysics for Geo-Energy and Geo-Resources*, *7*(2), 44. https://doi.org/10.1007/S40948-021-00228-7

Zhen, L. (2023). *Multiphysics—Poroelasticity and poromechanics*. https://multiphysics.us/poroelasticity.html

Zoback, M. D., & Kohli, A. H. (2019). *Unconventional reservoir geomechanics: Shale gas, tight oil, and induced seismicity* (pp. 1–484). https://doi.org/10.1017/9781316091869
Zou, C. (2013). *Unconventional petroleum geology*. https://doi.org/10.1016/C2011-0-06250-6
Zoveidavianpoor, M., Gharibi, A., & bin Jaafar, M. Z. (2018). Experimental characterization of a new high-strength ultra-lightweight composite proppant derived from renewable resources. *Journal of Petroleum Science and Engineering*, *170*, 1038–1047. https://doi.org/10.1016/j.petrol.2018.06.030

3 Drilling Engineering Aspects of Unconventional Resources

Aliyu Mohammed Kaura and Mesut Yurukcu

3.1 DESIGN AND PLANNING

Planning and design are distinct yet interdependent aspects of preparing for a drilling project. Planning involves steps from the initial drilling of a well until its plugging and abandonment (Finger & Blankenship, 2010). Many of these procedures are interconnected. Various factors influence a well's planning, including reservoir characteristics and fluid production rate. Any drilling operation requires careful planning. It necessitates the integration of engineering principles, corporate or special viewpoints, and experience-based aspects. Planning will cut costs and minimize the risks of damage to property from unexpected events. Although methodologies for well planning may differ across the drilling industry, the ultimate result should be a safe, cost-effective hole that meets the requirements for oil and gas production (Adams & Charrier, 1985).

A drilling plan should include and clearly state all actions required to complete the well, as well as the costs and timeframes, and provide adequate descriptions of individual tasks to define the order in which they must be completed. (A "critical path" technique clearly shows which processes must be sequential and which can be accomplished concurrently, which is often helpful. Any delay along the channel of sequential processes—the critical path—will delay the project's completion, but a delay in other operations may not). The drilling plan must be adaptable enough to accommodate unanticipated events or problems that may arise during the project, and a procedure must be clearly defined to identify the person responsible for plan modifications (Figure 3.1) (Adams & Charrier, 1985).

3.2 DESIGN

Well design defines the desired final status of a well. These are the features that are considered when planning and constructing a well. Design is crucial to the safe exploration and extraction of oil and gas resources. It is a "bottom-up" process

DOI: 10.1201/9781003319955-3

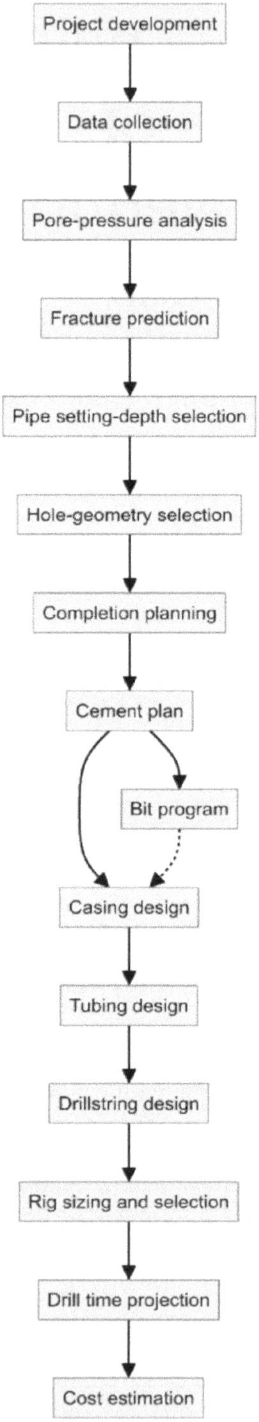

FIGURE 3.1 Flow path for well planning.

describing well depth, casing diameters, materials, thickness, reservoir production zone location, well dimensions, and pump settings. Once these are determined, other parameters such as wellbore diameter, completion methods, procedures, and drilling methods can be decided. The well-design process incorporates casing design, drilling fluids, bit and tool selection, and drilling techniques to manage drilling hazards. Other variables may be determined partly or entirely by local practices and the availability of materials.

Every well is designed (Figure 3.2), constructed, operated, maintained, and decommissioned with specific well integrity factors in mind. A preliminary design must be prepared to write specifications and obtain a bid; however, the design will probably change as more information about the actual hole conditions is obtained. Local, state, or federal agencies may specify certain requirements.

The overall objective of the design is to, among other things,

- Ensure that the pressure exerted by fluids in the subsurface formations is always safely controlled.
- Isolate distinct rock formations from one another to avoid fluid movement between them.
- Prevent the uncontrolled discharge of fluids from the well.
- Prevent the transmission of fluids from the surrounding environment into the borehole.

Well design always adheres to the principle that any potential leak channel must have at least two barriers (a primary and secondary barrier) in place. This indicates that if a primary barrier is breached, the secondary barrier will inhibit further leakage until the primary barrier is fixed (Mantle, 2014).

3.3 TYPES OF RIGS AND RIG REQUIREMENTS

A drilling rig is an integrated system of machines that drills wells, such as oil or water wells or holes for piling and other construction purposes into the earth's subsurface. In addition, it offers the ability to perform various additional tasks, such as logging and well testing. Modern drilling rigs are complex and necessitate highly trained and experienced personnel for effective operation. Incorrect rig selection can result in low penetration rates, formation damage due to inadequate solids control, and high ultimate well costs (Rig Worker, 2023).

A rig can be either stationary or moveable. These words describe how quickly the rig can exit a drilling site if needed. This classification is broad and fails to distinguish between different kinds of rigs. Rigs can be classified into two basic subcategories according to their capabilities and operating conditions: land (onshore) and marine (offshore). Both offshore and onshore rigs contribute significantly to worldwide production totals every day. Figure 3.3 summarizes the classification of rigs, although more rig types are available for marine operation. The types of rigs used in offshore drilling are distinct from the oil rigs on land. Offshore rigs must be able to endure the rigors of the ocean.

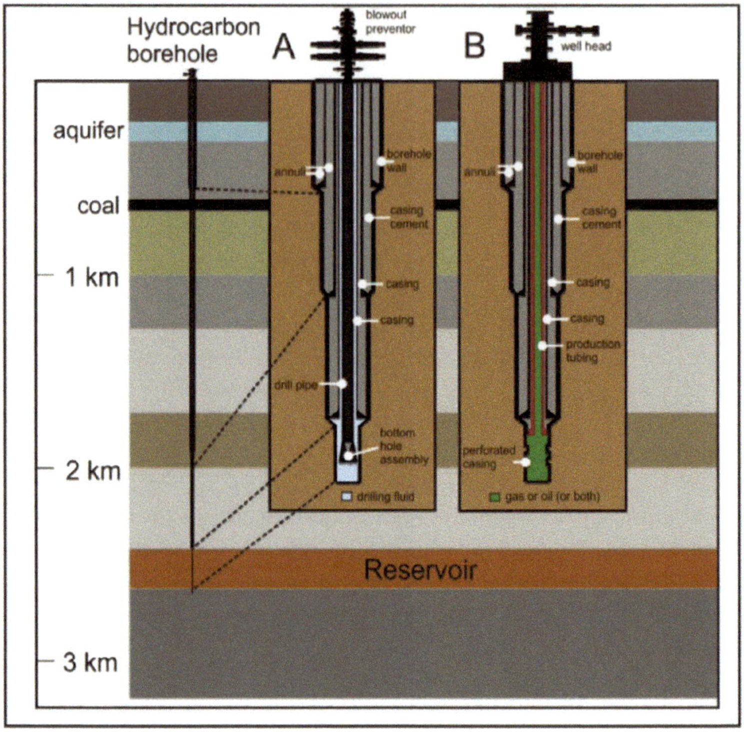

FIGURE 3.2 Schematic diagram of typical well design, showing (A) the structure of an exploration well and (B) a production well. Depths to which different casings are used vary according to the geology and pressure regime of the drill site. Well diameter is exaggerated to show sections more clearly.

Source: Davies et al., 2014

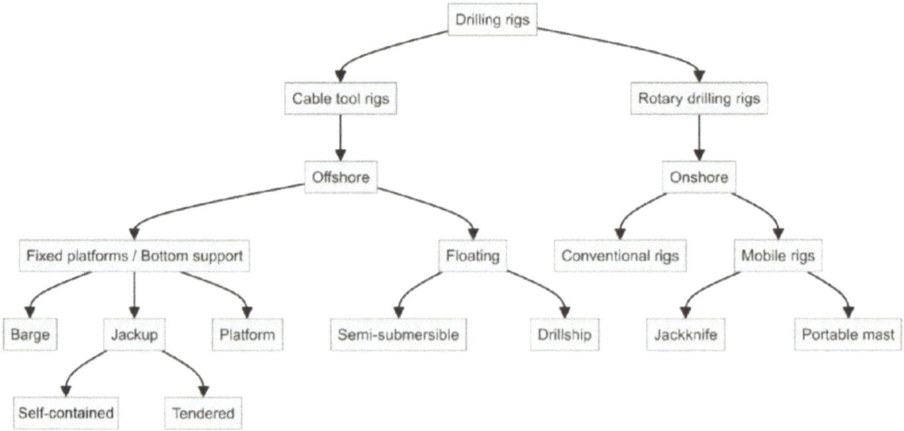

FIGURE 3.3 Classification of land and marine drilling rigs.

3.4 CABLE TOOL RIGS

Cable tool rigs (Figure 3.4) have played a significant role in the development of the drilling industry. They were the first drilling rigs utilized for hydrocarbon wells (Gregory King, 2020). The rigs have been widely used since the early 1800s and are still in use, although virtually all oil wells today are drilled using rotary rigs. Although not used very often today, cable rigs made up over 20% of all wells drilled in the United States in 1961 (Adams & Charrier, 1985). In the second part of the 19th century, shallow hydrocarbon wells in the Appalachian region of the US were drilled using cable tool rigs (Gregory King, 2020).

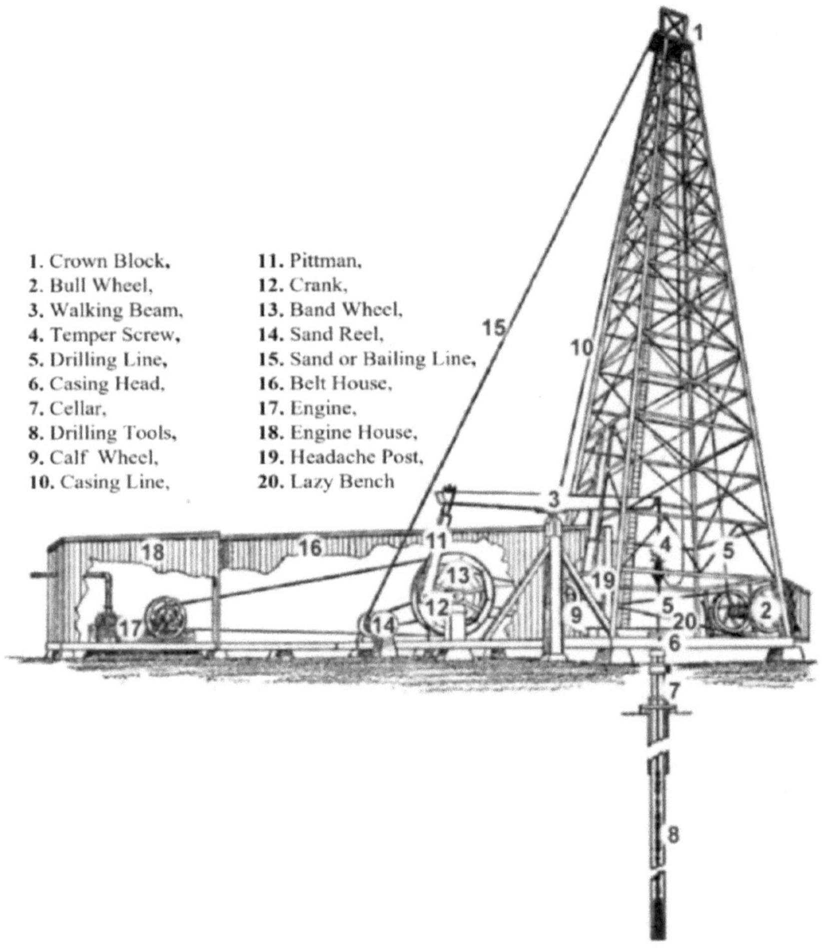

FIGURE 3.4 Cable rig.

Source: Gregory King, 2020

Drilling uses a cable tool rig by repeatedly lifting and dropping a heavy chisel drill bit on the rocks. The drill string comprises the bit, the drill stem, jars, and a rope socket connected to the drilling tools. The "tool" is a heavy (typically blunt) chisel bit, and the cable may consist of a simple manila rope or multiple steel strands (wireline) (Oilfieldteam, 2018).

3.5 ROTARY RIGS

Most drilling operations today use rotary drilling rigs (Figure 3.5). These rigs employ a drilling mechanism that applies torque (rotation) to the drill pipe or drill string with a drill bit connected to the bottom-hole assembly (BHA). As its names imply, the bottom-hole assembly is attached to the end of the drill string closest to the drilled formation (Gregory King, 2020). The drill string is the total length of drill pipe running from the drilling rig's rotary system to the wellbore's bottom-hole assembly. The BHA contains all the necessary equipment for drilling the current section of the wellbore.

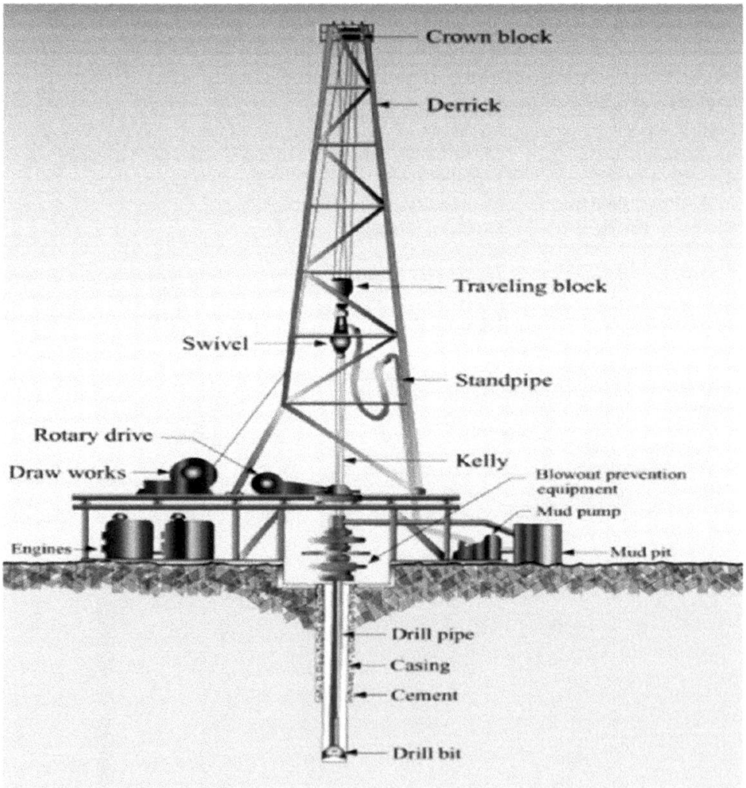

FIGURE 3.5 Rotary drilling rig.

Source: Khodja et al., 2010

This equipment may include drill collars, equipment for directional drilling, logging while drilling (LWD) and measurement while drilling (MWD) tools, etc.

A modern rotary drill rig system comprises multiple engines, power hoisting equipment that lifts and lowers the drill string, and rotating hardware that rotates the drill string and drill bit. These engines also power the circulating equipment that pumps fluids into the hole in order to lubricate the rotating drill string and drill bit. These liquids remove cuttings and regulate the pressure in the well to prevent blowouts.

Drilling using a rotary rig involves rotating a bit and exerting a downward force. Typically, the bit is rotated by revolving the entire drill string using a rotary table or a top drive at the surface, and the downward force is applied to the bit using drill collars, which are heavy, thick-walled pipes in the drill string located above the bit. While the bit cuts the rock at the bottom of the hole, the pump at the surface forces drilling fluids down the hole through the drill pipe to the bit. This fluid lubricates the bit and removes cuttings from the hole. The fluid with the cuttings then flows up through the casing annulus to the surface. Upon reaching the surface, the fluid is pumped back down the hole in a continuous cycle until the drilling process is stopped.

Generally, rotary rigs are categorized as onshore (land) or offshore (marine). Their primary design characteristics are portability, adaptability, and maximum operational depth. Components of onshore and offshore rigs are generally similar (Azar, 2021). The only notable variation when drilling in a maritime setting is the use of an extension pipe (drilling riser) between the rig floor and the seabed.

3.6 LAND OR ONSHORE RIGS

Land rigs are primarily used on land. Modern land rigs are built-in units and are skid mounted so they can be moved from one drilling site to another by truck or crane. Once onsite, the various rig parts are easily installed to drill the well. The derrick is designed to be easily assembled and dismantled; its roles are to carry loads suspended in the well and to provide space between the rig floor working area and the crown block, allowing for a specific length of the drill string to be created. The rigs can drill holes to a depth of more than 30,000 feet. Some light-duty drilling rigs resemble mobile cranes in design. Land rigs are classified into two types, depending on portability and maximum operational depth: conventional and mobile rigs.

3.7 MOBILE LAND RIGS

There are two types of mobile land rigs: jackknife and portable masts. A jackknife rig, also called a cantilever derrick, is a drilling rig with a mast resembling a jackknife rather than a conventional derrick. The rigs are assembled onsite, on the ground, and then hoisted using rig-hoisting equipment or drawworks to a vertical position. In contrast, a portable mast is typically installed as a single unit on wheeled trucks, transported to the construction site, and raised vertically by hydraulic pistons on the carrier unit. As with conventional derricks, the portable mast rig is supported by legs and hinged at the base. This rig is typically mounted on wheeled trucks or trailers that house the hoisting equipment, engines, and derrick as a single unit and is designed for medium-depth wells. Portable rigs are usually employed for workover operations and can drill to depths of approximately 10,000 feet (Rig Worker, 2023). When the rig must be relocated, it is lowered or laid horizontally and transported by truck.

3.8 CONVENTIONAL RIG

A conventional rotary rig is usually categorized as a component rig since the essential components are too large to move on a single truck bed. These rigs have depth ratings ranging from 6,000 to 35,000 feet (Gregory King, 2020; Rig Worker, 2023). The rigs are assembled onsite but dismantled and moved to another site after the well is completed. They can be used for workover operations throughout the well's lifecycle.

3.9 MARINE OR OFFSHORE RIGS

Due to increased global demand for fossil fuels, hydrocarbon exploration and production have been expanded to remote places such as offshore. Offshore drilling is performed from the space above water on platforms. The main design features for offshore rigs are portability and maximum water depth of operation (Khalifeh & Saasen, 2020b). Offshore rigs are classified broadly as floating or bottom support. (Figure 3.6).

3.9.1 Bottom-Supported Rig

A bottom-supported rig rests on the seafloor or pads built on the seafloor. This rig is categorized as a barge, jack-up, or platform rig. The selection of a particular rig

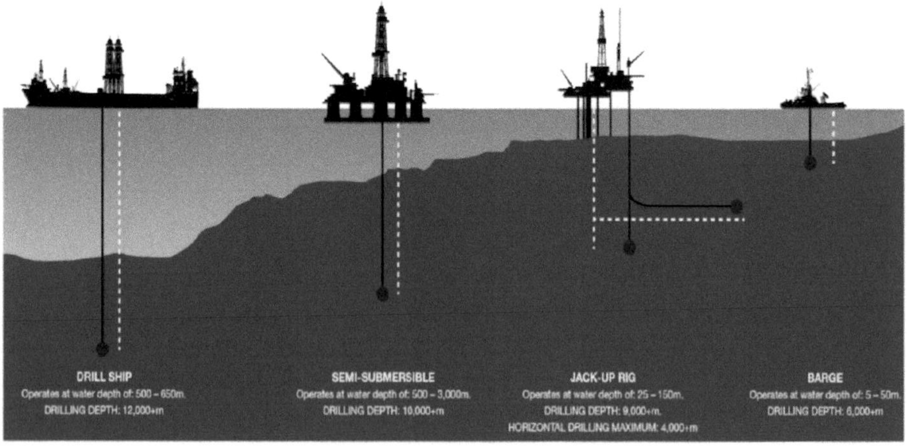

FIGURE 3.6 Types of offshore rigs and their water depth ranges.

Source: www.valiantoffshore.com

carrier depends on the operations to be performed and the water depth of the location where such operations are to be conducted.

3.9.2 Barge Rig

Barges are flat-bottomed boats designed to work in shallow waters, which are usually less than a feet in depth (Khalifeh & Saasen, 2020a). The rig floats to the drill site, and the lower hull sinks to the sea floor. The entire rig is built on the barge and transported to the area before being sunk by flooding the ship. Once drilling is completed, the water is pumped out of the barge, allowing it to be moved to the next location. As a result, the hull offers stability to the drilling platform. Barge rigs are not suitable for heavy seas or deep-water drilling. Built-in equipment on the boat serves as the housing for the operation (Morton-Thompson & Woods, 1993).

3.9.3 Platform Rig

Platform rigs are among of the most commonly used rigs for offshore drilling. As the name implies, this type of drilling rig is built on supports (a steel tubular structure anchored to the ocean floor to support the surface production equipment, living quarters, and drilling rig) (Figure 3.7). Oil platforms are typically utilized on the continental shelf but can also be used in lakes, inshore waters, and inland seas. Platform rigs can be employed during the development phase, exploiting an economically viable offshore field (Khalifeh & Saasen, 2020b).

FIGURE 3.7 Drilling platform.

Source: worldoceanreview.com

Platform rigs are either fully self-contained or tender supported (Adams & Charrier, 1985). If the production platform is sufficiently large, all drilling and personnel support equipment maintained on the rig is self-contained. If the platform's working space or load is limited, a floating tender supports people and auxiliary drilling equipment, including mud pumps and bulk chemical storage. Platform rigs typically have a shorter rig-up time than most Mobile Offshore Drilling Units (MODU) because neither a mooring system nor a dynamic positioning system is required (Khalifeh & Saasen, 2020b).

3.9.4 Jackup Rig

Jackup rigs (Figure 3.8) are specialized offshore drilling platforms designed to operate in shallow to medium water depths (Gregory King, 2020). One key advantage of jackup rigs is their ability to provide a stable drilling platform by elevating above the water using their retractable legs. This allows for safer and more efficient drilling operations (Gregory King, 2020).

The rotary rig on a jackup rig can either be a conventional Kelly drive rig or a top-drive rig. The jackup rig has a buoyant hull with numerous moveable legs to elevate it above the water's surface. Depending on the rig type, the rig is towed to the location and positioned over or near the well site. During the towing process, the legs are raised, and the barge is floating on the water (Adams & Charrier, 1985). The buoyant hull allows the platform to be transported to a drill site. Once on location, the legs are lowered to the bottom, and the platform is jacked up above the wave action utilizing sophisticated hydraulic jacking systems; the hull is also raised to the required elevation above the sea surface supported by the seabed. Once the legs reach the seabed,

FIGURE 3.8 A jackup rig.

Source: Gregory King, 2020

further extension allows the rig to be hoisted to the necessary height above the water level, and drilling preparations may begin (Gregory King, 2020).

One of the major benefits of jackup rigs is their ability to have a stable drilling platform by elevating above the water with retractable legs. This makes drilling operations safer and more efficient. Jackup rigs are commonly employed for exploratory, appraisal, development, and well intervention. They are adaptable in various environments, including coastal areas and offshore wind farms. On the seafloor, the legs are supported by either a spud can on each leg or a mat base connecting the legs (Adams & Charrier, 1985).

3.9.5 Submersible Rig

Submersible rigs are cutting-edge offshore drilling platforms that can be partially or completely submerged during the drilling process, providing stability and protection from harsh weather. The rig is typically used as a mobile offshore drilling unit (MODU) supported primarily by large pontoon-like structures submerged beneath the sea surface. The operating decks are elevated to around 100 feet or more above the pontoons by massive steel columns. Once in location, this structure is slowly flooded until it rests on the seafloor. As drilling is completed, the water is pumped out of the buoyancy tanks, and the vessel is refloated and towed to the next location. Submersibles operate in relatively shallow water environments (Gregory King, 2020).

3.9.6 Floaters

Floaters are drilling platforms that float on the water's surface. They require specialized equipment to stabilize and protect the drilling rig from waves. These rigs are mobile and have numerous features that make them advantageous for their jobs. These include (Gregory King, 2020):

- They are maintained in place by anchors or dynamic positioning.
- Motion compensators protect the drill string and riser from wave motion.
- The wellheads and blowout preventer (BOP) are located on the ocean floor and are linked to the rig by a riser to allow for mud circulation.

3.9.7 Drillships

Drillships are floating, self-propelled offshore drilling vessels (Figure 3.9) constructed with the ability to drill from the base. On a drill ship, the drilling apparatus is situated over a moon pool, a reinforced hole in the ship's hull whereby the drill string is raised and lowered. The moon pool (or "wet deck") permits drill pipe to run from the rotary system of the drilling rig through the ship's structure, past the hull into the water, to the sea bottom and the wellbore. The drill goes down and up through this moon pool in the middle of the drilling ship.

The drillship was designed for exploratory drilling of oil and gas wells in offshore locations. In recent years, many technological advancements have enabled the use of drillships for deep-water and ultra-deep-water sites. The drillship can also act as a platform to perform well maintenance or completion work such as casing and tubing, subsea tree installations, and well capping. Drillships usually have a large deck area for housing drilling equipment, storage, and crew accommodations. They may also have helipads for crew transportation and replenishment of supplies (Rig Worker, 2023).

3.9.8 Semisubmersible

The semisubmersible drilling unit is a specially designed vessel used exclusively for petroleum operations. The rigs can be used to drill wells inaccessible to jackup rigs (water depths greater than 400 feet). Semi-submersible rigs are buoyant and float during drilling operations, unlike jackup rigs, which rest on the seabed. The rig has pontoons and columns that help flood, which causes the unit to submerge in water to a predetermined depth. The necessary office space, limited residential space, and storage are reassembled on the deck. This rig is either self-propelled or towed to a drilling site and remains stationary at a well location via a series of anchors. Once onsite, water is utilized as ballast to partially flood the pontoons and columns, allowing them to drop below the water level. This increases the vessel's stability during drilling operations by adding mass and increasing the draft. Due to their modest cross-sectional area, the columns provide added stability to the vessel during drilling operations, as minimal wind, wave, and tidal energy are transmitted to them (Rig Worker, 2023).

3.10 REQUIREMENTS FOR DRILLING RIG SELECTION

In the past, footage rating was the basis for drilling rig selection. Heavier rigs were used to drill deeper wells, while lighter rigs drilled shallow wells (Adams & Charrier, 1985). For instance, a 20,000-foot rig would be chosen for a 19,000-foot prospect. This meant the rig could safely drill to a depth of 20,000 feet, carry out all related tasks necessary to drill, and complete a well at that depth (Rig Worker, 2023).

The main goal of rig selection is to pick the rig from the available ones that will most closely meet the requirements for drilling a usable hole at the lowest overall cost for the proposed well. All rig systems listed here are being evaluated as part of the selection process: power generation system, hoisting system, fluid circulation system, rotary system, well control system, and drilling data acquisition and monitoring system (Rig Worker, 2023).

Moreover, the quality of supervision and rig personnel must be paramount. Other factors in the selection of rigs include the following:

- Safety records
- Rig mobility and ease of handling
- Contractor dependability
- Contractual rates (footage, day, turnkey)
- Condition of all rig equipment

The minimum rig size standards should be established after completing the design requirements for the well's drilling programs and before requesting bids. Drilling programs that must be completed before rig selection include drilling fluid, drill string, drill bit, drilling fluid hydraulics, well control, data collecting and monitoring, and cementing/casing. There are numerous types of drilling rigs, and the selection criteria are determined from the drill sites' and well specifications, particularly diameter, depth, and casing design. Planning and engineering the well will determine the diameter, which is the fundamental criterion for determining whether the well is a "slim hole" or a standard well and, consequently, what type of rig will be utilized. The rig sizing is an essential component of the well planning procedure, but it must be accomplished after the actual well has been planned (Rig Worker, 2023).

3.11 CEMENTING

Cementing is an important operation in the development of oil and gas wells. It introduces cement to the annular space between the wellbore and casing or between two successive casing strings. Pumping cement behind the casing helps secure the casing, provides a barrier to keep the fluids flowing from or into the formation, and bonds the casing to the formation. Cement, in general, is mixed on the rig before being pumped into the well for the following reasons (Tao et al., 2020).

Numerous factors, such as the wellbore geometry, casing equipment, formation integrity, and drilling mud qualities, play a vital role in determining the right cement slurry design for well cementing (Hossain & Al-Majed, 2015). Well design is additionally influenced by the well's depth, geology, geothermal profile, pressure gradient, unusual pressure formations, and corrosive fluid. Cementing allows the drilling process to be completed safely and economically.

Unfortunately, because cementing is done towards the end of drilling, it is often performed in a rush to get a well up and running, and the time and commitment to get a decent cement job done are not always taken. As a result, much more time and money are spent fixing the cement job than would have been needed to do it right the first time (i.e., the initial cost). The cement program must be carefully planned for a well to be drilled successfully.

3.12 TYPES OF CEMENTING JOBS

Primary cementing and remedial cementing are the two main operations in well cementing. Primary cementing involves placing a cement sheath in the annulus between the casing and the formation. While remedial cementing occurs after primary cementing, engineers inject cement into strategic well locations for various purposes, including well repair and well abandonment.

3.12.1 Primary Cementing Job

Primary well cementing is the essential process during the drilling and completion of oil wells. It is the cement slurry injection into the void between the casing and formations exposed to the drilled wellbore. The cement slurry is then given time to solidify to create a strong, impermeable annular seal. The main objective of primary cementing in oil and gas wells is to provide zonal isolation, structural support, and corrosion resistance for the casing (Brandl et al., 2010).

There are numerous methods of primary cementing; multistage cementing, inner string cementing, and single-stage cementing are among the most common. This cementing process involves pumping a calculated volume of slurry into the casing after the pipe has been landed at the desired depth and displacing the slurry around the shoe and into the annulus in a circulating mode with another fluid (e.g., water, mud, or completion fluid) (Figure 3.9).

Multistage cementing usually involves pumping cement into the well in two or more separate stages, or batches, behind a casing string. This method is used in wells with critical fracture gradients or those requiring good cement jobs on long casing strings (Hossain & Al-Majed, 2015). Inner string cementing is another type of primary cementing procedure. In this case, the cementing is executed with a stinger passed from the drill pipe to the casing shoe and stubbed in it. Once the stringer has been engaged, drilling fluids are circulated to ensure no leaks allow fluids to flow from the drill pipe into the casing. The cement slurry is then pumped to the surface via the drill pipe, the casing shoe, and the annulus. Cementing the inner string is complete when surface cement reaches the cellar; at this point, the stinger is disconnected and pulled above the shoe. In such a circumstance, circulation through the drill pipe is required to flush the drill pipe of any cement.

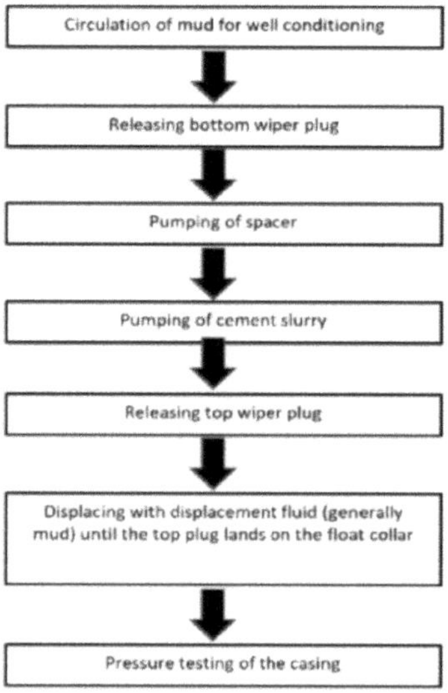

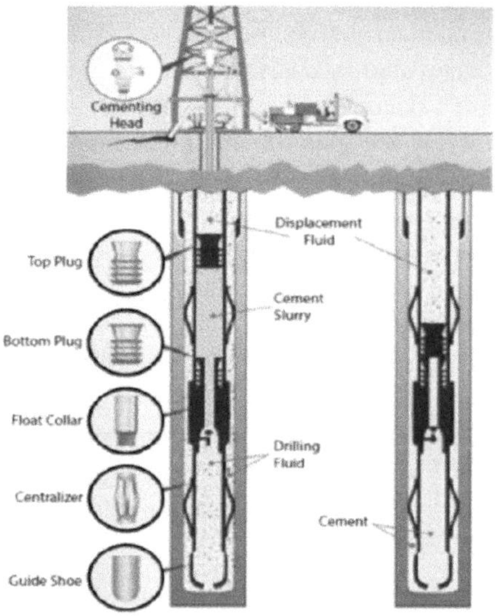

FIGURE 3.9 a) The procedure for single-stage cementing; b) Cementing a well.

Source: Hossain & Al-Majed, 2015 (courtesy MPG Petroleum)

Functions of primary cementing include:

- Providing structural support to the casing
- Sealing the annulus between the pipe and the formation to prevent fluid flow between zones and restrict fluid movement between formations and the surface
- Offering well control by weight and speedy curing after protective mud is displaced
- Preventing contamination of freshwater formations
- Keeping the casing's exterior from corrosion and the intermediate casing and liner pipe from torque and shock loads when drilling deeper
- Completely separating problematic zones behind the casing from formations to be drilled deeper
- Separating high-pressure formations beneath the casing from shallower zones behind the casing and isolate producing zones from water-bearing sands.

3.12.2 Squeeze Cementing Jobs

Squeeze cementing is a dehydration technique. A cement slurry is pumped to the problematic area or squeeze target through a wellbore. Isolating the area and applying pressure from the surface allows the slurry to be effectively forced into all voids. The slurry is designed to precisely fill a void in the wellbore, whether a small crack or micro-annuli, casing split or big vug, formation rock, or another type of void. Consequently, the design of the slurry and the rate of dehydration or fluid loss in it are crucial, and a poor design may not offer a complete fill and seal of the voids.

The main objectives of squeeze cementing are:

- Supplementing a faulty cementing zone behind the casing: Sometimes, void spaces or channels exist in the annulus between the casing and wellbore due to improper primary cement placement. A squeeze cementing job is usually carried out as a remedial action to reinforce such weak zones.
- Sealing off lost circulation zones. Cement can seal off fractures or porous geological formations a few centimeters below the well surface.
- Repairing casing failure. Cement slurry is directed and forced through casing leaks. A casing leak can develop due to corrosion caused by water, hydrogen sulfide gas, or other fluid.
- Isolating gas- or water-producing zones to maximize production from the completion interval.
- Restraining fluids escaping from abandoned zones.

3.12.3 High-Pressure Squeeze

This method requires the formation to be fractured before pumping cement slurry into the fractured region. The formation is first fractured by pumping fluids free of solids, such as water or brine, to prevent the creation of a filter cake. After fracturing the formation, the slurry is driven slowly through cracks by being put against broken zones. During cement squeezing, any backflow is prevented by releasing the pressure.

3.12.4 Low-Pressure Squeeze

The fracture gradient of the formation is not exceeded during the low-pressure squeeze. When pressure is applied to cement slurry against a formation, the cement's fluid content (filtrate) is forced into the rock. In contrast, the solid cement substance (filter cake) accumulates on the formation's surface (Hossain & Al-Majed, 2015). There are two procedures for low-pressure squeeze cementing: running and hesitation squeezes. In a running squeeze, the cement is pumped gradually and continuously until it reaches the desired final pressure. In the hesitation squeeze, the pumping is stopped at regular predetermined intervals to allow the slurry to dehydrate and form the necessary filter cake.

3.13 DRILLING SYSTEMS

A mud motor (or drilling motor) is a progressive cavity positive displacement (PCPD) pump attached to the drill string to provide extra power to the bit during drilling. It is situated just behind the bit within the drilling string. Mud motors employ the energy of the pressurized fluid to provide the necessary torque to drive the drill bit during the drilling process.

The key components of mud motors comprise a stator, rotor, transmission section, bearing assembly, and power section. The rotor is attached to the mandrel by a universal joint while the stator is connected to the drill string via threads. Bearings in the bearing assembly locate and guide the rotating mandrel while transferring axial and radial drilling loads between the mandrel and the drill string. The greater the number of stages of a motor, the greater the torque given to the drill bit but the slower its speed (drillingformulas, 2017; Trenchlesspedia, n.d.). Mud motors can also be configured with a bend, which enables them to be used for directional drilling.

3.14 ROTARY STEERABLE SYSTEMS

A rotary steerable system (RSS) is a novel form of technology used in directional drilling. It is a tool designed to drill directionally with continuous rotation from the surface, eliminating the need to slide a steerable motor (Schlumberger, 2023). The cutting-edge rotary steerable systems have minimal interaction with the borehole, thereby maintaining its quality. It enables full three-dimensional directional drilling control while drilling with continuous drill string rotation from the surface. (No "slide" drilling is necessary.) This capability needs a specialized BHA component above the bit to steer the well track in the desired direction, keeping the drilling trajectory's orientation regardless of the BHA and drill pipe's rotation above it.

There are three kinds of steering mechanisms for the rotary steering system to control the drilling direction of the drill bit: push-the-bit, point-the-bit and hybrid steering tools (Felczak et al., 2012; Li et al., 2020; Thaiprasert et al., 2016; Zhang et al., 2016). However, point-the-bit and push-the-bit systems are the most common in the industry.

Point-the-bit systems apply a principle similar to that of mud motor bend systems. The system has an internal mechanism that steers it by precisely tilting the bit in the direction the well path needs to be steered while maintaining rotation. Thus, the face of the drill bit points precisely in the required direction, and the bit does not experience any side loading (Schaa et al., 2000): that is, the point-the-bit type changes direction by bending the drill axis internally, which causes the drill bit to be headed in the desired direction (Figueredo, 2014). This operational principle has the advantage of allowing longer gauge bits to prevent the hole from spiraling. However, these systems are slower to respond to required trajectory adjustments, and their dogleg capacity is often inferior to a push-the-bit system.

Each point-the-bit RSS provider has different mechanisms to achieve this, but the two main principles are a system that utilizes a non-rotating collar or sleeve together with a specially designed stabilizer to grip the formation, which serves as an anchor for the system to know its pointing direction, and a system that utilizes a more complex counter-rotating system and universal joint. This system fully rotates outside the tool (Kim & Myung, 2017).

3.15 HYBRID ROTARY STEERABLE SYSTEM (RSS)

This hybrid RSS is a novel steering system that combines proven technology from two conventional types of RSS—point-the-bit and push-the-bit systems—to achieve better steerability (Table 3.1). That is, hybrid RSS exhibits the benefits of both point-the-bit and push-the-bit systems by using a united actuator (Figueredo, 2014). The hybrid system is realized through the use of hybrid pads, hydraulic cylinders, and a spherical joint (Kim & Myung, 2017).

TABLE 3.1
Summary of RSS Technologies

Company	RSS	Type of Offset Mechanism
Schlumberger	PowerDrive Archer	Hybrid
	Xceed	Point-the-bit
	PD X6	Push-the-bit
	ICE	Push-the-bit
Haliburton	Geopilot	Point-the-bit
Baker Hughes	Autotrac	Push-the-bit
Weatherford	Revolution	Point-the-bit
	Magnus	Push-the-bit
NOV	Vector	Push-the-bit
Terravici	Terravici	Point-the-bit
Gyrodata	Well Guide	Point-the-bit
APS	SureSteer	Push-the-bit
Sanvean	Scott	Push-the-bit
	D-Tech RST	Push-the-bit

Source: by Bryant, 2019; Li et al., 2020

In a conventional RSS that pushes the bit, external pads press against the wellbore wall to force the bit in the correct direction. The steerable motor employs an external bend, whereas the point-the-bit RSS employs an internal bend, which misaligns the tool about the borehole axis. However, the dogleg capability is significantly increased compared to either system alone by blending the two mechanisms. The new RSS includes a dependable electronic control unit and mechanical steering unit. All external components fully rotate with the drill string. In steering mode, electronics keep a rotary valve geostationary, diverting 4% to 5% of the mudflow to internal pads that push on the inside of a stabilizer sleeve instead of the borehole wall (JPT Staff, 2013). The sleeve is hinged at a universal joint, which allows it to pivot and direct the bit in the right direction. A mechanical strike ring controls the degree of deflection, which is held electronically in a definite direction for as long as necessary. Compared with other conventional steerable tools, hybrid RSS has many advantages: (1) the drilling string is rotating constantly, and lower friction is achieved; (2) wellbore trajectory can be adjusted in real time; and (3) higher drilling penetration is obtained. Effective and efficient downhole communication reduces human intervention in potentially dangerous situations.

3.15.1 Measurement while Drilling (MWD)

MWD systems permit the driller to collect and transfer data from the bottom of the hole to the surface without stopping routine drilling operations (Figure 3.10). The technique has become a crucial instrument for reservoir evaluation in the past decade. With the growth of the unconventional exploration industry, the usage of MWD has expanded because it enables the wellbore to be drilled in the desired

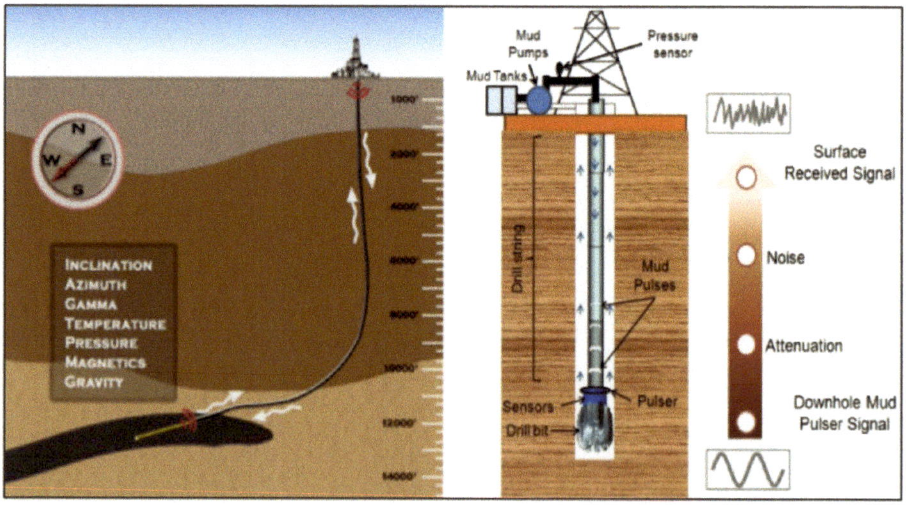

FIGURE 3.10 Measurement while drilling MWD system.

Source: globalspec.com

direction so specific reservoir formations can be accessed with precision (Ma et al., 2016). In general, MWD tools can conduct real-time directional surveys. They then transfer this information to the surface after measuring the inclination and azimuth of the wellbore at that spot using accelerometers and magnetometers.

MWD systems transmission technology works on the principle of mud pulse telemetry, which does not require a wireline. The system receives electrical power from a battery pack or a tiny turbine. In MWD systems, the hardware consists of sensors installed into a drill collar positioned near the bit and connected to the monitoring computer. The sensors and transmission equipment, located in a non-magnetic drill collar in the bottom-hole assembly, collect and transmit the data to the surface. The MWD operator receives real-time data from instruments housed in a module of the drill string's steering tool (Ma et al., 2016). The data is recorded in a solid-state memory and transmitted to a logic controller, which converts the data to binary. Typically, the data is transmitted to a pulser unit that varies the pressure of the drilling fluid within the drill pipe based on a mud pulse telemetry (MPT) code. Pressure transducers and computers on the surface identify positive, negative, and continuous sine wave pressure fluctuations and decode them for users (Short, 1993; Silva, 2006).

Positive: A valve is toggled to restrict mudflow in the drill pipe, resulting in a pressure increase at the well's surface. Data is encoded using pulse-position modulation line codes.

Negative: A valve is toggled to discharge drilling fluid from inside the drill pipe to the annulus, resulting in a noticeable decrease in surface pressure. With line codes or pulse-position modulation, data is encoded.

Continuous: A valve is gradually toggled to create sinusoidal pressure changes in the drilling fluid. Data is encoded with any digital modulation format, the most common being continuous phase modulation (Gravley & Corp, 1983; Zhantao Li et al., 2014).

3.15.2 Mud-Pulse Telemetry (MPT)

Mud-pulse telemetry (MPT) has been the global standard for real-time data delivery from MWD, in which a particular type of mud (drilling fluid) is pumped down the tubing and returned to the surface. Electronics at the bottom of the hole control a valve that may restrict and relax the flow of this mud, generating pressure pulses in the fluid circulating in and out of the hole. These pressure pulses are the fundamental element of mud-pulse telemetry. The downhole measurement technique uses pressure pulses in the mud system (Brandon et al., 1999). Typically, the readings are converted into a frequency- or amplitude-modulated pattern of mud pulses. The same telemetry technology is utilized to transmit surface-based commands. MPT technique was introduced to the oil and gas industry in the late 1970s (Berro & Reich, 2016). Since then, the technique has benefited from downhole data transmission technology, including the downhole mud-pulse generator and ground signal detection and extraction system (Su et al., 2011). Today, MPT is a popular, low-cost technology that delivers dependable, real-time, two-way data communication between the wellbore downhole system and the surface receiver (Liu et al., 2007). Proper well conditioning, navigation, and feedback control for real-time target relocation necessitate more advanced technology solutions to enhance drilling productivity, safety, and optimal scoping of the most productive hydrocarbon production zones. The availability of various telemetry system design enhancements has enabled the industry to overcome distinct obstacles posed by the transfer of information from the downhole to the surface through a variety of approaches and procedures.

MPT is a binary coding method of transmitting information through a flowing column of drilling mud (Silva, 2006). The drilling mud is pumped from the surface to the downhole BHA along the drill string. The mudflow wellbore telemetry device can selectively restrict the mud passages through the drill string to control and manipulate the pressure in the mud lines by operating a valve. At a point downhole, the pressure in the flowing mud column is periodically modulated by mechanical means, and the periodic pressure pulses that appear at the surface end of the mud column are detected by a pressure transducer conveniently located in the standpipe. The presence or absence of pulses or the relative duration of pulses in a sequence can convey information. Transducers on the surface detect and convert the pressure signal into a digital electrical signal through analog/digital (A/D) converters. This signal is then sent to a computer that will process and decode the signal through software specifically developed to recognize and treat these signals (Berro & Reich, 2016; Ivamberg et al., 2015; STS, 2020).

3.15.3 Bits

A drill bit is the cutting tool at the end of a drilling machine's drill string. They are used to create cylindrical holes, almost always of circular cross-section. Drill bits come in many types and sizes and have many uses (Figure 3.11). They are usually connected to a drill mechanism, which rotates them and provides torque and axial force. There are two primary types of drilling bits: rolling cutter bits and fixed cutter bits. At the same time, technological advancement has enabled the development of hybrid bits (Hossain & Al-Majed, 2015).

3.15.3.1 Roller Cone Bits

Roller cone bits are the most common type of bits used worldwide. They are also called rolling cutter bits, rock bits, or tri-cones. The great majority of roller cone bits today have three cones (Finger & Blankenship, 2010), with either milled steel teeth that are part of the cone itself or hard metal (usually tungsten carbide) teeth inserted

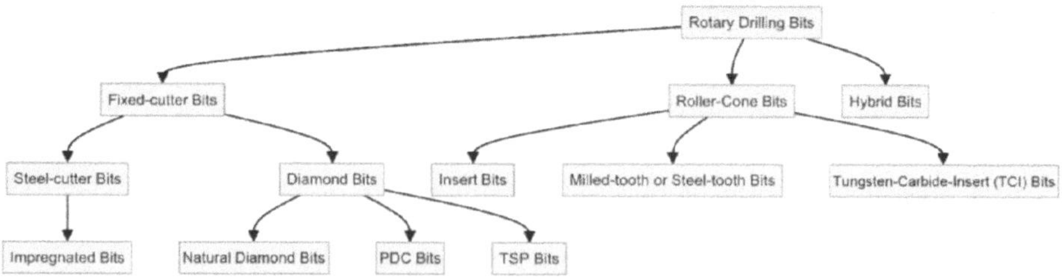

FIGURE 3.11 Drilling bits modified.

into the body of the steel cone. Milled-tooth bits are less expensive but are suited only for softer formations. Insert bits are used in medium to harder formations, with the size, shape, bearings, and number of inserts varied to fit the specific drilling conditions. Each cone can be independently rotated when the drill string turns the bit's body. The cutting action of the roller cone bit is predominantly done by grinding and chipping rocks (Hossain & Al-Majed, 2015).

3.15.3.2 Milled Tooth Bits

Milled tooth bits have steel teeth milled on a cone. When the bits are rotated, they carve or gouge formations out. The size and shape of the teeth vary depending on the formation. Milled tooth bits are available in three series; the selection is based on formation hardness (Azar, 2021). Series 1 apply in soft-to-medium formations, series 2 apply in medium-to-hard formations, and series 3 apply in hard-to-very-hard formations. In soft formations, the teeth should be long, slender, and widely spaced. These teeth will produce freshly broken cuttings from soft formations. However, the teeth should be short and closely spaced for hard formations to produce smaller, more rounded, crushed, and ground cuttings.

3.15.3.3 Insert Bits

The teeth in these bits are not milled into the cones; instead, tungsten carbide inserts are pressed into the cones. These techniques harden the bits, allowing them to last longer when drilling through hard rocks. The size and shape of the teeth also vary with the formation to be drilled; in soft formations, the teeth might be long and chisel shaped, but in hard formations, the teeth can be short and round (Azar, 2021). Insert bits are designed to drill into very hard formations. They cut rocks by chipping or crushing.

The International Association of Drilling Contractors (IADC) has established a standard for classifying roller cone bits. The IADC system categorizes milled and inserted teeth roller cone bits based on formation hardness, common mechanical characteristics, and manufacturer. The series numbers identify a category of drill bits for formations with the given characteristics. The numbers one through three represent milled-tooth bits, whereas four through eight refer to insert bits.

3.15.3.4 Fixed Cutter Bits

Fixed cutter bits, also called drag bits, are usually designed for soft formations such as sand, clay, or some soft rock (Hossain & Al-Majed, 2015). The cutting action of fixed cutter bit is by shearing formations rather than chipping or gouging as with rolling cutter bits (Hossain & Al-Majed, 2015; IADC, 2014). These bits do not have moving parts such as cones or bearings. The bit bodies and cutting elements are integrated. The components of bits consist of bit bodies made of steel or tungsten carbide matrix and fixed blades incorporating abrasion-resistant cutters. Fixed cutter bits are mainly diamond bits, steel-cutter bits, and impregnated bits.

3.15.3.5 Steel-Cutter Bits

Steel-cutter bits are drag bits containing steel-cutter components. Steel-cutter bits are superior to others in uniformly soft, homogeneous, unconsolidated formations. As the formation grows harder and more abrasive, the bit wear and drilling rate decrease significantly. Steel cutters are currently being replaced in all sectors by alternative types.

3.15.3.6 Diamond Drilling Bits

Due to their extreme hardness, diamonds have been cutting rock for centuries. During the bit's manufacturing, diamonds are embedded into the metal structure (typically a sintered or powdered carbide base matrix). The bit designer can create nearly unlimited combinations of bit shape, placement of hydraulic jetting ports, and diamond quantity and size (usually expressed as diamonds per carat).

A faster-drilling diamond bit has a shorter lifespan. If a bit is designed for a long life, it will typically drill more slowly. If a bit has a relatively high number of diamonds compared with other bits, it is considered "heavy set" and has higher durability. A "light-set" bit drills more aggressively but wears out more quickly because fewer diamonds are doing the work. The cutting action of the bit is produced by scraping away rock and drilling with a high-velocity plowing action that breaks the cementation between rock grains. This bit type is utilized in abrasive formations and performs best in hard, non-brittle formations. Natural diamond bits, polycrystalline diamond compact (PDC) bits, and thermally stable polycrystalline (TSP) bits are the three types of diamond bits.

3.15.3.7 The PDC Drilling Bits

Polycrystalline diamond compact (PDC) bits are the new generation of diamond bits. They comprise numerous tiny diamond crystals (polycrystalline diamonds) that are bonded together. Using a high-pressure and high-temperature technique, the PDC is bonded to a tungsten carbide bit body matrix or a tungsten carbide stud mounted in a steel bit body. This type of bit can be utilized on various formations, ranging from soft to hard. The cutter is self-sharpening because the sharp crystals are exposed continuously as each layer wears or disappears.

PDC bits drill commonly by shearing. A vertical penetrating force from the applied weight of the drill collar and a horizontal penetrating force from the rotating table are delivered to the cutters. The resultant force defines the cutter's thrust plane. The cuttings are then sheared off at an initial angle relative to the thrust plane that depends on the rock's strength. Recent advancements have been made in fixed cutter bit technology; although the PDC bits can drill almost any formation, if one plans to drill tough formations, using diamond bits has advantages.

3.15.3.8 TSP Drilling Bits

TSP bits comprise primarily thermally stable polycrystalline diamond (synthetic diamond) components. These elements are made entirely of triangular or cube-shaped pieces of micron-sized synthetic diamonds. TSP bits are manufactured similarly to PDC bits but tolerate much higher temperatures. TSP bits contain significantly larger cutting elements with a lower cutter density (fewer cutting edges per unit area) and greater primary cutter exposure than surface-set bits. In addition, synthetic diamond is the principal cutting medium in TSP bits, as opposed to natural diamond in surface-set bits.

3.15.3.9 Natural Diamond Bits

Natural diamond bits (Figure 3.12) drill through hard abrasive formations. Their cutting action is achieved by scraping away the rock (IADC, 2014). The face or crown of the bit consists of many diamonds set in the matrix body. The diamonds are arranged in a specific pattern and fused into a matrix on a steel body. This type of bit was developed in the mining industry (IADC, 2014).

Diamond bits differ structurally from roller cone bits. A matrix structure is covered in diamonds and contains waterways that lead from the bit's throat to its outer walls. To clean the cuttings and cool the diamonds, drilling fluid must flow across the face of the bit. Despite its high abrasion resistance, a diamond's sensitivity to shock and vibration necessitates extreme caution when employing a diamond bit. In addition, adequate fluid circulation across the face of the bit is required to prevent the diamonds and matrix material from overheating. Furthermore, it is necessary to prevent the bit face from being grimy with rock cuttings (bit balling). The primary advantages of these bits are their longevity, decreased number of run bits, and tripping time to the next casing point.

3.15.3.10 Hybrid Bits

Drill bit innovations impact drilling efficiency and cost. Due to advances in cutting material and bit design technology, polycrystalline diamond compact (PDC) bits are frequently utilized for their high rate of penetration (ROP) and wear resistance (Zhu et al., 2015, 2019). However, when employing a PDC bit in a directional well, the bit cuts irregularly due to the motor bending angle, and substantial torque fluctuations occur, resulting in instability of the bit's tool face (Yang et al., 2016, 2018). As a result, oil and gas drilling has adopted hybrid bit technology, a paradigm change in innovation that combines more than one bit design in a single patented design. The hybrid bit is designed to exploit the best qualities of each bit type, bridging the gap between them. With the cutting superiority and continuous scraping of diamond bits and the rock-crushing strength of roller cones, this repairable bit has proven its

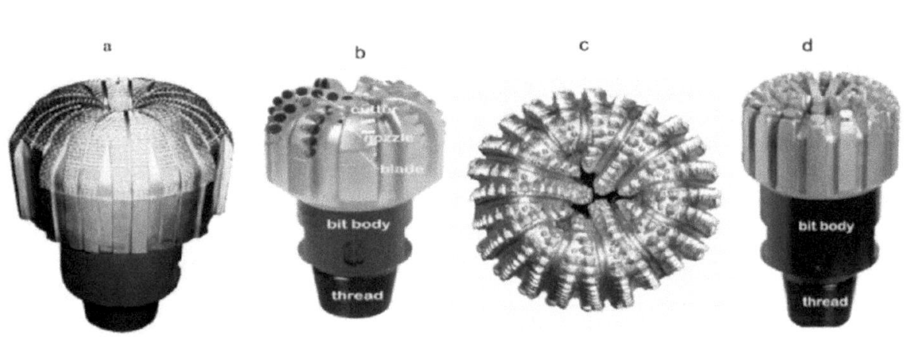

FIGURE 3.12 The fixed cutter bits: (a) natural diamond bit, (b) PDC drilling bit, (c) TSP drilling bit, and (d) impregnated bit.

Source: Niu et al., 2019

ability to endure highly interbedded formations with smooth drilling and excellent tool face control.

The major advantage of a hybrid bit over a conventional bit is that it can reduce the tripping time for a changing bit and improve the rate of penetration (ROP) by up to 25% with improved speed and durability without compromising stability; it also reduces drilling time, creates smoother drilling, and has remarkable torque management and precise steerability. The hybrid bit is an excellent directional bit for motor and rotary applications, as it offers enhanced buildup rate capabilities, muted torque response, and precise steerability on various bottom-hole assemblies (Hossain & Al-Majed, 2015). Some common types of hybrid bit technology are a combination of roller cone and fixed cutter bits, rotating and gouging insert bits, and PDC. See Adams and Charrier (1985), Drilling Course (2015), and Hossain and Al-Majed (2015) for more on drilling bits.

3.16 SELECTION OF DRILL BIT

Every oilfield target has distinct geological characteristics that create one-of-a-kind drilling situations. To optimize drilling, the operator must select drill bits with the optimal characteristics for the anticipated lithology. Along with drill pipes, MWD tools, stabilizers, reamers, and RSS, drill bits are crucial for the drilling system (or downhole motors). Choosing the appropriate bits for a well is a crucial decision that affects its total cost. Bit cost, expected parameters (e.g., turbines, rotary drilling, air operations), formation types and properties, mud systems, and rig cost must all be evaluated (Azar, 2021).

3.16.1 Drilling Fluids

One of the keys to success in drilling projects is the drilling fluids, circulated down the hole through the drill string and drilling bit and then up in the annular space towards the surface. They do this by combining density with any additional pressure acting on the fluid column to provide primary well control of subsurface pressures. The drilling fluids program is essential for operations and entails design, preparation, reactions with the formation, and surface treatment. It must be considered during well planning.

Drilling fluid systems are designed and prepared for optimal performance under anticipated wellbore conditions. An incomplete mud plan will cost the operator a lot of rig time, potentially making the difference between a successful and unsuccessful well (Adams & Charrier, 1985). Advances in drilling-fluid technology have made it feasible to adopt a cost-effective, purpose-specific system for each interval of the well-building process.

3.16.2 Types of Drilling Fluids

Drilling fluids come in many varieties. They consist of a base fluid, also called a solvent, and some additives (chemical agents) added to achieve specific properties. Although drilling fluids can be classified on other grounds, variety based on the type of base fluid is the most common and conventional. Major categories include fluids based on air, water, and oil. There are numerous sub-classes for each based on function, additives, or clay conditions (Oilfield Team, 2017).

Water-based drilling fluids (WBFs) have fresh or seawater as their continuous phase. Their constituents include mixtures of aqueous polymers, clay and brine, a pH control additive (hydroxyl source), and one or more organic polymers, surfactants, and deflocculants (Dareing, 2019). Three major groups of water-based fluids are inhibitive, noninhibitive, and polymer.

The OBFs and SBFs are also referred to as invert-emulsion systems, with an oil or synthetic base fluid serving as the continuous (or external) phase and brine as the internal phase. The use of oil-based muds necessitates environmental and safety precautions. Their primary applications are in high-temperature formations, water-sensitive shales, thick salt sections, lubricity-critical areas (such as extended-reach drilling), low-pore-pressure formations, and corrosive elements. Invert-emulsion systems have a greater unit cost than most water-based fluids, so they are frequently chosen when well conditions require dependable shale inhibition and excellent lubrication.

Synthetic fluids (SBF) are improved forms of oil-based mud designed to lessen the environmental impact of offshore drilling without compromising the cost effectiveness of oil-based systems. They are primarily made up of olefins, esters, and occasionally paraffin. SBFs have a low kinetic viscosity and can be used under low-pressure conditions. They still meet the stringent toxicity standards imposed by regulatory agencies today. Like conventional OBFs, SBFs can enhance lubricity in directional and horizontal wells and optimize the rate of penetration (ROP) as well as reducing wellbore instability issues, such as those induced by reactive shales.

Pneumatic drilling fluids are used when conventional drilling fluids cannot be due to loss of circulation or low reservoir pressures (Azar, 2021). The benefits of pneumatic drilling include increased penetration rate, longer drill bit life, enhanced control of loss circulation zones, and reduced formation damage. The drawbacks of dry air/natural gas include the possibility of downhole fire, corrosion, hole deviation, and hole erosion. Pneumatic fluids are classified into four types: dry air, mist, foam, and gasified mud.

3.16.3 Functions of Drilling Fluids

Drilling fluids are designed to perform a variety of functions. Although the list is lengthy and diverse, the following are the most essential functions:

- Controlling subsurface pressures to prevent formation fluids from returning to the wellbore
- Removing drill cuttings (debris of rock materials and mud) from beneath the bit and circulating them to the surface
- Cooling, cleaning, lubricating, and supporting the drill bit and its assembly during drilling and improving drilling rate
- Transmitting hydraulic energy to the bit and downhole tools
- Maintaining wellbore stability, mechanically and chemically

- Preventing or minimizing formation damage
- Reducing the weight of casing string and drill string while suspended in the wellbore
- Aiding in formation evaluation
- Minimizing corrosion of the drill string, casing, and tubing
- Minimizing the environmental impacts and hazards of drilling
- Sealing permeable formations, thereby reducing circulation loss
- Providing a completed wellbore that will produce hydrocarbons
- Forming a low permeability, thin, and tough filter cake across permeable formations, which prevents or reduces loss of circulation
- Supporting partial weight of drillstring or casing
- Ensuring maximum information from well

The appropriate drilling fluid is selected based on the type of formation being drilled and the various qualities of the drilling fluid. Due to varying conditions, not all functions will be performed on each well. The program for drilling fluids must be designed to meet the prospect well's highest-priority requirements. Sadly, these requirements may frequently be contradictory and impose stringent restrictions on the system. For instance, a low-solids system may be desired for increased drilling rates and minimal formation damage. However, oil muds or dispersed lignosulfonate systems may be easier to control when drilling high-pressure, high-activity shales at extreme temperatures. Drilling fluid performance is also affected by the drilling equipment being used. See Adams and Charrier (1985), Jamal (2007), and Warren (2022) for details on drilling fluids.

3.17 DRILLING FLUIDS PROPERTIES

3.17.1 Viscosity

Viscosity is one of the properties to be considered in drilling fluid. It is a general term used to define the internal friction a fluid generates when a force is applied to cause it to flow or how thick or thin it is. This internal friction is a result of the attraction between the molecules of a liquid and is related to shear stress (Khodja et al., 2010). This phenomenon is a measure of the combined effects of adhesion and cohesion on suspended particles and the liquid environment (Azar, 2021; Parate, 2021). The greater this resistance is, the higher will be the viscosity. The viscosity at the bit affects the penetration rate, which will be better when the viscosity is lower (Khodja et al., 2010). The viscosity in the annulus affects hole cleaning efficiency, and the viscosity in the pits influences the effectiveness of solids separation techniques.

3.17.2 Mud Weight or Density

Density is the mass per unit volume. During drilling operations, mud weight must be effectively managed and adjusted. If the mud weight is less than the required density for safe drilling, formation fluids can flow into the well, losing well control. If it exceeds the recommended level, it can lead to circulation loss.

Ideally, a mud weight as low as the weight of water is desired for optimum drilling rates and to minimize the chances of fracturing the formation. However, in practice, mud weights more than twice the weight of the water may be necessary to contain abnormal pressures or to stabilize unstable formations mechanically. Also, drilling muds with a higher density than the cuttings' density makes cleaning the wellbore easier and faster.

3.17.3 Filtrate and Filter Cake

The filter cake is created when mud solids deposit on the hole's walls. Fluid loss or filtration is the loss of mud filtrate into a permeable drilling formation. The primary objective of controlling fluid loss (filtrate) is to generate thin filter cakes while drilling through permeable formations to prevent excessive fluid loss. Loss of drilling fluids information can result in high water consumption and a thick filter cake, which can cause a tight hole, an increase in torque, and a stuck pipe.

3.17.4 Solids Content

Solids may have a high gravity (HGS) (e.g., barite and other materials used for weighing) or a low gravity (LGS) (e.g., clays, polymers, and bridging materials). High solid content drilling fluids (i.e., LGS) typically increase plastic viscosity and gel strength over low solid content drilling fluids (i.e., HGS), due to their thicker filter cakes and slower drilling rates (Jason, 2015). Consequently, low-solid muds are easier to control and offer faster ROPs in all areas of operation. Solids content affects drilling rate, flow characteristics, gel strengths, and overall mud stability (Parate, 2021). The optimum solids content and control are essential for superior mud performance.

3.17.5 Gel Strength

One of the most important properties to consider during the drilling process is gel strength. This denotes the pressure required to initiate flow after a static period, as well as the mud's ability to suspend cuttings when circulation is interrupted. During drilling shutdown, the mud becomes "thicker" the more it gels. Therefore, the higher the gel's viscosity, the greater the pump pressure required to recirculate it. In contrast, muds with weaker gel strengths are simpler to break, necessitating lower pump pressures to restore circulation.

Sufficient gel strength suspends drill cuttings and weighing materials throughout connections and other static conditions. Gel strength immediately affects surge and swabbing pressures while establishing connections, tripping pipe, or running casing. It also influences the pressure necessary to "break circulation" and the ease with which entrained gas or air can be released (IADC, 2000).

3.17.5.1 Factors to Be Considered when Choosing a Suitable Drilling Fluid

Drilling fluids are classified into numerous categories based on their composition and application. Cost, technical performance, and environmental impact are the three main criteria influencing the type of drilling fluid used for a specific well. However, the drilling fluid system for a particular well is chosen based on several parameters, including:

- The density required to control subsurface pressures
- Well type and hole size
- Characteristics of the formations to be drilled
- Anticipated temperature and pressure
- Completion type
- Common regional drilling problems
- Logistics: cost and quality
- Health, safety, and environmental considerations
- Well design
- Drilling rig
- Rock mechanics
- Casing program
- Makeup water
- Potential corrosion

3.18 PAD DRILLING TECHNOLOGY

Pad drilling, also known as multi-well pad drilling, is a technique that allows for drilling multiple wellbores from a single pad. This new technique enables rig operators to drill groups of wells more efficiently because improved rig mobility reduces the distance traveled from one well location to the next while reducing the overall surface footprint. A drilling pad is where the wellheads for several horizontally drilled wells are housed. The total number of wells and the related surface equipment are significant factors in deciding pad size.

Before the emergence of pad drilling, well-drilling operators applied traditional extraction methods, which involved drilling down a single well from a new pad—the location that houses the wellhead—for each new well (EIA, 2012). Moving a drilling rig between two well sites previously involved disassembling the rig and reassembling it at the new location, even if the new location was only a few yards away. This process can be time and labor intensive, not to mention costly to both the company and the environment.

Pad drilling technology allows many wells to be drilled from a single location. The wells are usually positioned in a straight line to minimize the amount of equipment that must be moved when repositioning the rig for the next well. After drilling one well, the fully assembled rig can be lifted and moved a few yards to the next well site utilizing hydraulic walking or skidding systems. Multi-well pad drilling has played a vital role in opening up capital-intensive tight formation oil plays. The technique originates from offshore drilling and gained notoriety with its onshore counterpart as recently as 2004 and became more widespread in 2010 and now accounts for about 75% of all new wells (EIA, 2012).

The advantage of pad drilling is that it enables producers to target a large area of subsurface resources with minimal surface disturbance. Concentrating the wellheads also reduces the costs of maintaining the resources above ground and transporting the output to market. Additionally, fewer access roads and trucks are required to transport the materials, thus minimizing the environmental impact. Pad drilling and other innovations continue to bend the cost curve for developing unconventional oil and gas resources. In the past, pad drilling comprised as few as two or three wells; however, today, operators are pushing the limits of shale field development and improving costs by utilizing superpads to drill as many as 24 to 40 wells from the same pad location. Multiwell pad drilling permits maximum reservoir penetration with minimal surface disturbance, which is essential in environmentally sensitive areas with limited infrastructure and mountainous regions (Perry, 2012).

3.19 CHALLENGES AND LATEST TECHNOLOGIES IN DRILLING SHALES

Wellbore instability is a major challenge in shale drilling and completion operations. Wellbore instability causes a lot of difficulties, time loss, and cost overruns while drilling. Clay in shale affects its physical properties and behavior when it is exposed to a drilling fluid (Khodja et al., 2010). The sloughing and swelling nature of shale formations encountered during drilling is the cause of wellbore instability. Swelling occurs due to an increase in the size of the silicate minerals, increasing the clay structure (layers of clay penetrated by water), leading to hole destabilization; however, the problem of shale sloughing is directly related to the adsorption of water from the drilling mud. Consequently, wellbore instability problems can be triggered and stuck pipe problems can be encountered due to tight spots or higher solids loading in the wellbore. This issue, if left unaddressed, leads to a high cost of drilling the hole and, more significantly, to other hole problems (pipe sticking, excessive solid buildup in the mud, and hole bridging), as well as the abandonment of the well due to its inability to reach the desired or anticipated pay depth.

Drilling operations most frequently utilize water-based mud. However, this method is ineffective when dealing with water-sensitive shale, which can result in shale hydration. Consequently, the wellbore's instability is compromised. Using synthetic-based mud (SBM) or oil-based mud (OBM) is an alternate method for handling this shale type. OBM is the best choice in terms of technical specifications. However, it is toxic and will cause environmental problems when released into terrestrial or marine environments. SBM is less risky than OBM.

Well operators lose billions of dollars annually, sometimes millions in a single well, due to drilling problems caused by time-dependent wellbore instability in clay-rich shale formations. Formation failure and borehole collapse cause stuck pipes, casing delays, loss of expensive tools in the hole, and undesirable sidetracks. Unstable hole conditions unnecessarily increase non-productive time, inflate equipment and materials costs, and impact the quality of critical formation evaluation data.

The latest technologies were invented to eliminate or reduce challenges in shale drilling. These technologies range from drilling techniques, drilling bits, drilling fluids, and BHA for steerable systems in directional drilling to downhole measurement techniques MWD and LWD. Improvements in drilling and completion technology for shale operations have increased the safe depth for drilling.

Recently, technologies such as real-time fluid tracking have helped improve drilling operations. Some prominent completion technologies include plug-and-perf, sliding sleeves, and intelligent, well-completion designs: e.g., interval control valves (ICV). Rotary steering systems play an even more vital role in unconventional oil and gas development, mostly shale gas or gas-tight resources. The most recent innovation in RSS is the hybrid steerable system, which combines point-the-bit and push-the-bit technology. The steerable rotary system, which continues to be at the center of today's oil drilling technology development, is crucial for exploiting unconventional oil and gas. The development of hybrid bits is another technological advancement that has helped overcome the difficulties of shale drilling. Due to its rough construction, this bit offers outstanding penetration in interbedded shale, sandstone, and hard shale formation.

3.20 HORIZONTAL DRILLING AND EVALUATION

A horizontal well is a multidirectional drilling technique in which a drill bit is steered to follow a horizontal path approximately 90 degrees from vertical through the reservoir rock (Lucas Downey, 2021; Schmidt, 2023). Typically, drilling a horizontal well can stimulate oil and natural gas reservoirs in ways that a vertical well cannot. Hydrocarbons can be extracted from previously unproductive zones when horizontal wells are coupled with hydraulic fracturing (Schmidt, 2023).

Traditionally, horizontal drilling was relatively unpopular in the oil and gas industry. The technique was introduced in the 1970s but became commercially viable during the late 1980s (EIA, 1993). HDD became more popular in the 1990s when Mitchell Energy commenced drilling horizontal wells in the Barnett shale and Fort Worth, Texas (Schmidt, 2023). With technological advancements, horizontal drilling is currently regarded as a viable reservoir-development technique, aiding oil and gas extraction worldwide.

Horizontal drilling starts by drilling a vertical well from the surface. When the drill bit reaches the target rock, the pipe is pulled out of the well, and a motor is attached to it. A flow of drilling mud down the drill pipe powers the motor, which rotates the bit without rotating the entire pipe (Leffler & Raymond, 2006). This enables the drill bit to create a route that differs from the drill pipe's orientation (King, 2016). The bit and pipe are lowered into the well, and the bit drills a path that bends from vertical to horizontal. Drilling resumes in a fully horizontal path once the required angle is obtained.

The success in drilling and completing horizontal wells has been attributed to significant technological breakthroughs, innovation in design approaches, effective teamwork, utilization of the learning process, effective planning, proper program implementation, real-time field monitoring of drilling data, and updating of program designs. Some reasons for drilling horizontal wells include (King, 2016):

- Ability to reach difficult targets. Some reservoirs are located in residential areas or parks where drilling is impossible. The reservoir can be reached by drilling down and then horizontally to drill underneath this area.
- Draining a broad area by branching off numerous times from one single main well reduces the surface footprint of a drilling operation.

- Increasing the well area, which permits more fluids to flow since more of the source rock is exposed to the well's edge.
- Enhancing the productivity of fractured reservoirs by intersecting the maximum number of fractures, increasing the oil and gas flow into the well.
- Sealing or relieving pressure in a well that has gotten out of control. A relief well can be drilled horizontally to intersect the original well and release pressure.

3.21 ACHIEVING DRILLING PRECISION AND EFFICIENCY

Drilling efficiency is an important cost-saving measure for achieving the optimal (high) penetration rate. In most instances, discussions relating to drilling efficiency have centered on the rate of penetration (ROP) (Khosravanian & Aadnøy, 2022b; Kosasih, 2018). As a result, ROP is either equated to drilling efficiency or seen as the parameter that establishes it (Mensa-Wilmot et al., 2010). Drilling for petroleum necessitates a complex process that depends on a variety of variables, including bit size, bit efficiency, torque, weight on bit (WOB), revolutions per minute (RPM), flow rate, mud rheology, and formation hardness (Khosravanian & Aadnøy, 2022a). This makes obtaining and maintaining a high ROP difficult, requiring more than just providing enough power to the drill string.

Selecting a suitable drill bit is one of the most fundamental ways a drilling engineer can enhance efficiency. As the principal piece of equipment conducting the mechanical task of rock drilling, the type of drill bit utilized is crucial to the overall effectiveness of the drilling operation. This drill bit has many designs, depending on the rock type. Choosing the right bit requires careful consideration since even a minor change might have an unforeseen and negative impact on the total expense and objectives of the well. Therefore, the drilling engineer must know how drill bit design and standard drilling objectives are related. Polycrystalline diamond bits and roller cone bits are today's most prevalent drill bits. Special drill bits are used for specialized operations such as rock coring. Bit design influences both the rate of penetration and the efficiency.

Drilling fluid (drilling mud) is essential to the rotary drilling system and is among the elements determining drilling efficiency. Most issues encountered during well drilling are directly or indirectly related to the mud. The successful completion of an oil and gas well and its cost are influenced to some degree by the properties of the drilling fluid. The drilling mud itself is not very expensive. However, the expense of making the right decision and maintaining the proper fluid quantity and quality during drilling operations increases significantly. Some of the most common drilling issues, such as rate of penetration, caving shale, stuck pipe, and loss of circulation, are directly related to the proper selection, properties, and quality of drilling mud. In addition, the mud affects the integrity of the formation and the subsequent production efficiency of the well.

3.22 LATEST ADVANCES, TECHNIQUES, AND METHODS

Technological innovation is a critical driver of economic growth in the oil and gas industry. The rapid progress of digital technology gives the industry a chance to utilize new capabilities offered by using data for decision-making and automation of technical and non-technical processes. So far, automation, edge analytics, real-time resource optimization, integrated technology platforms and a hybrid form of big data analytics are causing the digital shift in our industry (Lakhanpal & Samuel, 2018). The industry is already focusing on implementing automation, the Internet of Things (IoT), machine learning, artificial intelligence, etc.

Advances in well drilling and completion technologies have enabled the energy industry to access new oil and natural gas sources to fulfill an expanding global demand. They allow the identification of new resources, access to harsh or inaccessible places, and the development of challenging reservoirs previously unprofitable to extract. Through dramatic innovation and digital transformation, oil and gas firms increasingly focus on reducing costs and increasing recovery rates (Business Wire, 2018).

Over the years, significant advances in drilling technology development and utilization have occurred in oil and gas production. The future of drilling systems study by oil and gas companies provides an overview of emerging technology areas. Several emerging technologies can facilitate production and help comply with health, safety, and environmental issues and are cost effective for the oil and gas industry. The key emerging technologies developed by oil companies are:

- Automation and robotics technology for drilling systems
- Advanced measurement and instrumentation technology for drilling systems
- Advanced computing technology for drilling systems
- Enhanced communications technology for drilling systems
- Digital twin technology for drilling systems
- AR/VR technology for drilling systems
- Blockchain technology for drilling systems
- Additive manufacturing technology for drilling systems
- Advanced designs in bit technology and mud motors
- Advance in mud and solids control technology
- Advances in measurement while drilling (MWD) and other drilling technology

The future of drilling has received a reboot with the introduction of an innovative autonomous robotic drilling rig for unmanned land, offshore, and deep-water drilling operations. Drilling automation is the control of the drilling process by automatic means, minimizing the need for human intervention (Macpherson et al., 2013). Automating drilling systems aim to enhance productivity and quality, enhance employee safety, and effectively manage risk. Principal drivers of drilling automation include well complexity, data overload, efficiency, repetitive well manufacturing, access to limited expert resources, knowledge transfer resulting from the flight of qualified workers, and health, safety, and environmental concerns. With so many drivers and their potential economic benefits, it is not surprising that the sector has so many automation-related activities.

Artificial intelligence (AI) technology for drilling systems allows for improved drilling with less physical input from drillers (Nunoo, 2018). An adaptive drilling application uses AI to continuously optimize the drilling operations using machine-learning algorithms and a digital twin (Jeffery & Creegan, 2020). Automated parameter application allows the driller to concentrate on crucial tasks while adapting to changing conditions. Intelligent drilling optimization software functions as an adaptable autodriller. This innovative technology employs algorithms based on artificial intelligence to improve on-bottom drilling performance. Continuous learning capabilities enable the system to offer proactive drilling dysfunction mitigation while achieving a greater rate of penetration (ROP) and optimizing mechanical-specific energy. The system's advantage is that it relies less on humans in drilling, minimizing the risk of slow or wrong responses in drilling failure. During the field application, ROP increased by 61% while drilling efficiency (as measured by hours on the bottom) increased by 25% (Jeffery & Creegan, 2020).

Unconventional resources, such as shale gas and coalbed methane, are located horizontally beneath geological formations, in contrast to conventional resources. Directional drilling and hydraulic fracturing technologies are required to extract these resources. In directional drilling, the dogleg severity (DLS), which indicates how much the angle changes per 100 feet of drilling, is a major factor in terms of cost and time (Kim & Myung, 2017).

Rotary steerable technology has undergone a significant transformation in recent years and is quickly becoming a viable alternative to mud motors and, in some cases, standard practice. The trend of automation, pushed by the use of 3D visualization tools, has increased the potential of downhole tool technologies, thereby enhancing the performance of the drill bit and decreasing vibrations during drilling. The steer-at-bit is one of the new evolutionary alternatives to the conventional rotary steerable system (RSS). It is a rotary tool that utilizes internally directed pressure differentials to steer away from the drill bit face, delivering true "at-bit" geosteering (Teodoriu & Bello, 2021). The tool provides a robust, dependable, simple, and cost-effective direction based on successful initial testing. Automation enabled for precise steering and accurate well placement, this technique enables operators to reduce well time through faster drilling, dependable performance, and predictable outcomes. RSS with near-bit GR imaging facilitates accurate steer drilling in a Sichuan shale gas project (Červený & Pšenčík, 2011).

Another advance in rotary steerable technology is Schlumberger's introduction of PowerDrive Archer (Terrell, 2012). It is a high-build-rate rotary steerable system (RSS). The system can drill complex 3D trajectories and open hole side tracks from any inclination. It can also drill vertical and horizontal sections in one run without changing the bottom-hole assembly (BHA). The system, which is a hybrid steering, can combine both push-the-bit and point-the-bit technologies, allowing the tool to navigate in a way that was previously possible only with a downhole motor and to deliver high build rates for any deviation from vertical. The PowerDrive Archer RSS has been combined with new improvements in the SHARC high-abrasion-resistance PDC drill bits to reduce rig time significantly. SHARC bits are designed for superior performance in highly abrasive formations.

An automated well-control system is among the latest technologies to reduce well-control risks. Automated well-control solutions have successfully demonstrated that this system can support the driller, dramatically reducing exposure to risks from human factors (Teodoriu & Bello, 2021). The system can detect a fluid influx situation in a wellbore, decide based on criteria that favor a shut-in, and then conduct an initial well control routine that results in the well being safely shut down. This cutting-edge technology can reduce the magnitude of an influx compared to conventional methods (Woodroof, 2021).

Furthermore, machine intelligence improves oil and gas data analytics by providing algorithmic capabilities that complement employee performance, automate increasingly complicated workloads, and create "cognitive agents" that mimic human thinking and participation. Companies can use predictive oil and gas analytics, such as condition-based monitoring and variable analysis, to develop scenario-based simulations to predict likely future maintenance events and undertake any necessary repair before equipment damage. Predictive maintenance lowers costly reactive maintenance events and downtime and, when used in conjunction with a thorough maintenance plan, can shorten schedules for large-scale maintenance projects (Jeffery & Creegan, 2020).

The advanced technology in drill bits is the invention of hybrid drill bits. The technology represents a paradigm shift in innovation, combining roller cones and PDC bits into a single design. The result is a technology that exploits the best characteristics of each bit type while bridging the gap between them. With the cutting superiority, continuous scraping of diamond bits, and rock-crushing strength of roller cones, this repairable bit has demonstrated its ability to endure highly interbedded formations with smooth drilling and excellent tool face control.

REFERENCES

Adams, N., & Charrier, T. (1985). *Drilling engineering: A complete well planning approach* (p. 960). PennWell Publishing Co.

Azar, J. J. (2021). Drilling engineering. In *Drilling engineering*. PennWell Corporation. https://doi.org/10.1016/c2019-0-00943-0

Berro, M. J., & Reich, M. (2016, November). *Review of commercial telemetry systems for real time data transmission in boreholes*. www.researchgate.net/publication/309722398_Review_of_Commercial_Telemetry_Systems_for_Real_Time_Data_Transmission_in_Boreholes

Brandl, A., Cutler, J., Seholm, A., Sansil, M., & Braun, G. (2010). Cementing solutions for corrosive well environments. *Society of Petroleum Engineers—International Oil and Gas Conference and Exhibition in China 2010, IOGCEC*, *4*, 3104–3116. https://doi.org/10.2523/132228-ms

Brandon, T. L., Mintchev, M. P., & Tabler, H. (1999). Adaptive compensation of the mud pump noise in a measurement-while-drilling system. *SPE Journal*, *4*(2), 128–133. https://doi.org/10.2118/56852-PA

Bryant, T. (2019). *An operational comparison of push-the-bit rotary steerable tools* (pp. 1–12). APS Technology, Inc.

Business Wire. (2018). *The future of drilling systems (2018–2025): Emerging technologies to drive new efficiencies and business model shifts—ResearchAndMarkets.com | business wire*. Online Resources. www.businesswire.com/news/home/20180702005946/en/The-Future-of-Drilling-Systems-2018-2025-Emerging-Technologies-to-Drive-New-Efficiencies-and-Business-Model-Shifts-ResearchAndMarkets.com

Červený, V., & Pšenčík, I. (2011). Seismic, ray theory. *Encyclopedia of Earth Sciences Series, Part 5*, 1244–1258. https://doi.org/10.1007/978-90-481-8702-7_53

Dareing, D. W. (2019). *Oilwell drilling engineering*. https://doi.org/10.1115/1.861875

Davies, R. J., Almond, S., Ward, R. S., Jackson, R. B., Adams, C., Worrall, F., Herringshaw, L. G., Gluyas, J. G., & Whitehead, M. A. (2014). Oil and gas wells and their integrity: Implications for shale and unconventional resource exploitation. *Marine and Petroleum Geology*, *56*, 239–254. https://doi.org/10.1016/j.marpetgeo.2014.03.001

Drilling Course. (2015). *Introduction to drilling fluid*. Online Resources. www.drillingcourse.com/2015/12/drilling-fluids-functions-and-properties.html

drillingformulas. (2017). *Deviating the wellbore by positive displacement motor (Directional Drilling)*. Online Resources. www.drillingformulas.com/deviating-the-wellbore-by-positive-displacement-motor-directional-drilling/

EIA. (1993). *Drilling sideways—A review of horizontal well technology and its domestic application*. Energy Information Administration, DOE/EIA-TR-0565. https://www.eia.gov/petroleum/archive/tr0565.pdf

EIA. (2012). *Pad drilling and rig mobility lead to more efficient drilling*. www.eia.gov/todayinenergy/detail.php?id=7910

Felczak, E., Torre, A., Godwin, N. D., Mantle, K., & Naganathan, S. (2012). The best of both worlds—a hybrid rotary steerable system. *Oilfield Review*, *22*(4), 36–44.

Figueredo, C. (2014). Enhanced RSS technology pushes drilling envelope. *Journal of Petroleum Technology*, *66*(3), 32–35. https://doi.org/10.2118/0314-0032-jpt

Finger, J., & Blankenship, D. (2010, December). Handbook of best practices for geothermal drilling. *Sandia Report (SAND2010-6048)* . . ., 84, Figure 3.2. http://artikel-software.com/file/geothermal drilling handbook.pdf

Gravley, W., & Corp, D. (1983). Review of downhole measurement-while-drilling systems. *Journal of Petroleum Technology*, *35*(8), 1439–1445. https://doi.org/10.2118/10036-PA

Gregory King. (2020). *PNG 301 introduction to petroleum and natural gas engineering*. www.e-education.psu.edu/png301/node/904

Hossain, M. E., & Al-Majed, A. A. (2015). Fundamentals of sustainable drilling engineering. *Fundamentals of Sustainable Drilling Engineering*, 1–754. https://doi.org/10.1002/9781119100300

IADC. (2000). *IADC drilling manual*, 1–1463. International Association of Drilling Contractors. https://www.scribd.com/document/718269932/IADC-Vol-2-10-Well-Control-Equipment-and-Procedures

IADC. (2014). *IADC drilling manual*, 1–1463. International Association of Drilling Contractors. https://www.scribd.com/document/718269932/IADC-Vol-2-10-Well-Control-Equipment-and-Procedures

Ivamberg, N. de A. J., Antunes, P. D., Gonzalez, F. O. C., Yamachita, R. A., Nascimento, A., & Goncalves, J. L. (2015). A review of telemetry data transmission in unconventional petroleum environments focused on information density and reliability. *Journal of Software Engineering and Applications*, *8*(9), 455–462. https://doi.org/10.4236/jsea.2015.89043

Jamal, A. J. (2007). *Drilling engineering* (Vol. I). PennWell Corporation.

Jason, N. (2015). Selection and application of drilling fluids. *Drilling engineering and well completion* (pp. 0–14). http://dx.doi.org/10.13140/RG.2.1.3243.4722

Jeffery, C., & Creegan, A. (2020). Adaptive drilling application uses AI to enhance on-bottom drilling performance. *Journal of Petroleum Technology*, *72*(8), 45–47. https://doi.org/10.2118/0820-0045-jpt

JPT Staff. (2013). Hybrid rotary steerable system delivers higher build rates and smoother holes. *Journal of Petroleum Technology*, *65*(4), 32–34. https://doi.org/10.2118/0413-0032-JPT

Khalifeh, M., & Saasen, A. (2020a). *Different categories of working units* (Vol. 12, pp. 137–163). https://doi.org/10.1007/978-3-030-39970-2_5

Khalifeh, M., & Saasen, A. (2020b). *Introduction to permanent plug and abandonment of wells* (Vol. 12). https://doi.org/10.1007/978-3-030-39970-2_5

Khodja, M., Khodja-Saber, M., Paul, J., Cohaut, N., & Bergay, F. (2010). Drilling fluid technology: Performances and environmental considerations. *Products and Services; from R&D to Final Solutions*. https://doi.org/10.5772/10393

Khosravanian, R., & Aadnøy, B. S. (2022a). Mechanical specific energy and drilling efficiency. *Methods for Petroleum Well Optimization*, 193–247. https://doi.org/10.1016/b978-0-323-90231-1.00008-x

Khosravanian, R., & Aadnøy, B. S. (2022b). Mechanical specific energy and drilling efficiency. *Methods for Petroleum Well Optimization*, 193–247. https://doi.org/10.1016/B978-0-323-90231-1.00008-X

Kim, J., & Myung, H. (2017). Development of a novel hybrid-type rotary steerable system for directional drilling. *IEEE Access*, *5*, 24678–24687. https://doi.org/10.1109/ACCESS.2017.2768389

King, H. M. (2016). Directional and horizontal drilling in oil and gas wells. *Alternative Energy and Shale Gas Encyclopedia*, 695–696.

Klotz, C., Reckmann, H., Wassermann, I., Macpherson, J. D., Ortiz, J. A., & Brooks, A. G. (2007). *System and method for measurement while drilling telemetry*. US Patent (US20070189119A1).

Lakhanpal, V., & Samuel, R. (2018). Implementing blockchain technology in oil and gas industry: A review. *Proceedings—SPE annual technical conference and exhibition*, *2018-Septe*. https://doi.org/10.2118/191750-ms

Leffler, W., & Raymond, M. (2006). Oil and gas production in nontechnical language. *Choice Reviews Online*, *43*(8), 43-4690–43-4690. https://doi.org/10.5860/choice.43-4690

Li, F., Ma, X., & Tan, Y. (2020). Review of the development of rotary steerable systems. *Journal of Physics: Conference Series*, *1617*(1). https://doi.org/10.1088/1742-6596/1617/1/012085

Li, Z., Itakura, K.-I., & Ma, Y. (2014). Survey of measurement-while-drilling technology for small-diameter drilling machines. *Electronic Journal of Geotechnical Engineering*, *19*, 10267–10282.

Liu, X., Li, B., & Yue, Y. (2007). Transmission behavior of mud-pressure pulse along well bore. *Journal of Hydrodynamics*, *19*(2), 236–240. https://doi.org/10.1016/s1001-6058(07)60054-7

Lucas Downey. (2021). Horizontal well. *Investopedia*. www.investopedia.com/terms/h/horizontalwell.asp#:~:text=A horizontal well is a,and productive in recent years.

Ma, T., Chen, P., & Zhao, J. (2016). Overview on vertical and directional drilling technologies for the exploration and exploitation of deep petroleum resources. *Geomechanics and Geophysics for Geo-Energy and Geo-Resources*, *2*(4), 365–395. https://doi.org/10.1007/s40948-016-0038-y

Macpherson, J. D., De Wardt, J. P., Florence, F., Chapman, C. D., Zamora, M., Laing, M. L., & Iversen, F. P. (2013). Drilling systems automation: Current state, initiatives and potential impact. *Proceedings—SPE Annual Technical Conference and Exhibition*, *3*, 2229–2245. https://doi.org/10.2118/166263-ms

Mantle, K. (2014). Directional drilling practice. *Oilfield Review*, 25, 4, Schumberger.

Mensa-Wilmot, G., Harjadi, Y., Langdon, S., & Gagneaux, J. (2010). Drilling efficiency and rate of penetration—Definitions, influencing factors, relationships and value. *SPE/IADC Drilling Conference, Proceedings*, *1*, 578–590. https://doi.org/10.2118/128288-ms

Morton-Thompson, D., & Woods, A. M. (1993). *Development geology reference manual*. https://doi.org/10.1306/mth10573

Niu, S. W., Zheng, H. L., Yang, Y. X., Tong, X., Chen, L., Liu, Y., & Bao, Z. J. (2019). Experimental research on torque characteristic and weight distribution of the polycrystalline diamond compact–roller hybrid bit. *Advances in Mechanical Engineering*, *11*(5). https://doi.org/10.1177/1687814019849740

Nunoo, N. A. (2018). How artificial intelligence will benefit drilling. *JPT, Journal of Petroleum Technology*, *70*(5), 14–15.

Oilfield Team. (2017). *Drilling fluid classifications.* Online Resource. https://oilfieldteam.com/en/a/learning/Drilling-Fluid-Classifications

Oilfieldteam. (2018). *Cable-tool drilling.* Online Source. https://oilfieldteam.com/en/a/learning/Cable-tool-drilling-280218

Parate, N. B. (2021). *A review article on drilling fluids, types, properties and criterion for selection, 8.* www.jetir.org

Perry, M. J. (2012). *Advances in drilling technologies lower costs.* Online Source from CARPE DIEM. https://mjperry.blogspot.com/2012/09/advances-in-drilling-technologies-lower.html

Rig worker. (2023). *Rig sizing and selection—Drilling engineering—Rig worker.* www.rigworker.com/engineering-2/rig-sizing-and-selection-1.html

Schaa, S., Pafitis, D., & Guichemerre, E. (2000). Application of a point the bit rotary steerable system in directional drilling prototype well-bore profiles. *SPE/AAPG Western Regional Meetings*, 61–67. https://doi.org/10.2118/62519-ms

Schlumberger. (2023). *Rotary steerable system | Energy Glossary.* Online Resource. https://glossary.slb.com/en/terms/r/rotary_steerable_system

Schmidt, J. (2023). *Horizontal well.* Online Resources Corporate Finance Institute.

Short, J. A. "Jim." (1993). Introduction to directional and horizontal drilling. In *Drilling engineering problems and solutions* (p. 239). PennWell Publishing Co.

Silva, T. A. (2006). *MWD/LWD engineer guide. Grade 9.* Schlumberger.

STS. (2020). *Mud pulse telemetry: MWD data transmission with pressure sensors—STS Switzerland (EN).* www.stssensors.com/blog/2020/07/01/mud-pulse-telemetry-mwd-data-transmission-with-pressure-sensors/

Su, Y., Sheng, L., Li, L., Bian, H., Shi, R., & Corporation, P. (2011). High-data-rate measurement-while-drilling system for very deep wells. *American Association of Drilling Engineers*, 1–11. www.aade.org/technical-papers/

Tao, C., Kutchko, B. G., Rosenbaum, E., & Massoudi, M. (2020). A review of rheological modeling of cement slurry in oil well applications. *Energies*, *13*(3). https://doi.org/10.3390/en13030570

Teodoriu, C., & Bello, O. (2021). An outlook of drilling technologies and innovations: Present status and future trends. *Energies*, *14*(15). https://doi.org/10.3390/en14154499

Terrell, H. (2012). New drill bit technology: Greater durability, faster ROPs equal better economics. *World Oil*, *233*(10), 53–62.

Thaiprasert, N., Pensook, T., Tangcharoen, N., & Pattanapong, K. (2016). *Drilling performance improvement with rotary steerable system RSS for shallow soft formation in the gulf of Thailand.* https://doi.org/10.2523/iptc-18920-ms

Trenchlesspedia. (n.d.). *What is a Mud Motor?—Definition from trenchlesspedia.* Retrieved April 28, 2023, from www.trenchlesspedia.com/definition/2178/mud-motor

Vij, J., Saraiya, R., Saumya, S., Sarkar, S. K., & Majumdar, C. (2018). LWD as the absolute formation evaluation technology: Present-day capabilities, limitations, and future developments of LWD technology. *SPWLA 2nd Asia Pacific technical symposium, APTS 2018.* Paper Number: SPWLA-2018-1823.

Williamson, D. (2013). Drilling fluid basics. *Oilfield Review Spring*, 25, 1, Schlumberger.

Yang, Y., Zhang, C., Chen, L., & Liu, Y. (2016). Kinematic and bottom-hole pattern analysis of a composite drill bit of cross-scraping. *Proceedings of the Institution of Mechanical Engineers, Part C: Journal of Mechanical Engineering Science*, *231*(17), 3104–3117. https://doi.org/10.1177/0954406216642797

Yang, Y., Zhang, C., Lin, M., & Chen, L. (2018). Research on rock-breaking mechanism of cross-cutting PDC bit. *Journal of Petroleum Science and Engineering*, *161*, 657–666. https://doi.org/10.1016/J.PETROL.2017.11.034

Zhang, C., Zou, W., & Cheng, N. (2016). Overview of rotary steerable system and its control methods. *2016 IEEE international conference on mechatronics and automation, IEEE ICMA 2016*, 1559–1565. https://doi.org/10.1109/ICMA.2016.7558796

Zhu, H., Deng, J., Jin, X., Hu, L., & Luo, B. (2015). Hydraulic fracture initiation and propagation from wellbore with oriented perforation. *Rock Mechanics and Rock Engineering, 48*(2), 585–601. https://doi.org/10.1007/S00603-014-0608-7/METRICS

Zhu, H., Shen, J., & Zhang, F. (2019). A fracture conductivity model for channel fracturing and its implementation with Discrete Element Method. *Journal of Petroleum Science and Engineering, 172*, 149–161. https://doi.org/10.1016/J.PETROL.2018.09.054

4 Well Stimulation and Completion Aspects of Unconventional Resources

Ekrem Alagoz and Fatimah AlNasser

4.1 INTRODUCTION

Unconventional resources, such as shale gas, tight oil, and coalbed methane, have become increasingly important sources of energy in recent years. Unlike conventional reservoirs, unconventional resources are characterized by low permeability and complex geology, which makes them challenging and expensive to produce. Well stimulation and completion techniques have been key to unlocking the potential of unconventional resources, making their development economically viable.

Well stimulation involves creating artificial fractures in the formation to increase the flow of hydrocarbons while completion refers to the design and installation of production equipment in the wellbore to facilitate the flow of fluids. Well stimulation and completion techniques have evolved over the years, driven by advances in technology and increased knowledge of the reservoir characteristics. Recent advances have led to significant improvements in the efficiency and effectiveness of well stimulation and completion in unconventional resources, which have contributed to the growth of the industry.

In this chapter, we will discuss the definition and characteristics of unconventional resources, as well as the importance of well stimulation and completion in their development. We will also explore recent advances and techniques in well stimulation and completion that have improved the production of unconventional resources. Finally, we will consider the environmental considerations associated with well stimulation and completion and the future outlook for this important industry.

4.2 UNCONVENTIONAL RESOURCES AND THEIR CHARACTERISTICS

Unconventional resources are hydrocarbon deposits that are different from conventional resources in terms of their geological, physical, and chemical properties. The

DOI: 10.1201/9781003319955-4

most common types of unconventional resources include shale gas, tight oil, and coalbed methane (Figure 4.1), but there are other types as well. These resources have become increasingly important in recent years as conventional resources become harder to find and produce.

The characteristics of unconventional resources are distinct from those of conventional resources. For example, unconventional resources tend to have low permeability, which means that the rock formations containing the resources are not as permeable as those of conventional reservoirs. This makes it more difficult for the hydrocarbons to flow to the wellbore and requires more advanced well stimulation and completion techniques to improve production.

In addition, unconventional resources are often found in complex geological formations, which can make them more difficult and expensive to produce. The rocks containing these resources may be highly folded, faulted, or fractured, requiring sophisticated reservoir characterization and modeling to optimize production.

In this section, we will discuss the different types of unconventional resources, their geological and physical characteristics, and how these characteristics impact their development. We will also consider the challenges associated with producing unconventional resources and the importance of well stimulation and completion techniques in maximizing production. By understanding the characteristics of unconventional resources, we can better appreciate the challenges and opportunities associated with their development.

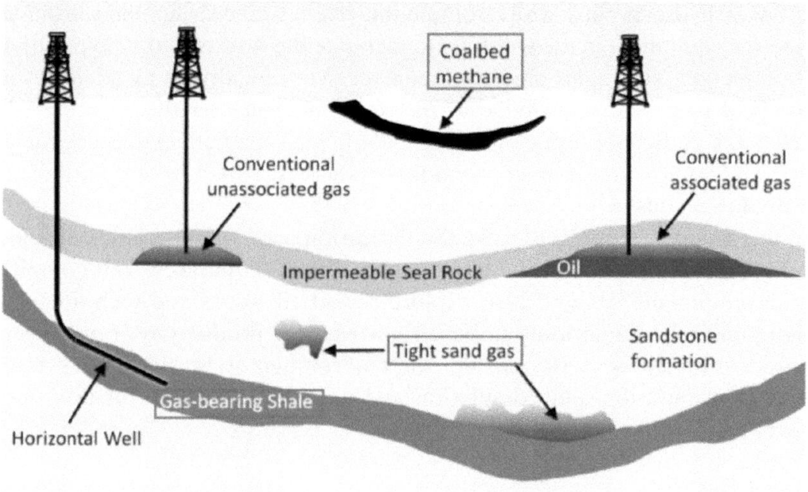

FIGURE 4.1 Conventional and unconventional resources, including gas-bearing shale, tight sand gas, and coal-bed methane.

Source: Ias, 2018, February 28

4.2.1 Types of Unconventional Resources

4.2.1.1 Shale Gas

Shale gas is a type of natural gas that is trapped within shale formations. Shale is a fine-grained sedimentary rock that is rich in organic matter, and the gas is generated from the thermal maturation of this organic matter (Figure 4.2). Unlike conventional natural gas reservoirs, which are typically found in permeable sandstones or limestones, shale gas is contained in the fine-grained shale itself. The shale must be fractured using hydraulic fracturing techniques to release the gas and allow it to flow to the wellbore.

The development of shale gas has revolutionized the energy industry, providing a vast new source of natural gas and contributing to energy security and independence. The United States has been a leader in shale gas production, with significant shale gas reserves located in various regions across the country. Other countries, such as China and Argentina, are also exploring their shale gas resources.

However, there are some challenges associated with the production of shale gas. The shale formations tend to have low permeability, which means that the gas is trapped in the rock and cannot flow to the wellbore without artificial stimulation. Additionally, shale gas development requires large volumes of water, which can be a concern in regions with limited water resources. There are also environmental concerns associated with hydraulic fracturing, including the risk of groundwater contamination and the release of greenhouse gases.

To optimize the production of shale gas, well stimulation and completion techniques must be carefully designed and executed. Hydraulic fracturing is the primary method used to stimulate shale formations, and advances in technology have enabled more efficient and effective stimulation of the reservoir (Figure 4.3). Completions

FIGURE 4.2 Shale rock sample.

Source: Anne, 2020

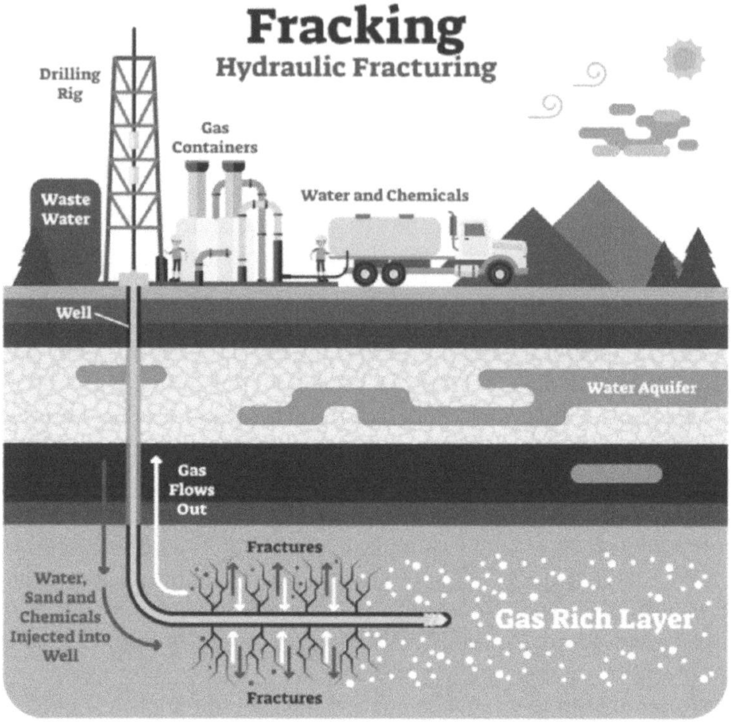

FIGURE 4.3 Hydraulic fracturing.

Source: Media, U. S. E., 2018

techniques such as multistage hydraulic fracturing have also been developed to improve production rates.

In summary, shale gas is a type of natural gas that is trapped in shale formations. While its development has revolutionized the energy industry, it presents unique challenges due to the low permeability of the shale and the large volumes of water required for production. Well stimulation and completion techniques have been critical to the successful development of shale gas resources, and ongoing research and development in these areas will continue to improve the efficiency and sustainability of shale gas production.

4.2.1.2 Tight Oil

Tight oil, also known as shale oil or oil shale, is a type of unconventional oil that is trapped in low-permeability rock formations, such as shale. Like shale gas, tight oil is produced through hydraulic fracturing, which involves injecting water, sand, and chemicals at high pressure into the rock formation to create fractures that allow the oil to flow to the wellbore.

The development of tight oil has transformed the energy landscape, providing a significant new source of oil and contributing to the energy independence of countries such as the United States. Tight oil resources are found in various regions around the world, including North America, South America, Europe, and Asia. The characteristics of tight oil reservoirs present some unique challenges for production. The rock formations tend to have low permeability, which means that the oil is trapped in the rock and cannot flow to the wellbore without hydraulic fracturing. In addition, tight oil reservoirs often have complex geological characteristics, such as folds, faults, and fractures, which require careful reservoir characterization and modeling to optimize production.

To maximize the production of tight oil, well stimulation and completion techniques must be carefully designed and executed. Advances in hydraulic fracturing technology, such as multistage hydraulic fracturing and refracturing, have improved the efficiency and effectiveness of well stimulation in tight oil reservoirs. In addition, innovations in well completion techniques, such as the use of smart well technologies and downhole sensors, have enabled better monitoring and control of the production process. However, the development of tight oil resources also raises some environmental and social concerns. The large volumes of water required for hydraulic fracturing can be a concern in regions with limited water resources, and there are also concerns about the disposal of wastewater and the release of greenhouse gases. In addition, the development of tight oil resources can have significant impacts on local communities and ecosystems.

In summary, tight oil is an important type of unconventional oil that is produced through hydraulic fracturing. Its development has transformed the energy landscape, but it presents unique challenges and raises environmental and social concerns. Well stimulation and completion techniques have been critical to the successful development of tight oil resources, and ongoing research and development in these areas will continue to improve the efficiency and sustainability of tight oil production.

4.2.1.3 Coalbed Methane

Coalbed methane (CBM) or coal seam gas (CSG) is an unconventional natural gas extracted from coalbeds (Figure 4.4). The gas is produced through the natural process of coalification as organic matter in the coal is transformed into methane gas by the heat and pressure of the surrounding rock (Pashin, 2015). CBM reservoirs are typically found in coal seams that are saturated with water. The gas is held in the coal seam by the pressure of the water, which must be removed to release the gas. To extract the gas, wells are drilled into the coal seam, and water is pumped out of the coal seam to reduce the pressure and release the gas (National Research Council (NRC), 2010).

The development of CBM resources has provided a significant new source of natural gas, particularly in countries such as Australia, the United States, and China. However, the production of CBM also presents some unique challenges. The coal seams tend to have low permeability, which means that the gas can be difficult to extract. In addition, the water that is produced with the gas can be high in salt and other impurities, which can be difficult to treat and dispose of (SEPA, 2010). To optimize the production of CBM, well stimulation and completion techniques must be carefully designed and executed. The primary method used to stimulate CBM

FIGURE 4.4 Coalbeds.

Source: Coal Bed Methane, 2023

reservoirs is depressurization, in which water is pumped out of the coal seam to reduce the pressure and release the gas. In addition, innovative completion techniques, such as the use of coiled tubing or underbalanced drilling, can be used to improve the efficiency of CBM production (Ghaderi et al., 2018). CBM production also raises some environmental and social concerns, particularly with respect to water resources. The large volumes of water required for CBM production can be a concern in regions with limited water resources, and the disposal of produced water can be challenging. In addition, there are concerns about the impact of CBM development on local communities and ecosystems (NRC), 2010).

In summary, coalbed methane is an important type of unconventional natural gas that is produced from coalbeds. Its development has provided a significant new source of natural gas, but it presents unique challenges and raises environmental and social concerns. Well stimulation and completion techniques are critical to the successful production of CBM, and ongoing research and development in these areas will continue to improve the efficiency and sustainability of CBM production.

4.2.2 Characteristics of Unconventional Resources

4.2.2.1 Low Permeability

Unconventional resources are characterized by low permeability, which refers to the ability of a fluid to flow through a rock formation (Zou et al., 2019). The low permeability of unconventional resources, such as shale gas and tight oil, presents a significant challenge to the production of these resources. Because the rock is so impermeable, it can be difficult to extract the gas or oil using conventional production methods.

To overcome this challenge, well stimulation techniques such as hydraulic fracturing are commonly used to create fractures in the rock and increase its permeability. Hydraulic fracturing involves injecting water, sand, and chemicals into the well at high pressure, which creates fractures in the rock and allows the gas or oil to flow more freely.

Recent research has focused on improving our understanding of the permeability of unconventional resources, as well as developing new techniques for enhancing the permeability of these resources. For example, some researchers have investigated the use of microorganisms to increase the permeability of shale formations (Zhang et al., 2019a) while others have explored the use of electromagnetic waves to stimulate the flow of oil in tight reservoirs (Alfi et al., 2021).

4.2.2.2 Complex Geology

Unconventional resources, such as shale gas and tight oil, are characterized by complex geology. These resources are often found in deep rock formations, which makes their exploration and extraction challenging (Figure 4.5). Additionally, the geological structures of unconventional resources are often more complex than those of conventional oil and gas reservoirs, which further complicates their exploration and development (Bashari, 2016). The complex geology of unconventional resources is due to the fact that they are often found in layered rock formations with varying mineral composition, porosity, and permeability. These rock formations can also contain natural fractures and faults that can impact the flow of fluids and the ability to extract resources (Mack, 2016).

The geological complexity of unconventional resources requires advanced exploration and production techniques to overcome these challenges. For example, hydraulic fracturing is used to create artificial fractures in the rock to increase permeability and enhance fluid flow (Hartsock, 2013). Additionally, 3D seismic imaging and other advanced technologies are used to gain a better understanding of the geological structures and properties of unconventional resources (Bashari, 2016).

In summary, the complex geology of unconventional resources presents significant challenges to their exploration and development. However, advances in technology and techniques are continually improving the industry's ability to extract these valuable resources.

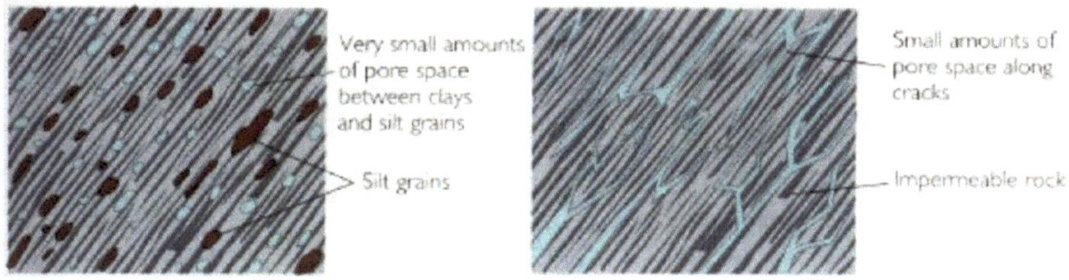

FIGURE 4.5 Unfractured shale (left) versus fractured shale (right).

Source: Water in the Ground, 2023

4.3 RECENT ADVANCES AND TECHNIQUES IN WELL STIMULATION FOR UNCONVENTIONAL RESOURCES

4.3.1 Advancements in Hydraulic Fracturing Technology

Hydraulic fracturing is a widely used technique for stimulating the production of unconventional resources such as shale gas and tight oil. Recent advancements in hydraulic fracturing technology have focused on improving the efficiency and effectiveness of this technique. One of the most significant advancements in hydraulic fracturing technology is the development of "zipper" fracturing (Figure 4.6). This technique involves simultaneously fracturing two parallel horizontal wells that are in close proximity to each other. By doing so, the fractures created in each well can intersect with each other, creating a network of fractures that can significantly increase the production of shale gas and tight oil. Zipper fracturing has been shown to increase the production of unconventional resources by as much as 30% (Zhang et al., 2020b).

Another important advancement in hydraulic fracturing technology is the use of gel-free fracturing fluids. Traditional hydraulic fracturing fluids contain a gelling agent that helps transport proppant into the fractures in the rock. However, these gels can be expensive and difficult to handle. Gel-free fracturing fluids, on the other hand, are easier to handle and can reduce the cost of hydraulic fracturing by as much as 30% (Bai et al., 2018). These fluids use alternative transport mechanisms, such as foam or slickwater, to carry proppant into the fractures. Other advancements in hydraulic fracturing technology include the use of microseismic monitoring to

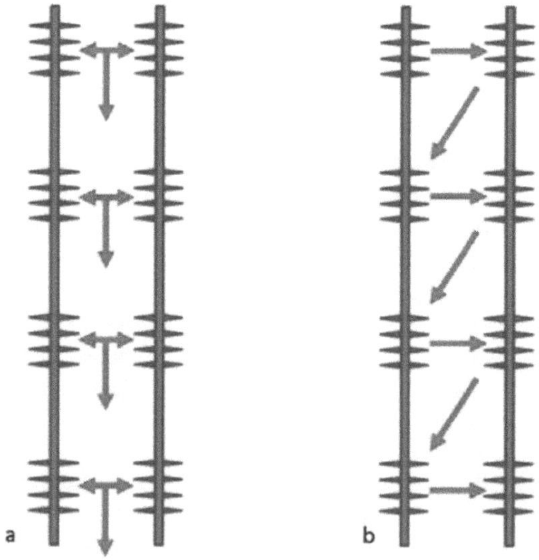

FIGURE 4.6 Simultaneous fracturing shale (a) versus zipper fracturing (b).

Source: Cuss et al., 2015

improve the accuracy of fracturing operations. Microseismic monitoring involves the placement of sensors in the ground near the wellbore to detect and map the fractures created during hydraulic fracturing. By analyzing this data, engineers can optimize the fracturing process to create more effective fractures and avoid creating fractures in unwanted locations. This technology has been used to improve the production of unconventional resources by as much as 15% (Maxwell et al., 2019).

Finally, the development of new proppant materials has also been an important advancement in hydraulic fracturing technology. Proppants are the solid particles that are placed in the fractures created during hydraulic fracturing to prop them open and allow oil or gas to flow through. New proppant materials, such as resin-coated sand and ceramic proppants, can increase the permeability of the fractures in the rock, leading to greater production of unconventional resources (Wang et al., 2018).

Overall, these recent advances in hydraulic fracturing technology have the potential to significantly improve the production of unconventional resources and reduce the cost of hydraulic fracturing operations.

4.3.1.1 Advancements in Fracturing Fluids

Traditional fracturing fluids consist of water, sand, and chemical additives. Recent advancements in fracturing fluids have focused on reducing water usage, increasing the effectiveness of the fracturing process, and reducing environmental impacts. Some of the recent advancements include the following:

Proppant-coated foam fluids: These fluids use a combination of proppants and foaming agents to create a lightweight, highly compressible fracturing fluid that can be pumped at low pressures. This technology can improve the efficiency of the fracturing process while reducing water usage and minimizing the risk of formation damage (Geng et al., 2020).

Self-breaking fracturing fluids: These fluids are designed to break down into harmless byproducts after they are pumped into the wellbore, reducing the environmental impact of fracturing operations. Self-breaking fluids can also reduce the risk of formation damage by minimizing the amount of residue left in the fracture (Chen et al., 2021).

Electrified fracturing fluids: Electrified fluids use a small electric current to control the flow of fracturing fluids and proppant, improving the accuracy of the fracturing process and reducing the risk of formation damage (Wang et al., 2021).

Alagoz and Sharma (2021) discuss the need to evaluate potential adverse interactions between shale and fracturing fluid additives, which can alter shale rock properties and lead to proppant embedment. The authors evaluate various frac fluids via a nuclear magnetic resonance (NMR) technique, which shows a strong correlation between the extent of proppant embedment and the extent of water imbibition into the shale as measured by NMR. The study also shows that certain fracturing fluids, such as a clay inhibitor with KCl, can significantly reduce water imbibition into shale, resulting in less embedment. This correlation between NMR-measured imbibition and proppant embedment can simplify the testing of different frac fluid additives for compatibility with shales, making this method very important.

4.3.1.2 Advancements in Proppants

Proppants are solid particles that are used to prop open the fractures created during hydraulic fracturing, allowing oil or gas to flow through. Recent advancements in proppant transport have focused on improving the effectiveness of the proppant and reducing the amount of proppant needed for fracturing operations. Some of the recent advancements include the following:

Dissolvable and degradable proppants: These proppants are designed to break down over time, reducing the risk of formation damage and minimizing the amount of proppant that needs to be produced (Zhang et al., 2020a).

Light-weight proppants: Light-weight proppants, such as resin-coated sand or ceramic beads, can be pumped at lower pressures and provide a more uniform proppant distribution in the fracture (Luo et al., 2018).

The experimental techniques developed by Alagoz et al. (28–30) are significant for understanding the mechanisms behind proppant embedment and decline in productivity in fractured wells. The unique methods that the authors developed to measure stress-dependent elastic and plastic deformation as well as time-dependent creep deformation in proppant embedment are crucial for improving field operations.

The authors' experimental setups used to tease apart the elastic, plastic, and creep deformation curves (presented in Figure 4.7) that are crucial for understanding the effects of proppant embedment in shale. However, the proppant pack in the field can be more complex than what can be replicated in the laboratory. Therefore, modifications of these techniques are necessary to study and tackle the real proppant pack problems that arise during field operations. In summary, experimental techniques (28–30) are vital for understanding the underlying mechanisms of proppant embedment in shale, and the setups used in their experiments are presented in Figure 4.7. However, further modifications of these methods are needed to tackle the more complex proppant pack problems encountered during field operations.

4.3.1.3 Advancements in Fracture Diagnostics

Fracture diagnostics are used to evaluate the effectiveness of the fracturing process and optimize future fracturing operations. Recent advancements in fracture diagnostics have focused on improving the accuracy and resolution of fracture mapping technologies. Recent advancements include the following:

Distributed acoustic sensing: Distributed acoustic sensing uses fiber optic cables to detect and locate microseismic events, providing detailed information about the location and extent of fractures created during the fracturing process (Bachari et al., 2019).

High-resolution microseismic imaging: High-resolution microseismic imaging uses advanced signal processing techniques to improve the resolution of microseismic data, allowing engineers to more accurately map the fractures created during the fracturing process (Ghorbani et al., 2020).

Downhole imaging: Downhole imaging technologies, such as X-ray CT scanners or acoustic imaging tools, can provide detailed information about the location and geometry of fractures, helping engineers optimize future fracturing operations (Ma et al., 2019).

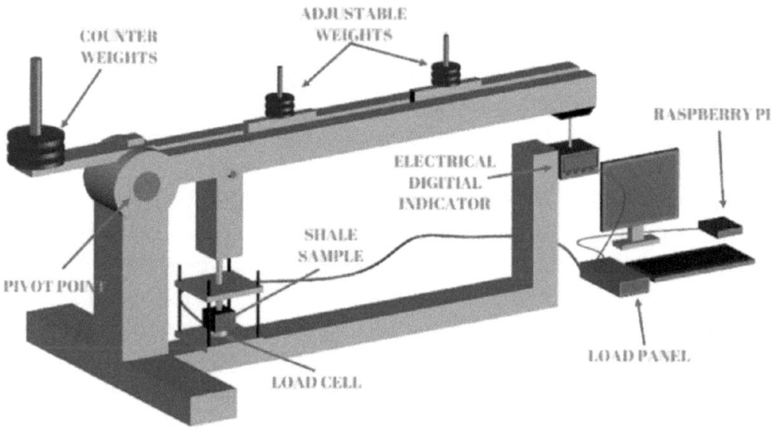

FIGURE 4.7 An AutoCAD view of the load-controlled test (not to scale).

Source: Alagoz & Sharma, 2021

4.3.2 Non-Hydraulic Fracturing Techniques

Non-hydraulic fracturing techniques have emerged as an alternative to conventional hydraulic fracturing to increase well productivity in unconventional reservoirs. These techniques aim to reduce the environmental footprint, cost, and operational challenges of hydraulic fracturing. This section provides an overview of the most commonly used non-hydraulic fracturing techniques, including electrically assisted hydraulic fracturing, pulse fracturing, and others.

4.3.2.1 Electrically Assisted Hydraulic Fracturing

Electrically assisted hydraulic fracturing (EAHF) is a non-hydraulic fracturing technique that uses electrical energy to enhance the fracturing process. EAHF involves the use of electrodes that are inserted into the formation to generate an electrical current. The electrical current heats the formation and reduces its viscosity, making it easier to fracture. EAHF has been shown to enhance the permeability of shale formations, increase fracture length, and reduce the amount of water required for fracturing.

According to a study by You et al. (2017), EAHF has been successfully applied in the Bakken formation, resulting in a 33% increase in well productivity compared to conventional hydraulic fracturing. Another study by Chen et al. (2017) reported that EAHF can significantly improve the production rate of tight gas reservoirs.

4.3.2.2 Pulse Fracturing

Pulse fracturing is a non-hydraulic fracturing technique that uses high-pressure gas to create a series of pressure pulses that fracture the formation. In pulse fracturing, a small amount of water is injected into the wellbore, followed by high-pressure gas that generates a series of pressure pulses. The pressure pulses create microfractures in the formation, which increase its permeability.

According to a study (Jia et al., 2019), pulse fracturing has been successfully applied in the Longmaxi formation in China, resulting in a significant increase in well productivity compared to conventional hydraulic fracturing. Another study (Zeng et al., 2020) reported that pulse fracturing can reduce water consumption by up to 90% compared to conventional hydraulic fracturing.

4.3.2.3 Other Non-Hydraulic Fracturing Techniques

Other non-hydraulic fracturing techniques include explosive fracturing, acid fracturing, and thermal fracturing. Explosive fracturing involves the use of explosives to create fractures in the formation. Acid fracturing involves the use of acid to dissolve the rock and create fractures. Thermal fracturing involves heating the formation to a high temperature to induce fracturing.

While these non-hydraulic fracturing techniques have shown promise in laboratory and field studies, they are not yet widely used in the industry due to various operational and technical challenges. Ongoing research and development in these areas are needed to improve the efficiency and sustainability of these techniques.

4.3.3 Reservoir Simulation and Modeling for Optimizing Stimulation Design

Reservoir simulation and modeling are important tools for optimizing stimulation design in oil and gas reservoirs. Simulation models can provide insights into the reservoir properties and the expected behavior of the reservoir under various conditions, such as different stimulation designs. These models can help identify the optimal stimulation design that can maximize the production of oil and gas while minimizing costs. One of the key benefits of reservoir simulation and modeling is the ability to test different stimulation scenarios before actual implementation. By using reservoir simulation models, engineers can test different hydraulic fracturing designs and predict their impact on the reservoir. This can help identify potential issues that may arise during the fracturing process and optimize the design to mitigate these issues.

Several studies have demonstrated the effectiveness of reservoir simulation and modeling in optimizing stimulation design. For example, a study by Salehi et al. (2019) used reservoir simulation to evaluate different hydraulic fracturing designs in a tight gas reservoir. The simulation results helped identify the optimal fracturing design that increased gas production while minimizing water usage.

Similarly, a study by Tang et al. (2020) used a reservoir simulation model to evaluate different fracturing designs in a shale gas reservoir. The simulation model was used to predict the impact of different parameters, such as the number of fracture stages and the spacing between them, on the well productivity. The results showed that an optimal fracturing design could be achieved by adjusting the fracture spacing and the number of stages.

In conclusion, reservoir simulation and modeling are essential tools for optimizing stimulation design in oil and gas reservoirs. These tools can help engineers predict the behavior of the reservoir under various conditions and optimize the fracturing design to maximize oil and gas production while minimizing costs and environmental impact.

4.4 COMPLETION ASPECTS FOR UNCONVENTIONAL RESOURCES

The completion of unconventional wells plays a crucial role in the development of these resources. Designing and optimizing completions for unconventional wells is necessary to ensure that the well is productive and economical. Completion components such as casing, tubing, and packers must be carefully selected and designed to withstand the high pressures and temperatures encountered in unconventional reservoirs. In particular, horizontal drilling and multistage hydraulic fracturing have become the norm in developing unconventional resources, allowing for the creation of long and highly stimulated intervals that maximize production. Wellbore integrity and well stimulation are equally important aspects of completion design, with the goal of achieving optimal reservoir contact and minimizing the interference of stress barriers. These completion aspects require careful consideration and planning to maximize the efficiency and economics of unconventional resource development.

4.4.1 Designing and Optimizing Completions for Unconventional Wells

Designing and optimizing completions for unconventional wells is a crucial aspect of unconventional resource development. The well completion design must be tailored to the specific reservoir and formation properties to maximize production. The key considerations for completion design include wellbore architecture, well trajectory, well spacing, completion type, number of stages, and frac fluid volumes (McClure & Mathews, 2016). There are several completion methods for unconventional wells, including plug-and-perf, sliding sleeve, and open hole (Figure 4.8). Researchers have investigated various factors that can impact the performance of completion designs, including the effect of stage spacing on well productivity, the impact of perforation clusters on fracture initiation, and the effect of completion type on reservoir drainage. Studies have shown that optimizing completion design can result in significant production increases. For example, a study by Anderson et al. (2017) found that increasing stage count and reducing stage spacing in plug-and-perf completions resulted in a significant increase in production in the Montney shale play. Similarly, a study by Gao et al. (2019) showed that using sliding sleeve completions with a longer open hole section and a larger number of frac stages increased well productivity in the Permian Basin.

4.4.2 Completion Components

The completion components of an unconventional well play a critical role in ensuring efficient production and preventing formation damage. In horizontal drilling, the casing is typically cemented in place to isolate the wellbore and prevent fluid migration between zones. The size and type of casing used can vary, depending on the specific well conditions and geological features. Tubing is then run inside the casing to transport the produced fluids to the surface. The selection of tubing size is based on the production rate and pressure, as well as the reservoir properties. Packers are used to isolate and seal off zones for selective stimulation and production. They can be either permanent or retrievable and can be set at various intervals along the wellbore.

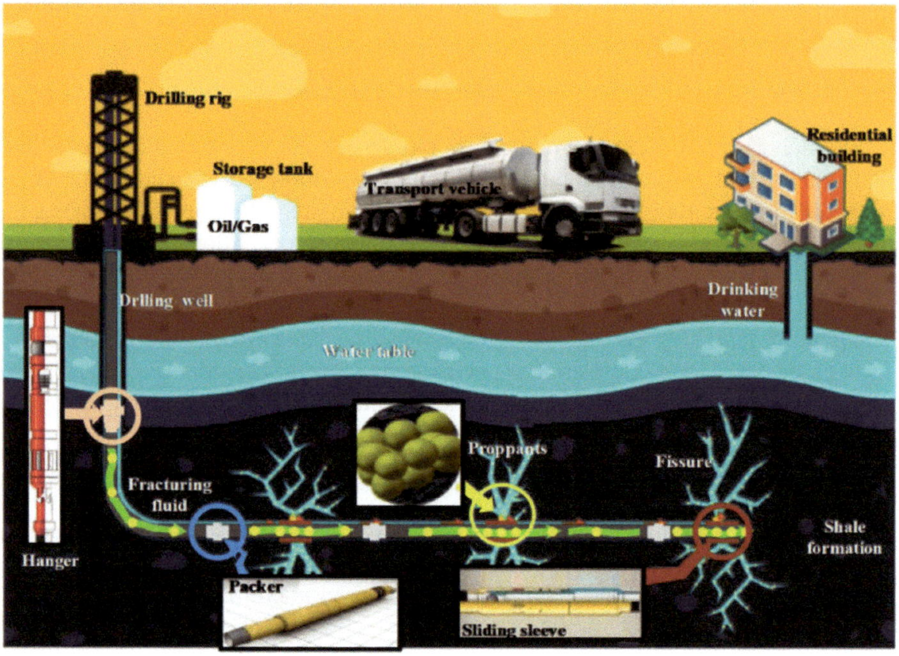

FIGURE 4.8 Sliding-sleeve completion.

Source: Guo et al., 2022

The use of specialized completion components can also enhance the performance of an unconventional well. For example, swellable packers can expand to create a tight seal against the casing, providing zonal isolation without the need for mechanical intervention (Figure 4.9). Open hole multistage fracturing systems allow for more efficient and precise stimulation by isolating individual stages and reducing the number of trips downhole. In addition, the use of dissolvable materials, such as plugs and balls, can eliminate the need for milling and reduce costs associated with post-fracturing operations.

Several studies have been conducted on the design and optimization of completion components for unconventional wells. Kuru et al. (2018) investigated the use of swellable packers in a tight gas reservoir and found that they provided effective zonal isolation and enhanced well productivity. Zhang et al. (2019a) examined the performance of dissolvable materials in hydraulic fracturing operations and found they could improve the efficiency of plug removal and reduce costs. In another study (Alklih et al., 2021), the use of open-hole multistage fracturing systems was shown to increase well productivity and decrease stimulation costs in a shale gas reservoir.

4.4.3 Horizontal Drilling and Multistage Hydraulic Fracturing

Horizontal drilling is a key aspect of completing unconventional wells. It involves drilling a wellbore horizontally through a reservoir, which increases the contact area between the wellbore and the formation, allowing for more efficient recovery of hydrocarbons (Figure 4.10).

Multistage hydraulic fracturing is also an important aspect of completion for unconventional wells, as it allows for the creation of multiple fractures along the

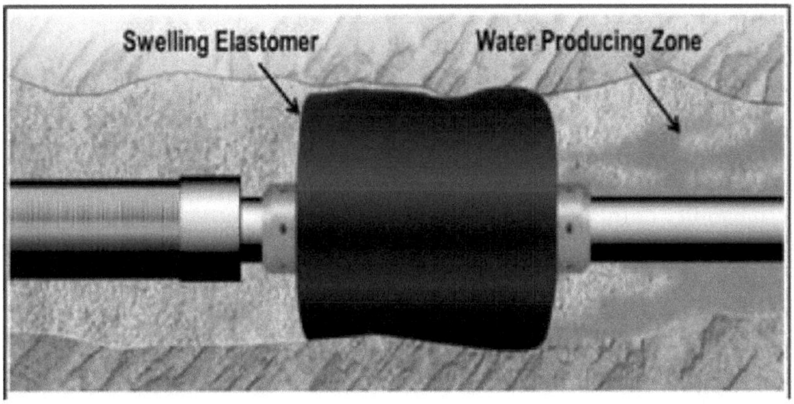

FIGURE 4.9 Swellable packers.

Source: Kuru et al., 2018

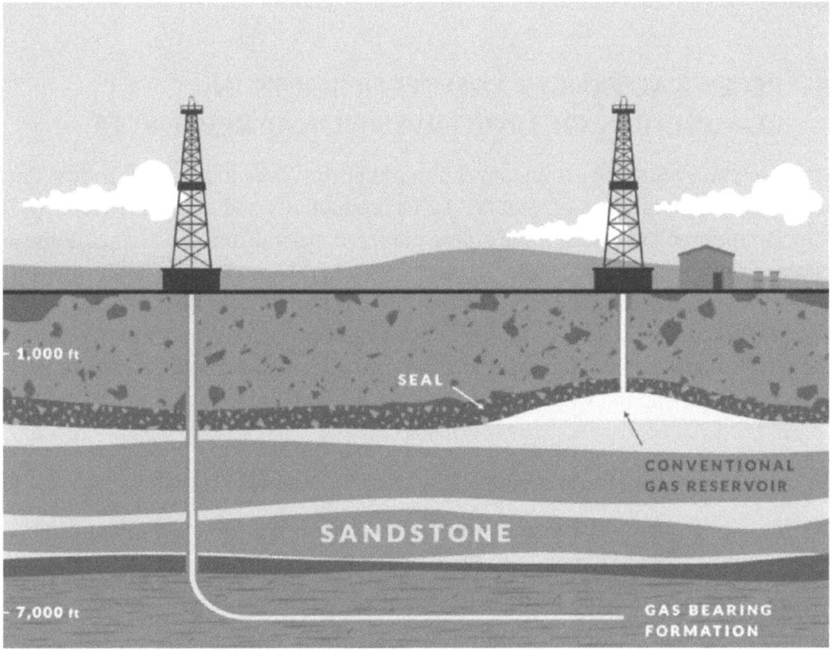

FIGURE 4.10 Horizontal versus vertical drilling.

Source: Rig Zone News, 2015

wellbore, which increases the amount of oil and gas that can be produced. The design of multistage hydraulic fracturing treatments involves the optimization of various parameters, including the number of stages, the spacing between stages, the size of the fractures, and the amount of proppant used.

Researchers have conducted numerous studies on the optimization of multistage hydraulic fracturing for unconventional resources. For instance, Li et al. (2018) investigated the effect of fracture spacing on production for horizontal wells in tight oil reservoirs. They found that smaller fracture spacing generally resulted in higher production rates, but the effect of fracture spacing on production was dependent on the rock properties of the reservoir. Similarly, Zhan et al. (2019) studied the impact of fracture spacing on the recovery of tight gas from a horizontal well. They found that the optimal fracture spacing for maximizing gas recovery varied depending on the permeability of the reservoir.

Other researchers have focused on the optimization of proppant placement in multistage hydraulic fracturing. Li et al. (2019) investigated the impact of proppant concentration on the conductivity of hydraulic fractures, finding that increasing proppant concentration resulted in greater fracture conductivity. In addition, Zhang et al. (2018) developed a numerical model to simulate proppant transport and placement during hydraulic fracturing treatments, allowing for the optimization of proppant placement for maximum production.

Overall, the design and optimization of horizontal drilling and multistage hydraulic fracturing are crucial aspects of completing unconventional wells, and research continues to explore the most effective techniques for enhancing production and recovery rates.

4.5 RECENT ADVANCES AND TECHNIQUES IN COMPLETION OF UNCONVENTIONAL RESOURCES

The completion phase of an unconventional well involves a series of critical decisions and actions that significantly affect the well's productivity and economic viability. Recent advances in completion techniques have played a crucial role in enhancing the overall success of unconventional resource development. In this section, we will discuss some of the latest techniques and advancements in completion technology, including multistage completion technology, new completion fluids and proppants, and wellbore monitoring and control. These advancements are helping unlock the vast potential of unconventional resources and paving the way for a brighter future for the oil and gas industry.

4.5.1 Advancements in Multistage Completion Technology

The multistage hydraulic fracturing technique is the cornerstone of unconventional resource development as it enables efficient stimulation of a large number of zones along the wellbore. The use of multistage completion technology has significantly increased in recent years due to its potential to enhance well performance and increase ultimate recovery. In this section, we will discuss some of the latest advancements in multistage completion technology that are revolutionizing the way we complete unconventional wells.

One of the latest advancements in multistage completion technology is the use of dissolvable fracturing balls (Tang et al., 2019). These balls are made of a dissolvable material that can be pumped downhole along with the proppant and fracturing fluid. Once the fracturing operation is complete, the balls dissolve, eliminating the need for costly and time-consuming milling operations to remove them. This technique not only saves time and money but also reduces the risk of damaging the formation during milling operations. The use of dissolvable fracturing balls has proven successful in several field trials, and its adoption is rapidly increasing in the industry.

Another advancement in multistage completion technology is the use of diverting agents (Hsieh et al., 2016) that help channel the fracturing fluid and proppant into unstimulated zones, increasing the efficiency of the stimulation process. Diverting agents such as fiber and degradable materials have been widely used in the industry, with promising results.

Additionally, advancements in perforating technology have enabled better communication between the wellbore and the reservoir, allowing for more effective stimulation of the targeted zones. New perforating gun systems that use shaped charges and electronic triggering systems have been developed, providing more precise and consistent perforations.

Furthermore, advances in modeling and simulation techniques have helped optimize the design of multistage completions, leading to more efficient and effective stimulation of the reservoir. These techniques (Yu et al., 2020) enable engineers to predict the fracture propagation and distribution and optimize the number, size, and spacing of the fracturing stages, resulting in better well performance.

Moreover, the use of smart completions, which allow real-time monitoring and control of the wellbore, has significantly improved the management of multistage completions. Smart completions (Zhao et al., 2021) enable the monitoring of reservoir parameters such as pressure, temperature, and flow rates, providing valuable data for reservoir management and optimization.

4.5.2 New Completion Fluids and Proppants

As the demand for unconventional resources increases, the development of new completion fluids and proppants has become essential to improve the overall efficiency and effectiveness of hydraulic fracturing. The selection of suitable completion fluids and proppants is crucial for optimizing fracture conductivity, reducing formation damage, and improving production rates. In recent years, several advancements have been made in the development of new completion fluids and proppants that are designed to address the specific challenges associated with unconventional reservoirs.

One significant advancement in completion fluids is the use of viscoelastic surfactant (VES) fluids. VES fluids have been shown to offer superior proppant-carrying capacity, better fracture cleanup, and reduced formation damage compared to conventional water-based fluids. Additionally, VES fluids have a lower friction pressure, which can lead to less pumping horsepower requirements and overall cost savings. The use of VES fluids has been successful in several field applications, including the Barnett shale and Eagle Ford shale formations.

Another advancement in completion fluids is the development of microemulsion-based fluids. These have been shown to provide improved proppant transport, reduced formation damage, and better fluid recovery compared to conventional water-based fluids. Microemulsion-based fluids have also been successful in reducing water usage in hydraulic fracturing operations as they can be formulated with brine instead of fresh water (Luo et al., 2018).

In addition to advancements in completion fluids, there have also been developments in proppant technology. One such advancement is the use of coated proppants. Coated proppants have a thin layer of coating material applied to their surface, which can improve their bonding to the formation and reduce proppant flowback. Coated proppants have been successfully used in several field applications, including the Haynesville shale and Marcellus shale formations (Luo et al., 2018).

Another advancement is the use of engineered proppants. These are designed to have specific size, shape, and surface properties to optimize the proppant pack conductivity and reduce the risk of proppant crushing. Engineered proppants have been shown to improve well productivity and reduce the amount of proppant required per well (Mayerhofer et al., 2006).

Overall, the development of new completion fluids and proppants has the potential to significantly improve the efficiency and effectiveness of hydraulic fracturing in unconventional reservoirs. As the industry continues to focus on optimizing completion designs, these advancements will play a crucial role in maximizing production rates and enhancing recovery.

4.5.3 Advances in Wellbore Monitoring and Control

Advances in wellbore monitoring and control technologies have significantly improved the safety and efficiency of oil and gas drilling operations. Monitoring the pressure, temperature, and other critical parameters in the wellbore is crucial to ensure the integrity of the well and to prevent costly accidents. In recent years, several technologies have been developed to improve wellbore monitoring and control, including fiber-optic sensors, downhole pressure gauges, and intelligent completions.

Fiber-optic sensors have been widely used in wellbore monitoring due to their high sensitivity, accuracy, and reliability. They can measure various parameters, such as temperature, pressure, strain, and acoustic signals, and provide real-time data for decision-making (Sun et al., 2017). Downhole pressure gauges are another crucial technology that enables continuous pressure monitoring in the wellbore. They can be used to detect changes in formation pressure and help prevent overpressuring of the wellbore, which can cause well blowouts (Wang et al., 2016).

Overall, these advances in wellbore monitoring and control technologies have improved the safety and efficiency of oil and gas drilling operations. By providing real-time data and enabling remote control, these technologies have reduced the risk of accidents and minimized downtime, resulting in significant cost savings for operators.

4.6 CASE STUDIES AND FUTURE OUTLOOK

This section presents several case studies and a future outlook in the field of oil and gas production. These case studies cover a range of topics, including the application of a stimulation workflow in a deep-water carbonate field, a training program for acid stimulation, an examination of fracture growth in different types of shale, an assessment of a novel hydraulic fracturing fluid for combined scale inhibition and hydraulic fracturing treatments, and a comparison of cemented plug-and-perf with open-hole ball-drop completion assemblies. These case studies offer insights into various aspects of oil and gas production and can serve as valuable references for those working in the industry. The future outlook section provides some thoughts on the direction of research and development in the field, highlighting potential areas for innovation and growth. Overall, this section aims to provide a comprehensive view of the latest advancements and trends in the field of oil and gas production.

4.6.1 Application of a Stimulation Workflow in the Deep-Water LAPA Pre-Salt Carbonate Field

LAPA Field is a pre-salt deep-water reservoir located in offshore Brazil. With water depths around 2,100 meters and long intervals varying from 100 to 250 meters, this field required a complex stimulation design. Factors that must be considered include permeability contrast of the pay zones, the need for mechanical or chemical diversion, and matrix acidizing method. Since no well stimulation vessel (WSV) was available, a vessel from an operator fleet—a multi-purpose field support vessel (FSV) equipped with ROV—was adapted for this operation.

4.6.1.1 Challenges

1. Completion challenges for effective stimulation: Reservoir was completed in an 8½-inch open hole with perforated liner. Stimulation with diesel emulsified acid (DEA) was done in three or four stages while chemical diversion was employed through a wash pipe with slick joints.
2. Stimulation fluids challenges: Stimulation fluid must be compatible with formation material and completion equipment. Additionally, it should propagate the wormholes to achieve the desired invasion ratio while preventing face dissolution. The stimulation fluid design for LAPA Field also included a corrosion inhibitor, a mud cake breaker, and a pipe pickle.
3. Pumping equipment challenges: Since the LAPA Field is located offshore, logistic challenges arise, including equipment availability and space limitation. The three layout options considered were:
   - Well stimulation vessel: eliminated due to high investment
   - Temporary well stimulation vessel: utilized for this operation
   - Rig-based plant: eliminated .due to space limitation

4.6.1.2 Completion Design

The selected LAPA Field wells had an 8½-inch open hole, and annular isolation was done with swell packers on three to four sections segmented based on the permeability profiles from the NMR log and segment length (40–70 meters).

The streamlined lower completion design allowed for safe and efficient well production prior to running upper completion. One service tool was used to perform two jobs (set the packer and perform the massive acid job) thus avoiding an extra run to

TABLE 4.1
The Pros and Cons of Each Layout Option

Aspect	Well Stimulation Vessel	Temporary Vessel	Rig-Based Plant
Pros	- High fluid volume - High hydraulic horsepower (HHP) - Can perform massive operations - A few persons onboard rig - Reduced rig time - Does not use rig space	- Multipurpose vessel - High fluid volume - High HHP - Can perform massive operations - A few persons onboard rig - Reduced rig time - Does not use rig space	- No need for dedicated supply vessel - No need for high- pressure connection hose hanger on rig - Faster mobilization \
Cons	- High capital expenditure - For long-term projects only - Need a high-pressure connection hose hanger on rig	- Need a dedicated supply vessel - Imposes logistics constraints - Need a high-pressure connection hose hanger on rig	- Low fluid volume - Increased rig time - HS&E concerns (spill & heavy lift) - Crane limitations - Many persons onboard rig - Some rigs require Zone 2 explosion-proof equipment

Source: Pedrosa et al., 2021

pull out of hole (POOH) the service tool prior to conducting the acid job. It was also modified to allow a 30-bpm pumping rate with minimum friction.

The wash pipe down to open hole toe included slick joint assemblies to seal on inverted seal subs positioned on perforated liner and provide mechanical diversion. Radio frequency identification remote-operated reservoir isolation device (RFID RO-RID) isolation valves were used to isolate the reservoir prior to upper completion installation after the conclusion of acid stimulation.

4.6.1.3 Stimulation Design

The biggest challenges for the stimulation design were the large acid volumes required and the need for diversion. The selected stimulation fluid was diesel emulsified acid (DEA), and placement was done by alternating stages of diversion by bull heading from the surface and injected into the open hole through a wash pipe with slick joints and perforated pup joints.

Single core flood testing was done with each stimulation fluid to test the worm holing efficiency and fluid incompatibility.

Laboratory tests: Tests conducted include compatibility, sludge creation, emulsion dissolution, iron precipitation, damage removal, performance, corrosion tests, etc.

Core flow test: Core flow tests were used to identify the optimum injection rate and acid concentration and to confirm fluid behavior. For this study, two acid concentrations (15% and 20%) and two plugs with different porosities (12.4% and 16%) were tested. The acid was injected until the breakthrough was observed either by pressure behavior or pH measured at the effluent side, then the plug was flushed with brine.

Results: See Figure 4.11: (A) Start pumping acid; (B) Viscous fluid contacts plug face causing pressure peak; C) Differential pressure drops due to acid action; D) Breakthrough observed and differential pressure gets close to zero; E) Brine flow and bleeding of differential pressure.

Both acid concentrations had the same volumetric breakthrough, hence 5% acid concentration difference will not affect wormhole creation capacity. Figure 4.12 demonstrates how emulsified acids present low-face dissolution.

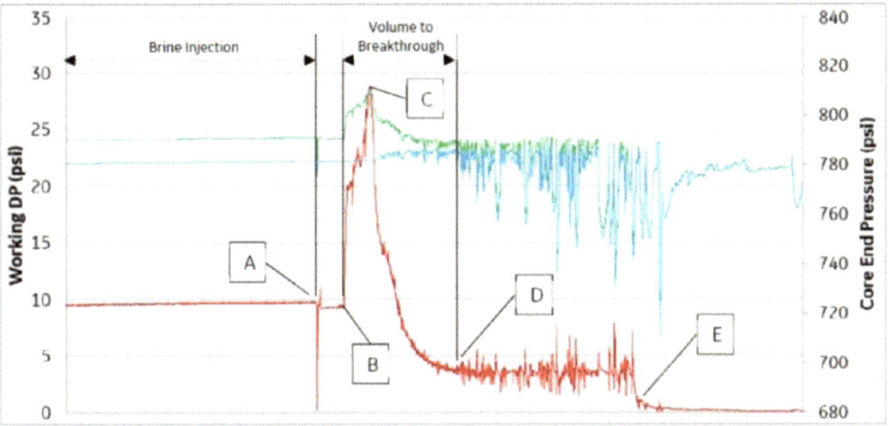

FIGURE 4.11 Emulsified acid test.

Source: Pedrosa et al., 2021

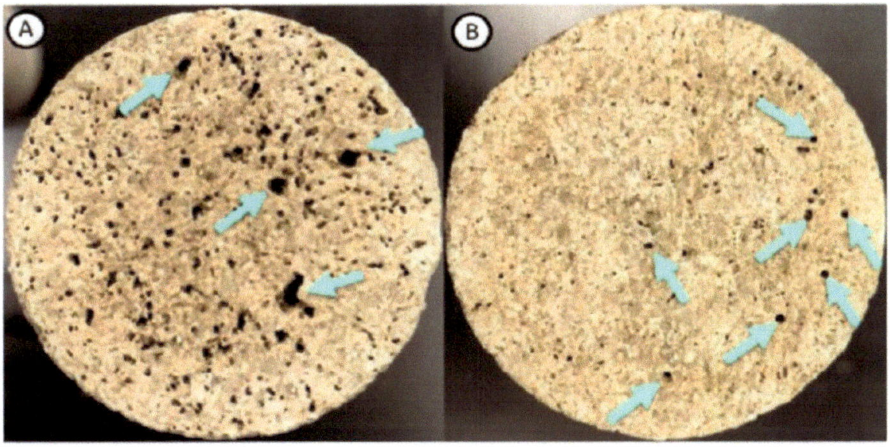

FIGURE 4.12 Plug pictures; inlet face is (A) and outlet face is (B).

Source: Pedrosa et al., 2021

Dual core flow test: This test was done to compare the viscoelastic, polymeric, and emulsified acid systems by analyzing the results of wormhole formation. The higher the vuggy fraction of porosity, the lower the pore volumes to breakthrough.

Results: Emulsified acid was the best system out of the three tested for the following reasons:

- Delay in reaction with carbonate
- Formation of extensive wormhole (Figure 4.13).

Software modeling and pumping schedule:
Software assumptions:

- Carbonate rock
- Steady-state models
- Incompressible fluids with lab-modeled pipe friction
- Matrix flow
- Vertical or horizontal IPR
- Wormhole model based on core flow measurements

Software outputs:

- Acid distribution in each zone or formation layer
- Wormhole propagation
- Skin factor as a function of time and well depth
- Injection rate, surface treating pressure, and bottom hole treating pressure as a function of time

4.6.1.4 Temporary Vessel Design

The vessel chosen for this project was a field support vessel (FSV) equipped with ROV. The onboard stimulation plant was distributed into the basic sub-areas: the contained area, the partial demobilization area, and the mezzanine deck.

Stimulation plant main equipment: See Figure 4.14.

- 1, 2, 3, and 4 are high pressure pumps.
- 5, 6, 7, and 8 are hydrochloric acid tanks.
- 9 is an acetic acid tank.
- 10 is a solvent tank.
- 11 is a batch mixer with electrical agitators.
- 12 is an on-the-fly mixing system.
- 13 is a blender and silo/
- 14 are buffer tanks, water, and brine.
- 15 is a high- pressure hose reel.
- 16 is the control cabin and chemical laboratory.
- 17 is the liquid additive tanks.
- 18 is the electronics control module.
- 19 is the liquid additive pumps.

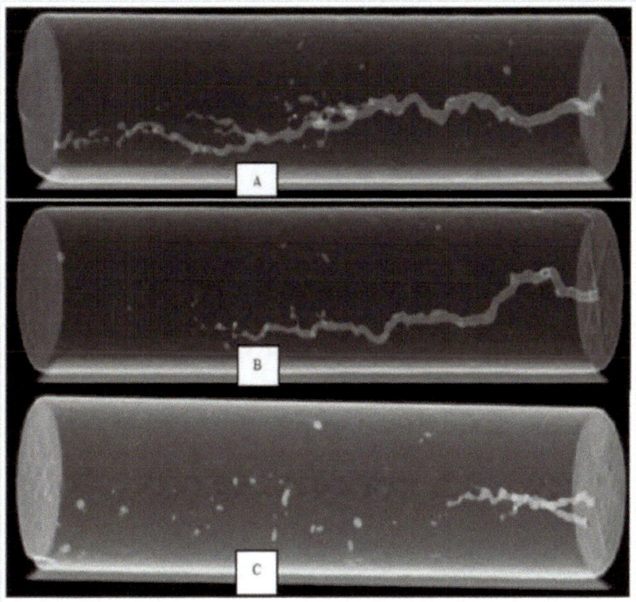

FIGURE 4.13 Emulsified acid (A), viscoelastic acid (B), and polymeric system (C).

Source: Pedrosa et al., 2021

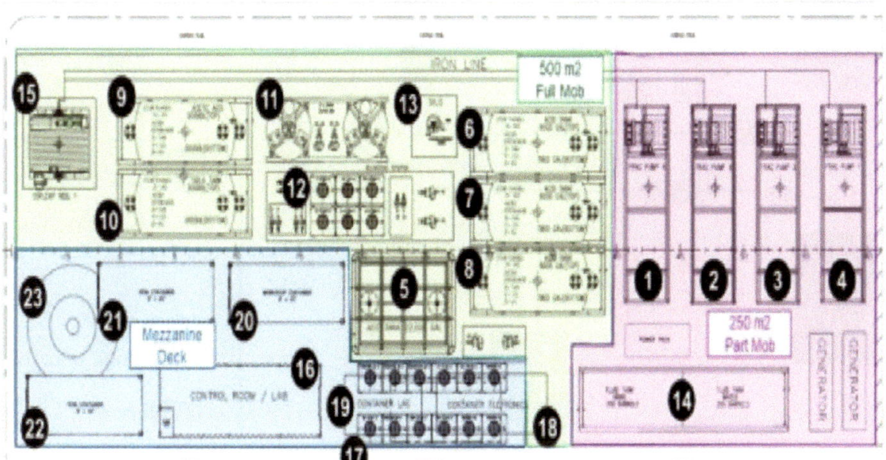

FIGURE 4.14 Schematic of stimulation plant.

Source: Pedrosa et al., 2021

- 20 is the workshop container.
- 21 and 22 are high-pressure iron containers.
- 23 is the transfer basket landing area.

Basic process flow: See Figure 4.15.

4.6.1.5 Execution

The design summarized earlier was used to successfully stimulate four LAPA Field wells; see Table 4.2.

Stimulation treatment performed in a production well: The job sequence is pipe pickle, mud cake breaker (MCB), and lastly, matrix acidizing (three stages). The stimulation fluids are summarized in Table 4.3, and the treatment results are shown in Figure 4.16.

4.6.2 Training Program for Acid Stimulation

Intervention operations, including light, heavy, mechanical, and hydraulic interventions, are regularly conducted on production and water-injection wells in order to diagnose downhole conditions or conduct mechanical repairs. A separate category of intervention operations with a lower frequency involves mainly stimulation.

In order to improve the results of a stimulation job, improved candidate selection, damage identification, and best practices must be implemented.

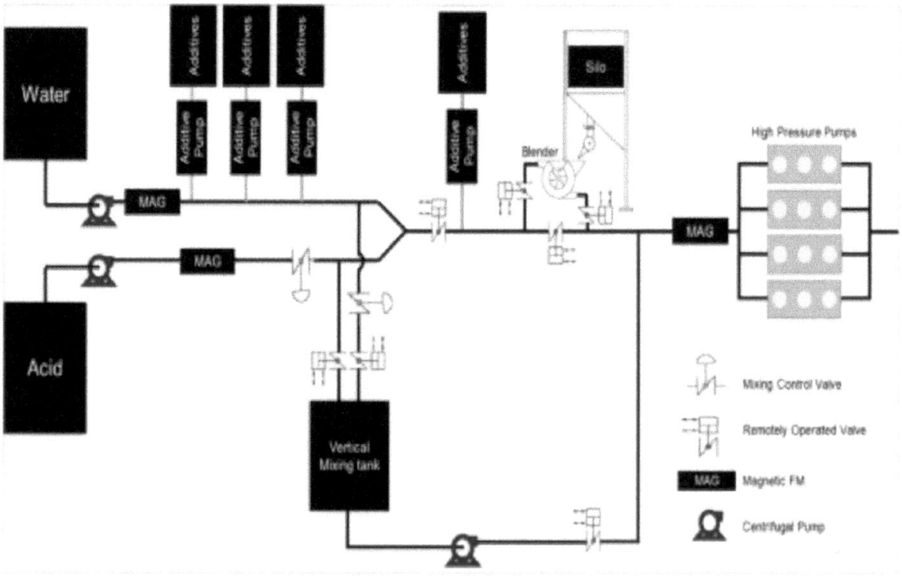

FIGURE 4.15 Basic flow process of stimulation plant.

Source: Pedrosa et al., 2021

TABLE 4.2
Well and Formation Data Set

	Well 01					Well 02					Well 03					Well 04				
		Top (m)		Bottom (m)			Top (m)		Bottom (m)			Top (m)		Bottom (m)			Top (m)		Bottom (m)	
		TVD	MD	TVD	MD		TVD	MD	TVD	MD		TVD	MD	TVD	MD		TVD	MD	TVD	MD
Zones	Zone 4	5143	5675	5190	5738	Zone 2	5216	5667	5262	5731	Zone 3	5162	5307	5203	5351	Zone 2	5214	5504	5248	5544
	Zone 3	5190	5738	5222	5780	Zone 1	5262	5731	5317	5804	Zone 2	5203	5351	5244	5402	Zone 1	5248	5544	5307	5614
	Zone 2	5222	5780	5253	5822						Zone 1	5244	5402	5312	5469					
	Zone 1	5253	5822	5327	5S96															

	Well 01	Well 02	Well 03	Well 04
Formation Fluid	Oil	Oil	Oil	Oil
BHST	72°C (0.025 °C/m)	72°C (0.025 °C/m)	72°C (0.025 °C/m)	72°C (0.025 °C/m)
BHSP	6,682 psi (1.25 psi/m)	6,454 psi (1.22 psi/m)	6,454 psi (1.22 psi/m)	6,454 psi (1.22 psi/m)
Fracture Gradient	10,600 psi (0.57 psi/ft)	10,600 psi (0.57 psi/ft)	10,600 psi (0.57 psi/ft)	10.600 psi (0.57 psi/ft)
Workstring	6 5/8" DP 41.13#/5.5" DP 24.7#/5.5" HWDP 65.14#	6 5/8" DP 41.13"/5.5" DP 24.7#/5.5" HWDP 65.14#	6 5/8" DP 41.13#/5.5" DP 24.7#/5.5" HWDP 65.14#	6 5/8" DP 41.13#/5.5" DP 24.7#/5.5" HWDP 65.14#
Wash Pipes	Perforated Wash Pipe 4" Hydrill 511	Perforated Wash Pipe 4" Hydrill 511	Perforated Wash Pipe 4" Hydrill 511	Perforated Wash Pipe 4" Hydrill 511
Liner Data	OD 7.297" * ID 5.791" (perforated)	OD 7.297" * ID 5.791" (perforated)	OD 7.297" * ID 5.791" (perforated)	OD 7.297" * ID 5.791" (perforated)
Open Hole Diameter	8.5"	8.5"	8.5"	8.5"
Open Hole Top	5,657 mMD	5,653 mMD	5,307 mMD	5,504 mMD
Open Hole Bottom	5,896 mMD	5,804 mMD	5,469 mMD	5,614 mMD
Open Hole Length	239 m	151 m	162 m	110 m

Source: Pedrosa et al., 2021

TABLE 4.3
Stage 1 Stimulation Fluids

Step	Fluid	Volume (bbl)
1	15% HCI System	70
2	Polymer-Based Acid System	100
3	20% HCI Emulsified System	260
4	Polymer-Based Acid System	100
5	15% HCI System	50
6	20% HCI Emulsified System	260
7	Polymer-Based Acid System	100
8	20% HCI Emulsified System	300
9	Displacement Brine	500

Source: Pedrosa et al., 2021

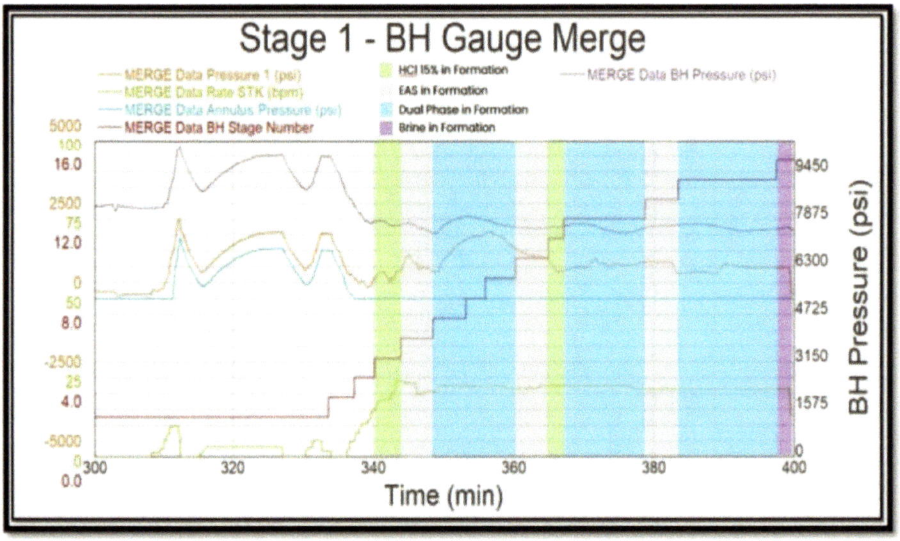

FIGURE 4.16 Stage 1 treatment graph.

Source: Pedrosa et al., 2021

4.6.2.1 Development of Simplified Workflows

Workflows and decision trees were developed to guide technical decisions (Figure 4.17). For example, student input was requested on whether the damage was acid soluble or insoluble. Then, based on the damage solubility, a damage mechanism could be selected, including an organic scale, water block, and barium or calcium sulfate scale.

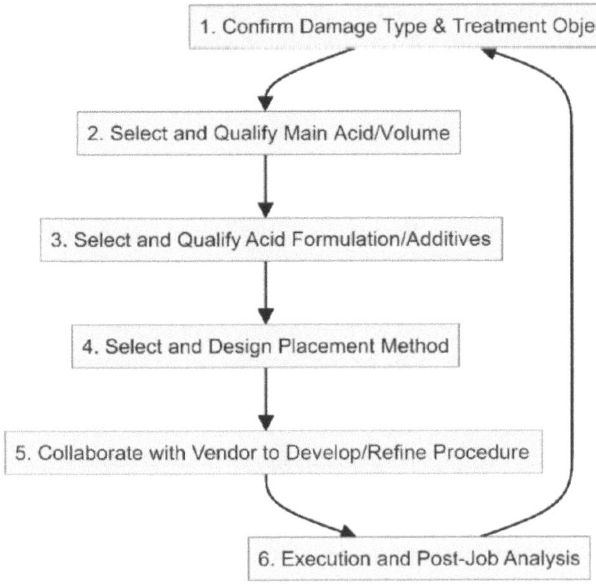

FIGURE 4.17 Practical stimulation design workflow.

This eased the process of appropriate fluid selection. After the optimal fluid system was identified, best practices were included for each selection.

4.6.2.2 Development of Improved Stimulation Design Parameters

The stimulation training module included instruction on selecting and designing the most appropriate stimulation fluids to dissolve damage, with special consideration for reservoir sensitivity. Downhole challenges, such as complicated completions and sand control completions, were also included in the acidizing fluid selection based on an understanding of the damage type associated.

4.6.2.3 Development of Operations/Practical Training Curriculum

After identifying damage mechanisms, the students' next step was to develop a reactive stimulation design that included matrix stimulation, scale soak, or wellbore cleanout. Each team presented their proposed stimulation and detailed best practices and all the additional data that would be needed to justify their selection. This was then peer reviewed and underwent a series of questions to discuss suitability and economic feasibility. The addition of practical exercises ensured that all the skills were successfully developed by the students with a practical and operational mindset.

4.6.3 Examining Fracture Growth in Low and High Brittle Shale

Khan et al. (2022) discusses the need for advancements in the technique to estimate the hydraulic fracturing potential of fracturing fluid in different shale formations.

The study aims to explore the fracture propagation and fracture initiation pressure under high confinement stress by performing hydraulic fracturing experiments on three different fields using different fracturing fluids: SC CO2, SC CO2 foam, and slick water. The study results show that SC CO2 foam injection produced complex fracture networks with high aperture and length, which can potentially enhance the stimulated reservoir volume.

In this experimental work, Eagle Ford, Wolfcamp, and Mancos shale cores were used. The cores were 38 mm in diameter and 176 mm in length and were laterally cut into two parts. One part was used for mineralogical and geomechanical characterization while the other part was used in hydraulic fracturing experiments. The samples for the experiments were prepared with perforations at angles of 0°, 45°, 60°, and 90° along the bedding plane. The sides of the samples were smoothed and straightened throughout the length with a lath machine, and several 'O' rings and epoxy resin were used to seal the nozzle in the perforation. A 10-mm-long hole inside the specimens was used as a fracture section, and precautions were taken during bonding to ensure connectivity between the bottom of the hole at the nozzle and the shale surface.

They described a hydraulic fracturing system that has various components that can be classified into four main parts: axial load, confinement flow line, surfactant, and CO_2 pumping section (Figure 4.18).

The system includes a fluid flow line with pumps, flow meters, pressure gauges, a flow control valve, and temperature meters, with a pressure range up to 9,000 psi. The system also includes a booster pump for CO_2 injection and a positive displacement pump for liquid injection with a pressure range up to 10,000 psi. The simulated cell of the system is made of high-strength stainless steel with dimensions of 45 mm (diameter) and 110 mm (length). The system is also equipped with a digital data-gathering system (Figure 4.19) and a SCADA system to collect the required dataset continuously.

They summarize the results as follows:

1. The effect of bedding angle on the pressure required to initiate fractures and the growth of those fractures in shale with varying levels of brittleness. Hydraulic fracturing was conducted on Wolfcamp shale at different bedding angles to observe the pressure and direction of fracture growth, as well as the overall fracture network. Results show that fractures occur primarily along the direction of the perforation, regardless of the bedding plane, but the pressure required to initiate fractures is strongly influenced by the bedding angle, with lower pressure at a 0° angle and higher pressure at higher angles (Figure 4.20).
2. The impact of bedding angle on the pressure required to initiate fractures and the growth of those fractures in the Eagle Ford shale at a constant perforation length. Results show that a high pressure is required to initiate fractures when using foam fracturing on the Eagle Ford shale. The study also found that when the bedding angle is 0°, fractures occur along the laminations, and when the bedding angle is 45°, the main propagation of the fracture is along the bedding plane (Figure 4.21).
3. The study also examines the impact of perforation length on the pressure required to initiate fractures and the overall fracture network in the

FIGURE 4.18 Hydraulic fracturing system: (a) front view, (b) Core holder with thrust pump and vertical loading cell, (c) confinement section, (d) pumping section for injection mediums.

Source: Khan et al., 2022

Mancos shale with bedding parallel to the perforation. Results show that a significant increase in the pressure required to initiate fractures is recorded when using short perforation length. However, this increase in pressure also results in a denser fracture network, as illustrated in Figure 4.22 (b).

4. The study examines the impact of slick water, supercritical carbon dioxide (SC CO_2), and CO2 foam on fracture growth and the pressure required to initiate fractures in shale with different levels of brittleness, under high confinement pressure. The results, presented in Figure 4.23, show that water

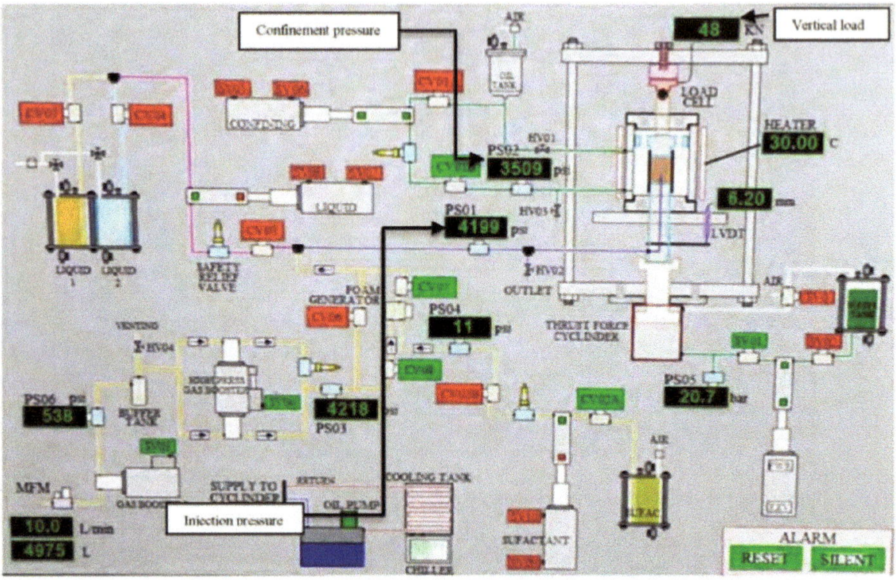

FIGURE 4.19 Hydraulic fracturing system flow diagram.

Source: Khan et al., 2022

fracturing creates small fractures that follow the laminations, and SC CO_2 only creates a single fracture in the direction of lamination. However, the use of SC CO_2 resulted in a significant decrease in the pressure required to initiate fractures. Using SC CO_2 foam resulted in a dense fracture network in different types of shale. The study shows that SC CO_2 foam is more effective than slick water and SC CO_2 alone in terms of imbibition rate and fracture propagation in low and high brittle shale under high confinement pressure.

5. The effect of foam fracturing on the Wolfcamp, Eagle Ford, and Mancos shale at different bedding angles was also studied. The results, presented in Figure 4.24, show that an open and dense fracture network is achieved in all types of shale using foam fracturing. The study also compares the pressure required to initiate fractures using slick water, supercritical carbon dioxide (SC CO_2), and SC CO_2 foam. Results show that water fracturing has a high fracture initiation pressure, SC CO_2 foam also has high fracture initiation pressure, but the injection of SC CO_2 only results in a significant drop in fracture initiation pressure. Overall, the study shows that a dense fracture network is achieved in different types of shale using SC CO_2 foam, and it can significantly increase the stimulated reservoir volume. Also, when using slick water, a consistent increase in the injection pressure is recorded while sudden fluctuations are recorded with the injection of SC CO_2. With the injection of SC CO_2, a sudden decrease in the injection pressure occurs due to the flow of SC CO_2 in the natural fracture.

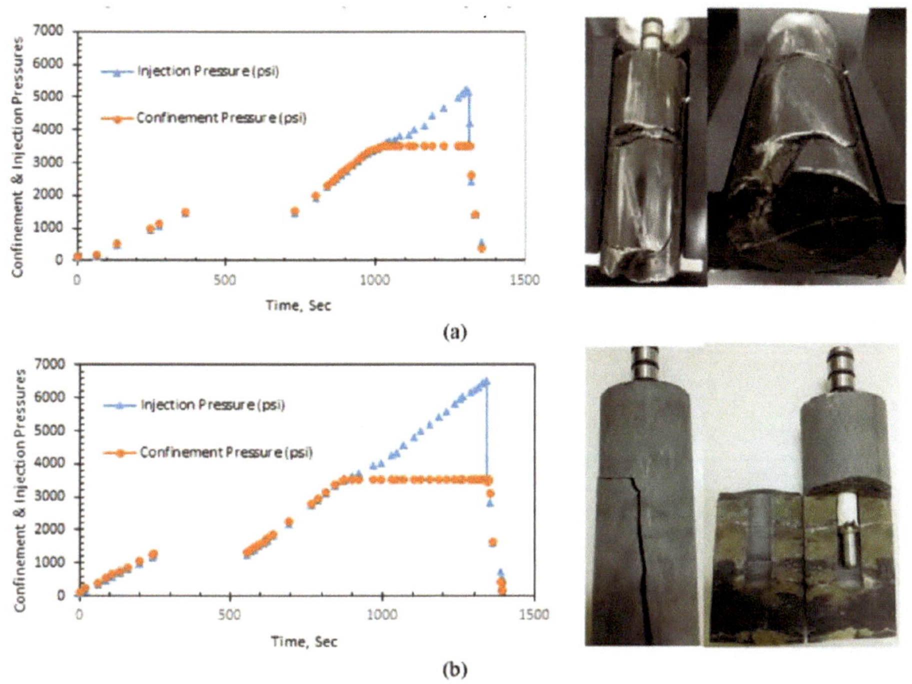

FIGURE 4.20 (a) Shale with bedding angle 0° and perforation length of 60 mm—fracture pressure is 5,239 psi; (b) Shale with bedding angle 60° and perforation length of 70 mm.

Source: Khan et al., 2022

4.6.4 Assessment of a Novel Hydraulic Fracturing Fluid for Combined Scale Inhibition and Hydraulic Fracturing Treatments: Three-Year Results From Russia

Levanyuk et al. (Ias, 2018, February 28) discuss the common problem of scale deposits in oil and gas wells, which can have detrimental effects on well production. Scale can form anywhere in the well production system (see Figure 4.25), including the formation, perforations, casing or tubular, and in or on the artificial lift equipment. Depending on the severity, scaling can stop production entirely. The authors describe several chemical and mechanical methods for removing scale deposits, but they note that the only solution to prevent scale deposition is through the injection of chemical inhibitors into the formation.

The authors (Levanyuk et al., 2012) then present approaches for improving the compatibility of a novel hydraulic fracturing fluid with a scale inhibitor to increase the average run life of electric submersible pumps (ESPs) in the Mayskoe and Snezhnoe oilfields. In these fields, the typical production system includes artificially lifted, stimulated wells placed in reservoirs where pressure maintenance is achieved by water flooding. Artificial lifting is typically accomplished through the use of ESPs. In reservoirs where produced fluids exhibit scaling tendencies, the life of the

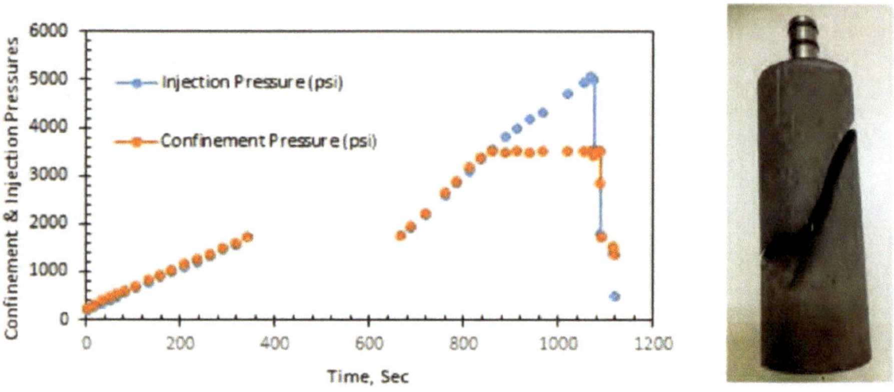

(a)

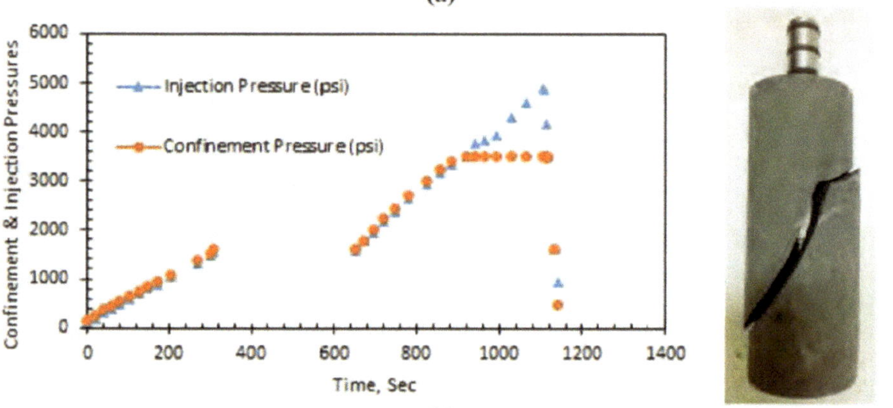

(b)

FIGURE 4.21 Fracture initiation pressure and fracture growth in shale along perforation: (a) Shale with bedding angle 0°—fracture initiation pressure of EF shale: 5,081psi @ 70 mm perforation; (b) Shale with bedding angle 45°—fracture initiation pressure of EF shale: 4,908psi @ 70 mm perforation.

Source: Khan et al., 2022

ESP can be significantly shortened by scale formation on the pump elements restricting rotation. By treating the formation with chemical inhibitors, the life of the ESP can be extended.

The authors (Levanyuk et al., 2012) describe a three-year campaign to combine scale inhibition with the hydraulic propped fracture. The novel fracturing fluid used in Russia allowed for a decrease in the amount of polymer in the fracturing fluid to 3.0 kg/m^3, which was beneficial for the inhibition of scale formation. However, the authors note that the use of a scale inhibitor and glazing agent can have negative impacts, which required several approaches to eliminate. For fluids with polymer loadings of 3.0 kg/m^3 to 3.24 kg/m^3, the authors proposed the use of a buffer and an additive that insignificantly increased delay time. For fluids with polymer loading

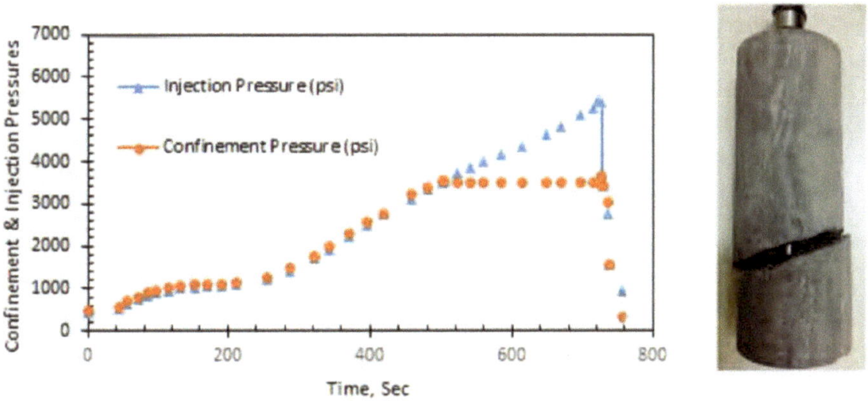

(a)

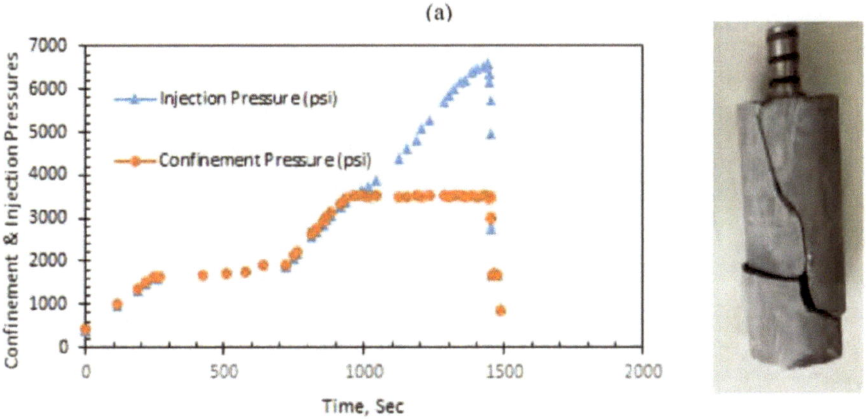

(b)

FIGURE 4.22 (a) Shale with bedding angle 0° and 70 mm perforation length—perforation: 70 mm; fracture pressure: 5,447 psi; (b) Shale with bedding angle 0° and 60 mm perforation length.

Source: Khan et al., 2022

of 3.60 kg/m^3, they proposed the use of a buffer, an additive that insignificantly increased delay time, and an additive that provided additional boron.

The implementation of a combined scale inhibition and fracturing treatment allowed for an increase in ESP run life. The authors report that the protection time exceeded one year. The success of the project was defined through five conclusions, including the confirmation of production problems caused by scale deposition and the implementation of a combined scale inhibition and fracturing treatment, which increased ESP run life and exceeded one year of protection time. Overall, the study presents a valuable approach for preventing and managing scale deposition in oil and gas wells.

FIGURE 4.23 (a) Hydraulic fracturing with water and fracture growth in Mancos shale with bedding angle 0° along perforation; (b) Hydraulic fracturing with SC CO_2 and fracture growth in Eagle Ford shale with bedding angle 0° along perforation; (c) Hydraulic fracturing with foam and fracture growth in Wolfcamp shale with bedding angle 90° along perforation.

Source: Khan et al., 2022

4.6.5 Field Examples of Comparing Cemented Plug-and-Perf With Open-Hole Ball Drop Completion Assemblies

Rahim et al. (2015) discuss the techniques used for fracturing horizontal wells in tight formations to maximize reservoir contact and optimize gas production. The multistage fracturing (MSF) technique is commonly used for this purpose and involves creating multiple fractures along the wellbore to increase contact area with the reservoir. The authors emphasize that the key to successful MSF treatment is generating independent fracture propagation and maintaining the bottom-hole treating

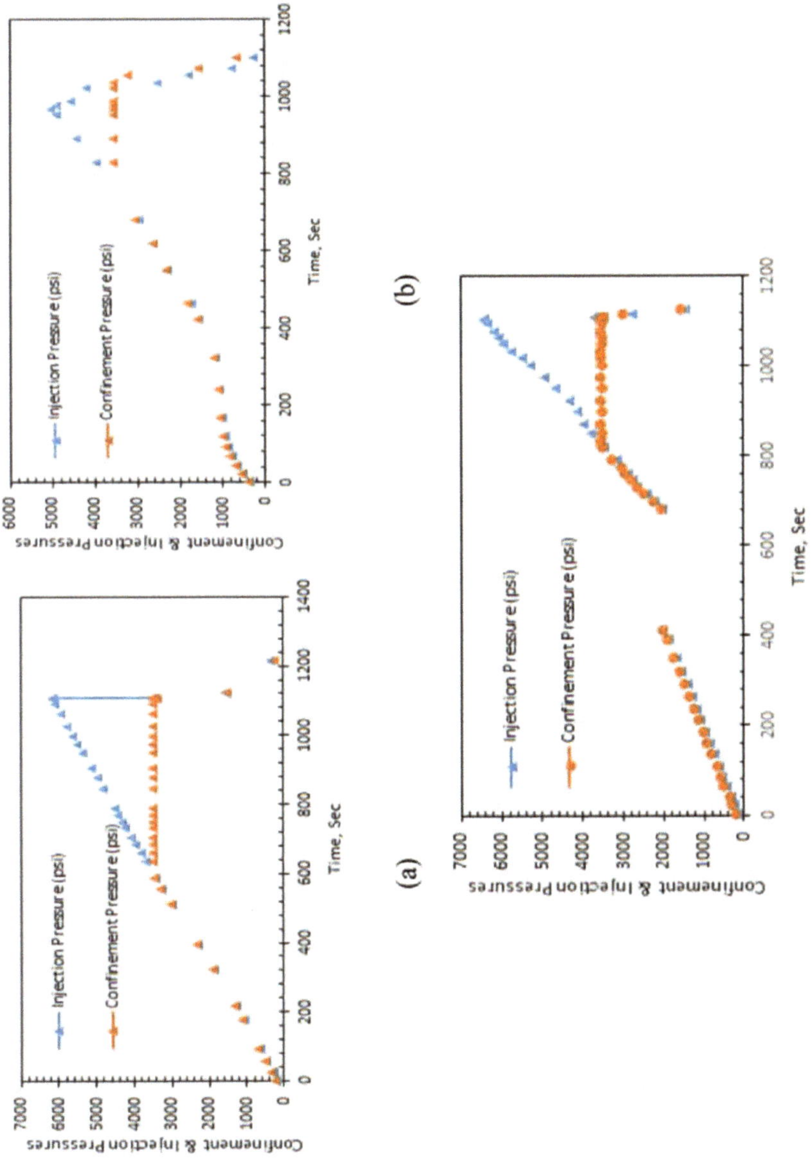

FIGURE 4.24 (a) Hydraulic fracturing with water and fracture initiation pressure in Mancos shale with bedding angle 0° along perforation; (b) Hydraulic fracturing with SC CO2 and fracture initiation pressure in Eagle Ford shale with bedding angle 0° along perforation; (c) Hydraulic fracturing with foam and fracture initiation pressure in Wolfcamp shale with bedding angle 90° along perforation.

Source: Khan et al., 2022

FIGURE 4.25 Scaling samples from ESP pumps.

Source: Levanyuk et al., 2012

pressure (BHTP) above the fracture closure stress (FCS) or the minimal horizontal stress (min) to ensure the fractures propagate as intended.

The authors compare two methods for fracturing horizontal wells: the open-hole multistage (OHMS) and cased hole plug-and-perforation completion (PPC). The OHMS completion method involves using swellable packers to isolate fracture stages in the open hole, which allow for direct gas flow from the reservoir into the wellbore. The PPC completion method, on the other hand, involves perforating reservoir intervals through a liner and then sealing each stage with cement. The cement provides an excellent seal and isolation between stages that cannot be achieved using either the mechanical or swellable OH packers used in OHMS. However, this method depends on manually perforating reservoir intervals, which may lead to misalignment of the perforation gun with the reservoir section and, consequently, reduced reservoir contact and gas production.

The authors found that the OHMS fracturing method is very effective in carbonate reservoirs when wells are drilled parallel to sigma-min and completed with swellable packers. This is because the chances of generating individual fractures are high, and the open lateral allows gas to flow into the wellbore directly, in addition to production through the induced fractures. The amount of gas flow through the OH section depends on the vertical permeability and how efficiently the near-wellbore section is etched and cleaned by the acid pumped. However, OHMS may not be suitable in reservoirs where generating independent fractures is challenging, and the risks of isolation packer failure are high.

In contrast, the CPPC completion method has proven promising, particularly in tight formations with low permeability values. This method involves perforating a single interval in multiple predetermined spots in clusters. The short perforated intervals create an orifice pressure drop at the entrance of the fracture during the treatment, which keeps the BHTP above min and the fracture open. With proper geomechanical modeling, a few perforation clusters can be optimally placed in one interval so that all can be treated by a single pumping stage, with individual fractures growing from each cluster and maximizing reservoir contact. The authors found that the CPPC completion method outperformed OHMS for a wide range of permeability values, providing additional cumulative production of up to 26% for a tight formation with 0.01 md permeability.

In conclusion, the authors suggest that the optimal completion and stimulation treatment will depend on the reservoir and well conditions. When generating independent fractures becomes challenging, and the risks of isolation packer failure are high, the CPPC technique should be the preferred choice. In reservoirs that have good permeability, the OH completions allow reservoirs to flow into the wellbore from intervals that are not necessarily fractured, thereby providing a better overall production rate for the well. In formations where a good zone is hard to detect and limited perforation has the chance to miss out on a potential producing interval, OHMS should be considered.

4.7 CONCLUSION

Unconventional resources like shale gas, tight oil, and coalbed methane are known for their low permeability and complex geology, making them extremely challenging to extract. However, new advances in well stimulation techniques helped overcome this challenge, including zipper fracturing, resin-coated sand, ceramic proppants, and modified fracturing fluids. Additionally, non-hydraulic fracturing techniques, such as electrically assisted hydraulic fracturing and pulse fracturing, were developed as a more economically viable alternative to conventional hydraulic fracturing. The completion aspect of unconventional wells also underwent great research and development in order to improve well productivity. This included optimizing the completion design for unconventional wells, utilizing innovative completion equipment, and employing techniques like horizontal drilling and multistage fracturing.

REFERENCES

Alagoz, E., & Sharma, M. M. (2021). Investigating shale-fluid interactions and its effect on proppant embedment using NMR techniques. *Paper ARMA 2021–1129, 55th US rock mechanics/geomechanics symposium held in Houston*, TX, 20–23 June. https://onepetro.org/ARMAUSRMS/proceedings-abstract/ARMA21/All-ARMA21/ARMA-2021-1129/467923

Alfi, M., Di Sario, M., & Florio, G. (2021). The use of electromagnetic waves for enhanced oil recovery in tight reservoirs: A review. *Journal of Petroleum Science and Engineering*, *199*, 108473.

Alklih, A. M., Al-Marhoun, M. A., Al-Qahtani, A. M., & Al-Muntasheri, G. A. (2021). Optimization of multistage hydraulic fracturing for shale gas reservoirs in Saudi Arabia. *Journal of Natural Gas Science and Engineering*, *93*, 103909. https://doi.org/10.1016/j.jngse.2021.103909

Anderson, D. M., Thompson, J. M., Popp, M., Behmanesh, H., & Liang, P. (2017, February). The impact of completion technique on well performance. *SPE Canada unconventional resources conference* (p. D011S002R007). SPE.

Bachari, N. E., Jones, D. A., & Bahi, L. (2019). Distributed acoustic sensing for hydraulic fracture monitoring. *Journal of Petroleum Science and Engineering*, *173*, 1169–1180.

Bai, B., Huh, C., & Ostadhassan, M. (2018). Gel-free fracturing fluids for hydraulic fracturing: A review. *Journal of Petroleum Science and Engineering*, *163*, 737–749.

Bashari, A. (2016). Unconventional oil and gas resources exploration and development in complex geological areas. *Journal of Natural Gas Science and Engineering*, *29*, 510–521.
Chen, J., Cui, B., Li, Q., Zhang, X., & Liu, C. (2017). Experimental study on electrically assisted fracturing of tight gas reservoirs. *Energies*, *10*(7), 985.
Chen, Z., Chen, Y., Li, J., Wang, Y., & Zhang, Z. (2021). Development and performance evaluation of a novel self-breaking fracturing fluid system for shale gas reservoirs. *Journal of Natural Gas Science and Engineering*, *94*, 103835.
Coal bed methane. Student Energy Coal Bed Methane Source. (n.d.). Retrieved February 25, 2023, from https://studentenergy.org/source/coal-bed-methane/
Cuss, R. J., Wiseall, A. C., Hennissen, J. A. I., Waters, C. N., Kemp, S. J., Ougier-Simonin, A., Holyoake, S., & Haslam, R. B. (2015). Hydraulic fracturing: A review of theory and field experience. *British geological survey open report*, OR/15/066. https://nora.nerc.ac.uk/id/eprint/514340/
Gao, B., Tan, Y., Huang, X., & Chen, Z. (2019). Optimization of completion design for horizontal wells with multistage fracturing in the Permian Basin. *SPE Journal*, *24*(5), 2523–2533. https://doi.org/10.2118/195714-PA
Geng, X., Liu, B., Yang, F., Li, L., Lu, Q., & Zhao, D. (2020). Experimental investigation of proppant-coated foam fluids for hydraulic fracturing in shale gas reservoirs. *Journal of Natural Gas Science and Engineering*, *78*, 103332.
Ghaderi, A., Amini, A., & Jamshidzadeh, M. (2018). A review on coalbed methane production and enhanced coalbed methane recovery. *Journal of Natural Gas Science and Engineering*, *55*, 311–326.
Ghorbani, A., Ameli, R., & Vamegh, M. (2020). High-resolution microseismic imaging in hydraulic fracture monitoring: A case study. *Journal of Petroleum Science and Engineering*, *184*, 106556.
Guo, Y., Zhang, M., Yang, H., Wang, D., Ramos, M. A., Hu, T. S., & Xu, Q. (2022). Friction challenge in hydraulic fracturing. *Lubricants*, *10*(2), 14. https://doi.org/10.3390/lubricants10020014
Hartsock, J. A. (2013). Hydraulic fracturing: History of an enduring technology. *Journal of Petroleum Technology*, *65*(12), 76–84.
Helmenstine, Anne Marie, PhD. (2020, August 2). What's the difference between shale and Slate Rock? *ThoughtCo*. Retrieved February 25, 2023, from www.thoughtco.com/shale-rock-4165848
Hsieh, T. Y., Wilke, J. P., & Abousleiman, Y. N. (2016). Stimulated reservoir volume enhancement with novel fiber-based diverters in multistage hydraulic fracturing. *Journal of Natural Gas Science and Engineering*, *33*, 824–835.
Ias, P. (2018, February 28). Unconventional gas resources: Shale gas & coalbed methane. *PMF IAS*. Retrieved February 25, 2023, from www.pmfias.com/shale-gas-coalbed-methane-reserves-india-world-extraction-shale-gas-hydro-fracturing-fracking/
Jia, K., Li, S., Li, X., Li, Y., & Bai, Y. (2019). A field experiment of pulse fracturing for shale gas in Longmaxi formation. *Journal of Natural Gas Science and Engineering*, *71*, 102985.
Khan, J. A., Padmanabhan, E., & Izhar Ul-Haq. (2022). Hydraulic fracturing to investigate impact of fracturing medium, bedding angle and perforation length on fracture growth in low and high brittle shale. *Paper presented at the offshore technology conference Asia*, Virtual and Kuala Lumpur, Malaysia, March. https://doi.org/10.4043/31531-MS
Kuru, E., Yasar, E., & Ozgen, S. (2018). Effective zonal isolation in hydraulic fracturing of a tight gas reservoir by swellable packers. *Journal of Natural Gas Science and Engineering*, *52*, 308–316. https://doi.org/10.1016/j.jngse.2017.12.026
Levanyuk, O., Overin, A., Sadykov, A., Parkhonyuk, S., Lungwitz, B., Enkababian, P., Klimov, A., Imperial, E., & Legeza, S. (2012). A 3-year results of application a combined scale inhibition and hydraulic fracturing treatments using a novel hydraulic fracturing fluid, Russia. *Paper presented at the SPE international conference on oilfield scale*, Aberdeen, UK, May. https://doi.org/10.2118/155243-MS

Li, J., Fan, Y., Wang, H., Cheng, L., & Liu, X. (2018). An experimental study of the impact of fracture spacing on tight oil production from horizontal wells. *Journal of Natural Gas Science and Engineering*, *49*, 240–250. https://doi.org/10.1016/j.jngse.2017.12.011

Li, Y., Zhang, X., Tang, H., Zhao, L., & Su, Y. (2019). An experimental study on the relationship between proppant concentration and hydraulic fracture conductivity in shale reservoirs. *Journal of Petroleum Science and Engineering*, *177*, 638–645. https://doi.org/10.1016/j.petrol.2019.02.026

Luo, J., Zhang, S., Huang, H., Li, B., & Xiao, Q. (2018). Study on the characteristics of lightweight proppant for hydraulic fracturing in tight sandstone reservoirs. *Journal of Natural Gas Science and Engineering*, *50*, 88–97.

Ma, X., Xu, H., Cui, Z., Zhao, D., & Zhang, X. (2019). Research on X-ray computed tomography and micro-seismic technique for hydraulic fracturing monitoring. *Journal of Natural Gas Science and Engineering*, *72*, 102959.

Mack, M. G. (2016). Unconventional geology: Exploring the complex and controversial world of shale gas. *Elements*, *12*(5), 327–331.

Maxwell, S. C., Prasad, M., & Mitchell, R. (2019). Microseismic monitoring and its role in unconventional reservoirs. *SPE Production & Operations*, *34*(2), 231–241.

Mayerhofer, M. J., Stutz, H. L., Davis, E. J., & Wolhart, S. L. (2006). Optimizing fracture stimulation using treatment-well tiltmeters and integrated fracture modeling. *SPE Production & Operations*, *21*(02), 222–229.

McClure, M. W., & Mathews, C. W. (2016). Completion design and hydraulic fracturing for maximizing well productivity in unconventional gas reservoirs. *Journal of Natural Gas Science and Engineering*, *34*, 881–896. https://doi.org/10.1016/j.jngse.2016.08.050

Media, U. S. E. (2018, December 27). 10 best performing tips for successful hydraulic fracturing. *Oilman Magazine*. Retrieved February 25, 2023, from https://oilmanmagazine.com/10-best-performing-tips-for-successful-hydraulic-fracturing/

National Research Council (NRC). (2010). *Management and effects of coalbed methane produced water in the western United States*. National Academies Press.

Pashin, J. C. (2015). *Coalbed methane: Principles and practices*. Elsevier.

Pedrosa, H. A., Colbert, F. C., Garcia, F. M., Gachet, R. F., Boldrini, A. C., De Souza, J. M., Carrara, M. O., Okullo, R. A., Belisario, J. H., Lacerda, L. A., & Goyallon, D. (2021). Stimulation workflow applied on LAPA pre-salt carbonate field deep-water Brazil. *Paper presented at the Abu Dhabi international petroleum exhibition & conference*, Abu Dhabi, November. https://doi.org/10.2118/207370-MS

Rahim, Z., Al-Kanaan, A., Al-Anazi, H., Kayumov, R., & Al-Jalal, Z. (2015). Comparing effectiveness between cemented plug and perf and open hole ball drop completion assemblies to ensure optimal multistage fracturing treatment and well performance—field examples. *Paper presented at the Abu Dhabi international petroleum exhibition and conference*, Abu Dhabi, November. https://doi.org/10.2118/177511-MS

Rig zone news. (2015, February 25). Horizontal drilling: How do they get it to go sideways? *Rigzone News*. Retrieved February 25, 2023, from https://rigzonenews.wordpress.com/2015/02/25/horizontal-drilling-how-do-they-get-it-to-go-sideways/

Salehi, S., Karimi, M., & Esmaili, S. (2019). Optimization of hydraulic fracturing design using reservoir simulation: A case study of a tight gas reservoir. *Journal of Petroleum Science and Engineering*, *178*, 1027–1039. https://doi.org/10.1016/j.petrol.2019.03.021

SEPA. (2010). *Coal bed methane development in Scotland: A review of best practice*. Scottish Environmental Protection Agency.

Sun, J., Murthy, S. S., & Liu, J. (2017). Fiber-optic sensors for downhole monitoring in oil and gas wells. *Sensors (Basel)*, *17*(6), 1242.

Tang, H., Xu, Y., Zhang, L., & Wang, Y. (2019). Experimental and numerical investigation of dissolvable fracturing balls in multistage fracturing. *Journal of Petroleum Science and Engineering*, *174*, 663–674.

Tang, J., Xu, Z., Chen, Y., Ren, X., Zhou, Y., & Luo, H. (2020). Optimization of shale gas hydraulic fracturing based on reservoir simulation. *Journal of Natural Gas Science and Engineering*, *83*, 103586. https://doi.org/10.1016/j.jngse.2020.103586

Wang, B., Sun, Y., & Tseng, C. H. (2016). Downhole pressure and temperature monitoring system for deepwater drilling operations. *IEEE Transactions on Industrial Electronics*, *63*(4), 2224–2233.

Wang, M., Chen, Y., Xu, L., & Fan, C. (2018). Research progress on proppants for hydraulic fracturing. *Journal of Petroleum Science and Engineering*, *169*, 679–692.

Water in the ground. (n.d.). Retrieved February 25, 2023, from www.csun.edu/~psk17793/ES9CP/ES9%20water_in_the_ground.htm

You, J., Zhang, C., & Wang, Q. (2017). Electrically assisted hydraulic fracturing of the Bakken formation: A feasibility study. *Journal of Natural Gas Science and Engineering*, *43*, 89–96.

Yu, Y., Li, Q., Chen, S., & Xie, W. (2020). Numerical simulation and optimization of multistage hydraulic fracturing in shale gas reservoirs. *Journal of Natural Gas Science and Engineering*, *83*, 103558.

Zeng, H., Wang, X., Zhang, J., Chen, H., & Zhang, W. (2020). A comparative study of pulse and conventional hydraulic fracturing in tight sandstone gas reservoirs. *Journal of Natural Gas Science and Engineering*, *77*, 103318. https://doi.org/10.1016/j.jngse.2020.103318

Zhang, H., Lu, H., Zheng, W., & Guo, Z. (2018). A numerical model for proppant transport and placement in hydraulic fracturing treatments. *Journal of Petroleum Science and Engineering*, *170*, 24–32. https://doi.org/10.1016/j.petrol.2018.04.049

Zhang, L., Zhang, C., Ma, J., Li, H., Liu, X., & Zhang, C. (2019a). Experimental study of dissolvable materials in hydraulic fracturing. *Journal of Natural Gas Science and Engineering*, *70*, 102982. https://doi.org/10.1016/j.jngse.2019.102982

Zhang, Y., Fan, M., Xie, Y., Xie, B., Xu, J., & Wang, L. (2020a). Degradable proppant and its application in hydraulic fracturing: A review. *Journal of Natural Gas Science and Engineering*, *83*, 103526.

Zhang, Y., Li, X., Liu, X., & Li, M. (2019b). Permeability enhancement in tight reservoirs: A review. *Journal of Petroleum Science and Engineering*, *181*, 106165.

Zhang, Y., Liu, S., Wang, S., & Zhang, Y. (2020b). Optimization of zipper fracturing parameters in shale gas reservoirs. *Journal of Natural Gas Science and Engineering*, *79*, 103313.

Zhang, Y., Zhan, Y., Zhou, J., Li, Y., Ma, J., & Wang, X. (2019c). The optimal spacing of hydraulic fractures for maximizing tight gas recovery from horizontal wells. *Journal of Petroleum Science and Engineering*, *174*, 1189–1199. https://doi.org/10.1016/j.petrol.2018.12.012

Zhao, H., Zhang, L., Chen, X., & Chen, H. (2021). Smart multistage completion technology for shale gas reservoirs. *Journal of Natural Gas Science and Engineering*, *93*, 103868.

Zou, C., Zhao, X., Du, X., & Hu, Y. (2019). Recent advances in research on shale gas permeability: A review. *Fuel*, *245*, 244–259.

5 Production Engineering Aspects and Artificial Lift in Unconventional Resources

Ekrem Alagoz, Emre Can Dundar, Mubarek Alpkiray, and Artur Davletshin

5.1 INTRODUCTION

The artificial lift method is used to raise oil from wells when the natural drive energy of the reservoir is not strong enough to push it to the surface. More than 50% of wells use artificial lift techniques. Production from unconventional reservoirs like shale and tight sand refers to production from very low-permeability formations. Special manipulations are required in such conditions to maintain efficient, economical production. Cenk Temizel et al. (2020) analyzed and described traditional and relatively new artificial lift methods and their optimization in an unconventional well.

Figure 5.1 shows the influence of the artificial lift (AL) system on the production rate during a well's lifespan.

Nowadays, the most popular AL systems are gas lift (40%), electrical submersible pumps (36%), sucker rod pumps (13%), plunger lift (7%) and jet pumps in 4% of unconventional wells (Figure 5.2). Each method has its advantages and disadvantages (Zhao et al., 2018).

Table 5.1 summarizes for the main limitation of the most common artificial lift systems described here. Choosing the most appropriate artificial lift method in every case is very important for the long-term profitability of producing oil and gas wells.

5.1.1 Unconventional Reservoirs

An unconventional reservoir is commonly defined as a reservoir with poor permeability consisting of ultra-tight source rock. Special recovery operations are required to enhance oil or gas production volumes in such oilfields. Also, specific issues related to the artificial lift need to be analyzed because it is one of the most important steps in gaining efficient production rates.

Oyewole (2016) summarized selection considerations for artificial lift system factors into four major considerations (Figure 5.3) and then validated them for

DOI: 10.1201/9781003319955-5

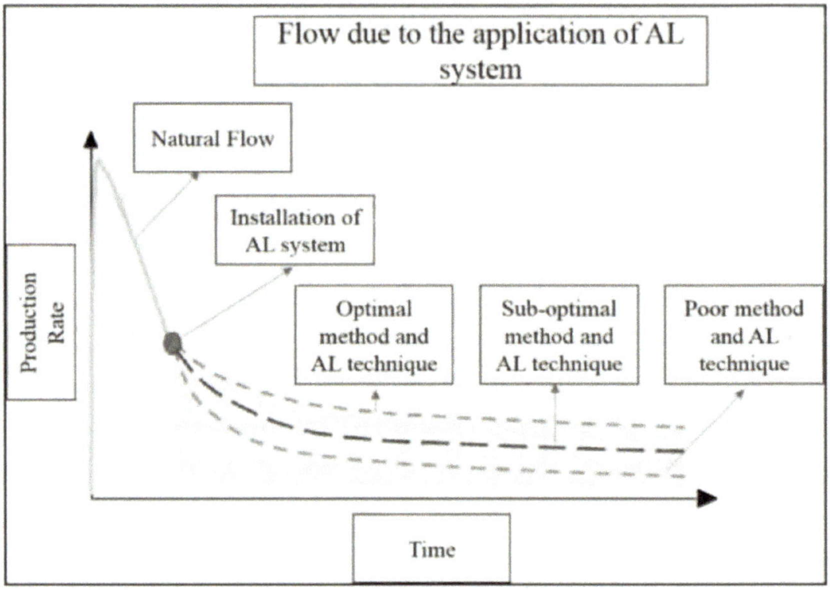

FIGURE 5.1 Impact of AL method on the rate of production.

Source: Pankaj et al., 2018

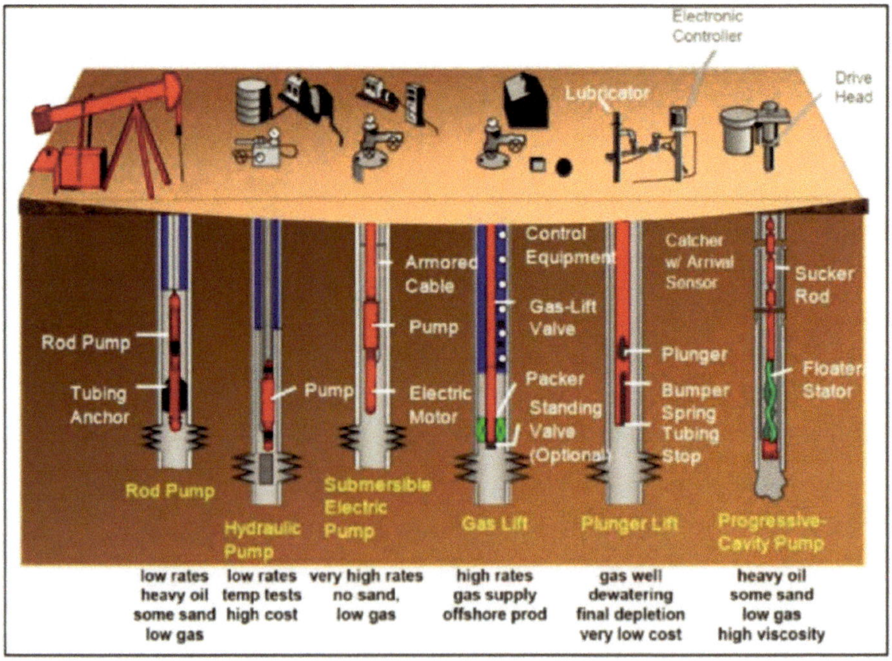

FIGURE 5.2 Various artificial lift systems.

Source: Temizel et al., 2020

TABLE 5.1
Comparison of Artificial Lift Systems

	Gas Lift	Plunger Lift	Sucker Road	ESP	PCP	Hydraulic Jet Pump
Maximum Temperature, Deg F	450	550	550	482	250	550
Maximum Depth, Ft	18,000	19,000	16,000	15,000	12,000	20,000
Fluid Gravity, API	> 15	> 15	> 8	> 10	8 < API < 40	> 6
Flow Rate, B/D	< 50,000	< 200	0–6.000	150–60,000	4500	< 35,000
Efficiency, %	10–30	N/A	45–60	35–60	50–75	10–30

Source: Artificial Lift Systems, 2020

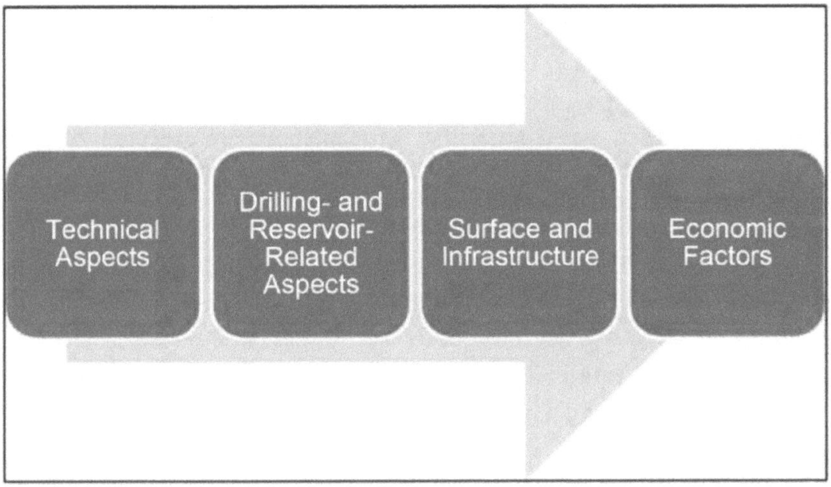

FIGURE 5.3 Considerations for artificial lift selection.

Source: Oyewole, 2016

unconventional reservoirs. The artificial lift selection process plays a crucial role in the long-term perspective, especially in unconventional reservoirs where different factors should be considered.

5.2 ARTIFICIAL LIFT METHOD SELECTION

5.2.1 Artificial Lift Method Selection for Mature Oilfields

The oil production rate of the wells may decrease during the development period as the reservoir pressure declines. When there is insufficient pressure in the reservoir to raise the produced fluids to the surface, artificial lift systems may be one of the main options to maintain the oil production rate, especially in the late-life period.

Darvish Sarvestani and Hadipour (2019) reviewed different artificial lift methods in the example field, which is located in southwest Iran near the Persian Gulf. Also, they determined three scenarios to predict the field performance by natural depletion, installing electrical submersible pumps, and implementing gas lift system. The selection of the best artificial lift method for oilfields depends on different parameters, the variety of which makes this procedure more difficult. These parameters are categorized in Figure 5.4.

The required information was gathered in the first stage of preliminary screening. Then, for a deeper analysis of all aspects, other important factors were considered (such as capital costs, operating costs, downhole equipment, and reliability). Authors compared and presented a summary of the relative advantages and disadvantages of gas lift systems, positive cavity pumps, electrical submersible pumps, and sucker rods. For each scenario (natural depletion, gas lift, and ESP), the inflow performance relationship and vertical-lift performance curves were generated and imported into the integrated model.

For accurate prediction of well pressure drop, an in-house code was developed for different wells and scenarios based on a modified Hagedron-Brown method. Also, to analyze the long-term efficiency of available options (GL and ESP), an integrated well and reservoir model was built, and results were compared (Figure 5.5).

Simulation showed that if no artificial lift methods are executed, only 13 million stock tank barrels (MMSTB) of oil will be produced in the next ten years. Additionally, it was concluded that GL may yield the highest cumulative oil production. More than 134 MMSTB of oil will be produced during the next ten years if a GL system is installed. In the third scenario, almost 112 MMSTB of oil will be produced if electrical submersible pumps are installed.

5.2.2 Screening Artificial Lift and Other Techniques for Liquid Unloading

Gas production in unconventional wells reduces much faster than that in conventional wells as production progresses. This results in a decline in reservoir pressure, leading to liquid loading problems in the well because reaching the dew point causes the condensing of the liquid in the gas phase. To decrease the negative economic impact of liquid loading, Chidirim Ejim and Jinjiang Xiao (2020) observed different artificial lift methods and techniques for use in unconventional gas well.

Artificial and non-artificial methods of lifting liquids from the well are mentioned in Ejim and Xiao (2020). Artificial lift methods are used to raise fluids to the surface of the wells when the reservoir energy is not sufficient to do so. Examples of artificial lift systems include gas lift, electrical submersible pumps, sucker rod pumps, and plungers. Non-artificial lift methods differ in that chemical, thermal, and fluid velocity may be used to maintain the production of hydrocarbon fluids to the surface. Examples of some these include soap sticks, foam-assisted lift or continuous chemical injection, liquid atomization, and wellbore heating.

To define the most applicable lift method, the specific unconventional gas well was selected by the authors. The target well had the typical decline in reservoir pressure, which led to liquid loading problems.

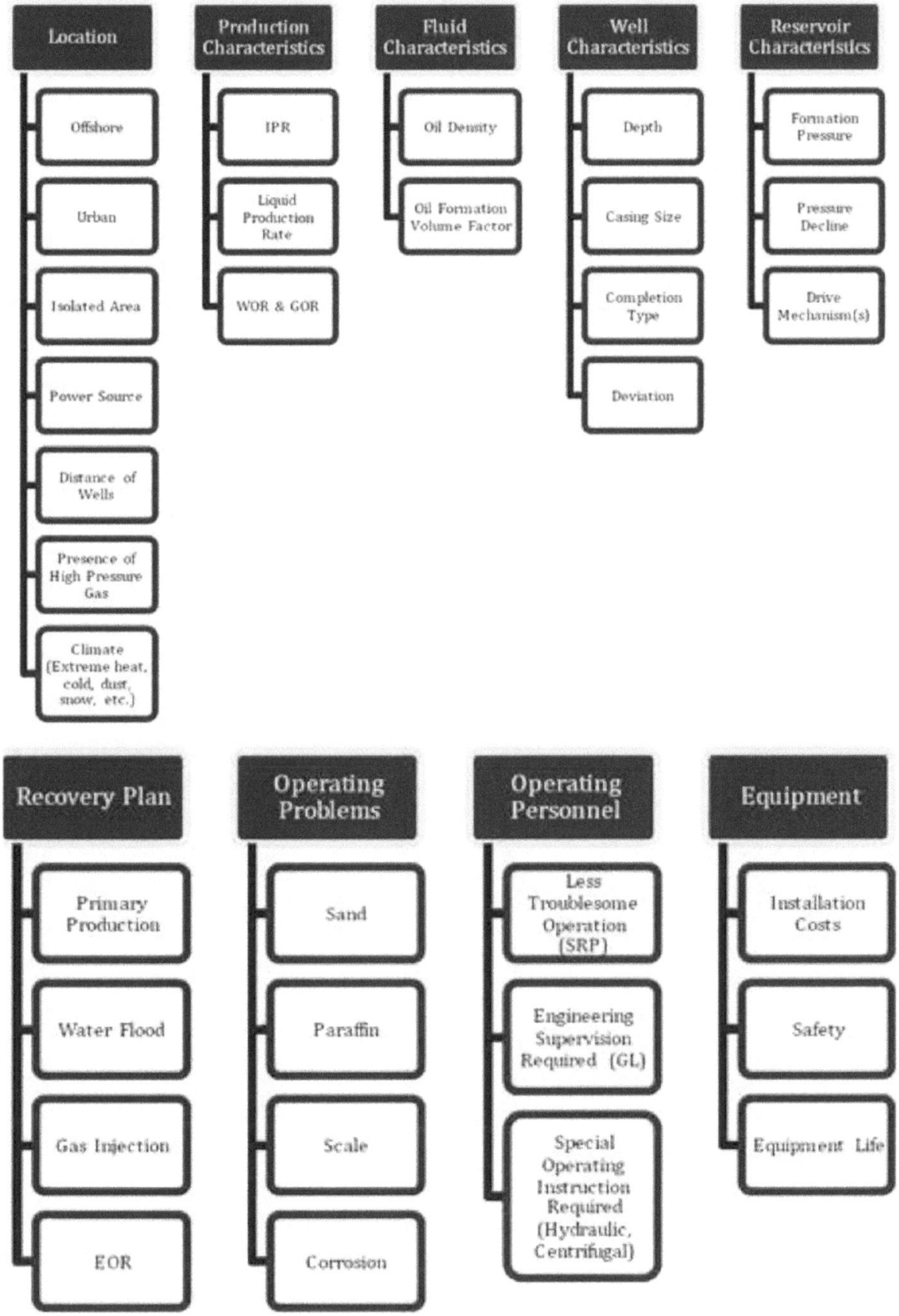

FIGURE 5.4 Influencing factors for proper selection of artificial lift method.

Source: Darvish Sarvestani & Hadipour, 2019

TABLE 5.2
Applicable Range of Depth and Production Rate for Different Artificial Lift Methods

a) based on production rate

Production Rate (bbl/day)	Possible Systems
> 20,000	GL or ESP
10,000–20.000	Any, except rod pumps
100–10,000	Any
< 100	Any, except ESP

b) based on installation depth

Depth (ft)	Possible Systems
> 15,000	Only hydraulic pumps
12,000–15,000	Any, except ESP and PCP
6,000–12,000	Any, except PCP
< 6,000	Any

Source: Artificial Lift Systems, 2020

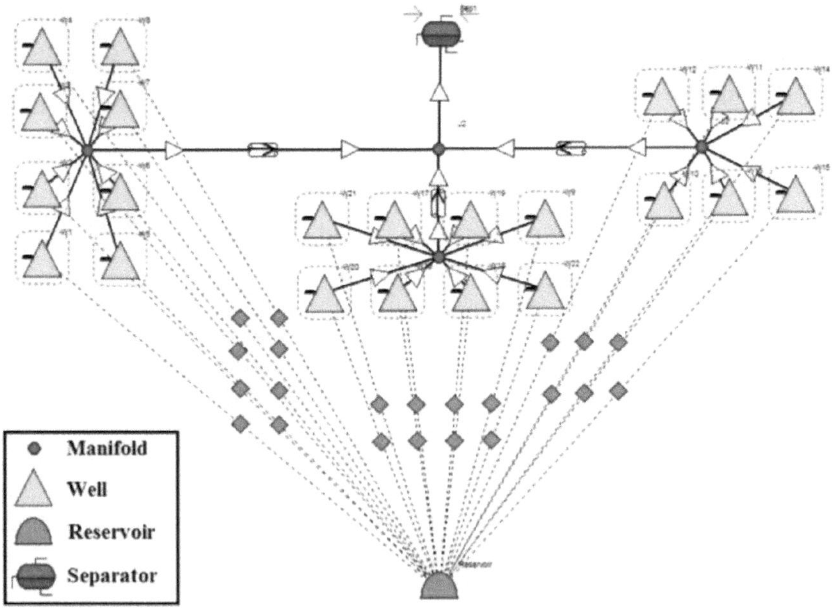

FIGURE 5.5 Schematic of integrated reservoir and well model.

Source: Darvish Sarvestani & Hadipour, 2019

Screening methodology similar to a Pugh matrix (Pugh, 1981) was used to narrow the number of options. This method is a criteria-based decision-making matrix in which every criterion is evaluated using a score of 1 to 5. Each criterion is based on the different well characteristics. The weightings of 1 to 5 have been selected arbitrarily to represent "very low" to "very high." For example, if the well depth is 10,000 feet, it was supposed that this requirement had a very high degree of importance (DOI) and was given a weighting of 5. For this analysis, the weightings are shown in Table 5.3.

A similar procedure is used to score the different alternatives (i.e., liquid unloading methods) as shown in Table 5.4. This is based on the characteristics of the liquid unloading method to the specific target unconventional gas well.

Then, the weighted sum for each alternative was calculated using criteria DOI and alternative score for criteria; after that, they were normalized to obtain a relative score. Table 5.5 shows the results of the screening analysis for the potential liquid unloading alternatives reviewed for the target unconventional gas well. They have been ranked from highest to lowest normalized weighted sum.

The analysis and screening enables ranking of the different liquid unloading methods, which is beneficial for field engineers and also aids in the initial planning process to optimize field assets for unconventional gas wells. Such a methodology is a first step

TABLE 5.3
Degree of Importance and Weightings

Degree of Importance (DOI)	Very High	High	Medium	Low	Very Low
Weighting	5	4	3	2	1

Source: Ejim & Xiao, 2020

TABLE 5.4
Alternatives Scoring against Criteria

	ALTERNATIVES DESCRIPTION/SCORES				
CRITERIA	**5**	**4**	**3**	**2**	**1**
Depth (10000 ft)					
Operating Temperature (280°F)					
Liquid Flow Rate (100–200 BPD)					
Casing Size (5 1/2 inch)	Excellent	Very Good	Good	Fair	Poor
Gas Handling					
Solids Handling					
Low CAPEX					
Low OPEX					
Technology Readiness Level (TRL)	Very High	High	Medium	Low	Very Low

Source: Ejim & Xiao, 2020

toward streamlining the number of potential alternatives for subsequent detailed analysis and final selection of an optimized solution to unload liquids in a target unconventional gas well. From Table 5.5, the five high-ranking liquid unloading methods were determined, which can be used to optimize gas production in unconventional wells.

5.2.3 Intelligent Decision-Making and Optimization of Artificial Lifting Based on MR Multiphase Flow Detection

At present, it is important to make decisions about artificial lifting parameter optimization and equipment in the petroleum industry according to consideration of multiphase flow in oil and gas wells and pipelines. Magnetic resonance (MR) can qualitatively and quantitatively evaluate reservoir fluids by acquiring information on molecular scale. This method makes it possible to simultaneously measure the flow rate and phase hold-up of multiphase flow in theory, but it is difficult to implement due to the effect of fluid motion in a hard-working environment. An article by Feng Deng et al., 2022 proposes a series of ideas for intelligent decision-making based on MR multiphase flow detection in measurement rooms, single wells, and oil and gas testing wells. Flow rate and phase hold-up of multiphase measurements are based on the measurement of the Carr-Purcell-Meiboom-Gill (CPMG) pulse sequence, assuming that the fluid is a mixture of composed components.

The oil-water ratio can be directly calculated using the areas of the oil and water peaks on the T2 spectrum shown in Figure 5.6. But the MR signal for gas is weak and can be interrupted by noises; therefore, a derivation method that can calculate the gas hold-up without direct measurement was proposed.

Nowadays, the most common ways of production metering are metering rooms/stations, single wells, oil testing, and gas testing, which use a more complicated set of tools and equipment than the conventional MR instruments used in laboratories. In

TABLE 5.5
Results Table

ALTERNATIVES	CRITERIA										
Criteria DOI Weighting (0 to 5)	5	5	5	5	5	3	4	5	5		
	Depth	Operating Temperature	Liquid Flow Rate	Casing Size	Gas Handling	Solids Handling	TRL	Low CAPEX	Low OPEX	WEIGHTED SUM	NORMALIZED WEIGHTED SUM
Multi-Stage PL	2	5	4	5	5	2	5	4	5	176	0.782
PAGL	3	5	4	4	5	4	5	3	3	167	0.742
Plunger Lift	1	5	3	5	5	2	5	5	4	166	0.738
GAPL	3	5	3	5	5	2	5	3	4	166	0.738
Continuous Gas Lift	5	5	5	2	5	5	5	2	2	165	0.733
Intermittent Gas Lift	5	5	4	4	5	3	5	2	2	164	0.729
Long-Stroke Sucker Rod Pump	3	5	5	5	2	2	5	2	5	161	0.716
Sucker Rod Pump	3	5	4	5	2	2	5	3	5	161	0.716
ESP	4	3	4	5	3	2	5	4	4	161	0.716
Hydraulic Jet Pump	5	5	5	3	2	3	5	3	3	159	0.707
RESP	3	5	4	5	3	3	3	2	5	156	0.693
Supersonic Nozzle	3	4	3	5	4	1	1	5	5	152	0.676
Diaphragm Insert Pump	2	4	2	5	4	4	3	4	4	149	0.662
Hydraulic Piston Pump	5	5	3	3	2	1	5	3	4	148	0.658
Top-Drive PCP	1	3	3	5	1	4	5	5	4	142	0.631
ESPCP	1	3	3	5	1	4	5	4	5	142	0.631
Foam-Assisted Lift	4	2	2	5	4	2	5	4	1	136	0.604
Wellbore Heating	3	5	3	5	4	3	1	3	1	133	0.591
Soap Sticks	5	1	1	5	4	2	5	4	1	131	0.582
SCS	4	4	4	4	5	1	1	1	2	127	0.564
Linear ESP	4	2	2	2	3	3	3	4	4	126	0.560

Source: Ejim & Xiao, 2020

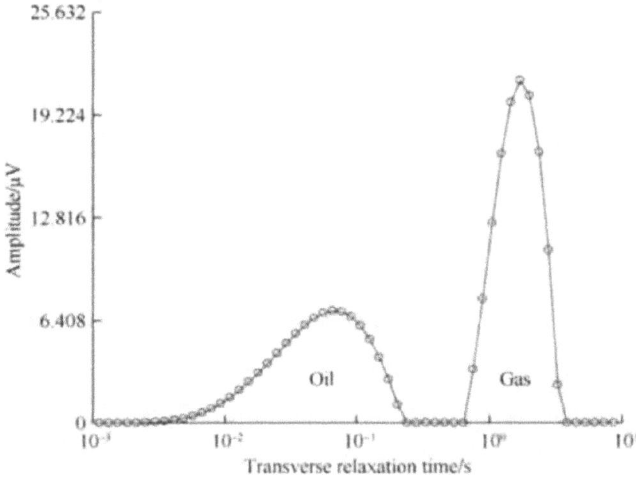

FIGURE 5.6 An example of multiphase flow T2 spectra.

Source: Feng Deng et al., 2022

the past two years, application tests using the MR multiphase flow have been realized in the different oilfield scenarios (Figure 5.7), and obtained actual field information.

One testing room can control multiple wells. However, it is recommended to develop the "one machine for multiple wells with virtual measurement" technology application in this case. In the single wells with continuous pumping, the changes in the flow rate vary during up and down direction. Real-time production data can be used to reflect the working conditions of the entire production system.

The intermittent pumping scenario represents the measurement scenario of intermittent liquid wells and unstable liquid-producing wells and requires a large dynamic measurement range for multiphase flowmeters. The flow data gathered can be used to guide the formulation of the system, making it more reasonable and realizing energy savings and consumption reduction.

Oil and gas testing is an important process for probing industrial oil and gas flow measurement, identifying the characteristics of oil and gas reservoirs, verifying the correctness of reservoir understanding and interpretation, evaluating reservoir development value, and determining the production system.

5.3 ARTIFICIAL LIFT ENCOUNTERED PROBLEMS AND SOLUTIONS

5.3.1 Sand Management Practices for Extended ESP Run Life

It is very important to analyze sand control practices during oilfield development because of its effect on the operation of electrical submersible pump (ESP) performance and its reliability. Xianglin Xu at el. (2022) described some study cases in which these practices were applied over the last ten years that had a positive effect on

FIGURE 5.7 MRMF application photos in the field: (a) Testing room; (b) Single well (continuous pumping); (c) Single well (intermittent pumping); (d) Well test.

Source: Feng Deng et al., 2022

the ESP run life. They analyzed them in the example of Tarapoa Block in Ecuador (Figure 5.8).

Tarapoa Block started commercial production in 1978; in 1996, an extensive development started with the field discovered on both sides of the Fanny-Dorine fault. Since the first tests, some preconditions were discovered that can induce sand production.

There have been different sand control practices applied to extend ESP performance in the Tarapoa Block since 2001 (Table 5.6).

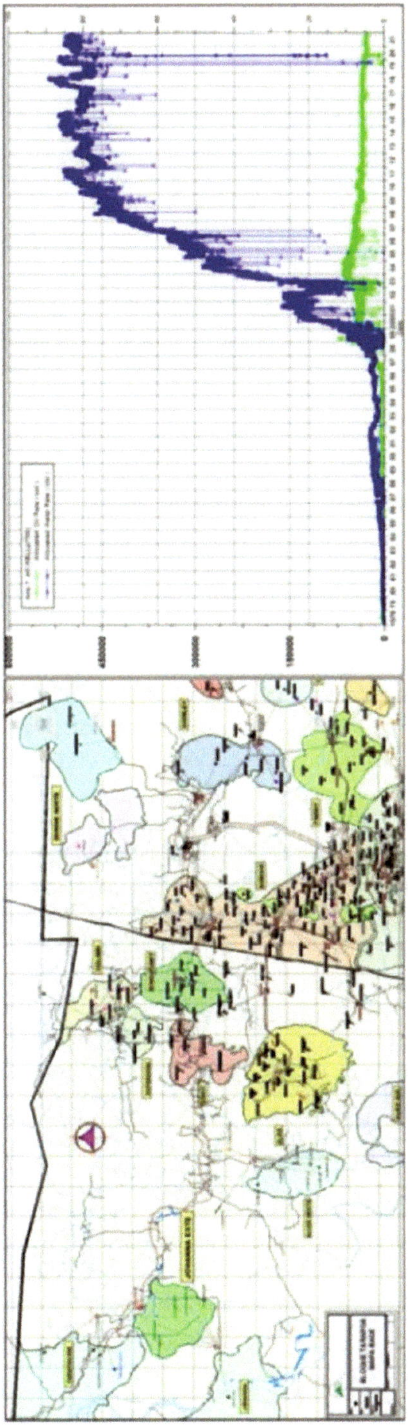

FIGURE 5.8 Tarapoa Block location and production history.

Source: Xu et al., 2022

TABLE 5.6
Sand Management Practices for Extended ESP Run Life in the Tarapoa Block

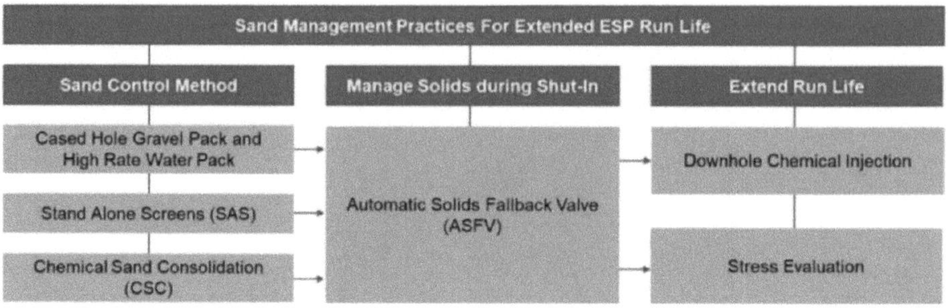

Source: Xu et al., 2022

Sand production is considered one of the important issues during the well production period. Sand control methods described in the article use gravel pack (GP), high rate water pack (HRWP), stand-alone screens (SAS), and chemical sand consolidation (CSC). The gravel pack method is widely used in the oil and gas industry and is an effective sand-control method in the early production phase, but this technique significantly reduced well productivity in the considered oilfield.

SAS are mostly used in horizontal wells, but, despite successful installation, some cases showed a low efficiency. CSC is the process of pumping chemical materials like resin consolidation fluid mix and had a good effect on the considered wells (Xu et al., 2022).

Automatic solids fallback valves (ASFV) provide protection to ESP by preventing sand deposition when the well is shut in. To control interaction between annulus and tubing, an ASFV valve is installed above ESP. During shut in, solids and fluid in the tubing string are diverted to the annulus rather than settling back in the pump (Figure 5.9).

Downhole chemical injection and stress evaluation are practices for extending run life. Diesel downhole injection is used to decrease viscosity influence on ESP equipment performance and to increase run life. This practice provides a positive result, leading to a longer production period.

Based on the analysis of the implemented sand management practices, Figure 5.10 shows the ESP performance for the past 20 years. As depicted in Figure 5.10, gravel packs were used in the beginning, but then this method was stopped due to the well productivity. Since 2006, the use of stand-alone screens (SAS) in horizontal wells and chemical sand control (CSC) in deviated wells changed the MTBF tendency, and in 2018, the use of ASFV also improved this tendency.

Implementing different sand control methods can significantly improve ESP performance and equipment reliability and, as a result, the oil production period.

5.3.2 Arresting Excessive Sand Production

The development of mature oilfields is ultimately characterized by increasing water cut and producing solids, which can lead to the complete stoppage of a well.

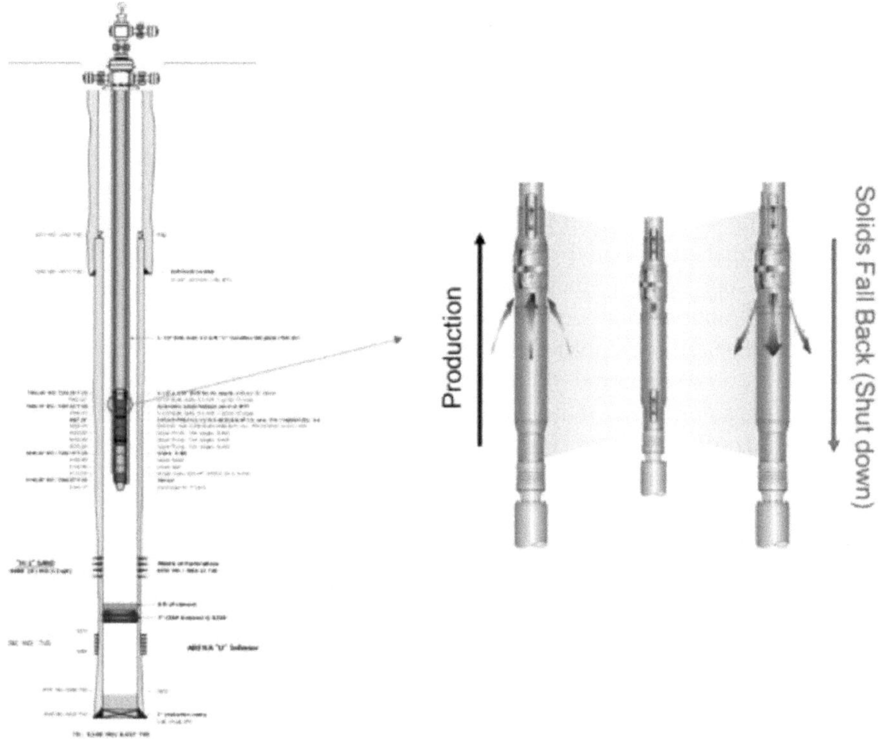

FIGURE 5.9 Automatic solids fallback (ASFV) valve.

Source: Xu et al., 2022

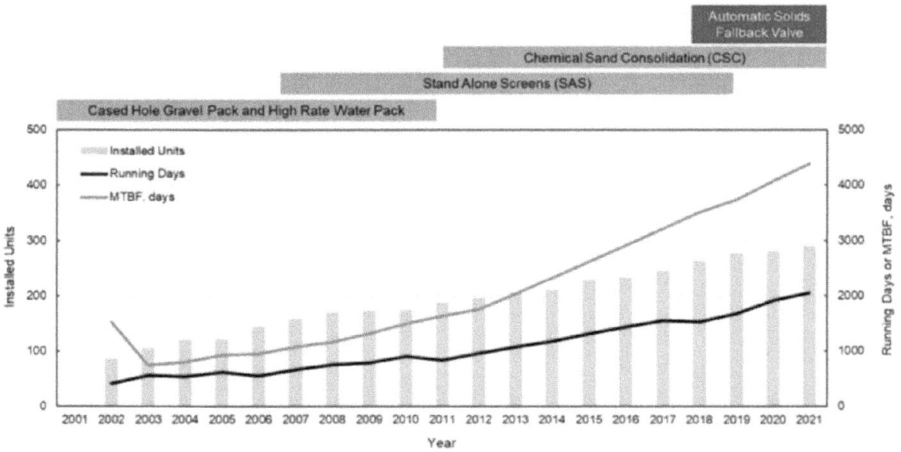

FIGURE 5.10 ESP Performance trends for 2001–2021.

Source: Xu et al., 2022

Therefore, in many cases, different practices are applied to extend the oil production period. As sand production is one of the important issues during the well production period, chemical sand consolidation (SCON) is considered helpful to solve this problem. Ako et al. (2021) described difficulties, lessons learned, and best practice to return the well to expected production levels.

ARAMU 037S was completed in August 1992 on the X4100 reservoir. Since the production began, it was observed that the interval was prone to sand production. The well was in production for 15 years, then it was closed due to extremely high sand production. Furthermore, in that period of time, a gas lift system was installed, which led to increasing sand production. Two unsuccessful chemical sand consolidation (SCON) interventions were conducted in May 2004 and February 2018. Figure 5.11 shows the measured sand production after the February 2018 intervention.

In 2012, reservoir saturation tool (RST) data was obtained in the well, and after analysis, it was concluded that an attic oil scope on the uppermost sand unit had a viable optimization potential (Figure 5.12).

After previous unsuccessful attempts, it was decided to use different approach to reduce the sand production in the well. This included analysis of the sand type, involving service providers, and a deeper understanding of the reasons for previous SCON fails. For further analysis in the laboratory, ditch cutting samples from the well perforation depth were taken. And an unconstrained compressive strength (UCS) test was conducted to determine the maximum force required to crush the consolidated sand sample. It provided extremely valuable information for sand evaluation and characterization.

Based on this analysis, a third treatment activity was executed in April 2019, involving a repeat sand cleanout and chemical consolidation on the existing perforation. It has since been producing way below the 10 pptb sand cut threshold at an average of 2 pptb (Figure 5.13). In conclusion, the determination of the UCS of a representative core sample helped define important indicators.

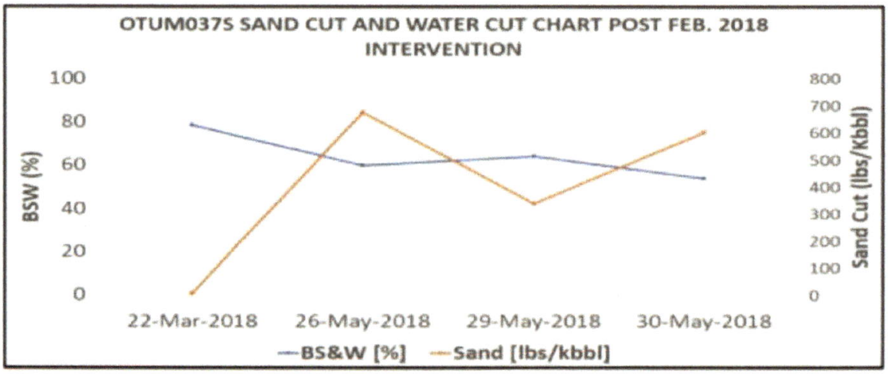

FIGURE 5.11 Water cut and sand cut sample plot.

Source: Ako et al., 2021

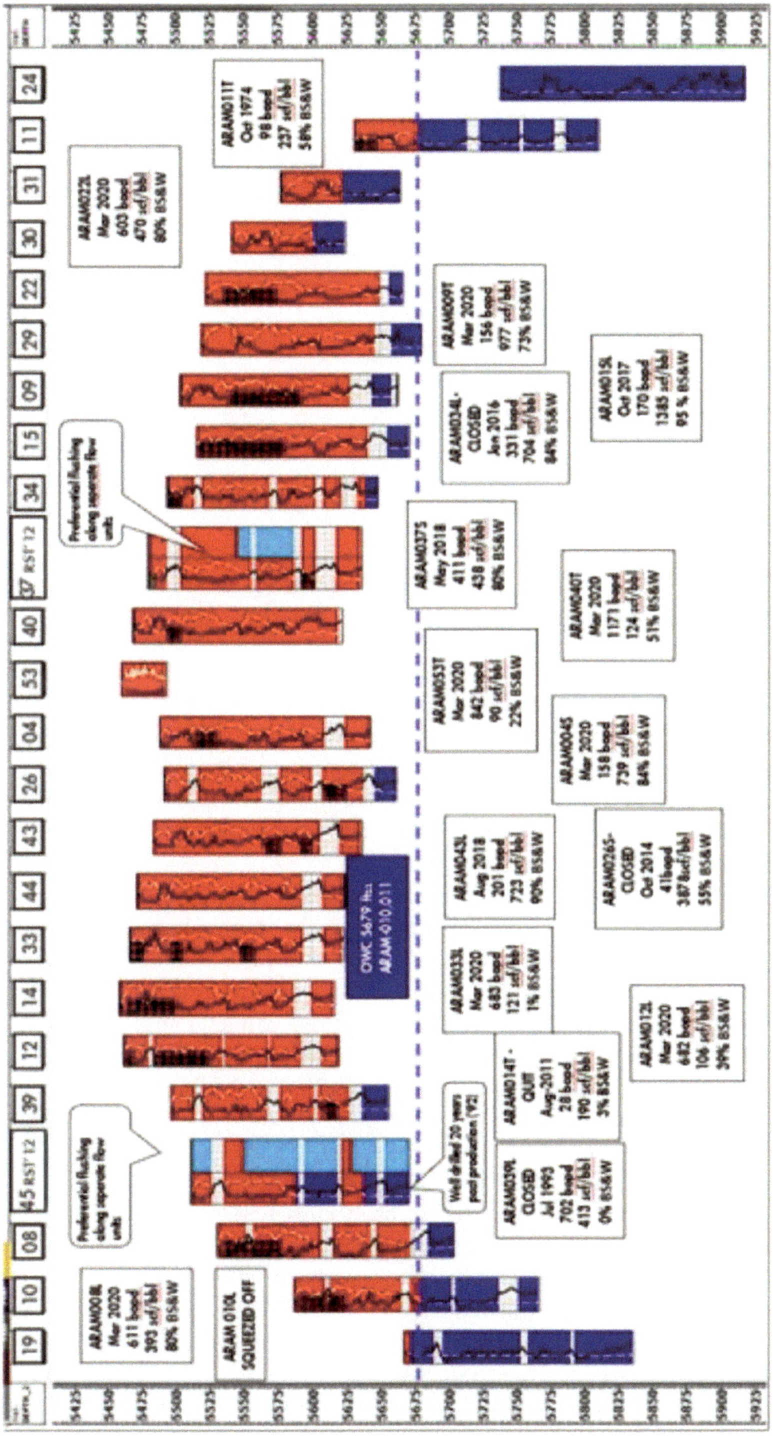

FIGURE 5.12 Fluid distribution plot.

Source: Ako et al., 2021

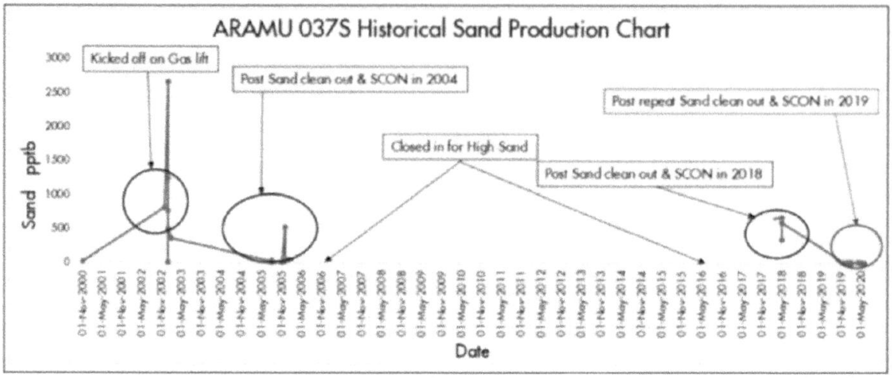

FIGURE 5.13 Historical sand production.

Source: Ako et al., 2021

5.3.3 Permanent Magnet Motor (PMM) as the Right Solution for ESP Challenges

In the Beleyam oilfield, an electric submersible pump (ESP) is used as the main artificial lift system. One of the most commonly issues encountered using ESP is electric failure, with varying oil production sustainability and cost. ESP is mainly based on conventional induction motor (IM), not to handle electric failures. One of the current designs being developed is a permanent magnet motor (PMM), eliminating motor obstacles. This type of motor has numerous pros, such as high efficiency, the power factor, and density (Gad et al., 2017).

Since ESP failures are classified as mechanical and electrical, there are requirements to provide solutions to reduce such failures. PMM is preferred as an effective solution to avoid problems. To determine the type of failure, a failure analysis report is implemented to minimize wells' downtime and operation costs. Failure report analysis shows that the motor short circuit is the predominant cause of electric failure. By elaborating short circuit, different sections of electric failure are shown in Figure 5.14. If well productivity is not proper, insufficient cooling of motors and low fluid velocity might be encountered.

5.3.3.1 Definition of a Permanent Magnet Motor

While PMM has almost all configuration properties with IM, it includes permanent magnets in the motor. Magnets can generate magnetic fields without requiring electrical energy. As this system gains popularity, it offers various advantages, including highly accurate positional sensing, maximized torque, and the elimination of slip. Since magnets produce the stator flux, high power is not required as it is with electric generation. Due to the increased efficiency and the synchronous operation, it might see decreased electrical losses and lower operating current. Therefore, the ESP PMM system will work at lower flow rates than conventional

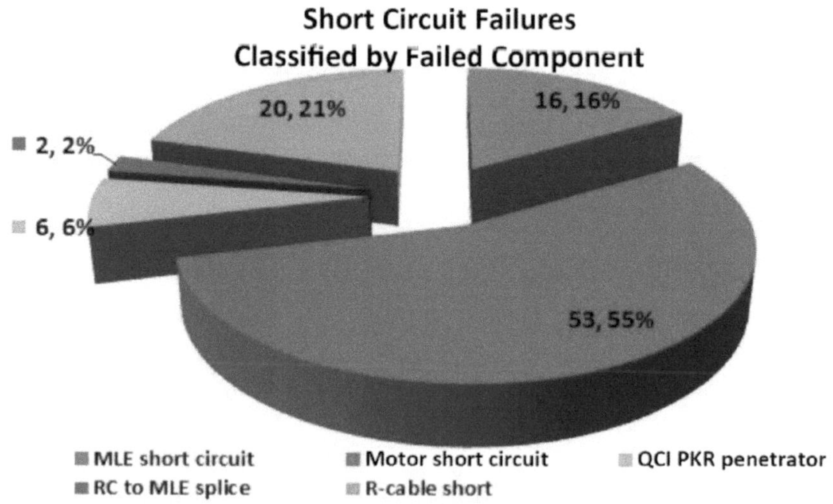

FIGURE 5.14 Classification of short circuits.

Source: Gad et al., 2017

ESP motors. Some of improvements with implementation of the ESP PMM system are as follows:

- Increased efficiency up to 93%
- Lowered power consumption 10 to 30%
- Constant torque
- Effects over power factor
- Decreased harmonic distortion
- Faster, more adaptive control,
- Operating rate expanded to 250–6,000 rpm.

Applying ESP PMM helps decrease failures based on Petrobel analysis, considered the main reasons for challenges on ESP motors. After utilizing PMM in Petrobel wells, electric failure rates were no longer encountered, as shown in Table 5.7.

5.3.4 Plunger Lift System Restores Production in Abandoned Gas Condensate Well

Gas condensate and mature gas wells are primarily based on high rates of gas production; however, liquid accumulation can result in shutdown or decrease in the well. To present an alternative approach to removing liquid, either a reciprocating rod lift (RRL) or a plunger lift system can be used. A plunger lift system is more cost effective, and it provides reliable maintenance for a naturally flowing well.

Gas and liquid systems are presented as multiphase flow conditions. Dewatering or liquid loading mechanisms are based on multiphase flow as shown in Figure 5.15.

TABLE 5.7
Implementation of PMM System in Different Wells

Well	Challenge	Run Days	Remarks
Well 1	Low recovery	695	Pulled due to well intervention (no electric failure)
Well 2	Low recovery	389	Still running
Well 3	Low recovery	72	Still running
Well 4	High GOR depleted well	246	Pulled due to well intervention (no electric failure)
Well 5	High GOR depleted well	194	Still running
Well 6	Obstructed well	163	Still running
Well 7	Obstructed well	41	Still running
Well 8	Obstructed well	47	Still running
Well 9	Obstructed well	43	Still running
Well 10	Obstructed high-potential well	45	Still running
Well 11	High viscous, tight emulsion, and high solid content fluid	469	Still running

Source: Gad et al., 2017

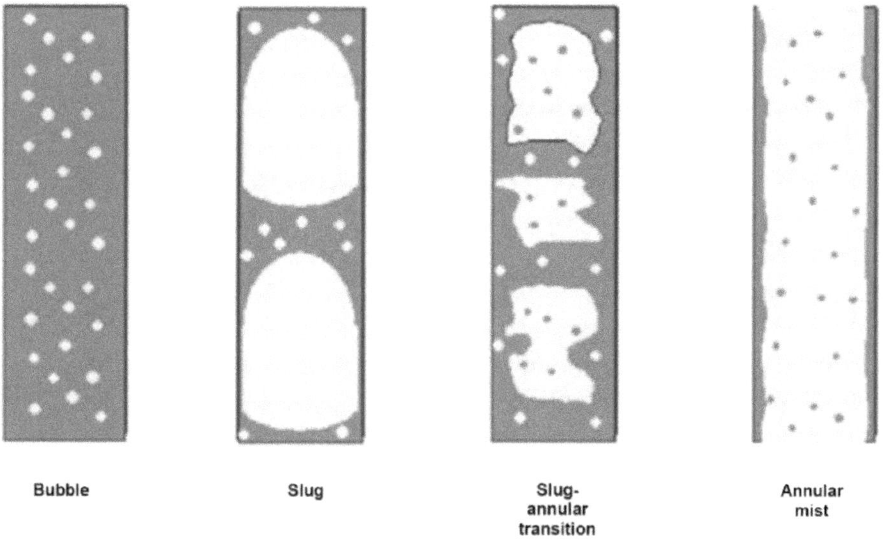

FIGURE 5.15 A multiphase flow illustration in a vertical well.

Source: Zhassulan, 2016

Considering well configuration, gas liquid ratio (GLR), and bottom-hole conditions may lead to multiphase flow in the well. While gas and liquid move up through the casing or tubing, well head pressure and flow state may alter due to the reduction in gas velocity. Liquid loading occurs due to the pressure difference between casing and tubing. (Casing is higher than tubing.)

In the early period of production, a mist flow can be encountered at the surface as proof of liquid loading in the bottom zone. If a cure is not applied, liquid accumulation may result in stopping production. Liquid loading means that the velocity of the gas is lower than critical gas velocity. The point of maximum flow leads to minimum speed, depending on BHP and gas compressibility. Once a low gas rate is monitored, gas velocity cannot be applied, resulting in common liquid accumulation in gas condensate wells. As long as production is not suspended, it is acceptable.

5.3.4.1 Determination of Liquid Loading

An empiric correlation was developed by Standing and Katz (1942) to analyze flowing wellhead pressure (FWHP) up to 1,000 psi. The critical velocity and critical gas rate equations are as follows;

$$V = \frac{5.321 \times (67 - 0.0031P)^{1/4}}{0.0031P^{1/4}} \tag{5.1}$$

$$q_{critical} = 10.155 \frac{\pi d^2 P \times (67 - 0.0031P)^{1/4}}{P^{1/2} \times (P + 460)} MMsf / day \tag{5.2}$$

To determine the presence of liquid loading, the operator can apply some of the techniques for using the Plunger lift. Gas production rate, well completions without packer, GLR, and load factor are considered to monitor the fluid accumulation. The use of this system decreases capital cost by more than three times compared to RRL lift system.

This study shows that using applications of the plunger lift system reduces costs and reliability with high GLR and liquid loading obstacles. It is important to consider accurate analyses of past and current production data. This approach requires more practical potential observations.

5.3.5 Dewatering Gas Well Successes with Sucker Rod Pumping System

Using a sucker rod pump in the oil and gas industry for dewatering a gas well provides high-rate production, cost, and efficiency advantages compared to other artificial lift systems. Adding a smart sensor to the pumping system helps monitor and control water production and gas volume calculations. Sucker rod pumping systems have several advantages. However, at higher depths, the pumping rate decreases rapidly.

The presence of liquid in a gas well is called liquid loading where the gas cannot remove the liquid from the wellbore. The accumulation of liquid reduces production, leads to slugging flow, and is time-consuming, eventually causing the wellbore to stop producing. A lower gas rate results in a larger tubing pressure gradient and higher formation pressure, leading to more liquid accumulation. Dewatering methods include ESP, PCP, beam pumping, hydraulic pumping, gas lift, velocity strings, compression systems, plungers, foaming, and injection systems. The variable slippage and two-stage pumps are used as types of sucker rod systems in the industry.

5.3.5.1 Case Studies

Some successful processes using sensor technology of unloading gas wells are analyzed as follows:

5.3.5.2 Well #3-36PM

The well suffered from gas locking during pump pulls. An approach was applied by running without pulling for 1½ years, resulting in increased gas production from 37 MCF to 96 MCF at 8 BWPD.

5.3.5.3 Well #1E

Inconsistent gas production was observed due to water buildup. A solution was based on 14 months running with steady gas production, and there was no gas lock as 95 MCF per day at 11 BWPD.

5.3.5.4 Well #2-1C

Five pulls were run through 6 months due to gas locking. To overcome this issue, the system was run for 14 months without pulling the pump, leading to an increase of gas production from 40 MCF to 60 MCF at 29 BWPD.

A proper run for a beam-pumping system is based on landing the tubing below the perforations to maximize gas separation. Additionally, applying a good pump design can assist in stopping gas pound problems. With proper analysis and design, beam pumping systems can operate in multiphase fluids, dewater gas wells and coal-bed methane wells, and increase the rate of fluid production. Installing smart sensor technology in the beam pumping system optimizes of operations and provides more accurate water and gas volumes production.

5.3.6 Well Performance Calculations for Artificial Lift Screening

Artificial lift techniques are a significant part of production operations in the oil industry. There are numerous types of these systems that have been used for decades. New technologies developed some screening criteria in the industry, such as rate, cost, reliability, flow assurance, solids handling, and well design. A variety of techniques may fit electric submersible pumps, gas lift, sucker rod pumps, progressing cavity pumps, jet pumps, and hydraulic submersible pumps. To select the appropriate techniques, nodal analysis is considered to observe the well performance during production of an artificial system based on power, flow capacity, and gas handling.

The selection of an artificial lift system is based on three stages of screening, forecasting, and economic analysis as shown in Figure 5.16. The screening criteria consider the fluid types, reservoir properties, and operating environment. The forecasting process depends on the filled size and complexity of the reservoir. Economic analysis evaluates the ultimate selection of lift the system. This analysis consists of capital and operational expenditures to estimate net present value (NPV).

Screening criteria evaluate hydraulic performance, well productivity, power requirements, and gas-handling ability. Since well performance impacts OPEX and APEX, hydraulic modeling software packages are used to predict fluid properties, operating environment, and completion design.

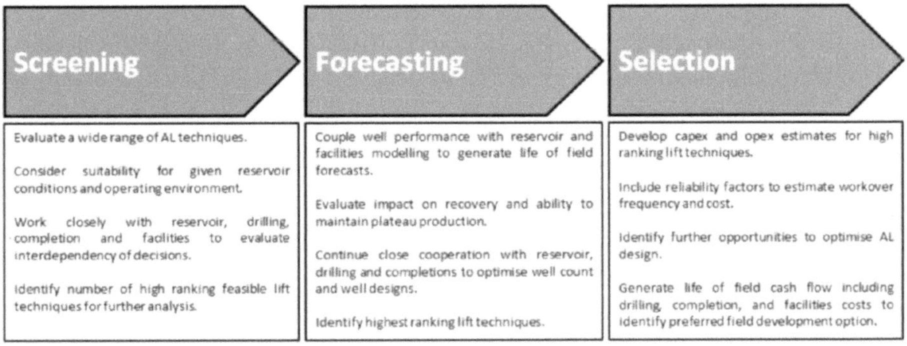

FIGURE 5.16 The selection of an artificial lift process.

Source: Kefford & Gaurav, 2016

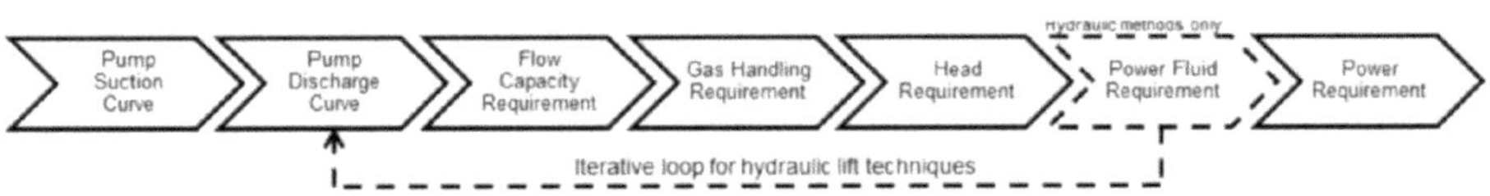

FIGURE 5.17 A typical of computation process.

Source: Kefford & Gaurav, 2016

The calculations and estimations are based on understanding the artificial lift equipment's capabilities and calculation of the well's artificial lift requirements. To understand the capability of lift technique, some criteria should be considered, such as fluid rate capacity, gas handling capacity, net positive suction head, head capacity, setting depth and dogleg severity, power capacity, and efficiency. The next step is based on determination of the actual well conditions for the future by considering maximum and target rates of reservoir pressure, inflow performance, water cut, and GOR. A typical calculation process is shown in Figure 5.17.

Software is used to provide fluid property tables, inflow performance curves, and tubing performance curves. The maximum achievable production rates are targeted using different tubing sizes and plot interpretations. Screening criteria and calculation processes are implemented to reach the optimum fit for purpose of artificial lift systems. The method is successfully applied to evaluate consistent and independent modeling. Calculations also enable detection of early development planning stages in a variety of field types.

5.4 DIGITALIZATION AND DATA SCIENCE

5.4.1 Data Science for Water Cut Analysis System by Utilizing ESP Sensor Data

Optimization and control of the critical parameters like water cut and oil production rate play significant roles in the field development. Appropriate estimation can be performed during the stable production of an oil well, but in an unstable well, it can be difficult. As most of the wells in the considered oilfield are equipped with electrical submersible pumps (ESP), Bogi Haryo Nugroho at el. (2021) suggest collecting data from their sensors and then analyze it using a data science approach to provide continuous control. There are five steps to developing a solution design (Figure 5.18).

The conceptual design process is generated using the obtain, scrub, explore, model, and interpret (OSEMN) framework (Nugroho et al., 2021), which covers every step of the data science project lifecycle from end to end, as shown in Figure 5.19.

Conceptual model test was done using Python and R programming languages, which are reliable and widespread data science tools. Figure 5.20 depicts informative data visualization.

Next, the proposed system solution was defined, which consists of three layers: source data, data science, and visual analytics (Figure 5.21).

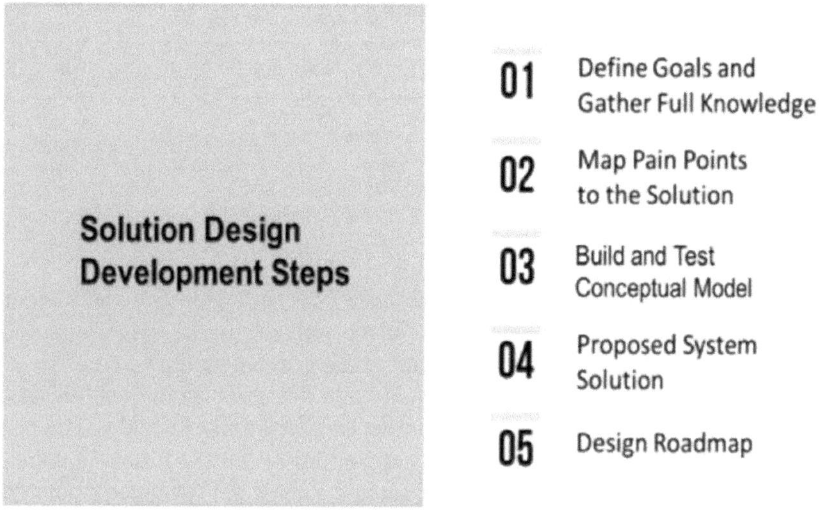

FIGURE 5.18 Solution design development steps.

Source: Nugroho et al., 2021

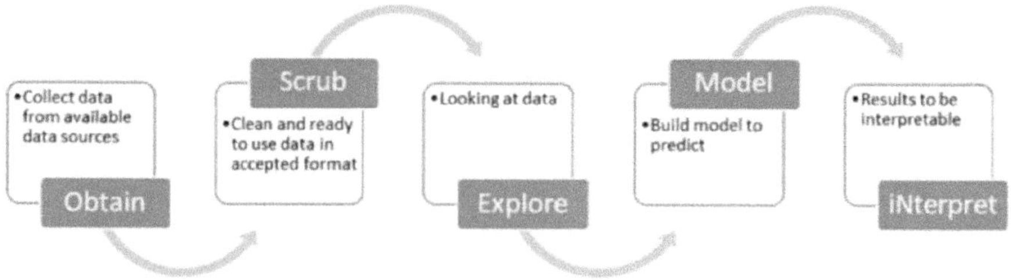

FIGURE 5.19 Data science process using OSEMN framework.

Source: Nugroho et al., 2021

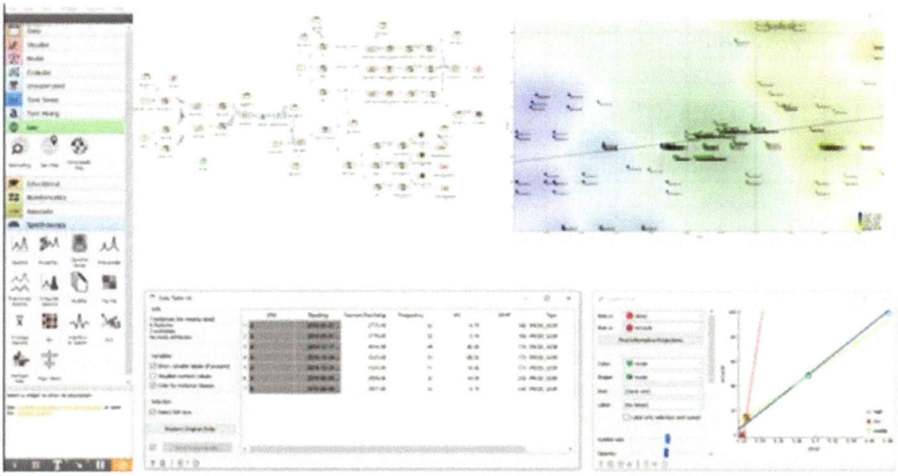

FIGURE 5.20 Conceptual model with exploratory analysis in Python environment.

Source: Nugroho et al., 2021

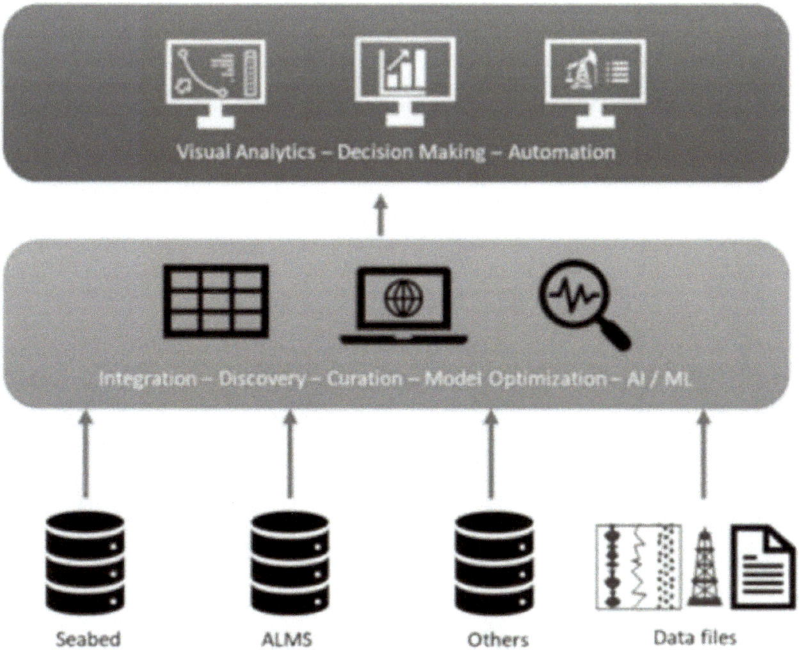

FIGURE 5.21 High-level solution summary.

Source: Nugroho et al., 2021

Finally, to accomplish the project goals, the authors determined a roadmap, which consisted of five phases (Figure 5.22).

5.4.2 Full Digital Intelligent Artificial Lift (DIAL) System Installation

Gas lift is an artificial lift method that uses an external source of high-pressure gas to help lift liquids up to the surface through the production tubing. Optimization methods have been evolving for a long time, and new approaches addressing increased efficiency and reliability have been implemented since early gas lift designs. Syahezat Ismail et al. (2018) described the implementation process of a digitally controlled in-well gas lift system called digital intelligent artificial lift (DIAL) on a pilot basis. As shown in Figure 5.23, the DIAL system consists of multiple digital gas lift units and is controlled by a surface control system (SCS), which provides broad remote lift management.

Technology evaluation included indicating economic objectives, selecting candidate wells, gas lift designing, planning risk-mitigation actions, and identifying success

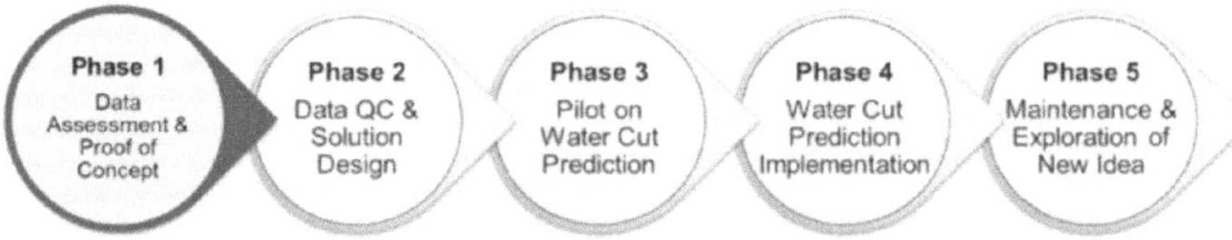

FIGURE 5.22 Roadmap of data science for water cut estimation/prediction.

Source: Nugroho et al., 2021

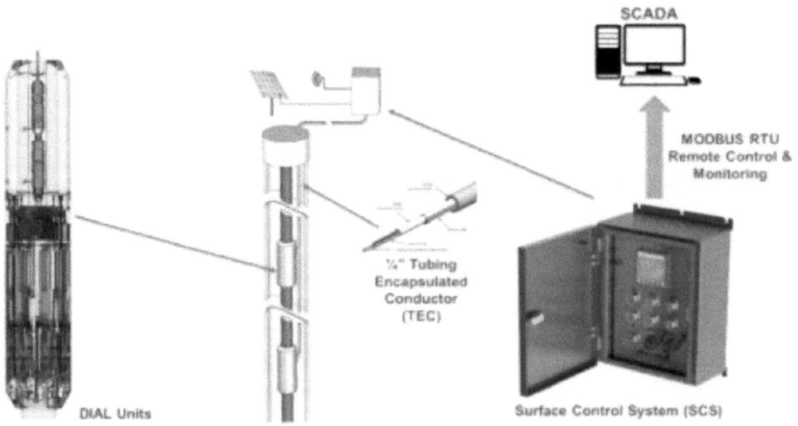

FIGURE 5.23 Digital gas lift system architecture.

Source: Syahezat Ismail et al., 2018

criteria. Revenue generation was compared in the case of DIAL implementation and legacy gas lift system, as well as APEX and OPEX rough estimation. The candidate well was chosen in consideration of several criteria, such as pressure, temperature, and location. Then, a gas lift design was made using simulation software from AppSmiths.

The first step of technology field implementation was integration testing. Systems integration and connection to DCS (Figure 5.24) allowed control and monitoring of the well's downhole parameters using a dashboard. Then training sessions were organized for participating staff.

Next, digital gas lift system equipment was installed (Figure 5.25), and during commissioning, a failure occurred in connection with one of the DIAL systems. This failure was further investigated, and a mitigation plan was proposed.

The left side of the dashboard (Figure 5.26) is used to visualize the status of downhole valves with a simple well schematic. The center has a real-time pressure versus depth graph, with pressure and temperature gradients below and above the point of injection. This facilitates production optimization decisions with a clear view of opportunities for deeper injection or change of injection rates. The right side tracks the history of all units with curves showing trends for casing pressure, tubing pressure, tubing temperature, and downhole gas injection rates.

However, during the implementation process, one of the DIALs showed that the sensor was not functioning correctly, as indicated by the fact that there was no change in temperature before and after the well started to be unloaded. It was later revealed that the lower connector of the DIAL had failed, with a similar failure mode as other DIAL. This was to be expected as all units were exposed to the same environment.

Even though this initial pilot was considered a "mixed success" for the digital gas lift implementation in the DL field, the operator is committed to continue evaluating and searching for innovative technology to be implemented in its operations worldwide. Based on the proposed mitigation plans provided, further candidate wells are currently being evaluated for digital gas lift implementation.

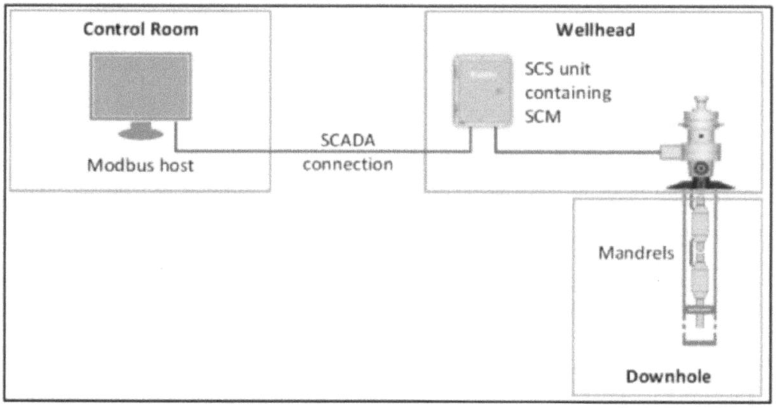

FIGURE 5.24 Systems integration and connection to DCS.

Source: Syahezat Ismail et al., 2018

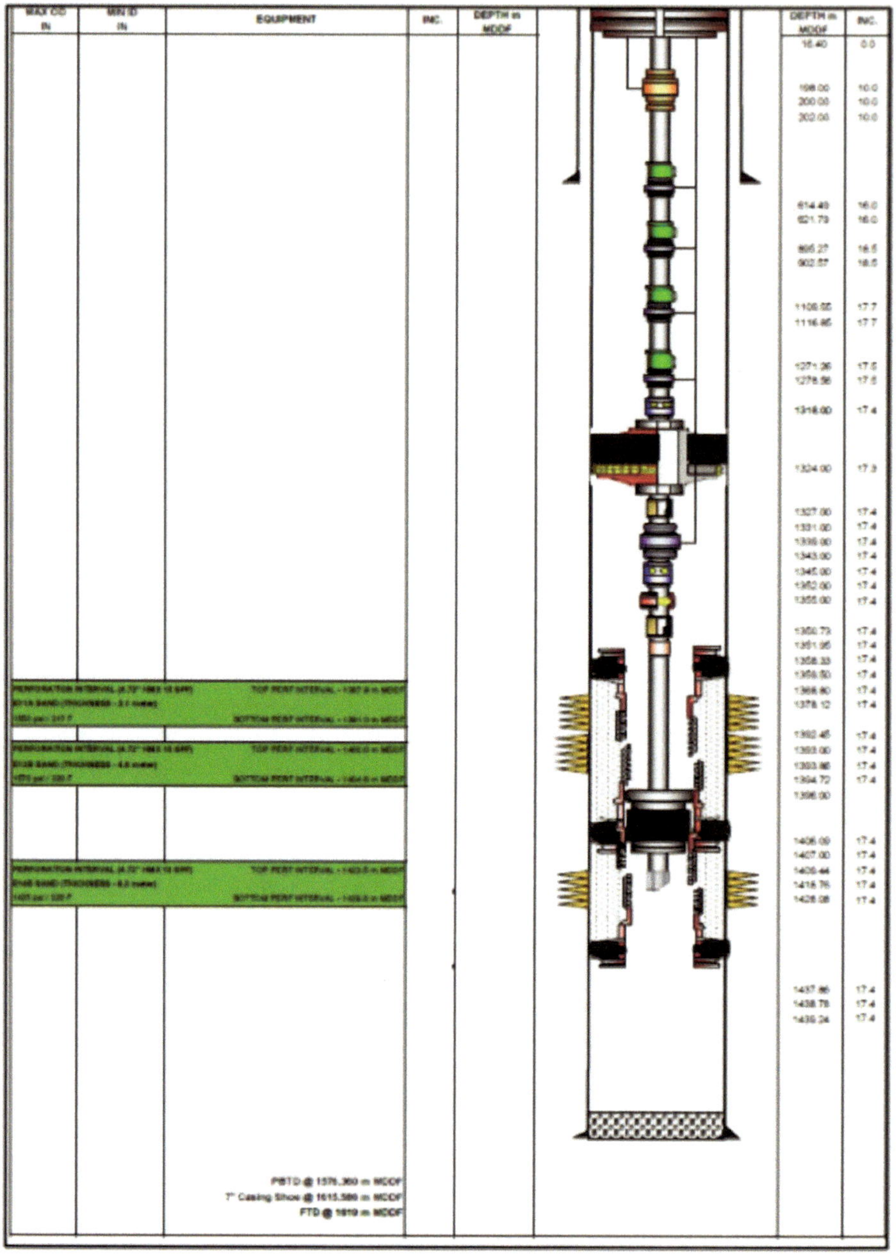

FIGURE 5.25 OP4 completion diagram.

Source: Syahezat Ismail et al., 2018

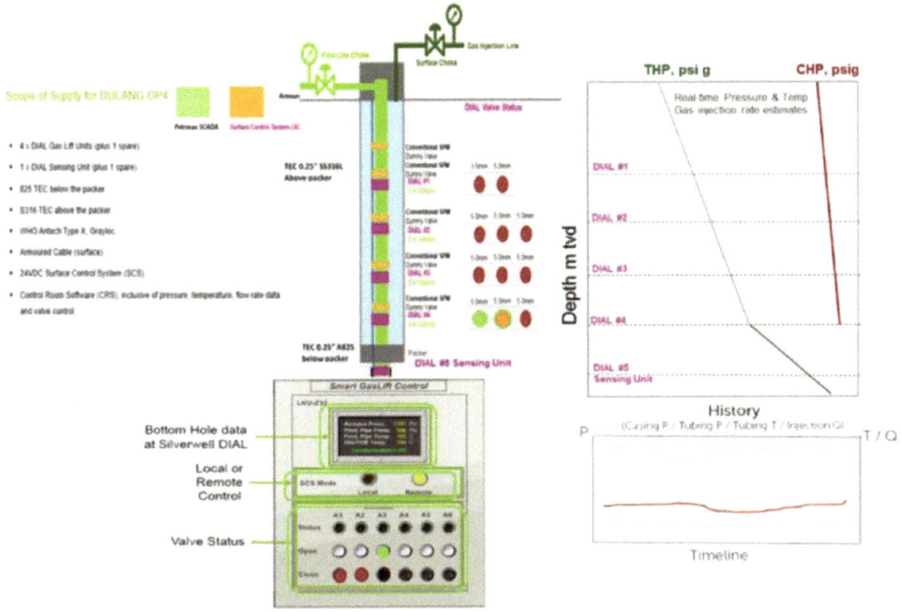

FIGURE 5.26 Dashboard/user interface.

Source: Syahezat Ismail et al., 2018

5.4.3 Gas Lift Optimization Using Artificial Neural Network

A selection of proper artificial lift methods is necessary to predict optimum and safe production for hydrocarbon reservoirs. One of the preferred artificial lift methods is the gas lift method, based on a high-pressure gas injection into the depth of fluid column, as shown in Figure 5.27. The injected gas enables the flow of the fluid with sufficient energy. The gas lift method is a preference not only on vertical wells but also on deviated wells. The aim of this method is to reduce the density of fluids at the bottom hole, resulting in the presence of less heavy fluid than in the original. There are two factors to consider with the gas lift method: the optimum position of injection depth and the gas injection rate.

5.4.3.1 Artificial Neural Network

To obtain maximum benefits from the reservoir while using the gas lift method, an artificial neuron network (ANN) is implemented in the filled operations. ANN is mainly based on the biological neuron network of the human body. A neuron network is interconnected that can be exerted to calculate various inputs and implement machine learning predictions. Two types of ANN can be used in machine learning predictions: static and dynamic. The static ANN cannot be run for the second attempt while the dynamic one can be optimized by a suitable algorithm, as illustrated in Figure 5.28.

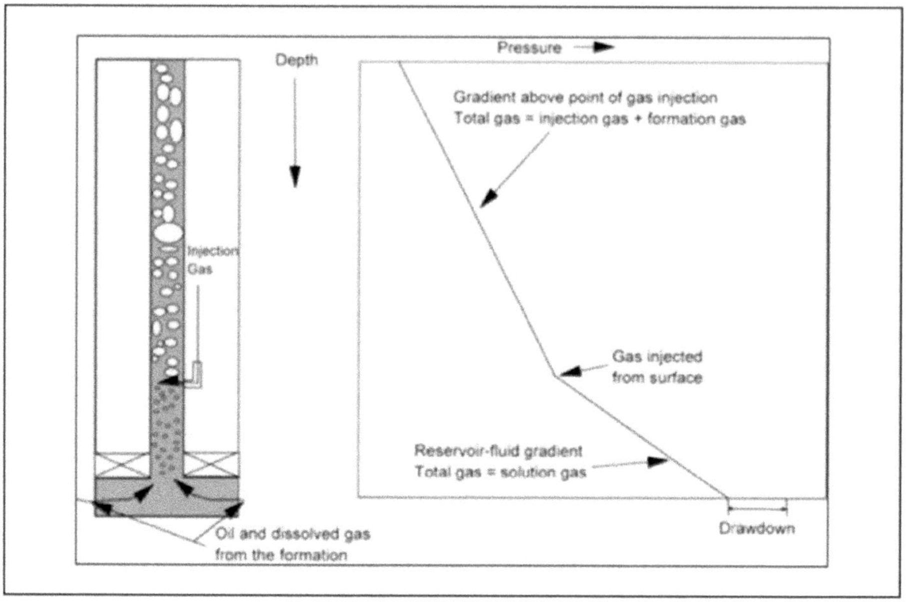

FIGURE 5.27 A schematic of gas lift design.

Source: Ranjan et al., 2015

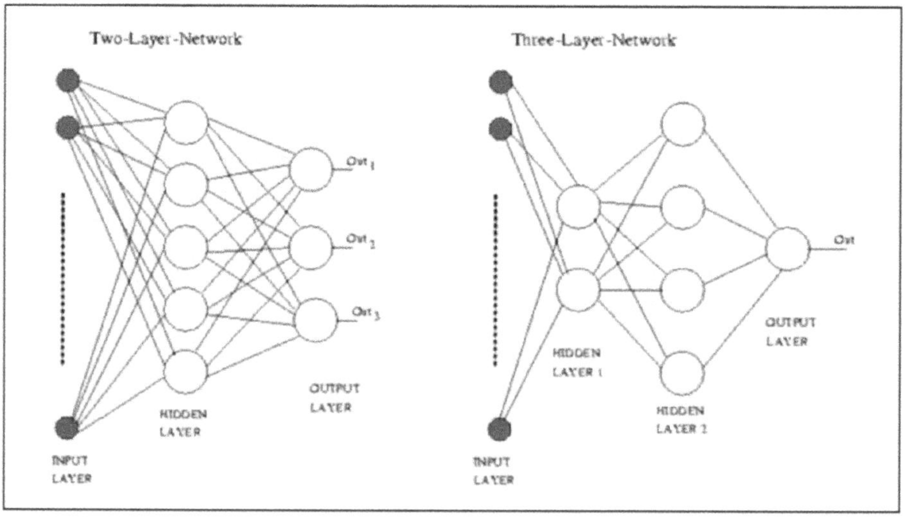

FIGURE 5.28 Multilayer network illustration.

Source: Ranjan et al., 2015

To find the optimum prediction for the gas lift method, there are variety of functions applying a trial-and-error basis, such as training function, transfer function, adaption learning function, number of hidden layers, and number of neurons in each layer, as shown in Figure 5.29.

A machine learning prediction is used to present two parameters for gas lift optimization. The injection rate is used to estimate the rate of oil production while the maximum rate of oil production is determined. The study shows that the simplicity of the network provides a better and faster calculation than previous approaches. This method concludes that time savings for engineers due to calculations of nodal analysis are eliminated, speeding up computations.

5.4.4 World's First Successful Dual String Installation of a Digital Intelligent Artificial Lift

Gas lift is one of the most common artificial lift systems to unload gas from a reservoir. Gas lift is separated by a two-phase or three-phase separator. Injection of gas reduces the density of the fluid column inside the production pipe, causing a low-flowing bottom-hole pressure and an increase in drawdown and production. To present efficiency conditions, the DIAL technology is applied to enhance production optimization and downhole monitoring of parameters. A successful application is used by PETRONAS on the B field in Malaysia.

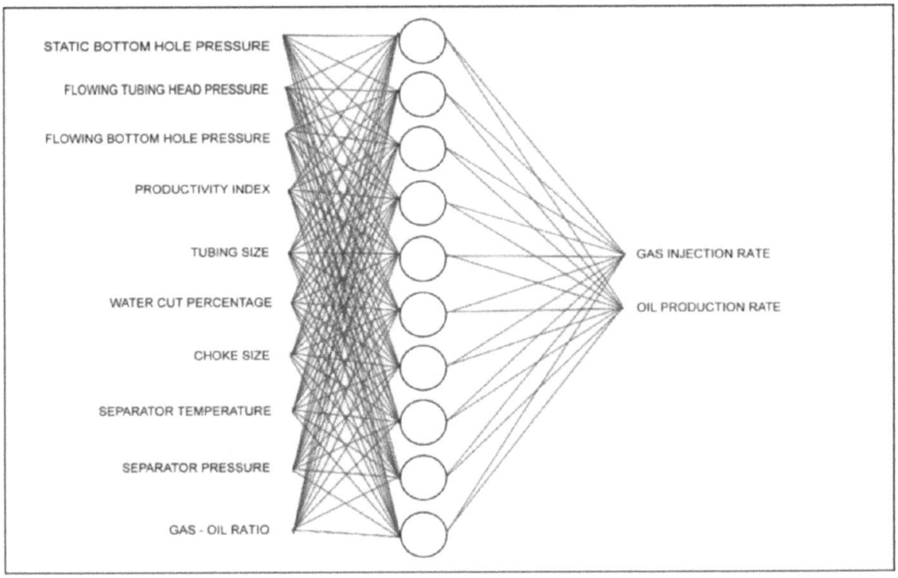

FIGURE 5.29 Gas lift optimum network architecture.

Source: Ranjan et al., 2015

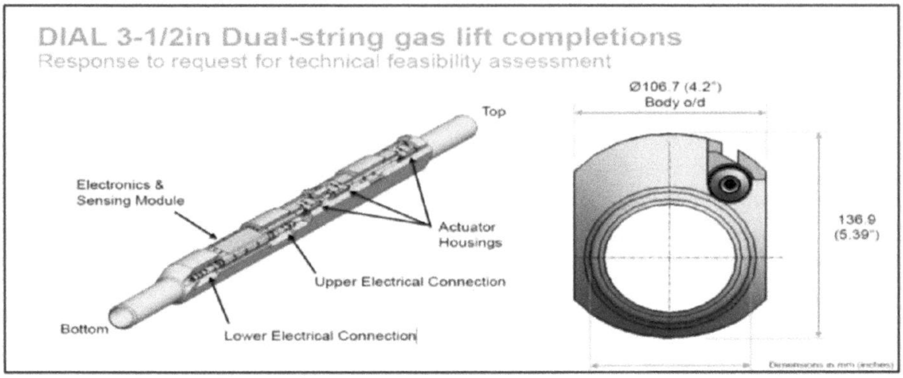

FIGURE 5.30 Proposed DIAL design for B field.

Source: Rozlan et al., 2022

The B field has the capacity to produce for approximately 50 years with 13 reservoir sequences. Used gas lift in this field exceeds 90%, with most strings flowing naturally in the early stages. A proposed DIAL system design is shown in Figure 5.30.

5.4.4.1 Digital Intelligent Artificial Lift (DIAL)

DIAL provides numerous benefits to control the system from the surface by electrical connection. It also eliminates some part of the gas lift system, such as side pocket mandrels, well intervention techniques, and wire line retrievals. The system can control up to six independent injection valves or orifices individually. Being able to control the rate of gas injection is another advantage of this technology, which includes a remote control with real-time analysis.

In Figure 5.31, a surface control system (SCS) box, MODUBUS connection, and SCADA communication system manage and monitor the system. The upper valve unit is used for well unloading, whereas the bottom units are operating units of orifices. Installed downhole sensors correspond to calculations of tubing and casing pressures and temperature values.

Installation of a DIAL system results in savings in either cost or non-productive time throughout controlling and monitoring of the system by PETRONAS and the service partner. DIAL technology introduces a fully digital and electronic system that provides real-time analysis of temperature and pressure. It can be deployed in highly inclined wells and allows for precise control of injection depth and rate.

5.5 OPTIMIZATION OF ARTIFICIAL LIFT

5.5.1 Optimum Well Completion for Marginal Field Development

Marginal field development relates to an oilfield that may not be worthy of developing at the present time. However, if technical or economic conditions change, such a field may become commercial. Therefore, it is very important to properly determine

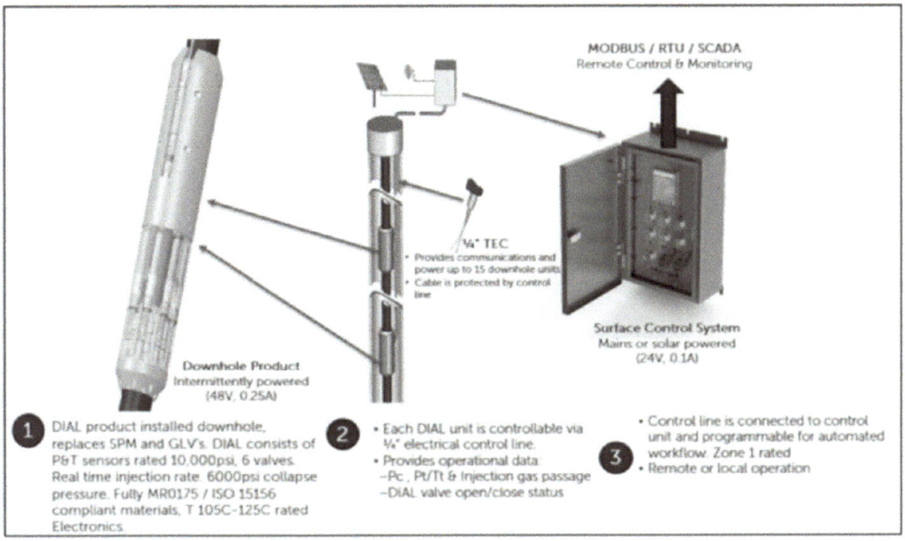

FIGURE 5.31 Surface and bottom-hole assembly design of DIAL system.

Source: Rozlan et al., 2022

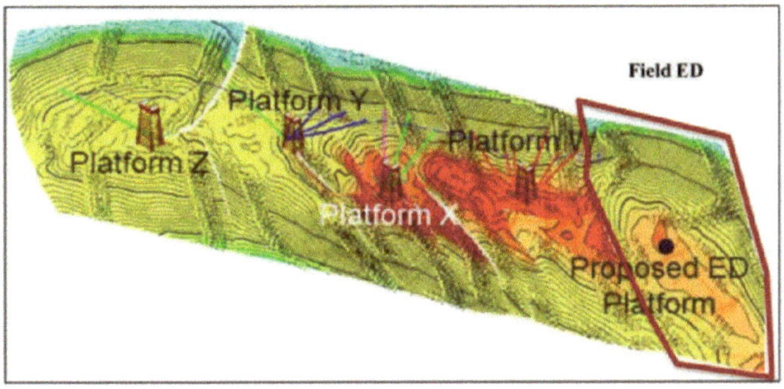

FIGURE 5.32 Location of field ED with nearby existing platform.

Source: Syarifah Puteh et al., 2022

all technological processes from drilling to well completion. In an article by Syarifah Puteh et al. (2022), the authors investigated eight completion strategies for the conditions of Malaysian oilfield ED (Figure 5.32), every of which has its own advantages and disadvantages.

The first step is to identify challenges this field can face. Analysis includes obtaining information regarding contaminant content (CO2, H2S), sand production issues, and pressure depletion. As was observed from the experience of adjacent field blocks and analyzing data from field ED itself, it is expected that this field will have aforementioned problems.

Completion types were evaluated based on the following factors:

1. Production flexibility
2. Early monetization
3. Operation complexity (drilling and completion)
4. Sand production management
5. Sand control failure probability
6. Associated risks
7. Stand-alone economics

Clearly, economic results were among the most important factors in decision-making. Net present value (NPV) is used as the economic criteria in each economic evaluation, and all possible risks were also taken into account. Table 5.8 shows the technical output for each completion strategy that has been identified.

Thus, a structured scoring system was applied by the authors (Syarifah Puteh et al., 2022) to give a clearer picture and lead the decision-making process. The key elements in the decision thought process are the associated cost of each option, the skin factor that leads to reserve prediction, and ultimately the net present value (NPV) indicator. These steps are summarized in Table 5.9. The traffic light method is used as the qualitative scoring system where:

- Circle represents lowest cost/least complexity/do-able operation—score of 5
- Triangle represents highest cost/highest complexity/unfeasible operation—score of 1
- Square represents the mid-cost/relatively simpler operation—score of 3

A process to determine optimal completion type and design will never give a single answer. While there are many well types and completion methods, it is very important to evaluate all decisive factors and options.

TABLE 5.8
Identified Completion Strategies

← Decreasing Cost

Subject	1 Monobore without san d control	2 Monobore with sand control	3 Cased & Perf	4 Cased & Perf with resin treatment	5 OHSAS	6 CHSAS	7 CHGP Circulating Pack Base Case	8 CHGP Frac Pack
Well Cost	- 70 %	- 60 %	- 50 %		- 50 %	- 30 %	-	+ 3 %
Skin values	3 (Drawdown Limit at 500psi)	5	3 (Drawdown Limit at 500psi)		5	5	15	3
Reserve (MMSTB)	+ 10 %						-	+ 10 %
Average Initial Production (STB/D)	+ 35 %	+ 40 %	+ 35 %		+ 40 %		-	+ 50 %

Source: Syarifah Puteh et al., 2022

TABLE 5.9
Pros, Cons and Risks Associated with Completion Strategies

Completion Concept	Monobore w/o sand control	Monobore with sand control	Cased & Perf	C&P with resin	OHSAS	CHSAS	CHGP	CHGP Frac Pack
Average Well Cost	- 50%	+ 3 %	- 20 %	- 30 %	- 15 %	- 10 %	*Base Case*	+ 25 %
Reversible Completion	■	▲	●	▲	●	●	▲	▲
Flexibility of Production (WSO, GSO etc.)	■ Straddle packer installation		●	●	■	●	●	●
Multiple Producing Zones	■ Less flexibility to manage multizonal production		■ Risk of sand accumulation	●	■ Unintended prod. from other zone	●	●	●
Early monetization	● All layers except for B02 & B40, has perf interval < 10m-TVD		●	●	■ Oil prod. hindered by unwanted fluid prod (diff zones)	●	●	●
Operation (drilling & completion) Complexity	●	■	●	●	■ # of swell packers	●	■ GP Operation	
Sand Production Management	Handle at surface	Handle at downhole and/or surface	Handle at surface	Handle at downhole and/or surface		Handle at downhole		
Years of Production*	N/A	N/A	> 20 years	N/A	> 8 years	> 8 years	> 15 years	
OPEX	▲ + 80 %	■ + 25 %	▲ + 80 %	■ + 80 %	■ + 25 %	■ + 20 %	*Base Case*	● 0
Chances of sand control failure*	No history	No history	N/A	N/A	No history	1 out of 4 strings	1 out of 10 strings	
Risk(s)	i. Sand build up in tubing ii. Erosion iii. Choke change frequency		i. Sand build up in tubing ii. Erosion iii. Choke change out frequency		i. Reserve is affected by prod of unwanted zones / fluid ii. Small particle management at surface	i. Small particle management at surface ii. Erosion	i. Lower production ii. Screen failure – proppant production (X20L & Y25S)	
Economic Results (NPV @ A%)	● + 35 %	■ + 20 %	● + 27 %	■ + 15 %	● + 27 %	● + 26 %	*Base Case*	■ + 20 %

** Based on adjacent fault block history*

Source: Syarifah Puteh et al., 2022

5.5.2 Exergoeconomic Optimization of Oil and Gas Production Systems

Today, fossil fuels comprise most of the primary global energy demand, and the energy system is one of the main sources of global greenhouse gas (GHG) emissions. In a study by Meziane Akchiche et al. (2020), the authors evaluate the overall energy efficiency and GHG emissions of a future offshore oil production system, implementing the exergy analysis with the help of a software platform.

The field development project discusses in this article consists of an offshore oil reservoir produced by three wells (W1, W2, and W3) (Figure 5.33). The performance of the oil production system was modeled and evaluated with three artificial lift and boosting techniques, including multiphase boosting, gas lift, and electrical submersible pumping (ESP) systems.

Nowadays, several engineering software platforms enable exergy calculations. Figure 5.34 shows the operating principle of numerical coupling that the authors developed for the calculation of physical, chemical, kinetic, and potential exergy.

To model and optimize the studied production system, classical and new approaches were used. The classic scheme of the software is based on the standard procedure and does not provide a high level of flexibility in the selection of optimization parameters.

Important information about the efficiency of three artificial lift and boosting techniques was obtained based on the exergy analysis.

5.5.3 Unlocking the Production Potential of Brown Fields through Gas Lift Optimization

Gas lift is a technique of artificial lift in which an external source of high-pressure gas is injected into the tubing/casing annulus to supplement the formation gas to lift well fluids by reducing the density of the fluids and lightening the formation fluid column. This project aims to increase the oil production rate by simulating the production system with different gas lift optimization scenarios using the production network model with a better approach to the complex interactions between

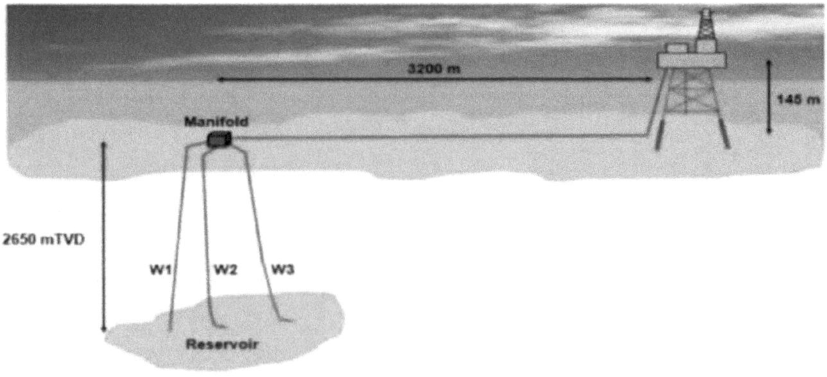

FIGURE 5.33 Scheme of the studied production system.

Source: Meziane Akchiche et al., 2020

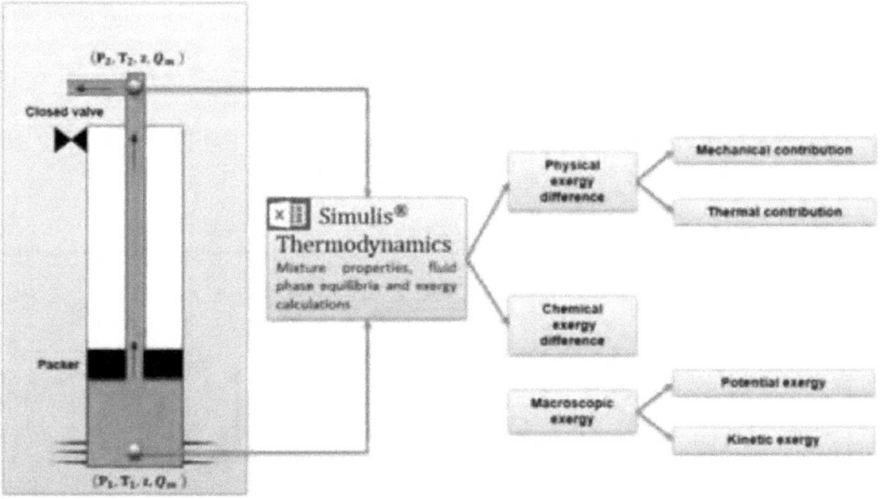

FIGURE 5.34 Nodal exergy analysis.

Source: Meziane Akchiche et al., 2020

reservoirs, wells, and pipelines. In this project, data were collected from 112 wells at the Zarzaitine field in North Africa, 95% of which are produced by gas lift system.

First, all well models were matched in to current conditions "well matching" by applying different techniques of model matching, depending on data availability. It is necessary to have complete and validated data to get a good match for each of the wells. For well model creation and calibration, if there is no way to get some required data, several educated assumptions can be made. For example, in this project:

It is assumed that the flowing bottom-hole pressure and reservoir pressure are the same to avoid mismatches due to hydrostatic levels. The optimal gas lift injection point is the deepest injection point.

- Flow gradient survey (FGS) data is provided by the company. This provides a pressure and temperature profile to decide which multiphase flow correlation should be used.

To match wells having FGS data, the following steps were used (Figure 5.35).

The gas lift's existing setup and injection systems are evaluated, and gas lift response evaluation and gas lift diagnostics are created for each well throughout the different separation centers.

5.5.3.1 Network Modeling

To understand wells' behavior before combining them, each separation model created a small cluster individually before integration into the network.

After integrating the wells in the network, the next step was to match field data for all the wells in the network. Figure 5.36 demonstrates that the network model is almost matched and ready for optimization.

FIGURE 5.35 Workflow to match wells with FGS data.

Source: Hammadi et al., 2020

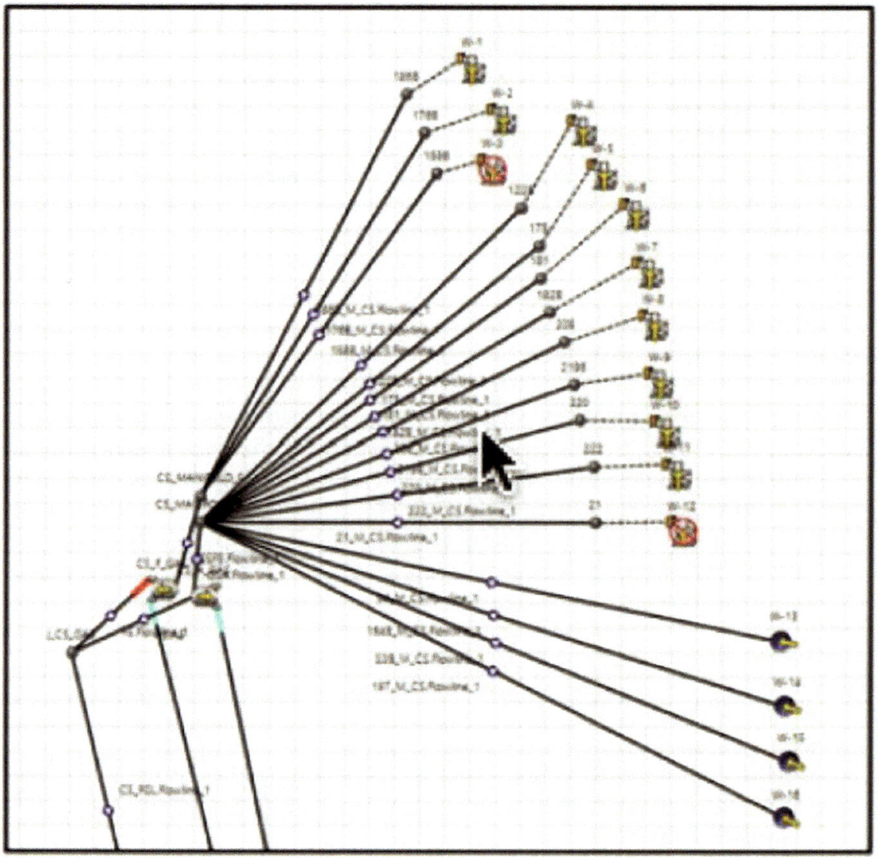

FIGURE 5.36 Separation center model (CS)

Source: Hammadi et al., 2020

After creating a network model to achieve an acceptable match with actual oil production at the C4 level, a network calibration and matching operation were done.

5.5.3.2 Gas Lift Optimization (GLO) Scenarios

Flow performance curves were constructed as a preprocessing step unless there was a substantial change in the wells, in which case all the curves had to be regenerated, which provides a quick visual comparison of the performance of all the wells. As a result of the GLO, by applying some local and global constraints, a 10% oil production increase was achieved.

5.6 CONCLUSION

In conclusion, the chapter provides in-depth insight into the production engineering aspects of artificial lift in unconventional resources. The selection and optimization of artificial lift methods are crucial for maximizing the production potential of oilfields, particularly mature fields. The use of digitalization and data science has revolutionized the approach to artificial lift, providing a more comprehensive and effective analysis of the performance of artificial lift systems. The intelligent decision-making process, based on MR multiphase flow detection, allows for a more accurate determination of the optimal artificial lift technique to be used in a specific field. The chapter also highlights various challenges faced with artificial lift, such as sand production and water cut analysis, and the innovative solutions that have been developed to address these challenges. The permanent magnet motor, plunger lift system, and sucker rod pumping system are just a few examples of the various techniques that have been developed to enhance the longevity and efficiency of artificial lift systems.

The importance of continued research and development in the field of artificial lift cannot be overstated. As the production of oil and gas from unconventional resources continues to grow, it is essential that we continue to develop and refine our understanding of the production engineering aspects of artificial lift. This will ensure that production outcomes are optimized, and the industry remains sustainable into the future. Overall, this chapter provides a comprehensive overview of the current state of artificial lift in unconventional resources and the ongoing efforts to improve and optimize this crucial aspect of oil and gas production.

REFERENCES

Akchiche, M., Beauquin, J.-L., Sochard, S., Serra, S., Reneaume, J.-M., & Stouffs, P. (2020). Exergoeconomic optimization of oil and gas production systems. *Paper presented at the SPE Europec*, Virtual, December. https://doi.org/10.2118/200607-MS

Ako, E., Nnanna, E., Somtochukwu, O., & Moradeke, A. (2021, August). Taming the monster: Arresting excessive sand production problem in ARAMU037. *SPE Nigeria annual international conference and exhibition* (p. D021S001R003). SPE.

Darvish Sarvestani, A., & Hadipour, A. (2019). Artificial lift method selection for mature oil fields: A case study. *Paper presented at the SPE annual caspian technical conference*, Baku, Azerbaijan, October. https://doi.org/10.2118/198424-MS

Deng, F., Chen, S., Chen, G., & Wang, M. (2022). Intelligent decision making and optimization of artificial lifting based on MR multi-phase flow detection. *Paper presented at the offshore technology conference Asia*, Kuala Lumpur, Malaysia, March. https://doi.org/10.4043/31349-MS

Ejim, C., & Xiao, J. (2020). Screening artificial lift and other techniques for liquid unloading in unconventional gas wells. *Paper presented at the Abu Dhabi international petroleum exhibition & conference*, Abu Dhabi, November. https://doi.org/10.2118/202653-MS

Gad, M., Fattah, M. A., Hassan, G., Zaied, S., Boccardi, A., Nady, A., & Sorour, A. (2017). Permanent magnet motor (PMM) as the right solution for electric submersible pump system challenges in Beleyem oilfields (case study). *Paper presented at the offshore mediterranean conference and exhibition*, Ravenna, Italy, March.

Hammadi, B., Agoudjil, K., Fahem, A., Tadjine, F., Benabdellah, M., Makhloufi, A., & Younsi, T. (2020). Unlocking the production potential of brown fields through gas lift optimization GLO. *Paper presented at the international petroleum technology conference*, Dhahran, Kingdom of Saudi Arabia, January. https://doi.org/10.2523/IPTC-19786-MS

Ias, P. (2018, February 28). Unconventional gas resources: Shale gas & coalbed methane. *PMF IAS*. Retrieved February 25, 2023, from www.pmfias.com/shale-gas-coalbed-methane-reserves-india-world-extraction-shale-gas-hydro-fracturing-fracking/

Kefford, P. A., & Gaurav, M. (2016). Well performance calculations for artificial lift screening. *Paper presented at the SPE annual technical conference and exhibition*, Dubai, September. https://doi.org/10.2118/181344-MS

Kuru, E., Yasar, E., & Ozgen, S. (2018). Effective zonal isolation in hydraulic fracturing of a tight gas reservoir by swellable packers. *Journal of Natural Gas Science and Engineering*, *52*, 308–316. https://doi.org/10.1016/j.jngse.2017.12.026

Maxwell, S. C., Prasad, M., & Mitchell, R. (2019). Microseismic monitoring and its role in unconventional reservoirs. *SPE Production & Operations*, *34*(2), 231–241.

Nugroho, B. H., Sumarto, B. A., Joenaedy, M. A., Al-Aradi, H. J., & Achmad, P. R. (2021). Data science for water cut analysis system by utilizing ESP sensor data, conceptual model, and its proposed solution. *Paper presented at the Abu Dhabi international petroleum exhibition & conference*, Abu Dhabi, November. https://doi.org/10.2118/207624-MS

Oyewole, P. (2016). Artificial lift selection strategy to maximize unconventional oil and gas assets value. *Paper presented at the SPE North America artificial lift conference and exhibition*. https://doi.org/10.2118/181233-ms

Pankaj, P., Patron, K. E., & Lu, H. (2018, September). Artificial lift selection and its applications for deep horizontal wells in unconventional reservoirs. *Unconventional resources technology conference, Houston, Texas, 23–25 July 2018* (pp. 3890–3910). Society of Exploration Geophysicists, American Association of Petroleum Geologists, Society of Petroleum Engineers.

Pugh, S. (1981). Concept selection: A method that works. *Proceedings of international conference on engineering design*, Heurista, Zürich, Switzerland, pp. 497–506.

Ranjan, A., Verma, S., & Singh, Y. (2015). Gas lift optimization using artificial neural network. *Paper presented at the SPE middle east oil & gas show and conference*, Manama, Bahrain, March. https://doi.org/10.2118/172610-MS

Rozlan, M. R., Ghazali, M., Haziq, M., Bakar, M. F., Ishak, M. F., Samuel, O. B., Misron, M. A. P., Ismall, M. S., Salleh, N. I., Rizwan, B., Amara, A. B., Faux, S., Ahmad A., Ahmad H., & Azis A. M. (2022). World's first successful dual string installation of a digital intelligent artificial lift DIAL—interventionless gas lift production optimization system, offshore Malaysia. *Paper presented at the offshore technology conference Asia, Virtual and Kuala Lumpur*, Malaysia, March. https://doi.org/10.4043/31563-MS

Standing, M. B., & Katz, D. L. (1942). Density of natural gases. *Transactions of the AIME*, *146*, 140–149.

Syarifah Puteh Mariah Syed Abd Rahim, M. F. A., Awang@Mohamed, Z., & Ahmad, M. Y. (2022). Defining the optimum well completion for marginal field development—An approach. *Paper presented at offshore technology conference Asia*, Kuala Lumpur, Malaysia, March. https://doi.org/10.4043/31552-MS

Temizel, C., Canbaz, C. H., Betancourt, D., Ozesen, A., Acar, C., & Krishna, S. (2020). A comprehensive review and optimization of artificial lift methods in unconventionals. *Paper presented at the SPE annual technical conference & exhibition*, Virtual, October.

Xu, X., Albuja, A., Izurieta, A., Wan, X., Yang, H., & Wang, Z. (2022). Sand management practices for extended ESP run life, a study case from a mature field in Ecuador. *Paper presented at the SPE Canadian energy technology conference*, Calgary, Alberta, Canada, March. https://doi.org/10.2118/208911-MS

Zhao, K., Tian, W., Li, X., & Bai, B. (2018). A physical model for liquid leakage flow rate during plunger lifting process in gas wells. *Journal of Natural Gas Science and Engineering*, *49*, 32–40.

Zhassulan, N. (2016). Plunger lift system restores production in previously abandoned gas condensate well. *SPE annual caspian technical conference* (p. SPE-182551). SPE.

6 Reservoir Engineering, Reservoir Management, and Surveillance in Unconventional Resources

Cenk Temizel, Cengiz Yegin, Jihoon Kim, Luigi Saputelli, and Ole Torsæter

6.1 STORAGE AND FLOW MECHANISMS

Unconventional reservoirs refer to sources of oil and gas that are more challenging to extract than conventional reservoirs. These reservoirs typically have low permeability, meaning that the movement of fluids, such as oil and gas, through the rock formations is restricted. This makes it more difficult for the fluids to flow naturally into wellbores and reach the surface.

To overcome these challenges, unconventional reservoirs require specialized storage and flow mechanisms. Two major techniques used are hydraulic fracturing and horizontal drilling. Hydraulic fracturing, also known as fracking, involves injecting a high-pressure fluid into the reservoir to create fractures or fissures in the rock formation. These fractures create pathways for oil and gas to flow through and be extracted. Fracturing fluids typically consist of water, chemicals, and proppants (such as sand) that help keep the fractures open, allowing for sustained production.

Horizontal drilling is another essential method used to improve the flow of oil and gas from unconventional reservoirs. Rather than drilling vertically, horizontal wells are drilled horizontally through the reservoir formations. This technique allows for greater access to unconventional reservoirs and increases the contact area between the wellbore and the reserves. By drilling horizontally, operators can access more reservoir rock and extract oil and gas from a larger volume. In addition to hydraulic fracturing and horizontal drilling, other storage and flow mechanisms may include well-stimulation techniques such as acidizing, water flooding, and gas injection. These methods are designed to enhance the recovery of oil and gas from unconventional reservoirs by altering the properties of the rock or improving fluid mobility.

Overall, the storage and flow mechanisms employed in unconventional reservoirs involve a combination of hydraulic fracturing, horizontal drilling, and various

DOI: 10.1201/9781003319955-6

other techniques to maximize the production of oil and gas from these challenging deposits.

6.2 WELL VARIABILITY ISSUES

Unconventional reservoirs can present several challenges when it comes to well variability. One key issue is the inherent heterogeneity of these reservoirs, which means that the properties and characteristics of the rock formations can vary significantly from one location to another.

This variability can affect multiple aspects of well performance. For example, it can affect the initial well productivity, with some areas yielding higher production rates while others may underperform. Understanding this variability is crucial for optimizing well placement and designing efficient stimulation techniques, such as hydraulic fracturing.

Another important factor contributing to well variability in unconventional reservoirs is the complex nature of reservoir fluid flow. Due to the presence of low-permeability rock, the flow of hydrocarbons (oil or gas) through the reservoir and into the wellbore can be highly challenging. Figure 6.1 illustrates the typical workflow process for unconventional reservoir exploitation and development.

Variations in reservoir pressure, saturation, and relative permeability can further complicate the production process and lead to varying well performance. In addition, unconventional reservoirs often exhibit a decline in well productivity over time, known as steep decline curves. This decline can be attributed to several factors, including the depletion of natural gas or oil resources, inefficient hydraulic fracturing, and uncertainties in reservoir characterization. Addressing these issues requires ongoing monitoring, data analysis, and advanced reservoir engineering techniques.

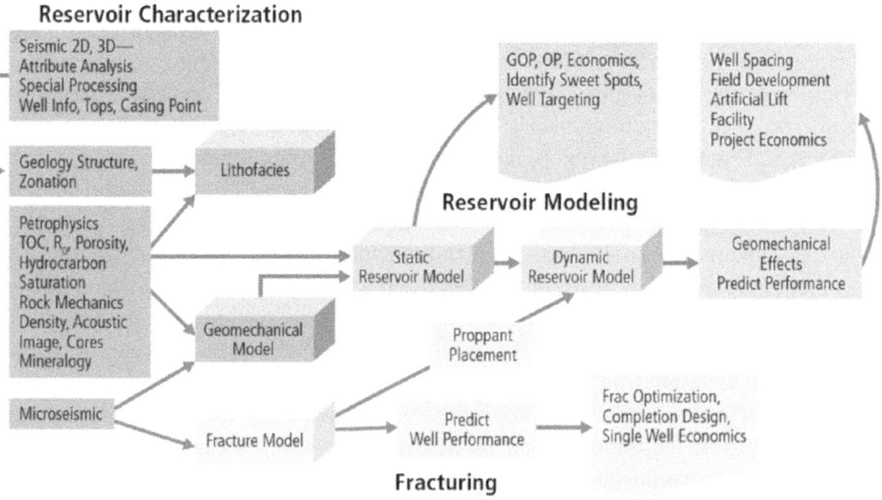

FIGURE 6.1 Workflow process and outline for unconventional resources exploitation and development.

To mitigate well variability issues, operators in unconventional reservoirs often rely on advanced technologies such as data analytics, machine learning, and reservoir simulation. These tools help in understanding and predicting reservoir behavior, optimizing well designs, and implementing effective production strategies. Continuous improvement in drilling techniques, completion methods, and reservoir characterization also plays a crucial role in managing well variability in unconventional reservoirs.

6.3 IMPACT OF FRACTURE CHARACTERISTICS

In recent years, unconventional energy sources including coalbed methane, tight oil, and shale gas have grown in importance. It is essential to comprehend how fracture features affect reservoir engineering, administration, and monitoring in order to effectively use these resources. Fractures are crucial for the effective development of unconventional reservoirs because they significantly increase the permeability and productivity of those reservoirs. Fractures greatly raise the formation's total permeability by supplying high-conductivity channels for fluid movement in low-permeability reservoirs (Thuwaini et al., 2009). These resources have significant hydrocarbon reserves that are ordinarily inaccessible, but the existence of natural fractures or produced hydraulic fractures can make such reserves accessible.

Performance of the well is directly influenced by the direction, density, and connectivity of fractures in the reservoir. To maximise production rates and final recovery, well placement and completion design may be improved with an understanding of the distribution and geometry of fractures. In unconventional reservoirs, fractures operate as conduits for the storage and transportation of fluid. The reservoir's reaction to production and injection activities is influenced by the fractures' capacity to hold and release enormous amounts of fluids. To anticipate production behavior and optimise field development methods, reservoir simulation models must accurately depict fractures. More accurate estimates of reservoir performance are possible, thanks to sophisticated numerical models that take fracture parameters into account.

The interplay between newly formed fractures and pre-existing natural fractures during hydraulic fracturing can have a big impact on how well the stimulation process works. To prevent unforeseen connections and potential interference between wells, hydraulic fracturing activities must be properly managed. Reservoir management is aided by continuous monitoring of fracture features using surveillance techniques such as microseismic monitoring, pressure transient analysis, and production recording. Making educated judgements about production and reservoir optimization is aided by understanding fracture behavior and their change over time (Demyanov et al., 2014). When assessing possible environmental effects, such as fluid migration to surrounding aquifers or surface water bodies, the orientation and size of cracks might affect the direction of fluid migration within the reservoir.

The economic feasibility of unconventional reservoirs is significantly influenced by the occurrence of productive and well-connected fractures. Operators can detect resource sweet spots and deploy resources more effectively by having a better understanding of the fracture network. Reservoir modeling, production forecasts, and reserve estimate are all made more uncertain by fracture characterization. To evaluate the range of probable outcomes and reduce uncertainty, adequate risk analysis and sensitivity studies are required. To optimize output, improve recuperation, and maximize economic advantages, it is crucial to comprehend fracture distribution, direction, connection, and behavior (Guerriero et al., 2013). Our capacity to tap into the enormous potential of unconventional resources will continue to grow with additional breakthroughs in fracture characterization and surveillance technology.

Proppants must be placed correctly into cracks during hydraulic fracturing operations in order to retain their conductivity. Engineers may create proppant placement techniques that maximize the contact area between the proppant and the fracture walls by understanding fracture characteristics, resulting in sustained production. Fracture interference in multi-well pad development can result from the interaction of fractures generated in several wells. Understanding these connections is essential for managing parent-child relationships and well-spaced parenting to prevent negative impacts on productivity (Gong & Rossen, 2017). Complex fracture networks are frequently seen in unconventional reservoirs as a result of stress-induced fractures, spontaneous fractures, or multistage hydraulic fracturing. Geomechanical modeling must accurately characterize these complexities in order to evaluate fracture stability and the possibility of induced seismicity.

The performance of improved oil recovery methods like gas injection and water flooding is also influenced by fracture features. Finding fracture-dominated flow routes enables more focused application of EOR techniques, increasing recovery rates. Fracture behavior can vary when reservoirs are depleted, impacting the reservoir's total pressure and fluid flow dynamics. Predicting reservoir performance throughout different production phases requires careful observation of pressure transients and fracture reactions. In unconventional reservoirs, cracks may eventually close as a result of a number of factors, including as stress redistribution, fluid imbibition, and mineral precipitation. Forecasting production and long-term reservoir management depends on an understanding of the behavior of fracture healing (Thuwaini et al., 2009).

Engineers can see and describe cracks in real time, thanks to advanced diagnostic tools, including electromagnetic imaging, tiltmeter surveys, and microseismic monitoring. During hydraulic fracturing operations and subsequent reservoir observation, these tools facilitate decision-making. Understanding fracture patterns can aid in identifying possible hazards of hydraulic fracturing, such as the potential for cracks to enter unwanted areas or serve as routes for fluid migration. Operators may assure ethical development of unconventional resources by proactively reducing environmental hazards. Unconventional wells have a decline behavior that is influenced by fracture features. Planning for well workovers or infill drilling may be done with more accuracy by analyzing the decline curves and comprehending the function that fractures play in production decline.

6.4 DRAINAGE VOLUME

In reservoir engineering, reservoir management, and the monitoring of unconventional resources, drainage volume is a crucial topic. It describes the area close to a wellbore where hydrocarbons can be generated or extracted effectively. In order to maximize the overall recovery from unconventional reservoirs, production tactics must be optimized, reserves must be estimated, and drainage volume must be understood. Calculating the drainage volume requires accurate reservoir characterization, which includes fracture characteristics, matrix permeability, porosity, and geomechanical parameters.

Engineers can define the area where hydrocarbons may be profitably produced from a well by having a thorough grasp of the geological qualities of the reservoir. For the most effective well location and spacing in unconventional resources, knowledge of the drainage volume is crucial (Britt & Smith, 2009). Engineers try to place wells so the drainage volume is covered as much as possible, maximizing total hydrocarbon recovery and minimizing interference between wells. By constructing a vast network of linked fractures, hydraulic fracturing designs are made to improve the drainage volume. Better production rates are made possible by well-planned fracturing procedures that maximize the contact area with the formation and permit enhanced connection with the reservoir.

For effective reservoir management, it is crucial to track the drainage volume over time. It is possible to spot variations in the drainage volume due to depletion and changes in reservoir conditions with the use of surveillance techniques like pressure transient analysis and production data analysis, allowing for well-informed judgements about production strategies and field development plans. Forecasting output is made more precise by estimating the drainage amount. Engineers can create accurate production estimates that may be used as decision-making tools by comprehending the reservoir's capacity for recovery and considering elements like pressure depletion and fluid flow dynamics (Williams et al., 2004).

The drainage volume's magnitude has a direct impact on how economically feasible it is to exploit unconventional resources. The drainage volume must be accurately estimated in order to make investment and field development planning decisions. In evaluating potential environmental effects, it is helpful to understand the drainage volume. Operators can put precautions in place to stop fluid movement into adjacent aquifers or other sensitive locations by determining the size of the drained area (Mohaghegh, 2011). Engineers can create more accurate simulations of reservoir behavior by including drainage volume data in reservoir models.

Pressure depletion in the drainage volume happens as hydrocarbons are extracted from unconventional reservoirs, which has an impact on fluid flow dynamics and well performance as a whole. Implementing efficient pressure management solutions, such as artificial lift or pressure maintenance techniques, to optimize production rates and ultimate recovery requires constant monitoring of pressure fluctuations and an understanding of pressure behavior inside the drainage volume (Yin, 2015). Due to intricate interactions between reservoir rock and hydraulic fractures, the fracture network inside the drainage volume may change throughout the course of a well's life. Making judgements on refracturing surgeries is influenced by knowledge of how fractures evolve over time. Operators can improve drainage connection and recover more hydrocarbons by stimulating new or existing fractures.

Finding sweet spots, or areas with the highest hydrocarbon contents and optimum production rates, depends on accurate drainage volume mapping. Geosteering techniques direct drilling within the most fruitful areas of the drainage volume using real-time data, such as well logs and seismic readings. A complete picture of reservoir behavior may be obtained by combining surveillance data, such as pressure, production, and reservoir modeling, with knowledge of the drainage volume (Britt & Smith, 2009). Operators may make well-informed decisions on well interventions, stimulation techniques, and reservoir management procedures thanks to this integrated approach. For efficient water management in the exploitation of unconventional resources, it is essential to comprehend the drainage volume.

Assessing the volume of incoming water and produced water movement enables responsible water recycling or disposal procedures, reducing possible environmental effects. Variations in the qualities of the rock may cause compartmentalization in some unconventional reservoirs, where different parts of the drainage volume exhibit different fluid flow characteristics. Designing effective production and injection methods requires locating these compartments and comprehending their relationship. Operators can assess the economic viability of various reservoir development scenarios and optimize production methods using knowledge of the drainage volume (Williams et al., 2004). Operators can maximize economic returns from unconventional resources by balancing drilling costs, stimulation costs, and anticipated reserves. The drainage volume should be well quantified to assist decrease reservoir modeling uncertainty.

6.5 STIMULATED RESERVOIR VOLUME (SRV) CHARACTERIZATION

Stimulated reservoir volume (SRV) characterization is an important aspect of understanding and optimizing production in unconventional reservoirs. It involves the assessment and quantitative analysis of the area around the wellbore that has been effectively stimulated by hydraulic fracturing. In unconventional reservoirs, such as shale or tight gas formations, the low-permeability nature of the rock makes it difficult for hydrocarbons to flow naturally. Hydraulic fracturing is commonly employed to create fractures and enhance permeability, allowing for the extraction of oil or gas. To characterize the stimulated reservoir volume, various methods and techniques are employed. These include the following:

Microseismic monitoring: This technique involves monitoring and analyzing the microseismic activity that occurs during the hydraulic fracturing process. By analyzing the location and magnitude of microseismic events, engineers can infer the extent and geometry of the stimulated volume.

Production analysis: By analyzing the production data from multiple wells in a field, engineers can observe the production behavior and infer the extent of the stimulated reservoir volume. This includes studying the production decline curves, flow rates, pressure data, and fluid composition.

Tracer tests: Tracer tests involve injecting a known tracer substance into the well during hydraulic fracturing and monitoring its movement through the reservoir. By analyzing the concentration and arrival time of the tracer at different wells, engineers can estimate the dimensions of the stimulated volume.

Pressure transient analysis: This technique involves analyzing the pressure response of the reservoir during production or interference tests. By studying the pressure data, engineers can estimate the extent of the stimulated reservoir volume and evaluate the reservoir properties.

The purpose of SRV characterization is to understand the effective stimulated area around the wellbore and assess the connectivity and productivity of the reservoir. This information is crucial for optimizing well placement, designing stimulation strategies, and improving reservoir management practices in unconventional reservoirs. Through detailed SRV characterization, operators can make informed decisions to maximize hydrocarbon recovery and optimize production economics.

6.6 RESERVOIR ROCK PROPERTIES AND RELATIVE PERMEABILITY IN MATRIX/FRACTURES

Reservoir rock properties and relative permeabilities play crucial roles in the development and production of unconventional oil and gas resources. In unconventional reservoirs, such as shale and tight formations, understanding the properties of the matrix and fractures is essential for effective reservoir characterization and production optimization. Reservoir rock properties refer to the physical and chemical characteristics of the rocks comprising the reservoir. This includes parameters like porosity, permeability, mineralogy, and organic content. These properties dictate the ability of the rock to store and transmit fluids, such as oil and gas.

In unconventional reservoirs, the matrix represents the low-permeability rock, typically shale, between the fractures. The matrix has significantly lower permeability than the fractures, and its properties influence fluid storage and flow within the reservoir. Fractures, on the other hand, are essential conduits for fluid flow in unconventional reservoirs. They can enhance reservoir permeability and facilitate better hydrocarbon extraction. The properties of fractures, including their density, length, and connectivity, directly impact fluid flow and production rates.

Relative permeability refers to the ability of different phases of fluid (e.g., oil, water, and gas) to flow through a porous medium. In unconventional reservoirs, determining the relative permeabilities of fluids in both the matrix and fractures is crucial for accurately predicting the reservoir behavior and optimizing production strategies. Understanding the reservoir rock properties and relative permeabilities in matrix and fractures helps in several ways. It aids in designing hydraulic fracturing operations, optimizing well placement, estimating production rates, and assessing the overall potential of an unconventional reservoir.

Different techniques are employed to evaluate these properties, including laboratory experiments, core analysis, well logging, and advanced imaging technologies. The data collected with these methods help reservoir engineers and geoscientists build accurate models and make informed decisions to maximize hydrocarbon recovery from unconventional reservoirs.

Relative permeability is a property that describes how easily fluids flow through a porous medium, such as a rock formation. In the context of matrix and fracture systems, different formulas are used to calculate relative permeability. For matrix permeability, the Corey-Brooks formula is often used. It relates the relative permeability of a fluid phase (oil, water, gas) to the saturation of that phase in the matrix. The formula is typically expressed as a function of the capillary pressure, which represents the resistance to fluid flow in the pores.

On the other hand, for fracture permeability, the formula can vary depending on the specific model used to describe the fluid flow in the fractures. Some common fracture models include the linear flow model, the cubic law model, and the Carman-Kozeny equation. It's important to note that these formulas are empirical and may vary depending on the specific rock properties, fluid characteristics, and the complexity of the system being studied. Additionally, there are different experimental methods to measure relative permeability, such as core analysis and numerical simulations.

6.7 PRESSURE, VOLUME, AND TEMPERATURE (PVT)

PVT stands for pressure, volume, and temperature, key properties that are important in the oil and gas industry to study the behavior of fluids in reservoirs. Unconventional reservoirs typically refer to reservoirs that have more challenging or complex characteristics than conventional reservoirs, such as tight oil and gas formations, shale gas, or coalbed methane.

In the context of oil and gas production, understanding the PVT properties of fluids in unconventional reservoirs is crucial for reservoir engineers, production engineers, and other professionals in the industry. PVT analysis helps determine the composition, phase behavior, and physical properties of the reservoir fluids, which, in turn, influence production strategies, well design, and reservoir management. By studying PVT properties, engineers can better predict fluid flow, estimate reserves, optimize production rates, select suitable recovery methods (such as hydraulic fracturing or enhanced oil recovery techniques), and make informed decisions regarding reservoir development and economic feasibility.

The PVT properties for compositional reservoir simulation can be determined by conducting laboratory experiments on fluid samples that accurately represent the reservoir conditions. However, in shale formations where the pore sizes are considerably smaller, it is possible that the availability of fluid molecules might be somewhat constrained. For example, a methane molecule is just one-fifth the size of a pore, with a diameter of 2 nm. In other words, the pore would be five times bigger than the methane molecule. In terms of pore size, there is a significant distinction between macroscale pores and shale nanopores.

6.8 PRESSURE TRANSIENT ANALYSES (PTA)

Pressure transient analyses, also known as pressure transient testing or well testing, are techniques used in the field of petroleum engineering to evaluate the behavior and characteristics of fluid flow in oil and gas reservoirs. These analyses involve monitoring pressure changes over time to gain insights into reservoir properties, storage capacity, productivity, and other key parameters. During a pressure transient analysis, data from well tests are collected and analyzed to understand how the pressure in the wellbore and reservoir changes in response to changing flow rates. These tests typically involve altering the flow rate or shutting in the well and monitoring the resulting pressure response.

The information obtained from pressure transient analyses helps reservoir engineers estimate important parameters such as reservoir permeability, skin factor (a measure of formation damage near the wellbore), and the presence of boundaries within the reservoir. This data plays a critical role in reservoir characterization, field development planning, and production optimization. Various analytical and numerical techniques are used to interpret the pressure data. Examples include the construction of pressure-depth plots, pressure derivative analysis, and numerical simulations to match the observed pressure behavior. Advanced well test analysis software and models are often used to aid in the interpretation process.

Pressure transient analysis is an essential tool for understanding and optimizing reservoir performance, ensuring efficient recovery of hydrocarbons, and maximizing production rates. By accurately characterizing reservoir properties, engineers can make more informed decisions regarding production strategies, wellbore placement, and reservoir management. Pressure transient analysis (PTA) is an important technique in evaluating the behavior of reservoirs, including unconventional resources. It involves analyzing pressure changes during well testing to gain insights into reservoir characteristics and performance.

6.9 TRANSIENT ANALYSES (RTA)

Transient analyses in unconventionals, such as in shale gas or tight oil reservoirs, are essential for understanding the behavior of these reservoirs over time. These analyses involve studying the flow dynamics and pressure changes during production or injection operations, considering the complex geomechanical and fluid properties of these unconventional reservoirs. Transient analyses aim to predict the behavior of the reservoir and optimize production or injection strategies by considering factors like fluid flow, reservoir heterogeneity, fracture networks, and stress redistribution effects. The method helps in determining well performance, estimating reserves, and evaluating the economic viability of these reservoirs.

Key aspects of transient analyses in unconventionals include analyzing pressure buildup or decline during production or injection, characterizing flow mechanisms and fluid flow regimes, assessing the impact of stimulated reservoir volume and fracture network connectivity, and optimizing well placement and completion designs.

Various numerical simulation techniques, such as numerical reservoir modeling, reservoir simulation, and fracture network modeling, are employed to perform these transient analyses. Additionally, combining the knowledge gained from field measurements, production data analysis, and laboratory experiments helps refine the understanding of the reservoir behavior and improve the accuracy of the analyses. By conducting transient analyses, operators can gain valuable insights into the behavior of unconventional reservoirs, optimize production or injection strategies, and make informed decisions regarding reservoir management and development plans.

6.10 GEOMECHANICS

Geomechanics in unconventional refers to the study of the mechanical behavior of rock formations in unconventional reservoirs, such as shale, tight gas, and coalbed methane. Unconventional reservoirs are characterized by low permeability, which makes the extraction of hydrocarbons challenging. Geomechanics plays a crucial role in understanding the behavior of these reservoirs and optimizing hydrocarbon recovery. In geomechanics, several factors are considered, such as rock properties, stress distribution, and the interaction between rock and fluid within the reservoir. These factors influence the success of hydraulic fracturing, which is a key technique used to enhance the permeability of unconventional reservoirs.

By analyzing the properties of the rock formations, engineers and geoscientists can determine the optimal design and placement of hydraulic fractures. They also assess the potential risks associated with the fracturing process, such as induced seismicity and wellbore stability issues. Geomechanics in unconventionals involves various techniques, including laboratory testing, numerical modeling, and real-time monitoring during drilling and fracturing operations. These methods help in understanding the stress regimes, rock deformation, and the potential for reservoir compaction or dilation. Overall, geomechanics in unconventionals is an essential field of study that enables engineers to optimize hydrocarbon extraction in unconventional reservoirs by understanding the behavior of the rock formations and their response to hydraulic fracturing techniques.

One of the main reasons completion optimization is important is because accurately predicting pore pressure plays a crucial role in this process. Pore pressure helps reduce the forces that are transmitted through the rock's structural framework (Couzens-Schultz et al., 2013; Xia & Michael, 2015). In Terzaghi's principle, the variation between stress and pore pressure is known as effective stress. This effectively describes the distinction between these two forces.

$$\sigma_e = \sigma_T - P_p \tag{6.1}$$

where σ_e is effective stress, σ_T is total stress, and P_p is pore pressure.

The deformation of a porous medium is influenced by the in situ stresses, which have varying magnitudes from different directions as well as the pore pressure. The amount of deformation experienced depends on the effective normal stress and shear stress in different orientations.

6.11 GEOCHEMISTRY

Geochemistry in unconventionals refers to the study of the chemical composition and reactions within unconventional reservoir rocks, such as shale, tight gas, and coalbed methane formations. It focuses on understanding the sources, migration, and abundance of hydrocarbons in these reservoirs. In geochemistry, researchers analyze the organic matter content of the rocks, as well as the composition of fluids, gases, and minerals present in the reservoirs. They also investigate the thermal maturation and transformation of organic matter into hydrocarbons, helping determine the potential for hydrocarbon generation and accumulation.

By studying the geochemical properties of unconventional reservoirs, scientists can assess the quality and quantity of hydrocarbon resources. They examine the organic richness, thermal maturity, and kerogen type to estimate the hydrocarbon potential of a reservoir. This information is crucial for identifying economically viable locations for drilling and production in unconventional plays.

Figure 6.2 shows how organic matter and volatile products change as we go deeper into the ground. On the left side, it demonstrates how potential source rock that is not yet thermally mature is buried and preserved. This preservation process is beneficial when anoxic (low oxygen) conditions are present during diagenesis. On the right side, we see that this burial and maturation of potential source rock can lead to the formation of effective source rock for conventional and unconventional petroleum accumulations. In the case of unconventional petroleum, it remains within the source rock, turning it into reservoir rock as well. Essentially, the diagram illustrates the stages of organic matter preservation, source rock maturation, and petroleum accumulation in both conventional and unconventional reservoirs (Peters et al., 2015).

Geochemical analyses also provide insights into the reservoir's history, including its depositional environment, burial history, and thermal evolution. This information helps in understanding the reservoir's geochemical characteristics and predicting its behavior during production. Understanding the geochemistry of unconventional reservoirs is vital for optimizing hydrocarbon recovery techniques. It allows engineers to design effective stimulation and production strategies, such as hydraulic fracturing and enhanced oil recovery methods. Overall, geochemistry plays a significant role in unraveling the hydrocarbon potential and behavior of unconventional reservoirs, aiding in decision-making processes related to exploration, development, and production activities.

6.12 NANOSCALE PORE-THROAT DIAMETER EFFECTS IN UNCONVENTIONALS

Nanoscale pore-throat diameter effects in unconventionals refer to the impact of the size of tiny interconnected channels, or pores, in unconventional reservoir rocks.

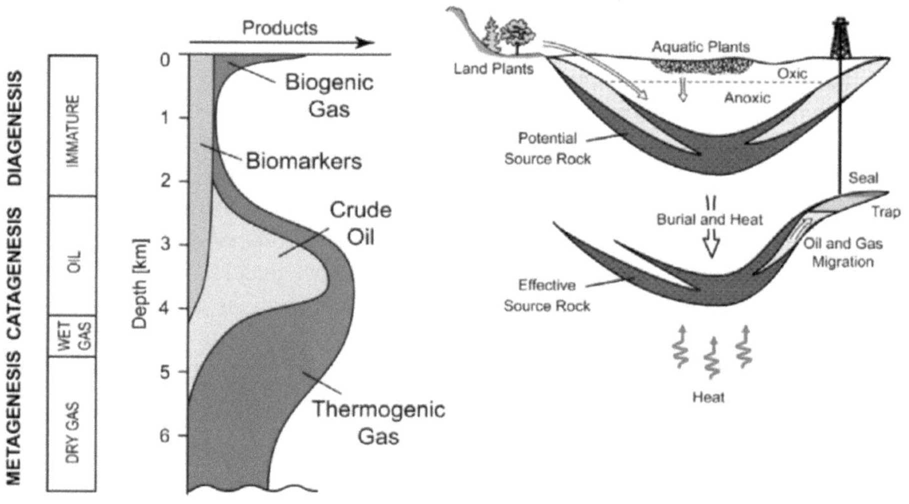

FIGURE 6.2 Three stages in the evolution of sedimentary organic matter.

Source: Peters et al., 2015

Unconventionals are oil and gas reservoirs trapped within low-permeability formations, such as shale. Understanding the behavior of these pores is crucial in determining how hydrocarbons flow through the rock matrix. At the nanoscale level, the size of the pore throats can significantly influence fluid flow. Smaller pore throats can limit the movement of hydrocarbons, impeding production rates and recovery. Conversely, larger pore throats provide more pathways for fluid migration, facilitating increased production.

The study of nanoscale pore-throat diameter effects in unconventionals involves characterizing the complex pore network, quantifying the size distribution, and investigating the relationship between pore sizes and fluid transport properties. This research is important for optimizing drilling and completion techniques, designing stimulation strategies, and enhancing hydrocarbon recovery from unconventional reservoirs.

Pore structure: Unconventional reservoirs are composed of fine-grained rocks with complex pore structures. The pore network consists of a combination of different pore sizes, including nanoscale throats that connect the pores. These throats play a critical role in determining fluid flow behaviour.

Fluid flow: Fluid flow in nanoscale pore throats is influenced by multiple factors, including surface forces, fluid viscosity, capillary pressure, and the wettability of the rock surface. The size and geometry of the pores and throats affect the efficiency of hydrocarbon transport through the rock matrix.

Knudsen diffusion: In very narrow nanoscale throats, a phenomenon called Knudsen diffusion becomes significant. Knudsen diffusion occurs when molecular collisions dominate over viscous flow, resulting in significantly reduced fluid flow rates. This diffusion mechanism becomes more prominent with decreasing pore-throat diameter.

Recovery efficiency: The understanding of nanoscale pore-throat diameter effects is crucial for optimizing recovery techniques. The presence of smaller pore throats can lead to lower hydrocarbon recovery due to restricted fluid flow and increased capillary trapping. Conversely, larger pore throats contribute to improved recovery rates by providing more pathways for fluid migration.

Characterization techniques: Researchers employ various techniques to characterize the nanoscale pore structure, such as scanning electron microscopy (SEM), focused ion beam (FIB), and advanced imaging techniques like nanoparticle tracking. These methods help in quantifying the size distribution, connectivity, and spatial arrangement of pores and throats. By gaining a deeper understanding of nanoscale pore-throat diameter effects, engineers and geoscientists can develop strategies to optimize drilling, completion, and stimulation techniques for unconventional reservoirs. This knowledge can enhance hydrocarbon recovery and improve the economics of unconventional resource development.

6.13 ROCK FLUID INTERACTIONS

Rock fluid interactions in unconventionals, also known as RFIs, are an important aspect of studying and understanding unconventional reservoirs. Unconventional reservoirs, such as shale formations, have unique properties that affect the interaction between rocks and fluids. RFIs play a crucial role in the production and recovery of hydrocarbons from these reservoirs. The study of RFIs involves investigating how different fluids interact with the rocks in unconventional reservoirs, including the adsorption, desorption, and flow of fluids within the rock matrix. Understanding these interactions is essential for optimizing production techniques, enhancing recovery factors, and mitigating potential issues like formation damage or wellbore instability.

Researchers and engineers study RFIs to analyze parameters such as mineralogy, pore structure, rock wettability, and fluid properties. By studying these factors, they can better understand how different fluids interact with the rock matrix, influencing permeability, porosity, and ultimately, hydrocarbon production. In recent years, numerous technological advancements, including hydraulic fracturing and horizontal drilling, have unlocked vast reserves of hydrocarbons from unconventional reservoirs. However, overcoming the challenges associated with RFIs remains crucial for efficient and sustainable production.

Researchers continue to explore new techniques and technologies to improve our understanding of RFIs. This knowledge can lead to the development of more advanced reservoir models, improved well designs, and optimized production strategies for unconventional reservoirs.

Adsorption and desorption: Unconventional reservoir rocks, such as shale, have high organic content and complex mineral compositions. These rocks have the ability to adsorb and desorb different types of fluids, including hydrocarbons and water. Understanding the adsorption and desorption characteristics of these rocks is important for estimating hydrocarbon reserves, assessing storage capacities, and designing effective production techniques.

Fluid flow in the rock matrix: In unconventional reservoirs, fluid flow occurs within the ultra-tight pores and fractures of the rock matrix. The pore size, connectivity, and distribution impact fluid movement within the reservoir. Researchers analyze the intricate flow mechanisms and permeability changes that occur during production to optimize recovery strategies and minimize production challenges.

Pore structure and wettability: The pore structure and wettability of the rocks significantly influence fluid-rock interactions. Pore size distribution, surface roughness, and connectivity affect the amount of fluid that can be stored and extracted from the rock. Additionally, the wettability of the rock surface plays a crucial role in determining how fluids spread and move within the pore system.

Fluid properties: Understanding the properties of the fluids that interact with the rocks is necessary for predicting their behavior in unconventional reservoirs. Different types of fluids, such as oil, gas, and water, have unique interactions with the rock matrix. Factors such as viscosity, density, and interfacial tension impact fluid flow, diffusion, and adsorption.

Formation damage and wellbore instability: RFIs can result in formation damage and wellbore instability issues, reducing well productivity. For instance, in hydraulic fracturing operations, the fluid-rock interaction can lead to proppant embedment, fines migration, or swelling/shrinking of the clay minerals. Researchers analyze these interactions to develop strategies to mitigate formation damage and ensure wellbore stability.

By studying these aspects of RFIs and continually improving our understanding, researchers and engineers can optimize production techniques, enhance hydrocarbon recovery, reduce operational risks, and develop sustainable approaches for exploiting unconventional reservoirs.

6.14 CONDENSATE BANKING AND MITIGATION, WORKFLOWS, METHODS

Condensate banking occurs in gas-condensate reservoirs when the pressure drops below the dew point, causing liquid hydrocarbons (condensate) to accumulate near the wellbore. This accumulation restricts the flow of gas and can significantly reduce gas production rates. Effective mitigation strategies are key to maintaining production levels and maximizing recovery.

Reservoir modeling plays a critical role in understanding the behavior of condensate banking. Engineers create numerical models that represent the reservoir's geology, fluid properties, and well configuration. These models simulate the flow of gas and liquid phases over time, helping predict the potential buildup of condensate and its impact on production rates.

Well testing is often performed to measure key reservoir properties and confirm the validity of the models. Pressure transient tests, such as buildup and drawdown tests, provide valuable data about the reservoir's behavior and characteristics. This information is then used to calibrate the reservoir models and improve their accuracy. Once the presence of condensate banking is confirmed, a variety of mitigation methods can be implemented.

Gas cycling involves injecting gas back into the reservoir, either through separate injection wells or by recycling gas produced from the same well. This helps maintain reservoir pressure and mobilize the accumulated condensate, facilitating improved gas production.

Increasing drawdown refers to lowering the pressure at the wellbore, which helps reduce the condensate saturation near the well. This can be achieved by adjusting production rates or implementing artificial lift techniques, such as gas lifts or electric submersible pumps.

Chemical treatments can be applied to reduce the interfacial tension between the gas and condensate, making it easier to displace and produce liquid hydrocarbons. Surfactants and foams are commonly used to modify the fluid properties and improve flow characteristics.

Reservoir stimulation techniques, such as hydraulic fracturing, can be employed to enhance the permeability of the reservoir. By creating fractures and increasing the network of flow paths, these methods improve the connectivity between the well and the reservoir, helping reduce the negative impact of condensate accumulation.

It's important to note that the specific methods and workflows employed for condensate banking mitigation depend on the unique characteristics of each reservoir. The geology of the formation, fluid properties, well completion design, and production strategy are all considered when developing mitigation strategies.

Continuous research and development efforts in the oil and gas industry aim to further refine the understanding of condensate banking and improve mitigation techniques. Advanced modeling approaches, enhanced characterization methods, and innovative chemical treatments are continuously explored to optimize production in unconventional reservoirs and ensure efficient recovery of both gas and condensate resources.

6.15 WORKFLOWS FOR LIQUID LOADING PREVENTION

Liquid loading prevention is an important aspect of unconventional oil and gas production. When the production rate declines, liquids such as water and condensate can accumulate in the wellbore, impeding the flow of gas and reducing productivity. To prevent liquid loading, various workflows can be implemented. Here are a few key steps:

Continuous monitoring: Regular monitoring of production rates, bottom-hole pressure, and fluid levels is crucial. This allows operators to detect early signs of liquid loading and take preventive measures promptly.

Artificial lift methods: Implementing artificial lift methods such as gas lift or plunger lift can help alleviate liquid loading. These techniques introduce gas into the wellbore to lift the accumulated liquids, allowing for enhanced gas production.

Velocity string: Installing a velocity string is another effective preventive measure. It reduces the diameter of the wellbore, increasing the gas flow velocity and preventing liquids from accumulating.

Chemical treatments: Chemical additives can be used to alter the surface tension of produced fluids, reducing the propensity for liquid to accumulate. This approach can enhance gas flow and delay the onset of liquid loading.

Production optimization: Optimizing production parameters such as choke sizes, tubing sizes, and operating pressures can help manage liquid loading. Balancing gas and liquid flow rates ensures efficient production and prevents liquid accumulation.

Wellbore hydraulics analysis: Conducting a comprehensive analysis of wellbore hydraulics can provide insights into fluid behavior and identify potential liquid loading scenarios. Advanced simulators can be used to model various production scenarios and aid in decision-making.

It's important to note that each unconventional reservoir is unique, so it's crucial to consider reservoir characteristics and historical production data to tailor liquid loading prevention workflows accordingly. Regular monitoring and proactive measures can help ensure sustained gas production and economic viability.

6.16 CHOKE OPTIMIZATION

Choke optimization in unconventional reservoirs is an important aspect of well production management. Unconventional reservoirs, such as shale gas and tight oil formations, have unique characteristics that require specific techniques for successful development and production. The choke is a device used in the production system that regulates the flow of fluids from the reservoir to the surface. It is typically located downhole, near the bottom of the well. Choke optimization refers to finding the optimal choke size and operating conditions to maximize production while minimizing potential issues such as sand production, liquid loading, or formation damage.

In unconventional reservoirs, choke optimization plays a crucial role in achieving optimal flow rates and maintaining reservoir performance. Several factors need to be considered when optimizing the choke, including reservoir properties, well completion design, and surface facilities. One common objective in choke optimization is to control the drawdown pressure in the reservoir. By adjusting the choke size, operators can manage the pressure differential across the reservoir and maintain the desired flow rate. Another objective is to prevent formation damage or sand production, which can occur due to excessive drawdown pressures or improper choke sizes.

Choke optimization in unconventionals often involves field monitoring and data analysis to understand the reservoir behavior and identify opportunities for optimization. Advanced technologies like real-time monitoring systems and production surveillance tools can provide valuable insights into the well performance, allowing operators to make informed decisions. Additionally, reservoir simulation models can be used to optimize choke settings by simulating different scenarios and predicting the impact on production. These models consider the reservoir properties, wellbore geometry, and fluid properties to simulate the flow behavior in the formation and provide valuable guidance for choke optimization strategies.

Overall, choke optimization in unconventional reservoirs is a multi-disciplinary approach that requires a thorough understanding of reservoir engineering, production operations, and data analysis techniques. By continuously optimizing the choke settings, operators can enhance production efficiency, improve ultimate recovery, and maximize the economic returns from unconventional resources.

6.17 WELL MANAGEMENT

Well management in unconventional reservoirs is a complex process that involves various aspects of planning, drilling, completion, and production. Unconventional reservoirs, such as shale gas and tight oil formations, require specific well-management strategies to achieve optimal production and maximize the recovery of hydrocarbons.

Here are some key considerations in well management for unconventional reservoirs:

Planning and design: Well planning in unconventional reservoirs involves assessing the reservoir characteristics, determining the optimal well locations, and designing the well trajectory. Factors such as reservoir depth, geomechanical properties, and the presence of natural fractures are considered during the planning phase.

Drilling techniques: Unconventional wells are often drilled horizontally or directionally to maximize contact with the reservoir. Advanced drilling techniques, such as extended-reach drilling or multi-lateral wells, may be employed to increase reservoir exposure and production potential.

Well completion: Completion techniques in unconventional reservoirs typically involve hydraulic fracturing, in which a high-pressure fluid is injected to create fractures in the reservoir. This enables the flow of hydrocarbons from the tight formation. The well completion design includes considerations such as fracture spacing, proppant volume, and fluid selection to optimize production.

Production optimization: Once the well is completed, continuous monitoring and optimization are necessary to maintain production rates and reservoir performance.

Techniques such as choke optimization and artificial lift methods may be employed to manage flow rates, prevent liquid loading or sand production, and enhance ultimate recovery.

Data analysis and monitoring: Well management in unconventional reservoirs relies heavily on data analysis and real-time monitoring. Advanced sensors, downhole gauges, and monitoring systems provide valuable insights into well performance, reservoir behavior, and hydraulic fracture effectiveness. Operators can use this data to make informed decisions and adjust management strategies accordingly.

Environmental considerations: In addition to production optimization, well management in unconventional reservoirs also involves managing environmental impacts. Measures are taken to minimize water usage, manage flowback fluids, and mitigate potential air emissions associated with production activities.

Overall, well management in unconventional reservoirs requires a comprehensive understanding of reservoir engineering, drilling and completions technology, and data analysis. By implementing effective well-management strategies, operators can maximize the potential of unconventional resources and improve the economic viability of these reservoirs.

6.18 RESERVOIR MANAGEMENT

Reservoir management in unconventional reservoirs is crucial for optimizing production and maximizing the recovery of hydrocarbons. Unconventional reservoirs, such as shale gas and tight oil formations, have unique characteristics that require specific approaches to reservoir management.

Here are some key aspects of reservoir management in unconventional reservoirs:

Reservoir characterization: Understanding the geological and reservoir properties is essential for effective reservoir management. This involves studying the rock composition, porosity, permeability, natural fracture networks, and other factors that influence fluid flow and hydrocarbon recovery. Advanced techniques like core analysis, well logging, and geophysical surveys are used to characterize the reservoir.

Reservoir modeling: Building accurate reservoir models is crucial for reservoir management in unconventional reservoirs. Reservoir simulation software is used to capture the complex behavior of the reservoir and predict how different production strategies may impact recovery. These models consider factors such as reservoir heterogeneity, natural fractures, and fluid properties to provide insights into the reservoir's behavior.

Production forecasting: Accurate production forecasting helps in planning and decision-making. By combining reservoir data with production history, operators can estimate future production rates and optimize field development strategies. Production forecasting models are continuously updated as new data becomes available, allowing for better reservoir management.

Hydraulic fracturing optimization: Hydraulic fracturing is a common method employed in unconventional reservoirs to enhance hydrocarbon recovery. Reservoir management involves optimizing the hydraulic fracturing design by selecting suitable fracturing fluids, proppant types, and injection parameters. It also involves monitoring fracture effectiveness with the help of pressure and microseismic data and adjusting fracture spacing and orientation accordingly.

Reservoir monitoring: Continuous reservoir monitoring is essential to identify changes in reservoir behavior and make informed decisions. Technologies like pressure gauges, production logging tools, and downhole sensors enable real-time monitoring of reservoir performance. Operators use this data to assess well productivity and identify opportunities for reservoir management optimization.

Enhanced recovery techniques: In addition to hydraulic fracturing, unconventional reservoirs may benefit from enhanced recovery techniques such as water flooding or gas injection. These methods can help improve sweep efficiency and increase hydrocarbon recovery. Reservoir management includes evaluating the feasibility and effectiveness of these techniques based on reservoir characteristics and economic considerations.

Reservoir surveillance: Surveillance techniques like production logging, chemical tracers, and downhole cameras are used to evaluate well performance and identify any issues such as wellbore damage or production anomalies. This information helps in detecting and addressing problems that may impact reservoir management.

By implementing effective reservoir management strategies, operators can optimize production, reduce costs, and maximize the recovery of hydrocarbons in unconventional reservoirs. Continuous monitoring and data analysis play crucial roles in adapting management strategies as the reservoir behavior evolves over time.

6.19 KPIS IN SHALE WELLS, IMPORTANT SURVEILLANCE METHODS AND PARAMETERS/KPIS, WORKFLOWS

In shale wells, key performance indicators (KPIs) play a crucial role in monitoring and evaluating the productivity and efficiency of operations. Some commonly tracked KPIs in shale wells include:

Production rate: This KPI measures the volume of oil or natural gas produced over a specific period. It helps assess the well's productivity and efficiency.

Decline curve analysis: By analyzing the production decline curve, operators can predict future production rates and estimate the economic viability of the well.

Drilling and completion costs: Monitoring the costs associated with drilling and completing shale wells is essential to evaluate operational efficiency and cost effectiveness.

Water usage: As water is a crucial component in hydraulic fracturing operations, tracking water usage helps ensure efficient water management practices and sustainability.

Regarding surveillance methods in unconventional wells, there are several techniques employed to monitor and optimize operations efficiently. Some important surveillance methods include:

Production monitoring: Continuously monitoring production rates, wellhead pressure, and other operational parameters helps operators detect any anomalies or performance issues.

Pressure transient analysis: This method involves analyzing pressure data collected during well testing to estimate reservoir properties, such as permeability and reservoir pressure, which helps in optimizing production.

Surveillance imaging techniques: Techniques like microseismic monitoring and downhole video surveillance aid in visualizing and analyzing the fracture network and overall reservoir behavior.

Real-time data analysis: By collecting and analyzing real-time data from downhole sensors, operators can identify trends, diagnose problems, and make informed decisions to optimize production.

Reservoir simulation: Using numerical models to simulate reservoir behavior helps operators understand how oil or gas flows within the reservoir. This provides valuable insights into maximizing recovery and minimizing operational risks.

It's worth noting that the specific surveillance methods and KPIs used may vary depending on the operator, reservoir characteristics, and specific unconventional play.

6.20 MONITORING/SURVEILLANCE TECHNOLOGIES

Various monitoring and surveillance technologies are employed in unconventional oil and gas operations to gather valuable data and optimize production. Here are some important technologies used in monitoring unconventional wells:

Fiber optic sensing: Fiber optic cables can be installed in or near wells to detect temperature, strain, pressure, and flow changes. This technology enables real-time monitoring and helps operators identify issues such as leaks, flow restrictions, or well integrity problems.

Production logging tools: These tools are used to assess downhole conditions by measuring parameters like fluid flow rates, pressure, temperature, and fluid composition at different depths. This data helps optimize production and diagnose any production issues.

Distributed acoustic sensing (DAS): DAS technology uses a fiber optic cable to detect acoustic signals generated by fluid flow within the wellbore. It provides real-time information on flow rates and well integrity and can even identify leakage or areas of fluid influx.

Pressure transient analysis tools: Specialized tools are employed to measure and analyze pressure transients during well testing, providing information about reservoir characteristics, stimulation effectiveness, and productivity.

Microseismic monitoring: This technique uses sensitive geophone arrays to detect and locate microseismic events occurring during hydraulic fracturing. It helps assess the effectiveness of stimulation and the geometry of the fracture network.

Satellite and aerial imagery: Remote sensing technologies, including satellite and aerial imagery, are used to monitor surface conditions such as vegetation growth, land subsidence, and environmental impacts related to unconventional oil and gas operations.

Internet of Things (IoT) and automation: IoT devices, sensors, and remote monitoring systems are increasingly being used to gather and transmit real-time data. Automated systems can monitor and control equipment, detect abnormalities, and improve operational efficiency.

These are just a few examples of the technologies used for monitoring and surveillance in unconventional oil and gas operations. Advancements in data analytics, artificial intelligence, and machine learning are also being leveraged to further enhance monitoring capabilities and optimize production in these challenging environments.

6.21 MICROSEISMIC

Microseismic monitoring in unconventional oil and gas reservoirs plays a crucial role in reservoir characterization and the evaluation of reservoir stimulation techniques, such as hydraulic fracturing. Microseismic monitoring involves the detection and analysis of very small seismic events generated during hydraulic fracturing operations.

By analyzing the microseismic data, engineers can gather information about the geometry and extent of the induced fractures within the reservoir. This information helps in optimizing well placement, designing effective hydraulic fracturing treatments, and assessing the potential for reservoir production.

Microseismic monitoring typically involves deploying an array of geophone sensors in and around the reservoir area. These sensors detect and record the small seismic events produced as a result of fracturing. The recorded data is then processed and analyzed to create a microseismic event catalog, which provides valuable insights into the fracturing processes.

This monitoring technique helps enhance the understanding of fracture propagation and distribution, which is crucial for efficient and cost-effective production from unconventional reservoirs. By assessing the success of hydraulic fracturing operations and identifying areas of enhanced fracture complexity, operators can optimize reservoir development strategies and increase production efficiency.

Overall, microseismic monitoring in unconventional reservoirs offers valuable information that helps operators make informed decisions during the development and production stages. It contributes to the efficient extraction of hydrocarbons and maximizes the recovery potential of these challenging resources.

6.21.1 SEISMOLOGY

Seismology plays a significant role in understanding the behavior of unconventional reservoirs and optimizing their production. The field of seismology involves studying the propagation of seismic waves through the Earth to gain insights into subsurface structures and properties.

In the context of unconventional reservoirs, seismology helps in several ways. First, it enables the imaging of subsurface layers and structures, allowing geoscientists and engineers to identify potential hydrocarbon-bearing formations. This information helps in the selection of drilling locations and the overall planning of exploration and production activities.

Seismic surveys can also be used to map the extent and distribution of fractures within the reservoir, which is particularly important for unconventional resources. By analyzing the seismic data, experts can infer the fracture network and understand how it influences fluid flow and production behavior.

Additionally, seismology can aid in monitoring and optimizing reservoir stimulation techniques, such as hydraulic fracturing. By deploying sensitive seismometers, engineers can capture and analyze the seismic signals generated during fracturing operations. This allows them to assess the effectiveness and efficiency of the stimulation, including fracture geometry and propagation.

Seismic data can also be used for time-lapse monitoring, in which repeated surveys are conducted to track changes in the reservoir over time. This provides valuable information on fluid movement, reservoir depletion, and potential production risks. Overall, seismology plays a crucial role in characterizing unconventional reservoirs, optimizing production strategies, and mitigating risks associated with reservoir development. It helps in understanding the subsurface dynamics, which is essential for efficient and sustainable extraction of hydrocarbons from these challenging resources.

6.21.2 Wave Types

In unconventional reservoirs, the understanding of wave types and their propagation characteristics is crucial for various purposes, including reservoir characterization and monitoring. Seismic waves can provide valuable information about the subsurface properties of the reservoir and help optimize production strategies.

During hydraulic fracturing operations, microseismic monitoring detects and analyzes small seismic events generated by the fracturing process. These events include the propagation of both P-waves and S-waves. The recorded seismic data aids in understanding the formation of induced fractures, their geometry, and the distribution of fractures within the reservoir.

P-waves and S-waves can be used to infer the mechanical properties of the reservoir rock and the presence of natural fractures. By analyzing the velocities and arrival times of these waves, geoscientists and engineers can gain insights into the rock's elastic properties, including its ability to store and transmit hydrocarbons.

Surface waves, such as Love waves and Rayleigh waves, are also relevant in unconventional reservoirs. Although they may not be as significant for fracture monitoring during hydraulic fracturing, they can provide valuable information on other aspects. For example, surface waves can be used in seismic surveys to map the subsurface structure, including potential hydrocarbon-bearing formations and natural fracture patterns.

Overall, understanding the behavior of different wave types in unconventional reservoirs allows for better reservoir characterization, optimization of production strategies, and assessment of the reservoir's response to hydraulic fracturing and other stimulation techniques. Seismic data gathered through the observation and analysis of these wave types helps optimize extraction techniques and improve the overall efficiency of hydrocarbon recovery.

6.21.3 Source Characterization

Source characterization in unconventionals refers to the process of understanding the properties and behavior of different reservoir rocks and fluids in unconventional resource plays. Unconventional resources, such as shale gas and tight oil, have complex pore structures and low permeability, making it challenging to extract these resources efficiently.

To characterize the source rock, various techniques are employed. These include core analysis, well logging, and geophysical data analysis. Core analysis involves studying rock samples obtained through drilling and analyzing their mineralogy, porosity, permeability, and organic content. Well logging involves measuring properties like formation resistivity, porosity, and rock density using downhole tools.

Geophysical data analysis techniques can provide valuable insights into the subsurface, such as seismic imaging, which helps in mapping the structure and identifying potential areas of hydrocarbon accumulation. Additionally, petrophysical analysis combines geophysical and well log data to estimate the properties of rocks and fluids.

Understanding the source rock properties is crucial in assessing the potential productivity of a reservoir and optimizing drilling and completion strategies. By characterizing the reservoir, engineers and geoscientists can better design hydraulic fracturing techniques and identify sweet spots within the formation that have the highest potential for hydrocarbon production.

Source characterization in unconventionals helps in making informed decisions regarding well placement, well spacing, and stimulation design, ultimately maximizing the recovery of hydrocarbons from these challenging reservoirs. It is an ongoing process that continues to evolve as technologies and methodologies improve.

6.21.4 Moment Tensor

The moment tensor is a mathematical representation of an earthquake source that describes the size, orientation, and type of seismic forces generated during an earthquake. In the context of unconventionals, moment tensor analysis can be used to understand the mechanisms of induced seismicity associated with hydraulic fracturing operations.

During hydraulic fracturing, rock formations are fractured by injecting high-pressure fluids into the reservoir. This process can induce seismic activity, albeit typically at a much smaller magnitude than natural earthquakes. By analyzing the moment tensor of these induced seismic events, scientists and engineers can gain insights into the orientation and characteristics of the fractures created during the hydraulic fracturing process.

Moment tensor analysis can help in evaluating the effectiveness of hydraulic fracturing operations by providing information about fracture connectivity, permeability, and the potential for production enhancement. It can also aid in assessing the potential for induced seismicity and understanding any associated risks. By monitoring and analyzing the moment tensor of induced seismic events in real time, operators can adjust their hydraulic fracturing operations to optimize well performance, mitigate seismic risks, and ensure the long-term viability of unconventional resource extraction.

6.21.5 Acquisition Designs

Microseismic acquisition designs play a crucial role in monitoring and optimizing hydraulic fracturing operations in unconventional oil plays. Microseismic monitoring involves recording and analyzing small seismic events induced by the fracturing process to understand the behavior of the fracture network, its dimensions, and propagation during stimulation.

The design of microseismic acquisition systems includes strategically placing arrays of geophones or seismometers in the subsurface to detect and record the microseismic events. These arrays are typically deployed in monitoring wells or through surface installations. Optimizing the placement and density of these sensors is important to ensure accurate imaging and interpretation of the fracture network. During hydraulic fracturing, the microseismic data collected is processed to generate maps or images that depict the location and geometry of the induced fractures. This information helps operators understand the effectiveness of their stimulation treatments, identify any potential issues such as fracture containment, and improve future well completion designs.

Microseismic acquisition designs also involve analyzing the microseismic data to estimate parameters such as fracture dimensions, stimulation efficiency, and the extent of fracture growth within the reservoir. This information is valuable in optimizing well spacing, stage spacing, and other completion strategies for future wells. In recent years, the industry has been exploring advancements in microseismic acquisition, such as the use of fiber-optic sensors for distributed sensing, which can provide higher-density data and more detailed information on fracture behaviors. Overall, microseismic acquisition designs in unconventional oil plays help operators gain insights into the effectiveness of hydraulic fracturing operations and improve reservoir performance by optimizing stimulation strategies and well designs.

6.21.6 Downhole Monitoring

Downhole monitoring in unconventional reservoirs refers to the practice of using various sensors and instruments to gather data from wells drilled in unconventional formations, such as shale plays. This monitoring helps operators gain a better understanding of the reservoir's behavior, optimize production, and mitigate risks.

In the context of unconventional reservoirs, downhole monitoring technologies play a crucial role in providing real-time data on well performance and reservoir conditions. These technologies allow operators to collect valuable information about parameters like pressure, temperature, flow rates, and fluid compositions at various depths.

With this data, operators can make informed decisions regarding well completion, fracturing techniques, and production strategies, ultimately maximizing the recovery of oil or gas from the reservoir. It also helps identify potential issues, such as wellbore integrity problems or fluid migration, allowing for timely intervention and remediation.

Downhole monitoring technologies commonly used in unconventional reservoirs include pressure and temperature gauges, distributed temperature sensing (DTS) systems, distributed acoustic sensing (DAS) systems, production logging tools, fiber optic sensors, and other wireless telemetry systems. These tools provide valuable insights into the reservoir's characteristics and help optimize well performance and hydrocarbon recovery. Overall, downhole monitoring in unconventional reservoirs has become essential for operators to optimize production, reduce costs, minimize environmental impact, and ensure the long-term sustainability of their operations. By continuously monitoring reservoir conditions and well behavior, operators can make data-driven decisions that improve overall operational efficiency and productivity.

6.21.7 Surface Monitoring

Microseismic surface monitoring in unconventionals is a technique used to gather information about the behavior of subsurface formations during unconventional oil and gas operations, such as hydraulic fracturing. It involves the detection and analysis of microseismic events, which are small seismic waves generated as a result of the fracturing process. By monitoring these microseismic events at the surface, operators can gain insights into the spatial distribution and extent of the fractures created during hydraulic fracturing. This information helps in optimizing well placement, designing future fracturing operations, and estimating reservoir properties.

During the monitoring process, multiple geophones or accelerometers are installed at the surface around the wellbore. These sensors detect the tiny vibrations caused by microseismic events and record their locations and magnitudes. The data collected is then processed and analyzed to create 3D images and models of the fracture network. Microseismic surface monitoring provides valuable information for assessing hydraulic fracturing efficiency, understanding fracture propagation, and evaluating reservoir stimulation. It helps operators make informed decisions to enhance production and reduce the environmental impact of unconventional oil and gas operations.

6.21.8 Shallow Buried Arrays

Shallow buried arrays in unconventional oil and gas refer to a specific method used for surveys and monitoring in the oil and gas industry. Typically, this technique involves the use of geophysical measurements to gather data on subsurface conditions and reservoir properties. Shallow buried arrays specifically focus on shallow depths, typically lying just a few meters below the surface.

In unconventional oil and gas operations, such as shale gas or tight oil extraction, understanding the geology and reservoir characteristics is crucial for effective resource exploitation. Shallow buried arrays can provide valuable insights into various aspects, including but not limited to:

Seismic monitoring: Seismic waves are used to create detailed images of the subsurface, allowing geologists to identify potential reservoirs and assess their properties. Shallow buried arrays can help capture seismic data from shallow depths, providing a more comprehensive understanding of the reservoir.

Microseismic monitoring: Microseismic monitoring involves identifying and analyzing small-scale seismic events that occur during hydraulic fracturing (fracking) operations. By utilizing shallow buried arrays, engineers can accurately locate and monitor these microseismic events, aiding in optimizing well placement and assessing the effectiveness of fracking processes.

Hydrocarbon detection: Shallow buried arrays can also be used to detect and monitor hydrocarbon accumulations near the surface. By measuring various geophysical attributes, such as electrical conductivity or electromagnetic signals, they can help identify potential oil or gas reservoirs in unconventional plays.

Overall, shallow buried arrays in unconventional oil and gas play a crucial role in understanding reservoir behavior, optimizing production techniques, and ensuring efficient resource extraction. By providing detailed subsurface data, these arrays contribute to the overall success and viability of unconventional oil and gas operations.

6.21.9 Microseismic Data Processing

Microseismic data processing plays a crucial role in unconventional oil and gas operations. It involves analyzing and interpreting the recorded microseismic events to gain insights into the subsurface behavior during hydraulic fracturing operations. By understanding the distribution and characteristics of these events, operators can optimize drilling and completion strategies to maximize production and reduce environmental impact. The first step in microseismic data processing is recording the microseismic events using an array of geophones or accelerometers. These sensors are placed in boreholes or on the surface to capture the induced seismicity during hydraulic fracturing. The recorded signals are then processed to remove noise and enhance the signal of interest. After preprocessing, the data is analyzed to identify and locate individual microseismic events. This involves applying time-picking algorithms and waveform correlation techniques to determine the arrival times and locations of the events. By triangulating the recorded signals from multiple sensors, the source locations can be estimated with reasonable accuracy.

Once the locations are determined, the data is further processed to obtain additional information about the subsurface behavior. This includes characterizing the magnitude and frequency content of the events, as well as determining their relationship to the hydraulic fracture geometry. This information can be used to optimize well placement and hydraulic fracturing designs for improved reservoir stimulation. In recent years, machine learning algorithms have been employed to automate and enhance microseismic data processing. These algorithms can quickly process large volumes of data and extract meaningful features that aid in reservoir characterization and production optimization. Overall, microseismic data processing is a valuable tool in the oil and gas industry, particularly for unconventional reservoirs. It allows operators to monitor and evaluate the effectiveness of hydraulic fracturing operations, leading to more efficient and sustainable production practices.

6.21.10 Velocity Modeling

The term "unconventional" refers to resources trapped in unconventional reservoirs like shale formations, tight sandstones, and coalbeds. These reservoirs have complex geological characteristics and require specialized techniques for accurate velocity modeling.

Velocity modeling involves estimating the speed of sound waves as they travel through the subsurface formations. This information is crucial for characterizing the subsurface and optimizing the drilling and production processes. In unconventional reservoirs, the complexity arises from the presence of natural fractures, variable lithology, and heterogeneity in rock properties.

To address these challenges, seismic surveys are conducted to collect data about the subsurface. Advanced algorithms and technologies are then applied to process the seismic data and create accurate velocity models. These models help generate images of the subsurface, identify potential hydrocarbon-bearing zones, and guide drilling decisions.

In unconventional oil and gas reservoirs, velocity modeling is particularly important for several reasons. First, accurate velocity models enhance the interpretation of seismic data, allowing geoscientists and engineers to identify potential sweet spots for drilling. Second, velocity models help in predicting and mitigating drilling hazards, such as the presence of hard minerals or unstable formations. They also aid in optimizing well placement and extraction techniques, improving production performance. As industry continues to evolve, researchers and engineers are constantly developing and refining velocity modeling techniques. This includes incorporating machine learning and artificial intelligence algorithms to improve accuracy and efficiency. The ultimate goal is to enhance the understanding of unconventional reservoirs and maximize oil and gas recovery while minimizing drilling risks and environmental impact.

6.21.11 Detection/Timing of Events

Detection and timing of events in unconventional oil and gas operations can involve various techniques and technologies. One common method is the use of wellhead pressure monitoring to detect changes in pressure, which can indicate operational events such as fracking, drilling, or production activities. Another approach is the analysis of microseismic data, which measures and maps the small seismic events that occur during hydraulic fracturing. This data can provide insight into the propagation and geometry of fractures, helping optimize extraction operations.

Timing is crucial in unconventional oil and gas operations, and advanced technologies like real-time monitoring systems and machine learning algorithms can assist in predicting and managing events. These technologies can continuously analyze multiple data sources, such as temperature, pressure, flow rates, and production rates, to identify potential issues or anomalies, allowing operators to take proactive measures to optimize production and mitigate risks.

It's worth noting that technologies and approaches for event detection and timing in unconventional oil and gas operations are continually evolving. With advancements in sensors, data analytics, and artificial intelligence, operators can gain greater insight into their operations, increasing efficiency, and reducing environmental impact.

6.21.12 Correlation Techniques

Correlation techniques play a crucial role in analyzing unconventional oil and gas reservoirs. These techniques help identify relationships and patterns between different parameters or variables, enabling a better understanding of the reservoir's behavior and optimization of production strategies. Some correlation techniques commonly used in unconventional oil and gas include:

Petrophysical correlations: These techniques establish relationships between petrophysical properties such as porosity, permeability, and lithology. By correlating these properties, engineers can estimate the reservoir's productivity, understand fluid flow behavior, and optimize completion strategies.

Geophysical correlations: These techniques involve integrating seismic data with well log data to identify geological features and reservoir characteristics. Correlations between seismic attributes and reservoir properties help in detailed characterization and mapping of unconventional reservoirs and aid in reservoir monitoring.

Production data correlations: These techniques involve analyzing production data, such as production rates and decline curves, to understand reservoir performance and optimize production strategies. By correlating production data with genetic factors like completion design, well spacing, and reservoir pressure, engineers can improve well performance and predict future production trends.

Reservoir performance correlations: These techniques establish relationships between performance parameters like well spacing, fluid viscosity, and ultimate recovery. By correlating these factors, engineers can optimize well spacing designs, assess deliverability, and estimate the impact of different completion techniques on reservoir performance.

Hydraulic fracture modeling correlations: These techniques involve integrating well and fracture treatment data to create accurate hydraulic fracture models. By correlating fracture modeling data with production data, engineers can understand how different fracture designs impact production and optimize future stimulation strategies.

Correlation techniques provide valuable insights into unconventional reservoirs, assisting in reservoir characterization, optimization of production strategies, and maximizing hydrocarbon recovery.

6.22 FIBER OPTIC MONITORING

Fiber optic monitoring technology has been gaining traction in the unconventional oil and gas industry in recent years. This advanced technique utilizes fiber optic cables to provide real-time data and insights into various aspects of oil and gas operations. One key application of fiber optic monitoring is in hydraulic fracturing, also known as fracking. By placing fiber optic cables in the wellbore, operators can monitor the entire fracturing process in detail. This includes monitoring the pressure and temperature changes, fluid flow rates, and propagation of fractures in the reservoir. Having access to such real-time information allows operators to make informed decisions and optimize their fracking operations for maximum efficiency and productivity.

Another application of fiber optic monitoring is in well integrity and production optimization. Fiber optic cables can be installed throughout the wellbore to continuously monitor the well's structural integrity, detecting any strain, pressure, or temperature changes that could indicate potential issues such as leaks or mechanical failures. By identifying these problems early on, operators can take corrective actions to prevent costly shutdowns or accidents.

Furthermore, fiber optic monitoring can help in reservoir characterization and production forecasting. Distributed acoustic sensing (DAS) and Distributed temperature sensing (DTS) technologies enable operators to measure acoustic and temperature variations, respectively, along the fiber optic cable. This data can be used to gain insights into reservoir behavior, such as fluid movement, flow rates, and temperature profiles. By analyzing this information, operators can make more accurate predictions about future production rates and optimize their extraction strategies accordingly. Overall, fiber optic monitoring technology offers valuable real-time data and insights, enabling operators to improve operational efficiency, enhance safety, and maximize production in unconventional oil and gas operations. As the industry continues to evolve, it is likely that we'll see further advancements in fiber optic monitoring techniques and their widespread adoption.

6.22.1 Distributed Temperature Sensing (DTS)

Distributed temperature sensing (DTS) is a technology commonly used in various industries, including unconventional oil and gas production. It allows for real-time monitoring and analysis of temperature variations along the entire length of a wellbore or pipeline. In unconventional oil and gas, such as shale gas or tight oil, DTS can provide valuable insights into the reservoir behavior and well performance. By measuring temperature variations, DTS systems can help operators identify flow and production patterns, detect anomalies, optimize well placement and completion designs, and even monitor hydraulic fracturing operations. This technology works by using a fiber optic cable installed along the wellbore or pipeline, which acts as a sensor. The cable is capable of measuring temperature changes along its length, which can then be analyzed to understand fluid movement, identify production zones, and assess well integrity. DTS is just one example of the many technologies leveraged in the unconventional oil and gas industry to enhance exploration and production operations. Its real-time monitoring capabilities can contribute to more efficient and effective resource extraction.

6.22.2 Distributed Acoustic Sensing (DAS)

DAS, which stands for distributed acoustic sensing, has proven to be a valuable technology in the oil and gas industry, particularly in unconventional oil and gas extraction. Unconventional resources, such as shale oil and gas, require efficient and cost-effective monitoring techniques. DAS allows for real-time monitoring and understanding of subsurface conditions during hydraulic fracturing operations, enabling operators to optimize the extraction process. With DAS, a fiber optic cable is installed along the wellbore, and by using laser pulses, it can convert the cable into an array of thousands of individual sensors. These sensors can pick up acoustic, seismic, and temperature signals, providing insights into reservoir behavior, fluid flow patterns, fracture propagation, and more.

One of the main advantages of DAS is its ability to provide continuous monitoring of the entire wellbore. Traditional monitoring techniques, such as downhole gauges or seismic sensors, may require multiple instruments at various depths, resulting in limited data and higher costs. DAS allows for more detailed and comprehensive data collection, enhancing reservoir characterization and production optimization. Furthermore, the flexibility and adaptability of DAS make it suitable for various unconventional extraction methods, such as hydraulic fracturing, carbon capture and storage (CCS), and geothermal energy production. It can also be utilized in unconventional drilling and well integrity monitoring to improve safety and reduce environmental risks.

Overall, DAS is an innovative technology that has great potential in the unconventional oil and gas industry. Its ability to provide real-time, detailed monitoring can help optimize production, improve reservoir management, and enhance environmental stewardship.

6.22.3 Hydraulic Fracture Monitoring

Hydraulic fracture monitoring (HFM) in unconventional reservoirs is an important aspect of the oil and gas industry. Unconventional reservoirs, such as shale formations, require hydraulic fracturing to extract hydrocarbons effectively. HFM techniques help operators optimize the fracking process and maximize production. There are several methods used for HFM, such as microseismic monitoring, tiltmeter arrays, pressure monitoring, and fiber optic sensing. Microseismic monitoring involves monitoring the seismic activity caused by the fracturing process. By analyzing microseismic events, operators can determine the extent and location of fractures in the reservoir. Tiltmeter arrays can measure the deformation of the rock near the wellbore, providing valuable insights into fracture propagation.

Pressure monitoring involves measuring the pressure changes during the fracking process, which helps operators understand the behavior of the fractures. Fiber optic sensing is a newer technology that uses fiber optic cables installed in the wellbore to detect strain and temperature changes. This method provides continuous, high-resolution monitoring and can offer valuable information about fracture growth in real time. By combining these monitoring techniques, operators can make informed decisions to optimize fracking operations, improve reservoir management, and enhance production efficiency. HFM in unconventional reservoirs has significantly advanced in recent years and continues to evolve as new technologies and techniques are developed.

6.22.4 Production Logging

Production logging in unconventional oil and gas refers to the practice of monitoring and analyzing data from wellbore measurements to gain insights into the production activities in unconventional oil and gas reservoirs. Unconventional reservoirs, such as shale and tight oil/gas formations, present unique challenges due to their complex and heterogeneous nature. Production logging techniques provide valuable information about fluid flow, well performance, and reservoir behavior, helping operators optimize production and enhance well performance in these challenging reservoirs. Production logging tools are deployed in the wellbore to measure various parameters such as flow rates, pressures, temperatures, and fluid properties at different points along the wellbore. By analyzing this data, operators can identify production challenges such as fluid channeling, completion issues, formation damage, or coning, which may impact production efficiency. This information can guide decision-making processes, enabling operators to optimize well design, completion techniques, and production strategies. In unconventional oil and gas operations, where hydraulic fracturing is common, production logging can also be used to evaluate the effectiveness of fracturing treatments. By monitoring fluid flow profiles and pressure differentials along the wellbore, operators can assess the distribution and extent of the created fractures, identify any communication or interference between different zones, and evaluate the overall effectiveness of the fracturing campaign.

Overall, production logging plays a critical role in understanding and optimizing production in unconventional oil and gas reservoirs. By leveraging data from downhole measurements, operators can make informed decisions to maximize production rates, improve recovery efficiency, and effectively manage these complex reservoirs.

6.23 TRACERS

Tracers play an important role in unconventional oil and gas operations, particularly in hydraulic fracturing. Tracers are chemical substances added to the fracturing fluid to track its movement and distribution within the reservoir. By incorporating tracers into the fracturing fluid, operators can gather valuable information about fluid flow pathways, fracture connectivity, and the interaction between different rock formations.

There are various types of tracers used in unconventional oil and gas operations, including radioactive tracers, chemical tracers, and nanotechnology-based tracers. Radioactive tracers, such as isotopes, emit radiation that can be detected and measured. Chemical tracers consist of unique chemical compounds that can be identified and quantified through laboratory analysis. Nanotechnology-based tracers utilize nanoparticles with distinct properties that can be detected and traced. During a hydraulic fracturing operation, tracers are injected along with the fracturing fluid into the reservoir. As the fluid flows through the created fractures, it carries the tracers along, allowing operators to determine the direction and extent of the fluid movement. By analyzing the concentration of the tracers at different points in the reservoir, operators can gain insights into the effectiveness of the fracturing treatment, the connectivity between fractures, and the overall performance of the well.

Tracers can be used to address various challenges in unconventional oil and gas operations. They can help identify preferential fluid flow paths, detect any communication or interference between adjacent wells, assess fracture height growth, and evaluate the impact of reservoir heterogeneity on production. This information is valuable for optimizing future fracturing designs, improving well spacing, and enhancing reservoir management strategies. Overall, tracers provide a valuable tool for understanding fluid behavior in unconventional oil and gas reservoirs. By tracking the movement of fracturing fluids, operators can make more informed decisions to optimize production and improve recovery efficiency.

6.23.1 Tracer Surveillance

Tracer surveillance is a technique commonly used in the oil and gas industry to monitor the movement and behavior of fluids in a reservoir. It involves injecting a specific chemical tracer into the reservoir and monitoring its movement over time. This can provide valuable insights into reservoir characterization and fluid flow patterns and can help optimize production.

When it comes to unconventional oil and gas, such as shale or tight reservoirs, tracer surveillance can be particularly useful. These reservoirs often have complex and heterogeneous characteristics, making it challenging to accurately understand and predict fluid flow behavior. By using tracer surveillance, operators can gain a better understanding of how fluids move within the reservoir, identifying key flow paths, preferential channels, and areas with bypassed resources.

The data obtained from tracer tests can be used to calibrate reservoir models, improve production techniques, and make informed decisions about well placement and stimulation strategies. It allows operators to evaluate the efficiency and effectiveness of different completion and hydraulic fracturing techniques, aiding in the optimization of recovery methods. Overall, tracer surveillance in unconventional oil and gas plays a vital role in maximizing recovery and reducing uncertainty. It enables operators to gain valuable insights into reservoir behavior and make informed decisions to enhance hydrocarbon production while minimizing environmental impact.

6.23.2 Inflow Tracer Monitoring

Inflow tracer monitoring is a technique used in unconventional oil and gas production to better understand reservoir behavior and optimize production. It involves the injection of a chemical tracer into one or more wells and subsequently monitoring the tracer's movement within the reservoir. The process starts with carefully selecting the tracer compound. Tracers are chosen based on their stability, detectability, and compatibility with the reservoir fluids. Once the tracer is injected, it follows the flow path of the reservoir fluids, providing valuable information about the underground processes.

Monitoring the tracer's movement is done by sampling production fluids at various wellheads and analyzing them for the presence of the tracer compound. This helps in determining the connectivity between wells, measuring fluid flow rates, and identifying any communication or interference between different zones or fractures. Inflow tracer monitoring can provide insights into reservoir behavior and help operators optimize production strategies. By understanding how fluids are flowing within the reservoir, companies can better target high-producing areas, optimize well completions, and make informed decisions about well spacing and stimulation techniques.

Overall, Inflow Tracer Monitoring is a valuable tool in unconventional oil and gas production as it enhances reservoir characterization and allows for more efficient and effective production.

In summary, tracer monitoring is a valuable tool in unconventional oil production. It helps operators optimize production strategies, understand reservoir characteristics, and improve reservoir management practices. The interpretation of tracer monitoring data involves analyzing breakthrough curves and utilizing mathematical models to extract valuable insights.

6.23.3 Polymer Tracer System

Polymer tracer systems are commonly used in unconventional oil production to enhance the understanding of fluid flow patterns and improve reservoir management. These tracers are chemical compounds that are added to the injected fluids, typically polymer solutions, to track their movement and behavior within the reservoir.

The use of polymer tracers in unconventional oil production offers several advantages. Firstly, they provide valuable information about fluid flow paths, reservoir connectivity, and sweep efficiency. By analyzing the concentration and distribution of polymer tracers, operators can gain insights into the efficiency of hydraulic fracturing procedures and the effectiveness of various stimulation techniques.

Polymer tracers are particularly useful in enhanced oil recovery (EOR) operations, where they can help optimize the injection and production strategies. By tracking the movement of polymer solutions, operators can identify preferential flow paths, evaluate reservoir heterogeneity, and design more effective sweep patterns to maximize oil recovery.

Interpreting polymer tracer data involves analyzing breakthrough curves and using mathematical models to simulate the fluid flow behavior within the reservoir. These models consider factors such as reservoir characteristics, fluid properties, and injection strategies to accurately interpret the tracer data and extract meaningful insights.

Overall, the use of polymer tracer systems in unconventional oil production is a valuable tool for reservoir management. It aids in optimizing production strategies, improving oil recovery, and enhancing overall field performance.

6.23.4 Inflow Interpretation, Quantitative Modeling

Inflow interpretation refers to the analysis and understanding of fluid flow into a wellbore or a reservoir. It involves studying the behavior of oil, gas, and water as they enter a wellbore or flow within the reservoir. Quantitative modeling, on the other hand, involves using mathematical and computational techniques to simulate and predict the behavior of the reservoir and its fluids. This can be extremely useful in the case of unconventional oil, which often involves complex geological formations and challenging extraction techniques. When it comes to unconventional oil, unconventional reservoirs like shale oil or tight oil formations present unique challenges due to their low permeability and complex rock structures. Therefore, accurate inflow interpretation and quantitative modeling are crucial for optimizing production and making informed decisions regarding drilling and production strategies.

These techniques can help operators understand the behavior of fluids within the reservoir, estimate reservoir properties like permeability and porosity, predict well performance, and optimize production strategies. By combining real-time data from the field, reservoir models, and sophisticated algorithms, engineers and geoscientists can gain insights into the reservoir's behavior and make more informed decisions.

In summary, inflow interpretation and quantitative modeling play important roles in understanding and optimizing the production of unconventional oil. They enable operators to make data-driven decisions that maximize recovery and minimize costs in these complex reservoirs.

6.24 RESERVOIR SIMULATION ASPECTS

Reservoir simulation plays a crucial role in understanding and optimizing the production of unconventional oil reservoirs. Unconventional oil reservoirs, such as shale oil and tight oil plays, have distinct characteristics that require special considerations in simulation.

First, the complex geology and heterogeneous nature of unconventional reservoirs pose challenges in accurately representing them in reservoir simulation models. Proper characterization of the reservoir properties such as porosity, permeability, and rock mechanics is critical for building reliable models. This involves incorporating data from well logs, core samples, and production tests to capture the reservoir's unique features.

Secondly, unconventional reservoirs often exhibit low permeability and significant variations in reservoir quality. Consequently, the flow behavior is often governed by mechanisms such as adsorption, desorption, and diffusion, in addition to the traditional viscous flow. Thus, advanced simulation techniques like dual-porosity and dual-permeability models or discrete fracture networks (DFN) are employed to capture the multiphase flow dynamics accurately.

Moreover, unconventional reservoirs are heavily influenced by hydraulic fracturing, which creates a network of fractures to enhance production. Accurate modeling of the complex fracture geometry and its interaction with the reservoir is essential for predicting well performance. Techniques like embedded discrete fracture modeling (EDFM) are employed to incorporate the fracture network into the simulation.

Lastly, unconventional reservoirs often require long-term development plans to ensure sustainable production. Reservoir simulation serves as a valuable tool in optimizing the hydraulic fracturing designs, well spacing, and production strategies. It helps evaluate the impact of various parameters on production, such as well interference, reservoir pressure depletion, and possible breakthrough of unwanted fluids.

Overall, reservoir simulation in unconventional oil plays a vital role in understanding the reservoir behavior, optimizing production, and making informed decisions for reservoir management. The continuous improvement of simulation techniques and incorporation of advanced models further enhance our understanding and ability to maximize recovery from unconventional oil reservoirs.

6.24.1 Numerical Model for Shale Gas

Developing a numerical model for shale gas reservoirs involves several key considerations. Shale gas reservoirs have unique characteristics that require specialized modeling techniques to accurately represent their behavior. Here are some aspects to consider when creating a numerical model for shale gas.

Geomechanics: Shale formations are highly brittle and sensitive to changes in stress. Incorporating geomechanical properties into the model, such as rock deformation, fracture propagation, and stress changes, is crucial for understanding how the reservoir will behave during production.

Porous media undergo dynamic changes during the extraction process in conventional reservoirs. As oil and gas flow and production take place, the reservoir's pore pressure and effective stress experience continuous fluctuations.

Adsorption and desorption: Shale gas is predominantly stored in the form of adsorbed gas on the surface of organic matter within the rock matrix. Modeling the adsorption and desorption processes accurately is essential to capture the behavior of gas production from shale reservoirs.

Dual-porosity and dual-permeability models: Shale formations typically have low matrix permeability and high fracture permeability. Utilizing dual-porosity and dual-permeability models helps represent the flow of gas both within the matrix and through the fracture network, enabling a more accurate simulation of the reservoir.

In simple terms, let's assume that each site has a greater chance of attracting a gas molecule than an adsorbed gas molecule attracting another one. Furthermore, if an adsorbed gas molecule is closer to a specific site, it has a higher likelihood of attracting another molecule than if it were farther away from that site.

Discrete fracture networks (DFN): Shale formations often contain natural fractures and can be further fractured through hydraulic fracturing. Incorporating discrete fracture networks into the model allows for a more realistic representation of the complex fracture systems and their interaction with the matrix during production.

The explicit description of large-scale natural/hydraulic fractures in reservoirs can include their spatial location, distribution characteristics, geometrical morphology, and other defining features. Additionally, the distribution of fractures at varying scales and their geometrical characteristics, physical parameters, and fluid flow behavior can be characterized using a discontinuous discrete function or equation. These non-continuous fractures, with various spatial locations and sizes, can collectively form the discrete fracture media.

Pore-level modeling: Shale gas recovery relies on gas desorption from nanopores in the rock matrix. Pore-level modeling techniques, such as Lattice Boltzmann or molecular dynamics simulations, can provide insights into the physics of gas flow and enable better understanding of the recovery potential.

History matching and uncertainty analysis: Calibration of the numerical model against field production data through history matching helps validate the model and reduce uncertainty. Performing uncertainty analyses, including sensitivity studies and stochastic simulations, provides a better understanding of the range of possible outcomes and helps in risk management.

It's worth noting that modeling shale gas reservoirs can be challenging due to the complexity of the rock properties, the presence of natural fractures, and the intricacies of hydraulic fracturing. Nonetheless, accurate numerical models serve as powerful tools for optimizing production strategies, evaluating well spacing, and making informed decisions for shale gas reservoir development and management.

6.24.2 Non-Darcy Flow Effect

Non-Darcy flow refers to fluid flow in porous media characterized by deviations from Darcy's law, which assumes that the flow is proportional to the pressure gradient. In unconventional oil and gas reservoirs, the non-Darcy flow effect can significantly impact fluid flow behavior due to the presence of small pore sizes, complex pore geometries, and large pressure differentials.

There are a few key aspects to consider regarding the non-Darcy flow effect in unconventional oil and gas reservoirs:

Forchheimer's equation: To account for the non-Darcy flow effect, Forchheimer's equation is often used as an extension of Darcy's law. It introduces an additional term related to the square of the fluid velocity to represent the inertial effects that become significant at high flow rates and in pores with small sizes.

$$-\frac{\partial P}{\partial x} = \frac{\mu}{k_{fh}} V + \beta \rho V^2 \tag{6.2}$$

where kfh is the Forchheimer permeability (m2), β is the inertial resistance (coefficient), non-Darcy flow coefficient, or Forchheimer coefficient (m-1).

Nonlinear flow behavior: The presence of non-Darcy flow can cause a nonlinear pressure drop through the reservoir. As the fluid velocity increases, the pressure drop may increase more steeply than predicted by Darcy's law. This behavior becomes more pronounced when flowing through extremely tight formations, such as shale, where the permeability is very low.

Knudsen diffusion: In unconventional reservoirs, the pore sizes can be so small that the flow regime transitions from continuum flow to Knudsen diffusion. Knudsen diffusion occurs when the mean free path of the gas molecules is comparable to or larger than the pore size. As a result, the gas molecules have collisions with the pore walls, which alter the flow behavior and decrease the permeability.

Nonlinear permeability: The permeability of unconventional reservoirs can exhibit nonlinear behavior due to the non-Darcy effect. This means that the permeability depends on the flow rate and pressure gradient. At high flow rates, the permeability can decrease, resulting in reduced fluid flow and increased resistance to flow.

Applicability of Darcy's law: While Darcy's law is widely used, it may not accurately represent the fluid flow behavior in unconventional reservoirs. Advanced models, such as the Forchheimer equation or modifications to the permeability models, are necessary to account for the non-Darcy flow effect and provide a more accurate representation of fluid flow behavior.

Understanding and quantifying the non-Darcy flow effect are crucial for reservoir engineers and simulation practitioners working with unconventional oil and gas reservoirs. Accurate models that incorporate the nonlinear behaviors and non-Darcy flow effects can help optimize production strategies, design hydraulic fracturing operations, and assess reservoir performance to maximize recovery from these challenging reservoirs.

6.24.3 Gas Desorption Effect

Gas desorption refers to the process of releasing gas from the organic matter in unconventional oil reservoirs. When oil is trapped in tight rock formations, it is often accompanied by significant amounts of adsorbed gas on the surface of organic material. Understanding and quantifying the gas desorption effect in unconventional oil reservoirs is essential as it can have a significant impact on reservoir behavior and production.

The gas desorption effect can influence key aspects of unconventional oil reservoirs:

Production behavior: During production, as pressure decreases, the gas desorption effect leads to the release of additional gas from the organic matter. This can enhance the overall production performance of the reservoir by contributing to the produced gas volume and changing the fluid composition.

Well performance: Gas desorption affects well productivity by altering the fluid flow properties in the reservoir. As the adsorbed gas is released, the effective permeability of the rock may increase, allowing for better flow of both oil and gas. Understanding the interaction between the desorption effect and well production is crucial for optimizing well designs and production strategies.

Reservoir pressure depletion: Gas desorption can contribute to pressure depletion in unconventional oil reservoirs. As the gas desorbs, it exerts additional pressure and helps maintain reservoir pressure, which is essential for sustaining production rates over the long term.

Fluid phase behavior: The presence of desorbed gas affects the fluid phase behavior in the reservoir. It can alter the saturation and distribution of oil and gas, impacting their relative mobility and flow rate. Accurate modeling of phase behavior, considering the desorption effect, is necessary for reliable reservoir simulation and forecasting.

Recovery factor: The gas desorption effect influences the overall recovery factor of unconventional oil reservoirs. The additional gas production resulting from desorption can enhance the overall hydrocarbon recovery from the reservoir. Understanding and quantifying the desorbed gas contribution to the total fluid production is crucial for accurate estimation of the recovery factor.

To account for the gas desorption effect, reservoir simulation models need to incorporate appropriate adsorption/desorption equations, such as Langmuir or BET isotherms. These equations provide a mathematical representation of the relationship between pressure, adsorbed gas content, and desorption behavior. The accuracy of the desorption model depends on reliable laboratory measurements using core samples and well logging data.

Overall, understanding and accounting for the gas desorption effect is imperative for accurate reservoir modeling, production optimization, and maximizing the recovery potential of unconventional oil reservoirs.

6.24.4 Geomechanics Effect

Geomechanics plays a crucial role in unconventional reservoirs, which are characterized by complex geomechanical properties. The geomechanics effects in these reservoirs are important to understand because they can significantly impact the production and exploration of hydrocarbons.

One prominent geomechanical effect is the interaction between the rock matrix and the hydraulic fractures. In unconventional reservoirs, hydraulic fracturing is utilized to create interconnected fractures in the rock, allowing for the extraction of oil or gas. Geomechanical considerations include assessing the stress distribution and its impact on fracture propagation, containment, and connectivity.

Another important geomechanical effect is reservoir compaction and subsidence, which can occur due to fluid extraction. As hydrocarbons are produced, the pore pressure decreases, and the overburden stress causes the formation to compact. This can lead to surface subsidence, which must be monitored and managed to prevent any adverse environmental or infrastructure effects.

Furthermore, geomechanical characteristics such as rock brittleness, natural fractures, and in situ stress distribution affect the production performance of unconventional reservoirs. Understanding these properties helps optimize drilling, completion, and stimulation techniques to maximize hydrocarbon recovery.

Geomechanical modeling and monitoring techniques, such as microseismic monitoring, are employed to assess reservoir behavior, fracture propagation, and stress distribution in real time. This enables operators to make informed decisions and adapt their production strategies accordingly.

Overall, the geomechanics effects in unconventional reservoirs are intricate and multifaceted. Understanding these effects is crucial for successful exploration and production, and ongoing research and technological advancements continue to push the boundaries of our understanding in this field.

6.24.5 Pressure-Dependent Fracture Conductivity

Pressure-dependent fracture conductivity is an important concept in the field of unconventional reservoir modeling. It refers to the change in conductivity of fractures in reservoir rocks as a result of pressure variations.

In unconventional reservoirs, such as shale or tight sands, hydraulic fracturing is essential to create paths for the flow of hydrocarbons. These fractures can have varying degrees of conductivity, which determine how easily the hydrocarbons can flow through them. However, the conductivity of fractures can change under different pressure conditions.

To model pressure-dependent fracture conductivity in unconventional reservoirs, engineers and geoscientists use different approaches. These approaches typically involve laboratory measurements, well tests, and numerical simulations. These techniques help in understanding how fracture conductivity changes with varying pressure conditions and aids in optimizing the design and production strategies for unconventional reservoirs.

By incorporating pressure-dependent fracture conductivity into modeling efforts, engineers can better predict and optimize the productivity of unconventional reservoirs. This understanding is crucial for optimizing drilling and completions strategies, enhancing production rates, and maximizing recovery from these challenging reservoirs.

6.24.6 Geomechanics Modeling

Geomechanics modeling in unconventional reservoirs is an essential aspect of understanding and optimizing the extraction of oil and gas resources. Unconventional reservoirs, such as shale and tight formations, have unique characteristics that require a specialized approach in geomechanics modeling.

In these reservoirs, rock properties, such as brittleness and natural fractures, play a crucial role in determining the production behavior of the reservoir. Geomechanics modeling helps in simulating the behavior of the rock under stress, predicting fracture propagation, estimating production rates, and optimizing well placement and design.

One common technique used in geomechanics modeling for unconventional reservoirs is hydraulic fracturing simulation. It involves injecting fluids at high pressure to create and propagate fractures within the rock, improving permeability and enhancing hydrocarbon extraction. Geomechanics models help in predicting the extent and direction of fracture growth, optimizing the design of the hydraulic fracturing treatment, and assessing the potential for unwanted fracture growth into nearby formations.

Geomechanical models also assist in understanding and mitigating potential geomechanical risks associated with unconventional reservoirs. These risks include subsidence, induced seismicity, and wellbore instability. By incorporating geomechanical properties into reservoir simulation models, engineers can better anticipate and manage these risks, improving production efficiency and operational safety.

Overall, geomechanics modeling is a critical tool in the development and production of hydrocarbons from unconventional reservoirs. It helps in optimizing reservoir performance, reducing operational risks, and maximizing resource recovery.

6.24.7 FMM Method Single- and Multi-Well Applications

The finite mobilization model (FMM) method is widely used in the oil and gas industry for analyzing unconventional reservoirs. It is commonly applied for both single-well and multi-well applications.

In single-well applications, the FMM method helps in estimating the production performance of a single well in a reservoir, considering factors such as fluid properties, well completion design, and reservoir characteristics. This information is crucial for making decisions regarding drilling, completion, and production strategies.

For multi-well applications, the FMM method is employed to assess the interactions between multiple wells in a reservoir. It takes into account the influence of neighboring wells on each other through factors such as interference, drainage area, and reservoir heterogeneity. This helps in optimizing well spacing, determining the optimal number of wells, and developing effective field development plans.

The FMM method is advantageous because it accounts for the complexity of unconventional reservoirs, which typically exhibit low permeability, complex geology, and varying fluid behavior. By combining numerical simulation and analytical models, the FMM method provides valuable insights into reservoir behavior and aids in optimizing production rates and recovery.

By utilizing the FMM method in both single- and multi-well applications, engineers and reservoir management teams can make informed decisions to maximize the production potential and economic viability of unconventional reservoirs.

6.24.8 Model Assumption and Fracture Discretization

In unconventional reservoirs, model assumption and fracture discretization play important roles in accurate modeling. Let's delve into these concepts.

Model assumption: When modeling unconventional reservoirs, certain assumptions are typically made to simplify the representation of the system. These assumptions can vary depending on specific conditions and the level of complexity desired in the model.

Some common assumptions include:

Homogeneous reservoir: This assumption assumes that the reservoir properties, such as porosity and permeability, remain constant throughout the reservoir. While this simplifies the model, it might not accurately capture the spatial heterogeneity present in the reservoir.

Isotropic reservoir: This assumption implies that the reservoir's properties are the same in all directions. In reality, unconventional reservoirs often exhibit anisotropic properties, meaning the permeability and other characteristics vary depending on the direction being considered.

Single fluid flow: This simplifying assumption assumes that only one fluid is flowing in the reservoir, neglecting interactions between different fluids, such as oil, gas, and water. Unconventional reservoirs often involve complex multiphase flow, which requires more sophisticated modeling techniques.

Fracture discretization: Fractures are critical in unconventional reservoirs as they enhance the flow of hydrocarbons by creating pathways. Discretization refers to the process of dividing fractures into discrete elements or grid cells to accurately represent their geometries.

Fracture discretization methods can be broadly classified into two categories:

Analytical methods: These methods use mathematical equations to describe the behavior of the fractures. Examples include the use of analytical solutions like Green's functions or the integral transform method. Analytical methods rely on simplifying assumptions and may not capture complex fracture geometries accurately.

Discrete methods: These methods divide the fractures into discrete elements and solve the governing equations for each discrete element. Examples of discrete methods are the discrete fracture network (DFN) approach and the discrete element method (DEM). Discrete methods can handle complex fracture networks but may require more computational resources.

Overall, accurate modeling of unconventional reservoirs requires carefully considering the assumptions made and appropriately discretizing fractures to capture the complex behavior of these reservoirs. Advanced modeling techniques and simulation tools are continuously being developed to improve the representation and understanding of these challenging reservoirs.

To calculate the original gas in place (OGIP) in an unconventional reservoir, you typically consider several factors. Here's a general approach for the calculation:

Determine the reservoir volume: The first step is to estimate the total volume of the reservoir that contains the hydrocarbon-bearing rocks. This can be done based on geological and geophysical data, such as seismic surveys.

Determine the porosity: Porosity refers to the percentage of pore space within the reservoir rocks. For unconventional reservoirs, such as shale or tight gas formations, porosity is typically low. You can estimate porosity from core samples or measurements from well logging data.

Determine the water saturation: Water saturation is the fraction of the pore space filled with water. For unconventional reservoirs, water saturation is generally low, as the rocks are relatively impermeable. This can also be estimated from well logging data.

Determine the gas saturation: Gas saturation refers to the fraction of the pore space filled with gas. It can be calculated using the porosity and water saturation values.

Determine the formation volume factor: The formation volume factor represents the ratio of gas volume at reservoir conditions to the volume at surface conditions. This factor accounts for changes in pressure and temperature. It can be determined through laboratory measurements or using correlations based on reservoir properties.

Estimate the initial reservoir pressure: The initial reservoir pressure can be estimated from well data, such as pressure measurements or pressure-transient analysis.

Calculate the original gas in place: To calculate the OGIP, you multiply the reservoir volume by the gas saturation and then multiply that by the formation volume factor. Finally, divide the result by the initial reservoir pressure.

Please note that there are various complexities involved in estimating OGIP accurately, and the steps outlined here provide a simplified approach. It's important to consult with reservoir engineers and use specialized software for a more precise calculation tailored to your specific reservoir conditions.

6.24.9 Latest Simulation Methods

Simulation methods for unconventional reservoirs have evolved significantly in recent years to address the complex nature of these reservoirs. Some of the latest simulation methods utilized in unconventional reservoirs include:

Discrete fracture Network (DFN) simulation: DFN simulation captures the complex network of fractures found in unconventional reservoirs. By explicitly modeling the discrete fractures and their connectivity, DFN simulations provide a more accurate representation of fluid flow and behavior in these reservoirs.

Dual porosity/dual permeability (DPDP) simulation: DPDP simulation is commonly used for dual media reservoirs, such as shale or tight gas formations. This approach involves modeling the matrix rock and the natural fractures as separate entities with distinct properties. DPDP simulations allow for a better understanding of the fluid flow mechanisms and the transfer of fluids between the matrix and fractures.

Geomechanical coupled simulation: Unconventional reservoirs are highly influenced by geomechanical effects, such as stress changes and induced fractures. Geomechanical coupled simulation integrates geomechanical models into reservoir simulations to capture the dynamic interplay between fluid flow and reservoir deformation. This approach helps in understanding the impact of hydraulic fracturing operations on reservoir behavior and optimizing production techniques.

Reduced order models (ROM): ROMs are simplified mathematical models that capture the essential features of complex reservoir simulation models. These models are computationally efficient and can be used to quickly evaluate multiple reservoir scenarios. ROMs help in rapid decision-making, such as optimizing well placement, evaluating completion strategies, and forecasting production profiles.

Uncertainty quantification (UQ): Unconventional reservoirs are characterized by significant geological uncertainties. UQ methods, such as Monte Carlo simulation and Latin hypercube sampling, are integrated with reservoir simulations to quantify the impact of these uncertainties on production forecasts. UQ techniques enable a better understanding of the range of potential outcomes and support robust decision-making processes.

These latest simulation methods in unconventional reservoirs aim to improve understanding, optimize recovery, and reduce operational risk. By incorporating complex geological and engineering aspects, these methods help engineers make informed decisions and enhance reservoir performance in these challenging settings.

6.24.10 History Matching

History matching in unconventional reservoirs refers to the process of calibrating reservoir models to match historical production data and fluid flow behavior in order to improve the predictive accuracy of the models. Traditional history-matching techniques have been primarily developed for conventional reservoirs but have been adapted and extended for unconventional reservoirs such as shale gas and tight oil formations.

In unconventional reservoirs, history matching can be challenging due to their complex geological and fluid flow behavior. These reservoirs often have low permeability, complex fracture networks, and heterogeneous matrix properties, making it difficult to accurately represent their behavior in reservoir models.

To perform history matching in unconventional reservoirs, various techniques can be employed. These may include adjusting parameters related to fracture conductivity, matrix permeability, reservoir geometry, and fluid properties to match observed production and pressure data. Advanced techniques such as ensemble-based methods, data assimilation, and probabilistic approaches are also being used to account for uncertainties and improve the reliability of history-matching results.

The goal of history matching in unconventional reservoirs is to obtain a calibrated reservoir model that accurately represents the behavior of the reservoir and provides reliable predictions for future production and reservoir performance. This process plays a crucial role in optimizing production strategies, designing stimulation techniques, and maximizing hydrocarbon recovery in unconventional reservoirs.

6.24.11 Complex and Planar Fractures

Complex and planar fractures play a significant role in the development of unconventional reservoirs. These types of fractures occur in tight formations, such as shale, where the rock is characterized by low permeability and porosity. Understanding the characteristics and behavior of these fractures is crucial for optimizing production in these reservoirs.

Complex fractures refer to fractures that exhibit intricate geometries, branching out into multiple pathways within the formation. These fractures can enhance the effective surface area available for fluid flow, thereby increasing production. However, their complex geometry poses challenges in accurately predicting and modeling fluid flow behavior.

On the other hand, planar fractures occur when a fracture propagates along a relatively straight plane within the formation. These fractures typically have high conductivity, allowing for efficient fluid flow. Planar fractures are desirable as they provide direct pathways for hydrocarbons to flow towards the production well. They are often created through hydraulic fracturing, where fluid is injected at high pressure to induce fractures within the formation.

Advances in technology and reservoir simulation have enabled engineers to better understand the behavior of complex and planar fractures. Numerical models and simulation tools can help predict and optimize fracture placement, geometry, and conductivity. Additionally, monitoring techniques like microseismicity and production data analysis can provide valuable insights into fracture performance and effectiveness.

Overall, the study and understanding of complex and planar fractures in unconventional reservoirs are vital for efficient and effective hydrocarbon recovery. Ongoing research and technological advancements continue to improve our understanding and optimization of fracture network development in these reservoirs.

6.25 REFRACTURING TO IMPROVE EUR

Refracturing is a technique used to enhance the estimated ultimate recovery (EUR) in unconventional reservoirs. Unconventional reservoirs, such as shale or tight formations, often exhibit low permeability, which complicates the efficient extraction of hydrocarbons.

Refracturing involves the process of restimulating existing fractures within the reservoir. This is typically achieved by injecting fluids, such as water or proppants, into the formation at high pressure. The injected fluids help reopen or extend existing fractures, which increases the connectivity between the reservoir and the wellbore.

By enhancing the connectivity, refracturing aims to improve the productivity of the well and ultimately increase the EUR. This technique is particularly useful in unconventional reservoirs where the initial hydraulic fracturing treatment may not have fully stimulated the formation or where the fractures have closed over time due to stress changes.

Refracturing can be performed multiple times throughout the life of a well, targeting different intervals within the reservoir. Operators may choose to refracture wells based on their performance decline, usually indicated by reduced production rates or declining pressure.

However, it's important to note that refracturing comes with its set of challenges. It requires careful evaluation of the reservoir's geomechanical properties, as well as selecting the appropriate fracturing fluids and proppants. Moreover, economic considerations play a significant role in determining the viability of refracturing operations.

Overall, refracturing is a technique that holds promise for improving the EUR in unconventional reservoirs. As technology continues to advance and understanding of reservoir behavior improves, it is likely to play an increasingly important role in maximizing hydrocarbon extraction from these challenging formations.

6.26 WATER MANAGEMENT

Water management in unconventional reservoirs plays a crucial role in optimizing the production and overall performance of these reservoirs. Unconventional reservoirs, such as shale and tight gas formations, have unique characteristics that require special considerations for effective water management.

One of the key challenges in water management for unconventional reservoirs is the large volumes of water required for hydraulic fracturing or fracking. Proper management of this water is important to minimize environmental impact and optimize resource usage. This involves sourcing water from sustainable and reliable sources, such as recycling produced water or using brackish water, to reduce the strain on freshwater sources.

Additionally, managing the flowback and produced water from the wells is essential to ensure safe disposal or treatment for reuse. Technologies such as water treatment systems and recycling facilities can be implemented to minimize the need for fresh water input and reduce costs associated with water disposal.

Another aspect of water management in unconventional reservoirs is flow assurance. Unconventional reservoirs often have complex geological structures, and it is important to monitor and manage water flow to prevent issues such as water coning, channeling, or preferential paths that can negatively impact production.

Implementing water monitoring systems and predictive analytics can help operators identify potential flow issues and make informed decisions to optimize water management strategies. This includes adjusting injection rates, monitoring reservoir pressure, and using diverters or chemicals to alter flow profiles.

Overall, an integrated approach to water management is essential for maximizing the productivity and efficiency of unconventional reservoirs while minimizing environmental impact. This involves considering water sourcing, treatment, reuse, and flow assurance throughout the life cycle of the reservoir.

6.27 LATEST ADVANCES, TECHNIQUES, AND METHODS

In recent years, there have been significant advancements in the techniques and methods used in unconventional oil and gas reservoirs. Here are some of the latest developments:

Improved hydraulic fracturing: Hydraulic fracturing, commonly known as fracking, has been crucial for extracting oil and gas from unconventional reservoirs. Over time, techniques have been refined to increase production rates and minimize environmental impact. Advanced fracturing fluids, optimized proppants, and more precise wellbore placement techniques have improved the efficiency and effectiveness of hydraulic fracturing.

Multi-well pad development: To enhance productivity and reduce surface footprint, operators have shifted towards multi-well pad development. This involves drilling multiple horizontal wells from a single pad, reducing cost and minimizing surface disturbance. Innovative drilling technologies such as pad drilling, walking rigs, and extended-reach drilling have made multi-well pad development more feasible.

Artificial intelligence and data analytics: With advances in computing power and data analytics, the use of artificial intelligence (AI) has gained momentum in unconventional reservoir development. AI algorithms help optimize well placement, stimulation, and production strategies by analyzing vast amounts of subsurface data. Predictive analytics can also improve decision-making for reservoir management, maximizing recovery.

Reservoir characterization: Unconventional reservoirs have complex geological formations, making accurate reservoir characterization crucial for optimizing production. Cutting-edge technologies such as 3D seismic imaging, microseismic monitoring, and advanced well logging tools provide valuable insights into rock properties, fracture networks, and fluid flow patterns. This information helps in predicting reservoir behavior and improving well placement.

Enhanced recovery techniques: As oil and gas extraction becomes more challenging in unconventional reservoirs, enhanced recovery techniques are being developed and deployed. These techniques include cyclic gas injection, chemical flooding, water-alternating-gas (WAG) injection, and carbon dioxide (CO_2) injection. By altering fluid properties and pressure conditions, these methods aim to improve reservoir sweep efficiency and increase recoverable reserves.

Environmental sustainability: As the industry acknowledges the importance of environmental sustainability, efforts are being made to minimize the environmental impact of unconventional reservoir development. Innovations such as waterless or reduced-water fracking, closed-loop drilling systems, and the use of renewable energy sources for operations are actively being explored to reduce water consumption, mitigate emissions, and promote responsible resource extraction.

These are just a few of the latest advances, techniques, and methods being employed in unconventional oil and gas reservoir development. As technology continues to evolve, it is expected that further advancements will improve efficiency, reduce costs, and optimize the extraction of these valuable resources.

REFERENCES

Britt, L. K., & Smith, M. B. (2009). Horizontal well completion, stimulation optimization, and risk mitigation. In *SPE eastern regional meeting, society of petroleum engineers*, Charleston, West Virginia.

Couzens-Schultz, B. A., Axon, A., Azbel, K., Haugland, M., Sarker, R., Tichelaar, B., Wiesneck, J. B., Wilheml, R., Zhang, J., & Zhang, Z. (2013). Pore pressure prediction in unconventional resources. *International petroleum technology conference*, Beijing, China, 26–28 March. IPTC paper #16849, 11 p.

Demyanov, V., Gopa, K., Arnold, D., & Elfeel, M. A. (2014). Production optimization under uncertainty in fractured reservoirs. *Proceedings of the ECMOR 2014, 14th European conference on the mathematics of oil recovery*, Catania, Italy, September 8–11.

Gong, J., & Rossen, W. (2017). Modeling flow in naturally fractured reservoirs: Effect of fracture aperture distribution on dominant sub-network for flow. *Petroleum Science*, *14*, 138–154.

Guerriero, V., Mazzoli, S., Iannace, A., Vitale, S., Carravetta, A., & Strauss, C. (2013). A permeability model for naturally fractured carbonate reservoirs. *Marine and Petroleum Geology*, *40*, 115–134.

Mohaghegh, S. D., Grujic, O., Zargari, S., & Kalantari, M. (2011). Modeling, history matching, forecasting and analysis of shale reservoirs performance using artificial intelligence. *SPE digital energy conference and exhibition.* Society of Petroleum Engineers, The Woodlands, TX.

Peters, K. E., Burnham, A. K., & Walters, C. C. (2015). Petroleum generation kinetics: Single versus multiple heating ramp open-system pyrolysis. *American Association of Petroleum Geologists Bulletin*, *99*, 591–616.

Thuwaini, J. S., Alwan, A. K., & Ranjan, A. (2009). Uncertainty analysis of a fractured carbonate reservoir. *Proceedings of the SPE middle east oil and gas show and conference*, Manama, Bahrain, 15–18 March. SPE paper #120722-MS.

Williams, G. J. J., Mansfield, M., MacDonald, D. G., & Bush, M. D. (2004). Top-down reservoir modelling. *SPE annual technical conference and exhibition*. Society of Petroleum Engineers, Houston, TX.

Xia, X., & Michael, G. E. (2015). Pore pressure in unconventional petroleum systems. In Y. Z. Ma, S. Holditch & J. J. Royer (Eds.), *Unconventional resource handbook: Evaluation and development*. Elsevier.

Yin, J., Xie, J., Datta-Gupta, A., & Hill, A. D. (2015). Improved characterization and performance prediction of shale gas wells by integrating stimulated reservoir volume and dynamic production data. *Journal of Petroleum Science and Engineering*, *127*, 124–136.

7 Recovery Enhancement in Unconventional Shale Resources

Amirsaman Rezaeyan, Son Tran, Hosein Rezvani, Ying Yu, Mohammadreza Keshavarz, and Leilei Zhang

7.1 INTRODUCTION

Unconventional shale resources hold an immense potential, with an estimated several thousand trillion cubic feet of endowment (Kim et al., 2000). Over the last decade, such reservoirs have dramatically changed how the oil industry operates, fundamentally altering global energy landscapes. The United States, a pioneer in this field, now produces over eight million barrels of oil each day from these previously inaccessible resources, contributing significantly to the global oil supply with a cumulative production of approximately ten billion barrels so far (Hoffman & Reichhardt, 2019; El-Samak et al., 2020). This remarkable achievement, once considered unattainable when the idea of extracting a million barrels a day from such challenging reservoirs seemed unlikely, has been realized thanks to significant advancements in drilling technologies. Shale and tight formations have been successfully explored and produced not only in the United States and North America but also in China (Jia et al., 2016; Wang & Chen, 2016). Accordingly, they became one of the major contributors to energy supplies. Although investigating flow and transport in unconventional reservoir rocks has made significant progress in the past decade, there is still a long way to go before we fully understand all mechanisms and key factors that control fluid flow and transport in such formations. The development of long horizontal wells combined with multistage hydraulic fracturing has made these previously unreachable goals achievable. The exploration and production success in the United States has showcased the substantial contribution these reservoirs can make to energy supplies. Unconventional reservoirs are diverse, encompassing organic-rich source rocks, tight oil reservoirs, and hybrid plays that combine elements of both source rocks and conventional reservoirs (Zoback & Kohli, 2019). Each category presents unique geological formations, necessitating customized approaches for optimal exploitation. Despite this progress and the vast potential oil volume in these unconventional formations, estimated to hold hundreds of billions of barrels per formation, the industry confronts a critical challenge: a rapid decline in production

 DOI: 10.1201/9781003319955-7

rates following initial extraction (Downey et al., 2011; Hoffman & Reichhardt, 2019). Wells typically exhibit high initial production rates, which quickly diminish, often leading to recovery factors below 10% (Clark, 2009; Hoffman & Reichhardt, 2019; King, 2014). This phenomenon underscores the imperative need for enhanced oil recovery (EOR) methods specifically designed for the peculiarities of unconventional reservoirs. While EOR techniques have long been established for conventional reservoirs, their adaptation to the unique conditions of unconventional formations is a relatively recent endeavor, necessitating a deeper understanding. This chapter explores the fundamentals of EOR methods suited for unconventional reservoirs. It explores the geological and petrophysical complexities that characterize these resources and examines how innovative EOR technologies, such as novel fracturing fluids, proppants, nanofluids, and gas injection strategies, can significantly enhance recovery rates. Coupled with technical insights and real-world case studies, we aim to provide a comprehensive perspective on the current and future potential of EOR techniques in unlocking the vast resources contained in unconventional reservoirs. Despite numerous applications in oil and gas exploration and the progress achieved, a complete understanding of flow and transport mechanisms in shales and tight sandstones, particularly across scales from pore to reservoir, remains elusive. This introduction sets the stage for an in-depth discussion on recent advances in unconventional reservoirs and the ongoing challenges in oil and gas exploration and production. Through comprehensive analysis and detailed discussion, we aim to equip readers with a thorough understanding of the evolving landscape of unconventional reservoir exploration and exploitation, paving the way for the detailed discussions that follow in the subsequent sections.

7.1.1 General Characteristics of Shale Reservoirs

Shale reservoirs are complex systems characterized by their significant hydrocarbon-bearing mudstone and shale compositions. These reservoirs play a pivotal role in the global energy supply, attributed to their extensive storage capacities and intricate pore interconnectivity. The presence of organic matter and the complexity of pore characteristics in shale reservoirs marked by abundant nanopores, natural fractures, and varying pore geometries are critical factors that influence the formation and recovery of shale oil (Feng et al., 2019; Pang et al., 2023). Nanoscale pores, predominant in these formations, alongside strong solid-liquid interactions in confined spaces, contribute to the nonlinear nature of fluid flow within the reservoirs (Wang et al., 2016; Garcez et al., 2023). Shale reservoirs can be broadly divided into two categories based on their heterogeneity: fracture type and matrix type. Fractured reservoirs feature extensive natural fractures that facilitate higher permeability and serve as primary channels for oil flow. In contrast, matrix-dominated reservoirs rely on matrix pores and microfractures for hydrocarbon storage and flow, with hydrocarbons existing in both free and adsorbed states within various pore types (Rezaeyan et al., 2022). The transport mechanisms within these nanopores are notably complex, with shale reservoirs also demonstrating sensitivity to stress. This sensitivity, heightened by the reservoirs' high clay mineral content and abundance of nanopores and fractures, can lead to significant changes in permeability under external forces, such as those encountered during the production process (Yan et al., 2021). Additionally, the compaction of reservoirs, alongside proppant embedding and crushing owing to fractures closing under production stress, further complicates the extraction process (Luo et al., 2018; Zhang & Emami-Meybodi, 2022). Addressing the challenges presented by these complex pore structures and stress sensitivities, numerous scholars have proposed various models to better understand and predict the transport mechanisms of oil and water within shale nanopores. These models take into account factors such as adsorption, diffusion, slip flow, oil viscosity, and the detailed pore structure and distribution, offering insights into the effective management of fluid transport in such constrained environments (Zhang et al., 2016; Cui et al., 2017; Qiao et al., 2020; Yu et al., 2021).

The mineral composition of shale significantly impacts its functional characteristics, with a rich array of clay minerals such as kaolinite, illite, chlorite, and mixed-layer illite-montmorillonite alongside feldspar, carbonates, organic matter, and quartz shaping reservoir properties (Rezaeyan et al., 2023, 2024b). This diverse mineralogy, coupled with shale's laminated and fissile nature, complicates not only the characterization of gas systems but also the identification of efficient pathways for hydrocarbon migration and recovery. The complex interaction of these minerals directly influences the reservoir's porosity and permeability, underscoring the critical role of mineralogy in shale's ability to store and transport fluids (Ulm et al., 2007; Aplin & Macquaker, 2011; Morley et al., 2018). Building on the foundational understanding of shale's mineralogy, the characterization of its pore systems, especially those on the nanometer scale, becomes essential for grasping the full scope of fluid transmissivity and storage capabilities within these reservoirs. These intricate pore systems, marked by significant surface areas and complex structures, don't just enable extensive gas absorption but also play a pivotal role in determining the overall porosity and permeability of the shale. These factors are crucial for the successful extraction of hydrocarbons, highlighting how the mineralogical composition directly shapes the reservoir's physical framework and its effectiveness in hydrocarbon recovery (Loucks & Ruppel, 2007; Chalmers & Bustin, 2012; Chen et al., 2017).

Key factors such as mineralogy, total organic content (TOC), rock thickness, permeability, gas in place, brittleness, pore pressure, and the thermal maturity of organic content are critical in assessing a shale gas reservoir's viability. The optimal interplay of these factors can significantly enhance production outcomes and improve the overall recovery factor of shale reservoirs, which is characteristically low for both shale gas and tight oil wells (Tang et al., 2012; Yu et al., 2014; Curtis, 2002; Loucks & Reed, 2007; Zou et al., 2012). Moreover, wettability and permeability emerge as vital elements affecting hydrocarbon recovery and reservoir performance. The wide spectrum of shale's wettability influences fluid trapping and recovery, while the detailed modeling of gas transport mechanisms, including slip flow, Knudsen diffusion, and surface diffusion, provides a deeper understanding of the complexities of shale's gas permeability (Rezaeyan et al., 2022). Such research highlights the complex nature of factors that govern hydrocarbon production rates and recovery efficiencies in unconventional reservoirs (Iglauer et al., 2015; Roshan et al., 2016; Feng et al., 2019; Zhang et al., 2019; Tahmasebi et al., 2020; Mehrabi et al.,

2021). The hydrocarbon production and recovery factor discussions highlight the industry's ongoing challenges in sustaining production rates and achieving substantial recovery factors. Despite the evolution of production methods and the adoption of EOR techniques such as gas huff-n-puff (HnP), continuous innovation is essential for optimizing recovery strategies and extending the productive lifespan of shale and tight oil reservoirs (Patzek et al., 2013; Speight, 2016; Zoback & Kohli, 2019).

7.1.2 Reservoir Rock, Fluid, and Geochemistry Aspects

Shale oil, as it is currently explored and developed, primarily exists in mudstone or shale formations that have reached the oil generation threshold. This type of oil is typically found in both free and adsorbed states within the formations, distinguished by high organic content and confined within nanoscale apertures. In contrast to tight oil, which may migrate short distances within tight sandstone or carbonate reservoirs before being trapped, shale oil remains predominantly within the organic-rich shale formations, undergoing minimal or no migration. This distinction is crucial as it implies that shale oil can be considered a type of tight oil accumulated directly in the source rock formations (Zou et al., 2012; Zhang et al., 2012).

The reservoir rocks for shale oil not only include mudstone and shale but may also contain interlayers of tight sandstone, carbonate, or volcanic rocks. However, the oil resources in such interlayers differ significantly from those in pure mudstone and shale in terms of their source-reservoir relationships, seepage mechanisms, and the factors controlling their accumulation and development. Specifically, shale oil reservoirs are characterized by their rich organic matter, complex mineral compositions, and the development of laminated structures, which together contribute to nanoscale pores. The oil in these reservoirs is primarily retained after hydrocarbon generation within the source rocks without significant migration (Zhang et al., 2012).

Organic matter plays a pivotal role in both the formation and the evaluation of shale oil reservoirs, serving as the primary source for generating hydrocarbons and also significantly contributing to reservoir porosity and permeability. Effective shale oil reservoir evaluation focuses on the abundance of organic carbon, with TOC thresholds often set at 2% for potential and 4% for high-quality reservoirs. The evaluation also considers the thickness of organic-rich shale layers, mineral composition, and the existence of fractures, which facilitate hydrocarbon movement and extraction.

Expanding on the organic geochemical aspects, the type and maturity of organic matter are crucial in determining the potential of a shale reservoir. Type I and type III kerogens, found respectively in continental and marine depositional environments, are assessed for their maturity levels using vitrinite reflectance (R_0), ideally ranging from 0.5% to 1.1%. These maturity levels are indicative of the thermal maturity stage of the organic matter, influencing the types and quantities of hydrocarbons that can be generated and retained within the shale matrix. Advanced geochemical analyses, such as pyrolysis and thermogravimetric analysis, are utilized to quantify the organic content and predict its behavior under thermal stress, thereby providing deeper insights into the kerogen types, their potential oil yields, and the temperatures at which these hydrocarbons are released. Moreover, the mineral composition of shale significantly influences its mechanical properties and, by extension, its suitability as a reservoir rock. A mix of brittle minerals such as quartz and calcite, which support the formation and maintenance of natural fractures, and clay minerals, which can complicate hydraulic fracturing owing to their pliability, affect the shale's brittleness—a critical factor during reservoir stimulation as it impacts the effectiveness of induced fracturing and subsequent hydrocarbon recovery. In summary, the "reservoir rock, fluid, and geochemistry aspects" of shale oil reservoirs encompass a broad spectrum of geological and chemical factors that must be meticulously analyzed to optimize hydrocarbon recovery. The complex interplay of organic matter content, mineral composition, and structural characteristics of the reservoir rock define the productivity and economic viability of shale oil extraction. This comprehensive evaluation helps delineate target, favorable, and non-viable reservoirs based on quantifiable geological and chemical criteria, setting the stage for customized exploration and development strategies to maximize efficiency and minimize environmental and financial risks.

7.1.3 EOR Mechanisms in Shale Reservoirs

EOR in shale reservoirs involves advanced techniques that go beyond conventional recovery methods, aiming to maximize oil extraction from formations typically characterized by ultra-low permeability. These methods have been rigorously evaluated through a combination of experimental research, simulations, and field pilot tests to determine their applicability and efficiency. Worldwide reported EOR studies of unconventional reservoirs further underscore the global scale and significance of these efforts in enhancing oil recovery, as can be seen at Figure 7.1.

7.1.3.1 Comprehensive Overview of EOR Techniques

EOR techniques in shale frequently involve gas flooding, cyclic gas injection or HnP, surfactant flooding, and low salinity water flooding. Notably, CO_2 injection is recognized as having the most potential for unconventional reservoirs, as demonstrated in numerous simulations and laboratory tests, which underscore its effectiveness in enhancing oil recovery. This is quantitatively supported by data presented in Figure 7.2, which shows that a significant percentage of studies favor CO_2 over other methods (Alfarge et al., 2017).

Despite this potential, field tests occasionally reveal that natural gas can outperform CO_2, indicating operational challenges and a substantial gap in our understanding of the physical and chemical interactions involved in CO_2 injection (Hoffman & Evans, 2016; Schmidt & Sekar, 2014). These insights have led to a preference for cyclic methods such as HnP over continuous injection, especially in formations where permeability does not support effective sweeping processes. The decision between cyclic and continuous EOR strategies is influenced by several key factors, including the permeability relative to injector-producer spacing and the diffusivity of the injected substance. High permeability can support continuous flooding, which is economically viable at certain spacing ratios. In contrast, in formations with lower permeability, cyclic strategies are more effective because of enhanced oil recovery during soaking periods and the high diffusivity of gases such as CO_2 and methane in shale matrices (Alfarge et al., 2017).

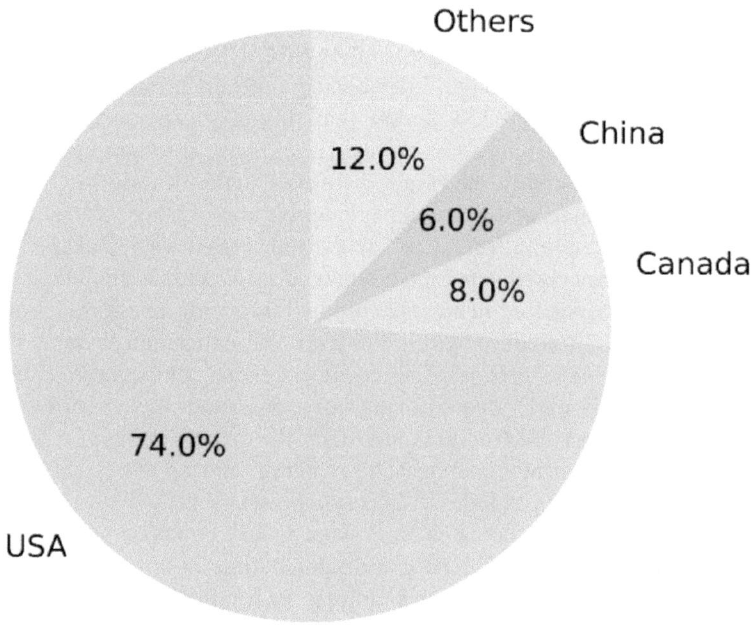

FIGURE 7.1 Worldwide reported EOR studies of unconventional reservoirs.

Source: Elturki & Imqam, 2020

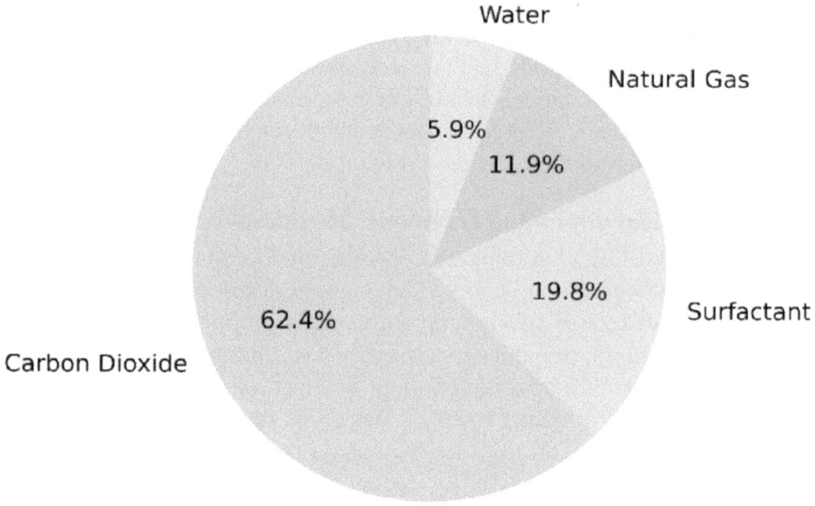

FIGURE 7.2 EOR methods with most potential for unconventional reservoirs.

Source: Alfarge et al., 2020a

7.1.3.2 Challenges in EOR Implementation

Despite advancements in technology, applying EOR methods in shale formations still presents significant challenges, particularly with injectivity and conformance. Injectivity concerns, though less severe than initially expected, require careful management of injection-induced fractures (IIFs). These fractures can develop even under low pressures owing to the fragile nature of shale formations. This phenomenon is well documented and necessitates ongoing monitoring to ensure that EOR processes do not worsen the formation's structural weaknesses (Baker et al., 2016). Moreover, conformance problems have emerged as a notable challenge in several EOR pilot projects conducted in unconventional reservoirs. These issues often manifest as uneven fluid distribution within the reservoir, which can severely impact the efficiency of the recovery process. The origins of these conformance problems are complex and not fully understood, raising questions about the roles of pre-existing natural fractures, hydraulic fractures, and IIFs. To diagnose these issues, Baker et al. (2016) suggested a strategic approach involving varying the injection rates and monitoring corresponding changes in injection pressure. If the relationship between injection pressure and rate deviates from what would be expected under diffusive Darcy law, it may indicate that IIFs are contributing to conformance problems. Additionally, completion interference among hydraulically fractured horizontal wells can also worsen these issues, complicating the effective application of EOR techniques (Schmidt & Sekar, 2014). Economic factors also play a crucial role in the implementation of EOR strategies. The costs associated with CO_2 capture and storage, for example, can significantly influence the overall feasibility of CO_2-EOR projects, impacting decision-making processes regarding which EOR methods to employ (Zhou et al., 2020). The success of EOR in unconventional reservoirs thus depends on a detailed understanding of geochemical interactions, reservoir characteristics, and the mechanics of fluid flow in nanoporous media. Continuous research and innovation are vital to navigate the physical limitations and economic barriers associated with EOR methods in these complex environments. As the field progresses, adapting EOR strategies to address both injectivity and conformance issues will be critical for enhancing recovery rates and ensuring the economic viability of shale oil extraction projects.

7.1.3.3 Future Directions and Economic Considerations

The evolution of EOR technologies continues to be shaped by both scientific advancements and economic factors. The ongoing development of more efficient, cost-effective EOR techniques is vital for making the most of existing oil reservoirs in an environmentally and economically sustainable manner. Consequently, the industry must remain adaptable to new scientific findings and market conditions, which directly influence strategic decisions in EOR applications. In summary, the effective application of EOR in shale requires an integrated approach that considers technical feasibility, economic viability, and environmental sustainability. The field is dynamic, with ongoing advancements promising to address current limitations and enhance the recovery rates from these challenging reservoirs.

7.2 HUFF-N-PUFF GAS INJECTION

HnP gas injection, also known as cyclic gas injection, emerges as a promising EOR method for boosting hydrocarbon recovery in unconventional reservoirs. The hypothesized sequence of events that occur during a HnP cycle in a tight oil reservoir, leading to gas transport into the matrix and subsequent oil production, is illustrated in Figure 7.3. Under initial conditions, both fractures and matrices exist at thermodynamic equilibrium, saturated with oil along with solution and/or free gas, depending on pressure and temperature. Typically, gas HnP is executed post primary production, so most of the leakoff fracturing water is expected to be cleaned up. During the injection phase, gas primarily channels through high-permeability fractures. A process of gas/oil mixing occurs near the fracture/matrix interface, resulting in the vaporizing of oil components into the gas phase and oil swelling. Throughout the early soaking period, advection and diffusion persist as the principal mechanisms of gas transportation. Progressing into the soaking period, the gas permeates deeper into the matrix, causing oil swelling via diffusive transport. During the pressure-depletion phase, the expansion of gas (free and solution gas) is believed to be the prevailing recovery mechanism (Ghanizadeh et al., 2021; Mahzari et al., 2021; Rezaeyan et al., 2024a; Song et al., 2023; Thomas et al., 2020; Tran et al., 2021).

7.2.1 Physics and Mechanisms in Shale

In the context of EOR, the capillary number (N_{cap}) serves as a critical parameter for understanding the mechanism by which pore-level oil trapping can be mitigated. When $N_{cap} > 1$, it indicates the dominance of viscous forces and the consequent reduction of interfacial tension between fluids within the pore space, thereby improving recovery potential.

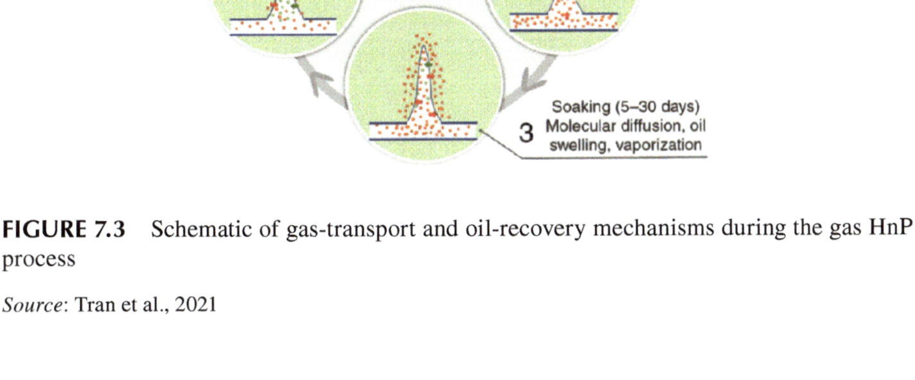

FIGURE 7.3 Schematic of gas-transport and oil-recovery mechanisms during the gas HnP process

Source: Tran et al., 2021

$$N_{\text{cap}} = \frac{u\mu}{\sigma \cos\theta} \tag{7.1}$$

where σ and θ, respectively, represent interfacial tension and contact angle, while u and μ are, respectively, Darcy velocity and viscosity of the oil. The goal of gas injection is to reduce σ and/or θ values, thereby increasing the dominance of viscous force and enhancing recovery.

A comprehensive understanding of oil-recovery mechanisms is vital for the optimization of the gas HnP process. The recovery mechanisms involved in a gas HnP process include gas expansion, reduction of oil viscosity, oil swelling, and vaporization of oil components to the gas phase (Tran et al., 2021; Carlsen et al., 2019; Alharthy et al., 2018), in addition to pore-volume compressibility (Davudov & Moghanloo, 2018). The transport of gas into the rock matrix is fundamentally dictated by the gradients of pressure and chemical potential. Moreover, there may be sufficient time for gas diffusion during the injection phase, which typically spans 20 to 90 days in HnP field pilots (Jacobs, 2019; Carlsen et al., 2019).

The diffusion process within porous media is significantly impeded by factors such as the tortuosity of the pores, the cross-sectional area available for diffusion, and potentially, the pore sizes (Grathwohl, 2012). Ullman and Aller (1982) proposed an equation for estimating the effective diffusion coefficient (D_{eff}), accounting for the tortuosity (τ) of the formation.

$$D_{\text{eff}} = \frac{D}{\tau^2} \tag{7.2}$$

Shen and Chen (2007) and Grathwohl (2012) proposed an empirical equation that relates tortuosity to the porosity ($\varnothing$) and the formation resistivity (F) as

$$\tau^2 = \left(F \times \varnothing\right)^n \tag{7.3}$$

where n is an adjustable empirical parameter.

Archie's law states that the formation resistivity can be expressed as $F = \frac{a}{\varnothing^m}$. Substituting F by Archie's law into Eq. 7.3 and thereafter substituting τ into Eq. 7.2 yields the D_{eff} by

$$D_{\text{eff}} = \frac{D}{\left(a \times \varnothing^{1-m}\right)^n} \tag{7.4}$$

Here, a and m are tortuosity constant and cementation factor within the porous media, respectively.

These parameters—a, n and m—are lithology dependent. Following the assumptions set out by Boudreau (1996), Shen and Chen (2007), Li et al. (2018) and Tran (2021), we can assume that a = n =1 and m = 2. Therefore, the D_{eff} can be estimated as

$$D_{\text{eff}} = \varnothing \times D \tag{7.5}$$

There remains, however, a debate on whether the mass transport of gas during the soaking period of an HnP process is dominated by advection or diffusion. The Peclét number (N_{Pe})—a ratio of convective to diffusive flows—can serve as a valuable tool in this context. The N_{Pe} for fluid flows that transport mass in the porous medium is (Hobbs & Ord, 2015)

$$N_{\text{Pe}} = \frac{u \times L}{D_{\text{eff}}} \tag{7.6}$$

where u, L, and D_{eff} are Darcy velocity, the length scale of the flow, and effective diffusion coefficient, respectively.

When $N_{\text{Pe}} < 1$, the mass transport is dominated by diffusion. Cronin et al. (2019) and Carlsen et al. (2019) suggested that the key mechanisms behind multi-contact miscibility in conventional gas-injection processes (e.g., vaporizing and/or condensing drive) are based on the advection-dominated transport within the matrix. They hypothesize that the transport of gas into the matrix during the soaking period of the HnP process in tight reservoirs may be entirely diffusion dominated ($N_{\text{Pe}} \approx 0$), thus implicating first-contact miscibility as the primary mechanism.

Tran et al. (2021) proposed an empirical equation to calculate the N_{Pe} utilizing the pressure-decay technique.

$$N_{\text{Pe}} = \frac{\sqrt{-\frac{k}{\mu} \times \frac{\partial p}{\partial t}} \times L}{\varnothing \times D} \tag{7.7}$$

where, $\frac{\partial p}{\partial t}$ is the pressure decline during the soaking period.

Following Tran et al. (2021)'s study, Doranehgard et al. (2021) proposed an analytical solution for determining D as

$$D = \frac{k\pi}{\omega_{s0}{}^2 \alpha^2 \mu} \frac{\left(p - p_{eq}\right)}{ln\left(\frac{t_{eq}}{t}\right)} \tag{7.8}$$

Here, ω_{so} represents the maximum value of the solvent mass fraction in the oil measured at the average pressure during the soaking process. The p_{eq} and t_{eq} represent equilibrium pressure and its corresponding time. The value of $\frac{\left(p-p_{eq}\right)}{ln\left(\frac{t_{eq}}{t}\right)}$ is determined using the slope of the straight line fitted to the plot of $(p\text{-}p_{eq})$ vs. $ln\left(\frac{t_{eq}}{t}\right)$.

During the soaking period, the total produced oil's mass (m) at a time t, can be calculated by

$$m = \frac{2A\sqrt{D\varnothing}\rho_o\sqrt{t}}{\sqrt{\pi}}\left(\frac{\omega_{s0}\alpha}{1+\omega_{s0}\alpha}\right) \tag{7.9}$$

In the early soaking period, the velocity is primarily controlled by the pressure gradient while the diffusion coefficient is driven by concentration gradient as the soaking process continues. The diffusion coefficient and velocity are controlled by the pressure gradient at the early soaking times and the concentration gradient when the soaking progresses. The composition of the injection gas plays a significant role in determining the extent of gas penetration in the matrix. This model also offers the potential to estimate swelled oil volume during the soaking period.

7.2.2 Effects of Pressure and Depletion Rate

The efficiency of HnP gas injection EOR is influenced by uncertain (geological) and operational parameters. It has been suggested that higher injection pressures could enhance gas dissolution and the swelling effect, subsequently leading to improved oil recovery. Tran et al. (2021) suggested that the optimization of gas transport into the matrix combined with an effective depletion strategy are crucial factors leading to high oil recovery.

In the field application, the efficiency of the gas HnP process may be enhanced by increasing injection pressure and/or enriching the injected gas with intermediate hydrocarbon components and/or CO_2. The preferable choice would be to increase injection pressure. First, it is feasible to inject gas at sufficiently high pressures to attain miscible conditions. Secondly, a higher injection pressure assists in spreading gas away from the matrix/fracture interface, thereby enlarging the surface area in contact with the gas. This fracturing-induced surface area substantially influences oil recovery via the gas expansion mechanism. However, reaching the MMP could present challenges as such pressure might surpass the capacity of the gas compressor, potentially leading to geomechanical complications such as refracturing of the formation. Given the constraints of injection pressure, enriching the injected gas is a practical alternative for enhancing oil recovery. Regarding the depletion strategy, choke-size control is necessary to optimize the rate of pressure depletion. If the pressure-depletion rate is too rapid, energy may be lost from the gas expansion too quickly, whereas a rate that is too slow may result in inadequate gas energy to drag the oil along its flow paths.

Figure 7.4 presents an approximation of total sensitivity indices, derived from the Sobol sensitivity analysis, illustrating the impact of different parameters on HnP performance (Hamdi et al., 2021). As shown in the figure, operational variables tend to have a more significant influence on HnP performance than uncertain (geological) variables. Nevertheless, certain uncertain variables such as the permeability multiplier, molecular diffusion coefficients, and fracture permeability can also considerably impact HnP performance.

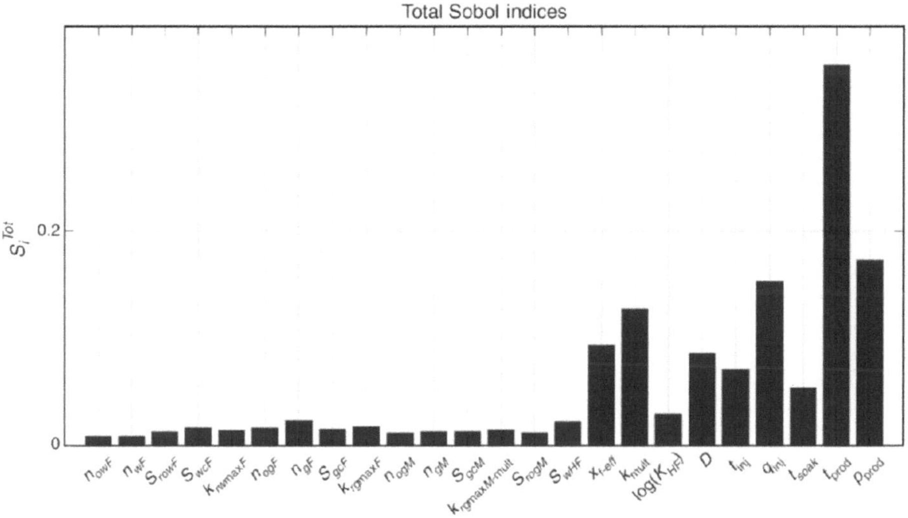

FIGURE 7.4 Effects of operational parameters and reservoir properties on total oil recovery by Sobol sensitivity indices.

Source: Hamdi et al., 2021

7.2.3 Effect of Soaking Time

The soaking period, a critical parameter in HnP gas injection, is defined as the duration during which the well is shut down to allow gas to permeate into the rock matrix. Sufficient soaking time facilitates enhanced gas diffusion and dissolution, leading to improved oil recovery (Yu et al., 2017). Nonetheless, lengthy soaking periods may not yield significant incremental recovery and may instead result in a negative return on investment owing to increased operational costs. Determination of optimal soaking time should be made considering the specific characteristics of the reservoir and any operational limitations.

According to Tran et al. (2021), significant gas penetration into the core plug was observed within the initial four days of the seven-day shut-in period. This suggests that an optimal shut-in period contributing to high oil recovery may be less than seven days. This observation aligns with the findings from field pilots, in which a shorter shut-in period (approximately five days) appears to facilitate better oil recovery than longer durations (Jacobs, 2019).

7.2.4 Injected Gas Composition

The utilization of associated gases for gas injection EOR and for the mitigation of gas flaring has gained increasing attention (Hoffman & Evans, 2016). It is noteworthy that methane, the primary constituent of natural gas, possesses a heat-trapping capacity that is 25 times greater than that of CO_2 (Dincer & Rosen, 2013). Investment in EOR designed to increase production such as utilizing produced gas, uneconomic oil wells, and existing gas compression/gathering facilities tends to be more economical than drilling new wells.

Achieving miscibility conditions depends on oil composition, injection-gas composition, pressure, and temperature conditions. In oilfield applications, it may be feasible to accomplish miscible conditions by adjusting the gas injection pressure and composition at the surface. The type of gas injected can have a significant bearing on the efficiency of the HnP process. For instance, CO_2 is highly efficacious owing to its higher solubility in oil, leading to the oil swelling viscosity reduction. Methane and nitrogen can also be used as injection gases, but their effectiveness is subject to several variables, such as reservoir conditions, the availability of gas, and cost considerations.

7.2.5 Gas Developed Miscibility

In a gas HnP process, achieving miscibility conditions is essential to improve oil recovery. These conditions may be achieved through either first-contact or multi-contact miscibility. First-contact miscibility occurs when oil and gas instantly form a homogeneous phase regardless of their mixing ratios at a particular pressure and temperature. Conversely, multi-contact miscibility necessitates repeated interactions and mass exchange between oil and injected gas. If one visualizes the porous medium as a series of mixing cells, the multi-contact miscibility can be classified into two processes: specifically, a vaporizing gas drive and a condensing gas drive (Orr, 2007). Ultimately, a comprehensive understanding of the miscibility conditions developed within the oil/gas systems under reservoir conditions allows for a more precise estimation of bottom-hole injection pressure, thus optimizing oil recovery.

Based on experimental and modeling studies involving Bakken and Permian oil samples, Adekunle and Hoffman (2014) and Liu et al. (2018) demonstrated that CO_2 establishes miscibility with Bakken oil at comparatively lower pressures than other hydrocarbon gases. Moreover, live-oil samples containing solution gas achieve a lower MMP than dead-oil samples (Adekunle & Hoffman, 2014).

Knowing the range of MMPs associated with varying degrees of enrichment aids operators in designing/optimizing the gas-compression system, reservoir-simulation-study sensitivity for the gas-injection EOR scenarios, and a long-term field development plan with consideration to the source of gas supply and gathering facility. Given the typical constraints on injection pressure imposed by the gas-compression system, miscibility conditions can be achieved by increasing the molar fraction of light hydrocarbon components and/or CO_2 in the injected gas, thereby reducing the MMP to an optimal level (Tran, 2021).

7.2.6 Asphaltene Precipitation and Deposition

During the HnP process, changes in pressure, temperature, and fluid composition can lead to the precipitation and deposition of asphaltene, which has the potential to impair near-wellbore permeability. The prudent management of the injection parameters and comprehensive understanding of the reservoir's asphaltene stability envelope can effectively counteract these issues (Gupta et al., 2019).

A significant challenge presented by CO_2 injection is the formation damage that can be incurred because of the incompatibility between the injected and reservoir fluids. According to Azin et al. (2022), CO_2 injection involves a higher level of asphaltene precipitation than both nitrogen and methane injections. The resulting asphaltene precipitation can decrease reservoir porosity and permeability through pore plugging, enhance the skin factor via formation damage, and obstruct surface facilities through plugging. In the case of carbonate reservoirs, CO_2 can chemically react with the reservoir rock, prompting the precipitation of calcite. Al-Maamari and Buckley (2000) reported that an excess of 4.6% asphaltene precipitation prompts a decrease in oil production owing to a shift in wettability from water-wet to oil-wet. Abedini and Torabi (2014) reported that the volume of asphaltene deposited during immiscible HnP is significantly lower than in miscible and near-miscible CO_2 HnP scenarios. Furthermore, Hasanzadeh et al. (2023) concluded that asphaltene precipitation can promote wettability alteration and reductions in permeability.

Several strategies can be used to reduce asphaltene precipitation during the CO_2 HnP process. One such approach involves the utilization of inhibitors to prevent asphaltene precipitation within reservoirs. Alternatively, CO_2 could be co-injected with solvents such as protic hexane-1-of toluene to deter or delay the deposition of asphaltenes.

7.2.7 Huff-n-Puff in Shale Oil

The experimental and simulation studies on HnP gas injection EOR, specifically utilizing various shale oil permeabilities ranging from nano-Darcy to milli-Darcy, have been well documented in literature. The most frequently used gases or solvents for injection include N_2, CO_2, CH_4, C_2H_6, and lean or rich gas mixtures. Sensitivity analyses are concurrently conducted with experiments and simulations to investigate the effects of operational parameters (injection pressure and rate, duration of injection, soaking time, number of cycles, and depressurization) on recovery performance. However, the economic viability of the HnP process in shale reservoirs depends on other factors such as gas availability, infrastructure costs, and reservoir characteristics.

Several studies reported that the phase behavior of a gas/oil system in nanopores of unconventional reservoirs deviates from the conventional ones. Teklu et al. (2014) studied the effects of confined pores on the MMP of CO_2 and a mixture of CO_2 and C_1 with Bakken oil. They reported that confinement-induced reduction in MMP becomes pronounced when the pore radius is less than 20 nm, a phenomenon attributable to shifts in both critical temperature and pressure. Additionally, Zhang et al. (2017) observed that high capillary pressures affect equilibrium liquid/vapor-phase compositions, resulting in a decrease the MMP of CO_2 and Bakken oil.

7.2.8 Huff-n-Puff in Shale Gas Condensate

The application of HnP gas injection can be extended to shale gas condensate reservoirs to enhance recovery. The injected gas helps minimizing condensate blockage near the wellbore, thereby enhancing gas production rates. Field trials and simulations have demonstrated the potential for increased recovery in these reservoirs.

In gas-condensate reservoirs, the fluid is primarily in the gaseous phase, and the initial reservoir pressure is typically above the dew-point pressure. During the production phase, the gas undergoes condensation into a liquid phase as pressure declines below the dew-point pressure. As the reservoir pressure drops below the dew-point pressure, the condensate accumulates near the wellbore, leading to a decrease in both gas relative permeability and gas production. This effect, referred to as retrograde condensation, heavily hinders the fluid flow from the reservoir to the wellbore. One prevalent method used to restore gas productivity after condensate blocking is to inject dry gas (Whitson & Brulé, 2000). The purpose is to increase or maintain reservoir pressure above the dew-point pressure, thereby revaporizing the condensate.

Overall, the injected gas could comprise methane, CO_2, N_2, or a mixture of gases. There are very few studies of gas injection in shale gas condensate reservoirs, primarily because of the difficulty of mitigating the condensate-blocking issue in low-permeability reservoirs. Further research is needed to optimize the process and understand the fundamental mechanisms fully.

7.2.9 Pre-Screening and Selection Criteria

Identifying suitable candidates for HnP gas injection in unconventional resources requires a thorough analysis of reservoir properties, fluid characteristics, and operational constraints. Important factors to consider include uncertain (reservoir properties) and operational variables. Geologic screening is a method used to evaluate types of reservoirs in terms of heterogeneity, connectivity, and other geological characteristics that play a pivotal role in risk management or correlate with process performance (Alvarado & Manrique, 2010). In resource assessments, analogous reservoirs—characterized by similar rock and fluid properties, reservoir conditions (depth, temperature, and pressure), and drive mechanisms—are typically more developed than the reservoir under consideration.

The availability of hydrocarbon (HC) gas for gas-injection EOR is better than that of other EOR injectants in both onshore and offshore environments. HC gas for injection can be sourced from produced gas, either from the reservoir itself or nearby field. Table 7.1 shows the proposed screening criteria for the HC gas and CO_2 miscible method. Kang et al. (2016) highlighted that for offshore fields, the availability of the injectant is a critical consideration, given its substantial contribution to the overall project cost. The recommended reservoir conditions for HC gas miscible operation, drawn from field statistics and simulation studies specific to offshore applications, are also applicable to onshore operations. The screening parameters for the CO_2 HnP process are comparable to those for the HC gas miscible process. Typically, the CO_2 miscible process is used in black oil with low viscosity conditions and oil saturation as low as 17%.

7.2.10 Laboratory and Field Design

To evaluate the feasibility and performance of HnP gas injection in a target reservoir, it is crucial to undertake both laboratory experiments and numerical simulations.

TABLE 7.1
Screening Criteria for the HC Gas Miscible Process

Screening Criteria by Oil and Reservoir Properties	Parameters for Gas HnP EOR	
	Hydrocarbon Gas	CO_2
Oil Viscosity (cP)	≤ 42	≤ 6
Oil Gravity (°API)	≥ 24	≥ 22
Oil Saturation (%)	≥ 25	≥ 17
Reservoir Pressure	≥ MMP	≥ MMP
Reservoir Permeability	Homogeneous preferred	Homogeneous preferred
Reservoir Net Thickness	Thin unless dipping	Thin unless dipping

Source: Kang et al., 2016

These investigations assist in establishing optimal operational parameters, which may include the composition of the gas, injection pressure, and soaking duration. Table 7.2 shows an overview of laboratory-based studies performed on tight hydrocarbon systems with permeability values extending down to the nano-Darcy range. The design of the field operation should consider factors such as well spacing, injection and production strategies, and the infrastructure required for handling gas. Implementing robust monitoring and control systems is also essential to maintain safe and effective HnP operations.

The optimal time to start the gas HnP EOR process is often a key question in planning reservoir development strategy. Injectivity can present challenges if injection commences early in the well life. However, injecting gas into a depleted reservoir would benefit from the total system compressibility, thereby leading to increased gas leakoff and, consequently, oil recovery through advective transport. In their simulation studies of CO_2 HnP application in Bakken shale, Sanchez-Rivera et al. (2015) and Chen and Gu (2017) reported that commencing the operation early in the well life does not fully harness the potential of repressurization (also referred to as gas expansion).

The preparation for a HnP gas-injection EOR includes the following phases:

1. Feasibility study
   - Conduct laboratory tests for miscibility, pressure-volume-temperature analysis, and core-flood experiments.
   - Identify potential pilot locations.
   - Carry out reservoir simulations of the pilot area to fine-tune operational parameters.
2. Facility design
   - Assess the production engineering design.
   - Engage in front-end engineering and design for the requisite facilities.
3. Construction and operation
   - Undertake the construction of the injection facility, including civil work, gas storage vessels, and pipelines.

- Conduct recompletions of existing production wells if injection gas is CO_2.
- Commission the injection facility, covering mechanical, electrical, and instrument aspects.
- Commence the HnP gas injection process, periodically collect samples of produced fluids, and undertake production logging runs for monitoring gas/oil contact and saturation levels.

7.2.11 Examples of Gas Huff-n-Puff Projects

Numerous successful HnP gas injection projects have been implemented in unconventional reservoirs across North America. Table 7.3 summarizes pilot performances of gas HnP in this region. Up to this point, some reported field pilots in the Bakken formation have yielded disappointing results, whereas several field pilots in the Eagle Ford formation have demonstrated promising performance. The unsuccessful tests in Bakken might be owing to the heterogeneous matrix and complex geology (Kurtoglu, 2013). Ongoing research and field trials will improve understanding of the underlying physics and mechanisms, as well as optimizing operational parameters.

7.3 GAS FLOODING

In unconventional reservoirs, the rapid decline in oil production rates coupled with low recovery factors (RF), typically between 5% and 10%, indicates a significant volume of oil remains trapped within the porous medium after primary recovery methods have been exhausted (Mohammed & Babadagli, 2016). EOR through fluid injection, a well-established technique in conventional reservoirs, employs predominantly water and gas as the primary injected fluids. Water injection is widely used in conventional reservoirs to maintain reservoir pressure and act as a displacement agent. However, in tight oil reservoirs, challenges such as reduced injectivity and productivity arise, owing to smaller pore and throat sizes, which severely diminish the overall efficacy of the development (Sheng, 2015a). Gas injection offers a more favorable alternative, owing to its easier injectability and minimal formation damage, making it a more suitable solution than water in tight reservoir settings (Sheng & Chen, 2014). Numerous studies have corroborated the superior performance of gas flooding over water flooding, particularly emphasizing its promise as an EOR method in tight reservoirs (Zhou et al., 2020). Gas injection enhances oil mobility by reducing its viscosity through the dissolution of injected gas, which can mix with oil to form a single phase through first or multiple contact mechanisms, predominantly utilizing CO_2 and N_2 (Barak & Oren, 1992; Du & Nojabaei, 2019; Kantzas et al., 1988). Despite the potential benefits, CO_2 injection poses significant economic uncertainties, owing to its cost and complexity, requiring extensive corrosion protection for pipelines, wellhead equipment, and tubular components (Alagorni et al., 2015). Furthermore, CO_2 is less efficient at restoring reservoir pressure than other gases such as N_2, which emerges as a more economical and effective option, especially when matrix permeability is below 0.3 mD (Wuensche, 1978; Clancy et al., 1985; Syed et al., 2011; Joslin et al., 2017). The application of gas injection in shale

TABLE 7.2
Summary of Recent Laboratory and Field Pilot Design on Cyclic Gas Injection

Formation	Porosity (%), Permeability (mD)	Pressure (psi), Temperature (°C)	Gas Type	Fluid System	Cycle No.	Soak Time (Hours)	Recovery (%)	Reference
Upper Bakken (USA)	4.4–5.4 and 0.008–0.1	6,200 and 110	CO_2	Bakken oil	3	24	11–43	Jin et al. (2017)
Lower Bakken (USA)	~3.8 and ~0.005	6,500 and 111	CO_2	Bakken oil	3	24	8–48	
Bakken (USA)	18.6–23.1 and 0.3–0.8	4,020 and 69	CO_2	Bakken oil	4–6	40–60	43–63	Song and Yang (2017)
Wolfcamp (USA)	7.5 and 0.0002	1,200–2,400 and 40	CO_2	Wolf camp oil	7	84	~35–70	Li et al. (2017b)
Wolfcamp (USA)	7.5 and 0.0003–0.0005	2,000 and 40	CO_2	Wolfcamp oil	6	78	~65	Li et al. (2017a)
			N_2		6	~145	35–50	
			CH_4		6	~145	~30–40	
Bakken (USA)	4.5–8.1 and 0.002–0.04	5,000 and 110	CO_2	Bakken oil	5	70–96	60–100	Hawthorne et al. (2013)
Eagle Ford (USA)	7.5 and 0.0003	2,000 and 40	CO_2	Wolfcamp oil	6	78	61	Li et al. (2019)
			N_2				33	
Baikouquan (China)	9.3 and 0.9	3,300 and 69	CO_2	Black oil	3	N/A	37	Du et al. (2020)

(Continued)

TABLE 7.2 (Continued)
Summary of Recent Laboratory and Field Pilot Design on Cyclic Gas Injection

Formation	Porosity (%), Permeability (mD)	Pressure (psi), Temperature (°C)	Gas Type	Fluid System	Cycle No.	Soak Time (Hours)	Recovery (%)	Reference
Eagle Ford (USA)	5.0–7.7 and ~0.0005	1,000 and 35	N_2	Eagle Ford oil	8	384–704	40–45	Wan et al. (2015)
Eagle Ford (USA)	9.7–10.6 and 0.0003–0.0004	1,000–5,000 and 23	N_2	Eagle Ford oil	12	12–288	18–25	Yu and Sheng (2017)
Middle Bakken (USA)	4.5–8.1 and 0.002–0.04	5,000 and 110	Lean gas	Bakken oil	N/A	24	95	Alharthy et al. (2018)
Haynesville Shale (USA)	8 and 0.0006–0.0009	8,000 and 132	Lean gas	Haynesville oil	7	230–350	35–50	Mahzari et al. (2021)
Montney/Duvernay (Canada)	6 and ~0.0008	3,100–5,200 and 60–92	Lean gas	Montney oil	7	~105	~35	Thomas et al. (2020)
Montney (Canada)	3.5 and 6E-5	2,000 and 50	Lean gas	Montney oil	1	168	29–35	Tran et al. (2021)
	3.5 and 6E-5		Natural gas	Montney oil	1	168	35–40	

Source: Ghanizadeh et al., 2021

TABLE 7.3
Summary of Recent Field Applications of Cyclic Gas Injection

Formation	Porosity (%) & Permeability (mD)	Gas Type	Fluid System	Pressure (psi) & Temperature (ºC)	Cycle No.	Soak Time (days)	Recovery Improvement (%)	Reference
Chattanooga (USA)	6.1 and 4.61E-06	CO_2	Gas condensate	434 and 62	1	120	800	Louk et al. (2017)
Eagle Ford (USA)	3–11 and 1E-6 to 5E-4	Natural gas	Black oil	6425 and 138	4	28–42	17–20	Hoffman (2018)
Eagle Ford (USA)	5.97 and 15.76E-4	Lean gas	Black oil	6,000 and 107	5	0	27.4	Orozco et al. (2018)
Bakken (USA)	5–7 and 0.0003–15	CO_2	Black oil	5,000–6200 and 70–132	1	30	Little or no improvement	Hoffman (2018)
Permian (USA)	7.76 and 71E-5	Natural gas	Black oil	3,300 and 69	12	0	29 and 82 for 2 wells	Ganjdanesh et al. (2020)
Montney (Canada)	6–11 and 0.001–0.007	Natural gas	Black oil	2,900 and 60	1	60	No improvement	Yassin et al. (2022)
Ellerlise (Canada)	7 and 0.02	Natural gas	Black oil	3,100 and 91	1	8	0.5–3	

or liquid-rich shale reservoirs remains largely experimental and limited to laboratory tests and simulation studies, with field applications yet to be implemented. Gas flooding involves continuous or water-alternating-gas (WAG) injection while HnP utilizes cyclic injection, in which gas produced during the production phase is reinjected into the well during subsequent cycles. Both methods contribute to cyclic gas injection, characterized by the repetitive injection of produced gas back into the reservoir, optimizing resource utilization and enhancing oil recovery (Monger & Coma, 1988; Haines & Monger, 1990; Torabi & Asghari, 2010). Despite the prominence of gas injection in conventional settings, extending it to unconventional shale reservoirs presents significant challenges. The effectiveness of EOR methods varies considerably with reservoir conditions. While CO_2 tops the list of potential EOR agents in unconventional settings based on lab studies, natural gases have shown better performance in pilot tests, revealing a gap between laboratory-scale results and field applications (Hoffman & Reichhardt, 2019; Schmidt & Sekar, 2014). This discrepancy likely stems from an incomplete understanding of the physical and chemical interactions of CO_2 in reservoir environments. The efficacy of gas injection methods in low-permeability shale reservoirs is further complicated by the unique properties of these formations. The ultra-low permeability typically encountered impedes the effectiveness of continuous flooding techniques, making the cyclic nature of HnP more appealing. The high diffusivity rates of miscible gases observed in laboratory conditions suggest that shorter, cyclic injections may facilitate better oil recovery in tight shales than continuous flooding. This process involves repeatedly injecting the produced gas back into the reservoir, thereby enhancing oil recovery while managing production costs and maintaining reservoir integrity. This background sets the stage for a deeper exploration of the physical mechanisms and comparative efficacy of these methods in subsequent sections.

7.3.1 Physics and Mechanisms of Gas Flooding in Shale

The physical mechanisms underlying gas flooding in shale reservoirs are distinct from those in conventional reservoirs, primarily because of the unique geological features and the properties of the injected gases. Sheng and Chen (2014) found that during gas flooding, the increase in oil production rates is primarily owing to the pressure maintenance rather than miscibility of the injected gas with the reservoir oil. This observation is crucial as it highlights the challenges of achieving miscibility in tight formations where natural fractures, though prevalent, are often poorly connected, limiting effective gas propagation from injectors to producers. In tight shale formations, the primary flow mechanism through which the gas interacts with the oil is existing fractures, rather than permeating the rock matrix as seen in more conventional settings (Hawthorne et al., 2013). This leads to scenarios in which gas may quickly break through to producers if not managed carefully. The HnP method, by allowing the injected gas to dissolve and swell the oil, can temporally mitigate this by enhancing miscibility under high-pressure conditions. However, the potential reproduction of injected gas during the subsequent puff phase remains a concern, necessitating further evaluation to determine the optimal EOR strategy between continuous gas flooding and HnP techniques (Wan et al., 2013; Sheng & Chen, 2014).

The application of CO_2 in shale gas EOR presents its own set of challenges and opportunities. Laboratory studies and field simulations indicate that the efficiency of CO_2 in EOR is significantly influenced by its molecular diffusion properties, which are markedly different in tight shale formations than in conventional reservoirs. For instance, Wang et al. (2010) demonstrated through simulation that CO_2 injection outperforms water flooding in the Bakken formation, attributing this to the superiority of CO_2 miscibility and diffusivity under the right conditions. Yet the molecular diffusion mechanism, while beneficial in smaller core samples, poses challenges when scaled to field operations, owing to varying natural fracture densities and the intrinsic low permeability of shale (Alfarge et al., 2020a). Comparatively, natural gas–based EOR has shown to outperform CO_2 at the field scale in pilot tests, suggesting that while laboratory studies often favor CO_2 owing to its molecular characteristics, natural gas may be more effective under actual reservoir conditions (Todd & Evans, 2016; Schmidt & Sekar, 2014). This disparity highlights the critical gap between laboratory-based predictions and field-level implementations, underscoring the need for enhanced understanding and optimization of gas properties and injection strategies tailored specifically to the complex environments of shale reservoirs. In summary, while gas flooding and HnP methods offer promising avenues for EOR in shale formations, their efficacy is highly dependent on the understanding of physical and chemical interactions within the reservoir. Continuous research and adaptation of these methods to the specific characteristics of shale formations are imperative to maximize recovery while mitigating the risks associated with gas breakthrough and poor miscibility.

7.3.2 Gas Flooding versus HnP

In the field of EOR in tight and shale reservoirs, two primary gas injection methods emerge: gas flooding and HnP. Each technique has its own set of mechanisms, advantages, and considerations, particularly in terms of reservoir permeability, operational complexity, and economic viability.

Gas flooding is a method by which gas is continuously injected through one or more wells to displace oil towards production wells. This technique has been shown to be effective in reservoirs with higher permeability, where the injection pressure can propagate adequately across the reservoir. Studies such as those by Yu et al. (2016) and Chen et al. (2020) have demonstrated the effectiveness of nitrogen and carbon dioxide gas flooding in improving oil recovery by creating a more uniform residual oil distribution within the matrix, especially as permeability increases. Liu et al. (2014) also confirmed the efficacy of CO_2 flooding in unconventional reservoirs, noting significant improvements in recovery factors. However, the efficiency of gas flooding diminishes in lower-permeability formations, where the pressure transmission becomes less effective. HnP, on the other hand, involves using the same well for both injection and production, performed cyclically. This method is particularly suited for formations with lower permeability. It involves three stages: injection (huff), shut-in for soaking, and production (puff), which allows for periodic recharging of formation pressure and has been shown to potentially increase oil recovery significantly (Sheng, 2015a; Su et al., 2015). The cyclic nature of HnP can better manage the challenges of poor matrix permeability and limited fracture connectivity that characterize many shale reservoirs. It leverages the high diffusivity rates of injected gases, such as CO_2, to enhance recovery, even under immiscible conditions.

The comparative performance of gas flooding versus HnP in field-scale studies reveals varied results, depending on the reservoir's physical characteristics. For instance, simulations and field tests have shown that gas flooding achieves higher recovery rates in reservoirs with matrix permeability above 0.01 millidarcies (mD), while HnP is more effective in ultra-tight formations with permeabilities lower than 0.001 mD (Tang & Sheng, 2022). Additionally, the economic analysis indicates that HnP may offer a better return on investment in higher-permeability formations than gas flooding, given the operational costs and recovery rates. The decision between continuous gas flooding and HnP should ideally be based on two critical factors: the ratio of reservoir permeability to injector-producer spacing and the diffusivity rate of the injected gas in the specific reservoir conditions (Alfarge et al., 2017). In reservoirs where the permeability is sufficient to allow an economic sweeping of oil from injector to producer, gas flooding might be preferable. Conversely, in tightly packed shale formations where traditional sweeping is less effective, the cyclic nature of HnP can capitalize on the high diffusivity of gases to enhance oil recovery more efficiently. In conclusion, while both gas flooding and HnP are viable EOR techniques, their applicability and effectiveness largely depend on the specific geological and operational characteristics of the reservoir. Understanding these intricacies is essential for optimizing recovery strategies and ensuring economic feasibility in the challenging environments of tight and shale oil reservoirs.

7.3.3 Prescreening and Selection Criteria

In ultra-low permeability reservoirs, injecting gas to sweep oil through the shale matrix and improve sweep efficiency is challenging. The complexity of fractures—including natural, hydraulic, and injection-induced fractures—coupled with the low injectivity in shale-oil reservoirs often leads to rapid gas breakthrough in gas flooding mode. Continuous gas flooding in shale reservoirs is nearly impractical owing to the ultra-low permeability of the shale matrix, except possibly through high-permeability channels such as fractures (Cronin et al., 2019).

Wood et al. (2008) presents a novel model for evaluating Gulf Coast reservoirs for CO_2 flooding and EOR, as well as CO_2 storage potential. This model, developed through a Box-Behnken experimental design, is notable for incorporating the effects of reservoir dip and providing quantitative estimates of both cumulative oil recovery and CO_2 storage potential. It employs ten critical dimensionless groups, including aspect ratio, dip angle group, mobility ratios, and various saturation levels. This approach offers a more comprehensive assessment than earlier models that focused solely on either qualitative evaluations or oil recovery. The model is applicable to both conventional and tight reservoirs, ranging from micro- to milli-Darcy permeability. Zhu et al. (2015) proposed an innovative fracture-to-fracture gas flooding scheme, which involves injecting gas into one fracture and producing fluids from an adjacent fracture within the same horizontal well. This method aims to overcome the challenge of high-pressure gradients in tight formations.

The prescreening and selection process for gas flooding projects in unconventional reservoirs often involves evaluating geological characteristics such as porosity, permeability, and connectivity within the shale matrix, including the presence of natural fractures.

Reservoir analogue identification is a method used when sufficient laboratory and field data are unavailable. It involves finding a field with similar reservoir properties and a successful EOR project to serve as a model. However, this approach is less popular in gas flooding EOR owing to the unique characteristics of each reservoir in the field pilot. The effectiveness of different gases (CO_2, N_2, or produced gas) in EOR varies significantly, depending on their interaction with the reservoir. CO_2 typically shows higher adsorption capacities in shale than methane and N_2, making it a crucial factor in both enhanced oil recovery and CO_2 sequestration. Nitrogen, with its high MMP and low dissolution in oil, appears to achieve better recovery in scenarios where immiscible displacement is sufficient. Lean gas flooding is considered in reservoirs where miscibility at existing reservoir pressures is feasible or where conditions favor gas expansion and displacement mechanisms.

7.3.4 Laboratory and Field Design

A full-field gas EOR process typically involves both simulation studies and pilot tests. In the laboratory, core samples from the reservoir are examined under controlled conditions. This examination aims to understand the interactions between various types of gases (such as C_1, enriched gas, N_2, and CO_2) and the reservoir oil/condensate as well as with the core samples themselves. Field design involves determining the optimal gas injection rate, pressure, and type, along with the development of monitoring and evaluation techniques. These techniques are essential to assess the effectiveness of the gas flooding process. Table 7.4 highlights recent advancements in the modeling and laboratory study of gas flooding in several unconventional reservoirs across North America.

7.3.5 Examples of Gas Flooding Projects

There have been relatively few worldwide projects in which gas flooding has been applied to unconventional reservoirs, primarily owing to the operational challenges and economic considerations regarding incremental recovery. These projects often utilize CO_2 or natural gas as the flooding agent.

7.3.5.1 North America

One of the earliest gas flooding projects in a tight oil reservoir was conducted at the Bakken formation in Saskatchewan (Schmidt & Sekar, 2014). This project involved immiscible continuous gas injection via a central horizontal injection well, interacting with nine horizontal production wells arranged north to south. The Bakken formation in this test area exhibited porosities of 9% to 10% and permeabilities ranging from 0.01 to 0.1 mD. Initially, the injection rate was set at 300 MSCF/d with an injection pressure of 500 psi. This rate was later increased to 1 MMSCF/d at 1,000 psi after adding compression capacity. Gas breakthrough was observed in two pattern wells, leading to a

TABLE 7.4
Summary of Recent Laboratory and Field Pilot Designs on Gas Flooding EOR

Method	Formation	Porosity (%) & Permeability (μD)	Pressure (psi) & Temperature (°C)	Gas Type	Fluid System	Recovery (%)	Reference
Modeling	Middle Bakken	7 and 1–100	6,200 and 110	CO_2	Bakken oil	0–30	Sheng and Chen (2014)
Modeling	Eagle Ford	0.1 and 6	6,500 and 111	CO_2	Eagle Ford oil	15	Zuloaga et al. (2017)
Modeling	Marcellus	2.5 and 0.5	870 and 65	CO_2	Marcellus gas	0–11	Godec et al. (2013)
Modeling	Barnett	6 and 0.1	2,100 and 49	CO_2	Barnett gas	9–12	Kim et al. (2015)
Laboratory	Eagle Ford	4.4–13.1 and 0.1–0.4	1,000 and 72–95	N_2	Wolfcamp oil	17–19	Yu et al. (2017)
Laboratory	Eagle Ford	0.1 and 6	9,500 and 154	C1/C2: 85/15	Condensate	4	Sheng (2015c)
Laboratory	Eagle Ford	0.1 and 6	4,000 and 93	C1	Condensate	5	Meng and Sheng (2016)
Laboratory	Eagle Ford	0.1 and 6	9,900 and 132	CO_2	Condensate	0	Sheng et al. (2016)

decrease in the oil production rate to 53 bbl/d. However, after workovers, oil rates consistently increased across all nine producers, with the total rate climbing to 295 bbl/d.

Another case study from the Bakken formation in North Dakota utilized natural gas flooding as the EOR method. In these reservoirs, with a permeability of about 1 mD, the application of natural gas flooding significantly improved oil production rates. Results indicated that continuous gas flooding could recover an additional 10% to 20% of oil, while the HnP protocol could recover 5% to 10% (Hoffman & Evans, 2016). Xu and Hoffman (2013) evaluated the Elm Coulee Field in Eastern Montana under CO_2 flooding with varying hydraulic fracture orientations. They discovered that transverse fractures yielded a higher oil recovery factor but had lower CO_2 utilization values than longitudinal fractures owing to breakthrough issues.

7.3.5.2 Worldwide

Jiang et al. (2008*) reported a CO2* flooding pilot in the Fuyang Reservoir in the Song-Fang-Dun Field, China. The reservoirs had permeabilities of 0.79 mD and oil viscosities of 6.6 cP. This project involved 14 injectors and 26 producers, with well spacings between 200 to 300 meters. Initially, the CO_2 injection rate was 22 tons/day, and the oil rate was 0.2 tons/day. When the producers responded to CO_2 injection, the oil rate increased to 0.6 tons/day. The resulting ratio of oil produced to CO_2 injected was 0.044 tons/ton, equivalent to 53.75 MSCF/bbl of oil. The performance of this CO_2 injection was hindered primarily by a low net-to-gross ratio (less than 0.5) and significant reservoir heterogeneity.

7.4 CHEMICAL AND NANOTECHNOLOGY METHODS

Since the early 2000s, advances in fracturing technology and increased oil and gas prices have driven the production of shale oil and gas to its record high in the United States, according to the US Energy Information Administration. In situ hydraulic fracturing and producing or mining, processing, and extracting are adopted for oil recovery for shale reservoirs. There are many chemicals and technologies involved in the hydraulic fracturing and production process. This section will discuss the chemicals and technologies used explicitly for shale reservoirs for oil recovery purposes.

Unconventional reservoir oil and gas recovery started in Texas Barnett shales in the 1990s. Beginning in the 2000s, it has bloomed, owing to advances in fracturing technology and increased oil and gas prices (Reynolds, 2020). The very low permeability of shale reservoirs requires specific production methods, such as hydraulic fracturing and producing or oil shale mining and processing (Zendehboudi & Bahadori, 2017). Various chemicals are engaged in shale reservoir hydraulic fracturing and producing, in which each chemical plays its role in the hydraulic fracturing fluid, a fluid with different bases (Table 7.5), combined with proppant and pumped into the shale formation to create fractures, untack the inaccessible hydrocarbons, and flow back with hydrocarbons. Depending on the reservoir, the fracturing fluid may include a combination of chemical additives, such as friction reducers and transport polymers, clay stabilizers, biocides, scale inhibitors, flowback aids, etc. (Ely & Herndon, 2019; Reynolds, 2020). Table 7.5 lists the chemical additives for hydraulic fracturing fluid and their functions, summarized based on Ely and Herndon's work (2019).

TABLE 7.5
Summary of Hydraulic Fracturing Fluid Base and Chemical Additives

Type of Fracturing Fluids	Features
Water-Based Fracturing Fluids	Mostly used, economical, easily controlled
1. Linear polymer solution	1. Water thickener, viscosifier
2. Crosslinked Fracturing Fluids	2. Tie polymer molecules together with crosslinkers to increase viscosity without increasing polymer usage
3. Viscoelastic Surfactant Fluids	3. Increase both viscosity and viscoelasticity
Oil-Based Fracturing Fluids	Avoid water-sensitive formation damage
Alcohol-Based Fracturing Fluids	Reduce water surface tension, remove water blocks
Emulsion Fracturing Fluids	Cost-effective, functional, high-friction pressure
Foam-Based Fracturing Fluids	High viscosity, moderate proppant-transport capability
Energized Fracturing Fluids	Entraining gas, cost effective but potential reservoir destruction effect
Chemical Additives	**Features**
Biocides	Kill bacteria
Breakers	Reduce fluid viscosity
Buffers	Control the pH
Surfactants	Lower surface tension
Clay stabilizers	Prevent clay swelling
KCl substitutes	Enhance clay swelling control effect of KCl and prevent fine migration
Fluid-loss additives	Improve fluid efficiently
Foamers	Foaming agents
Friction reducers	Reduce friction
Temperature stabilizers	Stabilize base chain polymer, fracturing fluid pH, and/or oxidizing agents to ensure temperature stability
Diverting agents	Divert flow of fracturing fluid

Source: Ely & Herndon, 2019

Tight oil reservoirs, including shale reservoirs, typically have an estimated ultimate recovery of less than 10% after hydraulic fracturing; thus, further recovery enhancement techniques are much needed for effective oil recovery (Liang et al., 2021; Milad et al., 2021). EOR mechanisms in shale reservoirs include (Wang et al., 2017):

- Gas and oil flow in nanometer pores
- Gas dissolution and diffusion through low permeability matrix
- Oil swelling
- Wettability alteration
- IFT reduction
- Fracture-matrix interaction

Here, chemical systems that improve oil recovery after fracturing or refracturing, whether chemicals serve as part of the hydraulic fracturing fluid or other formation stimulation fluids, and the chemical-associated techniques and mechanisms that contribute to the shale oil recovery process are taken into consideration and will be discussed in detail.

7.4.1 Chemicals

7.4.1.1 Surfactants/Solvents

Among the water-based fracturing fluid chemicals, such as polymers, the fracturing fluid thickener surfactants are more versatile and adopted as either the body of the fracturing fluid or as additives. As listed in Table 7.5, viscoelastic surfactants could improve the solution's viscosity and viscoelasticity (Yu et al., 2020), which helps carry the proppant efficiently. Meanwhile, as the surfactant, it could mobilize hydrocarbon by altering rock wettability and lowering interfacial tension, which are considered the primary mechanisms in shale oil recovery studies, thus improving oil recovery (Alvarez et al., 2023).

Studies have been performed for the selection of surfactants and also for the understanding of the mechanisms beneath. It has been found that when exposed to surfactant in spontaneous imbibition tests, 33% and 34% recoveries were achieved for Wolfcamp and Eagle Ford shale samples, compared to 19% and 14% of the surfactant absence base cases. Besides the interfacial tension decrease effect, preventing asphaltene precipitation might be another reason for the high recovery factor as some surfactants could linearly decrease the asphaltene precipitation with increased surfactant concentration. In addition, altering the oil-wet and mixed-wet to water-wet rock surface could also assist oil recovery (Gupta et al., 2020). Tu and Sheng (2019) performed a forced imbibition study with the assistance of surfactant in a high-pressure setting with a 12-hour soaking/12-hour depletion schedule for eight cycles. They obtained 38% recovery after eight cycles and concluded that wettability alteration is the dominating effect as interfacial tension was not prominently reduced. A lab-scale simulation was also set up, showing the soaking time is more critical than the depletion time. In addition, the surfactant itself dominates the recovery factor, while cyclic injection expedites the process (Tu & Sheng, 2019). In addition, Zelenev (2011) has performed work that showed no significant variation in surface energy determined from contact angle measurements of shales from different North American basins, which suggests that heterogeneity of shale surface chemistry could play only a secondary role in the development of shale-specific chemical treatments, and other factors should be considered first (Zelenev, 2011). Studies dealing with the adsorption of the surfactant to lower the cost have drawn attention. A natural nonionic surfactant extracted from Zyziphus Spina Christi leaves was tested with real shale-sandstone reservoir samples owing to its low cost. It has been found a higher concentration would lead to more surfactant adsorption until saturation point (Ahmadi & Shadizadeh, 2015).

7.4.1.2 Chemical Blends

Hydraulic fracturing fluid is naturally a chemical blend that includes base fluid, additives, and proppant. However, the production rate reduces by over 2/3 in the first two years. Researchers are looking for more chemical systems that improve oil recovery after fracturing or refracturing. Miller et al. (2019) tested a chemical blend of surfactants, a weak acid, a potential determining ion, and a solvent through spontaneous imbibition, and they found that the surfactant could change the wettability and interfacial tension, the solvent helped mix with oil and asphaltene, and the weak acid could dissolve part of the carbonate and improve permeability. Thus, they summarized that the synergy makes the chemical blend a good candidate for shale oil recovery (Miller et al., 2019). Zeng et al. tested a large number of surfactants (anionic, nonionic, and cationic) and their blends; at 125 °C, all aqueous stable surfactant blends changed the shale wettability from oil-wet to water-wet at both high and low salinities. Surfactant blends also improved the imbibition more than single surfactants (Zeng et al., 2018).

7.4.1.3 Nanoparticles

Nanoparticles for enhancing shale oil recovery have recently gained increasing attention. Table 7.6 summarizes nanomaterials and technologies adopted in hydraulic fracturing for unconventional oil and gas reservoir applications (Mao et al., 2022). Nanoparticles and related technologies that serve direct oil recovery purposes in shale reservoirs are discussed in this section, which comparably lacking studies. It has been found that aqueous-dispersed hydrophilic silicon dioxide nanoparticles induce less oil-wet wettability by significantly lowering the adhesion force and energy barrier to spontaneously detach oil molecules from clay and quartz mineral surfaces in shale reservoirs (Afekare et al., 2021). A nanoparticle-assisted CO_2 HnP-style field test was conducted on Eagle Ford shale (Zheng et al., 2020). The pilot test results showed a significant oil rate increase after the nanoparticle-assisted CO_2 was injected, and the lab work showed that nanoparticles lowered the interfacial tension between the water and oil and turned the wettability to a more water-wet state. Yet a CO_2-only field test was not performed as the baseline to demonstrate the nanoparticle effect. Scerbacova et al., on the other side, argued that nanoparticles do not outperform brine and cause pore blockage, which may eventually lead to critical damages; thus, nanoparticles are not recommended in shale reservoirs (Scerbacova et al., 2023). As for in situ oil shale exploration, it has been tested that in situ microwave heating in the presence of iron oxide nanoparticles at proper concentration, output power, and ultimate reaction temperature could lead to the highest oil yield and best oil quality (Yang et al., 2017). Tabatabaei et al. developed a graphite nanoplatelet-coated proppant that has more hydrophobic wettability and better conductivity and compressive strength, making it favorable as a flowback aid and indicating it may help oil recovery (Tabatabaei et al., 2019).

7.4.2 Mechanisms and Techniques

7.4.2.1 Enhanced Spontaneous Imbibition and Forced Imbibition with Wettability Alteration and IFT Changes

Spontaneous imbibition, in which the capillary action draws the wetting phase into the porous medium serves as the basic mechanism or technique for oil displacement (Morrow & Mason, 2001). The most-used expression for capillary pressure in a tube is listed as:

$$P_c = \frac{2\gamma\cos\theta}{r} \tag{7.10}$$

TABLE 7.6
Nanomaterials and Technologies Adopted in Hydraulic Fracturing for Unconventional Oil and Gas Reservoir Applications

Application Field	Mechanisms	Advantages
Fracturing Fluids	The spatial network structure of the fracturing fluid system is enhanced by adding nanoparticles to polymer solutions or cross-linked gel-fracturing fluids.	1. Improve the rheology of polymer-based fracturing fluid and its stability under high temperature and high pressure 2. Reduce the amount of polymer 3. Reduce the damage to the fractures' conductivity
	Additives such as nano drag reducer and drainage aid prepared based on nanomaterials can form a more stable solution system structure and reduce surface tension.	1. High viscosity and excellent sand suspension performance 2. Better temperature resistance, salt resistance, shear resistance, and drag reduction performance 3. Reduce water lock damage and help flowback after hydraulic fracturing
	Nanoparticles combine with VES micelles to form a tight spatial network structure through chemisorption and surface charge interactions.	1. Improve temperature and shear resistance and reduce fracturing fluids filtration loss 2. Enhance the sand-carrying performance of VES fracturing fluid
	Nanoparticles can be adsorbed on the interface with the synergistic effect of surfactants and form a stable adsorption layer on the surface of the foam film, which plays a role in stabilizing the foam film.	1. Effectively inhibit the coalescence effect of bubbles and stabilize the foam 2. Reduce the fracturing fluids filtration loss 3. Improve sand-carrying performance
Proppants	Nanoscale proppants were prepared by nanoparticles for supporting micro/nanofractures in the reservoir.	1. Prevent the closure of microfractures and effectively increase the fractures area 2. Improve the reservoir fractures network conductivity
	The surface of proppant matrix particles was coated with nanomaterials to improve its performance.	1. Improve proppant compressive strength and reduce the crushing of proppant 2. Maintain the hydraulic fractures' conductivity

(Continued)

TABLE 7.6 (Continued)
Nanomaterials and Technologies Adopted in Hydraulic Fracturing for Unconventional Oil and Gas Reservoir Applications

Application Field	Mechanisms	Advantages
Fracturing Flowback Fluid Treatment	Nanophotocatalysts have the ability to oxidize and decompose pollutants under light and deeply purify fracturing flowback fluids.	Improve the treatment efficiency and effect of fracturing flowback fluid, which can be recycled
	Utilize the nanofiltration characteristics to prepare separation membranes such as ultrafiltration and nanofiltration to separate pollutants in fracturing flowback fluids.	Effectively remove dissolved solids and organic pollutants
Degradable Fracturing Completion Tools Development	Fracturing completion tools made of degradable nanocomposites can be dissolved in fracturing flowback fluids.	Simplify operation procedures, improve operation efficiency and fracturing effect
Hydraulic Fracturing Monitoring and Fractures Mapping	Nanorobots and nanosensors can enter the micro/nanopores of unconventional reservoirs with the injected fluid to monitor hydraulic fracturing in real time.	Real-time acquisition of hydraulic fracture morphology parameters and accurate evaluation of fracturing effect
	Magnetic nanoparticles can be mixed with proppant and injected into the formation, and the location of the proppant can be detected by measuring the magnetic susceptibility to characterize the hydraulic fractures.	Precisely locate and track downhole proppant distribution and accurately map hydraulic fractures geometry

Source: Mao et al., 2022

where γ is the interfacial tension, θ is the contact angle, and r is the effective radius of the interface. During well completion, when fracturing fluid is introduced to the reservoir, if the external applied pressure (soaking pressure) at the fracturing place is right at the reservoir pressure, the displacement mechanism could still be characterized as spontaneous imbibition. However, in the real case, the induced pressure will most likely surpass the reservoir pressure: thus, forced imbibition occurs. After a certain amount of time soaking, the production begins once the well opens. Pressure control is critical for forced imbibition–induced oil recovery. It has been suggested in a forced imbibition study conducted for a range of applied pressure (1,000–5,000 psi) in unconventional sandstone, carbonate, and shale core plugs with a variety of wettability that a higher soaking pressure could lead to less oil recovery than a lower soaking pressure for the studied case; however, the water-wet state still dominates the oil recovery regardless of the soaking pressure (Tu & Sheng, 2020b). Low-salinity water could enter the matrix, owing to the osmotic pressure gradient, and displace oil from the matrix. By adding surfactants, not only could the wettability be altered to a more favorable condition but the IFT is also lowered for better oil movement (Liu & Sheng, 2019; Uzun et al., 2020).

Wettability alteration by chemicals is mainly induced by engaging surfactant into the hydraulic fracturing fluids during the multistage of the fracturing process to promote spontaneous imbibition, called enhanced spontaneous imbibition (Liu & Sheng, 2019; Neog & Schechter, 2016; Zeng et al., 2018). Various surfactants were studied in nano-Darcy permeability shale and could alter its wettability from intermediate-wet to water-wet and improve the oil recovery. It has also been found that the surfactants that alter the wettability significantly while slightly lowering the IFT promoted the highest oil recoveries (Neog & Schechter, 2016). Alvarez and Schechter (2016) observed that wettability can be altered from oil wet to systems favoring the fracturing fluid imbibition with anionic, nonionic, nonionic-cationic blended surfactants and nanofluids on Bakken shale cores that outperform water without additives (Alvarez & Schechter, 2016). They also confirmed that the oil recovery was consistent with the wettability alteration and IFT changes with additives. It seems wettability alteration and IFT changes complement each other, which leads to increased oil recovery. However, the dominating factor is still water wet instead of lowered IFT, which could be detrimental (Tu & Sheng, 2020a). It has also been tested that imbibed oil volume is correlated with porosity and permeability, with higher imbibed oil volume found in higher porosity and permeability cases (Habibi et al., 2019). Nevertheless, the spontaneous imbibition mechanism serves in different ways during the EOR process in shale reservoirs.

7.4.2.2 Fluid-Rock Interactions

Knowledge of chemical fluid-rock interactions is important for understanding the effect of a chemical for EOR purposes. Fluid-shale interaction during oxidative dissolution can dissolve unstable composition to enhance permeability. Li et al. performed a study with organic-rich shale collected from the Cambrian Shuijingtuo formation with three oxidants—hydrogen peroxide, sodium hypochlorite, and sodium persulfate (Li et al., 2020). Results showed that pyrite and organic matter can be oxidized with different oxidants. They also suggested that the acidic environment was beneficial to carbonate dissolution and that the alkaline environment was favorable to dolomite and tectosilicate minerals dissolution. Acid treatment could create mineral dissolution and secondary phase precipitation that greatly impact permeability alteration of the shale matrix and largely depend on the fluid pH and shale composition (Harrison et al., 2017). It takes only hours to days to neutralize in high-carbonate shale, while it takes weeks to months in low-carbonate shale (Li et al., 2020).

7.4.2.3 Chemical HnP

HnP, in which a recovery enhancement fluid is injected in a cyclic pattern, is a technique of interest for shale reservoir production. The enhancement fluid is usually a gas but could also be a liquid or a liquid mixed with gas (Scerbacova et al., 2023; Teklu et al., 2018). After each injection, there follows a soaking and a production process. For gas HnP, a kinetically slow, peripheral vaporizing gas drive mechanism was primarily responsible for recovery under gas injection for EOR in organic-rich shales (Tovar et al., 2018). For liquid HnP, Teklu et al. concluded that wettability alteration, capillary, and osmotic forces are the main mechanisms (Teklu et al., 2018). They highlighted that low-salinity water could improve oil production. Scerbacova et al. noticed the surfactant had no apparent effect using the HnP method compared to brine on recovery factors observed via computed tomography (CT) scan (Scerbacova et al., 2023). Teklu et al. claimed that surfactants in low-salinity water can significantly change wettability and decrease IFT, thus improving oil production. The HnP mode using low-salinity water followed by surfactant in low-salinity water was also recommended (Teklu et al., 2018). Low salinity seems to be the key to improved oil recovery regarding the contrast in the two literatures, but the role of reservoir heterogeneity is unknown.

Li et al. conducted 15 CO_2-Wolfcamp crude oil HnP experiments using three different Wolfcamp shale cores at pressures below and above the MMP. The result showed that below the MMP, the injection pressure had a significant effect on enhancing oil recovery while above MMP, the increased pressure further increased the oil recovery until the injection pressure was about 200 psi higher than the MMP. According to these experimental results, a higher injection pressure for higher oil recovery in shale oil reservoirs was suggested (Li et al., 2017a). It has also been concluded that nitrogen injection was less effective than CO_2 in organic-rich shale compared to conventional reservoirs (Tovar et al., 2018). Zeng et al. questioned coupling the chemical with CO_2 HnP as they found adding a chemical blend seemed to impede the oil recovery, owing to the shale sample heterogeneity (Zeng et al., 2019).

7.4.2.4 Surfactant-Assisted Gas Injection and Foam EOR

The low viscosity of gas makes gas injection a feasible EOR process in low- to ultra-low-permeability reservoirs. Surfactant-assisted gas injection can improve the oil recovery by (1) wettability alternation, (2) gas mobility control, (3) increase rock permeability, and (4) lowering IFT. The selection of surfactant or surfactant-chemical blend is critical during this process. By lowering the IFT and increasing the rock permeability, favorable imbibition can be stimulated. Foam injection is also found to provide gas mobility control to mitigate the gas channeling in the fractures (Díez et al., 2022).

Besides the mechanisms mentioned here, during CO_2 injection, surfactant can be added to lower the MMP to increase the oil-relative permeability (Almobarak et al., 2021).

In addition to modifying the surface and interfacial properties, the injection strategy also plays a critical role in ultimate recovery. Experimental study and reservoir simulation are insightful tools to optimize the production time, cycle number, etc. (Xiong & Sheng, 2021).

7.4.3 Laboratory Prescreening and Selection Criteria

In the laboratory studies, researchers mainly focused on answering three questions for chemical EOR in shale reservoirs: (1) What is the dominating mechanism? (2) What chemical performs the best? and (3) Do these apply to all shales or are they type specific? For these questions, these methods are generally used and have been performed in both atmospheric and elevated conditions:

1. Contact angle measurements
2. IFT measurements
3. Spontaneous imbibition experiments
4. HnP experiments.

Table 7.7 summarizes the laboratory prescreening methods and selection criteria in the featured literature.

7.4.4 Field Chemical and Nano EOR Projects

The major techniques for chemical-enhanced oil recovery include alkali/surfactant/polymer (ASP) combination flooding, foam flooding, and nanomaterials flooding, which aims at reducing the residual oil saturation by lowering the water-oil interfacial tension and increasing the volumetric sweep efficiency by reducing the water-oil mobility ratio (Ragab & Mansour, 2021). Depending on the maturity of the applied chemical flooding technique, many field pilots and commercial applications at different scales have been completed or are still in progress all over the world. According to the report from the International Energy Agency (IEA, 2018), 35 chemical flooding projects were in operation in 2017, which was much fewer than thermal recovery (120) and gas injection (166) projects.

Systematic investigation and application of ASP flooding have been developed for more than 30 years since its appearance in the 1970s (Foster, 1973). Different combinations such as alkali/surfactant (AS), alkali/polymer (AP), and surfactant/polymer (SP) were also tested to optimize recovery efficiency by incorporating the different strengths of each component (Yang et al., 2019). In China, ASP flooding has been extensively adopted with a total of 13 field pilots. In particular, the Daqing oilfield completed the indoor research and pilot test in 1994 and achieved industrialization in 2014, which proved that ASP flooding could improve the oil recovery by up to 30% with a cumulative oil increment of 21.57 million tons by May 2017 (Guo et al., 2016, 2017). The US has also conducted several pilot tests since 1987 in the West Kiehl field (Clark et al., 1993), the Cambridge field (Vargo et al., 2000), and the Tanner field (Pitts et al., 2006) in Wyoming. The reported crude oil recovery increment ranged from 12.8% to 43.4%, which proved that the technique of ASP flooding was economically feasible. Some ASP flooding projects have also been tested in Canada (Charest, 2013) and India (Jain et al., 2012).

Based on the gas phase component, foam flooding consists of N_2-foam flooding, air-foam flooding, CO_2-foam flooding, and hydrocarbon gas–foam flooding.

TABLE 7.7
Laboratory Prescreening Methods and Selection Criteria in Featured Literature

Reference	Recovery Enhancement Chemicals	Shale	Experiment	Findings
(Afekare et al., 2021)	Hydrophilic silicon dioxide nanoparticles	Tuscaloosa Marine	Zeta potential, atomic force microscope adsorption, scanning electron microscope micrographs	Nanoparticles promoted less oil-wet state by lowering the adhesion force and energy barrier to spontaneously detach oil molecules from clay and quartz mineral surfaces of shale; irreversible adsorption of silica nanoparticles.
(Alvarez & Schechter, 2016)	Anionic, nonionic, and blend surfactants Complex nanofluids	Bakken	Contact angle, IFT, spontaneous imbibition tests	Wettability could be altered from oil wet to systems favoring the fracturing fluid imbibition conducted with anionic, nonionic, and nonionic-cationic blended surfactants and nanofluids that outperform water without additives.
(Gupta et al., 2020)	Anionic, nonionic, and amphoteric surfactants; solvent limonene	Wolfcamp Eagle Ford	Contact angle, IFT, phase behavior, spontaneous imbibition, asphaltene precipitation	By adding surfactants, IFT decreased, wettability towards the more water-wet state, and the oil recovery significantly increased. Some surfactants served as suitable asphaltene inhibitors.
(Liang et al., 2021)	Diluted microemulsion with surfactants	Mahu	Aqueous stability test, emulsion tendency test, oil-water IFT measurement, shearing test	The adsorption of surfactant on the rock limited its enhancement effect on oil recovery. Diluted microemulsion effectively minimized the adsorption loss, enabled surfactants' penetration, and altered the wettability.
(Miller et al., 2019)	Anionic surfactant mixed with persulfate salt and various solvents	Permian Mancos	Spontaneous imbibition	Structurally diverse solvents tend to extract more oil from shales.

(Scerbacova et al., 2023)	Nanofluid booster Surfactant	N/A	Water/nanofluid flooding Surfactant HnP	Nanofluid agglomerates in porous media; the selected surfactant did not outperform brine.
(Teklu et al., 2018)	Low salinity brine Surfactant	Bakken	Spontaneous imbibition	Both low-salinity water and surfactant could improve oil production where imbibition is considered a slow process. Thus, HnP was recommended.
(Tu & Sheng, 2019)	Surfactant	Eagle Ford	Contact angle, IFT, cyclic surfactant injection	Surfactant cyclic injection was effective for oil recovery in shale; wettability alteration is the dominating factor that IFT changes.
(Zelenev, 2011)	Microemulsions	Marcellus Rhinestreet Dunkirk Horn River	Contact angle IFT	No significant variation in surface energy for shale samples from the different basins.
(Zeng et al., 2019)	CO_2 + Chemical blend	Eagle Ford Mancos	HnP experiments	Adding the chemical blend in addition to CO_2 seemed to impede the oil recovery.
(Zeng et al., 2018)	Low-salinity brine, high salinity brine, surfactants, surfactant blends	N/A	Aqueous stability test, IFT, contact angle, spontaneous imbibition, thermogravimetric analysis	Low-salinity brine changed the wettability to more water wet; high-salinity brine does not imbibe into the shales; adding a single surfactant to low-salinity brine did not outperform low-salinity brine, but adding surfactant blends worked better.

Many pilot studies and field tests have already been carried out and proved that foam flooding is a promising tool for enhanced oil recovery in terms of improving sweep profile, reducing gas breakthrough and interfacial tension, and increasing viscosity (Bello et al., 2023). The largest foam flooding in the oil industry is the foam-assisted WAG (FAWAG) in the Snorre field of Norway (Blaker et al., 2002; Skauge et al., 2002), which is one of the major oilfields in the North Sea. The oil production was initially driven by water injection and then followed by steam in 1992. The FAWAG was tested in the Upper Statfjord sandstone reservoir, with permeability ranging from 400 to 3,500 md. Based on laboratory core flooding experiments, alpha-olefin sulfonate with a carbon chain length of C_{14}/C_{16} was selected as the foaming agent. A surfactant alternating gas injection with two cycles was carried out. The added oil production was estimated at 250,000 m^3, while the procedure was operationally similar to WAG with few risks involved for well or reservoir damage. The N_2-based foam is one of the most widely used flooding techniques, owing to its relatively low cost, eco-friendliness, and stability (Afifi et al., 2021). A typical application is the pilot test reported (Jiang et al., 2017). Specifically, a nitrogen-assisted artificial foamy oil flooding was performed in the Tuha oilfield in China in 2016. In each cycle, 250,000 m^3 of gas and 250 m^3 of fluid were injected, which resulted in an average oil production increment of 7 tons while the water cut was reduced by 52%. On the other hand, the utilization of CO_2 foam for enhanced oil recovery has received a lot of attention recently as it naturally fits the target of carbon sequestration: i.e., reducing industrial contributions to global warming. Currently, only pilot tests have been conducted to evaluate its economic potential at different oilfields, such as the Salt Creek field in Wyoming (Patil et al., 2018) and the East Seminole field in West Texas (Alcorn et al., 2020).

Although nanomaterials have been widely utilized in different fields, including electronics, biomedicine, food, and textiles, the application of nanofluids as an efficient oil displacement agent for enhanced oil recovery still focuses on the experimental investigations to interpret its behavior under different reservoir conditions (Yakasai et al., 2021). Only a few pilot tests and field trials have been conducted, owing to its high cost and concern about its dispersion stability in aqueous solution (Liu et al., 2020). In the Jiangsu oilfield of China, for instance, a nano-poly silicon material was developed to overcome the high-pressure injection problem, particularly in low-permeability reservoirs, and was tested on many water injection wells in 2009 and 2010, which achieved a significant pressure reduction (from 24.8 to 5 MPa) and then satisfied the required daily water injection volume of 30 m^3 (Xiong et al., 2010). In two Colombian heavy oil fields, Castilla and Chichimene, oil-based nanofluid flooding was successfully performed in four wells by forcing a certain amount of nanofluid (from 86 to 200 barrels) within a radius of penetration of about three feet, which resulted in an instantaneous oil rate increment from 87 to 310 barrels per day and maintained this increased production gain for at least 174 days (Zabala et al., 2016). For the shale reservoir specifically, the nanoparticle-assisted CO_2 HnP field test was conducted for an Eagle Ford shale well (Zheng et al., 2020). This pilot test showed a significant oil rate increase after nanoparticles and CO_2 were injected, whereas the reservoir simulation suggests that by only injecting CO_2, the improved oil recovery would be much smaller.

7.5 AIR INJECTION

Air injection utilizes a readily available source—air—for enhanced oil recovery. Unlike other gas injections, this source-abundant method introduces the oxygen underground with high pressure, changes the oxidation-reduction (redox) condition of the reservoir, and causes the low-temperature or high-temperature oxidation (LTO or HTO) on a case-by-case basis. Owing to these oxidation reactions, thermal energy is generated, which leads to oil recovery through various mechanisms. Alfarge et al. have summarized these oil recovery mechanisms as shown in Table 7.8 (Alfarge et al., 2020b), where the mechanisms could synthetically serve the oil recovery.

7.5.1 Kinetic Parameters in Air Injection

Historically, air injection was considered a nitrogen gas flooding in simulation (Sakthikumar et al., 1995). Through time, the kinetic model was gradually preferred for high-pressure air injection, where oxidation behavior and chemical changes were presented in the simulation. There are two parts to the kinetic model: the first is to propose a set of pseudo-components based on experimental results; the second is to define their kinetic reactions during thermal transformation and the corresponding parameters: e.g., reaction activation energy (Bauman & Deo, 2011; Mukhina et al., 2021; Murzin & Salmi, 2005) with experimental results. The kinetic reactions (oxidation) could be studied via open-system pyrolysis, accelerating rate calorimetry, ramped-temperature oxidation, and pressurized differential scanning calorimetry experiments. After validating and modifying the reaction kinetic model with experimental results, the kinetic model could be used for high-pressure air injection reservoir simulation (Khakimova et al., 2019).

For kinetic parameters determination, in open-system pyrolysis, pyrolysis characteristics could be found with core samples; in pressurized differential scanning calorimetry, temperature ranges at significant heat-generation rates and activation energies at different heat-generation steps could be obtained with oil and core mixture; in accelerating rate calorimetry test, ignition temperature and temperature

TABLE 7.8
Mechanisms Involved in Air Injection Application

Process	Mechanism
Flue gas generation and steam generation	Gas displacement
Gas expansion	Pressurized reservoir
Heat generation	Reduced oil viscosity, thus better mobility
Gas dissolution	Oil swelling
Increased pressure	Miscible flooding
Induced microstructure changes	Better permeability
Bulldozing or pore-blocking effects	Redirect the flow for better sweep efficiency
Excessive heat, gases, and combustion products generation	Wettability and capillary pressure alteration

corresponding to maximum self-heating rate could be determined via oil and core mixture; and with ramped-temperature oxidation test, ignition temperature without oil saturation and temperature ranges over which oxygen uptake rates are high could be revealed with core samples (Khakimova et al., 2019).

7.5.2 Oxygen Addition

Oxygen additions involve adding oxygen to reservoir oil, primarily leading to the breaking of bonds rather than generating heat. This process results in the formation of partially oxygenated substances such as hydroperoxides, aldehydes, ketones, and acids, with minimal production of carbon oxides or water. These changes can deteriorate oil properties. A major concern is the potential progression of these reactions to a stage known as low-temperature oxidation (LTO) (Alfarge et al., 2020c).

7.5.3 Low-Temperature Oxidation Reactions

Low-temperature oxidation (LTO) refers to a type of reaction that breaks bonds and produces water and partially oxygenated hydrocarbons, such as carboxylic acids, aldehydes, and ketones. This reaction increases oil viscosity and decreases API gravity, making it less effective for oil mobilization. LTO generally occurs at temperatures between 150°C and 300°C. While LTO is less intense than high-temperature combustion, its role in enhancing oil recovery remains unclear, owing to its potential to increase oil viscosity, which can reduce recovery efficiency (Alfarge et al., 2020c).

7.5.4 Spontaneous Ignition

In spontaneous combustion reactions, some heat is produced, but this method is not the most effective for enhancing oil recovery. The high-temperature oxidation (HTO) reaction is considered the most effective, involving oxygen from injected air interacting with coke at temperatures above 343°C for light oils and between 400°C and 500°C for heavy oils. The main products of HTO are carbon dioxide, carbon monoxide, and water, and these reactions generate significant heat, making HTO highly effective at mobilizing oil. While the low-temperature oxidation (LTO) process can increase temperature, it may not significantly reduce viscosity and can increase it, owing to compounds such as carboxylic acids and ketones. Therefore, for air injection to be more effective than non-oxidation gas injection, combustion needs to occur, ensuring the process progresses to HTO (Alfarge et al., 2020c).

7.5.5 Prescreening and Selection Criteria

Air injection is considered a promising EOR approach in unconventional reservoirs. Shale oil reservoirs offer special opportunities and challenges for air injection because of their naturally fractured structure. However, conventional screening standards might not always be valid for unconventional resources; thus, the customization of evaluations must be considered. Prescreening and selection criteria for air injection in shale oil reservoirs usually entail a thorough evaluation to guarantee efficient EOR (Mukhina et al., 2021). The primary internal elements are its hydrocarbon composition, viscosity, API gravity, and unsaturated content (such as aromatic and asphaltene). Minerals, petrophysical characteristics, air injection rate, oil saturation, temperature, pressure, and geology are examples of external factors. These parameters are discussed in more details next.

7.5.5.1 Reservoir Depth and Thickness

The viability of air injection for EOR is largely dependent on reservoir depth. For implementation to be successful, certain requirements pertaining to reservoir depth are necessary. For air injection, a thick, continuous shale deposit is advantageous. However, a high air injection rate is frequently required in the very thick zone in order to ensure a stable oxidation process (Jia et al., 2012a; Rychlicki et al., 2011).

7.5.5.2 Geology

Given that shales usually have a high percentage of clay, air injection should be used because clay minerals have favorable catalytic effects on the oxidation of crude oil, which are beneficial for fuel deposition, oxygen consumption, and heat release (Vossoughi et al., 1983; Zhao et al., 2012). Quartz, feldspar (plagioclase and/or orthoclase), gypsum, calcite, and dolomite should not have any catalytic effects when it comes to oil oxidation. The only minerals in rock that have the ability to catalyze reactions are clays and other metal salts; the degree of catalyzing is determined by the relative amounts of these substances. A cutting that has very little or no clay in it should not have any catalytic properties. Different types of clays have varying levels of catalytic ability; smectite ranks top, illite ranks second, and chlorite and kaolinite rank third in terms of their capacity to oxidize crude oil. Air injection in shale should therefore be advantageous in terms of mineral composition (Jia et al., 2012b).

In addition, shale oil reservoirs usually come with dip angles, which help achieve high oil recovery. Prior to application, the shale oil reservoir's geological continuity and integrity must be considered.

7.5.5.3 Oil Properties

Density of reservoir oil is an important consideration when it comes to determining whether air injection is appropriate. To maximize recovery, different oil densities might need different injection procedure modifications. High content C_7–C_{15} hydrocarbons, which vary in mol fraction from 0.407 to 0.887, have been shown in recent studies to negatively impact crude oil LTO behavior. It takes a lot more activation energy to ignite a light oil. Conversely, oxygen intake benefits from the high-content, unsaturated, heavy hydrocarbons, which are mostly found in aromatics, resin, wax, and asphaltenes. These elements primarily control the oxygen addition reaction in LTO, which produces a range of different oxygenated hydrocarbons (Jia et al., 2014).

7.5.5.4 Pressure and Temperature

It is essential to determine reservoir pressure in order to inject air effectively. Prescreening entails assessing the pressure conditions to make sure the air injection procedure is compatible. The process of light oil oxidation can be accelerated by high pressure. As pressure increases in the low-temperature oxidation region of crude oil, more heat is released, resulting in a decrease in activation energy. Because of the decreased oxidation activity energy in an environment with high air pressure, exothermic activity appears to be found in a low-temperature oxidation range with an increase in pressure (Sarma & Das, 2009).

A protracted oil expulsion process is frequently observed in shale oil reserves. Conditions in reservoirs with high temperatures and excessive pressure are frequent. The liquid hydrocarbons are slightly overpressured if the formation pressure is between 200 and 450 bar, and the temperature is 65 to 140°C (Delaihdem, 2013; Mohan et al., 2013; Sheng et al., 2015; Tran, 2012). Auto-ignition in the light oil air-injection process might result from spontaneous LTO reactivity, as field experiences and project design suggest. This can happen in fields with relatively high reservoir temperatures (90–120°C) and pressures (Fassihi et al., 1997; JPT Staff, 1998, 1999). Shale oil reservoirs with high temperatures and overpressures should be ideal for the intended oxidation reaction; when air injection is used, spontaneous ignition might occur.

7.5.5.5 Petrophysics

The displacement efficiency of the injected air is influenced by porosity (Shah et al., 2010). A shale oil reservoir has porosity ranging from 3% to 9% on average with micro- to nano-Darcy matrix permeability. Shale's extremely low matrix permeability can have both favorable and unfavorable consequences. It seems like air injectivity would be lower in shale formations than in conventional reservoirs owing to their lower porosity. Shale does, however, share the common feature of having some potential natural-fracture development, which controls both air injectivity and petroleum fluid productivity. Tectonic activity, overburden pressure, and/or pressured liquid outflow are the causes of natural fractures (Berg & Gangi, 1999; Delaihdem, 2013). Thankfully, the high surface area offered by the ultra-low permeability matrix can encourage the oxidation of crude oil in porous media. Shale's surface area is orders of magnitude larger, owing to orders of magnitude smaller particle sizes. Crude oil oxidation activity energy will be reduced with higher heat release and potentially auto-ignition during the air injection process if tight shale has a notably large specific surface area.

7.5.6 Operational Parameters

Air injection optimization entails considering a number of operational factors. These factors affect air injection performance, and it is essential to analyze them for efficient use (Wang et al., 2022). With an emphasis on recovery elements and characteristics, experimental assessments investigate the viability and development efficacy of air injection in shale oil reservoirs (Chen et al., 2022). In unconventional reservoirs, air injection development has a greater recovery factor for light components, which affects the composition of residual oil (Vossoughi et al., 1983). In order to determine the project's viability, the feasibility is evaluated by contrasting past examples of successful air injection operations with conventional shale oil reservoirs (Jia & Sheng, 2017).

7.5.7 Examples of Air Injection Projects

Shale differs from other types of rocks, and each shale is unique. Marcelus and Bakken cannot duplicate what occurs at Eagle Ford. The Buffalo, Horse Creek, Coral Creek, Medicine Pole Hills Unit (MPHU) and Hackberry fields are the principal examples of past successful air injection initiatives (Clara et al., 1998; Gillham et al., 1998; Glandt et al., 1999; Gutierrez et al., 2009; Gutierrez et al., 2008; JPT Staff, 1998). Table 7.9 displays a review of previous air injection projects in shale oil reservoirs. The table reports the key parameters involved in air injection, which can be used as reference for future applications.

In general, high-temperature and high-pressure shale oil reservoirs can speed up light crude oil oxidation to release more heat. Catalytic oxidation can be carried out in the clay-rich tight shale with a high specific surface area, which helps mitigate early gas breakthrough. Sloped rock reservoirs also help in gravity drainage during air injection. In addition to the flow of water from the oil-filled matrix to the fracture, the heat generated by the air injection has the potential to dry the shale and induce thermal fracturing of the reservoir. This process frequently coexists with the reservoir's abrupt repressurization, which raises the possibility of shrinkage and overpressure fractures. It may be possible to extract additional light hydrocarbons using the air injection process of crude oil oxidation. Using air injection or air HnP injection as a means of upgrading oil-bearing shale in situ could be a different approach to producing hydrocarbon gases.

However, the overly mature shale oil has fewer unsaturates, primarily asphaltenes and aromatics. Fuel added during the air injection procedure will suffer from this. In order to provide a stable oxidation process and a thermal front, a large volume of air must be delivered into the extremely thick zone. Obstacles to achieving high injection rates include a limited number of accessible fractures and permeability of the nano-Darcy matrix. The nanometer diameter pore wall can be greatly affected by the molecular structure of the pore. The containment effect has the potential to significantly alter phase behavior in shale oil reservoirs, making it difficult to understand crude oil oxidation, evaporation, cracking, and combustion processes in nanopores. To improve air injection performance in shale oil reservoirs and reduce the pore confinement effect, more hydraulic and thermally generated cracks are required.

7.6 THERMAL EOR

Thermal EOR techniques have been increasingly utilized to improve hydrocarbon recovery in unconventional resources. These methods involve elevating reservoir temperature, thereby reducing oil viscosity and enhancing its mobility, which, in turn, facilitates improved recovery. The increased heat within the reservoir also reduces surface tension and increases the oil's effective permeability. A variety of thermal EOR techniques have been proposed and used, encompassing hot water injection, electrical heating, and electromagnetic heating. Furthermore, electrical heating can be subdivided into two main categories, depending on the frequency of the electrical current utilized: namely, low-frequency electrical heating (also referred to as resistive) and high-frequency electrical heating (typically referred to as electromagnetic heating).

TABLE 7.9
Important Air Injection Projects in Shale Oil Reservoirs with Their Typical Thermodynamic, Oil, and Reservoir Properties

	Reservoir				
Feature	**Eagle Ford**	**Marcellus**	**Barnett**	**Bakken**	**Wolfcamp**
Thermodynamic Parameters					
P/psi	6,400	3,500	3,700	4,600	4,600
T/°F	260	200	170	240	165
Oil Parameters					
Oil gravity/API	45	Higher than 40	38	40	45
Oil viscosity/cp	0.6	Smaller than 1	Smaller than 1	0.4	0.3
Reservoir Parameters					
Reservoir thickness/m	30–90	15–80	45–2,000	3–28	143
K/md	10^{-6} to 10^{-4}	10^{-6} to 10^{-3}	10^{-6} to 10^{-3}	10^{-3} to 1	10^{-6} to 10^{-4}
Φ/%	6	6	3 to 6	4 to 7	9
Clay/%	5 to 45	10 to 35	10 to 50	9 to 53	25
References	(Chaudhary et al., 2011; Delaihdem, 2013; Stegent et al., 2010)	(Bruner & Smosna, 2011)		(Cox et al., 2008; Sorensen & Terneus, 2011)	(Mohan et al., 2013)

7.6.1 Hot Water Injection

Hot water injection, a method involving the injection of heated water into the reservoir to increase temperature and reduce oil viscosity, has been extensively used in conventional reservoirs. This technique enhances oil flow properties by transferring heat to reservoir fluids, thereby facilitating their displacement.

The hot water HnP process has been investigated, owing to the difficulty in achieving displacement in shale reservoirs. The efficiency of hot water injection largely hinges on permeability enhancement attributed to mineral dissolution during the forced imbibition and shale cracking triggered by clay swelling after spontaneous imbibition (Wang et al., 2011) and changes in in situ stresses of the formation that reactivate pre-existing natural fractures or induce microfractures extending from the fracture surface in a tight matrix block (Fakcharoenphol et al., 2013).

Despite these benefits, the application of water flooding in shale and tight oil reservoirs with ultra-low permeability presents challenges as it can be difficult to implement any displacement process. Additionally, the application of this method faces several challenges, including managing heat loss from the low thermal conductivity of shale rocks and the potential alteration of the rock's wettability.

7.6.2 Electrical Heating

The frequency of the electric current determines the mode of electrical heating. Resistive heating involves the direct application of electrical energy to heat the reservoir. This method can be achieved through techniques such as resistive heating, in which an electrical current is passed through the reservoir rock, generating heat owing to the rock's resistivity. Electrical heating is an EOR process that works on the principle of transferring heat into the reservoir via electrical means. It involves the injection of electrical energy into the reservoir, heating the area and causing an increase in oil temperature. Heating the reservoir reduces the oil viscosity, making it easier for it to flow.

Electrical heating has shown potential for enhancing hydrocarbon recovery, particularly in heavy oil and bitumen-bearing formations (Bera & Babadagli, 2015). The electrical heating is controlled by water resistivity. When low-frequency currents flow through the reservoir in the presence of formation water, electrical energy is converted into heat. Figure 7.5-A illustrates a typical setup used for this application. The design was done for a vertical well with the electric cable running from the power source at the surface along the tubing and then connected to a series of electrodes placed across the perforation interval. A potential difference is applied across the two electrodes, and an electrical path through the formation is provided by the formation water. To maintain electrical conductivity, it is essential to keep the supply of water, especially around wellbore where heating can lead to a dry and sharp increase in resistivity.

7.6.3 Electromagnetic Heating

Using high-frequency electromagnets for wellbore heating was introduced more recently. The principle of this process is similar to the working microwave oven

(Figure 7.5-B). In microwave heating, microwaves interact with the water molecules present in the formation. These molecules are set in circulatory and oscillatory motion, generating frictional heat, which then may be transferred to its neighbors (mainly the matrix and the oil), ultimately elevating the temperature of the reservoir. The extent of the temperature elevation is determined by the amount of microwave energy absorbed by the fluids. The penetration depth of microwaves is usually small, less than 0.5 meters, but could be increased significantly if heat transport by water flow is allowed.

While electrical heating is a relatively new technique, advances in the past couple of years have allowed for testing by theoretical laboratories and field trial research worldwide. Jha and Chakma (1999) among many others, reported sandbox experiments and showed that the presence of higher salinity increases the conduction of the electromagnetic wave. Regarding the temperature dependence of electrical conductivity, Ucok et al. (1980) did a comprehensive experiment to assess geothermal brine resistivity and found that resistivity decreases with increasing temperature. They also observed a rapid drop in resistivity with a high salt concentration in geothermal brine.

Challenges include the need for specialized equipment, ensuring uniform heating of the reservoir, and managing the high energy consumption associated with these methods. The effects of EM waves on wettability in unconventional reservoirs (shales and even carbonates) require further clarification (Bera & Babadagli, 2015).

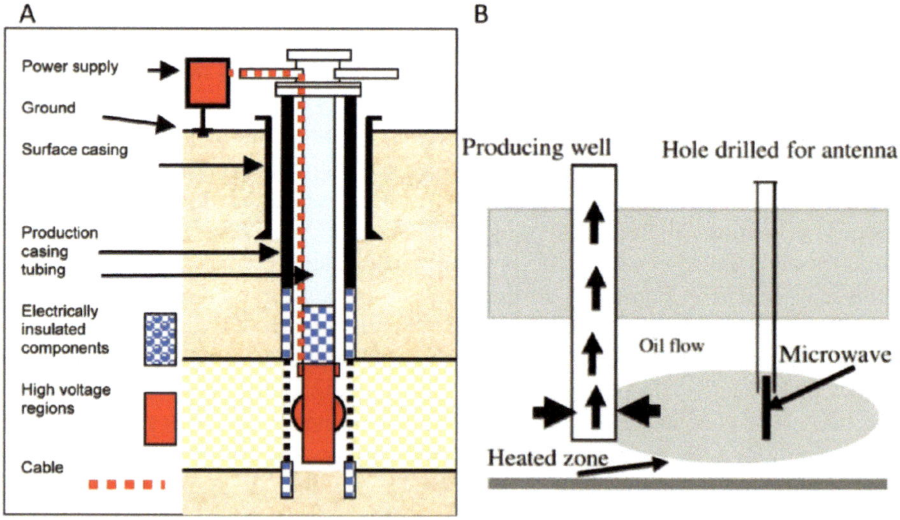

FIGURE 7.5 Schematic of (A) low-frequency electrical heating where the production pipe remains at ground potential and (B) electromagnetic heating mechanisms.

Source: Petrowiki, 2015; Vinsome et al., 1994 (A); Chhetri & Islam, 2008 (B)

7.6.4 Prescreening and Selection Criteria

Identifying suitable candidates for thermal EOR applications in unconventional shale resources requires a thorough evaluation of reservoir properties, fluid characteristics, and operational constraints. Reservoirs with relatively low permeability, high oil viscosity, and favorable thermal properties are considered prime candidates for thermal EOR. Additionally, reservoirs with good geological and geomechanical properties, such as minimal natural fractures and faults, are preferable to minimize heat loss and ensure efficient heat transfer. Table 7.10 presents screening criteria for selecting appropriate thermal EOR methods for unconventional reservoirs.

7.6.5 Laboratory and Field Design

To determine the viability and efficiency of thermal EOR techniques in unconventional reservoirs, laboratory experiments and numerical simulations are essential for optimizing operational parameters such as injection temperature, heating rate, and duration. Field designs should consider well spacing, injection strategies, production strategies, and heat management to optimize efficiency and minimize heat loss. Moreover, implementing robust monitoring and control systems is key to maintaining safe and effective thermal EOR operations. The following feasibility studies are reported in the literature regarding thermal EOR in unconventional reservoirs.

Egboga et al. (2017) conducted a simulation study investigating thermal stimulation's impact on enhanced oil recovery in shale reservoirs, utilizing electromagnetic

TABLE 7.10
Prescreening Criteria for Thermal EOR in Unconventional Reservoirs

Thermal EOR Methods	Screening Criteria
Hot water injection	• Compare energy intensity to produce one barrel of oil with other methods. • A homogeneous reservoir is preferred to avoid uneven heating and less efficient oil recovery. • Fractured carbonate reservoir with high water saturation.
Electrical heating	• Suitable for reservoirs with high permeability or fractures (Rehman & Meribout, 2012). • A homogeneous reservoir is preferred to avoid uneven heating and less efficient oil recovery. • Electrodes are prone to corrosion in high salinity formation brine. (Rehman & Meribout, 2012)
Electromagnetic heating	• Suitable for heavy oil reservoirs. • Favorable for offshore oilfields in terms of equipment compactness (Bientinesi et al., 2013). • Not suitable for water-flooded reservoirs (Rehman & Meribout, 2012).

heating at the wellbore. The research revealed that thermal stimulation for 1,000 days using a 371°C heater could potentially boost oil recovery economically from 7% to over 11.5%. It underscored thermal pressurization as the main mechanism for this improvement and noted that although kerogen decomposition doubled permeability, its effect on recovery was minimal, owing to limited connectivity between heated and fractured regions.

Askarova et al. (2020) explored thermal EOR methods in unconventional reservoirs through a new approach integrating autoclave experiments (hot water injection at reservoir pressure and temperature of 350°C) and numerical simulations. Key findings highlight the significance of exposure time and kinetic models in hydrocarbon transformation, the role of temperature in bitumen mobility, and the potential of the refined model in broader field applications. This technique, although requiring further refinement, yields critical parameters for large-scale field modeling.

Mukhina et al. (2021) conducted a comparative analysis of three EOR techniques in a Western Siberian oilfield with ultra-low permeability and high organic content. Using numerical simulations and laboratory experiments, the researchers tested surfactant solution injection, hydrocarbon gas injection, and hot water injection over eight years, revealing that hot water injection resulted in the highest oil recovery, yielding 26% more oil than the depletion mode. Despite certain drawbacks, such as heat loss and water blockage, the study concludes that hot water injection is the most effective EOR method for similar shale fields.

7.6.6 Examples of Thermal EOR Projects

Thermal EOR methods show potential for boosting hydrocarbon recovery in unconventional shale resources. As the oil and gas sector grapples with declining production and environmental issues, refining these techniques becomes critical. Although primarily deployed in conventional reservoirs, their viability in shale resources remains under investigation. These studies offer important insights, but further optimization is necessary for economic feasibility. Addressing challenges such as heat loss, reservoir damage, and the need for high-pressure, high-temperature equipment and energy-efficient heating methods is essential.

7.7 WATER FLOODING

Water flooding involves injecting water into an oil reservoir. It is often considered when the natural driving mechanisms are no longer able to produce oil, which usually happens when keeping the initial pressure is not possible anymore. Water can replace oil in pore spaces, pushing oil out and aiding its movement towards production wells.

Water flooding is an EOR technique that has been successfully applied to produce oil in conventional reservoirs and, more recently, in tight oil reservoirs. In projects involving water flooding, water is injected through injectors to activate production mechanisms that transport the reservoir oil in the direction of the producing wells. Water is often an inexpensive and readily available resource, making water flooding a cost-effective secondary recovery technique. Water can be sourced from nearby streams, rivers, oceans, or subsurface aquifers. Its accessibility and cost effectiveness could make it a possible choice in shale oil reservoirs (Chen, 2013).

Water can be injected into the oil column or bottom aquifer. It may be also injected into naturally fractured reservoirs. Water injection causes the oil zone to experience a bottom water drive, which forces the oil upward. Water injection used to be done in the reservoir's latter stages of life, but it is now done in the reservoir's early stages to prevent gas cap formation by pressure drop. Production is increased when water is injected earlier in the process because the water first has a tendency to compress the free gas cap before pushing the oil, necessitating the injection of much more water. When there is no gas drive or when water drive is insufficient, water injection is typically used. As a result, water injection is done when reservoir pressure is higher than saturation pressure in order to save money. Two purposes for water injection are to support the reservoir's pressure and to push oil away from the reservoir and towards an oil-producing well. The mobility ratio is the ratio of displacing fluid (i.e., water) to the displaced fluid (i.e., oil), which can determine the choice of injection water method. However, there are numerous drawbacks to the water injection, including (Chen, 2013):

- Damage to the formation may result from the interaction of injected water and formation water (scale formation).
- Surface and subsurface equipment corrosion.

The microscopic displacement efficiency is typically substantially below one in immiscible displacement processes such as water flooding. Depending on the rock wettability, some crude oil in the areas the displacing fluid contacts is trapped as solitary droplets, stringers, or pendular rings. In this situation, relative permeability to oil is almost nil, and subsequent flow simply avoids the trapped oil. Miscible displacement can be used to get around this oil recovery constraint (Chen, 2013).

7.7.1 Water Flooding in Unconventional Oil Reservoirs

Unconventional reservoirs have a significant amount of oil and gas resources across the world. Desires to exploit unconventional reservoirs, particularly shale gas and oil resources, was sparked by the high prices of oil and gas. As with tight gas, shale gas, and coalbed methane, tight oil can distribute over broad regions and build reserves without down-dip water, as opposed to traditional reservoirs, where oil is held in the up-dip parts above water-containing rocks (Chen, 2013). Shale reservoirs have been exploited using cutting-edge horizontal drilling and hydraulic fracturing techniques, but it is still unclear how traditional EOR methods such as gas and water flooding should function in these reservoirs.

Water flooding is frequently utilized because water injection is affordable and may be economically viable despite the modest final recoveries. Water flooding also has the advantage of being a low-risk method of recovering some extra oil while more sophisticated lab and pilot tests are being planned. Therefore, increasing oil recovery by flooding such reservoirs with water is a desirable objective (Chen, 2013). The efficiency of displacing oil with water depends on various factors, such as oil viscosity and rock characteristics. However, water flooding can effectively displace additional oil from the reservoir, increasing the overall recovery factor. It is especially beneficial in shale oil reservoirs where oil extraction through natural production methods may be challenging, owing to the low rock permeability. In certain oilfields, such as the Wilmington field in California, US, and the Ekofisk field in the North Sea, water flooding has been used to mitigate additional surface subsidence. These fields have experienced significant compaction of the reservoir rock, owing to the drawdown of reservoir pressure during primary production. By implementing water flooding and maintaining reservoir pressure, the compaction can be reduced.

In the context of shale oil reservoirs, water injections can be particularly beneficial. Shale oil extraction often involves hydraulic fracturing (fracking), which creates fractures in the rock formation to release trapped oil. However, over time, the production from these fractures tends to decline. Water injections can help maintain reservoir pressure, support the flow of oil from the shale formation, and improve overall oil recovery.

Through simulation studies, Chen observed no significant variations in oil production rates when water injection started after primary oil production. He pointed out that the transmission of injection fluid from injection well to producer is challenging, owing to the ultra-low permeability of shale reservoirs, and the production wells' poor response to water flooding also results in a low injection rate. During the water-flooding process (for 60 years both continuous and cyclic 5 years injection + 5 years shut-in), the oil rate can only be 0.7 to 0.8 bbl/day (compared to 0.6 bbl/day during primary stage), leading to an oil recovery factor of 11% to 12%. Using another strategy, they also conducted simulation by considering 70 years of water flooding from beginning and observed 11% oil recovery. He concluded that it is not required to inject water at the start of development, especially in a reservoir of this type with a high reservoir pressure and extremely poor permeability. Because of the extremely low porosity and permeability of shale reservoirs, it is difficult for injected fluids to move from the injection well to the production well, which results in low productivity and injectivity. The entire reservoir has a poor reaction to water injection (Chen, 2013). Jacobs reported production uplift through periodic injection of produced water back into a well with a soaking period of 30 days. Water spontaneously imbibes into the formation and displaces hydrocarbons (Jacobs, 2017). Bertoncello et al. reported that water leaks off through fracture surfaces and invades other regions of formation by viscous forces and imbibition capillary pressure, resulting in a high-water saturation region near the fracture surface. Lower water saturation and larger gas relative permeability are observed close to the fracture surface as a result of the imbibition process, which continues while the well is shut in and allows water to flow deeper into the water-wet lamina (Bertoncello et al., 2014).

It has been shown that injecting gas into a shale oil reservoir with ultra-low porosity and permeability and high oil viscosity has a greater influence on oil recovery than does injecting water via a fracture with high permeability. Injected water only helps keep the reservoir's pressure up, unlike miscible gas, which may lower oil viscosity. Based on the modeling findings of water and gas flooding, miscible gas injection can improve oil recovery in shale reservoirs more effectively. The solvent can be miscible with oil in addition to maintaining pressure, which reduces oil viscosity and results in a better sweep efficiency than water. Oil output from shale oil deposits is predicted to increase with the help of gas injection and an enhanced production and completion technique (Chen, 2013).

Poor sweep efficiency caused by incredibly low permeability leads to low productivity and low injectivity. As a result, the reservoir will receive a poor water injection. When implementing water injection in shale oil reservoirs, various factors need to be considered, including the source of injection water. Common sources include produced water (water extracted along with the oil), seawater, aquifer water, and river water. Each source may require treatment and filtration to remove impurities and ensure compatibility with the reservoir. Filtration systems, such as sand filters or self-cleaning back-flushable screen filters, are often used to clean the water before injection. De-oxygenation of the water is another critical step in the water injection process. Removing oxygen from the injected water is essential to prevent corrosion and other undesirable chemical reactions in the reservoir. Oxygen removal methods such as de-oxygenation towers are used to ensure the injected water is oxygen-free (SpA, 2017).

7.7.2 Low Salinity Water Flooding

Low salinity water flooding (LSWF) is an EOR technique to enhance the extraction of oil from reservoirs. This method involves injecting water with a lower salinity (ionic concentration) than that of the formation water in the reservoir. The goal of low salinity water flooding is to change the rock wettability and enhance oil recovery by modifying the interactions between the rock, water, and oil.

The process of low salinity water flooding is based on the understanding that the surface properties of reservoir rocks can be influenced by the injected water composition. When water with lower salinity is introduced into the reservoir, it can have several effects:

7.7.2.1 Wettability Alteration

Reservoir rocks typically exhibit a certain degree of wettability towards oil and water. Wettability refers to the preference of a solid surface to be in contact with one fluid (oil or water) over another. Low salinity water flooding can change the wettability of the rock surface, making it more water wet. This change in wettability can reduce the tendency of oil to adhere to the rock surface, making it easier to displace and recover.

7.7.2.2 Ion Exchange and Surface Charge Modification

The ions in the injected water can interact with the rock surface and exchange with ions present on the surface. This ion exchange can modify the surface charge of the rock, affecting the interactions between the rock, water, and oil. As a result, the mobility and displacement of oil within the rock can be enhanced.

7.7.2.3 Clay Swelling and Fines Migration

The lower salinity water can affect the swelling behavior of clays present in the reservoir rock. Clay minerals can swell or shrink based on the ionic concentration of the surrounding water. Changes in clay swelling can influence the pore structure of the rock, affecting fluid flow. Additionally, fines migration (movement of small particles within the rock matrix) can occur, owing to electrostatic interactions between the particles and the injected water.

7.7.2.4 Reservoir pH and Chemical Reactions

Altering the salinity can also influence the pH of the reservoir fluids, which, in turn, can impact the dissolution of minerals and other chemical reactions within the reservoir rock. These reactions can affect the permeability of the rock and the interactions between fluids and rock surfaces.

Since it takes advantage of the interaction between injected water and the surface characteristics of the reservoir rock to enhance oil recovery, low salinity water flooding is regarded as a promising EOR approach. However, the success of this method depends on various factors, including the specific reservoir conditions, the composition of the reservoir rock and fluids, and the design of the injection process. The technique is particularly effective when certain conditions are met, such as the presence of clay in sandstones, the presence of polar components in crude oil, and the presence of divalent ions such as Ca^{2+} in the water.

Shale oil reservoirs, characterized by low permeability, typically have low primary recovery factors, ranging from 3% to 10%. Therefore, implementing EOR methods becomes crucial to enhancing oil recovery from these formations. Owing to the limited permeability of unconventional shale reservoirs and the unrealistic pressure gradients needed to force the water through the reservoir matrix rock, traditional water-flooding techniques that rely on water displacing oil are not feasible in such reservoirs (Uzun et al., 2020). For shale oil reservoirs, low salinity water injection has been investigated as a viable EOR technique. The objective is to alter the fluid-rock interactions and improve oil recovery in these challenging formations. Low-salinity mechanisms in shale oil reservoirs include wettability alteration owing to alteration of surface charges, expansion of double layer, pH rise, multi-component ion exchange, fine migration, and rock dissolution. Combinations of several of these mechanisms are what promote favorable wettability changes, leading to increased oil recovery (Austad et al., 2010; Mahani et al., 2015). Capillary and osmotic forces are among the primary EOR mechanisms in shale, in addition to the potential low-salinity EOR mechanisms already described (Karimi et al., 2017; Li et al., 2016; Tuero et al., 2017).

Bui conducted experiments to determine how low salinity water injection affected oil recovery in unconventional reservoirs with lots of fluids. In making his assessment, he took into account variables including salinity, fluid type, shale swelling, and wettability modification. A coupled model incorporating mass transport and geomechanics was used to simulate the mass exchange between fractures and the shale matrix. The results suggest that osmosis and shale matrix swelling play significant roles in oil recovery while the density of hydraulic fractures also influences the outcome (Bui, 2021). Shale matrix swelling can lessen the permeability of the shale matrix and the cracks, which will result in less oil being recovered overall. Water injection is not advised for formations that have the potential to swell. For better oil recovery in shale oil resources, the modified-zipper pattern is advised to optimize operations (Bui, 2021).

The gap in chemical potential of two solutions when they are separated by a semipermeable membrane (permeable to solvent and impermeable to solute) is what causes osmosis, which is a thermodynamic force for solvent transport (ions). According to Uzun et al., a significant number of microfractures emerge when multistage hydraulic fracturing stimulates the shale reservoir and offers a favorable channel when water is later injected into the reservoir. If the salinity of this water is lower than that of the resident brine in the matrix pores, an osmotic pressure gradient will form between the microfractures and the matrix pores, allowing water to enter and forcing oil out. This osmotic pressure permits brine imbibition into the matrix in oil-wet shale reservoirs, which encourages countercurrent flow of oil into the fractures (Uzun et al., 2020). Shale reservoir cores can operate as semipermeable for osmotic force to have a significant impact on improving oil production from tight formations, according to Kurtoglu et al. (2014) and Fakcharoenphol et al. (2014). Furthermore, brine imbibition can reactivate sealed microfractures to increase system permeability by introducing chemical and physical interactions with the fracturing filing material (Zhou et al., 2016). Oil can function as a semipermeable membrane, according to Wen and Papadopoulos' research (2001). According to research by Sandengen and Arntzen (2013), Sandengen et al. (2016), and Fredriksen et al. (2017), an oil-water-rock system undergoes osmosis because the reservoir oil serves as a superb semipermeable membrane. These experiments demonstrated that the primary driving force for oil droplet ejection can be residual water expansion (swelling), owing to osmotic water molecule transfer from injected fluid into the in situ water through the oil phase as a semipermeable. Their investigation, which involved the microscopic observation of both strongly water- and oil-wet micromodels, provides evidence to corroborate their theory (silicon wafer). Torcuk et al. evaluated the effects of low-salinity brine on oil recovery in liquid-rich shale reservoirs and came to the conclusion that low-salinity brine invasion of the rock matrix occurs by osmosis rather than capillary force in oil-wet shale reservoirs (Torcuk et al., 2019). A novel technique for determining the osmotic pressure in samples of shale core was presented by Uzun and Kazemi in 2019. Both the oil recovery and the observed osmotic pressures were provided after the approach was applied to cores from the Niobrara and Codell formations in the DJ Basin (Uzun & Kazemi, 2019). Because of the presence of clays and their active interaction with salt ions, previous research also demonstrates that the application of low salinity water floods in shaly sandstone strata has not been entirely successful. Shales are predominantly made of fine-grained clays, such as illite, smectite (montmorillonite), kaolinite, and chlorite, are found as naturally occurring rocks. Several research projects have been conducted in recent years to assess the effect of clay minerals on the functionality of LSWF (RezaeiDoust et al., 2009; Sohrabi et al., 2017; Song & Kovscek, 2015). For instance, prior research has found a direct correlation between the presence of clays and a rise in oil recovery during low salinity water injection (Jerauld et al., 2008; Seccombe et al., 2008). Clay was found to improve oil recovery efficiency during laboratory experimentation, according to Rezaei Doust et al. (2009). However, Sohrabi et al. discovered that there is no correlation between oil production during LSWF and the presence of clays in the core samples (Sohrabi et al., 2017). A similar conclusion was reached by Farzaneh et al. (2017), who claimed that clays are not a necessary component for a successful flooding of sandstone reservoirs with low salinity. These findings might be explained by clays' charge on the surface, their affinity for water, and their interactions with salt ions (Lu, 1988). Nevertheless, the primary mechanisms of LSWF in reservoirs with substantial amounts of clay are still not fully understood (Safari et al., 2022). Chaturvedi et al. concluded that the addition of shale to sand packs can reduce the porosity by 34% on adding 25% shale while no permeability measurement was possible while 2 wt.% KCl lowered the effect of shale on sand pack features such that the impermeable sand pack showed significant permeability, finally resulting in increased oil recovery (Chaturvedi et al., 2019).

7.7.3 Hybrid Chemical-Assisted Water Flooding

EOR in shale reservoirs are challenging, owing to ultra-low permeability, low porosity, the presence of kerogen, matrix heterogeneity, and complex fracture network. Chemical additives have been targeted to increase the efficiency of water flooding (including low salinity water flooding) in shale oil reservoirs. The potential chemical additives include surfactants, polymers, alkalis, and more recently nanoparticles. However, most literature studies have used surfactants.

Surfactants can improve water imbibition, boost oil relative permeability, and lessen water obstruction at the matrix-fracture interface by lowering oil-water interfacial tension and the shale's wettability. The surfactants used in these treatments must be stable at the high temperatures typically seen in shale reservoirs. Higher aqueous stability is frequently obtained when two surfactants are combined than when they are used alone (Zeng et al., 2018). Zeng et al. investigated the water stability, contact angle, and spontaneous imbibition of several surfactants (anionic, nonionic, and cationic) and their mixtures.

Both high and low salinity brines at 125°C were shown to be stable for seven single surfactants and nine surfactant blends. In both low and high salinity, all aqueous stable blends transformed the wettability of oil-wet shale to predominantly water wet. Imbibition tests were conducted using five surfactant blends and seven single surfactants; 20% pore volume (PV) more water was absorbed thanks to surfactant solutions. The imbibition was improved more by surfactant blends than by single surfactants. In around three days, core imbibition peaked. In order to stimulate shale wells, surfactant blends may be utilized in low salinity fracturing or refracturing fluids.

Teklu et al. emphasised the possibility of producing oil and gas from shale reservoirs economically by utilizing horizontal wells and multistage hydraulic fracturing. They stated that chemical EOR must be carried out because these reservoirs often have low recovery coefficients. In liquid-rich shale reservoirs, they looked at the use of low salinity water and low salinity water plus surfactant as hydraulic fracturing and EOR fluids. The study found that these fluids can enhance oil recovery by modifying wettability, capillary pressure, and osmotic forces. This was discovered through imbibition trials. According to the findings, it is advised to use a HnP procedure that uses low salinity water followed by low salinity water plus surfactant to maximize production from liquid-rich shale resources while lowering costs. The number and size of natural cracks or laminations within the reservoir volume determines how effective these fluids are (Teklu et al., 2018). According to Wang et al., surfactant solutions can be used to encourage shale oil recovery. They discovered that surfactant solution spontaneously entered oil-saturated Pierre cores and preserved 41% of the original oil (Wang et al., 2011).

Spraberry tight cores (0.1 mD) from the Permian Basin in West Texas were used for IFT, contact angle measurements, and spontaneous imbibition tests by Mohan and Leonard. The imbibition investigations used formation brine (13.8%), ultra-low salinity brine (< 0.1%), and 0.1% anionic surfactant and formation brine solutions. Their findings showed that surfactant in high salinity brine and ultra-low salinity brine can both render wettability more water wet and boost spontaneous imbibition oil recovery (Mohan & Leonard, 2013). According to Mohan and Leonard (2013), spontaneous imbibition with formation brine resulted in a 15% to 18% oil recovery of OOIP and an incremental 15% to 40% oil recovery when ultra-low salinity and 0.1 wt.% anionic surfactant were combined with the formation brine (Mohan & Leonard, 2013). According to Mohan and Leonard (2013), improving oil recovery from the Spraberry siliceous tight formation in the Permian Basin will be possible with the injection of low concentration anionic surfactant mixed with generated formation water. On siliceous tight rocks (10 μD) at a reservoir temperature of 138°F and a formation brine of 132,000 ppm, Kathel and Mohanty (2013a, 2013b) reported similar outcomes. As imbibition fluids, several anionic and nonionic surfactants at 0.1 wt.% were combined with formation brine and formation brine that had been two-fold diluted. Additionally, they came to the conclusion through modeling that enhancing fracture density and wettability with the proper surfactant and brine combination can enhance oil recovery from the tight sandstone formation.

Alvarez and Schechter used anionic surfactant, a nonionic/cationic mixed surfactant, and fracturing fluid without surfactant to perform spontaneous imbibition on unconventional siliceous liquid-rich reservoir cores from Permian Field. Following nonionic/cationic and fracturing fluid without surfactants, they observed that anionic surfactant produced the most oil after ten days of spontaneous imbibition trials. They also detailed how Permian shale cores treated with wettability-altering surfactants changed in terms of contact angle, zeta potential, and oil-water interfacial tension. They believed that the best surfactant for the shale samples they utilized was an anionic one. Additionally, they conducted spontaneous imbibition studies on core samples under the supervision of a computerized tomography (CT) scanner, and the outcomes demonstrated that the anionic surfactant could restore up to 40.6% of the initial oil saturation in the core (Alvarez & Schechter, 2017).

Wang et al. scaled up the results of surfactant imbibition rates to a field scale in fractured shale deposits using analytical models. The Bakken reservoir cores utilized for the upscaling trials have porosities between 4% and 8% and permeabilities between 0.0015 and 0.096 mD. They came to the conclusion that owing to the fact that surfactant imbibition proceeds very slowly in very tight shales and may take months to years to reach into 3 × 3 × 3 m^3 or larger tight matrix blocks, the presence and size of natural fractures linked to the main induced hydraulic fractures are crucial for practical applicability at the field scale (Wang et al., 2016).

According to Uzun et al., the addition of a simple surfactant or wettability-altering chemical not only improves the imbibition of water into the matrix but also creates a low IFT environment that helps the oil move out of the micro- and macrofractures more easily and improves oil recovery from the matrix. A multi-component mass transport model that incorporates the advective and diffusive movement of water molecules was constructed and utilized to match the experimental data in order to scale laboratory results and observations to the field settings. The findings of numerical modeling and core imbibition showed that adding a diluted surfactant to low salinity brine improves oil output (Uzun et al., 2020).

He et al. examined the surfactant adsorption in the Barnett shale and proposed that polyelectrolyte agents be utilized to lessen the surfactant adsorption in unconventional shale formations in order to increase oil production (He et al., 2015). Alvarez et al. performed core flooding and described the impact of adding a surfactant. Wolfcamp cores showed reduced oil-water interfacial tension, which boosted oil recovery (Alvarez et al., 2017). The effectiveness of the surfactant HnP for unconventional reservoirs was researched by Shuler et al. To increase the initial production, they suggested adding surfactants to the hydraulic fracturing fluid (Shuler et al., 2016).

7.8 LATEST ADVANCES, TECHNIQUES, AND METHODS

7.8.1 Electric Methods

In tight oil reservoirs, electric techniques have been investigated as viable for EOR. There have been studies on a number of electric techniques, including:

1. Electromagnetic heating: This process uses electromagnetic waves to warm the oil reservoir and reduce oil viscosity, allowing the oil to flow more easily. The electromagnetic waves can be produced by downhole or surface-mounted antennas (Sahni et al., 2000).
2. Electrokinetics: By applying an electric field to the reservoir, electrokinetics can mobilize the oil and boost its flow. Combining this technology with other strategies, such as hydraulic fracturing, is possible (Zhang et al., 2023).
3. Electro-osmosis: This technique can improve oil displacement and increase recovery by pumping water or other fluids through the reservoir with the help of an applied electric field (Skomedal et al., 2021).
4. Electrothermal: This technique includes heating the oil electrically to make it less viscous and improve flow (Bera & Babadagli, 2015).

7.8.2 Waves Methods

Acoustic waves in particular have been researched as potential EOR methods for tight oil reserves. The following are some of the wave-based techniques investigated for EOR from shale oil reservoirs:

1. Acoustic stimulation: The reservoir can be stimulated with acoustic waves, which can enhance the permeability and fracture the rock, facilitating oil flow. The effectiveness of this technique can be increased by combining it with hydraulic fracturing (Li et al., 2019).
2. Resonance-based stimulation: By applying a particular frequency of acoustic waves to the reservoir, resonance-based stimulation can cause the formation of new cracks and enhance the permeability of the reservoir by causing resonance in the rock (Agarwal & Kudapa, 2023).
3. Seismic stimulation: Explosions or tremors can produce seismic waves that can be used to stimulate the reservoir and boost oil recovery. More research is necessary for this technique, which is still in the experimental stage (Li et al., 2019).
4. Electromagnetic waves: As was already indicated, electromagnetic waves can also be used to heat the reservoir and reduce the oil's viscosity, allowing it to flow more easily (Chen et al., 2018).

7.8.3 Laser Methods

EOR from shale oil reservoirs has been investigated using laser techniques as a potential technology. The following are some of the laser-based techniques for EOR from shale oil reserves that have been researched:

1. Laser-induced breakdown spectroscopy (LIBS): In LIBS, laser pulses are used to generate plasma on the reservoir's surface that can provide details about the makeup and characteristics of the rock and oil. This knowledge can help increase oil recovery and optimize EOR methods (Han et al., 2019).
2. Laser-assisted gravity drainage (LAGD): LAGD uses laser heating to lower the oil viscosity and enhance its flow toward production wells. Although this approach is still in its infancy, laboratory tests have yielded encouraging results.
3. Laser-assisted hydraulic fracturing (LAHF): By creating microfractures in the reservoir using laser pulses, LAHF can improve oil flow and increase permeability. Although this approach has produced encouraging results in laboratory tests, more investigation is required to assess its utility in real-world settings.

Overall, it should be highlighted that the development of electric-, wave-, and laser-based EOR from shale oil reservoirs is still in its infancy, and further study is necessary before they can be widely used. However, these methods offer hope for enhancing oil recovery and lessening the negative effects of oil production on the environment.

REFERENCES

Abedini, A., & Torabi, F. (2014). Oil recovery performance of immiscible and miscible CO2 huff-and-puff processes. *Energy & Fuels*, *28*, 774–784.

Adekunle, O., & Hoffman, B. T. (2014). Minimum miscibility pressure studies in the Bakken. *SPE improved oil recovery symposium*, Tulsa, OK. SPE-169077-MS.

Afekare, D., Garno, J., & Rao, D. (2021, April). Enhancing oil recovery using silica nanoparticles: Nanoscale wettability alteration effects and implications for shale oil recovery. *Journal of Petroleum Science and Engineering, 203*, 108897. https://doi.org/10.1016/j.petrol.2021.108897

Afifi, H. R., Mohammadi, S., Mirzaei Derazi, A., Moradi, S., Mahmoudi Alemi, F., Hamed Mahvelati, E., & Fouladi Hossein Abad, K. (2021). A comprehensive review on critical affecting parameters on foam stability and recent advancements for foam-based EOR scenario. *Journal of Molecular Liquids*, 116808. https://doi.org/10.1016/j.molliq.2021.116808

Agarwal, M., & Kudapa, V. K. (2023). Plasma based fracking in unconventional shale–A review. *Materials Today: Proceedings, 72*, 2791–2795.

Ahmadi, M. A., & Shadizadeh, S. R. (2015). Experimental investigation of a natural surfactant adsorption on shale-sandstone reservoir rocks: Static and dynamic conditions. *Fuel, 159*, 15–26. https://doi.org/10.1016/j.fuel.2015.06.035

Alagorni, A. H., Yaacob, Z. B., & Nour, A. H. (2015). An overview of oil production stages: Enhanced oil recovery techniques and nitrogen injection. *International Journal of Environmental Science and Development, 6*(9), 693.

Alcorn, Z. P., Føyen, T., Zhang, L., Karakas, M., Biswal, S. L., Hirasaki, G., & Graue, A. (2020, August 30). CO2 foam field pilot design and initial results. *Day 1 Mon*, 31 August. https://doi.org/10.2118/200450-MS

Alfarge, D., Wei, M., & Bai, B. (2017). IOR methods in unconventional reservoirs of North America: Comprehensive review. *Day 3 Tue*, 25 April, D031S008R005. https://doi.org/10.2118/185640-MS

Alfarge, D., Wei, M., & Bai, B. (2020a). *Air injection in shale and tight oil reservoirs* (pp. 87–111). https://doi.org/10.1016/B978-0-12-818343-4.00005-X

Alfarge, D., Wei, M., & Bai, B. (2020b). Chapter 5—Air injection in shale and tight oil reservoirs. In D. Alfarge, M. Wei, B. Bai, B. Bai & Z. Chen (Eds.), *Developments in petroleum science* (Vol. 67, pp. 87–111). Elsevier. https://doi.org/10.1016/B978-0-12-818343-4.00005-X

Alfarge, D., Wei, M., & Bai, B. (2020c). *Fundamentals of enhanced oil recovery methods for unconventional oil reservoirs*. Elsevier.

Alharthy, N., Teklu, T., Kazemi, H., Graves, R., Hawthorne, S., Braunberger, J., & Kurtoglu, B. (2018). Enhanced oil recovery in liquid-rich shale reservoirs: Laboratory to field. *SPE Reservoir Evaluation & Engineering, 21*(1), SPE-175034-PA.

Al-Maamari, R. S., & Buckley, J. S. (2000). Asphaltene precipitation and alteration of wetting: Can wettability change during oil production? *Paper presented at the SPE/DOE improved oil recovery symposium*, Tulsa, OK. SPE-59292-MS.

Almobarak, M., Wu, Z., Zhou, D., Fan, K., Liu, Y., & Xie, Q. (2021). A review of chemical-assisted minimum miscibility pressure reduction in CO2 injection for enhanced oil recovery. *Petroleum, 7*(3), 245–253. https://doi.org/10.1016/j.petlm.2021.01.001

Alvarado, V., & Manrique, E. (2010, July 30). *Enhanced oil recovery: Field planning and development strategies*. Gulf Professional Publishing—Technology & Engineering.

Alvarez, J. O., & Schechter, D. S. (2016, April 11). Altering wettability in Bakken shale by surfactant additives and potential of improving oil recovery during injection of completion fluids. *All Days*. https://doi.org/10.2118/179688-MS

Alvarez, J. O., & Schechter, D. S. (2017). Wettability alteration and spontaneous imbibition in unconventional liquid reservoirs by surfactant additives. *SPE Reservoir Evaluation & Engineering, 20*, 107–117.

Alvarez, J. O., Tagavifar, M., Doorwar, S., Vo, D. T., Malik, T., Zhou, D., Strickler, J., Mathis, L. R., Neal, S., & Americas, C. (2023). *Design of chemical EOR in unconventional reservoirs*, 1–25. https://doi.org/10.15530/urtec-2023-3870505

Alvarez, J., Tovar, F., & Schechter, D. (2017). Improving oil recovery in unconventional liquid reservoirs by soaking-flowback production schedule with surfactant additives. *SPE liquids-rich basins conference*, North America, D021S006R002.

Aplin, A. C., & Macquaker, J. H. S. (2011). Mudstone diversity: Origin and implications for source, seal, and reservoir properties in petroleum systems. *AAPG Bulletin*, *95*(12), 2031–2059. https://doi.org/10.1306/03281110162

Askarova, A., Turakhanov, A., Mukhina, E., Cheremisin, A., & Cheremisin, A. (2020). Unconventional reservoirs: Methodological approaches for thermal EOR simulation. *Unconventional resources technology conference*, Virtual. URTeC, 2112.

Austad, T., RezaeiDoust, A., & Puntervold, T. (2010). Chemical mechanism of low salinity water flooding in sandstone reservoirs. *SPE improved oil recovery symposium*. SPE-129767-MS, https://doi.org/10.2118/129767-MS

Azin, R., Izadpanahi, A., & Mohamadi-Baghmolaei, M. (2022). *Gas recycling, fundamentals and practical aspects of gas injection*. Springer.

Baker, R., Dieva, R., Jobling, R., & Lok, C. (2016). The myths of waterfloods, EOR floods and how to optimize real injection schemes. *All Days*, SPE-179536-MS. https://doi.org/10.2118/179536-MS

Barak, Y., & Oren, M. (1992). Enhanced binding of a 95 kDa protein to p53 in cells undergoing p53-mediated growth arrest. *The EMBO Journal*, *11*(6), 2115–2121. https://doi.org/10.1002/j.1460-2075.1992.tb05270.x

Bauman, J. H., & Deo, M. D. (2011). Parameter space reduction and sensitivity analysis in complex thermal subsurface production processes. *Energy & Fuels*, *25*(1), 251–259. https://doi.org/10.1021/ef101225g

Bello, A., Ivanova, A., & Cheremisin, A. (2023). Foam EOR as an optimization technique for gas EOR: A comprehensive review of laboratory and field implementations. *Energies*, *16*(2), 972. https://doi.org/10.3390/en16020972

Bera, A., & Babadagli, T. (2015). Status of electromagnetic heating for enhanced heavy oil/bitumen recovery and future prospects: A review. *Applied Energy*, *151*(2015), 206–226.

Berg, R. R., & Gangi, A. F. (1999). Primary migration by oil-generation microfracturing in low-permeability source rocks: Application to the Austin Chalk, Texas. *AAPG Bulletin*, *83*, 727–756.

Bertoncello, A., Wallace, J., Blyton, C., Honarpour, M., & Kabir, C. S. (2014). Imbibition and water blockage in unconventional reservoirs: Well-management implications during flowback and early production. *SPE Reservoir Evaluation & Engineering*, *17*, 497–506.

Bientinesi, M., Petarca, L., Cerutti, A., Bandinelli, M., De Simoni, M., Manotti, M., & Maddinelli, G. (2013). A radiofrequency/microwave heating method for thermal heavy oil recovery based on a novel tight-shell conceptual design. *Journal of Petroleum Science and Engineering*, *107*, 18–30.

Blaker, T., Aarra, M. G., Skauge, A., Rasmussen, L., Celius, H. K., Martinsen, H. A., & Vassenden, F. (2002). Foam for gas mobility control in the Snorre field: The FAWAG project. *SPE Reservoir Evaluation & Engineering*, *5*(4), 317–323. https://doi.org/10.2118/78824-PA

Boudreau, B. P. (1996). The diffusive tortuosity of fine-grained unlithified sediments. *Geochimica et Cosmochimica Acta*, *60*(16), 3139–3142.

Bruner, K. R., & Smosna, R. A. (2011). *A comparative study of the Mississippian Barnett shale, Fort Worth basin, and Devonian Marcellus shale, Appalachian basin*. US Department of Energy.

Bui, B. T. (2021). Evaluation of low salinity water injection improved oil recovery in liquid-rich unconventional reservoirs. *Rock Mechanics and Rock Engineering, 54*, 3541–3567. https://doi.org/10.1007/s00603-021-02444-9

Carlsen, M. L., Whitson, C. H., Dahouk, M. M., Younus, B., Yusra, I., Kerr, E., Nohavitza, J., Thuesen, M., Drozd, J. H., Ambrose, R., & Mydland, S. (2019). Compositional tracking of a HnP project in the eagle ford. *SPE/AAPG/SEG unconventional resources technology conference*, Denver, CO, 22–24 July. URTEC-2019-539-MS.

Chalmers, G. R. L., & Bustin, R. M. (2012). Geological evaluation of Halfway–Doig–Montney hybrid gas shale–tight gas reservoir, northeastern British Columbia. *Marine and Petroleum Geology, 38*(1), 53–72. https://doi.org/10.1016/j.marpetgeo.2012.08.004

Charest, M. (2013). Alkaline-surfactant-polymer (ASP) flooding in Alberta: Small amounts of the right chemicals can make a big difference. *Canadian Discovery Digest.* https://digest.canadiandiscovery.com/article/6030

Chaturvedi, K. R., Singh, A. K., & Sharma, T. (2019). Impact of shale on properties and oil recovery potential of sandstone formation for low-salinity waterflooding applications. *Asia-Pacific Journal of Chemical Engineering, 14*, e2352. https://doi.org/10.1002/apj.2352

Chaudhary, A. S., Ehlig-Economides, C., & Wattenbarger, R. (2011). Shale oil production performance from a stimulated reservoir volume. *SPE annual technical conference and exhibition.* SPE-147596-MS, https://doi.org/10.2118/147596-MS

Chen, C., & Gu, M. (2017). Investigation of cyclic CO2 huff-and-puff recovery in shale oil reservoirs using reservoir simulation and sensitivity analysis. *Fuel, 188*, 102–111.

Chen, C., Tang, X., Qin, M., Zhou, R., Ding, Z., Lian, G., Qi, H., Chen, X., Liu, Z., & Li, Y. (2022). Experimental evaluation of shale oil development effectiveness by air injection. *Energies, 15*, 9513.

Chen, H., Li, B., Duncan, I., Elkhider, M., & Liu, X. (2020). Empirical correlations for prediction of minimum miscible pressure and near-miscible pressure interval for oil and CO2 systems. *Fuel, 278*, 118272. https://doi.org/10.1016/j.fuel.2020.118272

Chen, J.-H., Georgi, D. T., & Liu, H.-H. (2018). Electromagnetic thermal stimulation of shale reservoirs for petroleum production. *Journal of Natural Gas Science and Engineering, 59*, 183–192.

Chen, K. (2013). *Evaluation of EOR potential by gas and water flooding in shale oil reservoirs.* M.S. Thesis, Texas Tech University.

Chen, S., Zhao, W., Ouyang, Y., Zeng, Q., Yang, Q., Hou, H., Gai, S., Bao, S., & Li, X. (2017). Prediction of sweet spots in shale reservoir based on geophysical well logging and 3D seismic data: A case study of lower Silurian Longmaxi formation in W4 block, Sichuan Basin, China. *Energy Exploration & Exploitation, 35*(2), 147–171. https://doi.org/10.1177/0144598716679961

Chhetri, A. B., & Islam, M. R. (2008). A critical review of electromagnetic heating for enhanced oil recovery. *Petroleum Science and Technology, 26*(14), 1619–1631.

Chopra, S., & Michelena, R. J. (2011). Introduction to this special section: Reservoir characterization. *The Leading Edge January*, 35–37.

Clancy, J. P., Gilchrist, R. E., Cheng, L. H. K., & Bywater, D. R. (1985). Analysis of nitrogen-injection projects to develop screening guides and offshore design criteria. *Journal of Petroleum Technology, 37*(6), 1097–1104. https://doi.org/10.2118/11902-PA

Clara, C., Zelenko, V., Schirmer, P., & Wolter, T. (1998). Appraisal of the Horse Creek Air injection project performance. *Abu Dhabi international petroleum exhibition and conference*, p. SPE-49519-MS. https://doi.org/10.2118/49519-MS

Clark, A. J. (2009). Determination of recovery factor in the Bakken Formation, Mountrail County, ND. *All Days*, SPE-133719-STU. https://doi.org/10.2118/133719-STU

Clark, S. R., Pitts, M. J., & Smith, S. M. (1993). Design and application of an alkaline-surfactant-polymer recovery system for the west Kiehl field. *SPE Advanced Technology Series*, *1*(1), 172–179. https://doi.org/10.2118/17538-PA

Cox, S. A., Cook, D. M., Dunek, K., Daniels, R., Jump, C., & Barree, B. (2008). Unconventional resource play evaluation: A look at the Bakken Shale Play of North Dakota. *SPE unconventional reservoirs conference*. SPE-114171-MS, https://doi.org/10.2118/114171-MS

Cronin, M., Meybodi, H. E., & Johns, R. T. (2019). Diffusion-dominated proxy model for solvent injection in ultratight oil reservoirs. *SPE Journal*, *24*(2), SPE-190305-PA.

Cui, J., Sang, Q., Li, Y., Yin, C., Li, Y., & Dong, M. (2017). Liquid permeability of organic nanopores in shale: Calculation and analysis. *Fuel*, *202*, 426–434. https://doi.org/10.1016/j.fuel.2017.04.057

Curtis, J. B. (2002). Fractured shale-gas systems. *AAPG Bulletin*, *86*. https://doi.org/10.1306/61EEDDBE-173E-11D7-8645000102C1865D

Davudov, D., & Moghanloo, R. G. (2018). Impact of pore compressibility and connectivity loss on shale permeability. *International Journal of Coal Geology*, *187*, 98–113.

Delaihdem, D. K. (2013). *Decline curve analysis and enhanced shale oil recovery based on Eagle Ford Shale data*. University of Alaska Fairbanks.

Díez, K., Ocampo, A., Restrepo, A., Patiño, J., Rayo, J., Ayala, D., & Rueda, L. (2022, April 18). A novel gas dispersible foam technology can improve the efficiency of gas injection processes for IOR-EOR operations in unconventional reservoirs. *Day 1 Mon*, 25 April. https://doi.org/10.2118/209381-MS

Dincer, I., & Rosen, M. A. (2013). *Exergy: Energy, environment and sustainable development* (2nd ed., Chapter 4). Elsevier.

Doranehgard, M. H., Tran, S., & Dehghanpour, H. (2021). Modeling of natural-gas diffusion in oil-saturated tight porous media. *Fuel*, *300*, 120999.

Downey, M. W., Garvin, J., Lagomarsino, R. C., & Nicklin, D. F. (2011). *Quick look determination of oil-in-place in oil shale resource plays*. Eagle Ford Shale: American Association of Petroleum Geologists (AAPG) Search and Discovery Article No. 40764.

Du, D., Pu, W., Jin, F., & Liu, R. (2020). Experimental study on EOR by CO2 HnP and CO2 flooding in tight conglomerate reservoirs with pore scale. *Chemical Engineering Research and Design*, *156*, 425–432.

Du, F., & Nojabaei, B. (2019). A review of gas injection in shale reservoirs: Enhanced oil/gas recovery approaches and greenhouse gas control. *Energies*, *12*(12), 2355. https://doi.org/10.3390/en12122355

Egboga, N. U., Mohanty, K. K., & Balhoff, M. T. (2017). A feasibility study of thermal stimulation in unconventional shale reservoirs. *Journal of Petroleum Science and Engineering*, *154*, 576–588.

El-Samak, A. A., Ponnamma, D., Hassan, M. K., Ammar, A., Adham, S., Al-Maadeed, M. A. A., & Karim, A. (2020). Designing flexible and porous fibrous membranes for oil water separation—A review of recent developments. *Polymer Reviews*, *60*(4), 671–716. https://doi.org/10.1080/15583724.2020.1714651

Elturki, M., & Imqam, A. (2020). *Application of enhanced oil recovery methods in unconventional reservoirs: A review and data analysis*. https://doi.org/ARMA 20–1081

Ely, J. W., & Herndon, R. A. (2019). Fracturing fluids and additives. In *Hydraulic fracturing: Fundamentals and advancements* (pp. 165–197). Society of Petroleum Engineers Richardson. https://doi.org/10.2118/9781613997192-06

Fakcharoenphol, P., Charoenwongsa, S., Kazemi, H., & Wu, Y. S. (2013). The effect of water- induced stress to enhance hydrocarbon recovery in shale reservoirs. *SPEJ*, *18*(5), 897–909.

Fakcharoenphol, P., Kurtoglu, B., Kazemi, H., Charoenwongsa, S., & Wu, Y.-S. (2014). The effect of osmotic pressure on improve oil recovery from fractured shale formations. *SPE unconventional resources conference*. SPE-168998-MS, https://doi.org/10.2118/168998-MS

Farzaneh, S. A., Carnegie, A., Sohrabi, M., Mills, J. R., Facanha, J. M., & Sellers, B. (2017). A case study of oil recovery improvement by low salinity water injection. *Abu Dhabi international petroleum exhibition and conference*. SPE-188512-MS, https://doi.org/10.2118/188512-MS

Fassihi, M., Yannimaras, D., & Kumar, V. (1997). Estimation of recovery factor in light-oil air-injection projects. *SPE Reservoir Engineering*, *12*, 173–178.

Feng, R., Chen, S., Bryant, S., & Liu, J. (2019). Stress-dependent permeability measurement techniques for unconventional gas reservoirs: Review, evaluation, and application. *Fuel*, *256*, 115987. https://doi.org/10.1016/j.fuel.2019.115987

Foster, W. R. (1973). A low-tension waterflooding process. *Journal of Petroleum Technology*, *25*(2), 205–210. https://doi.org/10.2118/3803-PA

Fredriksen, S., Rognmo, A., Sandengen, K., & Fernø, M. (2017). Wettability effects on osmosis as an oil-mobilization mechanism during low-salinity waterflooding. *Petrophysics*, *58*, 28–35.

Ganjdanesh, R., Eltahan, E., Sepehrnoori, K., Drozd, H., & Ambrose, R. (2020). A field pilot of HnP gas injection for enhanced oil recovery in Permian Basin. *SPE annual technical conference & exhibition*, Virtual. SPE-201622-MS.

Garcez, J., Zhang, M., & Ayala, L. F. (2023). Modeling of anomalous gas transport in heterogeneous unconventional reservoirs using a nonlinear generalized diffusivity equation. *Fuel*, *340*, 127519. https://doi.org/10.1016/j.fuel.2023.127519

Ghanizadeh, A., Song, C., Hamdi, H., & Clarkson, C. R. (2021). Experimental and computational evaluation of cyclic solvent injection in fractured tight hydrocarbon reservoirs. *Scientific Reports*, *11*, 9497. https://doi.org/10.1038/s41598-021-88247-y

Gillham, T., Cerveny, B., Fornea, M., & Bassiouni, Z. (1998). Low cost IOR: An update on the W. Hackberry air injection project. *SPE improved oil recovery conference*, p. SPE-39642-MS. https://doi.org/10.2118/39642-MS

Glandt, C., Pieterson, R., Dombrowski, A., & Balzarini, M. (1999). Coral creek field study: A comprehensive assessment of the potential of high-pressure air injection in a mature waterflood project. *SPE Oklahoma city oil and gas symposium/production and operations symposium*, p. SPE-52198-MS. https://doi.org/10.2118/52198-MS

Godec, M., Koperna, G., Petrusak, R., & Oudinot, A. (2013). Potential for enhanced gas recovery and CO2 storage in the Marcellus shale in the Eastern United States. *International Journal of Coal Geology*, *118*, 95–104.

Grathwohl, P. (2012). *Diffusion in natural porous media contaminant transport, sorption/desorption and dissolution kinetics*. Springer US. Springer Science+Business Media New York.

Guo, H., Li, Y., Gu, Y., Wang, F., & Yuliang, Z. (2016, April 11). Comparison of strong alkali and weak alkali ASP flooding pilot tests in Daqing oilfield. *All Days*. https://doi.org/10.2118/179661-MS

Guo, H., Ma, R., & Kong, D. (2017, October 17). Success and lessons learned from ASP flooding field tests in China. *Day 2 Wed*, 18 October. https://doi.org/10.2118/186931-MS

Gupta, I., Rai, C. S., & Sondergeld, C. H. (2020). Impact of surfactants on hydrocarbon mobility in shales. *SPE Reservoir Evaluation & Engineering*, *23*(3), 1105–1117. https://doi.org/10.2118/201110-PA

Gupta, S., Kamari, A., & Mahani, H. (2019). Application of HnP gas injection for EOR in shale oil reservoirs: A review. *Journal of Petroleum Science and Engineering*, *180*, 1044–1058.

Gutierrez, D., Miller, R. J., Taylor, A. R., Thies, B., & Kumar, V. (2009). Buffalo field high-pressure-air-injection projects: Technical performance and operational challenges. *SPE Reservoir Evaluation & Engineering, 12*, 542–550.

Gutierrez, D., Taylor, A. R., Kumar, V., Ursenbach, M. G., Moore, R. G., & Mehta, S. A. (2008). Recovery factors in high-pressure air injection projects revisited. *SPE Reservoir Evaluation & Engineering, 11*, 1097–1106.

Habibi, A., Javaheri, A., & Dehghanpour, H. (2019). Tight rock wettability and its relationship with petrophysical properties. In *Petrophysical characterization and fluids transport in unconventional reservoirs.* Elsevier Inc. https://doi.org/10.1016/B978-0-12-816698-7.00008-5

Haines, H. K., & Monger, T. G. (1990). A laboratory study of natural gas huff 'n' puff. *All Days*, SPE-21576-MS. https://doi.org/10.2118/21576-MS

Hamdi, H., Clarkson, C. R., Esmail, A., & Sousa, M. C. (2021). Optimizing the huff 'n' puff gas injection performance in shale reservoirs considering the uncertainty: A DuVernay shale example. *SPE Reservoir Evaluation & Engineering, 24*(1), 219–237. SPE-195438-PA.

Han, H., Dang, S., Acosta, J., Fu, J., Sondergeld, C., & Rai, C. (2019). X-ray fluorescence and laser-induced breakdown spectroscopy for advanced rock elemental analysis. *Unconventional resources technology conference*, Denver, CO, 22–24 July, pp. 3755–3766.

Harrison, A. L., Jew, A. D., Dustin, M. K., Thomas, D. L., Joe-Wong, C. M., Bargar, J. R., Johnson, N., Brown, G. E., & Maher, K. (2017). Element release and reaction-induced porosity alteration during shale-hydraulic fracturing fluid interactions. *Applied Geochemistry, 82*, 47–62. https://doi.org/10.1016/j.apgeochem.2017.05.001

Hasanzadeh, M., Izadpanahi, A., & Naghizadeh, A. (2023). Chapter 5—Carbon dioxide HnP. In *Enhanced oil recovery series, gas injection methods* (pp. 171–198). Gulf Professional Publishing.

Hawthorne, S. B., Gorecki, C. D., Sorensen, J. A., Steadman, E. N., Harju, J. A., & Melzer, S. (2013). Hydrocarbon mobilization mechanisms from upper, middle, and lower Bakken reservoir rocks exposed to CO2. *Paper presented at the SPE unconventional resources conference Canada*, 5–7 November, Canada. SPE-167200-MS.

He, K., Yue, Z., Fan, C., & Xu, L. (2015). Minimizing surfactant adsorption using polyelectrolyte based sacrificial agent: A way to optimize surfactant performance in unconventional formations. *SPE international conference on oilfield chemistry*, SPE-173750-MS, https://doi.org/10.2118/173750-MS

Hobbs, B., & Ord, A. (2015). *Structural geology: The mechanics of deforming metamorphic rocks.* Imprint Elsevier.

Hoffman, B. T. (2018). HnP gas injection pilot projects in the eagle ford. *Paper presented at the SPE Canada unconventional resources conference*, Calgary, Alberta, Canada. SPE-189816-MS.

Hoffman, B. T., & Evans, J. (2016). Improved oil recovery IOR pilot projects in the Bakken formation. *SPE-180270-MS paper presented at the SPE low perm symposium held in Denver*, CO, 5–6 May. Society of Petroleum Engineers.

Hoffman, B. T., & Reichhardt, D. (2019). Quantitative evaluation of recovery mechanisms for HnP gas injection in unconventional reservoirs. *Proceedings of the 7th unconventional resources technology conference. Unconventional resources technology conference*, Denver, CO. https://doi.org/10.15530/urtec-2019–147

IEA. (2018). *Number of EOR projects in operation globally* (pp. 1971–2017). International Energy Agency.

Iglauer, S., Pentland, C. H., & Busch, A. (2015). CO 2 wettability of seal and reservoir rocks and the implications for carbon geo-sequestration. *Water Resources Research*, *51*(1), 729–774. https://doi.org/10.1002/2014WR015553

Jacobs, T. (2017). Optimism and activity rising in the Vaca Muerta. *Journal of Petroleum Technology*, *69*, 34–38.

Jacobs, T. (2019). Shale EOR delivers, so why won't the sector go big? *JPT*, *71*(5). SPE-0519-0037-JPT.

Jain, A. K., Dhawan, A. K., & Misra, T. R. (2012, March 28). ASP flood pilot in Jhalora (K-IV)—A case study. *All Days*. https://doi.org/10.2118/153667-MS

Jerauld, G. R., Lin, C., Webb, K. J., & Seccombe, J. C. (2008). Modeling low-salinity waterflooding. *SPE Reservoir Evaluation & Engineering*, *11*, 1000–1012.

Jha, K. N., & Chakma, A. (1999). Heavy-oil recovery from thin pay zones by electromagnetic heating. *Energy Sources*, *21*(1–2), 63–73.

Jia, C., Zheng, M., & Zhang, Y. (2016). Some key issues on the unconventional petroleum systems. *Petroleum Research*, *1*(2), 113–122. https://doi.org/10.1016/S2096-2495(17)30036-4

Jia, H., & Sheng, J. J. (2017). Discussion of the feasibility of air injection for enhanced oil recovery in shale oil reservoirs. *Petroleum*, *3*, 249–257.

Jia, H., Ni, J., Pu, W., Yue, P., Jiang, H., & Yang, J. (2014). New view on the oxidation mechanisms of crude oil through combined thermal analysis methods: TG/DTA and oxidation tube experiment. *Journal of Thermal Analysis and Calorimetry*, *118*, 1707–1714.

Jia, H., Yuan, C.-D., Zhang, Y.-C., Peng, H., Zhong, D., & Zhao, J.-Z. (2012a). Recent progress of high-pressure air injection (HPAI) process in light oil reservoir: Laboratory investigation and field application. *SPE heavy oil conference Canada*. SPE-156974-MS, https://doi.org/10.2118/156974-MS

Jia, H., Zhao, J.-Z., Pu, W.-F., Liao, R., & Wang, L.-L. (2012b). The influence of clay minerals types on the oxidation thermokinetics of crude oil. *Energy Sources, Part A: Recovery, Utilization, and Environmental Effects*, *34*, 877–886.

Jiang, H.-F., Lei, Y.Z., Xiong, X., Yan, L.P., Pi, W.F., Li, X.J., Yu, C.H., 2008. A CO2 immiscible displacement experimental study aiming at Fuyang extra-low permeability layer at the Daqing Changheng peripheral. *Geoscience*, *22*(4), 659–663.

Jiang, Y., Wang, B., Li, Q., Zhao, J., Zhang, Y., & Zheng, H. (2017, November 13). Mechanisms and EOR potential of nitrogen assisted artificial foamy oil flooding. *Day 2 Tue*, 14 November. https://doi.org/10.2118/188172-MS

Jin, L., Hawthorne, S., Sorensen, J., Pekot, L., Kurz, B., Smith, S., Heebink, L., Bosshart, N., Torres, J., Dalkhaa, C., Gorecki, C., Steadman, E., & John, H. (2017). Extraction of oil from the Bakken shales with supercritical CO2. *Paper presented at the unconventional resources technology conference held in Austin, TX*. URTeC, 2671596.

Joslin, K., Ghedan, S. G., Abraham, A. M., & Pathak, V. (2017). EOR in tight reservoirs, technical and economical feasibility. *Paper presented at the SPE unconventional resources conference*, Calgary, Alberta, Canada. SPE-185037-MS.

JPT Staff. (1998). Horse creek air-injection project: An overview. *Journal of Petroleum Technology*, *50*, 78–80.

JPT Staff. (1999). Air injection for IOR from light-oil reservoirs: Oxidation rate and displacement studies. *Journal of Petroleum Technology*, *51*, 46–48.

Kang, P.-S., Lim, J.-S., & Huh, C. (2016). Screening criteria and considerations of offshore enhanced oil recovery. *Energies*, *9*, 1–18.

Kantzas, A., Chatzis, I., & Dullien, F. A. L. (1988). Enhanced oil recovery by inert gas injection. *All Days*, SPE-17379-MS. https://doi.org/10.2118/17379-MS

Karimi, S., Kazemi, H., & Simpson, G. A. (2017). Capillary pressure, fluid distribution, and oil recovery in preserved middle Bakken cores. *SPE Oklahoma city oil and gas symposium*. SPE-185095-MS, https://doi.org/10.2118/185095-MS

Kathel, P., & Mohanty, K. (2013a). Wettability alteration in a tight oil reservoir. *Energy & Fuels, 27*, 6460–6468.

Kathel, P., & Mohanty, K. (2013b). EOR in tight oil reservoirs through wettability alteration. *SPE annual technical conference and exhibition.* SPE-166281-MS, https://doi.org/10.2118/166281-MS

Khakimova, L., Bondarenko, T., Cheremisin, A., Myasnikov, A., & Varfolomeev, M. (2019). High pressure air injection kinetic model for Bazhenov Shale Formation based on a set of oxidation studies. *Journal of Petroleum Science and Engineering, 172*, 1120–1132. https://doi.org/10.1016/j.petrol.2018.09.021

Kim, E. M., Tinker, S. W., Fisher, W. L., & Ruppel, S. C. (2000). *Benefit/cost analysis of GRI's gas supply research initiative: Economics of a gas supply research trust fund to increase gas production of Federal lands.* The University of Texas at Austin. Bureau of Economic Geology.

Kim, T. H., Park, S. S., & Lee, K. S. (2015). Modeling of CO2 injection considering multicomponent transport and geotechnical effect in shale gas reservoirs. *Paper presented at the SPE/IATMI Asia pacific oil & gas conference and exhibition*, Nusa Dua, Bali, Indonesia. SPE-176174-MS.

King, G. E. (2014). 60 years of multi-fractured vertical, deviated and horizontal wells: What have we learned? *All Days*, SPE-170952-MS. https://doi.org/10.2118/170952-MS

Kurtoglu, B. (2013). *Integrated reservoir characterization and modeling in support of enhanced oil recovery for Bakken* [Ph.D. dissertation, Department of Petroleum Engineering at the Colorado School of Mines].

Kurtoglu, B., Kazemi, H., Rosen, R., Mickelson, W., & Kosanke, T. (2014). A rock and fluid study of middle Bakken formation: Key to enhanced oil recovery. *SPE/CSUR unconventional resources conference*, Canada.

Li, L., Sheng, J. J., & Xu, J. (2017a). Gas selection for HnP EOR in shale oil reservoirs based upon experimental and numerical study. *Paper presented at the SPE unconventional resources conference*, Calgary, Alberta, Canada. SPE-185066-MS.

Li, L., Tan, J., Wood, D. A., Zhao, Z., Becker, D., Lyu, Q., Shu, B., & Chen, H. (2019). A review of the current status of induced seismicity monitoring for hydraulic fracturing in unconventional tight oil and gas reservoirs. *Fuel, 242*, 195–210.

Li, L., Zhang, Y., & Sheng, J. J. (2017b). Effect of the injection pressure on enhancing oil recovery in shale cores during the CO2 HnP process when it is above and below the minimum miscibility pressure. *Energy Fuels, 31*(4), 3856–3867.

Li, Q., Jew, A. D., Brown, G. E., Bargar, J. R., & Maher, K. (2020). Reactive transport modeling of shale–fluid interactions after imbibition of fracturing fluids. *Energy & Fuels, 34*(5), 5511–5523. https://doi.org/10.1021/acs.energyfuels.9b04542

Li, S., Qiao, C., Zhang, C., & Li, Z. (2018). Determination of diffusion coefficients of supercritical CO2 under tight oil reservoir conditions with pressure-decay method. *Journal of CO2 Utilization, 24*, 430–443.

Li, X., Abass, H., Teklu, T. W., & Cui, Q. (2016). A shale matrix imbibition model–interplay between capillary pressure and osmotic pressure. *SPE annual technical conference and exhibition.* SPE-181407-MS, https://doi.org/10.2118/181407-MS

Liang, T., Zhao, X., Yuan, S., Zhu, J., Liang, X., Li, X., & Zhou, F. (2021). Surfactant-EOR in tight oil reservoirs: Current status and a systematic surfactant screening method with field experiments. *Journal of Petroleum Science and Engineering, 196*, 108097. https://doi.org/10.1016/j.petrol.2020.108097

Liu, J., & Sheng, J. J. (2019). Experimental investigation of surfactant enhanced spontaneous imbibition in Chinese shale oil reservoirs using NMR tests. *Journal of Industrial and Engineering Chemistry, 72*, 414–422. https://doi.org/10.1016/j.jiec.2018.12.044

Liu, S., Sahni, V., Tan, J., Beckett, D., & Vo, T. (2018). Laboratory investigation of EOR techniques for organic rich shales in the Permian basin. *Paper presented at SPE/AAPG/*

SEG unconventional resources technology conference, Houston, TX, 23–25 July. URTEC-2890074-MS.

Liu, X., Godbole, A., Lu, C., Michal, G., & Venton, P. (2014). Source strength and dispersion of CO2 releases from high-pressure pipelines: CFD model using real gas equation of state. *Applied Energy, 126*, 56–68. https://doi.org/10.1016/j.apenergy.2014.03.073

Liu, Z., Liang, Y., Wang, Q., Guo, Y., Gao, M., Wang, Z., & Liu, W. (2020). Status and progress of worldwide EOR field applications. *Journal of Petroleum Science and Engineering, 193*, 107449. https://doi.org/10.1016/j.petrol.2020.107449

Loucks, R. G., & Ruppel, S. C. (2007). Mississippian Barnett shale: Lithofacies and depositional setting of a deep-water shale-gas succession in the Fort Worth Basin, Texas. *American Association of Petroleum Geologists Bulletin, 91*(4), 579–601. https://doi.org/10.1306/11020606059.

Louk, K., Ripepi, N., Luxbacher, K., Gilliland, E., Tang, X., Keles, C., Schlosser, C., Diminick, E., Keim, S., Amante, J., & Michael, K. (2017). Monitoring CO2 storage and enhanced gas recovery in unconventional shale reservoirs: Results from the Morgan County, Tennessee injection test. *Journal of Natural Gas Science and Engineering, 45*, 11–25.

Lu, C.-F. (1988). A new technique for the evaluation of shale stability in the presence of polymeric drilling fluid. *SPE Production Engineering, 3*, 366–374.

Luo, G., Tian, Y., Bychina, M., & Ehlig-Economides, C. (2018). Production optimization using machine learning in Bakken shale. *Proceedings of the 6th unconventional resources technology conference. Unconventional resources technology conference*, Houston, TX. https://doi.org/10.15530/urtec-2018-2902505

Mahani, H., Keya, A. L., Berg, S., Bartels, W.-B., Nasralla, R., & Rossen, W. R. (2015). Insights into the mechanism of wettability alteration by low-salinity flooding (LSF) in carbonates. *Energy & Fuels, 29*, 1352–1367. https://doi.org/10.1021/ef5023847

Mahzari, P., Mitchell, T. M., Jones, A. P., Oelkers, E. H., Striolo, A., Iacoviello, F., Shearing, P. R., & Ernesto Juri, J. (2021). Novel laboratory investigation of HnP gas injection for shale oils under realistic reservoir conditions. *Fuel, 284*, 118950. https://doi.org/10.1016/j.fuel.2020.118950

Mao, Z., Cheng, L., Liu, D., Li, T., Zhao, J., & Yang, Q. (2022). Nanomaterials and technology applications for hydraulic fracturing of unconventional oil and gas reservoirs: A state-of-the-art review of recent advances and perspectives. *ACS Omega, 7*(34), 29543–29570. https://doi.org/10.1021/acsomega.2c02897

Mehrabi, M., Pei, Y., Haddad, M., Javadpour, F., & Sepehrnoori, K. (2021). Quasi-static fracture height growth in laminated reservoirs: Impacts of stress and toughness barriers, horizontal well landing depth, and fracturing fluid density. *Proceedings of the 9th unconventional resources technology conference. Unconventional resources technology conference*, Houston, TX. https://doi.org/10.15530/urtec-2021-5723

Meng, X., & Sheng, J. J. (2016). Optimization of HnP gas injection in a shale gas condensate reservoir. *Journal of Unconventional Oil and Gas Resources, 16*, 34–44.

Milad, M., Junin, R., Sidek, A., Imqam, A., & Tarhuni, M. (2021). HnP technology for enhanced oil recovery in shale/tight oil reservoirs: Progress, gaps, and perspectives. *Energy & Fuels, 35*(21), 17279–17333. https://doi.org/10.1021/acs.energyfuels.1c02561

Miller, C., Zeng, T., & Mohanty, K. (2019, September 23). Evaluation of chemical blends for shale EOR. *Day 2 Tue*, 1 October. https://doi.org/10.2118/195819-MS

Mohammed, M. A., & Babadagli, T. (2016). Experimental investigation of wettability alteration in oil-wet reservoirs containing heavy oil. *SPE Reservoir Evaluation & Engineering, 19*(4), 633–644. https://doi.org/10.2118/170034-PA

Mohan, K., & Leonard, P. A. (2013). Evaluating EOR techniques in the Spraberry. *SPE/AAPG/SEG unconventional resources technology conference*. URTEC-1579528-MS, https://doi.org/10.1190/urtec2013-188

Mohan, K., Scott, K. D., Monson, G. D., & Leonard, P. A. (2013). A systematic approach to understanding well performance in unconventional reservoirs: A Wolfcamp case study. *SPE/AAPG/SEG unconventional resources technology conference*, URTEC-1579514-MS. https://doi.org/10.1190/urtec2013-051

Monger, T. G., & Coma, J. M. (1988). A laboratory and field evaluation of the CO2 huff 'n' puff process for light-oil recovery. *SPE Reservoir Engineering*, *3*(4), 1168–1176. https://doi.org/10.2118/15501-PA

Morley, C. K., Von Hagke, C., Hansberry, R., Collins, A., Kanitpanyacharoen, W., & King, R. (2018). Review of major shale-dominated detachment and thrust characteristics in the diagenetic zone: Part II, rock mechanics and microscopic scale. *Earth-Science Reviews*, *176*, 19–50. https://doi.org/10.1016/j.earscirev.2017.09.015

Morrow, N. R., & Mason, G. (2001). Recovery of oil by spontaneous imbibition. *Current Opinion in Colloid & Interface Science*, *6*(4), 321–337. https://doi.org/10.1016/S1359-0294(01)00100-5

Mukhina, E., Cheremisin, A., Khakimova, L., Garipova, A., Dvoretskaya, E., Zvada, M., Kalacheva, D., Prochukhan, K., Kasyanenko, A., & Cheremisin, A. (2021). Enhanced oil recovery method selection for shale oil based on numerical simulations. *ACS Omega*, *6*(37), 23731–23741. https://doi.org/10.1021/acsomega.1c01779

Murzin, D., & Salmi, T. (2005). Elementary reactions. In *Catalytic kinetics* (pp. 73–109). Elsevier. https://doi.org/10.1016/B978-044451605-3/50003-4

Neog, A., & Schechter, D. S. (2016, April 11). Investigation of surfactant induced wettability alteration in Wolfcamp shale for hydraulic fracturing and EOR applications. *All Days*. https://doi.org/10.2118/179600-MS

Orozco, D., Fragoso, A., Selvan, K., & Aguilera, R. (2018). Eagle ford huff-and-puff gas injection pilot: Comparison of reservoir simulation, material balance and real performance of the pilot well. *SPE Reservoir Evaluation & Engineering*, *23*(1), 247–260. SPE-191575-PA.

Orr, F. M. Jr. (2007). *Theory of gas injection processes*. Tie-line Publications.

Pang, X., Li, M., Li, B., Wang, T., Hui, S., Liu, Y., Liu, G., Hu, T., Xu, T., Jiang, F., Pang, H., & Wang, C. (2023). Main controlling factors and movability evaluation of continental shale oil. *Earth-Science Reviews*, *243*, 104472. https://doi.org/10.1016/j.earscirev.2023.104472

Patil, P. D., Knight, T., Katiyar, A., Vanderwal, P., Scherlin, J., Rozowski, P., Ibrahim, M., Sridhar, G. B., & Nguyen, Q. P. (2018, April 14). CO2 foam field pilot test in sandstone reservoir: Complete analysis of foam pilot response. *Day 4 Tue*, 17 April. https://doi.org/10.2118/190312-MS

Patzek, T. W., Male, F., & Marder, M. (2013). Gas production in the Barnett Shale obeys a simple scaling theory. *Proceedings of the National Academy of Sciences*, *110*(49), 19731–19736. https://doi.org/10.1073/pnas.1313380110

PetroWiki. (2015). *Electromagnetic heating of oil*. Society of Petroleum Engineers.

Pitts, M. J., Dowling, P., Wyatt, K., Surkalo, H., & Adams, C. (2006, April 22). Alkaline-surfactant-polymer flood of the tanner field. *All Days*. https://doi.org/10.2118/100004-MS

Qiao, J., Zeng, J., Ma, Y., Jiang, S., Feng, S., & Hu, H. (2020). Effects of mineralogy on pore structure and fluid flow capacity of deeply buried sandstone reservoirs with a case study in the Junggar Basin. *Journal of Petroleum Science and Engineering*, *189*, 106986. https://doi.org/10.1016/j.petrol.2020.106986

Ragab, A., & Mansour, E. (2021). Enhanced oil recovery: Chemical flooding. In *Geophysics and Ocean waves studies*. IntechOpen. https://doi.org/10.5772/intechopen.90335

Rehman, M., & Meribout, M. (2012). Conventional versus electrical enhanced oil recovery: A review. *Journal of Petroleum Exploration and Production Technology*, *2*, 157–167.

Reynolds, M. A. (2020). A technical playbook for chemicals and additives used in the hydraulic fracturing of shales. *Energy and Fuels*, *34*(12), 15106–15125. https://doi.org/10.1021/acs.energyfuels.0c02527

RezaeiDoust, A., Puntervold, T., Strand, S., & Austad, T. (2009). Smart water as wettability modifier in carbonate and sandstone: A discussion of similarities/differences in the chemical mechanisms. *Energy & Fuels*, *23*, 4479–4485. https://doi.org/10.1021/ef900185q

Rezaeyan, A., Humphrey-Kugbega, K., Tran, S., Younis, A., Song, C., Ghanizadeh, A., Hamdi, H., & Clarkson, C. R. (2024a). Experimental investigation of cyclic CO2 injection and storage in unconventional reservoirs under reservoir conditions. *SPE/AAPG/SEG unconventional resources technology conference (URTeC)*. https://doi.org/10.15530/urtec-2024-4034665

Rezaeyan, A., Kampman, N., Pipich, V., Barnsley, L. C., Rother, G., Magill, C., Ma, J., & Busch, A. (2023). Evolution of pore structure in organic-lean and organic-rich mudrocks. *Energy & Fuels*, *37*, 16446–16460. https://doi.org/10.1021/acs.energyfuels.3c02180

Rezaeyan, A., Kampman, N., Pipich, V., Barnsley, L. C., Rother, G., Magill, C., Ma, J., & Busch, A. (2024b). Compaction and clay content control mudrock porosity. *Energy*, *289*, 129966. https://doi.org/10.1016/j.energy.2023.129966

Rezaeyan, A., Pipich, V., Ma, J., Leu, L., Seemann, T., Rother, G., Barnsley, L. C., & Busch, A. (2022). Predicting fluid flow regime, permeability, and diffusivity in mudrocks from multiscale pore characterisation. *Transport in Porous Media*, *141*, 201–229. https://doi.org/10.1007/s11242-021-01717-9

Roshan, H., Al-Yaseri, A. Z., Sarmadivaleh, M., & Iglauer, S. (2016). On wettability of shale rocks. *Journal of Colloid and Interface Science*, *475*, 104–111. https://doi.org/10.1016/j.jcis.2016.04.041

Rychlicki, S., Stopa, J., Uliasz-Misiak, B., & Zawisza, L. (2011). Screening criteria for selecting oil reservoirs for CO2 enhanced oil recovery. *Gospodarka Surowcami Mineralnymi-mineral Resources Management*, 125–139–125–139.

Safari, M., Rahimi, A., Gholami, R., Permana, A., & Siaw Khur, W. (2022). Underlying mechanisms of shale wettability alteration by low salinity water injection (LSWI). *Journal of Dispersion Science and Technology*, *43*, 33–41. https://doi.org/10.1080/01932691.2020.1813156

Sahni, A., Kumar, M., & Knapp, R. B. (2000). Electromagnetic heating methods for heavy oil reservoirs. *SPE western regional meeting*, SPE-62550-MS. https://digital.library.unt.edu/ark:/67531/metadc719772/m2/1/high_res_d/790586.pdf

Sakthikumar, S., Madaoui, K., & Chastang, J. (1995, March 11). An investigation of the feasibility of air injection into a waterflooded light oil reservoir. *All Days*. https://doi.org/10.2118/29806-MS

Sanchez-Rivera, D., Mohanty, K., & Balhoff, M. (2015). Reservoir simulation and optimization of huff-and-puff operations in the Bakken Shale. *Fuel*, *147*, 82–94.

Sandengen, K., & Arntzen, O. (2013). Osmosis during low salinity water flooding. *IOR 2013-17th European symposium on improved oil recovery*, p. cp-342-00015. https://doi.org/10.3997/2214-4609.20142608

Sandengen, K., Kristoffersen, A., Melhuus, K., & Jøsang, L. O. (2016). Osmosis as mechanism for low-salinity enhanced oil recovery. *Spe Journal*, *21*, 1227–1235.

Sarma, H., & Das, S. (2009). Air injection potential in Kenmore oilfield in Eromanga Basin, Australia: A screening study through thermogravimetric and calorimetric analyses. *SPE middle east oil and gas show and conference*, SPE-120595-MS. https://doi.org/10.2118/120595-MS

Scerbacova, A., Mukhina, E., Bakulin, D., Burukhin, A., Ivanova, A., Cheremisin, A., Spivakova, M., Ushakova, A., & Cheremisin, A. (2023). Water- and surfactant-based HnP injection into unconventional liquid hydrocarbon reservoirs: Experimental and modeling study. *Energy & Fuels*. https://doi.org/10.1021/acs.energyfuels.3c01344

Schmidt, M., & Sekar, B. K. (2014). *Innovative unconventional 2EOR-A Light EOR an unconventional tertiary recovery approach to an unconventional Bakken reservoir in Southeast Saskatchewan*. World Petroleum Congress.

Seccombe, J., Lager, A., Webb, K., Jerauld, G., & Fueg, E. (2008). Improving waterflood recovery: LoSalTM EOR field evaluation. *Society of petroleum engineers: Proceedings of SPE symposium on improved oil recovery*. SPE-113480, https://doi.org/10.2118/113480-MS

Shah, A., Fishwick, R., Wood, J., Leeke, G., Rigby, S., & Greaves, M. (2010). A review of novel techniques for heavy oil and bitumen extraction and upgrading. *Energy & Environmental Science*, *3*, 700–714. https://doi.org/10.1039/B918960B

Shen, L., & Chen, Z. (2007). Critical review of the impact of tortuosity on diffusion. *Chemical Engineering Science*, *62*, 3748–3755.

Sheng, J. J. (2015a). *Enhanced oil recovery field case studies*. Gulf Professional Publishing.

Sheng, J. J. (2015b). Enhanced oil recovery in shale reservoirs by gas injection. *Journal of Natural Gas Science and Engineering*, *22*, 252–259. https://doi.org/10.1016/j.jngse.2014.12.002

Sheng, J. J. (2015c). Increase liquid oil production by HnP of produced gas in shale gas condensate reservoirs. *Journal of Unconventional Oil and Gas Resources*, *11*, 19–26.

Sheng, J. J., & Chen, K. (2014). Evaluation of the EOR potential of gas and water injection in shale oil reservoirs. *Journal of Unconventional Oil and Gas Resources*, *5*, 1–9. https://doi.org/10.1016/j.juogr.2013.12.001

Sheng, J. J., Cook, T., Barnes, W., Mody, F., Watson, M., Porter, M., & Viswanathan, H. (2015). Screening of the EOR potential of a Wolfcamp shale oil reservoir. *ARMA US rock mechanics/geomechanics symposium*, ARMA-2015-438. https://onepetro.org/ARMAUSRMS/proceedings-abstract/ARMA15/ARMA15/ARMA-2015-438/65865

Sheng, J. J., Mody, F., Griffith, P. J., & Barnes, W. N. (2016). Potential to increase condensate oil production by HnP gas injection in a shale condensate reservoir. *Journal of Natural Gas Science and Engineering*, *28*, 46–51.

Shuler, P. J., Lu, Z., Ma, Q., & Tang, Y. (2016). Surfactant HnP application potentials for unconventional reservoirs. *SPE improved oil recovery conference*, SPE-179667-MS. https://doi.org/10.2118/179667-MS

Skauge, A., Aarra, M. G., Surguchev, L., Martinsen, H. A., & Rasmussen, L. (2002, April 13). Foam-assisted WAG: Experience from the Snorre field. *All Days*. https://doi.org/10.2118/75157-MS

Skomedal, E., Park, J., Soldal, M., Smith, H., & Mokkelbost, S. (2021). Electro-osmosis to enhance shale barrier. *ARMA US rock mechanics/geomechanics symposium*, ARMA-2021-1329. https://onepetro.org/ARMAUSRMS/proceedings-abstract/ARMA21/ARMA21/ARMA-2021-1329/467997

Sohrabi, M., Mahzari, P., Farzaneh, S. A., Mills, J. R., Tsolis, P., & Ireland, S. (2017). Novel insights into mechanisms of oil recovery by use of low-salinity-water injection. *Spe Journal*, *22*, 407–416.

Song, C., Ghanizadeh, A., Younis, A., & Clarkson, C. R. (2023). Experimental evaluation of the impact of induced fractures on cyclic solvent injection performance in low-permeability hydrocarbon reservoirs. *Fuel*, *354*, 129327. https://doi.org/10.1016/j.fuel.2023.129327

Song, C., & Yang, D. (2017). Experimental and numerical evaluation of CO2 HnP processes in Bakken formation. *Fuel*, *190*, 145–162.

Song, W., & Kovscek, A. R. (2015). Functionalization of micromodels with kaolinite for investigation of low salinity oil-recovery processes. *Lab on a Chip*, *15*, 3314–3325.

Sorensen, J., & Terneus, J. (2011). *Subtask 1.2-evaluation of key factors affecting successful oil production in the Bakken formation*. North Dakota.

SpA, A. I. (2017). Oxygen removal from injection seawater in offshore platforms: Vacuum towers vs membrane deaeration technology. *Impiantestica Italiana*, 28–36.

Speight, J. G. (2016). *Deep shale oil and gas* (1st ed.). Elsevier.

Stegent, N. A., Wagner, A. L., Mullen, J., & Borstmayer, R. E. (2010). Engineering a successful fracture-stimulation treatment in the Eagle Ford shale: Tight gas completions. *SPE unconventional resources conference/gas technology symposium*. SPE-136183, https://doi.org/10.2118/136183-MS

Su, Y., Ren, L., Meng, F., Xu, C., & Wang, W. (2015). Theoretical analysis of the mechanism of fracture network propagation with stimulated reservoir volume (SRV) fracturing in tight oil reservoirs. *PLoS ONE*, *10*(5), e0125319. https://doi.org/10.1371/journal.pone.0125319

Syed, A., Shi, J.-Q., & Durucan, S. (2011). Permeability and injectivity improvement in CO2 enhanced coalbed methane recovery: Thermal stimulation of the near wellbore region. *Energy Procedia*, *4*, 2137–2143. https://doi.org/10.1016/j.egypro.2011.02.098

Tabatabaei, M., Dahi Taleghani, A., Cai, Y., Yang Santos, L., & Alem, N. (2019, September 23). Using nanoparticles coating to enhance proppant functions to achieve sustainable production. *Day 1 Mon*, 30 September. https://doi.org/10.2118/196067-MS

Tahmasebi, P., Javadpour, F., & Enayati, S. F. (2020). Digital rock techniques to study shale permeability: A mini-review. *Energy & Fuels*, *34*(12), 15672–15685. https://doi.org/10.1021/acs.energyfuels.0c03397

Tang, W., & Sheng, J. J. (2022). HnP gas injection or gas flooding in tight oil reservoirs? *Journal of Petroleum Science and Engineering*, *208*, 109725. https://doi.org/10.1016/j.petrol.2021.109725

Tang, Y., Xing, Y., Li, L. Z., & Jiang, S. X. (2012). Influence factors and evaluation methods of the gas shale fracability. *Earth Science Frontiers*, *19*(5), 356–363 (in Chinese with English abstract).

Teklu, T. W., Alharthy, N., Kazemi, H., Xiaolong, Y., Graves, R. M., & AlSumaiti, A. M. (2014). Phase behavior and minimum miscibility pressure in nanopores. *SPE Reservoir Evaluation and Engineering*, *17*(3).

Teklu, T. W., Li, X., Zhou, Z., Alharthy, N., Wang, L., & Abass, H. (2018). Low-salinity water and surfactants for hydraulic fracturing and EOR of shales. *Journal of Petroleum Science and Engineering*, *162*, 367–377. https://doi.org/10.1016/j.petrol.2017.12.057

Thomas, F. B., Piwowar, M., Noroozi, M., Gibb, W., Marin, J., & Zhang, H. (2020). Experimental measurements of Montney and Duvernay gas-cycling enhanced oil recovery GCEOR. *Paper presented at the SPE Canada unconventional resources conference*, Virtual. SPE-199994-MS. https://doi.org/10.2118/199994-ms

Todd, H. B., & Evans, J. G. (2016). Improved oil recovery IOR pilot projects in the Bakken formation. *All Days*, SPE-180270-MS. https://doi.org/10.2118/180270-MS

Torabi, F., & Asghari, K. (2010). Effect of connate water saturation, oil viscosity and matrix permeability on rate of gravity drainage during immiscible and miscible displacement tests in matrix-fracture experimental model. *Journal of Canadian Petroleum Technology*, *49*(11), 61–68. https://doi.org/10.2118/141295-PA

Torcuk, M. A., Uzun, O., Padin, A., & Kazemi, H. (2019). Impact of chemical osmosis on brine imbibition and hydrocarbon recovery in liquid-rich shale reservoirs. *SPE annual technical conference and exhibition*. SPE-195909. https://doi.org/10.2118/195909-MS

Tovar, F. D., Barrufet, M. A., & Schechter, D. S. (2018). Gas injection for EOR in organic rich shales. Part II: Mechanisms of recovery. *Proceedings of the 6th unconventional resources technology conference*. https://doi.org/10.15530/urtec-2018-2903026

Tran, S. (2021). *Evaluation of enhanced oil recovery application in tight-oil formations* [Ph.D. dissertation, Department of Civil and Environmental Engineering at the University of Alberta].

Tran, S., Yassin, M. R., Eghbali, S., Doranehgard, M. H., & Dehghanpour, H. (2021). Quantifying oil-recovery mechanisms during natural-gas huff 'n' puff experiments on ultratight core plugs. *SPE Journal, 26*, 498–514. https://doi.org/10.2118/200341-pa

Tran, T. (2012). *Bakken shale oil production trends*. Texas A & M University.

Tu, J., & Sheng, J. (2019). Experimental and numerical study of shale oil EOR by surfactant additives in fracturing fluid. *Proceedings of the 7th unconventional resources technology conference*. https://doi.org/10.15530/urtec-2019-295

Tu, J., & Sheng, J. J. (2020a). Experimental and numerical study of surfactant solution spontaneous imbibition in shale oil reservoirs. *Journal of the Taiwan Institute of Chemical Engineers, 106*, 169–182. https://doi.org/10.1016/j.jtice.2019.11.003

Tu, J., & Sheng, J. J. (2020b). Effect of pressure on imbibition in shale oil reservoirs with wettability considered. *Energy & Fuels, 34*(4), 4260–4272. https://doi.org/10.1021/acs.energyfuels.0c00034

Tuero, F., Crotti, M., & Labayen, I. (2017). Water imbibition EOR proposal for shale oil scenarios. *SPE Latin America and Caribbean petroleum engineering conference*. SPE-185560, https://doi.org/10.2118/185560-MS

Ucok, H., Ershaghi, I., & Olhoeft, G. R. (1980). Electrical resistivity of geothermal brines. *Journal of Petroleum Technology, 32*(4), 717–727. SPE-7878-PA.

Ullman, W. J., & Aller, R. C. (1982). Diffusion coefficients in near shore marine sediments. *Limnology Oceanography, 27*, 552–556.

Ulm, F., Vandamme, M., Bobko, C., Alberto Ortega, J., Tai, K., & Ortiz, C. (2007). Statistical indentation techniques for hydrated nanocomposites: Concrete, bone, and shale. *Journal of the American Ceramic Society, 90*(9), 2677–2692. https://doi.org/10.1111/j.1551-2916.2007.02012.x

Uzun, O., & Kazemi, H. (2019). Assessment of improved oil recovery by osmotic pressure in unconventional reservoirs: Application to Niobrara chalk and Codell sandstone. *Fuel, 306*, 121270, https://doi.org/10.1016/j.fuel.2021.121270

Uzun, O., Torcuk, M. A., & Kazemi, H. (2020). Enhanced oil recovery for liquid-rich unconventional shale reservoirs using low-salinity surfactant and wettability altering dilute surfactants. *SPE improved oil recovery conference*. https://doi.org/10.2118/200346-ms

Vargo, J., Turner, J., Vergnani, B., Pitts, M. J., Wyatt, K., Surkalo, H., & Patterson, D. (2000). Alkaline-surfactant-polymer flooding of the Cambridge Minnelusa field. *SPE Reservoir Evaluation & Engineering, 3*(6), 552–558. https://doi.org/10.2118/68285-PA

Vinsome, K., McGee, B. C. W., & Vermeulen, F. E. (1994). Electrical heating. *Journal of Canadian Petroleum Technology, 33*(4). PETSOC-94-04-04.

Vossoughi, S., Willhite, G., El Shoubary, Y., & Bartlett, G. (1983). Study of the clay effect on crude oil combustion by thermogravimetry and differential scanning calorimetry. *Journal of Thermal Analysis, 27*, 17–36.

Wan, T., Sheng, J. J., & Soliman, M. Y. (2013). Evaluate EOR potential in fractured shale oil reservoirs by cyclic gas injection. *Unconventional resources technology conference*, Denver, CO, 12–14 August, 1845–1854. https://doi.org/10.1190/urtec2013-187

Wan, T., Yu, Y., & Sheng, J. J. (2015). Experimental and numerical study of the EOR potential in liquid-rich shales by cyclic gas injection. *Journal of Unconventional Oil and Gas Resources, 12*, 56–67.

Wang, D., Butler, R., Liu, H., & Ahmed, S. (2011). Flow-rate behavior and imbibition in shale. *SPE Reservoir Evaluation & Engineering, 14*, 505–512.

Wang, D., Zhang, J., Butler, R., & Olatunji, K. (2016). Scaling laboratory-data surfactant-imbibition rates to the field in fractured-shale formations. *SPE Reservoir Evaluation & Engineering, 19*, 440–449.

Wang, L. K., Aulenbach, D. B., & Shammas, N. K. (2010). Treatment of seafood processing wastewater. In L. K. Wang, N. K. Shammas, W. A. Selke & D. B. Aulenbach (Eds.), *Flotation technology* (pp. 567–592). Humana Press. https://doi.org/10.1007/978-1-60327-133-2_17

Wang, L., Tian, Y., Yu, X., Wang, C., Yao, B., Wang, S., Winterfeld, P. H., Wang, X., Yang, Z., Wang, Y., Cui, J., & Wu, Y.-S. (2017). Advances in improved/enhanced oil recovery technologies for tight and shale reservoirs. *Fuel*, *210*, 425–445. https://doi.org/10.1016/j.fuel.2017.08.095

Wang, S., & Chen, S. (2016). A comprehensive evaluation of well completion and production performance in Bakken shale using data-driven approaches. *Paper presented at the SPE Asia Pacific hydraulic fracturing conference*, 24–26 August. Society of Petroleum Engineers. https://doi.org/10.2118/181803-MS

Wang, Y., Hu, J., Xie, W., & Zhang, Y. (2022). Optimization and analysis of CO2 HnP process in shale oil reservoirs using response surface methodology (RSM). *Geofluids*, *2022*, 5927853. https://doi.org/10.1155/2022/5927853

Wen, L., & Papadopoulos, K. D. (2001). Effects of osmotic pressure on water transport in W1/O/W2 emulsions. *Journal of Colloid and Interface Science*, *235*, 398–404.

Whitson, C. H., & Brulé, M. R. (2000). *Phase behavior.* Monograph Series, Society of Petroleum Engineer.

Wood, D. J., Lake, L. W., Johns, R. T., & Nunez, V. (2008). A screening model for CO2 flooding and storage in gulf coast reservoirs based on dimensionless groups. *SPE Reservoir Evaluation and Engineering*, *11*(3), 513–520. SPE-100021-PA.

Wuensche, R. (1978). Nitrogen injection for enhanced oil recovery. *Annual technical meeting*, Calgary, Alberta. https://doi.org/10.2118/78-29-44

Xiong, J., Li, Q., Tao, J., He, Y., Zhang, J., & Liu, J. (2010). Nano-injection technology application and understanding in injection wells. *Oil Drilling & Production Technology*, *32*(4), 78–80.

Xiong, X., & Sheng, J. J. (2021, August 26). An optimized experimental investigation of foam-assisted N2 HnP enhanced oil recovery in fractured shale cores. *All Days*. https://doi.org/10.2118/208431-MS

Xu, T., & Hoffman, T. (2013). Hydraulic fracture orientation for miscible gas injection EOR in unconventional oil reservoirs. *Society of Petroleum Engineers*. https://doi.org/10.1190/URTEC2013-189

Yakasai, F., Jaafar, M. Z., Bandyopadhyay, S., & Agi, A. (2021). Current developments and future outlook in nanofluid flooding: A comprehensive review of various parameters influencing oil recovery mechanisms. *Journal of Industrial and Engineering Chemistry*, *93*, 138–162. https://doi.org/10.1016/j.jiec.2020.10.017

Yan, G., Qin, Z., Marsh, S., Grebby, S., Mou, Y., Song, L., & Zhang, C. (2021). Nanopore size distribution heterogeneity of organic-rich shale reservoirs using multifractal analysis and its influence on porosity–permeability variation. *Energy & Fuels*, *35*(17), 13700–13711. https://doi.org/10.1021/acs.energyfuels.1c01654

Yang, P., Li, Z., Xia, B., Yuan, Y., Huang, Q., Liu, W., & Cheng, C. (2019). *Comprehensive review of Alkaline–Surfactant–Polymer (ASP)-enhanced oil recovery (EOR)* (pp. 858–872). https://doi.org/10.1007/978-981-10-7560-5_79

Yang, Z., Zhu, J., Li, X., Luo, D., Qi, S., & Jia, M. (2017). Experimental investigation of the transformation of oil shale with fracturing fluids under microwave heating in the presence of nanoparticles. *Energy & Fuels*, *31*(10), 10348–10357. https://doi.org/10.1021/acs.energyfuels.7b00908

Yassin, M. R., Alinejad, A., Asl, T. S., & Dehghanpour, H. (2022). Unconventional well shut-in and reopening: Multiphase gas-oil interactions and their consequences on well performance. *Journal of Petroleum Science and Engineering*, *215*, 110613.

Yu, H., Xu, H., Fan, J., Zhu, Y.-B., Wang, F., & Wu, H. (2021). Transport of shale gas in microporous/nanoporous media: Molecular to pore-scale simulations. *Energy & Fuels*, *35*(2), 911–943. https://doi.org/10.1021/acs.energyfuels.0c03276

Yu, W., Lashgari, H. R., & Sepehrnoori, K. (2014). Simulation study of CO2 HnP process in Bakken tight oil reservoirs. *All Days*, SPE-169575-MS. https://doi.org/10.2118/169575-MS

Yu, W., Lashgari, H. R., & Sepehrnoori, K. (2017). Optimizing HnP gas injection scenario in shale reservoirs. *Journal of Petroleum Science and Engineering*, *149*, 361–372.

Yu, Y., García, B. F., & Saraji, S. (2020). Surfactant viscoelasticity as a key parameter to improve supercritical CO2 foam stability/foamability and performance in porous media. *Journal of Non-Newtonian Fluid Mechanics*, *282*, 104311. https://doi.org/10.1016/j.jnnfm.2020.104311

Yu, Y., Li, L., & Sheng, J. J. (2017). A comparative experimental study of gas injection in shale plugs by flooding and HnP processes. *Journal of Natural Gas Science and Engineering*, *38*, 195–202.

Yu, Y., Meng, X., & Sheng, J. J. (2016). Experimental and numerical evaluation of the potential of improving oil recovery from shale plugs by nitrogen gas flooding. *Journal of Unconventional Oil and Gas Resources*, *15*, 56–65. https://doi.org/10.1016/j.juogr.2016.05.003

Yu, Y., & Sheng, J. J. (2017). A comparative experimental study of IOR potential in fractured shale reservoirs by cyclic water and nitrogen gas injection. *Journal of Petroleum Science and Engineering*, *149*, 844–850.

Zabala, R., Franco, C. A., & Cortés, F. B. (2016, April 11). Application of nanofluids for improving oil mobility in heavy oil and extra-heavy oil: A field test. *All Days*. https://doi.org/10.2118/179677-MS

Zelenev, A. S. (2011, April 11). Surface energy of North American shales and its role in interaction of shale with surfactants and microemulsions. *All Days*. https://doi.org/10.2118/141459-MS

Zendehboudi, S., & Bahadori, A. (2017). Production methods in shale oil reservoirs. In *Shale oil and gas handbook*. https://doi.org/10.1016/b978-0-12-802100-2.00008-3

Zeng, T., Miller, C. S., & Mohanty, K. K. (2018). Application of surfactants in shale chemical EOR at high temperatures. *SPE improved oil recovery conference*. SPE-190318, https://doi.org/10.2118/190318-MS

Zeng, T., Miller, C. S., & Mohanty, K. K. (2019). Chemical blend-CO2 HnP for enhanced oil recovery in shales. *Proceedings of the 7th unconventional resources technology conference*. https://doi.org/10.15530/urtec-2019-362

Zhang, F., Adel, I. A., Saputra, I. W., Chen, W., & Schechter, D. S. (2019). Numerical investigation to understand the mechanisms of CO2 EOR in unconventional liquid reservoirs. *Day 2 Tue*, 1 October, D021S021R007. https://doi.org/10.2118/196019-MS

Zhang, F., & Emami-Meybodi, H. (2022). A type-curve method for two-phase flowback analysis in hydraulically fractured hydrocarbon reservoirs. *Journal of Petroleum Science and Engineering*, *209*, 109912. https://doi.org/10.1016/j.petrol.2021.109912

Zhang, J. C., Lin, L. M., & Li, Y. X. (2012). Classification and evaluation of shale oil. *Earth Science Frontiers*, *19*(5), 322–331 (in Chinese with English abstract).

Zhang, W., Ning, Z., Wang, Q., Cheng, Z., Lyu, C., Wang, Y., Liu, Z., & Wang, H. (2023). Experimental study on the effects of an electric field on the pore characterization in anode, middle and cathode regions of tight sandstone samples. *Geoenergy Science and Engineering*, *223*, 211500.

Zhang, X.-S., Wang, H.-J., Ma, F., Sun, X.-C., Zhang, Y., & Song, Z.-H. (2016). Classification and characteristics of tight oil plays. *Petroleum Science*, *13*(1), 18–33. https://doi.org/10.1007/s12182-015-0075-0

Zhang, Y., Lashgari, H. R., Di, Y., et al. (2017). Capillary pressure effect on phase behavior of CO2/hydrocarbons in unconventional reservoirs. *Fuel*, *197*, 575–582.

Zhao, J.-Z., Jia, H., Pu, W.-F., Wang, L.-L., & Peng, H. (2012). Sensitivity studies on the oxidation behavior of crude oil in porous media. *Energy & Fuels*, *26*, 6815–6823.

Zheng, S., Sharma, M., Watts, R., & Ahmad, Y. (2020, August 30). A nanoparticle assisted CO2 HnP field test in the eagle ford shale. *Day 1 Mon*, 31 August. https://doi.org/10.2118/200375-MS

Zhou, W., Xin, C., Chen, S., Yu, Q., & Wang, K. (2020). Polymer-enhanced foam flooding for improving heavy oil recovery in thin reservoirs. *Energy & Fuels*, *34*(4), 4116–4128. https://doi.org/10.1021/acs.energyfuels.9b04298

Zhou, Z., Abass, H., Li, X., & Teklu, T. (2016). Experimental investigation of the effect of imbibition on shale permeability during hydraulic fracturing. *Journal of Natural Gas Science and Engineering*, *29*, 413–430.

Zhu, P., Balhoff, M. T., & Mohanty, K. K. (2015). Simulation of fracture-to-fracture gas injection in an oil-rich shale. *Paper presented at the SPE annual technical conference and exhibition*, Houston, TX. SPE-175131-MS.

Zoback, M. D., & Kohli, A. H. (2019). *Unconventional reservoir geomechanics: Shale gas, tight oil, and induced seismicity* (1st ed.). Cambridge University Press. https://doi.org/10.1017/9781316091869

Zou, C., Zhu, R., Liu, K., Su, K., Bai, B., Zhang, X., Yuan, X., & Wang, J. (2012). Tight gas sandstone reservoirs in China: Characteristics and recognition criteria. *Journal of Petroleum Science and Engineering*, *88–89*, 82–91. https://doi.org/10.1016/j.petrol.2012.02.001.

Zuloaga, P., Yu, W., Miao, J., & Sepehrnoori, K. (2017). Performance evaluation of CO2 HnP and continuous CO2 injection in tight oil reservoirs. *Energy*, *134*, 181–192.

8 Production Forecasting in Unconventional Resources

Ekrem Alagoz, Yaser Alghawi, Mubarek Alpkiray, and Emre Can Dundar

8.1 INTRODUCTION TO UNCONVENTIONAL RESERVOIRS

Unconventional reservoirs present unique challenges for production forecasting. Most of the methods applied in conventional reservoirs are not valid due to the different mechanisms and flow regimes encountered in unconventional reservoirs, as well as the invalidity of many of the fundamental assumptions upon which conventional reservoir analysis techniques are based.

Unconventional reservoirs include, but are not limited to (Clarkson, 2021):

- Low-permeability oil and gas reservoirs (0.001–< 0.1 md matrix permeability)
- Coalbed methane (CBM) reservoirs (< 0.001 md matrix permeability)
- Shale gas and oil reservoirs (< 0.001 md matrix permeability)

The first and third types are commonly developed using multi-fractured horizontal wells (MFHWs, Figure 8.1) due to the low matrix permeability. Whereas production from conventional reservoirs is characterized mainly by a short period of transient radial flow followed by a boundary-dominated flow (BDF) regime, production from unconventional shale plays usually is characterized by a long period of transient linear flow (TLF), followed by transitional flow and finally by fracture interference in the stimulated reservoir volume (SRV) region. Most classical pressure transient analysis techniques are not applicable to such reservoirs because of the very long time it takes to reach stabilized conditions, which makes it impractical to shut in wells. Hence, the classical decline curve analysis techniques, alongside rate transient analysis (RTA) and numerical simulation, are the most common methods for analyzing and forecasting production from unconventional reservoirs.

8.1.1 Chapter Structure

Section 2 provides a review of the commonly used decline curve analysis methods and their advantages and disadvantages when applied in unconventional reservoirs.

DOI: 10.1201/9781003319955-8

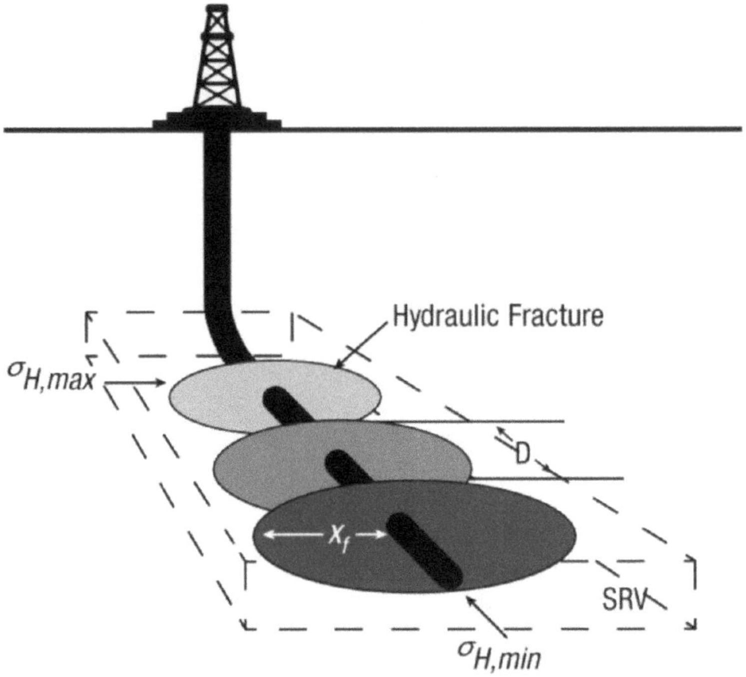

FIGURE 8.1 Multi-fractured horizontal well schematic.

Source: Economides, 2013

Section 3 focuses on rate transient analysis methods and provides a recommended workflow at the end. Section 4 introduces the most common unconventional reservoir numerical simulation models and their applications.

8.2 DECLINE CURVE ANALYSIS

8.2.1 Arps Decline Models

Arps (1945) reviewed the development of decline curve analysis and presented a classification of decline curves based on their "loss ratios," alongside a series of decline charts for straight line extrapolation for both rate-time and rate-cumulative curves. Through empirical relationships, the following definitions were made:

Decline rate, D:

$$D = -\frac{\frac{dq}{dt}}{q} \tag{8.1}$$

Parameter b, change in loss ratio with time:

$$b = \frac{d\left(\frac{1}{D}\right)}{dt} \tag{8.2}$$

Through the integration of *b*, the following formula for Arps's decline model was presented:

$$q = \frac{q_i}{\left(1 + bD_i t\right)^{1/b}} \tag{8.3}$$

Integration of flow rate over time interval yields cumulative production.

$$Q = \int_0^t q\,dt \tag{8.4}$$

Most of the wells that Arps analyzed had a constant value of *b*. Three distinct decline models were developed based on the value of *b*. They are presented in the following sections.

8.2.2 Exponential Decline Model

The exponential decline is defined by a constant decline rate ($b = 0$), which results in the following formulas for flow rate and cumulative production:

$$q = q_i e^{-D_i t} \tag{8.5}$$

$$Q = \frac{q_i - q}{D_i} \tag{8.6}$$

The exponential decline can be characterized by a straight line on either the semi-log plot of rate versus time (Figure 8.2), or the Cartesian plot of rate versus cumulative production (Figure 8.3).

8.2.3 Harmonic Decline Model

The harmonic decline is a special case of Arps's decline in which $b = 1$. Rate and cumulative production can be calculated by:

$$q = \frac{q_i}{1 + D_i t} \tag{8.7}$$

$$Q = \frac{q_i}{D_i} \ln \frac{q_i}{q} \tag{8.8}$$

A semi-log scale plot of rate versus cumulative production would produce a straight line for the harmonic decline case (Figure 8.4).

8.2.4 Hyperbolic Decline Model

Hyperbolic decline is defined by $0 < b < 1$:

$$q = \frac{q_i}{\left(1 + bD_i t\right)^{1/b}} \tag{8.9}$$

$$Q = \frac{q_i^b}{(1-b)D_i}\left(q_i^{1-b} - q^{1-b}\right) \tag{8.10}$$

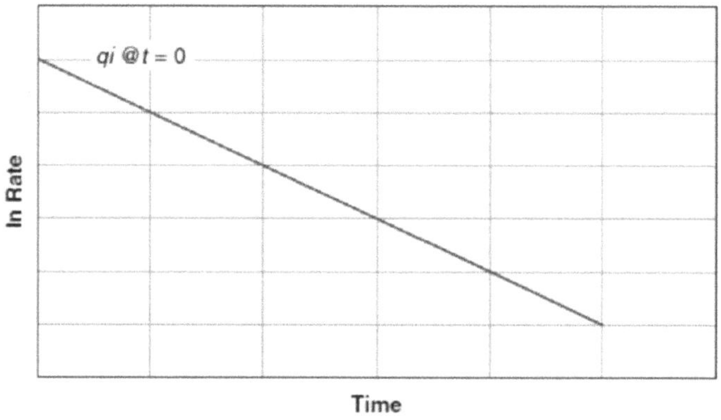

FIGURE 8.2 Semi-log plot of rate versus time, exponential decline.

Source: Poston et al., 2019

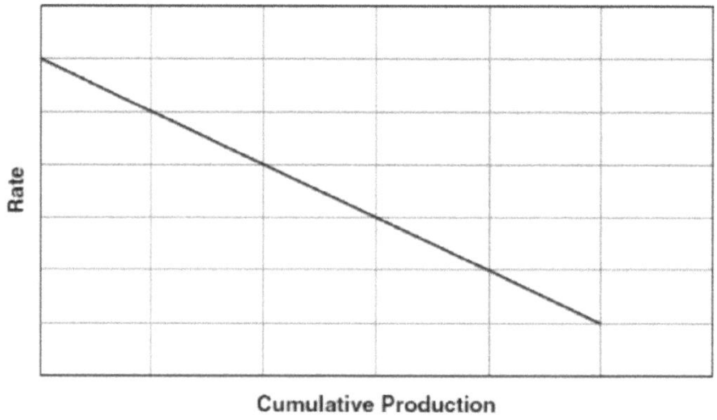

FIGURE 8.3 Cartesian plot of rate versus cumulative production, exponential decline.

Source: Poston et al., 2019

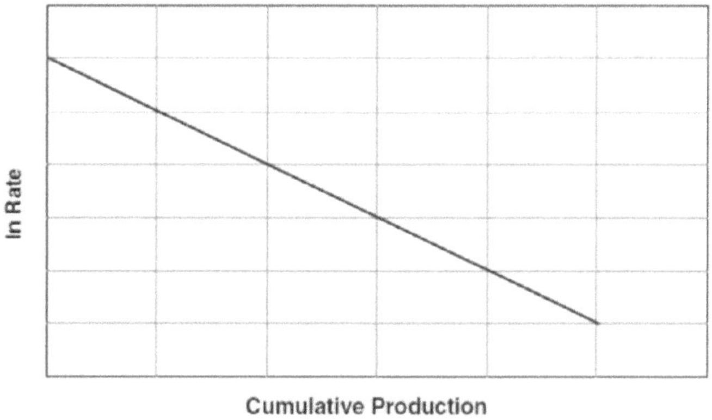

FIGURE 8.4 Semi-log plot of rate versus cumulative production, harmonic decline.

Source: Poston et al., 2019

8.2.5 Problems Associated with b > 1 Values

The original Arps decline models were developed for wells in conventional reservoirs with sufficient production time to reach boundary dominated flow regime. Typical values of b-factor for transient linear and transient bilinear flow regimes are 2 and 4, respectively. Numerous operators have used the Arps model with b > 1 to forecast production from unconventional reservoirs over the whole lifespan of the wells. This leads to physically impossible EUR estimations as:

$$\lim_{t\to\infty} Q(t) = \lim_{q\to 0} \frac{q_i^b}{D_i(1-b)}\left(q_i^{1-b} - q^{1-b}\right) = \infty \tag{8.11}$$

Although it is possible to have b = 2 or b = 4 during transient flow regimes, there must be a flow regime with b < 1 that succeeds them. It is strongly advised to not use b > 1 for the last segment in any forecast from unconventional reservoirs for EUR estimation.

8.2.6 Fetkovich Type Curves

Fetkovich (1980) developed a set of type curves for constant bottom-hole pressure production based on the dimensionless rate q_{Dd} and dimensionless time t_{Dd} variables. Their formulas for oil wells are:

$$q_{Dd} = \frac{141.2 q \mu B \left[\ln\left(\frac{r_e}{r_{wa}}\right) - \frac{1}{2}\right]}{kh\left(p_i - p_{wf}\right)} \tag{8.12}$$

$$t_{Dd} = \frac{0.01266kt}{\phi\mu c r_{wa}^2 \left[\left(\frac{r_e}{r_{wa}}\right)^2 - 1\right]\left(\frac{r_e}{r_{wa}} - \frac{1}{2}\right)} \tag{8.13}$$

$$r_{wa} = r_w r^{-s} \tag{8.14}$$

The type curves (Figure 8.5) show two distinctive flow regimes: transient flow (left) and boundary-dominated flow (right). The boundary-dominated flow stems represent different Arps b-values. While the original Fetkovich type curves were designed to model transient radial flow and boundary dominated flow regimes, Chen and Teufel (2000) modified the type curves to add a transient linear flow stem (b = 2, Figure 8.6 (a)). Eleiott et al. (2019) further modified the type curves, adding transient flow b stems ranging from 1 to 4 (Figure 8.6 (b)).

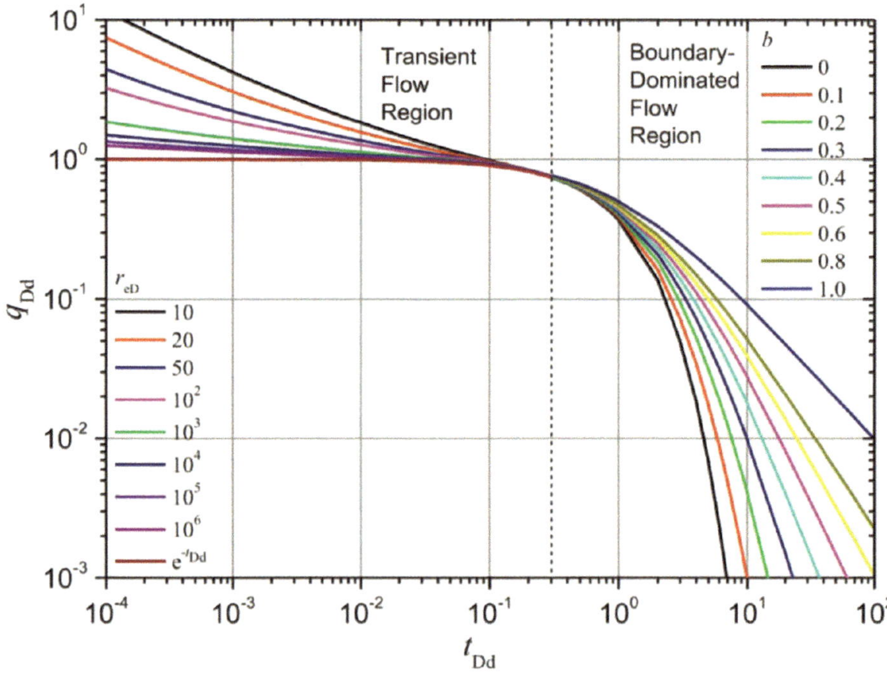

FIGURE 8.5 Fetkovich dimensionless flow rate type curves.

Source: Fetkovich, 1980

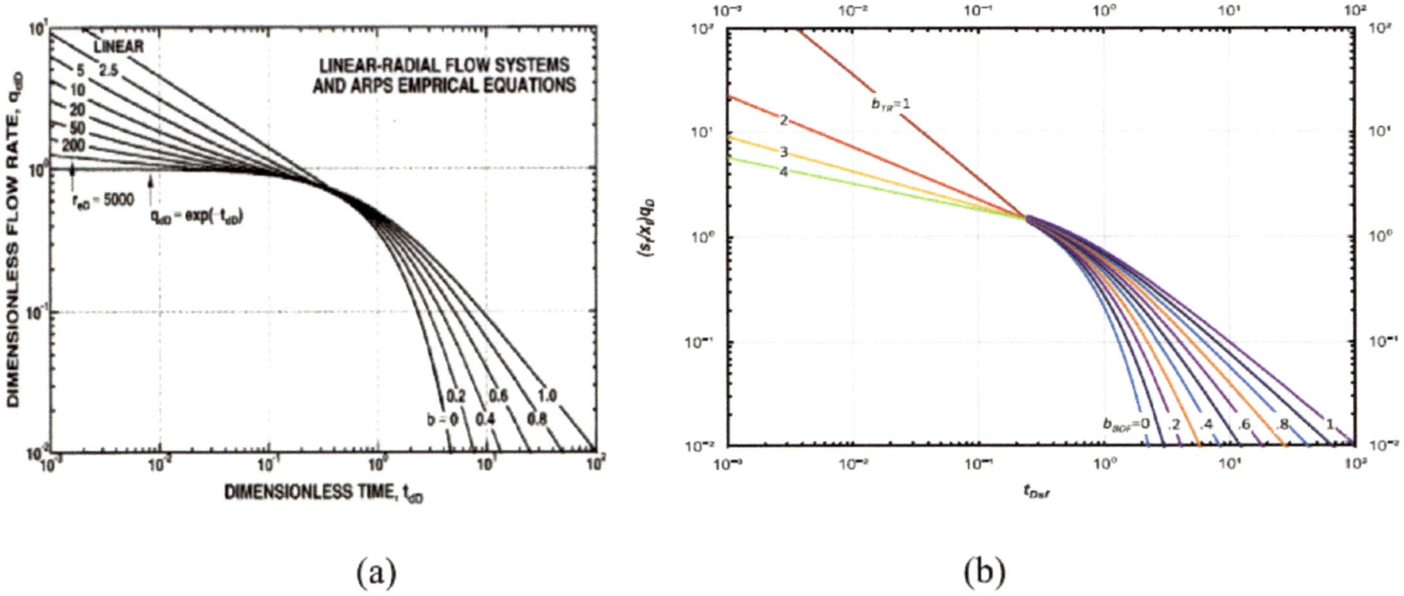

FIGURE 8.6 (a) Chen and Teufel modification of Fetkovich type curve; (b) Eleiott's modification of Fetkovitch type curve.

Source: Lee, 2022a

8.2.7 Power-Law Exponential Decline Model

Ilk et al. (2008) presented a non-hyperbolic model to estimate unconventional reserves. Their formulation is based on modified parameters D and b. These are given by:

$$D(t) = D_\infty + \hat{D}_i n t^{n-1} \tag{8.15}$$

$$b(t) = \frac{-\hat{D}_i (n-1) n t^n}{\left(D_\infty + \hat{D}_i n t^n\right)^2} \tag{8.16}$$

Using these definitions, the power-law exponential decline formula is given as:

$$q(t) = \hat{q}_i \exp\left[-D_\infty t - \hat{D}_i t^n\right] \tag{8.17}$$

where D_∞ controls late time behavior and can be set to 0 for production data that does not exhibit boundary-dominated flow. Since the last equation cannot be integrated analytically, numerical methods have to be used to calculate cumulative production. Figure 8.7 shows a comparison of the behavior of the power-law exponential with Arps hyperbolic decline model.

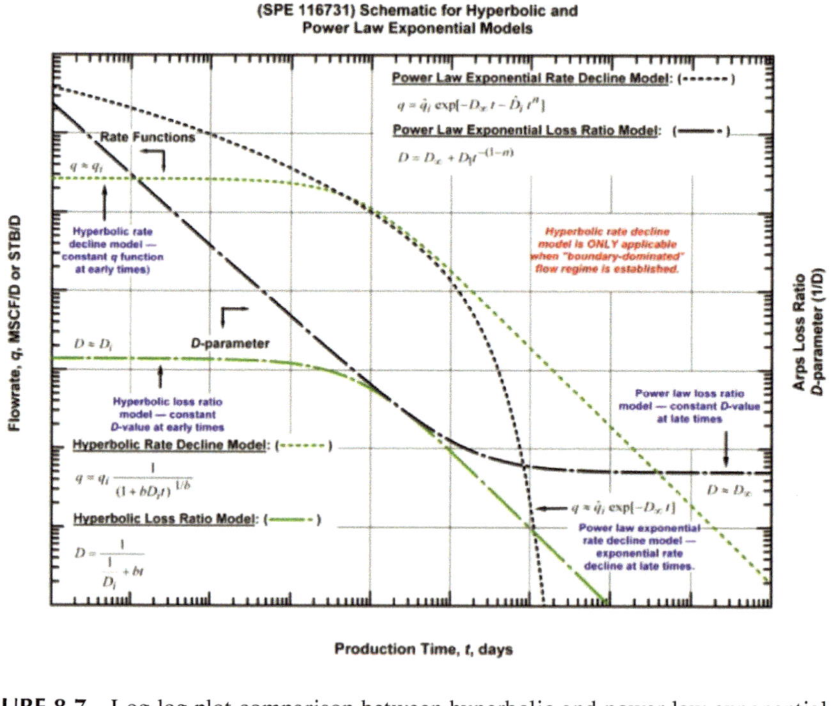

FIGURE 8.7 Log-log plot comparison between hyperbolic and power-law exponential rate decline and loss ratio models.

Source: Ilk et al., 2008

8.2.8 Stretched Exponential Decline Model

Valkó and Lee (2010) introduced the stretched exponential decline curve model (SEPD) for improved production forecasts in unconventional gas wells. This model is based on the idea that the actual production decline is determined by a multitude of individual contributing volumes in exponential decay, yet with a specific distribution of characteristic time constants. The distribution is characterized by two parameters (n,τ). The following formulas have been introduced by the authors to describe the model:

$$\text{Governing differential equation: } \frac{dq}{dt} = -n\left(\frac{t}{\tau}\right)^n \frac{q}{t} \tag{8.18}$$

$$\text{Rate as a function of time: } q = q_0 exp\left[-\left(\frac{t}{\tau}\right)^n\right] \tag{8.19}$$

Cumulative production as a function of time:

$$Q = \frac{q_0\tau}{n}\left\{\Gamma\left[\frac{1}{n}\right] - \Gamma\left[\frac{1}{n},\left(\frac{t}{\tau}\right)^n\right]\right\} \tag{8.20}$$

$$\text{Estimated ultimate recovery (EUR): } EUR = \frac{q_0\tau}{n}\Gamma\left[\frac{1}{n}\right] \tag{8.21}$$

$$\text{Recovery potential: } R_p = 1 - \frac{Q}{EUR} = \frac{1}{\Gamma\left[\frac{1}{n}\right]}\Gamma\left[\frac{1}{n}, -\ln\frac{q}{q_0}\right] \tag{8.22}$$

For positive values of n,τ and q_0, this model yields finite EUR values, in contrast to the Arps model, which produces unbounded and physically impossible estimates of EUR for $b \geq 1$.

Yu et al. (2013) noted that SEPD will most likely result in the underestimation of EUR in reservoirs with permeability values in the range of 0.0001 through 0.1 mD. Therefore, they provided a modified SEPD (YM-SEPD) method to overcome this limitation. The authors introduced a new specialized plot to find the relevant parameters and analyzed hundreds of wells from various unconventional basins. They reported that reliable EURs and production profiles can be predicted by employing two to three years of production history and recommended coupling their method with Duong's method for analyzing wells with less than two years of production history.

The specialized straight-line plot is based on the equation:

$$\ln\left[\frac{q_0}{q(t)}\right] = \tau^{-n}t^n \tag{8.23}$$

and its purpose is to find the parameters n and τ, as well as to eliminate production data that does not follow the SEPD model. Figure 8.8 shows an example of the specialized plot for a dry gas well.

The authors provided several synthetic and actual production data examples for dry gas, wet gas, retrograde gas, volatile oil, and black oil fluid systems. Figures 8.9 and 8.10 show production and production history and the material balance dime diagnostic plot, respectively, for the dry gas well example for which the specialized plot was presented in Figure 8.8. The permeability of the reservoir is 0.05 mD. The reported relative errors in EUR estimations are presented in Table 8.1.

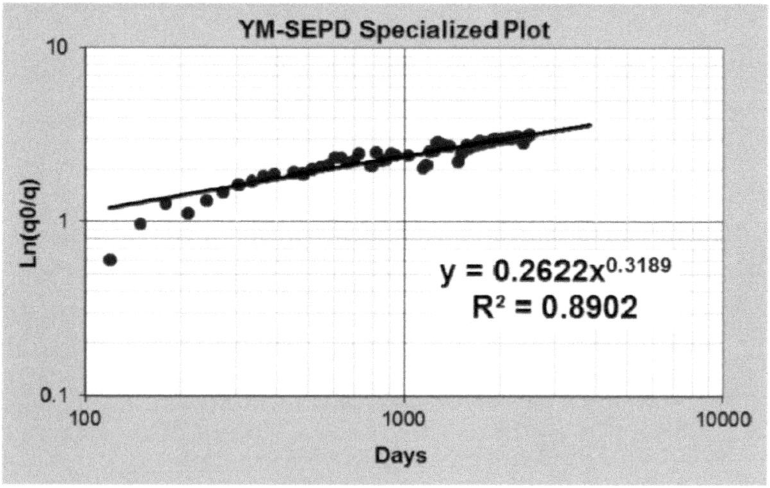

FIGURE 8.8 YM-SEPD specialized plot for example dry gas well.

Source: Yu et al., 2013

TABLE 8.1
YM-SEPD Results for Example Dry Gas Well

Data Available Used (Years)	% of EUR Remaining	EUR Relative Error (%)
~1	69.0	−57.9
~1 + pseudo rate (Duong)	69.0	−30.6
~2	58.7	−20.7
~2 + pseudo rate (Duong)	58.7	−11.0
~3	52.3	−9.6
~4	46.9	−9.0
All Data	33.7	−1.2

Source: Yu et al., 2013

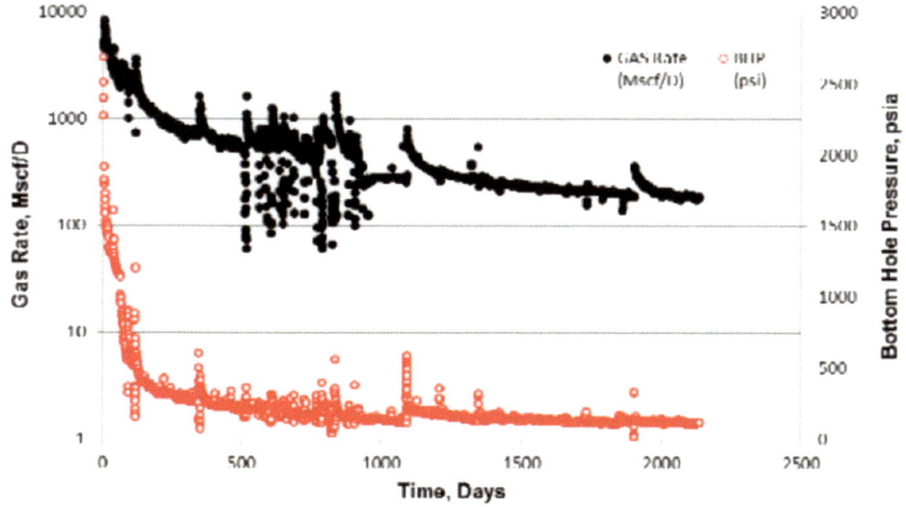

FIGURE 8.9 Production and pressure history for example dry gas well.

Source: Yu et al., 2013

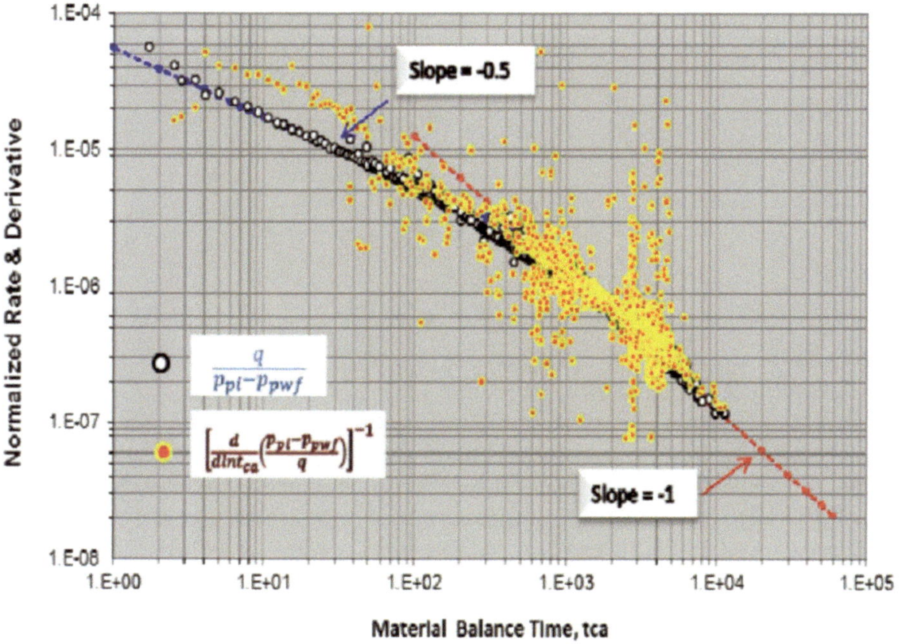

FIGURE 8.10 Diagnostic plot of flow regimes for example dry gas well.

Source: Yu et al., 2013

8.2.9 Duong's Model

Duong (2010) developed and tested a new model for tight/shale reservoir decline curve analysis. The empirically derived model is based on long-term linear flow in a large number of wells in tight and shale gas reservoirs.

For a well with a prolonged linear or bilinear flow regime, the gas flow rate can be estimated by:

$q = q_1 t^{-n}$ where $n = 1/2$ for linear flow, $n = 1/4$ for bilinear flow, q_1 is the initial flow rate.

Cumulative production is $Q = \int_0^t q\,dt = q_1 \frac{t^{1-n}}{(1-n)}$ (8.24)

which results in $\frac{q}{Q} = \frac{(1-n)}{t}$ (8.25)

The author tested the last equation on field data from various shale gas plays. Plots yielded a log-log straight line with a negative slope $-m$ and an intercept of a. Thus, the relationship becomes:

$$\frac{q}{Q} = a t^{-m} \tag{8.26}$$

m always has a positive value and is greater than 1 for shale reservoirs, whereas a value less than 1 might indicate a conventional tight well. Equations for q and G_p are presented here:

$$q = q_1 t^{-m} e^{\frac{a}{1-m}\left(t^{1-m}-1\right)} + q_\infty \tag{8.27}$$

$$Q = \frac{q_1}{a} e^{\frac{a}{1-m}\left(t^{1-m}-1\right)} \tag{8.28}$$

$$EUR = \frac{q_{final}}{a} t_{final}^{m} \tag{8.29}$$

Based on the developed model, a set of type curves has been presented in a dimensionless form. They are shown in Figure 8.12.

Several field examples were presented, including a retrograde gas case, and a Bakken oil case. The ranges of a and m for the retrograde case were 1.02–1.67 and 1.04–1.21, respectively, whereas for the Bakken case, a and m values were 1.7894 and 1.1477, respectively. Plots for both cases are shown in Figures 8.13 and 8.14.

The author presented the following step-by-step workflow for individual well analysis:

1. Data check, correction of rates based on average pressures
2. a and m determination using $\frac{q}{G_p}$ versus time log-log plot
3. q_1 determination using a plot of flow rate against $t(a, m)$
4. Reserve estimation using minimum economic rate

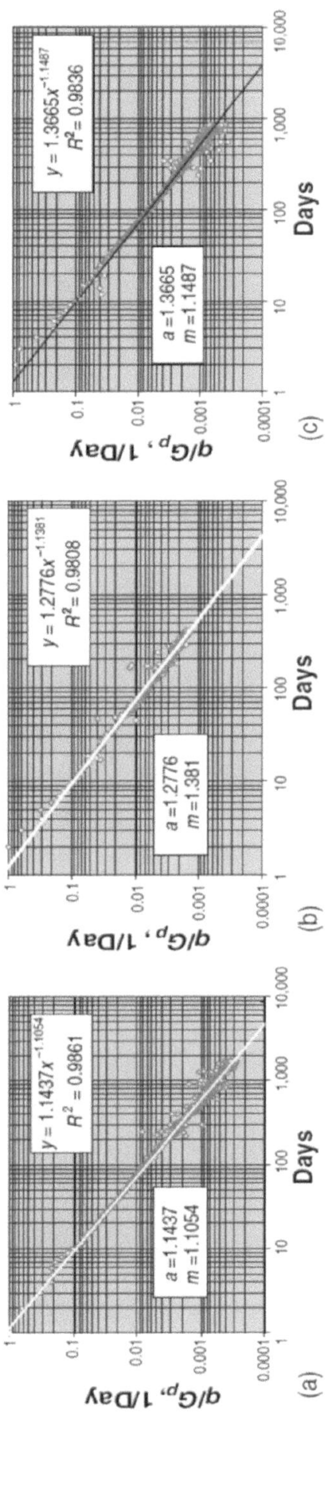

FIGURE 8.11 A log-log plot of q/Gp vs. t for various shale-gas wells with > 5 years of production.

Source: Duong, 2010

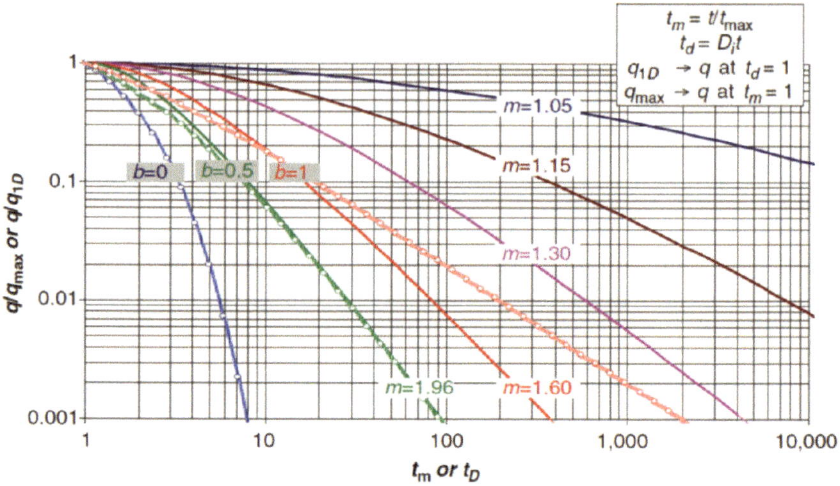

FIGURE 8.12 Comparison between Arps's and Duong's type curves for dimensionless rate and time.

Source: Duong, 2010

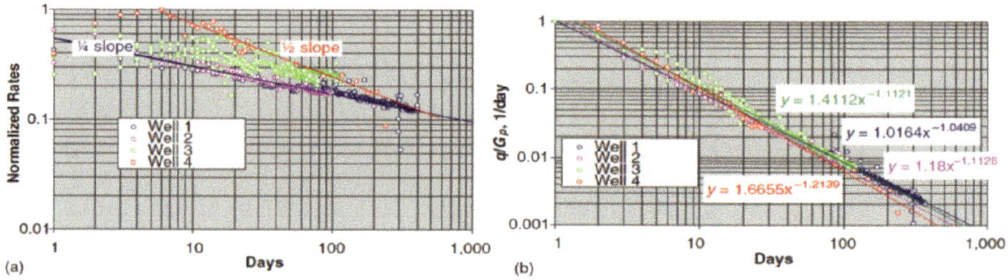

FIGURE 8.13 Condensate/shale-gas example. Flow regimes varying from linear to bilinear were attributed to potential liquid dropout in the fracture.

Source: Duong, 2010

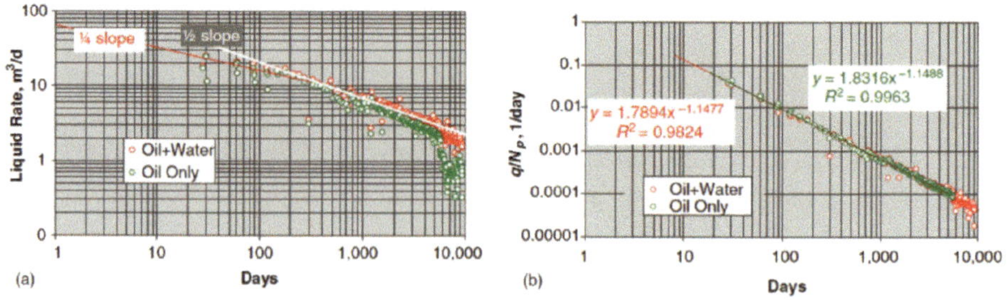

FIGURE 8.14 Bakken oil example of fractured shale with 26 years of monthly production data.

Source: Duong, 2010

Joshi and Lee (2013) proposed a modification to the Duong model to better fit data in the boundary-dominated flow regime. They observed that a non-zero value of q_∞ can produce unrealistic results, particularly for historical production data in the range of 6 to 12 months. The modification includes forcing the regressed line through the origin ($q_\infty = 0$) for wells with < 18 months of production data, which leads to the minimum error in remaining production estimates (Figure 8.15).

Also, the authors suggested switching to an Arps BDF model after a specific rate or time to account for this flow regime. A b-value of 0.4 was used by the authors, whereas they stated that finding the right value for the decline rate at the time of the switch needs more research. A D_{switch} value of 5% was used by the authors, and improved match results were observed after using the switched version (Figure 8.16).

8.2.10 Logistic Growth Decline Model

Clark et al. (2011) presented a new decline model on the basis of logistic growth. The model is based on the principle of population growth that is possible to a certain size and incorporates known physical oil and gas quantities into the forecast to constrain reserve estimated to a reasonable amount. The production rate and cumulative production can be calculated by:

$$q = \frac{Knbt^{n-1}}{\left(a + t^n\right)^2} \tag{8.30}$$

$$Q = \frac{Kt^n}{\left(a + t^n\right)^2} \tag{8.31}$$

8.2.11 Multi-Segment Arps Decline Model

The multi-segment Arps model is based on the idea that unique flow regimes require unique Arps segments with b-values that honor the flow regime. Lee (2021) presented a comprehensive theoretical and practical basis for applying multi-segment Arps decline models for multi-fractured horizontal wells (MFHWs) based on the common flow regimes typically encountered in these wells.

The typical flow regimes observed as MFHWs often include (1) an increasing rate period (ramp-up) in which the well produces back some of the fracture fluid, causing a steady decline in bottom-hole pressure (BHP) before stabilizing; (2) a transient flow period after the ramp-up; (3) a transition flow regime following the first influence of no-flow boundaries (interference between adjacent fractures); and (4) boundary-dominated flow (BDF), in which the well is draining the entire simulated reservoir volume. Figure 8.17 shows a production history that includes all but the first of the mentioned flow regimes.

Transient linear flow: A close approximation of transient flow is common in hydraulically fractured wells producing from low and ultra-low permeability reservoirs. It will span the time range starting at the end of early fracture fluid cleanup

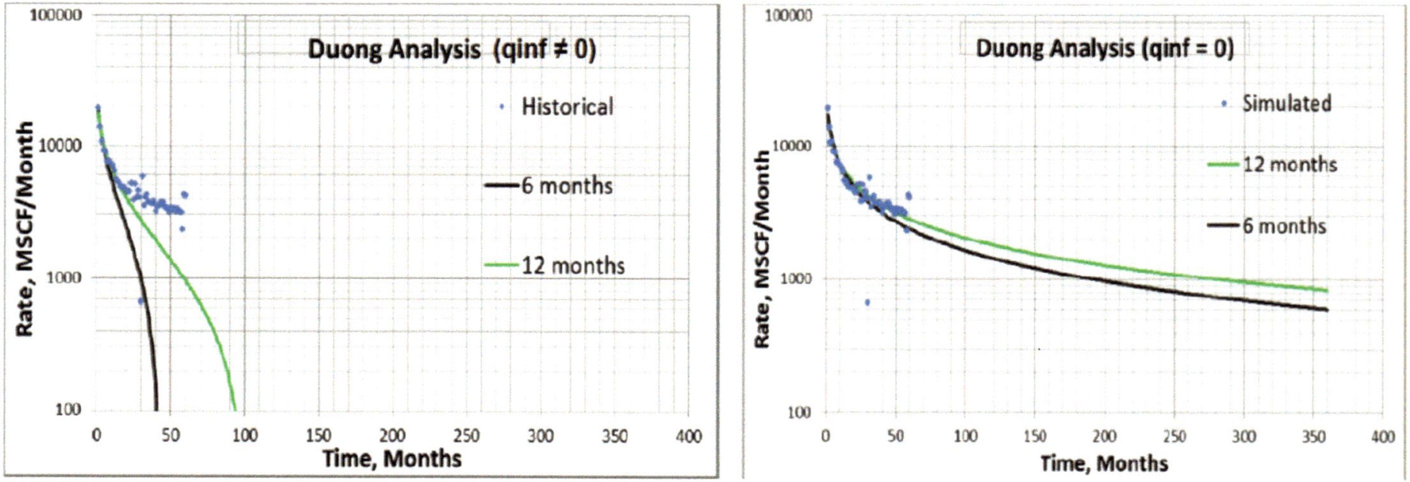

FIGURE 8.15 Comparison between the original and modified versions of Duong's model.

Source: Joshi & Lee, 2013

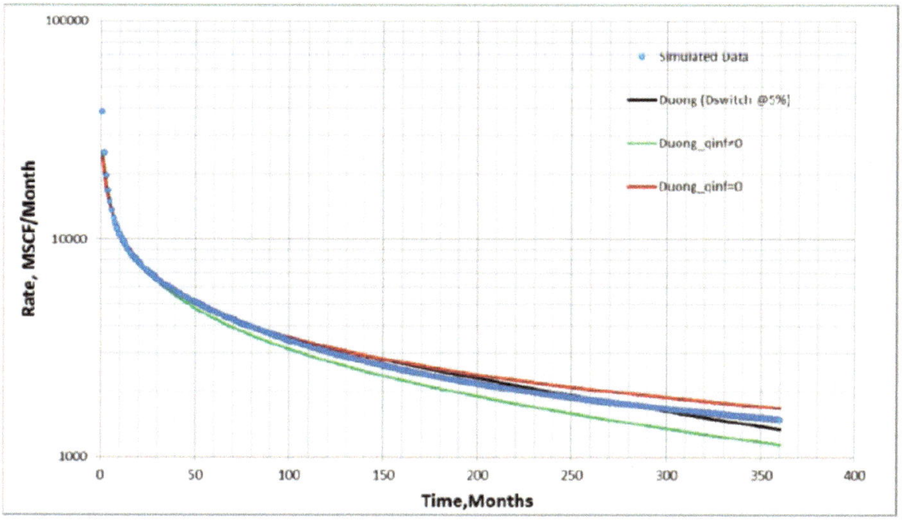

FIGURE 8.16 Comparison of the original, modified, and modified with D_{switch} = 5% versions of the Duong model for a Barnett shale simulation.

Source: Joshi & Lee, 2013

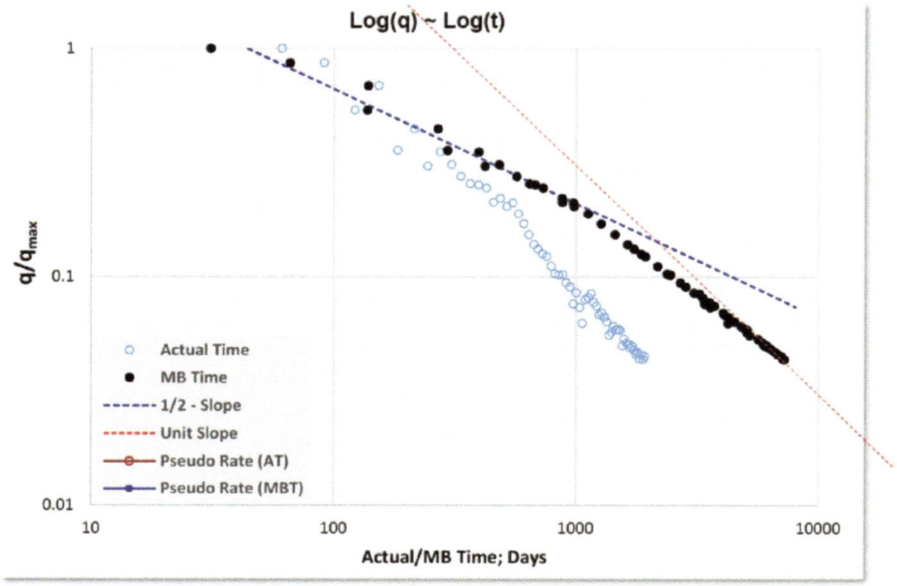

FIGURE 8.17 Production profile with plots of normalized rate versus actual time and material balance time.

Source: Lee, 2021

and ending with interference between adjacent fractures for dimensionless fracture conductivity, F_{CD}, greater than a threshold value.

$$F_{CD} = \frac{k_f w_f}{k x_f} > 300 \tag{8.32}$$

This threshold level is almost always exceeded in ultra-low permeability reservoirs. The modeling equation for transient linear flow for variable rate production at or near constant BHP is:

$$\frac{p_i - p_{wf}}{q} = 25.54 \frac{B}{A_f}\left(\frac{\mu t}{k\phi c_t}\right)^{0.5} + 141.2 \frac{B\mu}{kh} S_f = a_1\sqrt{t} + b_1 S_f \tag{8.33}$$

where A_f is the cross-sectional area perpendicular to flow and S_f is the fracture skin. From the equation, we can see that a log-log plot of $\frac{p_i - p_{wf}}{q} vst$ will exhibit a straight line with a slope of 0.5 at this flow regime. The corresponding value of Arps b factor will be (1/0.5) = 2. This behavior will continue until adjacent fractures begin to interfere with each other.

Transient bilinear flow: For fractures with lower conductivity ($F_{CD} < 10-50$), which is unlikely in ultra-low permeability reservoirs, the transient bilinear flow may occur. For wells producing at or near constant BHP:

$$\frac{p_i - p_{wf}}{q} = 44.1 \frac{B\mu}{h\left(w_f k_f\right)^{0.5}}\left(\frac{t}{\phi\mu k c_t}\right)^{0.25} + 141.2 \frac{B\mu}{kh} S_f = a_2 t^{0.25} + b_1 S_f \tag{8.34}$$

This indicates that a log-log plot of q^{vst} will yield a straight line with a slope of 0.25, resulting in an Arps b factor value of 4.

Boundary-dominated flow: A general form for the boundary-dominated flow equation (constant rate production at pseudo–steady state (PSS) flow regime) is:

$$\frac{p_i - p_{wf}}{q} = a_3 t + b_3 \tag{8.35}$$

This implies that a log-log plot of $\frac{p_i - p_{wf}}{q} vst$ will result in a straight line with a slope of 1. Whereas this implies an Arps b-factor of unity, this is misleading because the PSS equation is applicable for constant rate production only, which is highly unlikely in practice. The actual value of b in PSS could be anything between 0 and 1.

Material balance time (MBT), applications, and limitations: Material balance time is defined as the cumulative production divided by the instantaneous rate. MBT has a valuable property as it transforms a variable-rate production profile into an equivalent constant-rate production profile. This transformation is especially important in the identification of BDF. For constant BHP production, log-log rate versus time does not exhibit a characteristic or constant slope starting at BDF onset. However, plotting versus MBT would yield a unit slope at the BDF while still maintaining the characteristic ½ slope during transient linear flow (Figure 8.18).

While a material balance time plot is useful for flow-regime identification, it has its limitations. For early times, the cumulative production may be affected by

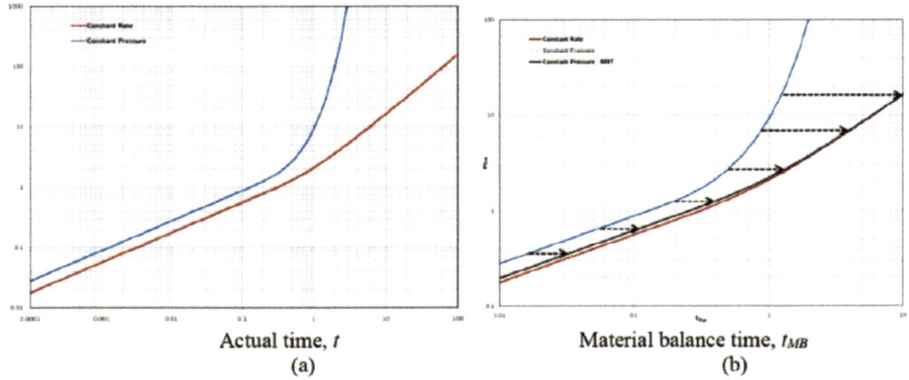

FIGURE 8.18 (a) Reciprocal rate versus time for constant rate and constant pressure production from a hydraulically fractured well; (b) Reciprocal rate versus t_{MB} for both constant rate and constant pressure.

Source: Lee, 2021

fracture fluid cleanup, decreasing BHP, or significant fracture skin. These effects will have permanent effects on cumulative production, and some time will be needed for these errors to become unimportant. For this reason, the log-log q versus t plot is more reliable for identifying flow regimes at early times (Figure 8.19 (a)). Also, operational difficulties at late times, such as liquid loading for gas wells or slugging for oil wells, would introduce a "zig-zag" effect in late-time data points, which, in turn, would form a false straight line with a unit slope, yielding an erroneous impression of BDF (Figure 8.19 (b)).

8.2.12 Comparison of Models

Each of the presented methods has certain advantages and disadvantages when applied to production forecasting in unconventional reservoirs (Temizel et al., 2020; Lee, 2022b). Table 8.2 presents a comparison of several of the decline models commonly used in unconventional wells forecasting.

8.3 RATE TRANSIENT ANALYSIS METHODS

8.3.1 Straight Line Analysis Methods

8.3.1.1 Bilinear Flow Analysis

Dimensionless fracture conductivity F_{cD} is defined as the ratio of the fracture conductivity (product of fracture permeability and fracture width) to the product of matrix permeability and fracture half-length.

$$F_{cD} = \frac{k_f w_f}{k x_f} \tag{8.36}$$

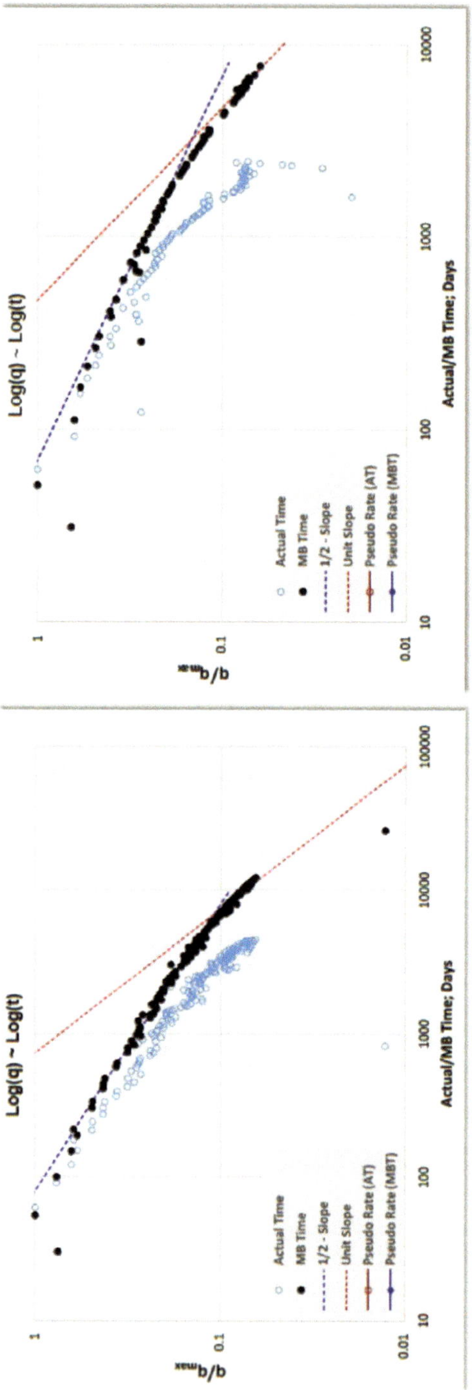

FIGURE 8.19 (a) Log-log plot of rate versus time and t_{MB} for a Barnett shale gas well; (b) Log-log plot of rate versus time and t_{MB} for a Bakken oil well.

Source: Lee, 2021

TABLE 8.2
Comparison of Decline Models for Unconventional Reservoirs

	Model					
Criteria	**Arps**	**Modified Arps**	**SEDM**	**PLE**	**Duong**	**Modified Duong**
Reasonable for unconventional reservoirs	No	Maybe	Yes with enough history	Possible	Yes, but switch to BDF model is required	Yes
Reasonable forecasts with < 1 year data	No	No	No	No	Yes	Yes
Valid during transient flow	No	No	Yes	Yes	Yes	Yes
Valid for BDF	Maybe	Maybe	No	Yes	No	Yes
Need to change parameters with longer history	Yes	Yes	No	No	No	No
Easy to use and couple with economics software	Yes	Yes	Some, not all	No	No	No

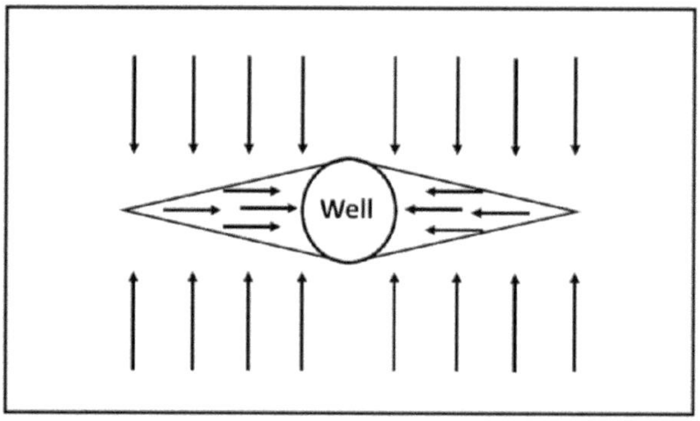

FIGURE 8.20 Bilinear flow schematic, after Cinco-Ley and Samaniego-V.

Source: Cinco-Ley & Samaniego-V, 1981

For $F_{cD} < 50$ (Wattenbarger et al., 1998), the pressure drop along the fracture is significant, and transient bilinear flow regime is exhibited (Figure 8.20). Analysis of this flow regime can provide an estimate of fracture conductivity if the matrix permeability is known. The quarter-root of the time plot $\left(\frac{p_i - p_{wf}}{q}\right)$, rate-normalized pressure (RNP) versus quarter-root of time (Figure 8.21) and the bilinear flow plot, which uses bilinear superposition time instead (Figure 8.22), are used to analyze data in this flow regime.

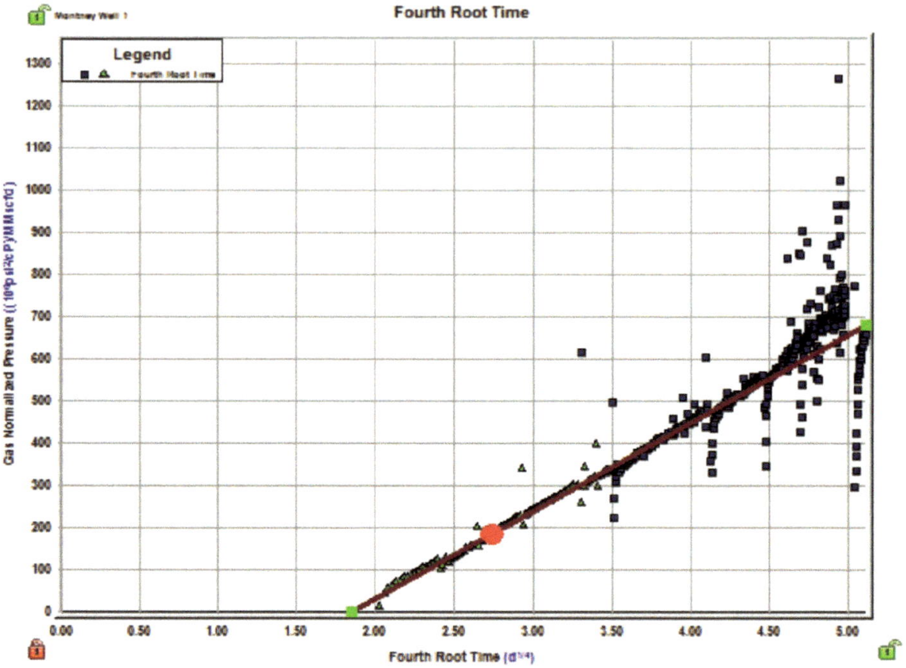

FIGURE 8.21 Quarter-root time plot, gas case.

Source: Unconventional Reservoir Plots, 2020

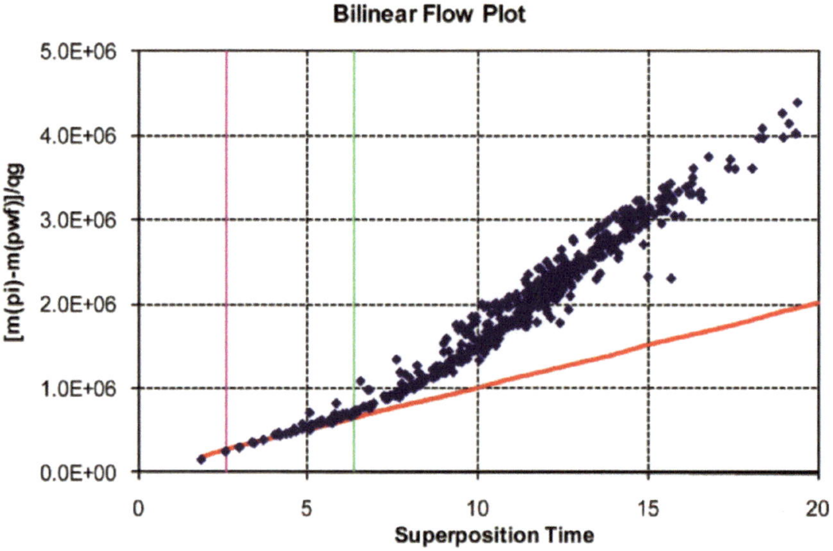

FIGURE 8.22 Bilinear flow plot, gas case.

Source: Clarkson & Beierle, 2011

The bilinear flow plot gives

$$RNP = mt_{BLS} + b \tag{8.37}$$

where the intercept b is often assumed to equal zero, and the slope m is used to evaluate fracture conductivity using the following formulas (Clarkson, 2021):

$$\text{Oil wells: } \sqrt{k_f w_f}\, k^{1/4} = \frac{44.13\mu_{oi} B_{oi}}{mh} \left(\frac{1}{\phi \mu_o c} \right)_i^{1/4} \tag{8.38}$$

$$\text{Gas wells: } \sqrt{k_f w_f}\, k^{1/4} = \frac{443.98T}{mh} \left(\frac{1}{\phi \mu_g c} \right)_i^{1/4} \tag{8.39}$$

Bilinear superposition time is defined as:

$$t_{BLS} = \sum\nolimits_{j=1}^{n} \frac{\left(q_j - q_{j-1}\right)}{q_n} \left(t - t_{j-1}\right)^{1/4} \tag{8.40}$$

8.3.1.2 Linear Flow Analysis

Fractured horizontal wells typically exhibit formation linear flow regime early in the life of the well. In contrast to bilinear flow, pressure drop along the fracture is negligible, and the flow is characterized by a linear component from the matrix to fractures (Figure 8.23). A key parameter from analyzing this data is an estimate of the product of the matrix surface area draining into the fracture system and the square root of permeability, $A\sqrt{k}$. The square root-time plot, $\frac{p_i - p_{wf}}{q}$ (rate-normalized pressure, RNP) versus $\sqrt{t}$ for oil wells and $\frac{m(p_i) - m(p_{wf})}{q}$ versus $\sqrt{t}$ for gas wells (Figure 8.24), will produce a straight line during linear flow. The linear flow plot uses linear superposition time instead (Figure 8.25).

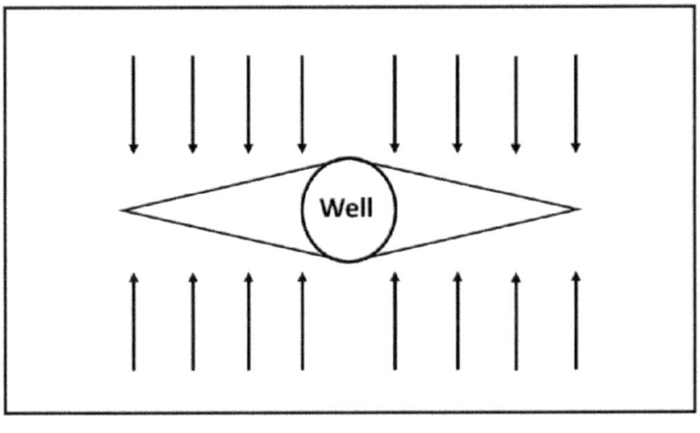

FIGURE 8.23 Linear flow schematic.

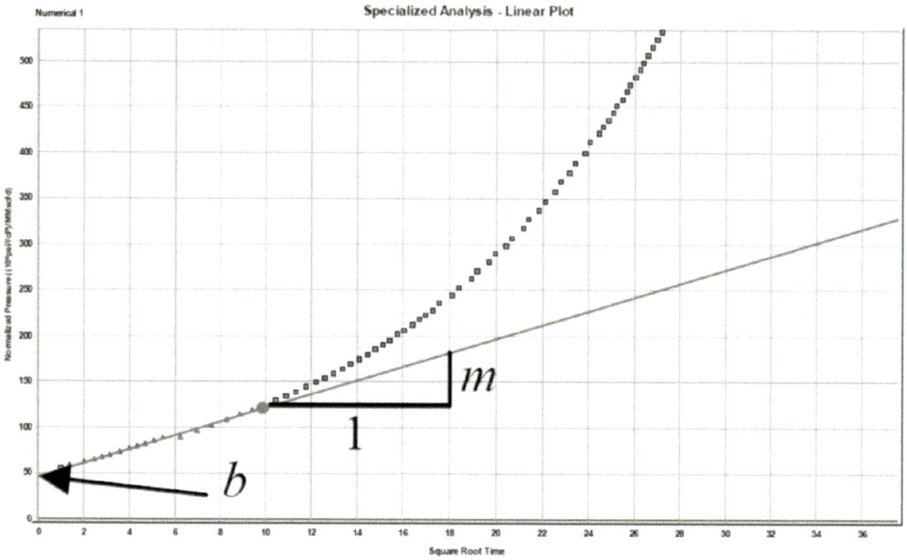

FIGURE 8.24 Square root-time plot.

Source: Anderson et al., 2010

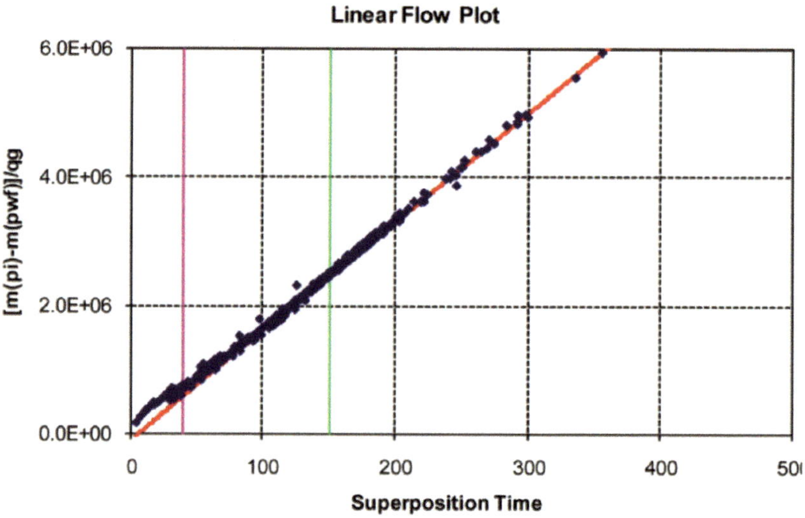

FIGURE 8.25 Linear flow plot, gas case.

Source: Clarkson, 2021

Similarly to the bilinear flow case,

$$RNP = mt_{LS} + b \tag{8.41}$$

is used to characterize the straight line portion of the plot, where the slope m is used to calculate the parameter $A\sqrt{k}$, which is also called the linear flow parameter (LFP), using the following formulas (Clarkson, 2021):

$$\text{Oil wells: } A\sqrt{k} = \frac{16.256\mu_{oi}B_{oi}}{m}\left(\frac{1}{\phi\mu_o c}\right)_i^{1/2} \tag{8.42}$$

$$\text{Gas wells: } A\sqrt{k} = \frac{163.72T}{m}\left(\frac{1}{\phi\mu_g c}\right)_i^{1/2} \tag{8.43}$$

The linear superposition time is defined as:

$$t_{LS} = \sum_{j=1}^{n} \frac{\left(q_j - q_{j-1}\right)}{q_n}\left(t - t_{j-1}\right)^{1/2} \tag{8.44}$$

The intercept b should be zero. A positive intercept indicates fracture skin and the presence of finite conductivity fracture system.

8.3.1.3 End of Linear Flow Analysis

The time at the end of linear flow t_{elf} marks the beginning period for which conservative reserve estimations can be made. Based on the slope of the square root-time plot, the following formulas for oil initially in place, N, and gas initially in place, G, are presented by Wattenbarger (Wattenbarger et al., 1998):

$$N = \frac{0.724S_{oi}}{mB_{oi}c_{ti}}\sqrt{t_{elf}} \tag{8.45}$$

$$G = f_{CP}\frac{40.95TS_{gi}}{mB_{gi}\mu_{gi}c_{ti}}\sqrt{t_{elf}} \tag{8.46}$$

where f_{CP} is a drawdown correction factor.

8.3.1.4 Boundary-Dominated Flow Analysis

Flowing material balance (FMB) method: Mattar and McNeil (1998) introduced the method of flowing material balance to analyze production data and flowing pressures as an alternative to the classical material balance, which requires average shut-in pressures. The method is applicable to variable rate/variable pressure production cases, which makes it extremely useful for rate-transient analysis.

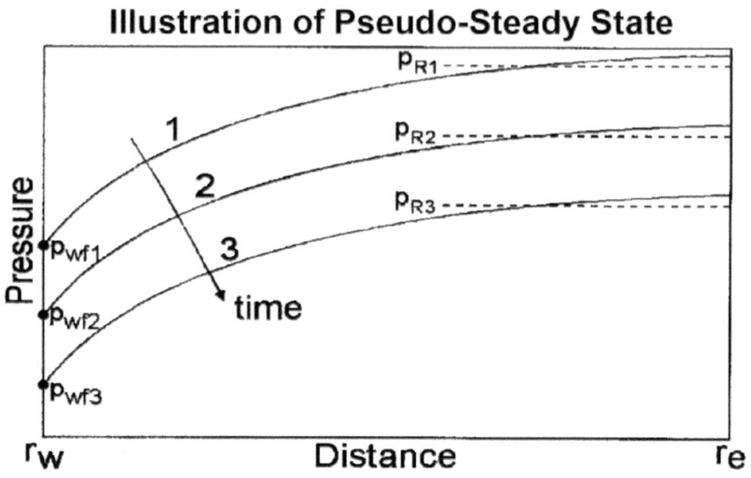

FIGURE 8.26 Illustration of pseudo–steady state.

Source: Mattar & McNeil, 1998

For pseudo–steady state flow (Figure 8.26),

$$p_i - p_{wf} = \frac{qt}{cN} + b_{ss}q \tag{8.47}$$

Material balance equation:

$$p_i - \bar{p}_R = \frac{Q}{cN} \tag{8.48}$$

Combining equations yields $\bar{p}_R - p_{wf} = b_{ss}q$, which can be rearranged as $\bar{p}_R = p_{wf} + b_{ss}q$.

A plot of $(\frac{p_i - p_{wf}}{q})$ or $(\frac{q}{p_i - p_{wf}})$ versus $(\frac{Q}{q})$ should produce a straight line if boundary-dominated flow regime is encountered. The intercept of the straight line is b_{ss} (Figure 8.27). Several FMB formulations are available in the literature. One simple formula that utilizes pressure-normalized rate (PNR) is:

$$\text{Oil Wells: } PNR = \frac{q_o}{p_i - p_{wf}} = m\frac{S_{oi}N_p}{c_t\left[p_i - p_{wf}\right]} + b \tag{8.49}$$

An example FMB plot is shown in Figure 8.28.

Hence, a plot of PNR versus $\frac{S_{oi}N_p}{c_t\left[p_i - p_{wf}\right]}$, which is referred to as normalized cumulative production, would produce a straight line in the boundary-dominated flow portion of the data with a slope m and an intercept b. We can calculate the original oil in place and permeability as follows:

$$N = \frac{b}{-m} \tag{8.50}$$

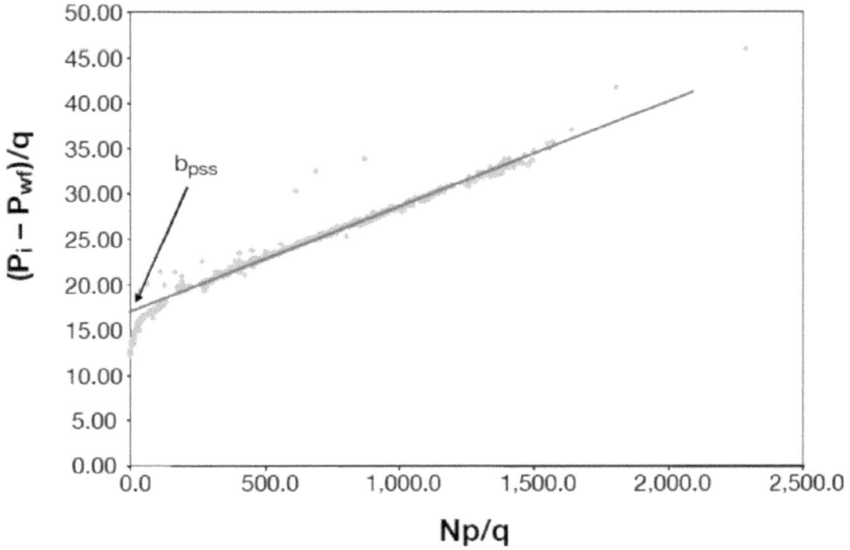

FIGURE 8.27 Determination of b_{pss} value.

Source: Mattar et al., 2006

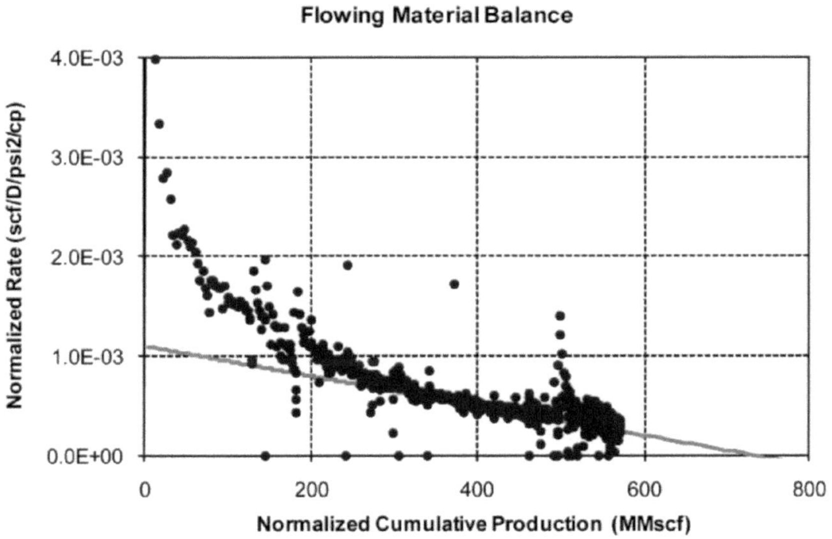

FIGURE 8.28 Flowing material balance plot.

Source: Tabatabaie et al.

$$k = \frac{141.2B_{oi}\mu_{oi}}{h}\left[\frac{1}{2}\ln\left(\frac{4A}{e^{\gamma}C_A r_{wa}^2}\right)\right]b \tag{8.51}$$

where r_{wa} is the apparent wellbore radius, $\gamma = 0.5772$ is Euler's constant, A is the drainage area of the well, and C_A is the Dietz shape factor. Adopting equation 8.55 to the case of a fracture extending to the boundaries of a rectangular reservoir, we get the following formula, which can be derived from Wattenbarger's work (Wattenbarger et al., 1998):

$$k = \frac{141.2B_{oi}\mu_{oi}}{h}\left[\frac{2}{\pi}\left(\frac{y_e}{x_e}\right)+s\right]b \tag{8.52}$$

For gas wells, pseudo-pressure is used instead of pressure, which results in the following formulas:

$$PNR = \frac{q_g}{m(p_i)-m(p_{wf})} = m\left(\left[\frac{m(p_i)-m(\bar{p})}{m(p_i)-m(p_{wf})}\right]G\right)+b \tag{8.53}$$

Similarly to the oil case,

$$G = \frac{b}{-m} \tag{8.54}$$

$$k = \frac{1.417T}{h}\left[\frac{1}{2}\ln\left(\frac{4A}{e^{\gamma}C_A r_{wa}^2}\right)\right]b \tag{8.55}$$

$$k = \frac{141.2B_{oi}\mu_{oi}}{h}\left[\frac{2}{\pi}\left(\frac{y_e}{x_e}\right)+s\right]b \tag{8.56}$$

Shahamat and Clarkson-Multi-Well (2017) improved on the method, using pseudo-pressure for two-phase oil and gas flow, and adding water production/injection and multi-well effects into the analysis. The proposed workflow is summarized in the following steps:

1. Assume values for oil, gas, and water initially in place (OOIP, OGIP, and OWIP, respectively).
2. Calculate average reservoir pressures using cumulative productions, in-place volumes, and a minimization routing in combination with the following equation:

$$\Phi_o\frac{N_p}{N}+\Phi_g\frac{G_p}{G}+\Phi_w\frac{W_p}{W}=1 \tag{8.57}$$

Additionally, calculate average oil, gas, and water saturations at each time, using the following formulas:

$$\bar{S}_o = \frac{\bar{B}_o\left[\left(1-\frac{N_p}{N}\right)-\left(1-\frac{G_p}{G}\right)\left(\frac{G}{N}\right)\bar{R}_v\right]}{\left(1-\frac{W_{ps}}{W}\right)\left(\frac{W}{N}\right)\bar{B}_w\left[1-\bar{R}_s\bar{R}_v\right]+\left(1-\frac{G_p}{G}\right)\left(\frac{G}{N}\right)\left[\bar{B}_g-\bar{R}_v\bar{B}_o\right]+\left(1-\frac{N_p}{N}\right)\left[\bar{B}_o-\bar{R}_s\bar{B}_g\right]} \tag{8.58}$$

$$\bar{S}_g = \frac{\bar{B}_g\left[\left(1-\frac{G_p}{G}\right)\left(\frac{G}{N}\right)-\left(1-\frac{N_p}{N}\right)\bar{R}_s\right]}{\left(1-\frac{W_{ps}}{W}\right)\left(\frac{W}{N}\right)\bar{B}_w\left[1-\bar{R}_s\bar{R}_v\right]+\left(1-\frac{G_p}{G}\right)\left(\frac{G}{N}\right)\left[\bar{B}_g-\bar{R}_v\bar{B}_o\right]+\left(1-\frac{N_p}{N}\right)\left[\bar{B}_o-\bar{R}_s\bar{B}_g\right]} \tag{8.59}$$

$$\bar{S}_w = \frac{\left(1-\frac{W_{ps}}{W}\right)\left(\frac{W}{N}\right)\bar{B}_w\left[1-\bar{R}_s\bar{R}_v\right]}{\left(1-\frac{W_{ps}}{W}\right)\left(\frac{W}{N}\right)\bar{B}_w\left[1-\bar{R}_s\bar{R}_v\right]+\left(1-\frac{G_p}{G}\right)\left(\frac{G}{N}\right)\left[\bar{B}_g-\bar{R}_v\bar{B}_o\right]+\left(1-\frac{N_p}{N}\right)\left[\bar{B}_o-\bar{R}_s\bar{B}_g\right]} \tag{8.60}$$

3. Depending on the values of average pressure at the current time step and using guidance from Table 8.3, calculate pseudo-pressure differences for oil and gas phases. Pressure is sufficient for the water case.
4. Construct a plot of normalized rate versus normalized cumulative production for the three phases.
5. Fit a straight line through the late segment of the three plots and extrapolate to zero y-intercept. This calculates initial in-place fluid volumes.
6. Compare the calculated in-place volumes with those from the previous steps. If the difference is outside the acceptable tolerance, iterate through steps 2–6 until convergence is achieved.

The authors provided several examples to validate their method, one of which was to test a multi-well oil reservoir with injection case. Figure 8.29 shows the pressure distribution of the numerical simulation case after nine years of production. Figure 8.30 shows the application of multiphase FMB analysis for both oil and gas phases. The authors reported good agreement with the results of the simulation model.

8.3.1.5 Straight Line Methods Comparison

Table 8.4 presents a comparison of some of the common straight line rate transient methods used in analyzing unconventional reservoirs (Clarkson, 2021).

TABLE 8.3
Summary of the Parameters and Recommended Methods Used to Calculate Pseudo-Pressures in Multiphase FMB

$\alpha_1 = \int_{p_{wf}}^{p_{avg}} \left(\frac{k_{ro}}{\mu_o B_o} + \frac{k_{rg}}{\mu_g B_g} R_v \right) dp$	$\alpha_1 = \int_{p_{wf}}^{p_{avg}} \left(\frac{k_{ro}}{\mu_o B_o} R_s + \frac{k_{rg}}{\mu_g B_g} \right) dp$	Steady-State Model
$\alpha_2 = \int_{p_{avg}}^{p_b} \left(\frac{k_{ro}}{\mu_o B_o} \right) dp$	$\alpha_2 = \int_{p_{avg}}^{p_b} \left(\frac{k_{rg}}{\mu_g B_g} \right) dp$	Tank Model
$\alpha_3 = \int_{p_b}^{p_i} \left(\frac{k_{ro}(S_{wi})}{\mu_o B_o} \right) dp$	$\alpha_3 = \int_{p_b}^{p_i} \left(\frac{k_{rg}(S_{wi})}{\mu_g B_g} \right) dp$	
$\alpha_4 = \int_{p_{avg}}^{p_i} \left(\frac{k_{ro}(S_{wi})}{\mu_o B_o} \right) dp$	$\alpha_4 = \int_{p_{avg}}^{p_i} \left(\frac{k_{rg}(S_{wi})}{\mu_g B_g} \right) dp$	
$\alpha_5 = \int_{p_{wf}}^{p_b} \left(\frac{k_{ro}}{\mu_o B_o} + \frac{k_{rg}}{\mu_g B_g} R_v \right) dp$	$\alpha_5 = \int_{p_{wf}}^{p_b} \left(\frac{k_{ro}}{\mu_o B_o} R_s + \frac{k_{rg}}{\mu_g B_g} \right) dp$	Steady-State Model

$$p_{pi} - p_{pwf} = \begin{cases} \mu_i B_i (\alpha_3 + \alpha_5) & \text{If } p_{avg} > p_b \\ \mu_i B_i (\alpha_1 + \alpha_2 + \alpha_3) & \text{If } p_{avg} \le p_b \end{cases}$$

$$p_{pi} - p_{pavg} = \begin{cases} \mu_i B_i (\alpha_4) & \text{If } p_{avg} > p_b \\ \mu_i B_i (\alpha_2 + \alpha_3) & \text{If } p_{avg} \le p_b \end{cases}$$

Source: Shahamat & Clarkson-Multi-Well, 2017

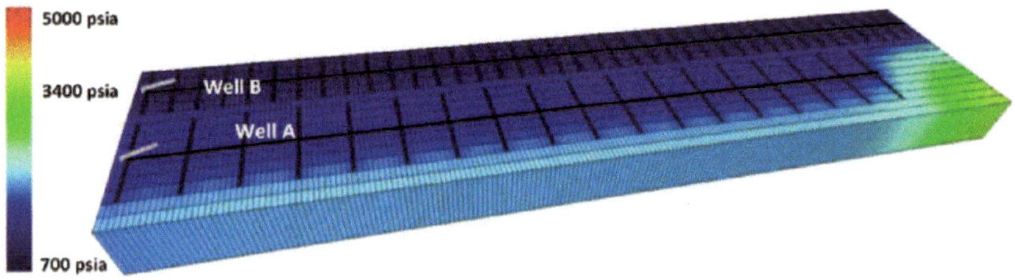

FIGURE 8.29 Pressure distribution of a two-well simulation case, nine years of production.

Source: Shahamat & Clarkson-Multi-Well, 2017

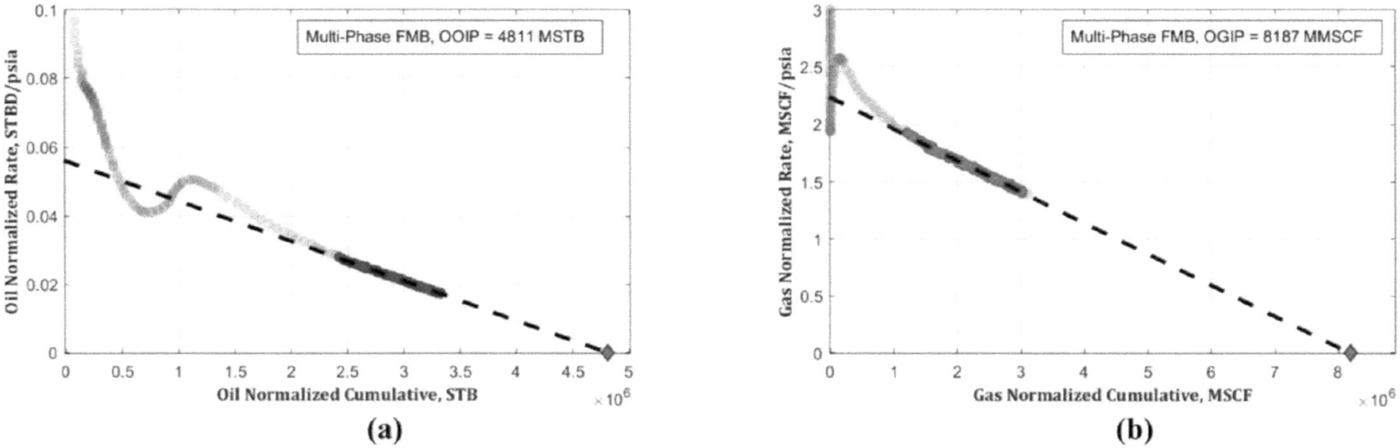

FIGURE 8.30 Analysis of (a) oil and (b) gas production using multi-well, multiphase FMB

Source: Shahamat & Clarkson-Multi-Well, 2017

TABLE 8.4
Straight Line Methods Comparison

Method	Flow Regime Signature	Specialized Plot	Extracted Properties
Bilinear Flow Analysis	¼ slope	RNP vs t_{BLS}	$\sqrt{k_f w_f} k^{1/4}$
Linear Flow Analysis	½ slope	RNP vs t_{LS}	$x_f\sqrt{k}, A_f\sqrt{k}$
End of Linear Flow Analysis	End of ½ slope	RNP vs $\sqrt{t}$	N, G
Flowing Material Balance		PNR vs N_{pN}	N, G, K

8.3.2 Type Curve Methods

8.3.2.1 Wattenbarger Type Curves

Wattenbarger et al. (1998) provided solutions and type curves for analyzing production from tight gas wells. The solutions are valid for both constant rate and constant pressure cases. Based on the assumption of a hydraulically fractured well in the center of a rectangular reservoir where the fracture extends to the boundaries (Figure 8.31), the authors presented the following solution for the constant rate case:

$$p_D = \frac{\pi}{2}\left(\frac{y_e}{x_f}\right)\left[\frac{1}{3}+\left(\frac{x_f}{y_e}\right)^2 t_{Dxf}\right]-\frac{2}{\pi^2}\left(\frac{y_e}{x_f}\right)\sum_{n-1}^{\infty}\frac{1}{n^2}exp\left[-n^2\pi^2\left(\frac{x_f}{y_e}\right)^2 t_{Dxf}\right] \tag{8.61}$$

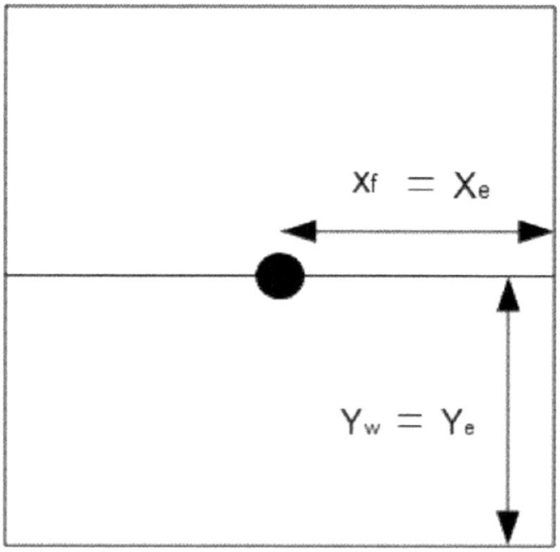

FIGURE 8.31 A hydraulically fractured well in a rectangular reservoir,

Source: Wattenbarger et al., 1998

where for oil wells:

$$p_D = \frac{kh\left(p_i - p_{wf}\right)}{141.2qB\mu} \tag{8.62}$$

and for gas wells:

$$p_D = \frac{kh\left(m\left(p_i\right) - m\left(p_{wf}\right)\right)}{1422Tq_g} \tag{8.63}$$

$$t_{Dxf} = \frac{0.00633kt}{\phi\mu c_t x_f^2} \tag{8.64}$$

For the constant bottom-hole pressure, the authors presented the following solution:

$$\frac{1}{q_D} = \frac{\frac{\pi}{4}\left(\frac{y_e}{x_f}\right)}{\sum_{n=1}^{\infty} exp\left[-\frac{(2n-1)^2\pi^2}{4}t_{Dye}\right]} \tag{8.65}$$

where y_e is the distance from the fracture to the boundary in the perpendicular direction to the fracture.

$$t_{Dye} = \frac{0.0002637kt}{\phi\mu c_t y_e^2} = \left(\frac{x_f}{y_e}\right)^2 t_{Dxf} \tag{8.66}$$

Based on the value of t_{Dye}, the equation can be replaced by two equations: one for early time and the other for late time.

$$\text{Early time } (t_{Dye} < 0.25)\text{:} \quad \frac{1}{q_D} = \frac{\pi}{2}\sqrt{\pi t_{Dxf}} \tag{8.67}$$

$$\text{Late time } (t_{Dye} > 1.25)\text{:} \quad \frac{1}{q_D} = \frac{\pi}{4}\left(\frac{y_e}{x_f}\right) exp\left[\frac{\pi^2}{4}t_{Dye}\right] \tag{8.68}$$

For $0.25 < t_{Dye} < 1.25$, the full equation must be used.

Figure 8.32 shows Wattenbarger's linear type curves.

8.3.2.2 Pratikno-Blasingame Type Curves

Pratikno et al. (2003) generated type curves for vertical wells with finite conductivity fractures in bounded low permeability reservoirs. The authors used a wide range of

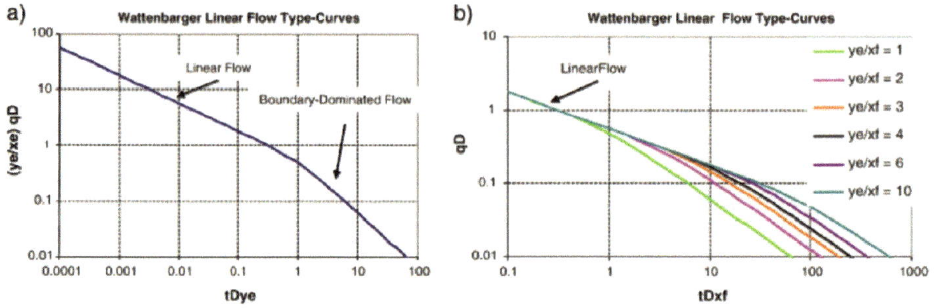

FIGURE 8.32 Wattenbarger linear flow type curves.

Source: Wattenbarger et al., 1998

dimensionless fracture conductivity values F_{cD} and dimensionless drainage radius r_{eD}, where $r_{eD} = \frac{r_e}{x_f}$ (8.69).

The authors defined q_D, t_D for oil wells as:

$$q_D = \frac{141.2qB\mu}{kh\left(p_i - p_{wf}\right)} \tag{8.70}$$

$$t_{DA} = \frac{0.00633k}{\phi\mu cA}t \tag{8.71}$$

For gas wells:

$$q_D = \frac{1422Tq_g}{kh\left(m\left(p_i\right) - m\left(p_{wf}\right)\right)} \tag{8.72}$$

$$t_{DA} = \frac{0.00633k}{\left(\phi\mu c\right)_i A}t_{MBa} \tag{8.73}$$

where t_{MBa} is material balance pseudo-time.
The type curves variables q_{Dd}, q_{Dd} can be calculated by:

$$q_{Dd} = q_D b_{D_{pss}} \tag{8.74}$$

$$t_{Dd} = \frac{2\pi}{b_{D_{pss}}}t_{DA} \tag{8.75}$$

Where the parameter $b_{D_{pss}}$ is a dimensionless pseudo–steady state parameter that is a function of F_{cD} and r_{eD} during bilinear flow and is constant during pseudo–steady state flow. The authors provided both analytical solutions and an empirical correlation to calculate $b_{D_{pss}}$. Figure 8.33 shows an example type curve set for $F_{cD} = 0.1$.

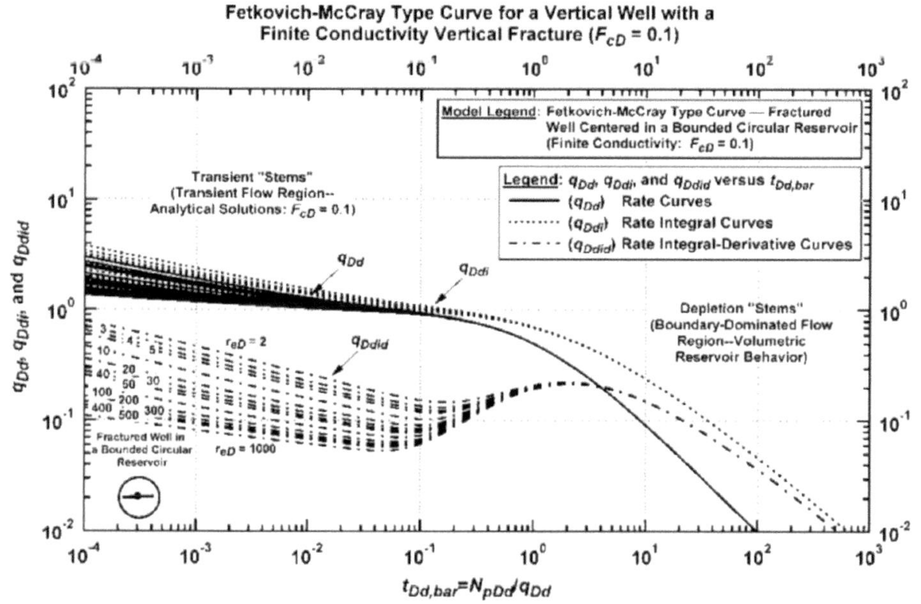

FIGURE 8.33 Pratikno-Blasingame type curve, F_{CD}=0.1.

Source: Pratikno et al., 2003

8.3.2.3 Agarwal-Gardner Type Curves

Agarwal et al. (1998) presented type curves for production data analysis. Their work is an extension of the work done by Fetkovich, Palacio, and Blasingame. The dimensionless flow rate and inverse normalized pressure derivative curves (Figure 8.34) are generated using the following formulas for oil wells:

$$t_c = \frac{Q}{q}$$

$$t_{DA} = \frac{0.00633kt_c}{\pi\phi\mu c r_e^2} \tag{8.76}$$

$$q_D = \frac{141.2qB\mu}{kh\left(p_i - p_{wf}\right)} \tag{8.77}$$

Similar formulas are available for gas wells using pseudo-pressure and material balance pseudo-time, as shown in the previous methods.

8.3.2.4 Nobakht-Clarkson Type Curves

Nobakht et al. (2013) presented a type curve set for the analysis of production from MFHWs with planar fractures. Whereas Wattenbarger's type curves assumed fully

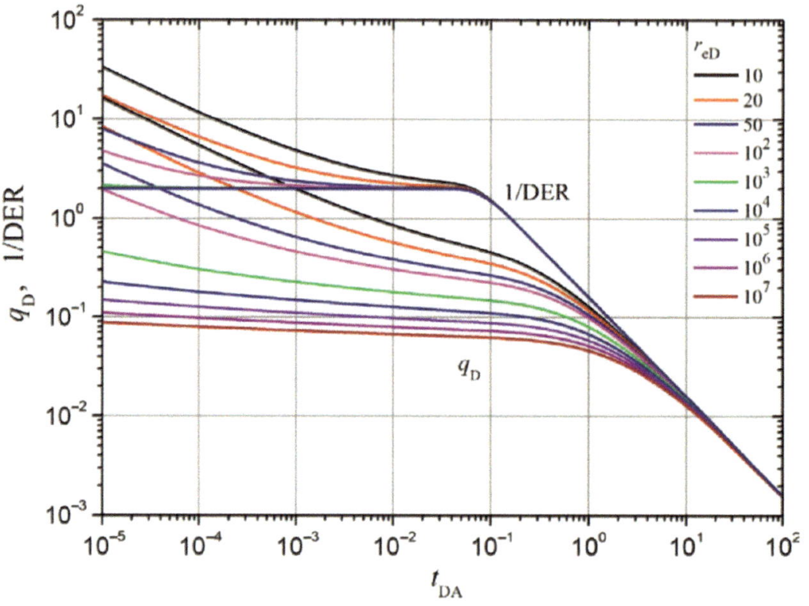

FIGURE 8.34 Agarwal-Gardner dimensionless flow rate qD and inverse normalized pressure derivative (1/DER) type curves.

Source: Agarwal et al., 1998

extended fractures to the boundaries of the reservoir, Nobakht-Clarkson type curves consider flow beyond fracture tips during inter-fracture flow (Figure 8.35).

The dimensionless variables for the y- and x-axes for the type curves are given by:

$$q_{DM} = q_D \left(\frac{x_e}{y_e}\right)\left(\frac{y_e}{x_f}\right)^2 \tag{8.78}$$

$$t_{DM} = t_{D_{xf}} \left(\frac{y_e}{x_e}\right)^2 \left(\frac{x_f}{y_e}\right)^4 \tag{8.79}$$

Contrary to Wattenbarger's nomenclature, y_e denotes well spacing, and x_e denotes fracture spacing. q_D has the same definition as Wattenbarger's. The authors used the total rate divided by the total number of fractures to calculate q_D, which leads to a slightly different $t_{D_{xf}}$ formulation:

$$t_{D_{xf}} = \frac{0.00633k}{\phi \mu c x_f^2} t \tag{8.80}$$

Here x_f describes the individual fracture half-length, not the total fracture half-length. The same pseudo-pressure and material balance pseudo-time conversions that were

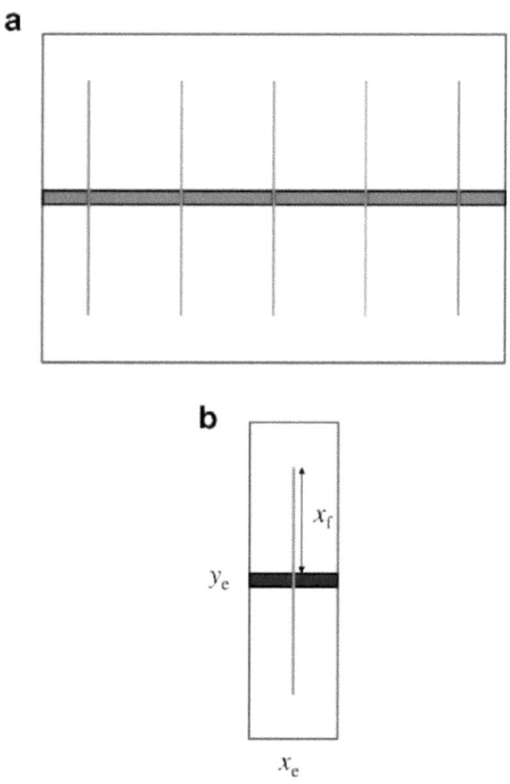

FIGURE 8.35 (a) conceptual MHFW model; (b) Geometry used to generate type curves.

Source: Nobakht et al., 2013

used in the previous type curves are needed here as well. Figure 8.36 shows an example Nobakht-Clarkson type curve set.

8.3.2.5 Compound Linear Flow Type Curves

Liang et al. (2012) developed a type curve set for MFHWs that takes into account inter-fracture formation transient linear flow, which is referred to as primary flow by the authors; a transition flow regime; and inter-well formation transient linear flow regime, which is referred to as compound linear flow by the authors.

In the development of the type curves, the authors modeled transient linear flow and the semi-log derivative using the short-term infinite-acting constant rate solution of Wattenbarger et al. (1998):

$$p_D = \sqrt{\pi t_{D_{xf}}} \tag{8.81}$$

$$p_D' = \frac{1}{2}\sqrt{\pi t_{D_{xf}}} \tag{8.82}$$

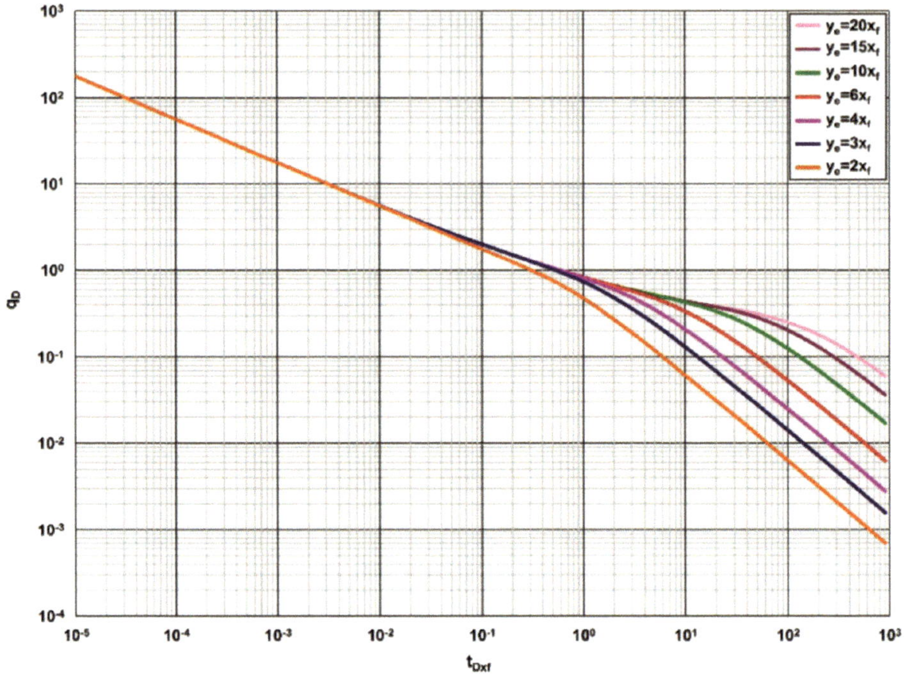

FIGURE 8.36 Nobakht-Clarkson type curve for y_e/x_e= 1.

Source: Nobakht et al., 2013

The compound linear flow regime and the derivative can be modeled by:

$$p_D = \frac{2x_f}{x_e}\sqrt{\pi t_{D_{xf}}} + s' \tag{8.83}$$

$$p_D' = \frac{x_f}{x_e}\sqrt{\pi t_{D_{xf}}} \tag{8.84}$$

Figures 8.37 and 8.38 show pressure and derivative compound linear flow type curves, respectively.

8.3.2.6 Type Curve Methods Comparison

Table 8.5 presents a comparison of some of the common type curve methods used in analyzing unconventional reservoirs (Clarkson, 2021).

8.3.3 Suggested RTA Workflow

Clarkson (2013) presented an integrated workflow that combines analytical and empirical production analysis techniques to characterize reservoir and hydraulic fractures, determine reserves, and forecast production. The workflow consists of the following eight steps and is summarized in Figure 8.39.

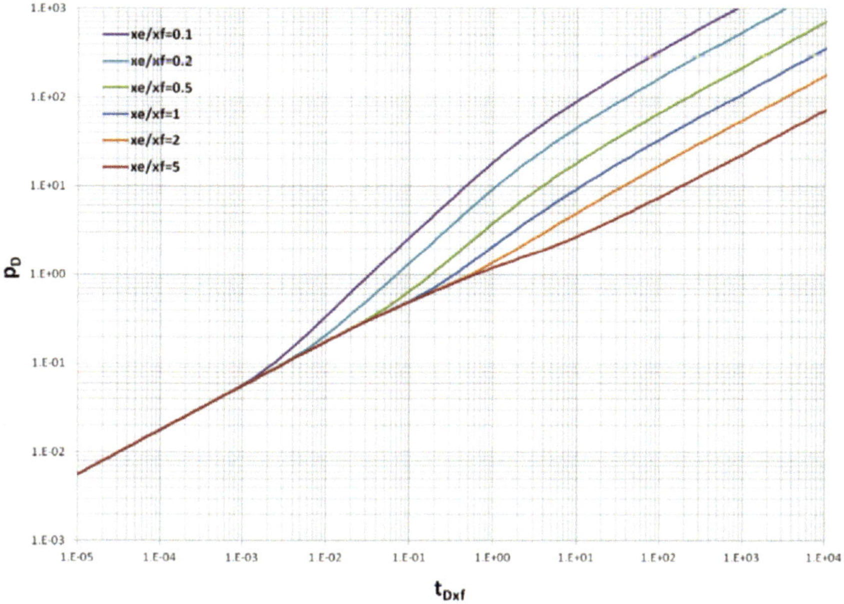

FIGURE 8.37 Dimensionless pressure type curves for primary, transitional, and compound linear flow.

Source: Liang, 2012

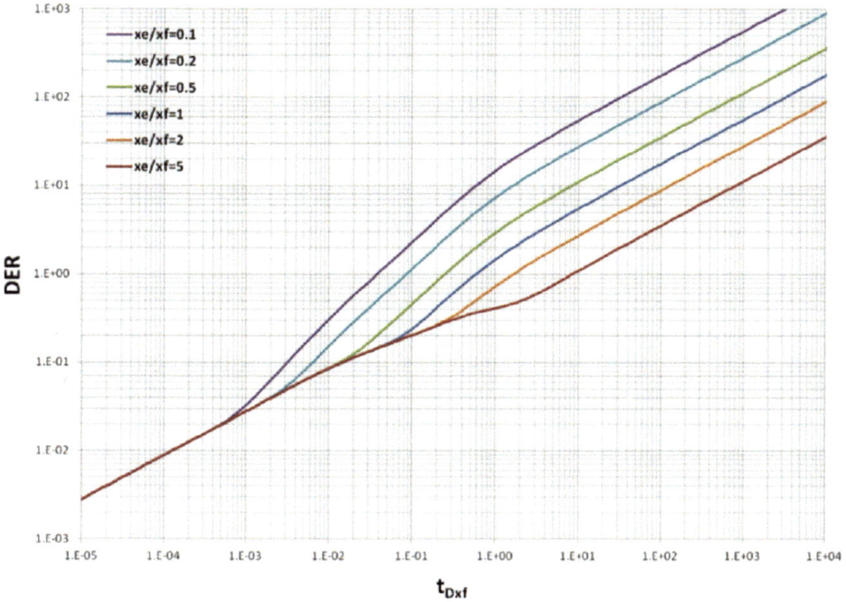

FIGURE 8.38 Derivative type curves for primary, transitional, and compound linear flow.

Source: Liang, 2012

TABLE 8.5
Type Curve Methods Comparison

Method	Dimensionless Variables	Extracted Properties
Wattenbarger	q_D, t_{Dy_e}	Linear Flow Parameter, Drainage Area, $L_w\sqrt{k}$
Pratikno-Blasingame	q_D, t_{DA}	kh, x_f, N, G
Agarwal-Gardner	$q_D, \frac{1}{DER}, t_{DA}$	kh, x_f, A, N, G
Nobakht-Clarkson	q_D, t_{Dx_f}	Linear Flow Parameter, Drainage Area
Compound Linear Flow	p_D, t_{Dx_f}	kh, x_f, x_e

Step 1: Assess data viability

Prior to performing the analysis, a careful review of production data and well history is required. Relevant reservoir, completion, and PVT data should be collected.

Step 2: Check for data correlation

Data should be checked for rates and pressures correlation. (They should have a similar trend.)

Step 3: Preliminary diagnosis

At this step, anomalous or uncorrelated data must be filtered out. Also, the influence of operational upsets, such as liquid loading, should be identified and filtered out if necessary.

Step 4: Identify flow regimes

The author described this step as "the most critical, and often most difficult" step in production analysis as errors will propagate through the rest of the analysis.

Step 5: Perform straight-line analysis

In this step, the flow periods (that correspond to the specific flow regimes identified in step 4) are selected, and flow regime-specific data are analyzed using specialty plots. Reservoir and hydraulic fracture information can be extracted from the analysis based on the flow regime, reservoir type, and fracture geometry.

Step 6: Perform type-curve analysis

Using the guidance of the interpretation of geological, completion, and surveillance data, a model can be identified using the sequence of flow regimes identified in step 4. Type-curve set that match the identified model can be selected, and data can be matched to the type curve subsequently. Reservoir and hydraulic fracture property data can be extracted from the type-curve match and be compared with those obtained from straight-line analysis.

Step 7: Perform forecast with model

This step involves history matching the dynamic data, adjustment of uncertain parameters, and well production forecast to abandonment limit.

Step 8: Fit empirical model to forecast

The last step is fitting empirical relationships, such as multi-segment Arps, power-law exponential, etc., to the forecast generated by the analytical models.

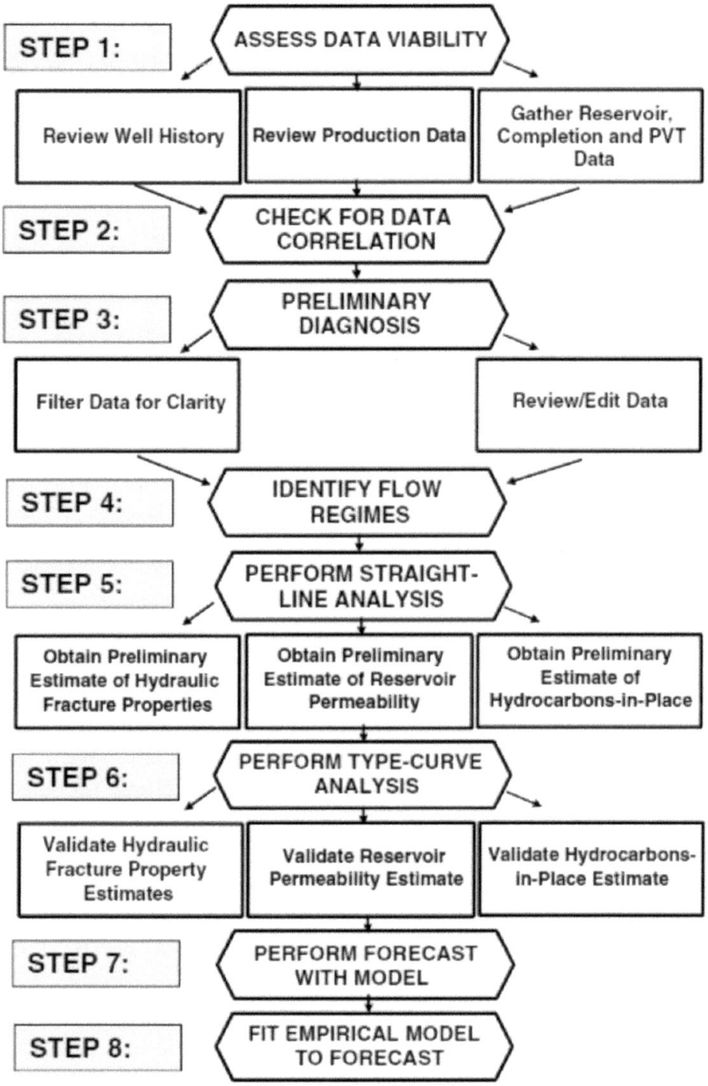

FIGURE 8.39 Workflow for production data analysis.

Source: Guo & Sankaran, 2022

8.3.3.1 Workflow Examples

8.3.3.1.1 Case 1: Simulated Example—Shale Gas Well

The model geometry is shown in Figure 8.40 (vertical well with an infinite-conductivity hydraulic fracture or a single fracture stage of a multi-fractured horizontal well). Reservoir permeability is 0.1 md, and the reservoir contains both free and adsorbed gas.

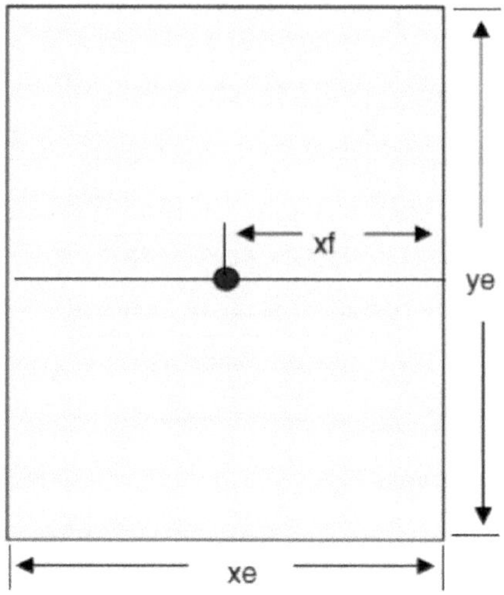

FIGURE 8.40 Case 1 geometry.

Source: Clarkson, 2013

Because this is a simulated case, steps 1 through 3 were skipped. Starting with flow-regime identification (Figure 8.41), transient linear flow and boundary-dominated flow were identified as the dominant flow regimes. Next was the flow-regime analysis (Figure 8.43). Using the square root of the time plot and linear flow plot, $\sqrt{kx_f}$ was obtained. Estimation of OGIP can be obtained from both the intercept of a straight line on the flowing material balance (FMB) plot and from the square root of the time plot.

Next is type-curve analysis. Type curves selected (Figure 8.45) correspond to reservoir type and geometry, as well as fracture geometry. Using a slight adjustment of x_f, a good match with the type curve was made.

The parameters derived from steps 5 and 6 were used as inputs to an analytical simulator, which was used to history match the data and generate a forecast (Figure 8.44). Afterward, an empirical model was fit to the analytical model forecast data (step 8, Figure 8.46). Using a two-segment Arps model, a good match was achieved, providing confidence that the empirical forecast will not lead to errors in reserve estimations.

8.3.3.1.2 Case 2: Field Example—Multi-Fractured Horizontal Well in Shale Gas Reservoir

This well is producing from the Barnett shale. It was stimulated in six stages, using five clusters per stage. Steps 1 through 3 were performed with minimum filtering. Rates and pressures were correlated.

Step 4 (Figure 8.42) was performed, resulting in the observance of linear flow, followed possibly by fracture interference. Next, step 5 was conducted (Figure 8.47)

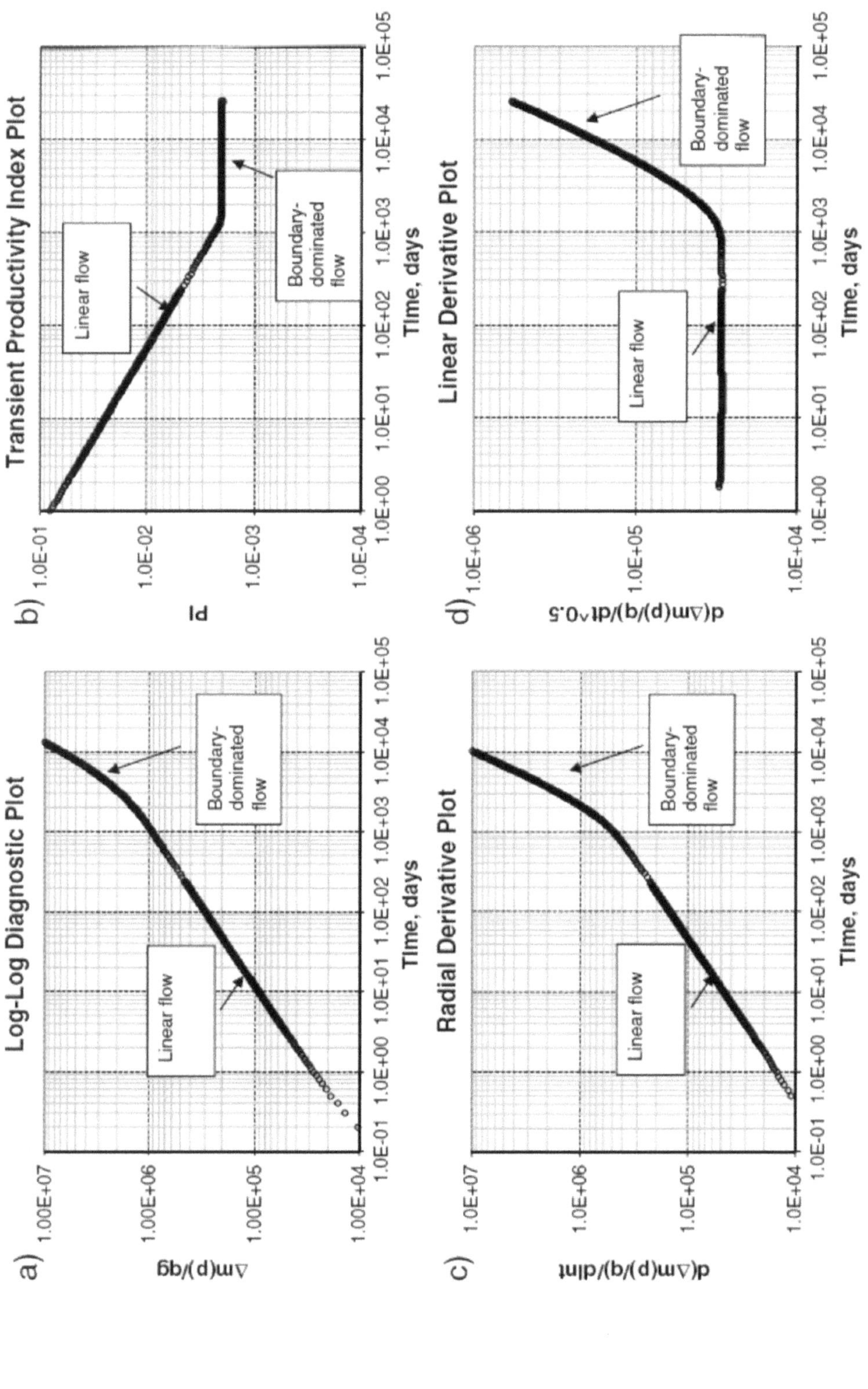

FIGURE 8.41 Flow-regime identification for case 1.

Source: Clarkson, 2013

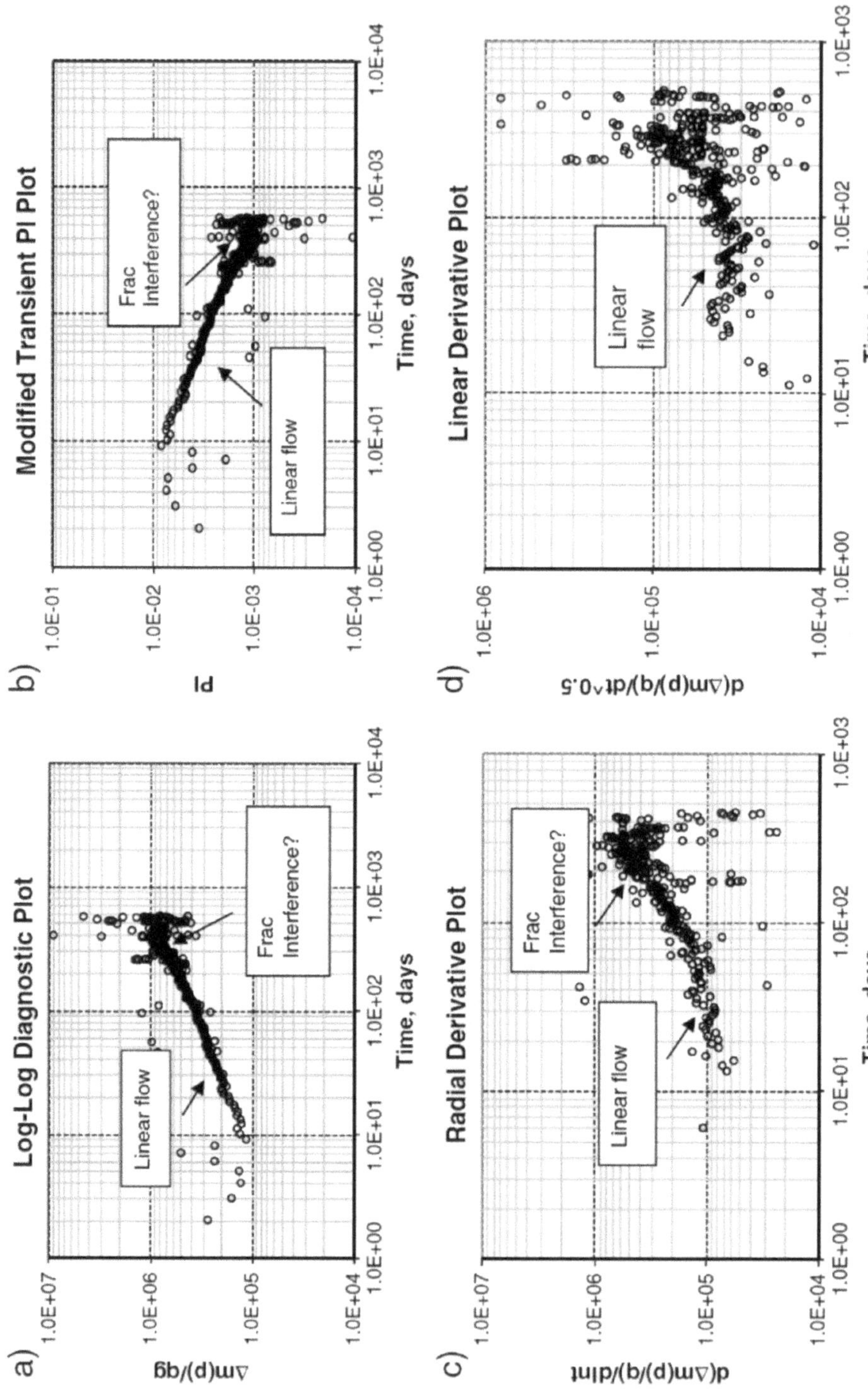

FIGURE 8.42 Flow-regime identification for case 2.

Source: Clarkson, 2013

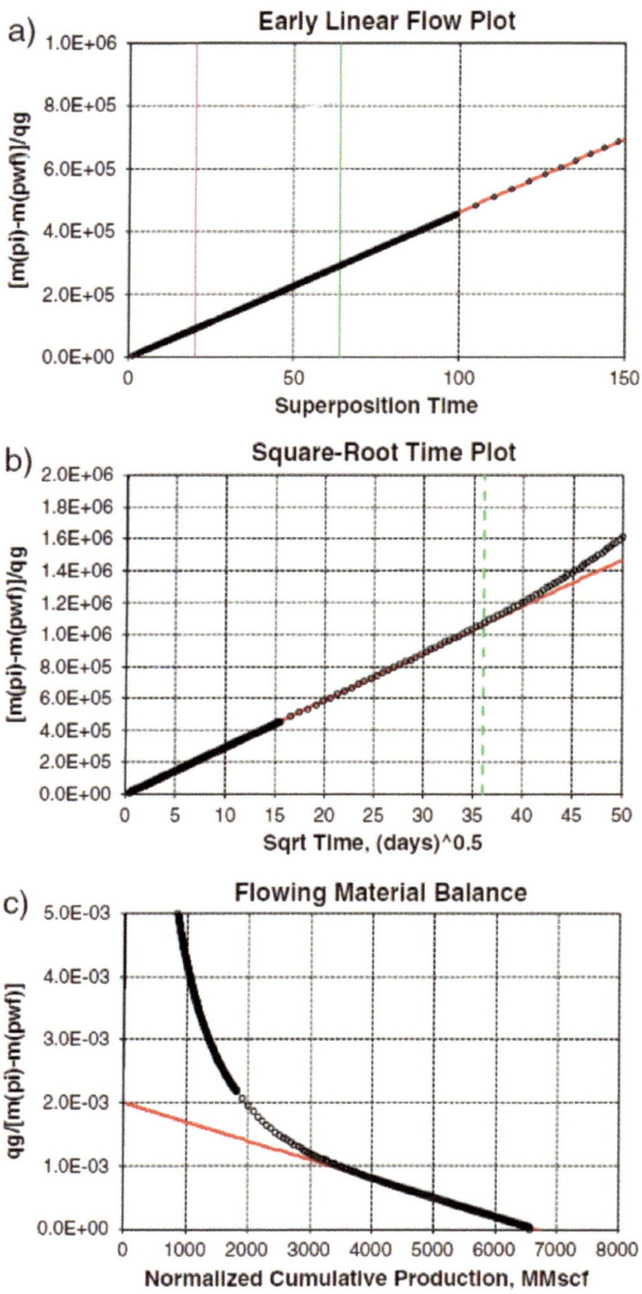

FIGURE 8.43 Flow regime analysis for case 1.

Source: Clarkson, 2013

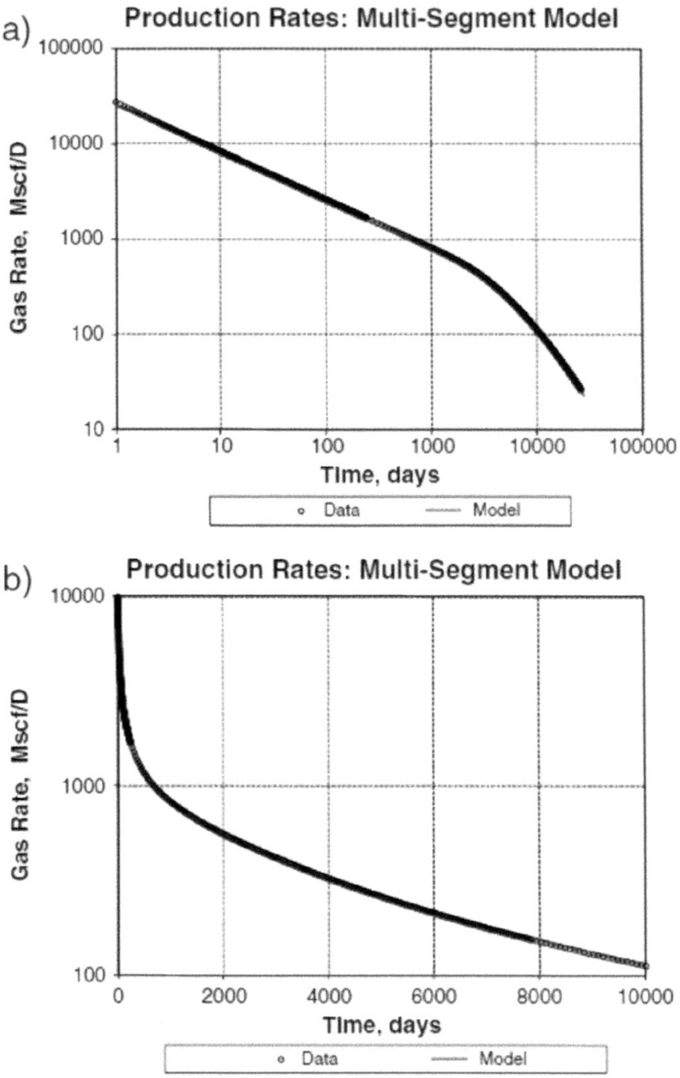

FIGURE 8.44 Analytical model match for case 1.

Source: Clarkson, 2013

to estimate permeability and fracture half-length. Good agreement for both calculations was found. FMB plot was used to calculate a conservative estimated OGIP value of 1.5 bscf.

Next (step 6, Figure 8.49), Wattenbarger bounded linear flow type curves were used to match the data using properties derived from the previous step. A small adjustment to the total fracture half-length was performed to achieve the match.

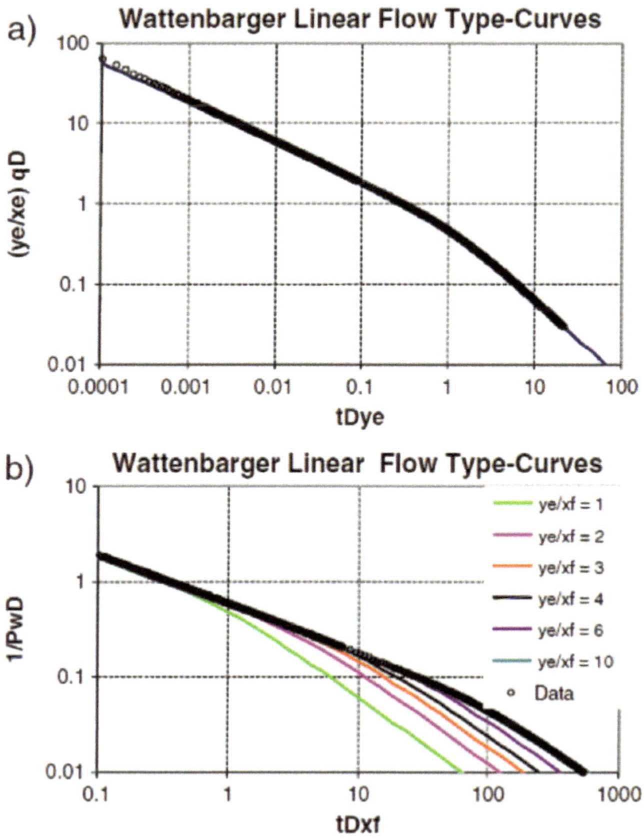

FIGURE 8.45 Wattenbarger type curve match for case 1.

Source: Clarkson, 2013

Step 7 (Figure 8.48) is the analytical forecasting. The same steps were used as for Case 1. Finally, step 8 (Figure 8.50) is fitting an empirical model to the analytical forecast. The two-segment Arps method was used again. A good match was achieved, with the EUR estimates from the analytical forecast and the empirical model forecast agreeing with 4%.

8.4 NUMERICAL SIMULATION METHODS

8.4.1 Complex Geometry Models

The analytical methods discussed in Section 3 have many simplifying assumptions that do not necessarily hold true for all applications. Hence, numerical models have been developed to provide more realistic modeling of the systems. One of the assumptions is uniform transverse fractures with equal lengths. Off-center fractures

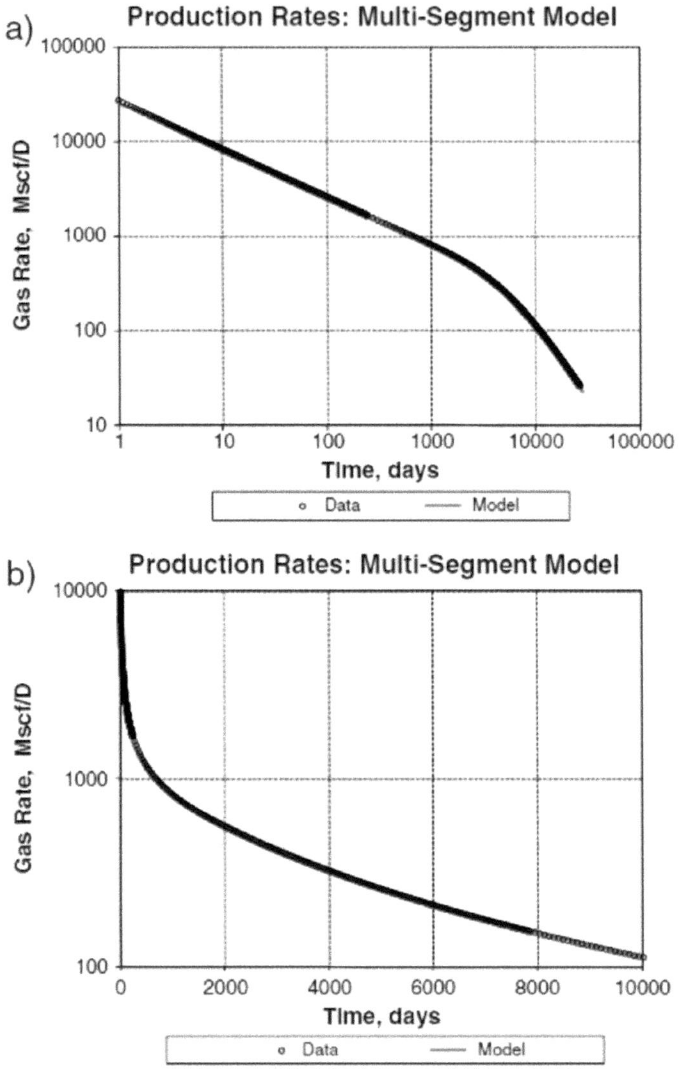

FIGURE 8.46 Two-segment Arps hyperbolic decline model for case 1.

Source: Clarkson, 2013

and uneven spacing along the well (Figure 8.51) are commonly present and can be modeled numerically using local grid refinement (LGR) as shown in Figure 8.52.

8.4.2 Discrete Fracture Network (DFN) Models

Discrete fracture network models treat the reservoir as a network of pre-existing natural fractures that may have undergone stimulation. DFN models, coupled with

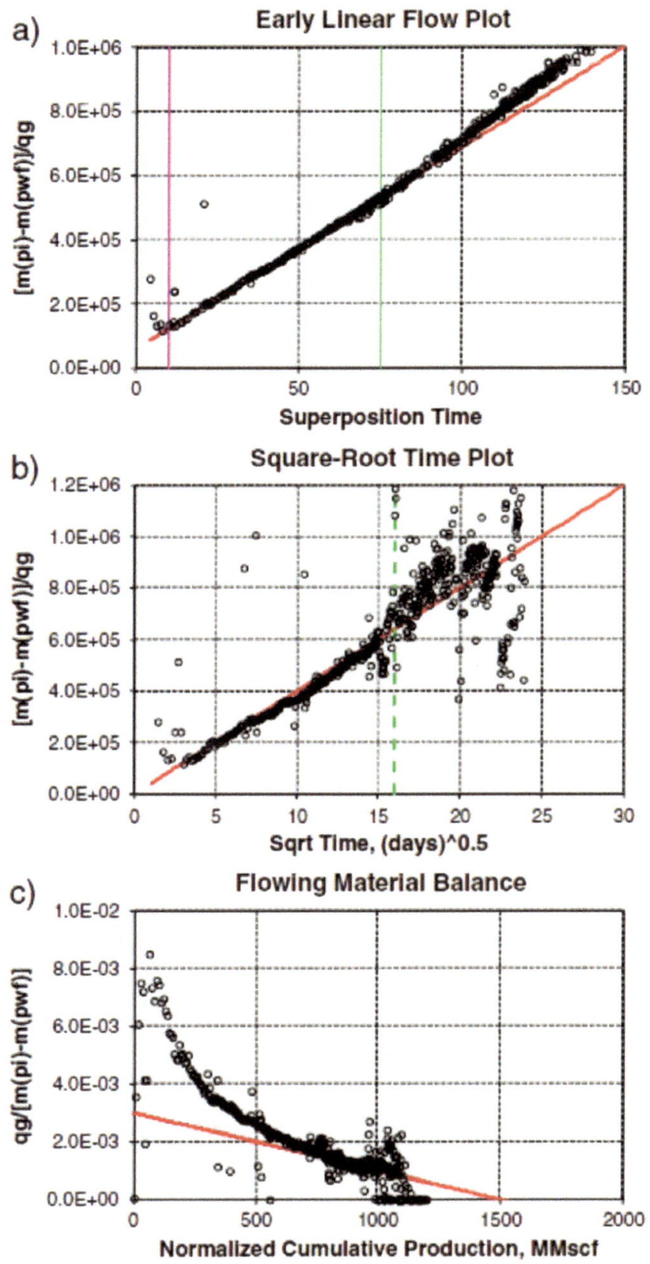

FIGURE 8.47 Flow regime analysis for case 2.

Source: Clarkson, 2013

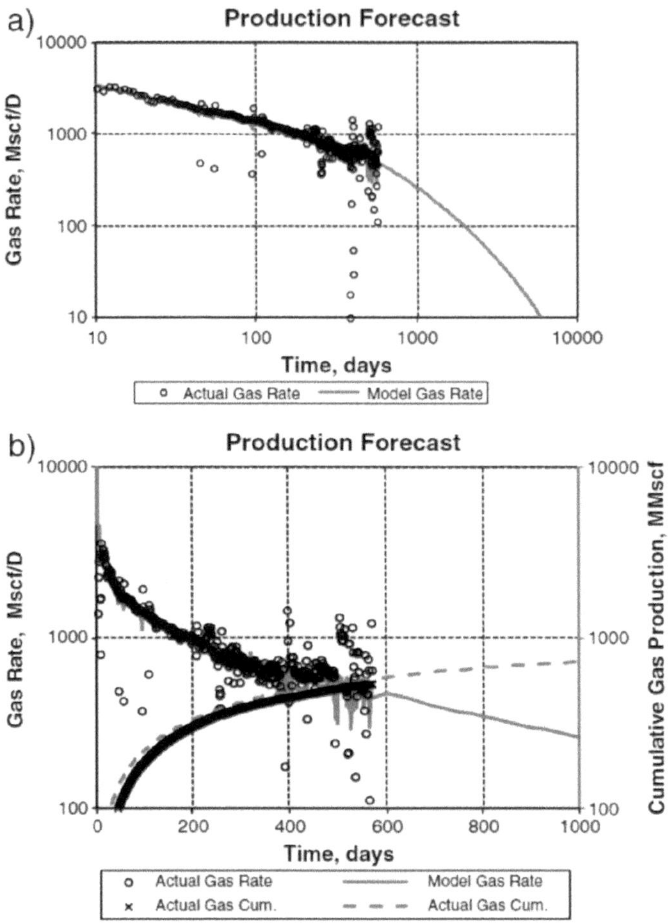

FIGURE 8.48 Analytical model match of case 2.

Source: Clarkson, 2013

complex fracture geometry, provide a robust way to model such complex geometries with more detailed treatment (Figure 8.53).

8.4.3 Hybrid Analytical-Numerical Models

Ayoub et al. (2018) presented a hybrid analytical-numerical model for unconventional reservoirs. They used a mathematical model based on complex fracture systems and multiphase flow as a concept of pseudo-pressure and pseudo-time. The gridding scheme is based on the work of Bennett et al. (1986). Figure 8.54 shows the Bennett gridding scheme, and Figure 8.55 shows the use of a regular tight grid scheme. The

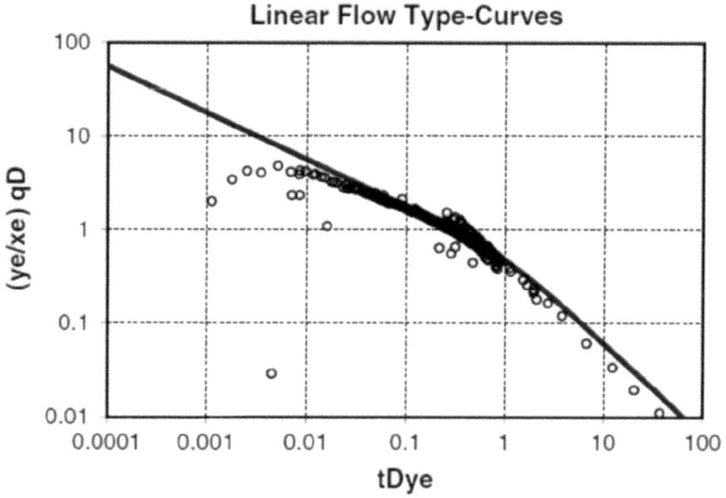

FIGURE 8.49 Type curve analysis for case 2.

Source: Clarkson, 2013

authors compared standard published solutions with the numerical model results. A perfect match was observed (Figure 8.56).

As for the regular tight gridding scheme, early-time data did not match perfectly with published solutions, especially for high-permeability reservoirs (Figure 8.57). A better match was observed in the late-time region of the data.

The Bennett grid system was selected because it provided better matching results. The hybrid model was tested against Cinco-Ley and Gringarten type curves based on infinite and finite fracture conductivity. Figure 8.58 shows that this model provides an excellent match to the models.

The authors validated their model using reservoir and completion data summarized in Tables 8.6 and 8.7, respectively. A 3,000-foot horizontal well with 21 planar fractures was drilled. Fractures have a half-length of 200 feet and infinite conductivity.

The authors generated simulation runs to compare oil and gas production rates and cumulative volume results between the hybrid and the numerical model. Results are shown in Figures 8.59 and 8.60. The change of flow regimes from linear to radial is captured by the model as Figure 8.61 shows.

8.5 DATA-DRIVEN METHODS

8.5.1 Decline Curve Analysis Using Machine Learning

Temizel et al. (2020) presented a comparison between traditional decline curve analysis and several machine learning methods used for unconventional tight shale reservoir production forecasting. The algorithms used by the authors are artificial

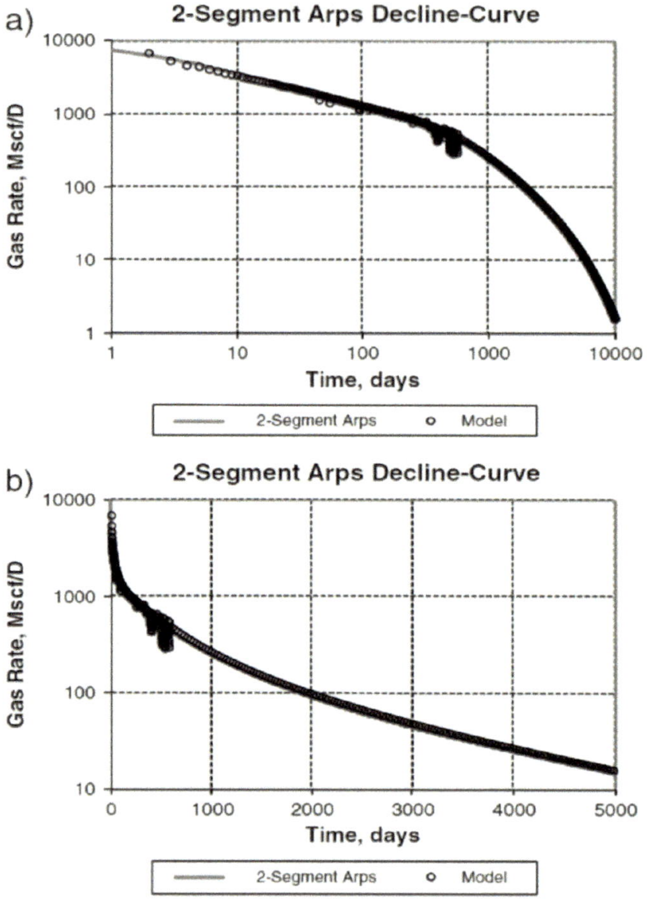

FIGURE 8.50 Two-segment Arps model match for case 2.

Source: Clarkson, 2013

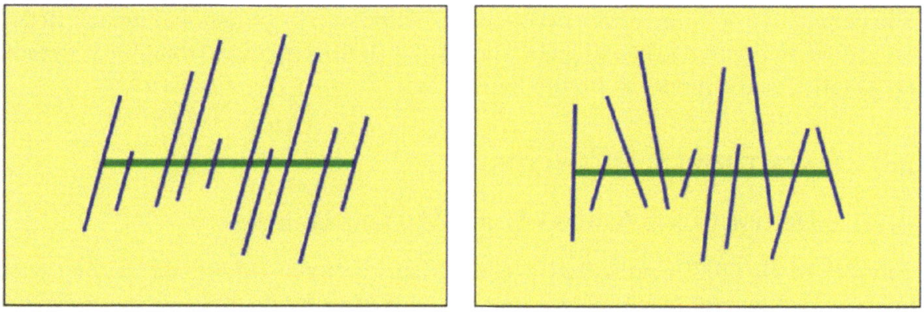

FIGURE 8.51 Off-center fractures and arbitrary fracture angles.

Source: Ilk, 2016

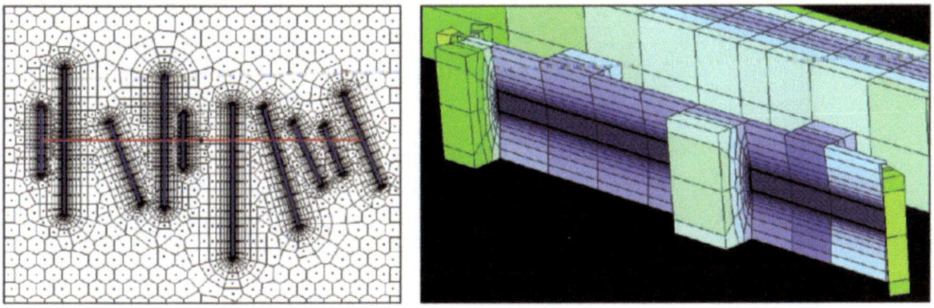

FIGURE 8.52 Grid pattern representation (left) and 3D refinement for partially penetrating fractures.

Source: Ilk, 2016

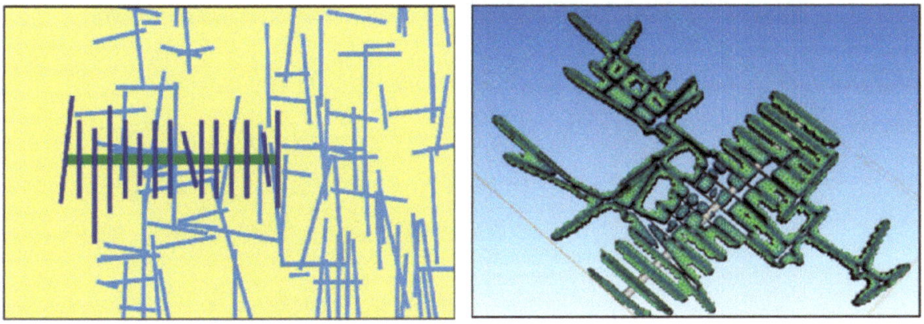

FIGURE 8.53 Numerical DFN representation.

Source: Ilk, 2016

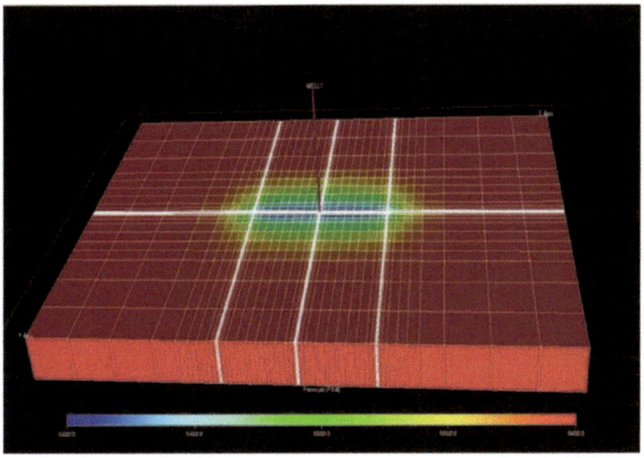

FIGURE 8.54 Vertical well fracturing based on Bennett et al. gridding.

Source: Ayoub et al., 2018

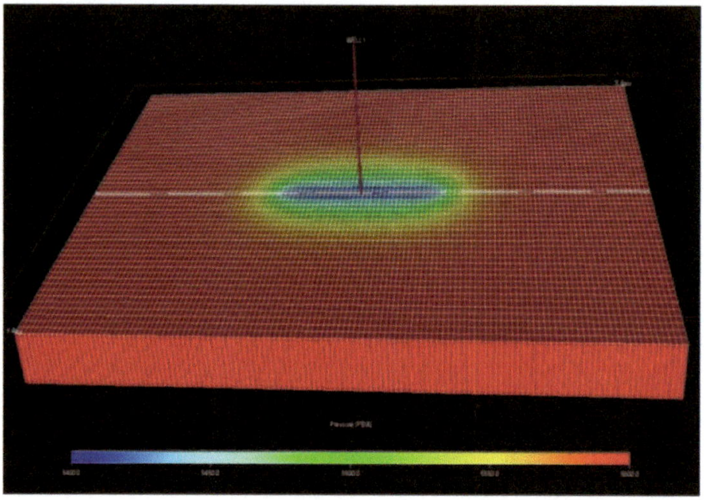

FIGURE 8.55 Vertical well fractured based on regular tight grid scheme.

Source: Ayoub et al., 2018

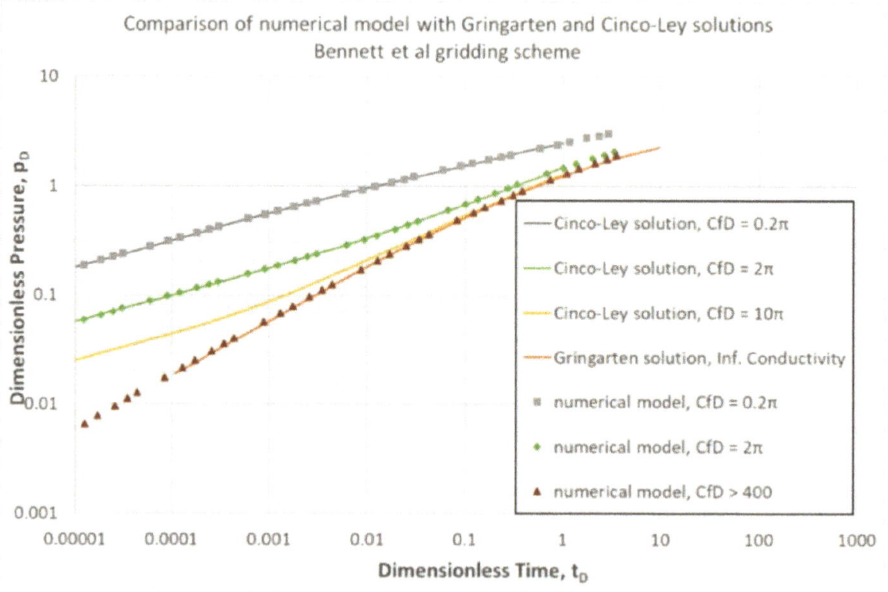

FIGURE 8.56 Validation of numerical approach based on Bennett et al. gridding scheme.

Source: Ayoub et al., 2018

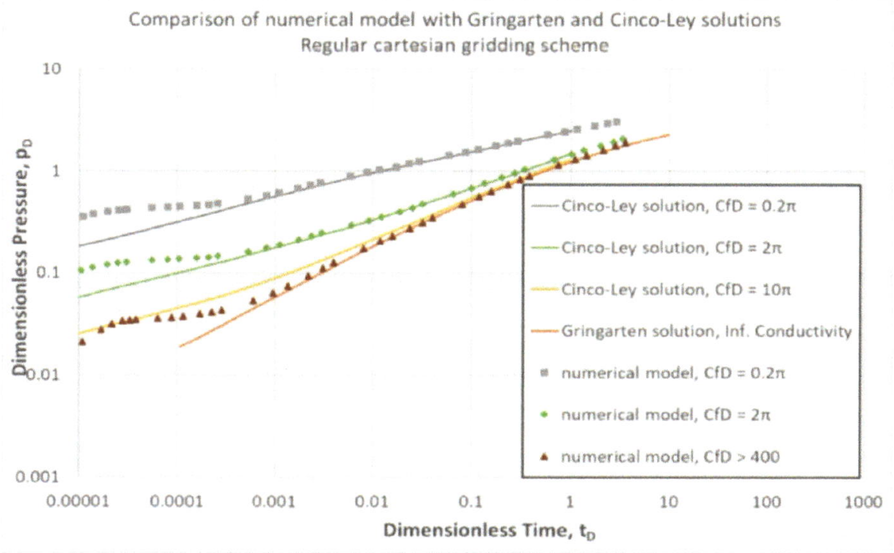

FIGURE 8.57 Validation of numerical approach based on a regular tight gridding scheme.

Source: Ayoub et al., 2018

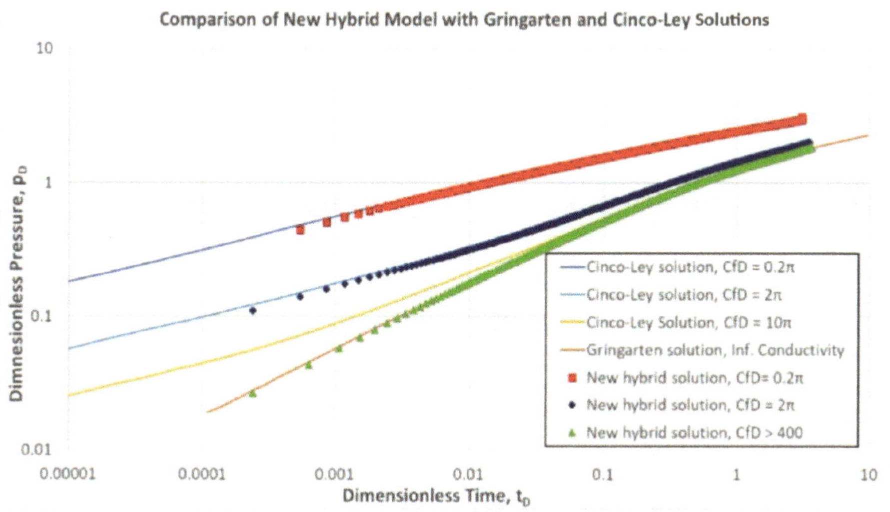

FIGURE 8.58 Validation of hybrid model based on published studies.

Source: Ayoub et al., 2018

TABLE 8.6
The Well Reservoir Properties

Reservoir top, ft	6,644
Reservoir length, ft	4,000
Reservoir width, ft	1,400
Reservoir height/net pay, ft	100
Reservoir permeability, md	0.005
Reservoir porosity	8.50%
Reservoir pressure, psi	5,600
Rock compressibility, psi−1	5.31E-06
Reservoir temperature, °F	200
Water saturation	20%

TABLE 8.7
Completion Data

Horizontal well depth, ft	6,694
Horizontal well length, ft	3,200
Number of orthogonal hydr. fractures	21
Effective fracture spacing, ft	150
Individual fracture half-length, ft	200
Fracture height, ft	100
Reservoir extent from end perfs, ft	500
Reservoir extent from fracture tips, ft	500
Dimensionless fracture conductivity	500

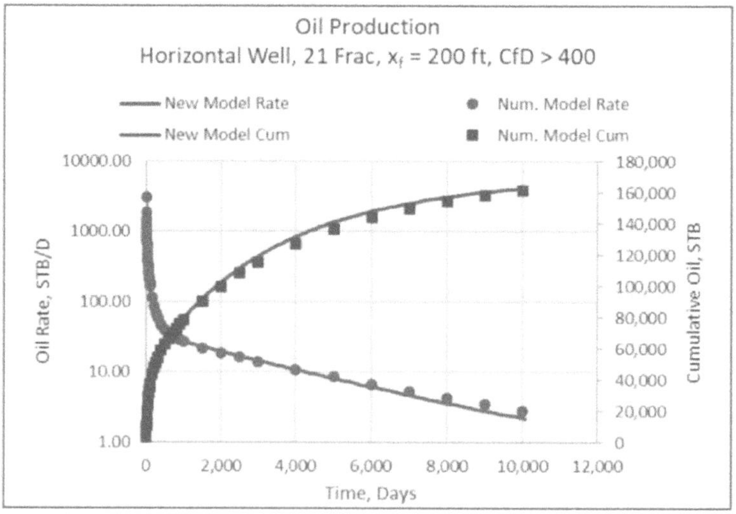

FIGURE 8.59 Oil production comparison.

Source: Ayoub et al., 2018

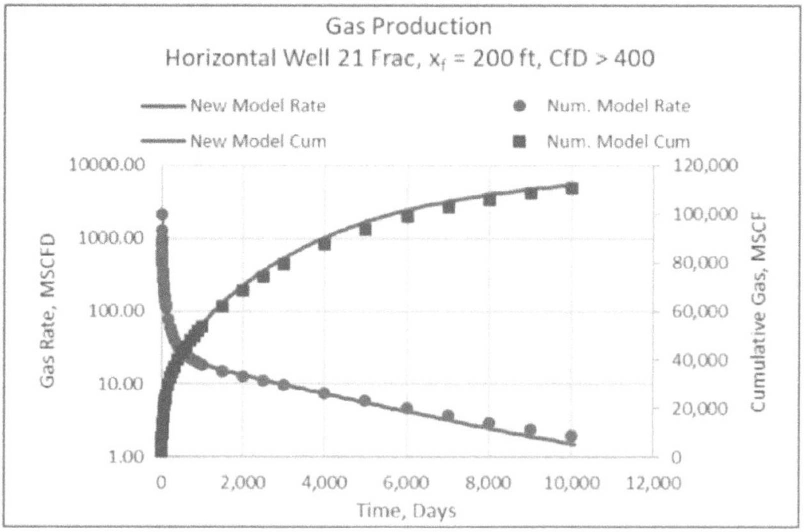

FIGURE 8.60 Gas production comparison.

Source: Ayoub et al., 2018

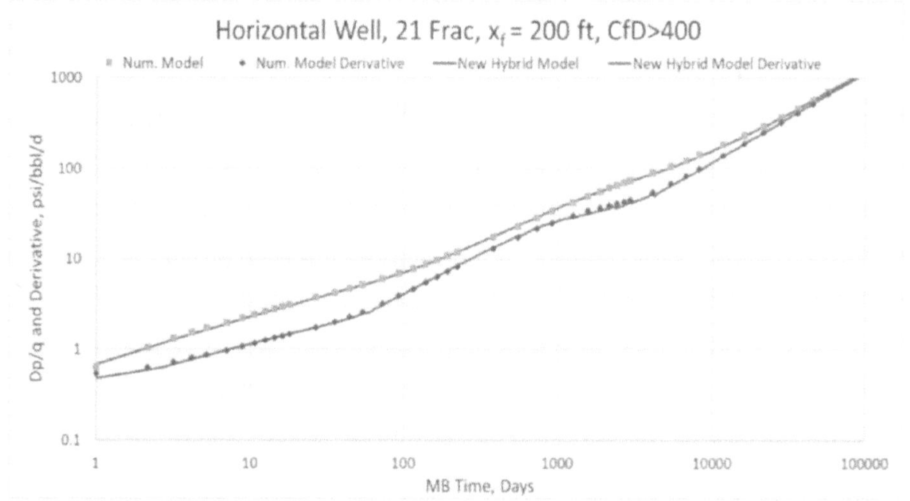

FIGURE 8.61 Log-log diagnostic plot based on 21 planar fractures.

Source: Ayoub et al., 2018

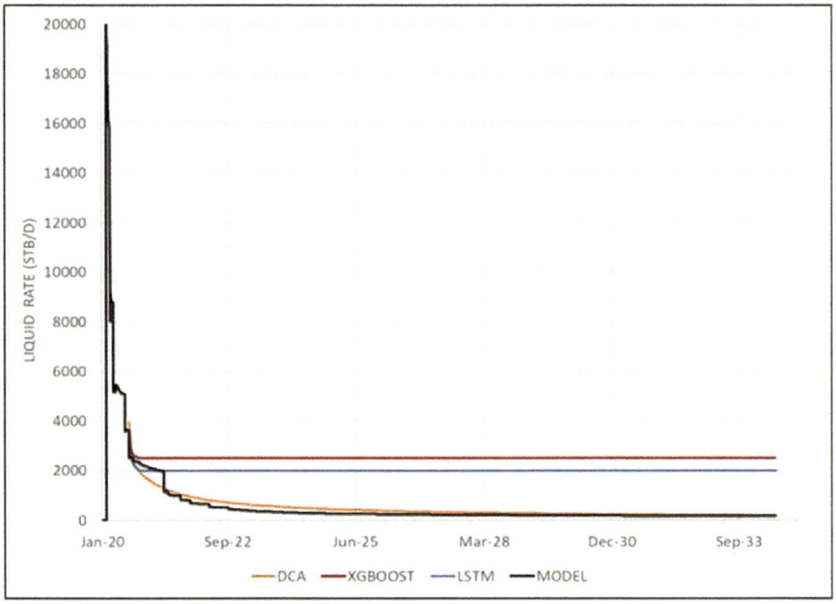

FIGURE 8.62 Liquid rate comparison (training time: 180 days).

Source: Temizel et al., 2020

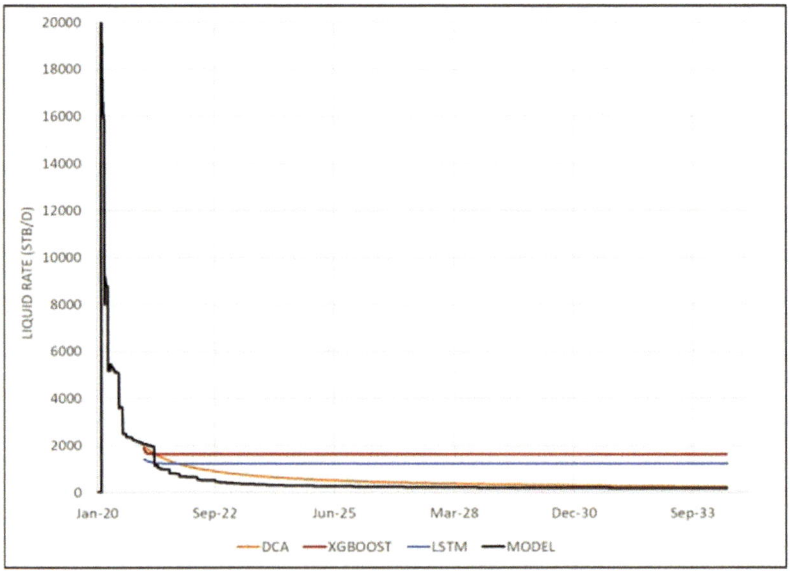

FIGURE 8.63 Liquid rate comparison (training time: 360 days).

Source: Temizel et al., 2020

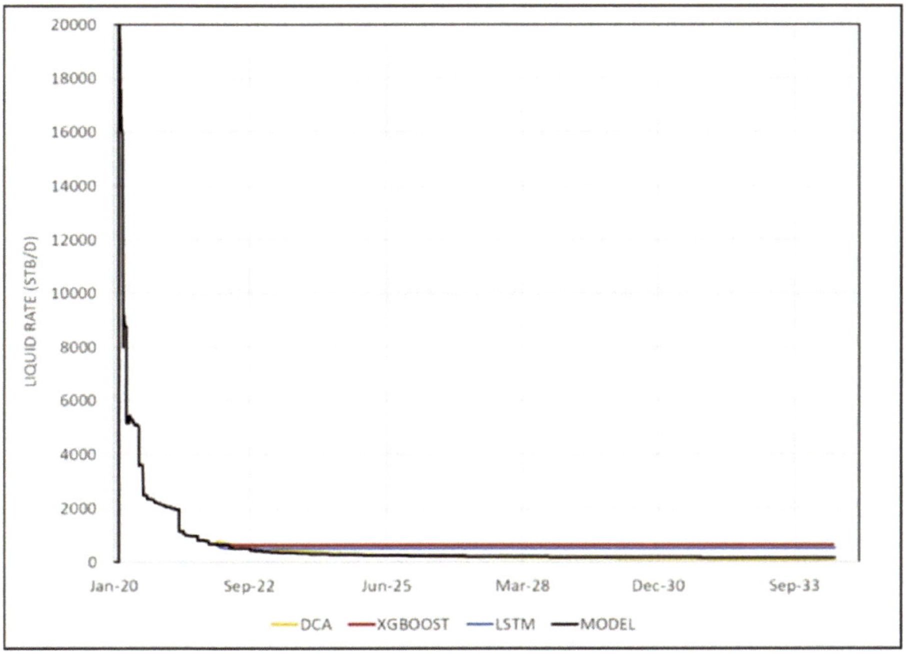

FIGURE 8.64 Liquid rate comparison (training time: 720 days).

Source: Temizel et al., 2020

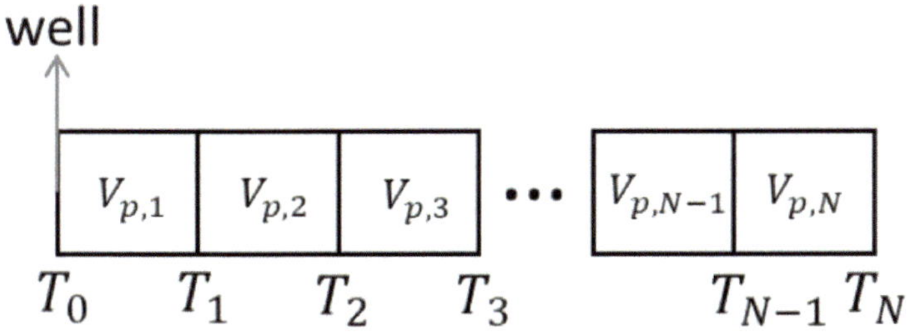

FIGURE 8.65 1D simulation of RGNet grid.

Source: Guo & Sankaran, 2022

neural networks (ANN) utilizing long-short term memory architecture (LSTM) and decision-tree boosting algorithm XGBoost. The authors provided a comparison for volatile oil, light oil, condensate, and dry gas cases. Figures 8.62, 8.63, and 8.64 show the volatile oil case. The results show that machine learning methods can provide good results if trained for a long time, but traditional DCA methods outperform machine learning methods when the training term is short.

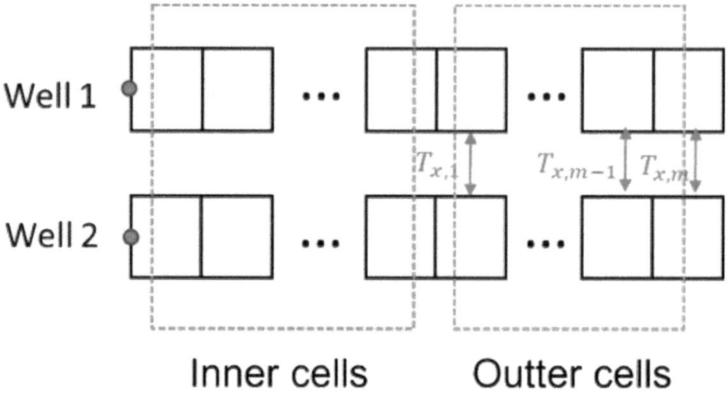

FIGURE 8.66 Simulation of two-well model.

Source: Guo & Sankaran, 2022

8.5.2 Graph Network-Based Unconventional Reservoir Modeling

Guo and Sankaran (2022) presented the use of the reservoir graph network (RGNet) to model unconventional reservoirs on the basis of reduced physics to reduce model complexity. They utilized 3D reservoir parameters as 1D grid blocks covering interconnecting wells. The model is based on history-matching observations from field data such as initial saturations, relative permeabilities, pore volumes, well indices, and transmissibility.

Figure 8.65 shows the structure of the RGNet grid. The characterization of the model is used transmissibility (T), well index (T_0), and pore volume (V_p).

The proposed model has multi-well capability; hence, the grid blocks can be divided into inner and outer cells (Figure 8.66). While inner cells constitute drainage volumes, communication between wells is represented by outer cells, which retain flowing property implementation.

Multiple data assimilation (ES-MDA) was used as the basis for the history-matching algorithm in the study. The authors presented two examples from the US onshore basin, one of which covers three wells while the other has a single-well case. The first case is based on one year of production of gas and water modeling with 50 inner cells. Gas production rate, bottom-hole flowing pressure, and water production rate are the observations (Figure 8.67). No initial oil saturation was ass umed.

Simulation results of history matching and drainage volume characterization are shown in Figures 8.68 and 8.69. The reasonable similarity between BHP and gas and water rates was reported by the authors. Future prediction for two years is shown in Figure 8.69, with uncertainty analysis incorporated into the model.

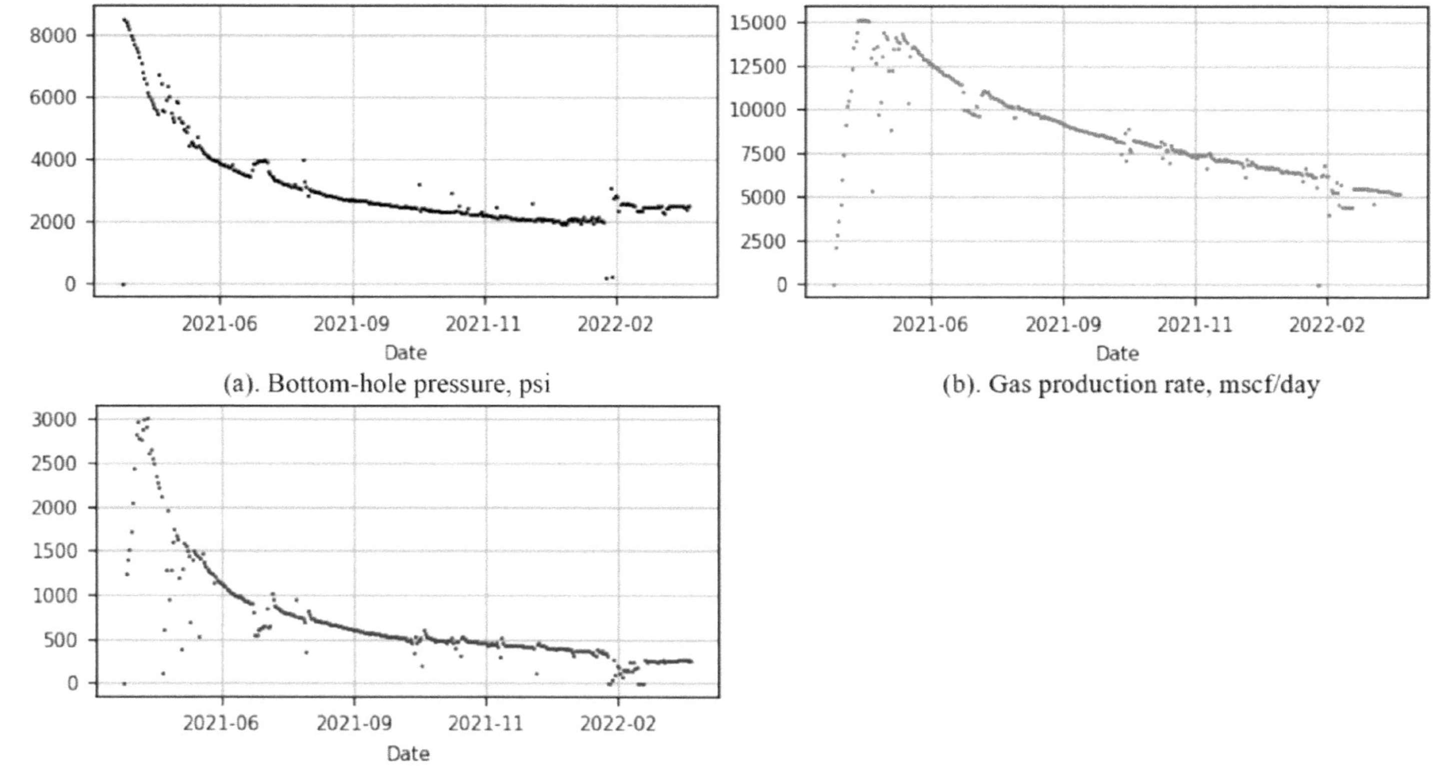

FIGURE 8.67 Case 1 production history data: (a) bottom-hole pressure, (b) gas production rate, and (c) water production rate.

Source: Guo & Sankaran, 2022

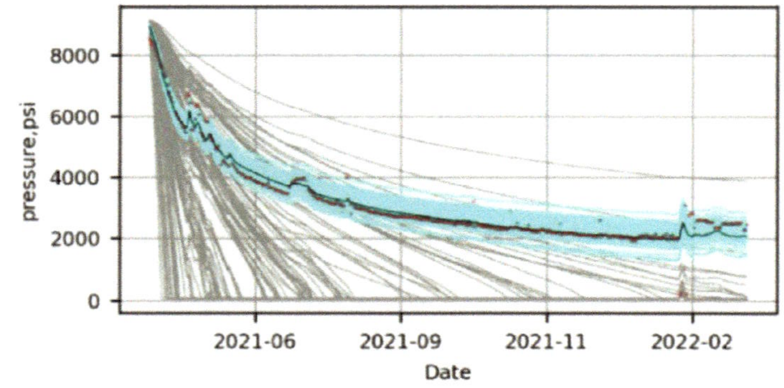

(a). Downhole pressures, psi

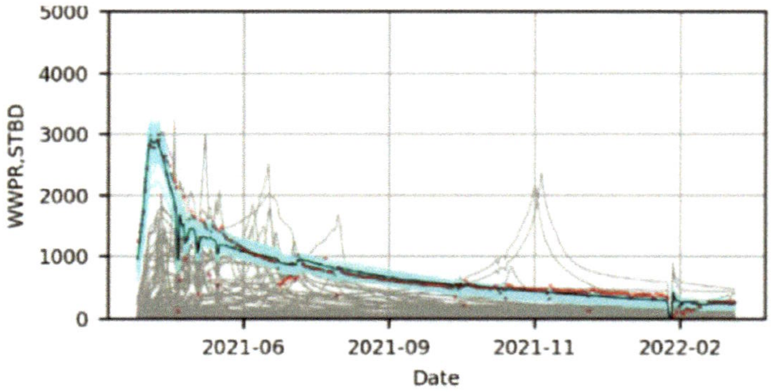

(b). Gas production rates, mscf/day

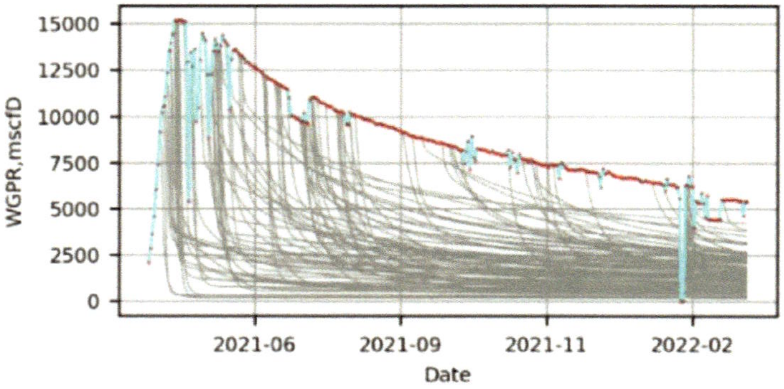

FIGURE 8.68 History-matching results: red dots—observed data, (thin) gray lines—prior ensembles, (thick) light blue lines—posterior ensembles.

Source: Guo & Sankaran, 2022

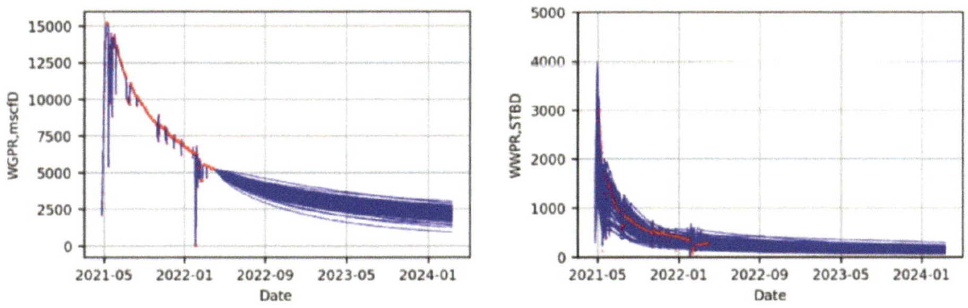

FIGURE 8.69 Results based on history-matching model: red (or dark gray when grayscaled) = data, blue (or light gray when grayscaled) = simulated).

Source: Guo & Sankaran, 2022

REFERENCES

Agarwal, R. G., Gardner, D. C., Kleinsteiber, S. W., & Fussell, D. D. (1998). Analyzing well production data using combined type curve and decline curve analysis concepts. *SPE annual technical conference and exhibition*. SPE-49222, https://doi.org/10.2118/49222-MS

Anderson, D. M., Nobakht, M., Moghadam, S., & Mattar, L. (2010, February). (2010). Analysis of production data from fractured shale gas wells. *SPE unconventional gas conference*. SPE-131787, https://doi.org/10.2118/131787-MS

Arps, J. J. (1945). Analysis of decline curves. *Transactions of the AIME*, *160*(1), 228–247.

Ayoub, J. A., Blakey, B., Krishnamurthy, S., & Thambynayagam, R. K. M. (2018). Applications of a novel hybrid model for unconventional reservoirs. *SPE/AAPG/SEG unconventional resources technology conference*. URTEC-2902834, https://doi.org/10.15530/URTEC-2018-2902834

Bennett, C. O., Reynolds, A. C., Raghavan, R., & Elbel, J. L. (1986). Performance of finite-conductivity, vertically fractured wells in single-layer reservoirs. *SPE Formation Evaluation*, *1*(4), 399–412.

Chen, H.-Y., & Teufel, L. W. (2000). A new rate-time type curve for analysis of tight-gas linear and radial flows. *SPE annual technical conference and exhibition*. SPE-63094, https://doi.org/10.2118/63094-MS

Cinco-Ley, H., & Samaniego-V, F. (1981). Transient pressure analysis for fractured wells. *Journal of Petroleum Technology*, *33*(9), 1749–1766.

Clark, A. J., Lake, L. W., & Patzek, T. W. (2011). Production forecasting with logistic growth models. *SPE annual technical conference and exhibition*. SPE-144790, https://doi.org/10.2118/144790-MS

Clarkson, C. R. (2013a). Production data analysis of unconventional gas wells: Review of theory and best practices. *International Journal of Coal Geology*, *109–110*, 101–146.

Clarkson, C. R. (2013b). Production data analysis of unconventional gas wells: Workflow. *International Journal of Coal Geology*, *109–110*, 147–157.

Clarkson, C. R. (2021). *Unconventional reservoir rate-transient analysis*. Gulf Professional Publishing.

Clarkson, C. R., & Beierle, J. J. (2011). Integration of microseismic and other post-fracture surveillance with production analysis: A tight gas study. *Journal of Natural Gas Science and Engineering*, *3*(2), 382–401.

Duong, A. N. (2010). An unconventional rate decline approach for tight and fracture-dominated gas wells. *Canadian unconventional resources and international petroleum conference*. SPE-137748, https://doi.org/10.2118/137748-MS

Economides, M. J. (2013). *Petroleum production systems*. Pearson Education.

Eleiott, A., Lee, W. J., & Moridis, N. (2019). Modified Fetkovich type curve enhances type well construction for horizontal wells with multiple fractures. *SPE/AAPG/SEG unconventional resources technology conference*. URTEC-2019-471, https://doi.org/10.15530/urtec-2019-471

Fetkovich, M. J. (1980). Decline curve analysis using type curves. *Journal of Petroleum Technology, 32*(6), 1065–1077.

Guo, Z., & Sankaran, S. (2022). Modeling unconventional reservoirs using a graph network model. *SPE/AAPG/SEG unconventional resources technology conference*. URTEC-3718999, https://doi.org/10.15530/urtec-2022-3718999

Ilk, D., Rushing, J. A., Perego, A. D., & Blasingame, T. A. (2008). Exponential vs. Hyperbolic decline in tight gas sands—understanding the origin and implications for reserve estimates using Arps's decline curves. *SPE annual technical conference and exhibition*. SPE-116731, https://doi.org/10.2118/116731-MS

Ilk, D., Blasingame, T. A., Houzé, O. (2016). Practical Considerations for Well Performance Analysis and Forecasting in Shale Plays, The Open Petroleum Journal, 9, (Suppl-1, M7), 107-136. doi: 10.2174/18748341016090100107.

Joshi, K., & Lee, J. (2013). Comparison of various deterministic forecasting techniques in shale gas reservoirs. *SPE hydraulic fracturing technology conference*. SPE-163870, https://doi.org/10.2118/163870-MS

Lee, J. (2021). Establishing the basis for multi-segment Arps decline models. *SPE/AAPG/SEG unconventional resources technology conference*. URTEC-2021-5393, https://doi.org/10.15530/urtec-2021-5393

Lee, W. J. (2022a). *Deterministic reserves evaluation course notes*. Texas A&M University.

Lee, W. J. (2022b). *Why do we use multi-segment arps decline models for forecasting in unconventional reservoirs*? Springer Nature Singapore.

Liang, P., Thompson, J. M., & Mattar, L. (2012). Importance of the transition period to compound linear flow in unconventional reservoirs. *SPE Canadian unconventional resources conference*. SPE-162646, https://doi.org/10.2118/162646-MS

Mattar, L., & McNeil, R. (1998). The "flowing" gas material balance. *Journal of Canadian Petroleum Technology, 37*(2).

Nobakht, M., Clarkson, C. R., & Kaviani, D. (2013). New type curves for analyzing horizontal well with multiple fractures in shale gas reservoirs. *Journal of Natural Gas Science and Engineering, 10*, 99–112.

Poston, S. W., Laprea-Bigott, M., & Poe, B. D. Jr. (2019). *Analysis of oil and gas production performance*. Society of Petroleum Engineers.

Pratikno, H., Rushing, J. A., & Blasingame, T. A. (2003). Decline curve analysis using type curves—fractured wells. *SPE annual technical conference and exhibition*. SPE-84287, https://doi.org/10.2118/84287-MS

Shahamat, M. S., and Clarkson, C. R., (2017), Multi-Well, Multi-Phase Flowing Material Balance, *SPE Unconventional Resources Conference*, SPE-185052, doi: https://doi.org/10.2118/185052-MS

Tabatabaie, S. H., Behmanesh, H., & Mattar, L. (2022). Using the flowing material balance model to determine which wells out of a group of wells belong to the same common pool. *SPE Reservoir Evaluation & Engineering, 25*(4), 719–729.

Temizel, C., Hakan, C. C., Saracoglu, O., Putra, D., Baser, A., Erfando, T., Krishna, S., & Saputelli, L. (2020). Production forecasting in shale reservoirs through conventional DCA and machine/deep learning methods. *Unconventional resources technology conference*, Virtual, 20–22, pp. 4843–4894.

Valkó, P. P., & Lee, W. J. (2010). A better way to forecast production from unconventional gas wells. *SPE annual technical conference and exhibition*. SPE-134231-MS, https://doi.org/10.2118/134231-MS

Wattenbarger, R. A., El-Banbi, A. H., Villegas, M. E., & Maggard, J. B. (1998). Production analysis of linear flow into fractured tight gas wells. *SPE rocky mountain regional/low-permeability reservoirs symposium*. SPE-39931, https://doi.org/10.2118/39931-MS

Yu, S., Lee, W. J., Miocevic, D. J., Li, D., & Harris, S. (2013). Estimating proved reserves in tight/shale wells using the modified SEPD method. *SPE annual technical conference and exhibition*. SPE-166198, https://doi.org/10.2118/166198-MS

9 Reserves and Resources Evaluation in Unconventional Resources

Ekrem Alagoz and Baris Tali

9.1 RESERVES AND RESOURCES IN SHALES

9.1.1 What Are Reserves and Resources?

The terms "resources" and "reserves" are usually used to quantify and qualify the presence of a mineral, hydrocarbon, or any other valuable material in a particular volume. The two distinct terms are usually confused and sometimes interchangeably used. It is essential to make the definitions clear to avoid incorrect usage.

By definition, a resource includes the amount of geologic commodity that presents in both discovered and undiscovered deposits. Reserves are a subgroup of a resource that has been discovered and defined within a known size and can be produced at a profit. The economic potential of shale reserves can be evaluated by aggregating works of different disciplines, such as geology, drilling, reservoir engineering, production engineering, etc. For a project to be considered reserves, a high probability of commercial development is essential.

9.1.2 Why Are Reserves and Resource Estimates Important?

It is extremely important to make reserves and resources estimates intelligently to understand the long-term economics and production forecasting of shale gas production. Companies and countries plan for future energy demand and cash flows based on the reserves and resources estimations. The estimate calculations include lots of uncertain parameters such as reservoir properties, production technology, the existence of a hydrocarbon system, etc. By conducting well testing, numerical simulations, and production analyses, the uncertainty of parameters is decreased to a certain level.

Cook (2009) presented a resource evaluation of the Bakken formation of North Dakota. The study includes a wide range of modeling methods to reduce uncertainty in resource play developments. The technical evaluation of the Bakken shale involved the following evaluations: geology of the Bakken, the historical development of the Bakken, well level analysis, the multi-tool method to address uniqueness, conceptual

DOI: 10.1201/9781003319955-9

model, descriptive reservoir models, well testing, production analysis, and decline curve analysis. They emphasized that the key development of a resource play is integrating all the related disciplines by use of available data.

Nascimento et al. (2018) used the Petroleum Resource Management System (PRMS) concept along with a 12-step methodology to decrease the uncertainty for a successive plan in exploration and production projects (Figure 9.1).

The representation of the PRMS method is shown in Figure 9.2. A 12-step methodology is a useful tool for decision analysis of petroleum field developments.

Lee and Sidle (2010) analyzed procedures and identified the techniques for gas reserve estimation in resource plays. The study emphasized the lack of complete understanding of fluid flow in shale gas formations because of different flow mechanisms like adsorbed gas on organic matter. The strengths and disadvantages of forecasting methods for resource plays were discussed in detail. Volumetric, material balance, analogy, decline curves, history matching, and type curves are the most widely used methods. Decline curve analysis was noted as providing valuable information about reserve estimation.

9.1.3 Regulators

The oil and gas companies require a contract with governments to legally carry out exploration and production activities.

Nascimento et al. (2018) discussed contractual and economic issues related to exploration and production in Brazil. Different contractual and fiscal regimes and their limits were presented in detail. The exploration and development of oil and gas reservoirs in Brazil are issued by Constitutional Amendment 9/1995, which includes a contract between Union and private/state companies to perform exploration and production activities. The petroleum law (law 9478) was established in 1997. Petrobras has a contract with Union for five billion barrels equivalent oil for exploration of the pre-salt. The limits of Brazilian contracts are given in Table 9.1.

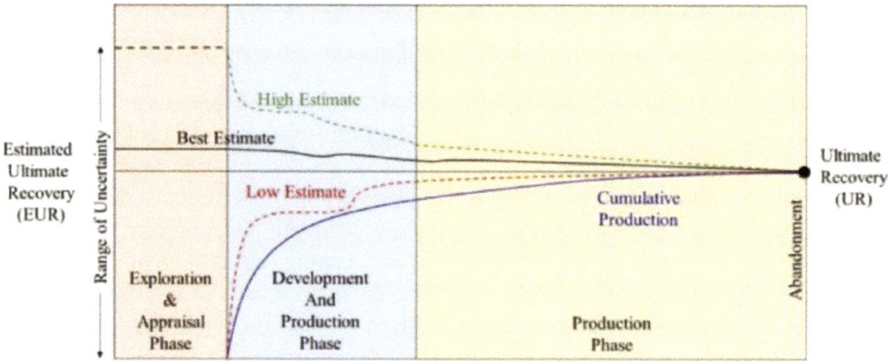

FIGURE 9.1 Uncertainty in exploration and production projects.

Source: Nascimento et al., 2018

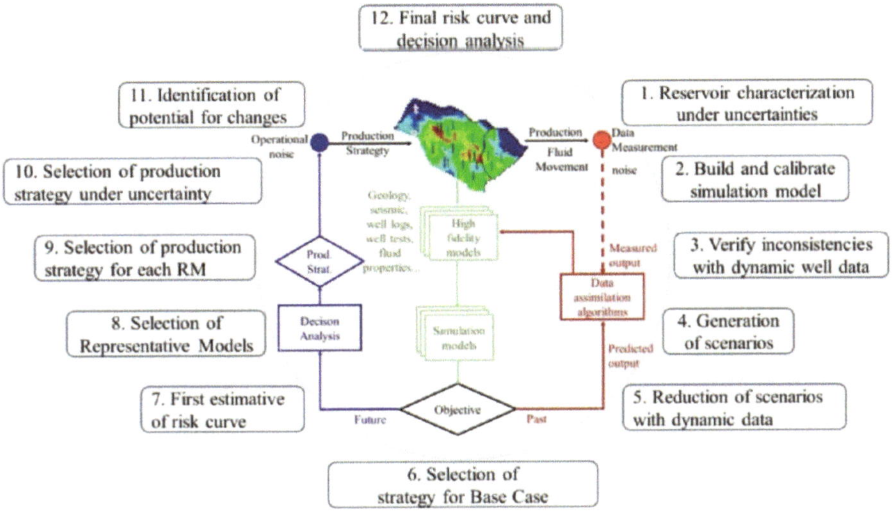

FIGURE 9.2 Schematic representation of the PRMS method.

Source: Nascimento et al., 2018

TABLE 9.1
Time Limits in Brazilian Contracts

	Limits	Renewal
Concession	Time: 5 + 27 years	Yes
Transfer of Rights/ Onerous Assignment	Volume: 5 Billion boe Tempo: 40 + 5 years	No
Prod. Sharing Contracts	Time: 35 years	No
Surplus of Onerous Assignment	Time: 35 years	Same as PSC

9.2 GOVERNMENTS IOCS AND NOCS

Petroleum is one of the largest industries in the world. There is a competition between governments to attract oil and gas developers for a homogeneous product, which requires intervention. International oil companies (IOCs) effectively develop oil and gas production with a good technical knowledge base and superior capital backing. On the other hand, national oil companies (NOCs) try to become self-sufficient and independent in developing oil and gas projects. Nowadays, NOCs and IOCs cooperate to increase the effectiveness of exploration, production, and asset development. Governments are not comfortable with the privatization of oilfields, and they try to exhibit resource nationalism by limiting the global trend of field privatization. Although joint ventures and alliances maximize profits, there is also a consideration of the political value of the assets from the point of view of the government.

Rudra (2013) proposed a conceptual framework for the technical complexity of the fields from the point of political economy in which NOCs have the exploration and production rights (Figure 9.3). The control variables are oil prices, fiscal regimes, and non-commercial objectives. The multilevel model used in the structural equation model includes government involvement and NOC resource capability. The objective of the study was to understand how the type and alignment of NOC resources play a role in the structure of a potential NOC-IOC alliance.

Radical changes have been observed in the relationship between NOCs and IOCs over the last two decades. The level of resource nationalism determines the power of NOCs and IOCs in a country. The knowledge base of IOCs can be incorporated with NOCs' ability to successfully execute a petroleum project. NOCs have increased resource activities to manage countries' resources, and they also decide which IOCs should participate in the projects. Governments usually try to extend the activities of NOCs beyond their borders, which means using NOCs as IOCs. These are called quasi-IOCs (Ledesma, 2009). Governments take risks to expand the responsibilities of NOCs. Ledesma (2009) published the roles of NOCs from Malaysia, Russia, Qatar, Algeria, Egypt, Indonesia, and Nigeria and their activities in Figure 9.4.

9.2.1 Banks and Other Lending Institutions

Banks and other financing companies give credit facilities to the oil and gas developers by direct lending, buying, or financial leasing of movable und immovable property. By definition, this credit cannot be given to banks, investment houses, insurance companies, or other financial institutions (SEC. 3(a)). The SEC provides the authority to companies to provide credit to consumers. The consumers are granted loans with reasonable interest rates and charges.

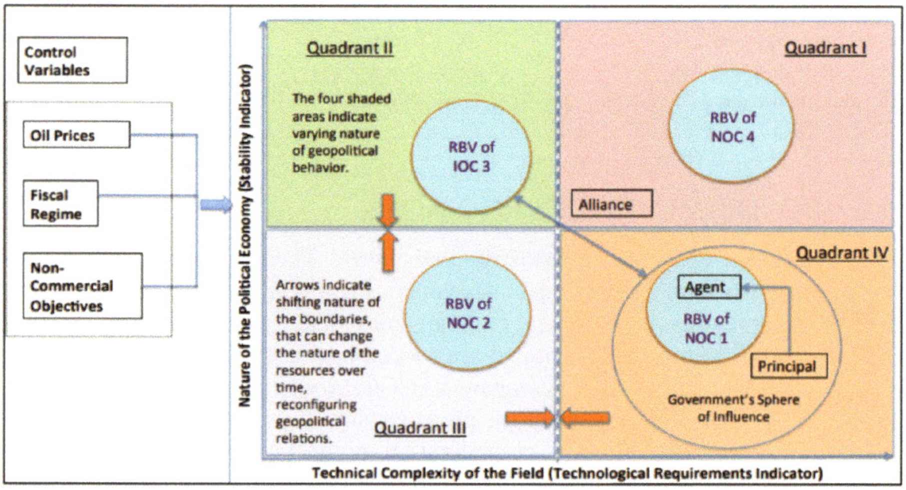

FIGURE 9.3 Conceptual framework for the technical complexity of the field.

Source: Rudra, 2013

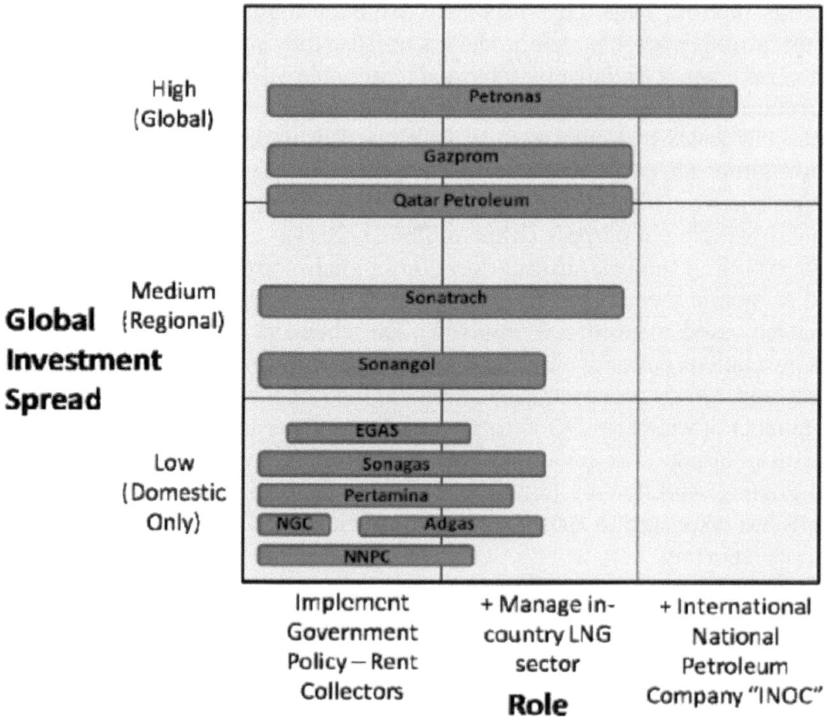

FIGURE 9.4 An illustration of NOC activities.

Source: Ledesma, 2009

9.2.2 What Do We Mean by Reserves Booking?

Reserves are booked when there is a reasonable expectation that the required infrastructures for transporting those reserves will be installed. A reserve can be booked if the following conditions are met:

- SEC proved criteria
- The development plan included in the budget plan
- Market is available

A market must be established for a reserve to be booked. If the market is not continuous, only that portion of produced gas should be booked as proved reserves (Wagner, 2014).

9.2.3 Concepts of Rights, Risk, and Reward

The volatility in oil and gas markets has a significant effect on shale gas development projects. The COVID-19 pandemic outbreak caused a shock effect on oil and gas consumption. Thus, oil prices have exhibited high volatility. Lower oil prices are not favorable for shale gas producers because they consider all the operation and production costs at a relatively stable and reasonable oil price level. However, the law governing companies' contracts can deal with unexpected situations. To illustrate, French law states that contracts can be revised if unforeseen situations disturb the initial equilibrium of the contract. Similarly, Egyptian law recognizes the unforeseen circumstances doctrine; New York law and Texas law include doctrines of impossibility and frustration of purpose (Blum & Zhong, 2021).

Determining the best drilling location of shale formations is a critical step for good economic metrics of shale gas projects. Statistical evaluation approaches are effectively used to minimize this risk. The chemical and physical properties of shale formations determine the success of the drilling and well completion program. Chorn et al. (2013) proposed a geostatistical workflow using standard logs of a limited number of wellbores. With the proposed workflow, a quantitative risk assessment of drilling targets was achieved. Thus, a higher rate of return was obtained with high-grading workflow by minimizing risks and maximizing rewards. Companies usually use exposure to risk in conjunction with the rights they assign to support reserves reporting.

9.2.4 Entitlement under Tax Royalty and PSC Contracts

The type of contractual relations provides how the NOCs, IOCs, and governments work together. Petroleum's fiscal regime has two main divisions: concessionary and contractual-based systems.

Concessionary is also known as royalty tax, and contractual is known as production sharing contracts (PSCs) and service agreements (SAs).

Types of agreements for some regions of the world are shown in Table 9.2:

The entitlement interest in a resource project is known as a royalty. Unlike working interest, royalty is free and clear on the sots and expenses of development and production to the royalty owner. Resource owners commonly retain royalties when they are granted rights for field production and development. The royalty owner recognizes only net revenue interest quantities. In some cases, the production taxes taken by governments are called royalties. These payments depend on various terms such as cost recovery, quantities produced, production rates, prices, and profits.

Royalty tax can be specified by the following features:

- Oil companies have the right to hydrocarbon exploration.
- If a commercial discovery is made, IOC has the right to produce from the field.
- IOC can title to its share-gross production less the royalty during production.
- Entitlement—gross production less royalty—at the wellhead.
- If the royalty is 10%, IOC can lift 90% of production. If it is paid in cash, then the IOC can lift 100% of production.

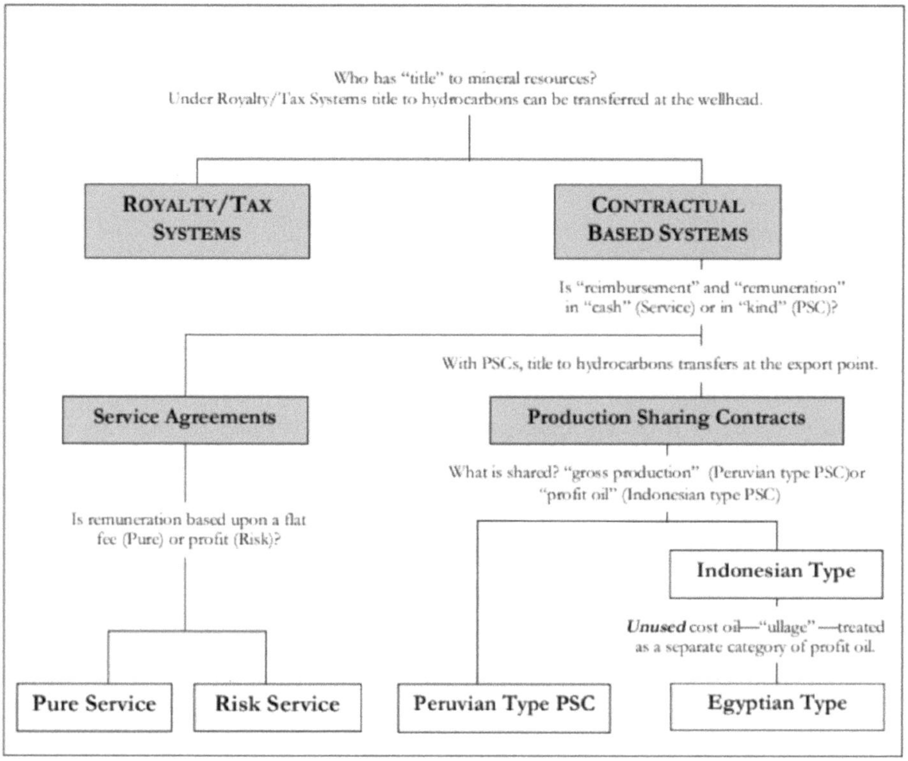

FIGURE 9.5 Classification of petroleum fiscal regimes.

Source: Johnston, 2017

TABLE 9.2
Types of Agreements in Different Regions of the World

Region	Type of Agreement
Latin America and the Middle East	Service agreements
Africa and FSU	Rate-of-return (ROR) features
Africa	Cost recovery limits based on net production
Former British Colonies	Work program bidding
FSU	PSA terminology (vs. PSC)
W, Africa	"Cost stop" terminology (vs. cost recovery limit)
Middle East	Taxes "in lieu"

Source: Johnston, 2017

Some characteristic features of production sharing contracts are as follows:

- Title to the hydrocarbons remained with the state
- Cost recovery limit
- Taxes paid
- Profit oil split
- Purchased equipment
- Company entitlement

9.2.5 Working Interest and Net Revenue Interest

The minimum share of gross revenue that the government takes in a given period of a field is called the effective royalty rate (ERR). Access to gross revenue (AGR) is the maximum share of revenues a company can take relative to their working interest. ERR is complementary to AGR, which is very important from the perspective of the international oil company. The net revenue interest refers to the income while the working interest means the expenses of a project.

9.2.6 Production-Sharing Contracts, Entitlement, and Tax Barrels

Contractors have to provide the technical and capital expertise needed for exploration and production in production-sharing contracts. An illustration of a production-sharing contract between the government and a contractor is shown in Figure 9.6.

Since a small share of production is needed for payback of investments, the contractor's entitlement to production usually decreases with increasing prices. It is a

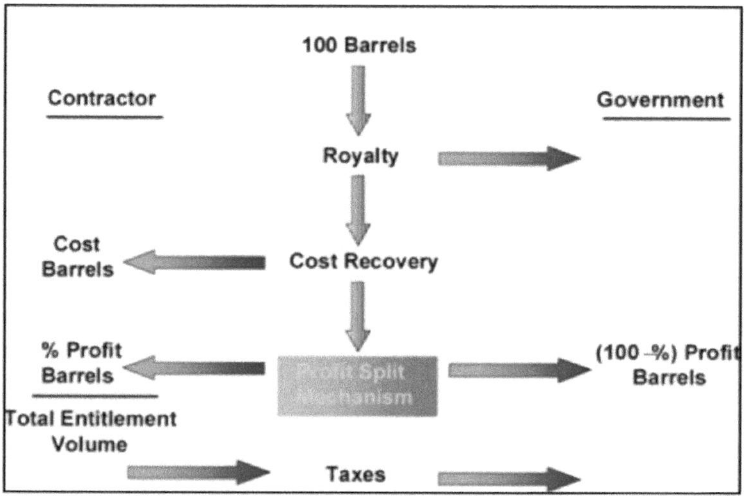

FIGURE 9.6 An example of a production-sharing contract.

Source: McMichael & Young, 2001

common procedure in this agreement to add terms that reduce entitlement as the production rate increases. The schematic distribution of a yearly project production between contractor and government is shown in Figure 9.7.

9.2.7 Risk Service Agreements

In risk service agreements, contractors take a defined revenue share rather than a production share. It is similar to a production-sharing mechanism, except for contractor payment. Capital and technical expertise are required for risk service agreements as well as production-sharing contracts. An illustration of a risk service contract is shown in Figure 9.8. This type of agreement is usually used for improved recovery operations.

9.2.8 Formal Reporting to Regulatory Authorities

Oil and gas companies are regulated by the Federal Energy Regulatory Commission (FERC). However, specific oil and gas components are issued by a number of other federal offices. Government agencies in the US are responsible for regulating the oil and gas regulations in their states. Some of the important agencies in the United States are as follows:

- California Division of Oil, Gas, and Geothermal Resources
- Texas Railroad Commission
- Louisiana Department of Natural Resources Oil and Gas

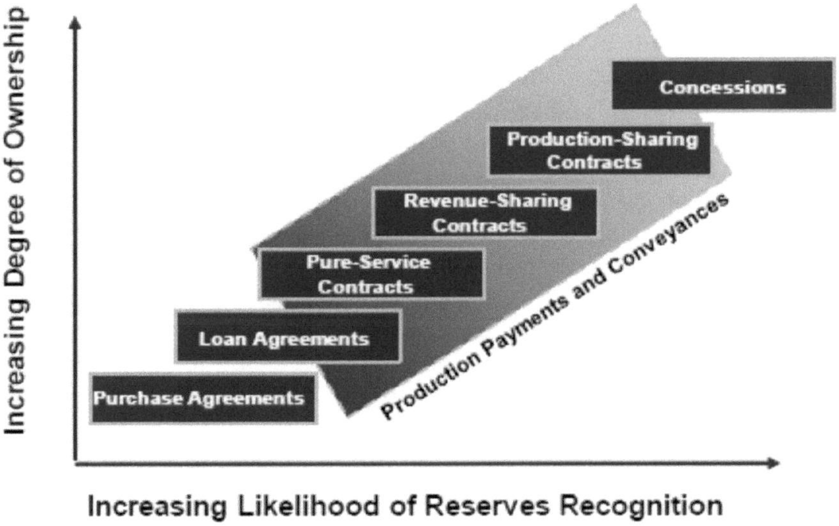

FIGURE 9.7 Spectrum of petroleum fiscal systems.

Source: Ampem-Darkoh, 2019

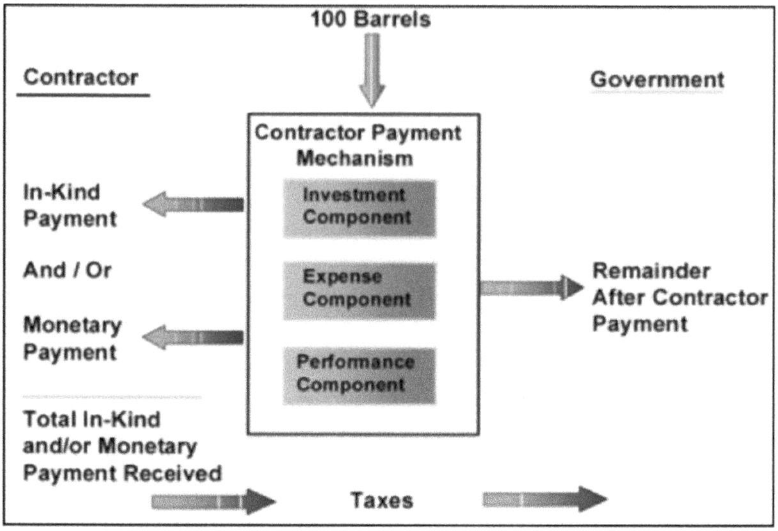

FIGURE 9.8 An illustration of risk service contract.

Source: McMichael & Young, 2001

- Alaska Oil and Gas Conservation Commission
- Oklahoma Corporation Commission Oil and Gas Division

Although it is difficult to estimate the future production of shale gas wells and tight gas reservoirs, resource owners have to forecast these reserves annually, and the results are reported to regulatory agencies and other stakeholders. To illustrate, companies that want to participate in the US capital market must report the reserve estimates to the US SEC. According to the SEC, the disclosed reserve has to be estimated based on reliable technology and uncertainty.

Scott and Blasingame (2014) overviewed various requirements for reporting oil and gas resources specified by market rules, regulators, accounting standards, and technical societies supported by the governments. Apart from national oil companies (NOCs), most oil and gas companies are listed publicly. They are usually required to report to investors using the guidelines. There is a contractual bond between companies and investors that the stock exchange publishes. There are also government legislations for the stock exchange. Companies that breach the stock exchange and reporting obligations can be penalized by the regulators empowered by the legislation. Oil and gas companies should fully inform investors who buy or sell the shares. Reserves and resources are used as a direct indication of investment decisions for average-size investors. Rules and requirements for some of the large, commonly known stock exchanges are given in Table 9.3.

Ampem-Darkoh (2019) examined detailed international agreements to grant rights to explore, develop, and produce hydrocarbons. A broad spectrum was covered with variation in fiscal terms. Specific agreements and terms with different fiscal

TABLE 9.3
Reserves and Resources Requirements of the Markets

Stock Exchange	Key Rule(s)	Specific Section(s)	Comment
ASX (Australia)	Section 5 Additional Reporting on Mining and Exploration Activities and Chapter 19 Definitions.	Hydrocarbon Reports 5.15, 5.16, and 5.17. Proved, probable and possible defined.	Rules provided by ASX.
TSX (Canada)	No specific requirements in listing rules.		Rules provided by the Canadian Securities Administrators.
LSE Main Market (UK)	No specific requirements in listing rules.		A number of rules provided by external regulators and authorities.
LSE AIM Market (UK)	Aim Guidance Note for Mining and Oil & Gas Companies—2009.	Definition of Standard and Appendix 3.	Rules provided by LSE and a number of rules provided by external regulators and authorities.
NYSE (USA)	No specific requirements in listing rules.	-	Rules provided by the US Securities and Exchange Commission.
SGX Main Market (Singapore)	No specific requirements in listing rules.	-	-
SGX Catalist Market (Singapore)	Catalist Rules 705(7), 1204 (23), Practice Note 4C, Appendix 7D.	The parts pertaining to oil and gas and not minerals.	Rules provided by SGX.
HKEx (Hong Kong)	Chapter 18 and Appendix 18	18.32 and 18.33.	Rules provided by HKEx
-	-	Proved, probable and possible defined.	-

Source: Scott & Blasingame, 2014

regimes were evaluated in the study. External and internal reporting of reserves were also discussed by considering the government's regulations. They aimed to increase understanding of PRMS 2011 and entitlement to production under modern fiscal agreements. The study emphasized the definitions, terms used for reserves by different institutes, companies, host governments, and home governments. However, PRMS reserves and resource categorization clearly state the terms and requirements for all the interested government regulatory agencies. National regulatory bodies have published regulations and standards for reporting oil and gas reserves in their countries. However, with the publication of PRMS, they mostly adopted its regulations. The format of reports required information, and description of the categories of reserves. The concept of economic interest was also used to report reserves. The Financial Accounting Standards Board (FASB) is used to determine if an economic interest exists.

9.2.9 The US Securities and Exchange Commission and the Financial Accounting Standards Board

The US Securities and Exchange Commission (SEC) was established after the Great Depression in the US. It was created to structure setting accounting standards. The establishment of FASB was encouraged by the SEC for private standard-setting bodies. The SEC works with private organizations very closely.

The Financial Accounting Standard Board (FASB) was created because of the need to reform the Accounting Principles Board, created in 1959.

9.2.10 Definitions of Reserves and Resources

Ross (1997) defined resources as the summation of the produced volume of hydrocarbon and the volume contained in the subsurface. The Canadian Institute of Mining, Metallurgy, and Petroleum (CIM) defined resources more specifically by adding time, quantities, discoveries, and accumulation terms. According to CIM, resources are "The total quantities of hydrocarbon and related substances that are estimated, at a particular time, contained in, or that have been produced from known accumulations, plus estimated quantities in accumulation yet to be discovered."

Ross (1997) proposed four key criteria to constitute a basis for reserve definitions:

- Discovered
- Recoverable
- Commercial
- Remaining

A resource classification tree was constructed based on these terms (Figure 9.9).

9.2.11 Overview and History of Reserves and Resources Definitions

The establishment of reserve and resources definitions started in the 1960s. The first SPE Oil and Gas Reserves Committee (OGRC) had an appointment in 1962 with cooperative industry organizations to inform and educate the public about the importance of the Petroleum Resources Management System (PRSM). The World Petroleum Council (WPC), the American Association of Petroleum Geologists (AAPG), and the Society of Petroleum Evaluation Engineers (SPEE) are the most effective organizations in the oil and gas industry. It is widely known that a standard of petroleum reserves and resources definitions started in 1997 with the joint publication of SPE and WPC. In 2000, AAPG also joined SPE and WPC to develop definitions of petroleum reserves and resources. By the inclusion of SPEE in 2007, PRMS 2007 was released. The Arab oil embargo of 1973 culminated in the Energy Policy and Conservation Act of 1975 (EPCA). Previously, the US government created the Financial Accounting Standards Board (FASB) in 1973 for the private sector to establish standards of financial reporting. The FASB's members are from academia, industry, and US-based firms in the financial community. The SEC was developed to take a larger role in reserves reporting and quantification with US oil supplies. The SEC definitions were for proved oil and gas reserves in 1975 (Harrell, 2008).

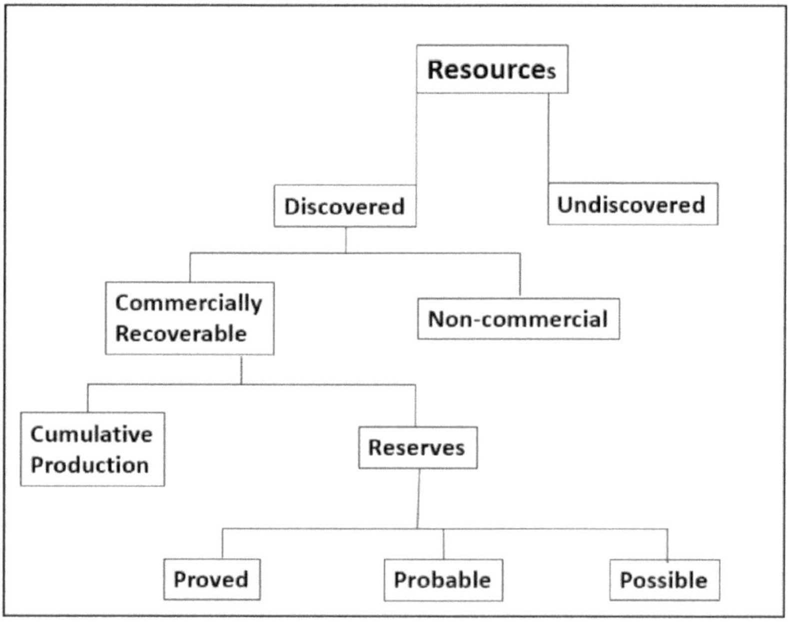

FIGURE 9.9 Resource classification tree.

Source: Ross, 1997

9.2.12 Reserves (Proved, Probable, Possible, 1P, 2P, and 3P) Criteria and Contingent Resources

Oil and gas projects are shaped based on reserve estimation. Stochastic reserve estimation is a widely used approach to assess assets by providing their chances of occurring. However, deterministic reserve estimation may also be compared with the stochastic approach if the best-estimate philosophy is applied. Ross (1997) published a study on the philosophy of reserve estimation. The study focuses on the resource classification system to understand the distinction between risk and uncertainty and establishing a basis for reserve definitions.

Ogilvie and Keyser (2004) recommended the use of geophysical technology for classifying resources and reserves. The standard reference definition for proved reserves, contingent resources, and possible reserves in the United States can be found in Security and Exchange Commission Regulation S-X (17 CFR 210.4-10-11/88). The SX regulation states that the lower proved reserve is defined with the lowest known structural occurrence. However, geophysical studies are now effectively used to define the structure area with reasonable uncertainty. The terminology of commercial reserves and contingent, as shown in Figure 9.10, includes the term "contingent resources," which include P4, P5, and P6. However, most of the time, they are used as proved resources (P4), probable resources (P5), and possible resources (P5). P1, P2, and P3 refer to proved reserves, probable reserves, and possible reserves, respectively.

Etherington and Ritter (2007) published a document on approved reserves/resource definitions by the Society of Petroleum Engineers (SPE), the American of Petroleum Geologists (AAPG), and the Society of Petroleum Evaluation Engineers (SPEE). The approved petroleum resource management (PRSM) is shown in Figure 9.11. The revised PRMS provides better classification guidelines using commercial evaluation processes. It was aimed to appropriate all types of petroleum deposits without considering their characteristics, extraction methods, etc. Respectively, 1C, 2C, and 3C represent low, middle and, high contingent resources. Proved, probable, and possible are used for commercial reserves amounts and labeled as 1P, 2P, and 3P, respectively.

PRMS has been accepted as a standard reference for reserve/resource evaluation since 2007. The SPE Oil and Gas Reserves Committee (ORGC) periodically updates the PRMS based on recent technology and commercial arrangements. Diluzio (2018) released the 2018 PRMS update, which was also approved by the SPE board of directors (Figure 9.12). The improvement of PRMS and retained properties are mentioned in the study in detail. Sub-classes of contingent resources (C1, C2, and C3) are updated to define development from unclarified separately. Technology is a criterion for reserve classification. Technical and commercial conditions are applied during EUR estimations.

The contingent resource quantities are important for project developments; 1C is too small to commit a project, and 2C is viable. The contingent quantities (1C, 2C, and 3C) are removed as all the requirements for satisfying reserves are assigned. Thus, the terms 1C, 2C, and 3C are directly transferred to 1P, 2P, and 3P reserve estimations. In the industry, there are no standard estimates for contingent resources; however, the common estimates suggest an 80% probability of proceeding to development (Society of Petroleum Engineers, 2011).

Moridis et al. (2019) defined the proper order of movements from prospective resources to contingent resources and reserves. The proposed methodology describes the "causes of change in classification" and the meaning of these changes when classifying resources. From prospective resources to reserves, updated PRMS and COGEH documents (2008) were revisited, and important terms, such as discovery,

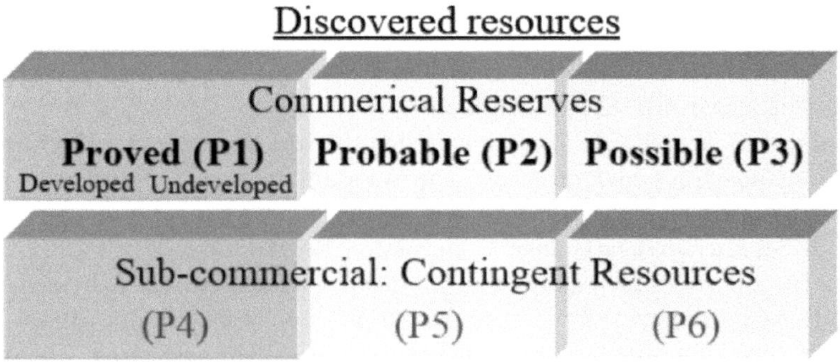

FIGURE 9.10 6P categorization system, including contingent resources and commercial reserves, adopted from the SPE, AAPG, and WPC.

Source: Ogilvie & Keyser, 2004

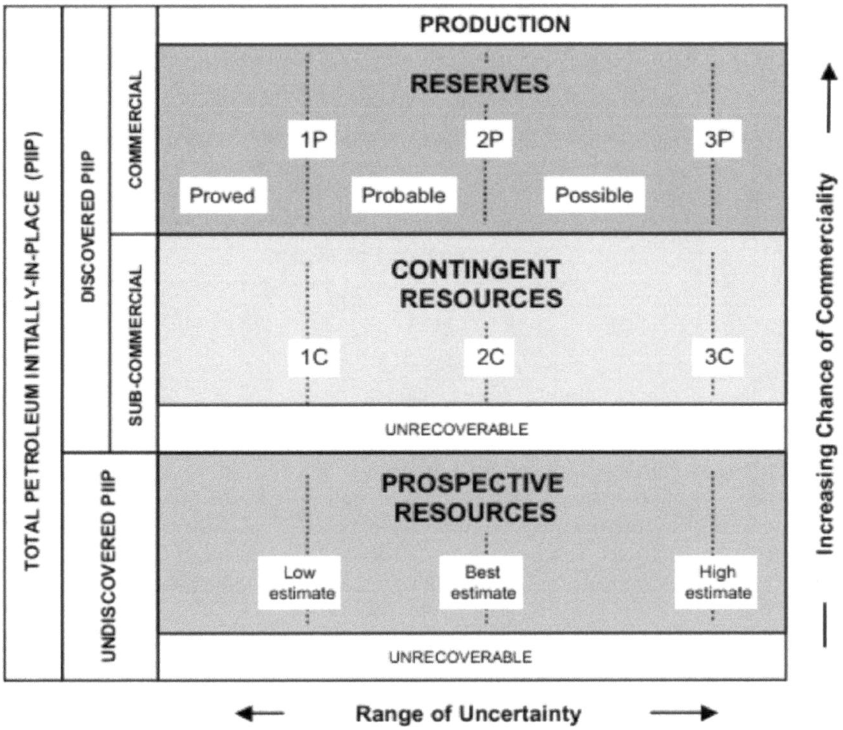

FIGURE 9.11 Petroleum resource management (approved).

Source: Etherington & Ritter, 2007

commerciality, and technology, were addressed. The study showed that straight movements from one class to another could not be applied for the movement from prospective resources to contingent resources to reserves. There are certain criteria that must be met to move between classes. The definitions and sub-classes published by PRMS (2018) are given in Table 9.4.

The movements between classes were based on commerciality and uncertainty. The workflow of the proposed methodology is as follows:

- Movements for undiscovered resources to become discovered are defined.
- Progression in the chance of commerciality is defined in project maturity sub-classes within the classification of contingent resources.
- The elements of the testing stage of technology are defined, and the criteria for technology to become "established technology" are described.
- Different contingencies and movement through each contingency are defined.

A visual representation of a process that can be used to move from prospective resources to contingent resources is shown in Figure 9.13.

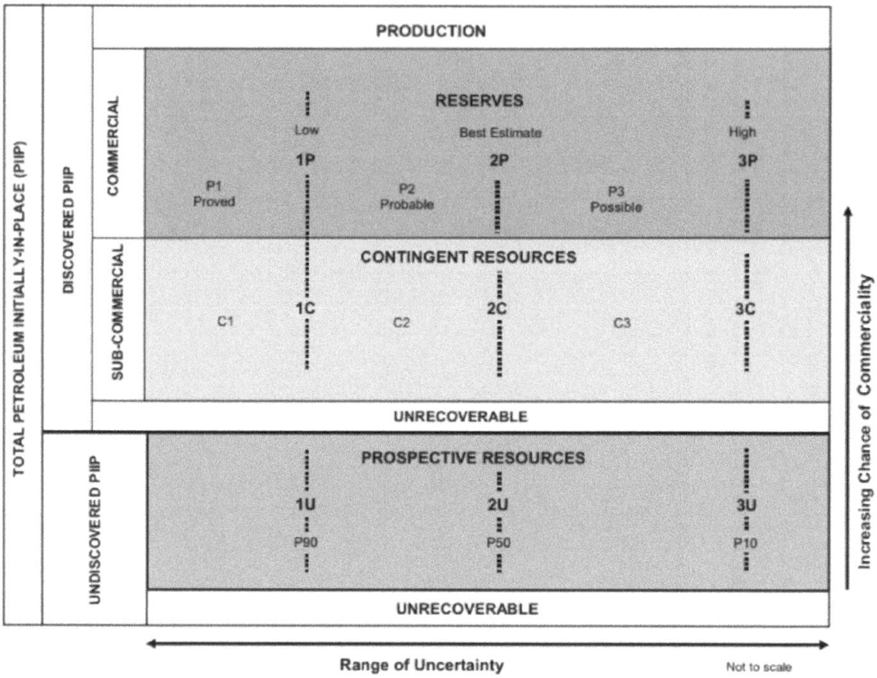

FIGURE 9.12 Framework of updated PRMS classification.

Source: Diluzio, 2018

Galvao et al. (2012) proposed a correlation-based probabilistic methodology for aggregating conventional and unconventional resources. The proposed methodology considers PRMS (2007) and the rules in the SEC (2009). The Monte Carlo simulation was performed for probabilistic aggregation. The study started with gathering information for uncertain parameters used in the simulations. Then, probabilistic simulations were conducted with reasonable risks. The flow diagram of information is shown in Figure 9.14.

The correlation coefficient between variables is used to estimate aggregated reserves, also known as the portfolio effect. Using the correlation coefficients for each project in Table 9.5, an aggregated reserve curve was constructed (Figure 9.15).

Badia et al. (2021) developed a corporate database for reporting reserve estimations to external authorities like the US Securities and Exchange Commission (SEC). The database provides approved reference data about reserve and resource estimates. It also has the ability to export data to other systems. Numerous reports are available in this system, including geological reports, license, reservoir, field, summary level, etc.

Zolotukhin (2000) proposed a novel approach to determining resources and reserves based on fuzzy mathematics. Fuzzy mathematics is good at handling

TABLE 9.4
Definitions of the Sub-Classes of Reserves, Contingent Resources, and Prospective Resources

Class	Sub-Class	Definition	PRMS
Reserves	On Production	A development project currently producing or capable of producing.	p. 31
	Approved for Development	All necessary approvals have been obtained, capital funds have been committed, and implementation of the development project is ready or is underway.	p. 31
	Justified for Development	The implementation of the development project is justified based on reasonable forecasted commercial conditions at the time of reporting.	p. 32
Contingent Resources	Development Pending	A discovered accumulation in which project activities are ongoing to justify commercial development in the foreseeable future.	p. 32
	Development on Hold	A discovered accumulation in which project activities are on hold, and/or justification as a commercial development may be subject to significant delay.	p. 32
	Development Unclarified	A discovered accumulation in which project activities are under evaluation, and justification as a commercial development is unknown based on available information	p. 32
	Development Not Viable	A discovered accumulation in which project activities are on hold, and/or justification as a commercial development may be subject to significant delay.	p. 32
Prospective Resources	Play	A project associated with a prospective trend of potential prospects that requires more data acquisition and/or evaluation to define specific leads or prospects.	p. 46
	Lead	A project associated with a potential accumulation that is currently poorly defined and requires more data acquisition and/or evaluation to be classified as a prospect.	p. 44
	Prospect	A project associated with an undrilled potential accumulation that is sufficiently well defined to represent a viable drilling target.	p. 47

Source: PRMS, 2018

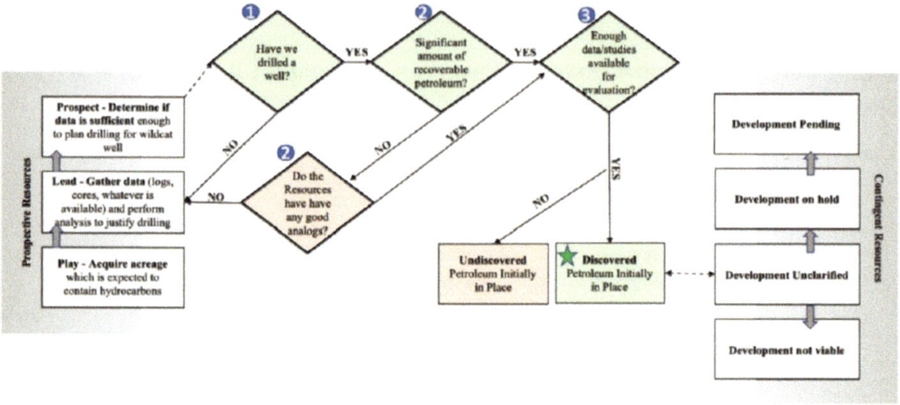

FIGURE 9.13 An illustration of movement from undiscovered to discovered resources.

Source: Moridis et al., 2019

uncertainties in a scattered environment. Fuzzy math was used to introduce uncertainty in deterministic reserve estimation. Fuzzy numbers handled the correlation between porosity and permeability (Figure 9.16).

Chan et al. proposed a process to evaluate unconventional resources. Special problems associated with the estimation and evaluation of shale gas and bitumen were summarized. To define the problems, it is important first to clarify the petroleum-resource triangle (Figure 9.17). As is apparent from Figure 9.17, the developed extraction technology is needed for unconventional hydrocarbon resources. The amount of hydrocarbon accumulation in unconventional reservoirs is much higher than conventional resources. The authors emphasized that recoverable resource quantities must allocate PRMS categories using the same methodology for traditional resources. A typical evaluation process starts with estimates of original-in-place volumes, and then, recoverable quantities are determined based on current technology and development programs. The PRMS does not require testing results or flowing behavior of hydrocarbon to be classified as discovery. Logging, sampling, or other evidence of hydrocarbon may be good enough for PRMS to classify them as "discovered." The discovered hydrocarbon technically has production potential. However, there are many identified recoverable accumulations that do not meet commercial criteria. Sub-classes of PRMS project maturity are shown in Figure 9.18. Chan et al. noted that the main challenge of evaluating unconventional resources is defining sufficient scale and recovery rates to be used for commercial projects.

Karacaer and Onur conducted a study on analytical probabilistic reserve estimation with a volumetric approach and resource aggregation. The uncertainty of the volumetric method was handled with the analytical uncertainty propagation method, which is based on Taylor-series expansion around the mean values of input parameters. The results were found comparable with the Monte Carlo method. Probabilistic aggregation of reserves was also studied for projects with more than one reservoir.

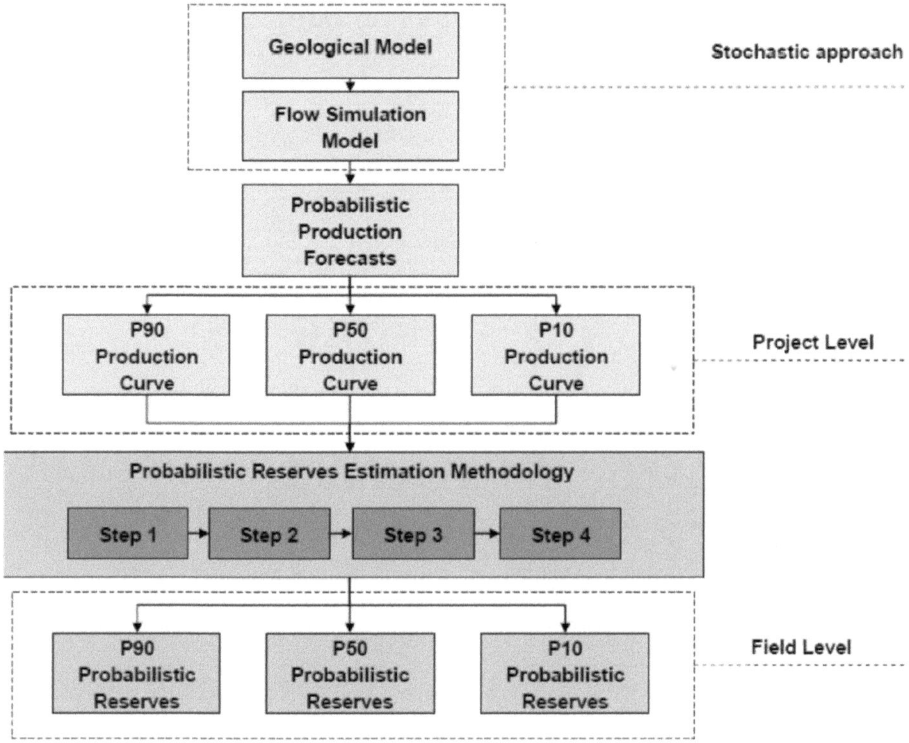

FIGURE 9.14 Information flow diagram.

Source: Galvao et al., 2012

TABLE 9.5
Correlation Coefficients

	A	B	C	D	E	F	G	H	I	J
A	1									
B	0.65	1								
C	0.82	0.64	1							
D	0.55	0.50	0.55	1						
E	0.61	0.63	0.63	0.67	1					
F	0.63	0.62	0.66	0.74	0.61	1				
G	0.55	0.65	0.71	0.67	0.52	0.60	1			
H	0.64	0.64	0.65	0.66	0.67	0.62	0.73	1		
I	0.57	0.63	0.67	0.68	0.64	0.59	0.56	0.65	1	
J	0.58	0.69	0.69	0.72	0.63	0.59	0.51	0.62	0.71	1

Source: Galvao et al., 2012

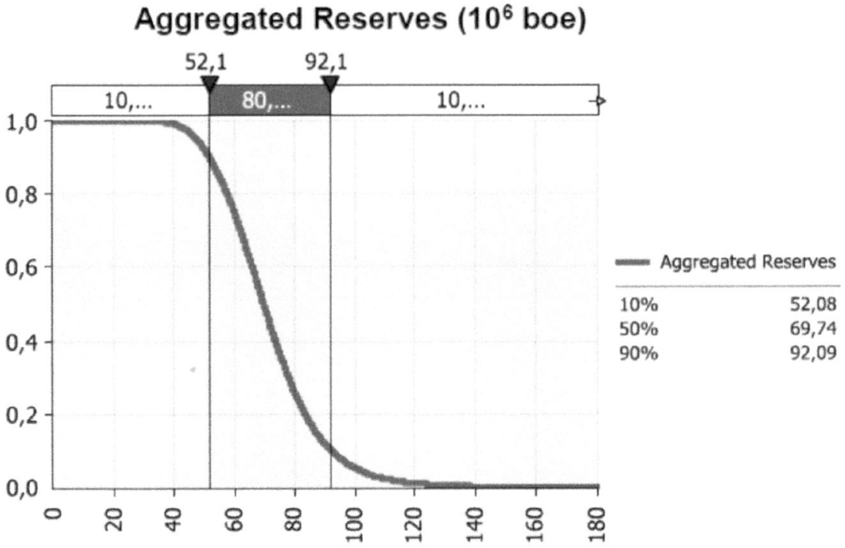

FIGURE 9.15 Aggregated reserves.

Source: Galvao et al., 2012

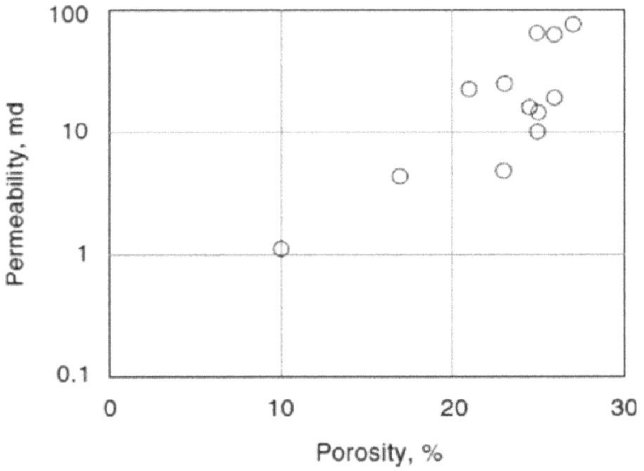

FIGURE 9.16 Porosity-permeability cross-plot.

Source: Zolotukhin, 2000

The uncertainty sorting method was proposed to determine correlation coefficients for multiple resources. A simple arithmetic sum of reserves may lead to underestimation of the P90 and significant overestimation of the P10 compared with probabilistic sums. The log-normal distribution model was used for aggregation of different numbers for synthetic fields. The skewness coefficient was found to change between 1.0 and 1.8 (Figure 9.19).

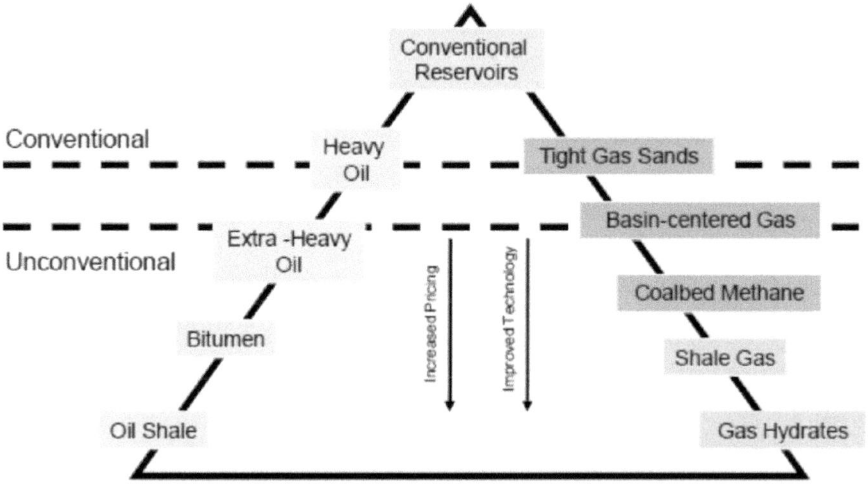

FIGURE 9.17 Petroleum-resources triangle.

Source: Chan et al., 2010

Coll (2019) applied probabilistic and deterministic approaches for reserves and resources evaluation. Deterministic and probabilistic methods are the two important commonly used methods for resources and reserves estimation, and they are often considered mutually exclusive. The probabilistic method covers a wide range of uncertainty on key reservoir properties with given confidence intervals. However, the deterministic method uses the best estimates for each reservoir parameter, and it does not provide a confidence level of uncertainty. The study focused on probabilistic methods compliant with the known regulatory institutes such as the SEC and the SPE PRMS. Since most of the model parameters include certain uncertainty, probability distribution functions are used for all the input parameters. Monte Carlo method, experimental design method, global optimization method, evolutionary strategy, and genetic algorithms are some of the important probabilistic methods assessed in the study. Figure 9.20 shows the workflow to be used for probabilistic purposes in pre-production fields. First of all, input parameters are defined with their uncertainty ranges. Sensitivity studies are then performed for reservoir simulation to identify the key parameters. History match and risk analysis of the fields are constructed. Lastly, the cumulative probability distribution of the response parameter R is estimated, and P90, P50, and P10 are defined.

9.2.13 Prospective Resources

Prospective resources include the chances of hydrocarbon discovery using the probability of accumulation components such as trap, hydrocarbon resources, and migration. Uncertain parameters are handled with distribution functions to reveal the potential of recoverable quantities for a discovered resource. Generally, prospective resources are not published externally, and they are internally documented.

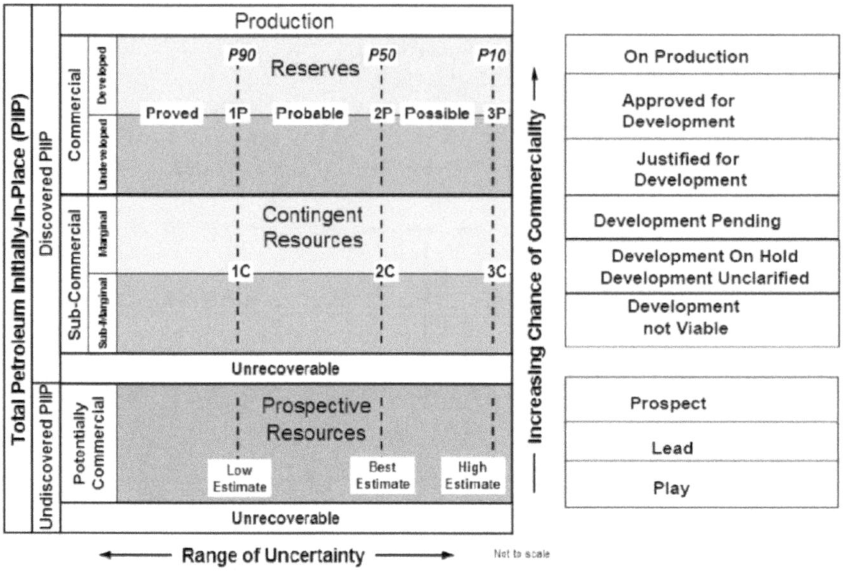

FIGURE 9.18 PRMS reserves and resources classification with project maturity sub-classes.

Source: Chan et al., 2010

9.2.14 Concept of a Project

The concept of shale gas projects is to increase the amount of gas that can be economically produced from a given area while minimizing the project cost. The objective of projects is to maximize profit, which is achieved by drilling multiple horizontal wells from a single drilling pad and creating vertical fractures with multistage frac operations. Created vertical fractures are planar surfaces that are thousands of square feet in size. Wells with high formation contact areas generally provide very large gas productivity.

The SPE Petroleum Resource Management System (2007) was a project-based system. The estimation and categorization of reserves were based on project definition and its treatment play. Morales and Lee (2015) studied probabilistic reserves addition within a project. They noted that a wrongly defined project must be considered improper risk assessment of resources estimation. SPE-PRMS specifies the type of aggregation to be used in a project with probabilistic addition. However, most companies still have not adopted their systems for best practices. Morales and Lee (2015) used a liquefied natural gas (LNG) project to describe a simple reserve estimation project. The project included regulations, recommendations, and existing standards for joint-venture project partners. Proved reserves (P90) are usually desirable for commitment purposes of LNG projects. Underestimation or overestimation of results may lead to loss of opportunities or reflecting less uncertainty. Therefore, the requirements of regulatory bodies such as the SEC and ESMA are essential for reserve estimations.

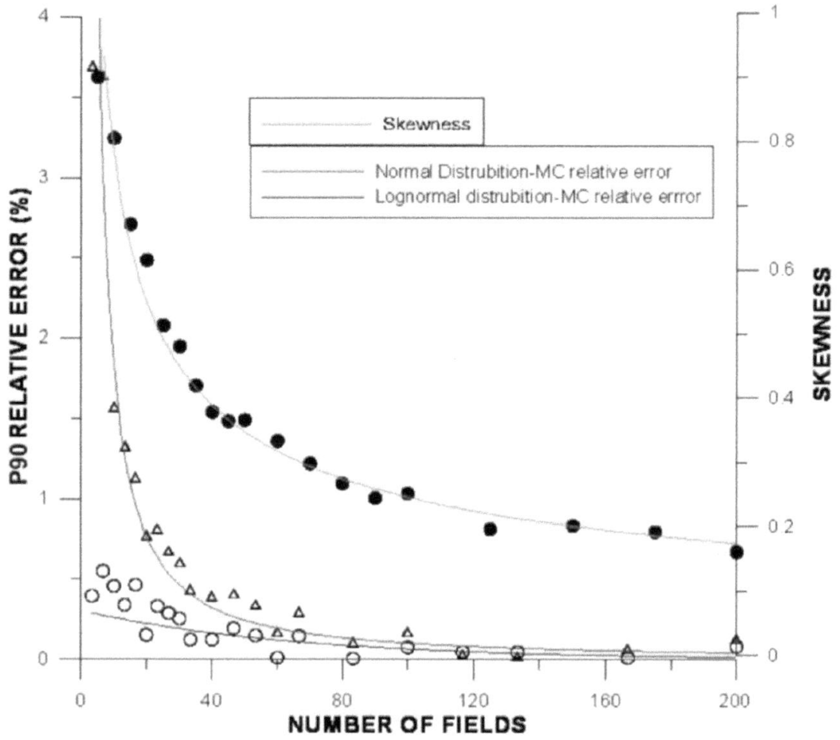

FIGURE 9.19 Aggregation of a different number of random synthetic fields.

Source: Karacaer & Onur, 2012

PRMS (2018) notes that the concept of a project-based classification can be clarified by analyzing the evaluation elements of net recoverable resources (Figure 9.21).

A project may include the development of a reservoir, well, or small field production. Each level of the project has a certain cash-flow schedule. The schedules are shaped based on technical, economic, and contracts of projects that constitute these schedules. Unique contractual rights may be assigned for each license area, including fiscal terms. One license area may include more than one reservoir or one reservoir produced by different license holders. Thus, one can say that the projects are the primary elements of resource classification, and the recoverable quantities are derived from projects. PRMS (2018) stated that there should be a range for estimated recoverable quantities derived from projects.

Technically feasible projects are not necessarily commercial. The economic viability of projects is determined based on a forecast of conditions such as prices, capital costs, marketing, rates, and cumulative production.

9.2.15 Methodologies

Hydrocarbon accumulation and storage capacity within a formation should first be properly characterized by conducting geological, geophysical, and formation

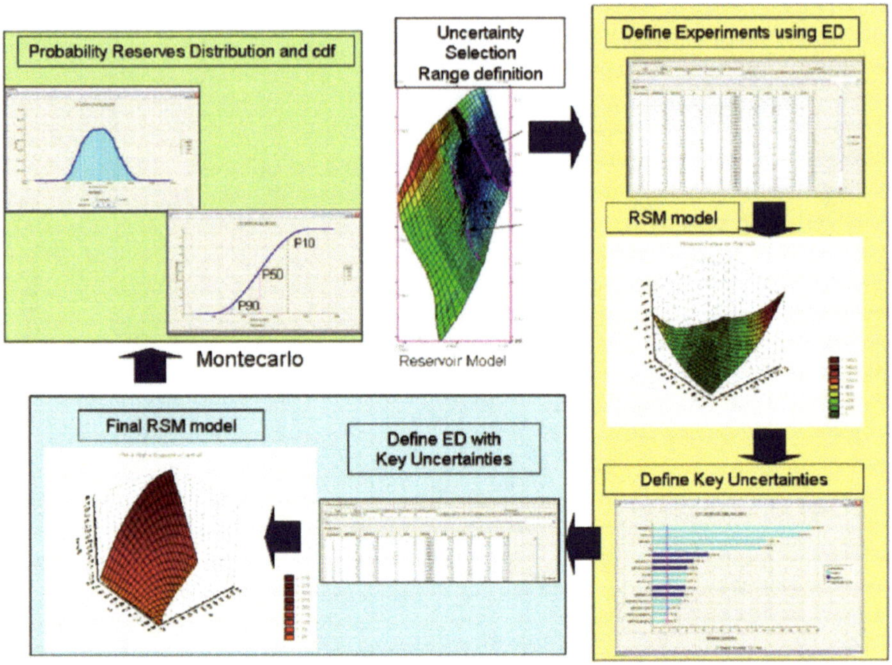

FIGURE 9.20 General probabilistic workflow to be used for pre-production fields.

Source: Coll, 2019

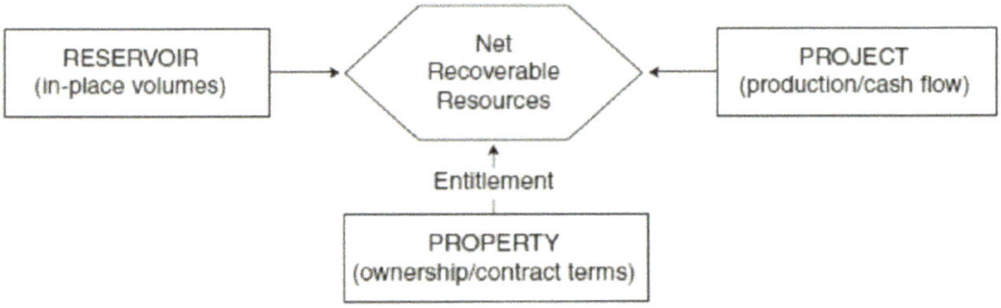

FIGURE 9.21 Resources evaluation.

Source: PRMS, 2018

evaluations. The most proper characterization of a resource can be made with production data. History matching of production data with representative numerical models or analytic models enables us to forecast the production of the particular formations. The characterization process can be carried out using different data sources such as drilling cuttings, production data, well logging, outcrops, and core data.

After reservoir characterization, the extent and thickness of the formations are plotted for a given field area. If the field is mature, then offset wells are analyzed carefully to understand the production behavior of older wells under different production strategies. However, in wildcat fields, assumptions are made by creating an analogy with fields that have similar geological and reservoir properties and well completion design in the world. The most applied methods of forecasting hydrocarbon production are:

9.2.15.1 Material Balance

In the material balance method, a plot of p/z versus cumulative gas production is usually applied for conventional gas reservoirs to determine gas in place. Different assumptions are made in the material balance method. To illustrate, wells are assumed to produce from the same drainage area, and stabilized flow is assumed. However, in reality, the drainage area of wells usually changes because of production rates and wells' interference. Therefore, this method is less applicable to individual wells. In unconventional reservoirs, the drainage area of the well may not be reached in the lifetime of wells. Thus, the p/z plot method requires assumptions for applicability of material balance in unconventional reservoirs.

Gherabati (2018) estimated the drainage volume of multi-fracture unconventional oil wells using the material-balance method. They aimed to optimize well spacing and completions of oil shale wells. In shale wells, the stimulated rock volume is determined through microseismic and proppant tracer logs. However, the study states that microseismic overestimates and tracer logs underestimate the stimulated reservoir volume. To address this issue, the authors used a modified material balance method, which is based on production data and fluid properties. Oil, water, and gas rates of more than 100 wells in the Midland Basin were used to calculate drainage volumes. The analyses showed that higher proppant concentration provides more drainage volume, as shown in Figure 9.22, and overlapping drainage volumes obtained with completions less than 500 feet in length. The main drawback of this method is that it is applicable only in oil wells.

Duarte et al. (2014) performed a material balance method for shale gas reservoirs to maximize the final recovery by preventing premature closing of fractures. Sensitivity studies were performed for original gas in place stored in the fractures. To determine the time at which the matrix's contribution starts, a p/Z versus Gp plot was constructed (Figure 9.23). This critical time is assumed to be the point at which fractures to start to close. Thus, knowing critical Gp is essential for estimating adsorbed gas calculations.

The study concluded the following:

- More sorption capacity of rocks provides greater original gas in place (OGIP).
- The material balance method was effectively used to calculate OGIP for shale wells.
- Conventional MBE leads to overestimation when desorption manifestation is ignored.

Orozco et al. (2020) estimated huff-n-puff gas injection in shale oil reservoirs using the material balance method. Generally, multi-porosity effective oil compressibility is used to perform material balance in undersaturated oil reservoirs. However, it is

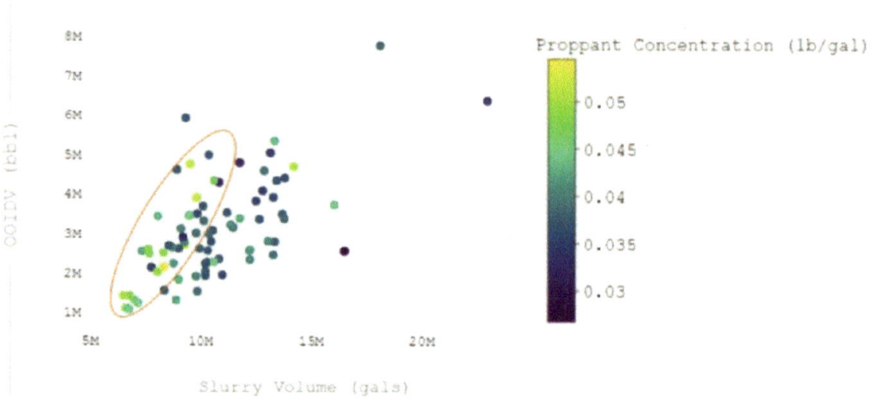

FIGURE 9.22 Proppant concentration versus drainage volume.

Source: Gherabati, 2018

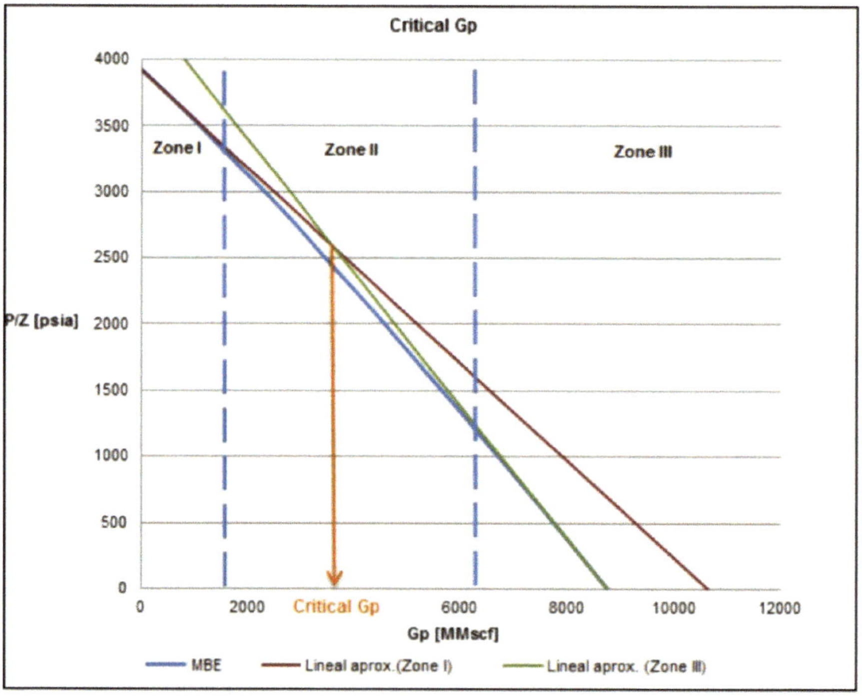

FIGURE 9.23 Critical zones for fracture (zone I), matrix (zone II), and desorption (zone III) manifestation.

Source: Duarte et al., 2014

solved at each pressure step with a numerical approximation for saturated oil reservoirs. In huff-n-puff gas injection, the average reservoir pressure is calculated after the injection period. The injected gas volume makes the MBE nonlinear. The proposed method was applied in Eagle Ford shale wells, and reasonable matches were obtained with field data (Figure 9.24).

The proposed new MBE is believed to make quick and reasonable forecasts for oil shale wells with huff-n-puff gas injection.

Firanda (2011) modified material balance equations for coalbed methane by adding new terms such as connate water expansion, gas expansion, moisture expansion, and rock compaction. The study also proposed a modified p/Z^* versus Gp straight-line analysis, which is believed to address the derivation of average water saturation.

Fatlawi et al. (2016) developed a multi-tank material balance method to better represent the behavior of tight gas reservoirs. A computational tool was employed to handle the numerical model using the false position iterative method. The proposed method was applied to the Australian tight gas field for ultimate recovery and gas-in-place calculations. The study subdivides the reservoir into small compartments, and a material balance equation is applied to each compartment. The rock and fluid properties of the blocks are assumed to be the same. However, the block sizes are different from each other. Fluid withdrawn from each block causes pressure decline, and thus, fluid flow between blocks takes place from thin and permeable layers. The proposed tool was found to be simple and practical to apply in tight gas reservoirs for gas-in-place calculations and ultimate estimated recovery.

The influence of adsorption hysteresis must not be neglected in shale gas calculations because it has significant impacts on the results. He et al. (2019) proposed an advanced MBE for shale gas reservoirs to be used for more effective field development. They performed experiments on shale core samples taken from Fuling shale of China. Methane was used as the adsorbent. The experiment was carried out at a constant temperature. Isothermal adsorption/desorption curves were obtained from the experiments (Figure 9.25).

9.2.15.2 Volumetric Method

The volumetric method uses formation dimensions and rock and fluid properties to calculate the amount of gas in place and the recoverable amount of gas. In unconventional reservoirs such as gas methane, tight gas, and shale gas, it is difficult to estimate gas in place using the volumetric method. The reason is that there are uncertainties about the extended area of the formations and reservoir thickness. Gas adsorption approaches may be applicable to determine the amount of gas adsorbed on organic matter; however, pore volume calculation includes uncertain parameters. Another important unknown value is recovery rates in unconventional resources, which directly affect the producible gas amount. The ratio of adsorbed gas to free gas is also questionable in unconventionals. Well logs do not directly provide gas saturation in shale gas reservoirs.

9.2.15.3 Analytic Models

Analytical models are developed to describe the production behavior of shale reservoirs. Simulation of complex fractures network and physics controlling gas flow from these fractures take too much time and effort to history match gas production from

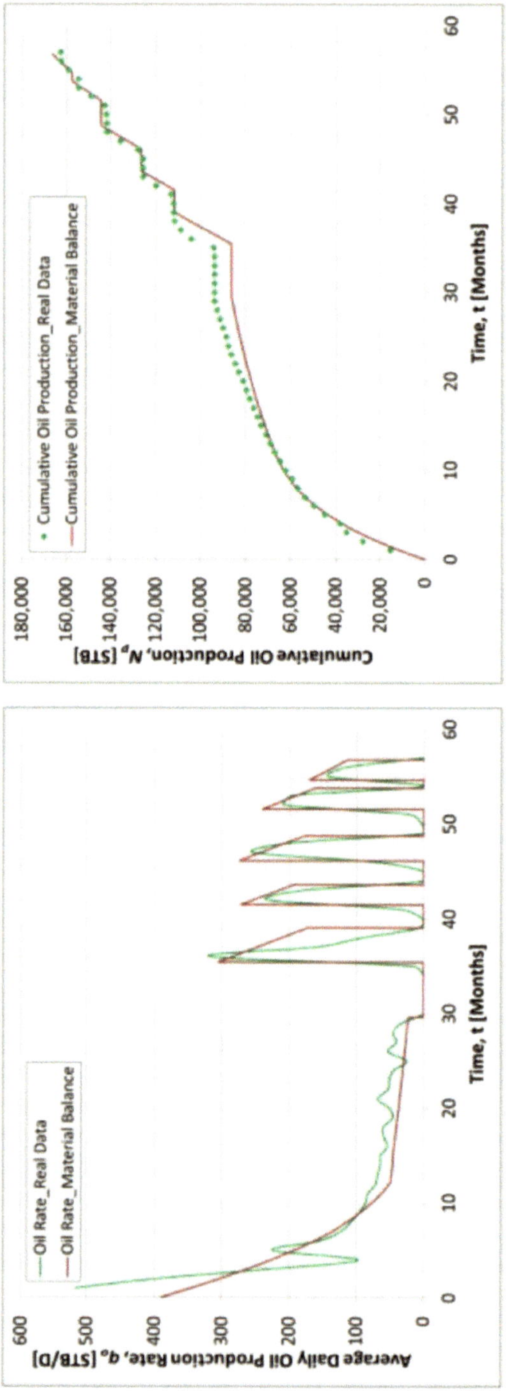

FIGURE 9.24 History matches between real data and MBE.

Source: Orozco et al., 2020

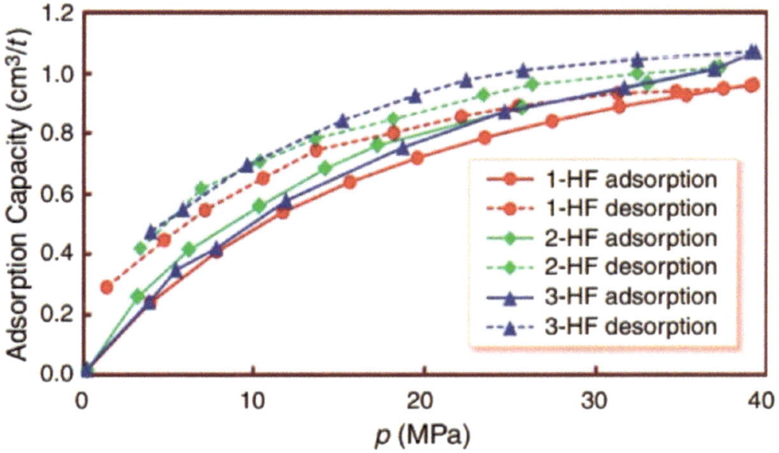

FIGURE 9.25 Isothermal adsorption/desorption curves for shale core samples from China.

Source: He et al., 2019

the wells. Lack of full understanding of desorption and flow mechanisms in shale gas production has made analytical models necessary.

Al-Rbeawi and Owayed (2019) performed a temporal analytical analysis of diffusion flow in fractured, unconventional reservoirs. The objective of the study was to develop analytical models for different flow regimes that may occur in a well lifetime. Figure 9.26 shows the wellbore pressure obtained from an analytical and numerical simulation. Excellent matches were determined between the two methods for different flow regimes.

The following results were obtained from the study:

- Decline rate behavior and pressure distribution are highly affected by anomalous diffusion flow mechanisms.
- Pressure distribution is more significantly influenced in early and intermediate production times. In the boundary-dominated flow regime, this effect was relatively minor.
- Anomalous diffusion flow mechanisms provide better performance than diffusion flow mechanisms in the reservoir.

Stalgorova and Mattar (2013) presented an analytical model for unconventional reservoirs with multiple fractures in horizontal wells. The model subdivides the reservoir into five regions, and it is an extension of trilinear flow simulation. This technique was reported to be more effective for generating production forecasts of reserve evaluation because the effect of the surrounding non-stimulated region was also included in the model for long-term production. The schematic configuration of the five-region model is shown in Figure 9.27 and Figure 9.28.

The results of the five-region model were compared with a 2D finite-difference numerical simulation. An excellent agreement was obtained between the two models.

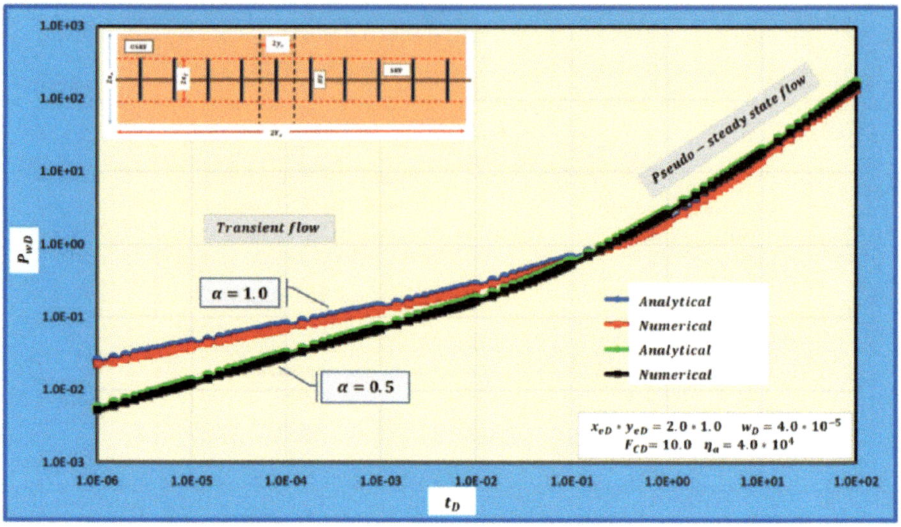

FIGURE 9.26 Comparison of analytical and numerical models.

Source: Al-Rbeawi & Owayed, 2019

Kabir et al. (2011) presented a semi-analytical approach for estimating EUR in unconventional reservoirs. The study aimed to understand the intrinsic production behavior in unconventional reservoirs. The proposed model considers concentric compressibility elements and is analogous to electric capacitors in series. Thus, production from conceptual geo-bodies is obtained with the continuity equation. The material balance of a series of blocks was used to formulate the primary from unconventional reservoirs. Multivariate regression was employed to minimize the error between model results and the production history of wells. The models were tested in an unconventional synthetic reservoir, and matches were obtained for both rate and cumulative production (Figure 9.29).

Jia et al. (2020) proposed a semi-analytical model to simulate production from a complex fracture network in unconventional reservoirs. The proposed model treats the fractures explicitly rather than considering them in a dual-porosity medium. To simulate the flow process analytically, they adopted the real domain and superposition principle to the model. Star-delta transformation was used for flux redistribution. The semi-analytical model was validated with the ECLIPSE simulator (Figure 9.30). A field example from Barnett shale was presented in the study.

Molina and Zeidouni (2017) investigated the application of a nonlinear analytical model for unconventional multi-fractured horizontal wells. The proposed model used a modified five-region model to account for pressure-dependent rock and fluid pressures. The model was tested in several cases studied against numerical simulation. The nonlinear diffusion processes in liquid-rich shales for pressure-dependent rock and fluid properties were reported in exponential form.

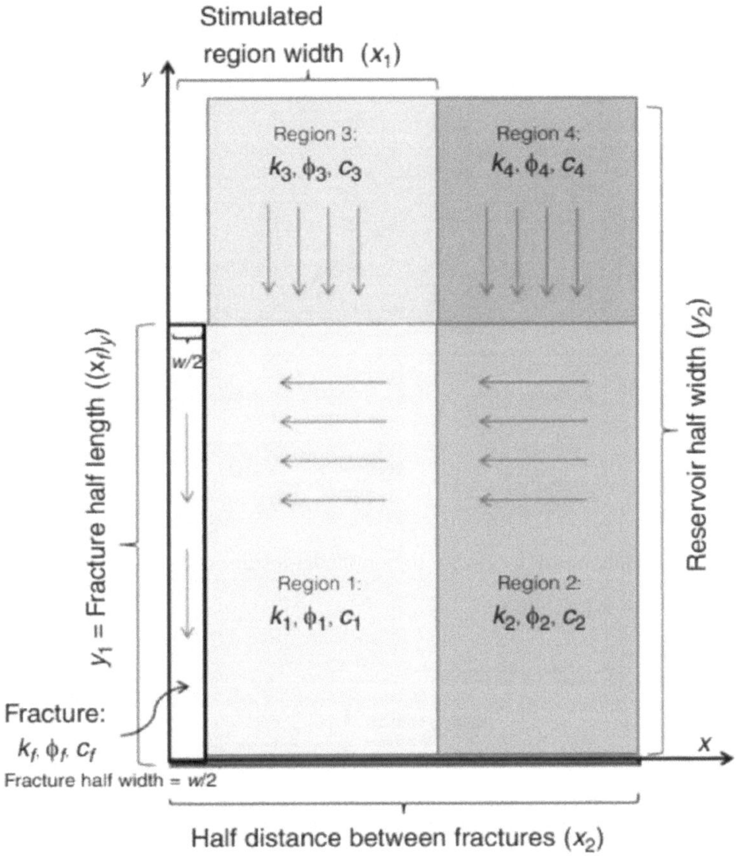

FIGURE 9.27 A view of one-quarter of the fracture in a five-region model.

Source: Stalgorova & Mattar, 2013

Qiu and Li (2020) proposed a new analytical model for EUR of unconventional reservoirs. The model considers the simultaneous matrix-fracture flow. The quadrilinear flow model (QLFM) is a widely used technique for simulating flow between matrix and microfracture and macrofracture. However, numerical inversion and Laplace transform of QLFM is relatively complicated. To simplify the problem, they subdivide the matrix flow into two parts: one feeding the microfracture and the other the microfracture. The study used ordinary differential equations to solve the four partial differential equations obtained as the result of the simplification process. To validate the model results, they first calculated the productivity index. Then, the transmissibility between fractures and two sub-matrixes was found. After that, the pore volume of microfractures and macrofractures and two sub-matrixes were calculated. Finally, permeability was obtained using the Darcy equation. A case study of a shale oil well from Canadian Cordillera was presented for production analysis (Figure 9.31).

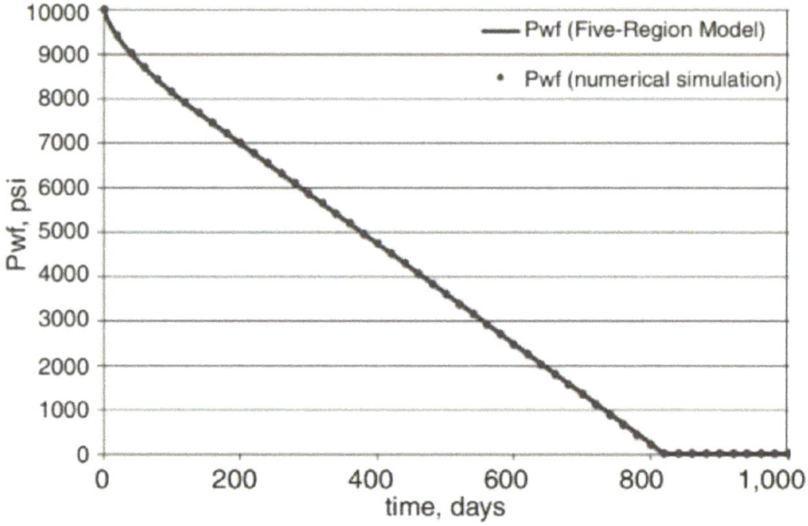

FIGURE 9.28 Matches between five-region model results and numerical model results.

Source: Stalgorova & Mattar, 2013

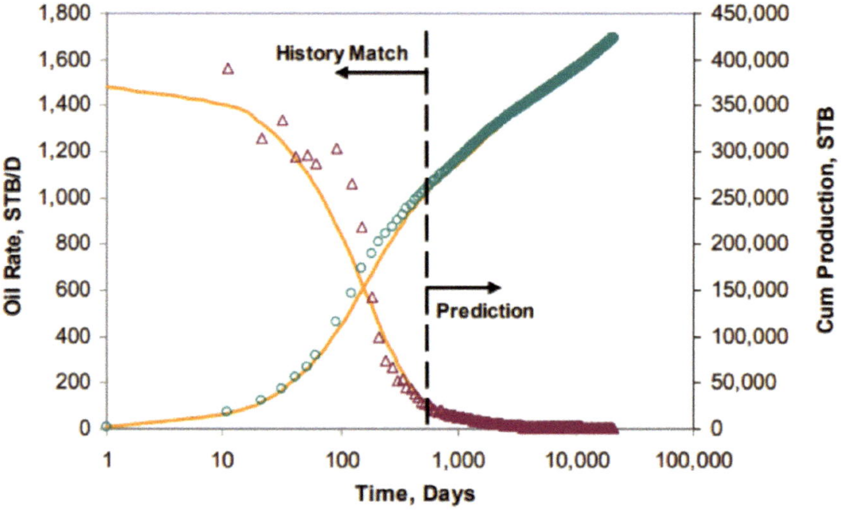

FIGURE 9.29 Matches between synthetic reservoir and model results.

Source: Kabir et al., 2011

9.2.15.4 Numerical Models

A large amount of data is required to construct a representative numerical reservoir model. It is difficult to simulate the behavior of fluid flow in extremely tiny samples. The most widely used simulation techniques are the Native-Stoke equations and the

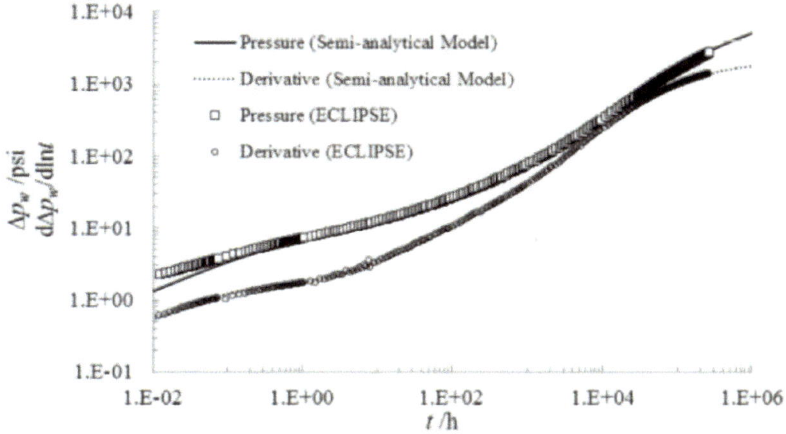

FIGURE 9.30 Comparison of the semi-analytical model and numerical model results.

Source: Jia et al., 2020

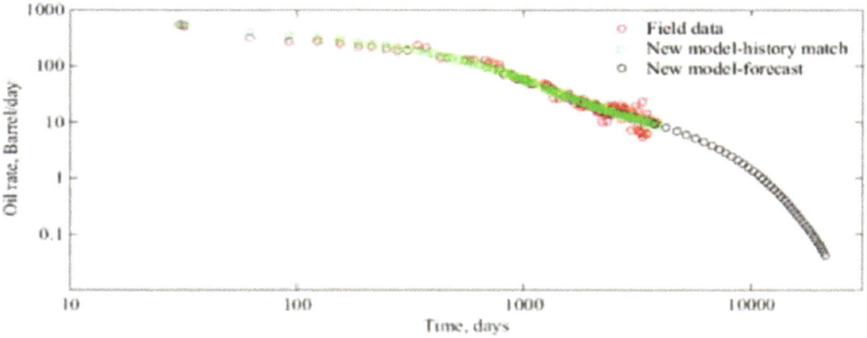

FIGURE 9.31 Good history matches obtained with field data and the new model.

Source: Qiu & Li, 2020

Lattice Boltzmann method (LBM). Numerical models are now capable of better simulating tight gas and shale reservoirs by embedding the Lattice Boltzmann method. The Lattice Boltzmann method uses the bottom-up technique, and it starts to simulate flow mechanism at the molecular levels within very tiny porous media. Once a representative model is constructed, new scenarios or production strategies can be forecasted. In unconventional gas reservoirs, the gas flow is mainly accepted under transitional or slip flow based on the pressure, type of gas molecules, and pore sizes.

Moghaddam and Jamiolahmady (2017) studied slip flow in unconventional shale rocks using the Lattice Boltzmann method. Flow mechanisms and rock-fluid interactions in micro- or nanopores are important phenomena for understanding the performance of unconventional reservoirs. Lattice Boltzmann is a widely accepted method for simulating slip flow mechanisms in shale rocks. The study used the two-relaxation time-based approach to simulate slip flow in a single channel and 2D–3D flow media (Figure 9.32). They compared the simulation results with experimental data obtained from several steady-state and unsteady-state gas permeability measurements with core-flooding experiments—the core-flooding experiments conducted on Eagle Ford, Pierre, and Barnett shale rocks.

The study concludes the following:

- In a single channel, the gas slippage could not be captured with LBM. This is because underestimated permeability enhancement occurred with the diffusive reflection boundary.
- D3Q19 LBM was successfully used to simulate gas flow in a simplified 3D porous medium, and the results were in agreement with the experiment's results.
- Overestimated permeability enhancement was found for all the models available in the literature for Knudsen number higher than 0.1

Civan studied the effect of porosity, tortuosity, and intrinsic permeability on the Klinkenberg slippage factor, rarefaction coefficient, and apparent gas permeability in tight formations. The study presented a correlation of the rarefaction coefficient, derivation of apparent gas permeability, and correlation of the Klinkenberg gas slippage factor.

Javadpour (2009) used atomic force microscopy (AFM) to capture images of nanopores in mudrocks. The study presented a formulation for gas flow in tiny pores using slippage gas glow and Knudson diffusion. An apparent permeability term was introduced to quantify the complexity of flow in nanopores (Figure 9.33).

The AFM confirmed the existence of nanopores in the mudrock strata. Using the apparent permeability term, it was found that permeability is not just a function of the solid matrix, but the gas type and operating conditions are also very effective for gas flow in the fine pores.

Darabi et al. (2012) conducted a study on formulating pressure depending on the permeability function by considering apparent permeability function. The study assumes that Knudson diffusion and slip flow mainly affect the overall flow in porous media. The study showed that the Knudsen diffusion term increased with increasing pore size. However, slip flow becomes more effective than Knudsen diffusion as the pore size increases. For large-size pores, APF converges to 1, and Darcy permeability becomes useful. However, with tiny pores, the flow behavior deviates from Darcy permeability, and the need for advective slip flow and Knudsen diffusion coefficient becomes more considerable.

Deng et al. (2014) proposed a multiscale seepage model considering slippage, diffusion, and desorption effects in shale gas reservoirs. They used the elliptical flow model coupling matrix and fractures to reveal the effects of sorption and poromechanical response on the permeability. Besides, the study aimed to numerically show the influence of fracture conductivity, fracture penetration ratio, and the status of gas on gas productivity. In the seepage model, the authors presented the Knudsen

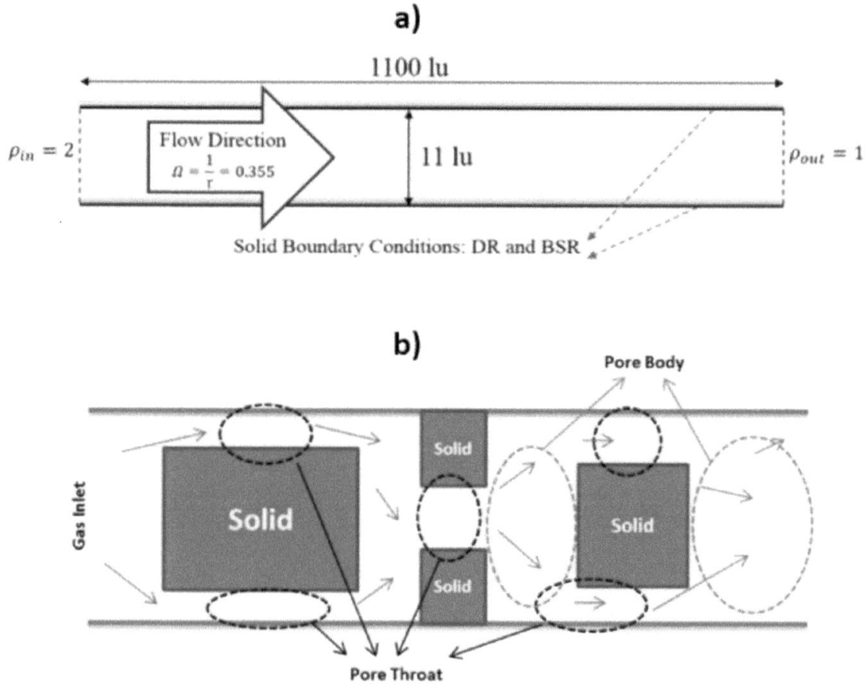

FIGURE 9.32 a) Schematic of micro-channel and b) schematic of pore-throat system.

Source: Moghaddam & Jamiolahmady, 2017

number as a function of pressure with different pore sizes ranging from 10 nm to 50 nm and different flow regimes (Figure 9.34). As the pore size is larger than 50 um, the continuum model can be used for simulation. However, in pore sizes less than 50 um, the gas flow is either in slippage flow or transition flow regime.

Guo et al. (2015) performed a study on gas flow in shale gas reservoirs. The study included analyses of nitrogen flow through nano-membranes. The discrepancy between apparent permeability and intrinsic permeability was observed. To characterize gas flow in nanopores, a new mathematical model was proposed based on Knudsen diffusion and advection. The simulation results were verified with experimental data (Figure 9.35).

Fathi et al. (2012) proposed a correction to Klinkenberg slip theory for gas flow in nanopores. They used Lattice Boltzmann simulation to show apparent permeability values of nanoscale capillaries could be more significant than Klinkenberg slip theory. The kinetic effects of gas molecules through collisions are found as the main differences between the two methods. This causes dramatic changes in the fluid flow velocity profile. A double-slip Klinkenberg equation was proposed. To quantify the molecular streaming effect, the Klinkenberg chart was modified (Figure 9.36).

Fathi and Akkutlu (2013) proposed a new Lattice Boltzmann method that simulates shale gas transport in kerogen. The adsorbed-phase transport was

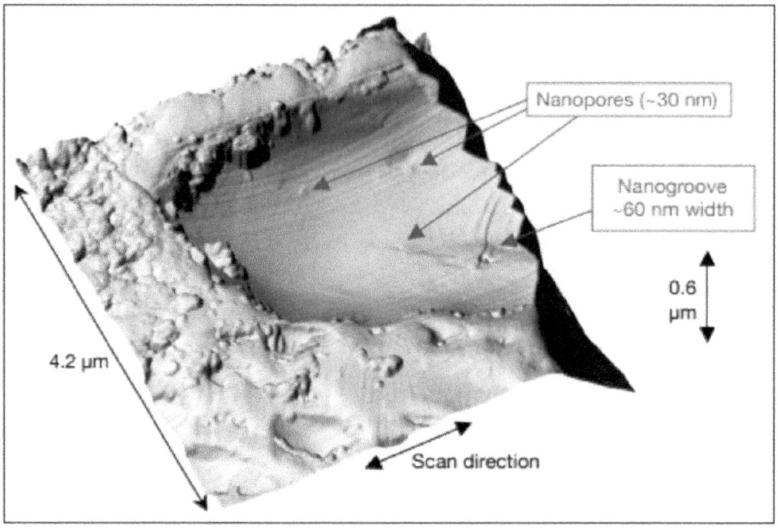

FIGURE 9.33 Atomic force microscopy image of nanopores in a fine-grained mudrock.

Source: Javadpour, 2009

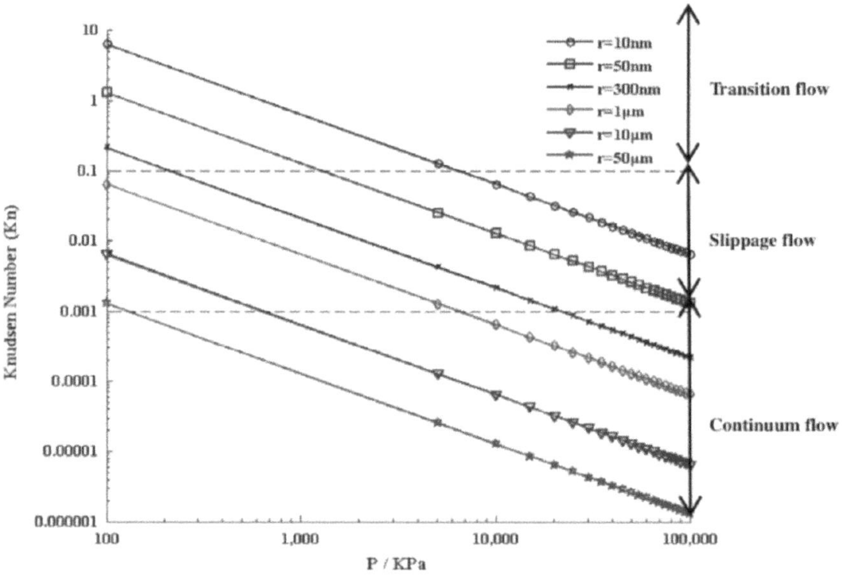

FIGURE 9.34 Relationship between Knudsen number and pressure.

Source: Deng et al., 2014

FIGURE 9.35 SEM images of experimental membranes with pore sizes 20 nm, 55 nm, and 100 nm.

Source: Guo et al., 2015

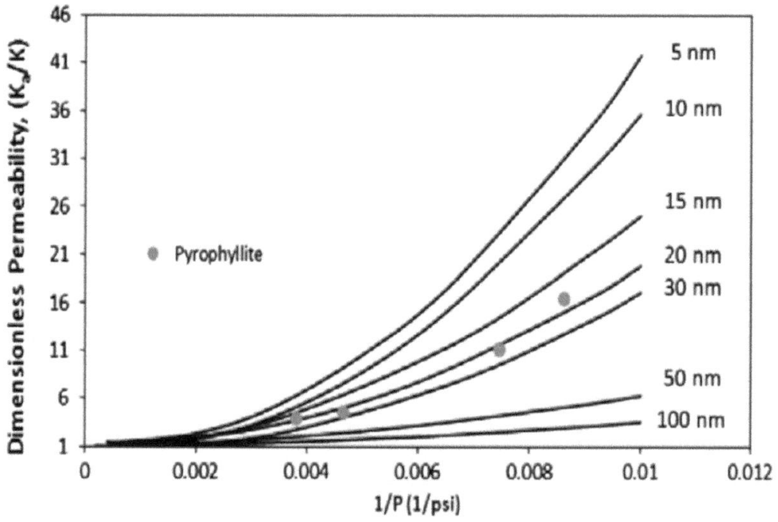

FIGURE 9.36 A modified double-slip Klinkenberg chart for steady-state gas flow in nano-capillaries.

Source: Fathi et al., 2012

introduced as a wall movement, and at each time step, the speed of movement was calculated using local pressure and gradient. Then, the effect of pore size and adsorbed gas was investigated. It was found that as the Knudsen number increases, the flow regime becomes diffusive rather than advective transport at shrunken capillaries.

Zhang et al. (2015) investigated the micro-flow of shale gas using the Lattice Boltzmann method. They also studied the influence of mass flux in nanoscale pores with analytical models, which are based on the capillary model (Figure 9.37).

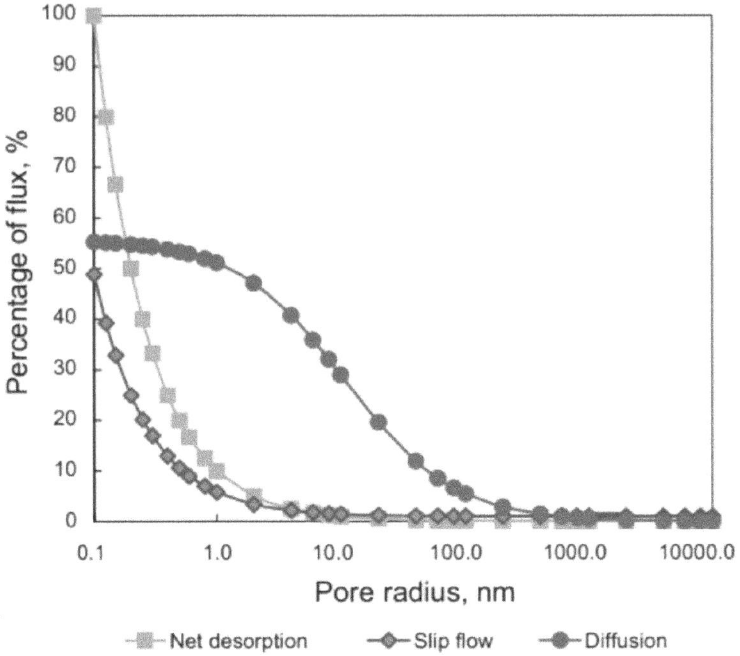

FIGURE 9.37 Mass flux ratio for diffusion, slip flow, and net desorption.

Source: Zhang et al., 2015

The study emphasized the effects of temperature, pressure, diffusion, slip flow, and net desorption on gas transport in shale strata. To estimate the matrix permeability of the complex pore structure, the Lattice Boltzmann method was employed.

Ding et al. (2014) performed a numerical reservoir simulation and mathematical model for unconventional gas reservoirs. The flow behavior in microfractures and primary hydraulic fractures in stimulated reservoir volume (SRV) was considered in the study. They quantified the impacts of Klinkenberg, adsorption/desorption, and geomechanics on flow behavior. To simulate hydraulic fractures, the study used a single-porosity model. Microfractures in the system were considered as a part of the matrix medium, and the existence of microfractures was assumed to increase effective matrix permeability. As an example, a multi-fractured system was assumed with the size of 250 feet in the x-direction, 1,000 feet in the y-direction, and 50 feet in the z-direction (Figure 9.38). Different sizes of coarse grid were used to understand its effect on the CPU time. Very similar CPU times were found for Coarsel (fracture cell one foot wide) and Coarse2 (fracture cell 0.005 feet wide) grids.

Ye et al. (2020) proposed a fully integrated reservoir simulation method for unlocking unconventional reservoirs. The study includes transient rate analysis, geomodeling, hydraulic fracture modeling, and reservoir simulation. To demonstrate the study, case studies from a few fields, including a dry gas field and a black oil field, were presented. The workflow of the study is shown in Figure 9.39.

Satisfying results were reported from the field examples with the integrated reservoir management method. With this method, the uncertainty of unconventional reservoirs was also reduced by using geological and geomechanical properties.

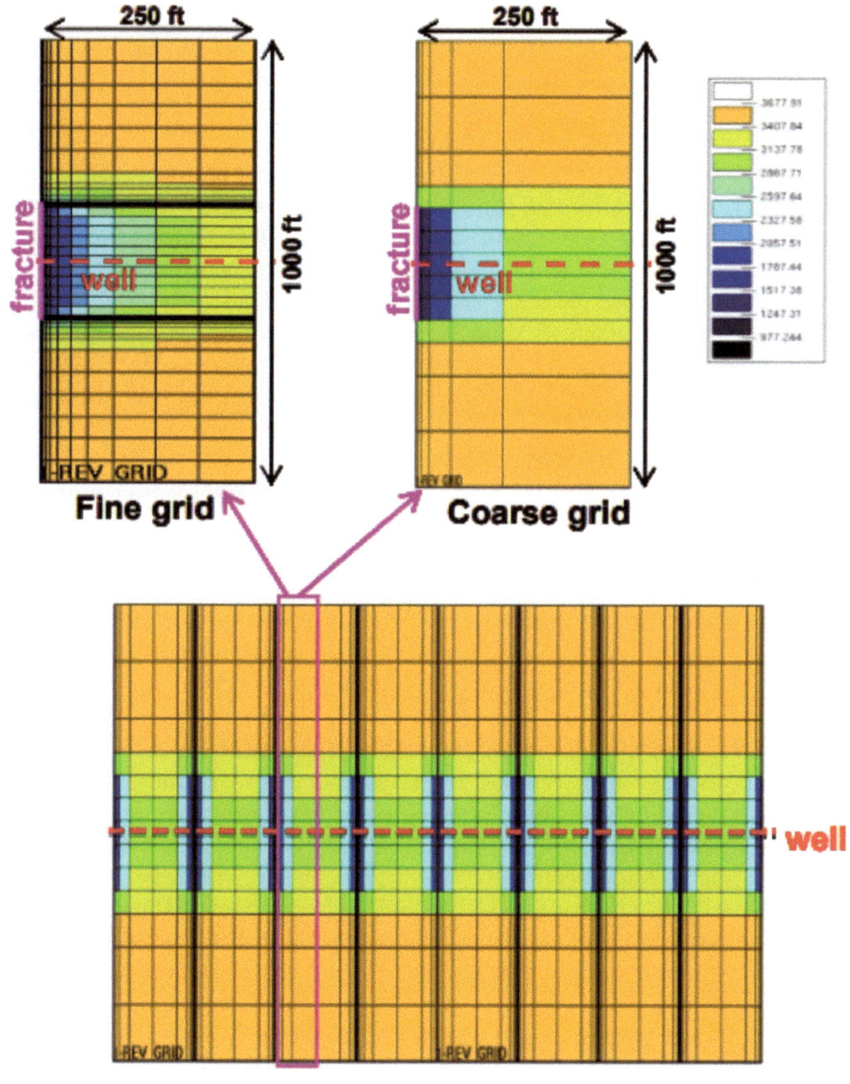

FIGURE 9.38 Hydraulic fractures simulation.

Source: Ding et al., 2014

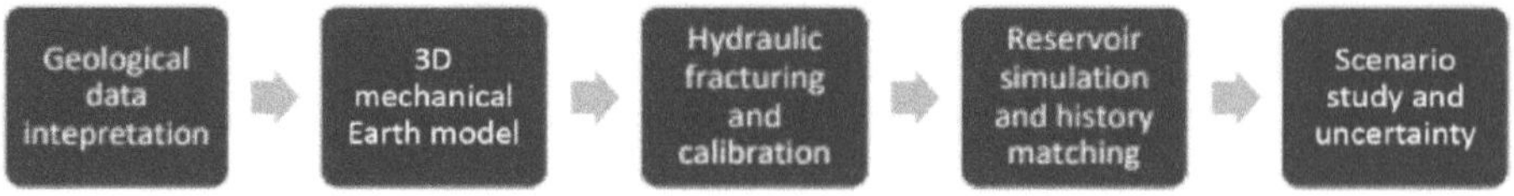

FIGURE 9.39 Workflow for an integrated modeling approach.

Source: Ye et al., 2020

Cipolla et al. (2010) presented a study on the impacts of gas desorption on ultimate gas recovery in shale reservoirs. The study also focused on the impact of changing closure stress distribution on gas recovery. Examples from the microseismic map of Barnett shale were presented (Figure 9.40).

The study concluded with the following remarks:

- Adsorbed gas constitutes 40% to 50% of total gas in place in shale reservoirs.
- Because of very low permeability and high bottom-hole flowing pressure, gas desorption may not significantly contribute to shale gas recovery.
- 5% to 15% of ultimate recovery gas is from desorbed gas.
- Shale reservoirs with a high Young's modulus experience minimal impacts from stress dependence.
- More complex hydraulic fracturing may improve the recovery of adsorbed gas.

Negara et al. (2015) investigated natural gas flow in shale reservoirs using numerical reservoir simulation. By considering anisotropy in shale reservoirs, a dual-porosity dual-permeability model was developed through the full-tensor apparent permeability. The full-tensor apparent permeability was handled with multipoint flux approximation. Calculated undetermined coefficients from pressure fields generate the solution domain for the problem. The model results were compared with the isotropic model to understand the differences between them. The effect of anisotropy on gas production of shale reservoirs was investigated for both matrix and fracture systems. Figure 9.41 shows the anisotropic cases with different degrees to understand its impact on the pressure field distribution. It was found that as we get closer to the wellbore, the pressure gradient increases. Rapid pressure decline occurred in the fracture boundary and in the matrix until they converged with the bottom hole flowing pressure. However, the pressure decline was less significant in anisotropy of 0 degrees and isotropic cases. The numerical models used Knudsen diffusion, slip flow, and gas adsorption-desorption mechanisms. The study concluded that anisotropy, along with pressure gradient, dominates the direction and magnitude of the flow. It was stated that incorporating anisotropy with flow mechanisms in shale formations is essential to maximize gas production from an unconventional gas field.

9.2.15.5 Machine Learning Models

Performance-based, geological-driven, and physics-constrained models are widely used machine learning techniques to simulate and forecast shale gas production.

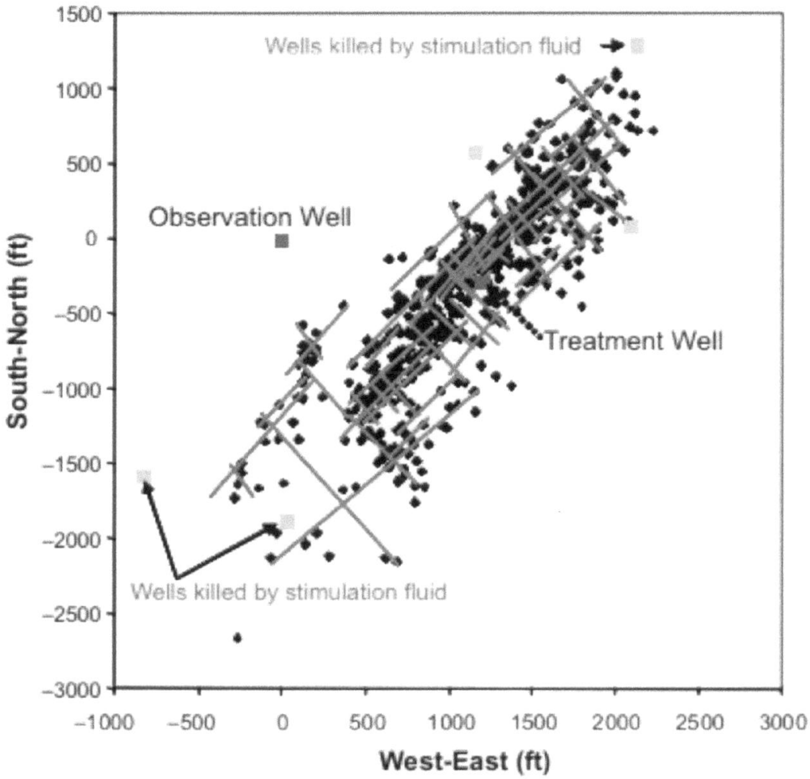

FIGURE 9.40 Microseismic data from Barnett shale well.

Source: Cipolla et al., 2010

Performance-based ML methods such as random forest regression, neural networks, and deconvolution regression are adopted to overcome the complex production system of shale reservoirs. Xue et al. (2021) studied forecasting shale gas production using a data-driven method based on multi-objective random forest regression. They proposed the data-driven model as an alternative to the physics-based shale gas production prediction. The technique uses geological and hydraulic fracturing parameters as input. Once the hyper-parameters are determined, the performance of the model is evaluated with the R squared (R^2). They noted that the initial peak production rate is the most important parameter for increasing the model's prediction accuracy. Xue et al. (2021) compared the result of multi-objective random forest with multiple output regression chains. It was reported that the former method has much more accurate results than the latter approach. Pan et al. (2019) proposed a novel data-driven deconvolution algorithm to analyze production data in unconventional reservoirs. A rigorous regression was applied on the superposition equations to overcome pressure curve fitting, sensitive to noise in traditional deconvolution approaches. They validated the proposed algorithm with two synthetic cases: drawdown pressure

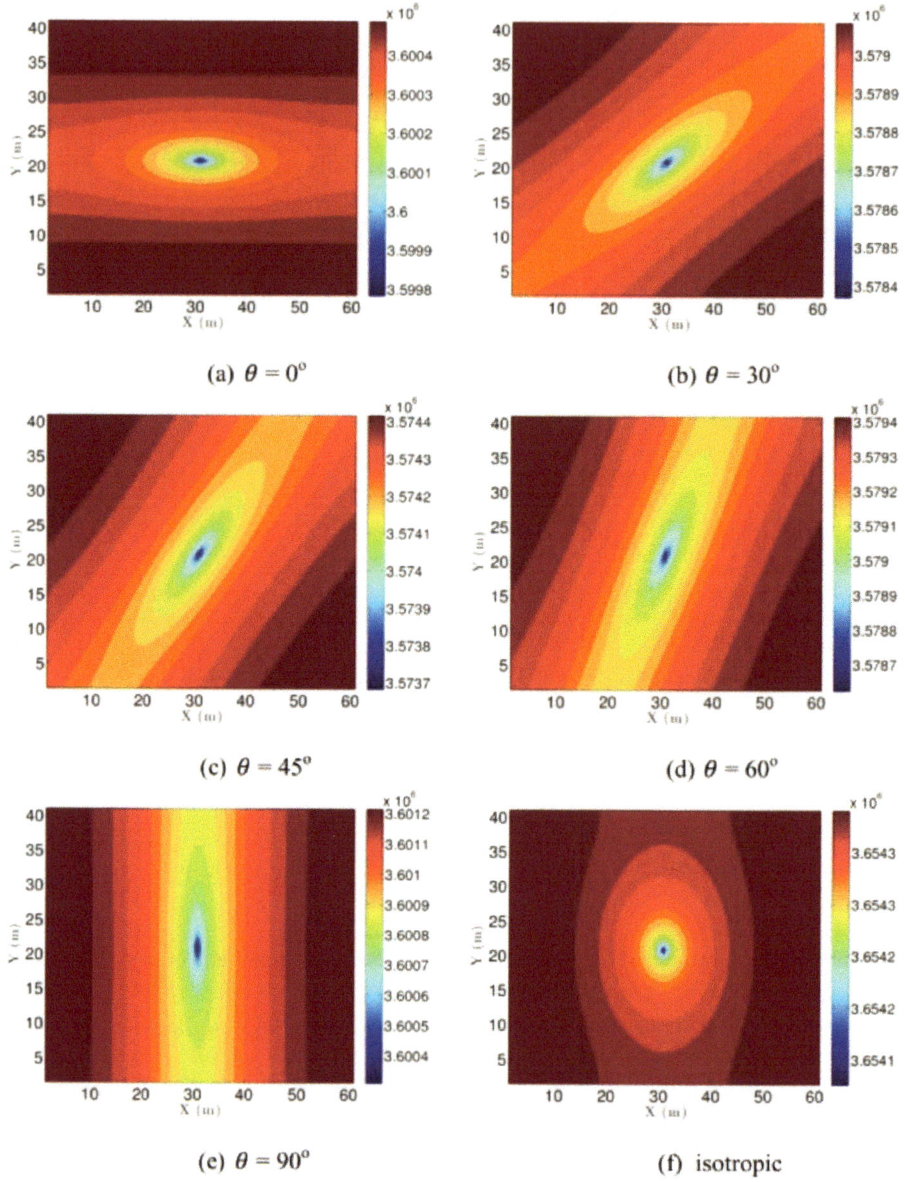

FIGURE 9.41 Matrix pressure after three years of production.

Source: Negara et al., 2015

response and a horizontal well with single and multiple transverse fractures. Gupta et al. (2014) used data mining and time series analysis for forecasting production in unconventional resources. They used a neural network method using historical production data of the wells. This method discovers meaningful relationships between datasets. The data-driven model is proposed to be fast, simple, and easy to apply in unconventional resources.

The geological-driven method uses completion and physical reservoir parameters in a machine learning model to create a relationship between production data and reservoir parameters. Cunningham et al. (2012) used multiple linear regression to simulate EUR of horizontal wells in Marcellus shale. Multiple linear regression has the capability of combining various parameters and design aspects for comparison against productivity. They used a regression software package based on the least-squares method. Different parameters were included in the study, such as perforation design, lateral design, fracturing design, and formation characteristics. The study showed that treated lateral length has a near-linear relationship with EUR. Friis and Erlandson (2015) presented geology-driven EUR prediction for multistage hydraulically fractured horizontal wells in Eagle Ford shale using deep learning. The study includes more than 200,000 geological data points and more than 800 wells. To determine a relationship between geology and the EUR, a deep neural network is employed. The proposed method is only applicable when the field has been producing for some time. The geological parameters used as inputs are porosity, thickness, bulk density, water saturation, total organic carbon, and brittleness. Gaurav (2017) investigated EUR of horizontal shale wells using machine learning, pattern recognition, statistical analysis, and geology. The critical well parameters for better productivity are identified with machine learning. Pattern recognition is performed to check the quality of machine learning. The method provided more EUR for deeper Wolfcamp wells as compared with average type curve matching. Liang et al. (2019) performed a study on big data application for correlation between EUR and petrophysics/production parameters in shale reservoirs. Data regression and nonlinear multivariate interpolation analysis were carried out to determine the correlation between EUR and petrophysics/production for 4,067 wells in Eagle Ford shale. The sensitivity studies showed that the most sensitive parameters are total organic carbon (TOC), resource density, and vitrinite reflectance equivalent (VRE). Effective lateral length, proppant weight, and true vertical depth had positive effects on EUR. It was also noted that no correlation was identified between EUR and fracturing stage and perforation per cluster. Stephenson and Coflin (2015) introduced a study on structural stage spacing to provide frac guidelines in faulted rock for EUR. The chemical tracer data showed that the wells are in good communication across the faults. Implementation of structural spacing was found to increase the EUR of wells.

The physics-constrained method is applicable to overcome nonlinearity and high dimensionality of the multiscale nature of fluid flow in shale reservoirs. Artun presented physics-reduced data-driven models to forecast water-flood performance. Capacitance-resistance models were used to incorporate reservoir physics with some assumptions. Pattern-based forecasting methodology was used to quantify connectivity between wells. Flow barriers and high-permeability faults are defined to interpret fluid movement in an artificial reservoir. Male et al. (2018) presented a study on production forecasting by using data analytics. Long-term production profiles of Bakken wells are evaluated to understand the influences of completion design on productivity. The evaluation is performed with a physics-based approach. To test the predictability of completion parameters, a tree-based machine learning approach was used. The parameters affecting EUR (completion type, length, fluid type, fluid pumped, total proppant) are analyzed with gradient boosting regressors. The study showed that the terminal decline rate increases by upwards of 10% in modern completions multi-wells pad. Lower terminal decline rates were found for early wells with modern completions. Pan et al. (2019) performed a study on forecasting unconventional reservoirs using Laplacian echo state networks (LESN). They introduced a physics-assisted analytics workflow to accelerate the analysis of historical production data. The proposed method (Laplacian Eigenmaps coupled echo-state network) overcomes a common issue in traditional echo-state network, an ill-posed, complex time series when training sets less than the number of neurons. The other advantage of the proposed method is honoring transient flow physics. With a trained model, reservoir characterization and EUR can be possible. Matches of one-year production between the rate of constant pressure data and the LIESN model are shown in Figure 9.42.

Li et al. (2020) applied machine learning models to overcome the weaknesses of traditional decline curve analysis. The study aimed to increase the accuracy of DCA models by considering the following aspects:

- Capturing reservoir signals
- Considering the production of neighboring wells in the same geologic formation
- Well completion, well spacing, the geology and geography of wells

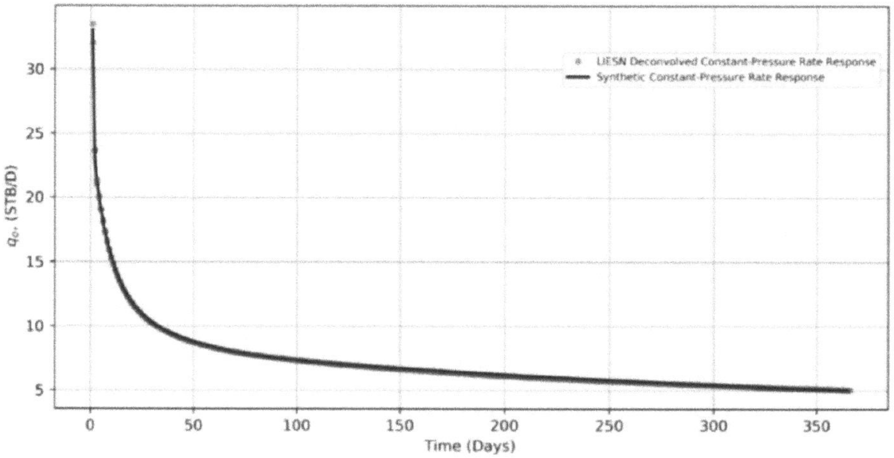

FIGURE 9.42 Comparison of LIESN deconvolved rate responses.

Source: Pan et al., 2019

They populated the DCA models with the aforementioned properties using machine learning methods such as support vector regression, random forest, etc. A novel method, dynamic production rescaling (DPR), was developed to decrease the mean error and skewness of machine learning DCA models. The DPR reduced the error by 15% to 30% and the error of traditional DCA models was reduced by 30% to 60% with the machine learning models (Figure 9.43).

Han et al. (2019) used machine learning and decline curve models to forecast production for shale gas wells in transient flow regimes. DCA models are good at transient flow; however, they usually deviate from the actual production data under boundary-dominated flow behavior. The study used an artificial neural network (ANN) to predict the future production of the wells under a transient flow regime with the DCA method. Well completion data, reservoir properties, fracturing factors, and reservoir management data were used as inputs for the ANN model. A set consisting of 8,560 individual pieces of data from 150 wells was used for training. Clustering analysis was performed to increase the accuracy of the ANN model. The error between the DCA model and ANN model was less than 8%.

Zhou and Lascaud (2019) developed a flexible, scalable, and integrated machine learning framework for the optimization of unconventional resources. The proposed method uses a large volume of heterogeneous data. Different types of data were related to estimating the production performance of wells. Figure 9.44 shows the major categories of data used for the system. The data were transformed into a more meaningful form. The framework was extensively applied for technical evaluations such as optimization of well completion and spacing, evaluation of new formation, PUD reserve booking, etc.

Zhan et al. (2019) developed a new methodology to forecast oil production from unconventional resources. The study is based on long short-term memory (LSTM) to handle the production forecasting problems. As compared with the modified hyperbolic decline model, LSTM can handle nonlinear events more dynamically. More

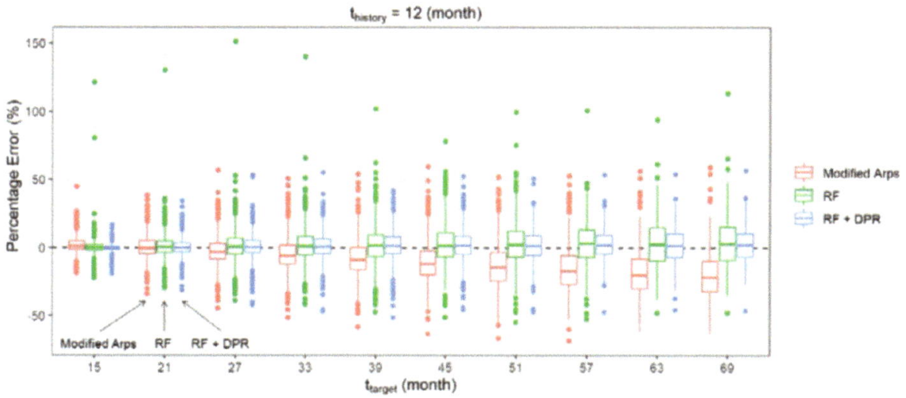

FIGURE 9.43 Percentage error for modified DCA and ML DCA with and without DPR.

Source: Li et al., 2020

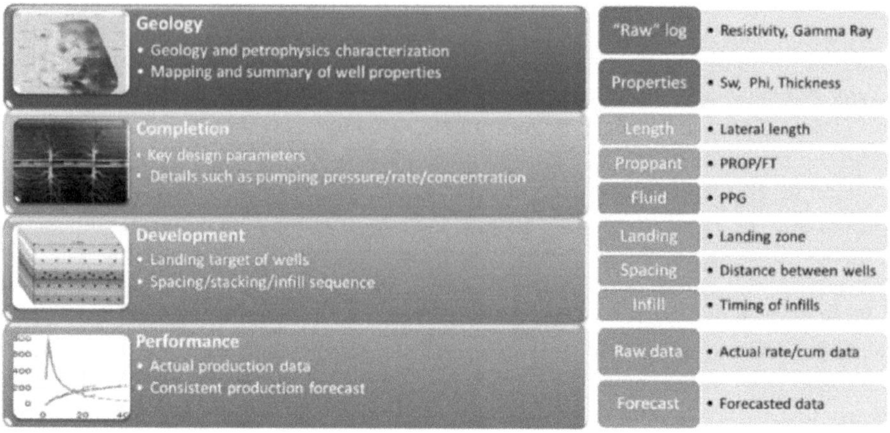

FIGURE 9.44 The data in the integrated machine learning framework.

Source: Zhou & Lascaud, 2019

than 300 onshore unconventional wells with more than two years of production were included in the study, and less than 0.2% error was obtained between the predicted and actual data. The proposed method is based on recurrent neural networks (RNN) to predict the future. The models were trained 3 months of production data, and 21 months of production were used for validation. The LSTM method improved the forecasting accuracy as compared with hyperbolic decline curve analysis (Figure 9.45).

Tandon (2019) used supervised learning techniques to improve the analysis of stimulated reservoir volumes in unconventional reservoirs. The study aimed to provide fracture design and nonlinear models for sweet spot analysis in unconventional resources. A fracture simulator was combined with a simulation of fluid flow in discrete fractures—different fracturing designs obtained by changing the geomechanic properties of reservoir rock in fracturing simulator. Short-term stimulated rock volume was created, and its performance was quantified. Formation geomechanic properties were then related to SRV using support vector machines, decision trees, and random forests. The parameters were then ranked based on their influence on the SRV. The study showed that unpropped fracture conductivity significantly affects the success of hydraulic fracturing treatments.

Liang and Zhao (2019) investigated the application of machine learning to estimate EUR of unconventional reservoirs. The most important independent variables, such as petrophysics and completion parameters, were ranked based on the importance of their influence on EUR using random forest regression (Figure 9.46). Twenty-five independent variables, including production/engineering and petrophysics data from 4,000 wells in Eagle Ford shale, were summarized for analysis. The study was arranged in three parts. In the first part, multivariate linear regression was employed to forecast EUR. In the second stage, a spatial multivariate regression model was performed to understand if spatial variables have a reasonable influence on the multivariate regression model. In the last step, the reliability of RFR models was tested with spatially auto-correlated data.

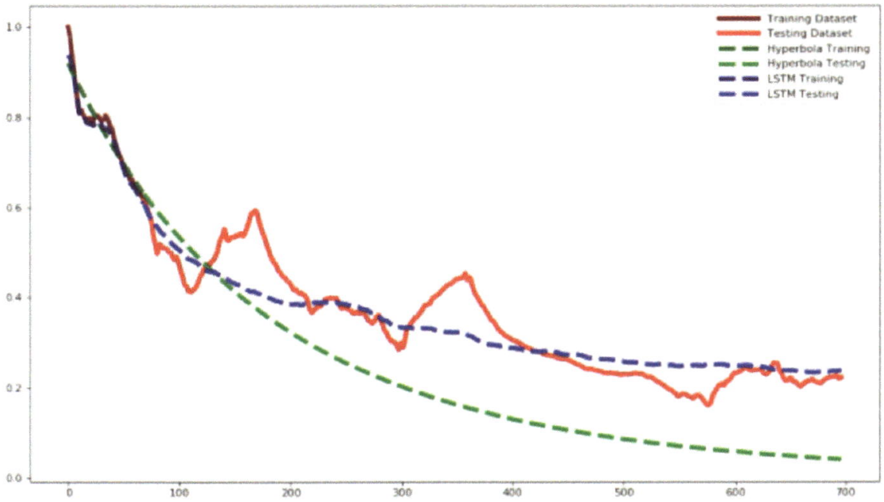

FIGURE 9.45 Oil rate versus cumulative production.

Source: Zhan et al., 2019

Mousavi et al. (2020) presented a study on optimization of well placement using machine learning methods. The study compared the experimental-based approach to a central composite design with the xGBoost machine learning method. The aim of the study was the net present value of the operational life of the studied reservoir. Both approaches were applied to three different cases:

- Production from a homogeneous reservoir
- Production from a heterogeneous reservoir
- Water flooding into a heterogeneous reservoir

The workflow of the study is shown in Figure 9.47.

In the first and second scenarios, both approaches showed very good results, and the results were quite comparable. However, in the third case, xGBoost showed more accurate results than the CCD method (Figure 9.48).

Zhao et al. (2020) tried to improve the prediction of estimated ultimate recovery in unconventional formations using the multi-variant relationship of geological and engineering parameters in machine learning. Multiscale machine learning was applied to a big dataset from Eagle Ford shale. More than 30 geological, petrophysical, and drilling/completion features from 4,067 wells were first checked for quality control. After that, the K-nearest neighbor algorithm and bagging technique were employed to estimate the spatial trend of EUR across the studied area. The workflow of the study used is shown in Figure 9.49.

Rahmanifard et al. (2020) predicted the performance of unconventional wells using machine learning methods. The study aimed to use the capability of big data analytics to develop a model for a typical unconventional gas formation, the Montney

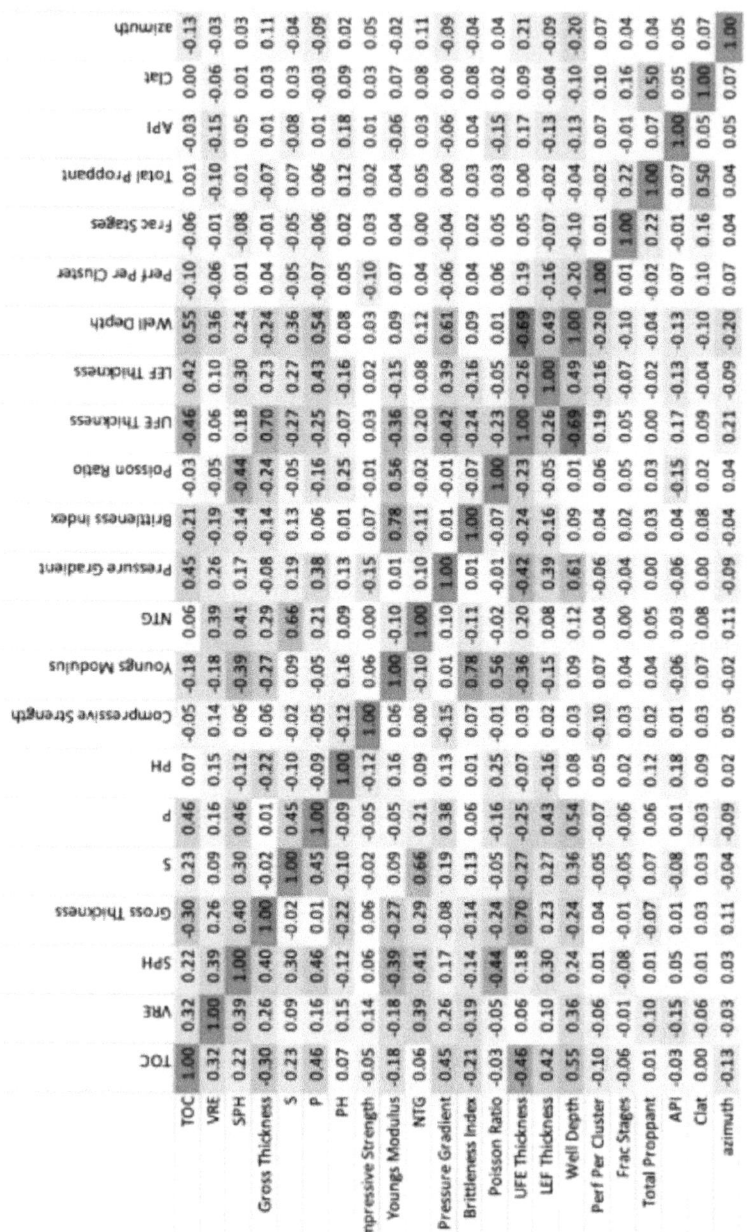

FIGURE 9.46 Independent variable correlation.

Source: Liang & Zhao, 2019

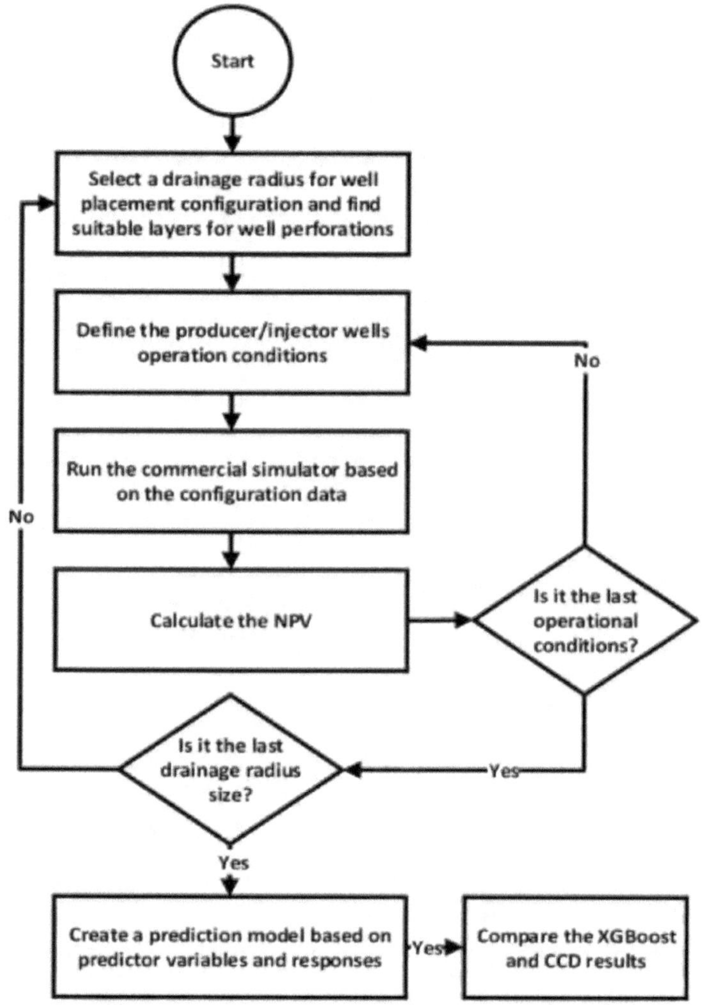

FIGURE 9.47 Workflow of the study.

Source: Mousavi et al., 2020

formation in Canada. A database with 16 features from 603 horizontal wells from the Montney formation was first created. After that, the relationship between various parameters and gas production rate was identified. In the last step, an artificial neural network was carried out for a comparative study to determine the best solution for the estimation of the gas production rate. The workflow of the study is shown in Figure 9.50. Root mean square error, mean absolute error, correlation coefficient, and mean relative error were the benchmarks for the comparison. The study concluded that the average gas production rate is significantly influenced by well geometry, completion interval, and completion design parameters.

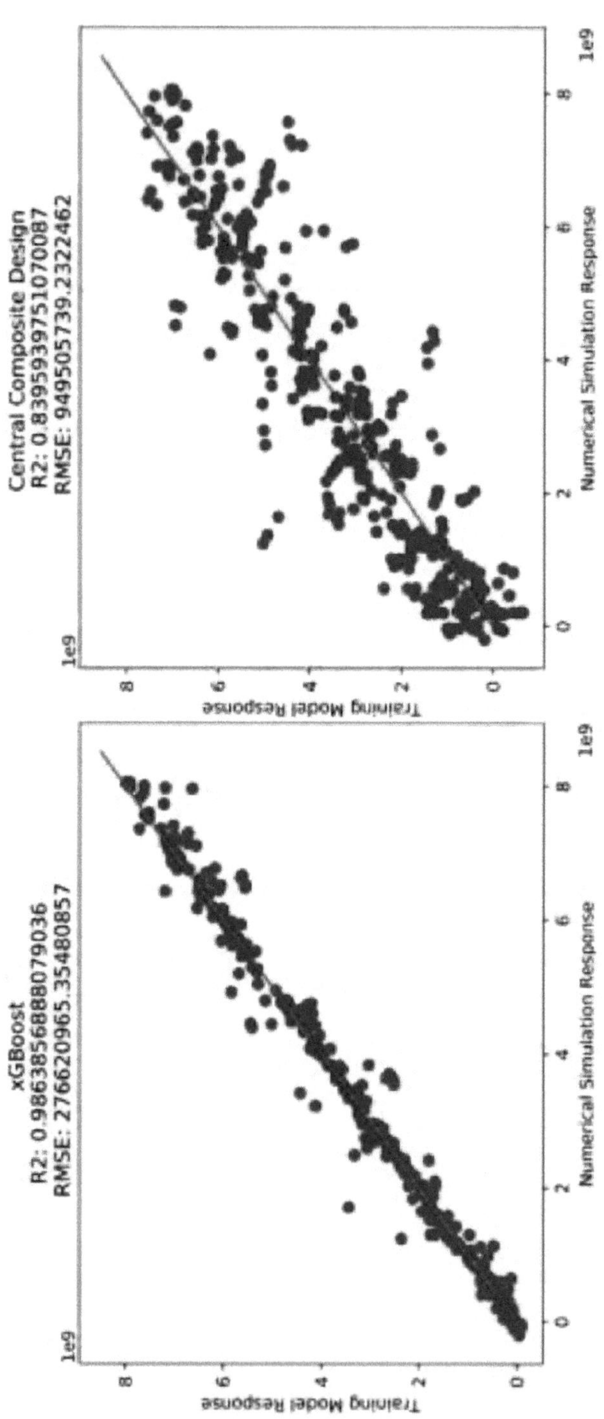

FIGURE 9.48 Cross-validation of xGBoost and CCD models.

Source: Mousavi et al., 2020

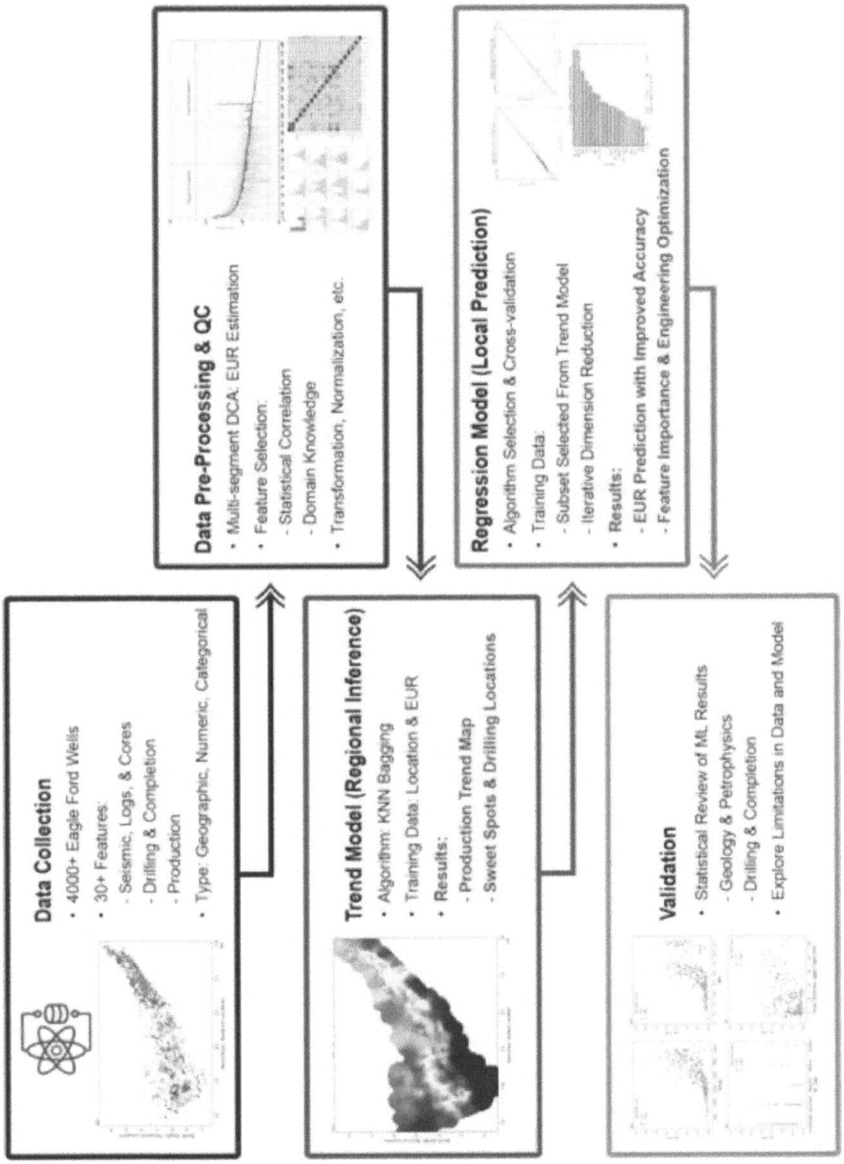

FIGURE 9.49 Workflow for the machine learning-based EUR prediction.

Source: Zhao et al., 2020

9.2.15.6 Decline Curve Analysis

Decline curve analysis is one of the most widely used methods for estimating the ultimate recovery of shale gas reservoirs. It is simply a curve-matching method based on empirical parameters determined from historical matches of production. Arps's decline curves (1945), exponential, hyperbolic, and harmonic, are the originally proposed curves for decline curve analysis. The hyperbolic decline curve was found a better match with shale gas production. Arps's curve is applicable when the following conditions are maintained:

- Boundary-dominated flow
- Constant bottom-hole pressure
- Unchanging drainage area
- Same skin factor

Cheng et al. (2008) presented a study about improving reserves estimates using DCA in multi-layer gas wells. They investigated the in-depth decline behavior of multi-layer gas wells by analyzing the characteristic decline behavior of simulated data. In commingled production wells, the flow regime is complex, and abnormal decline behavior can be observed in multi-layer gas wells. The abnormal behavior (b > 1) may lead to the wrong estimation of the EUR of wells. The study emphasized that the b-value significantly affected the calculated EUR of wells. It was concluded that the b-value is not constant, and it decreases as the reservoir is exploited under constant bottom-hole pressure. Fulford and Blasingame (2013) investigated the relation of transient hyperbolic with the time-rate performance of shale wells. The study aimed to develop a formulation for the early-time production profile of shale wells. The logistic transition of the b-parameter was used to describe the change between binary states in a confined space. The proposed model was applied in several shale reservoirs and the results compared against power-law exponential and modified hyperbolic models. Tugan and Weijermars (2020) performed a new three-segment decline curve analysis method to improve production forecasting from tight gas wells by describing the flow regime transition with variation in b-sigmoids. Synthetic production data was generated using a commercial reservoir simulator by changing fracture parameters. The synthetic data was then used in diagnostic log-rate-log plots to identify four flow regimes. The characteristic slope of each flow regime was attributed to a typical flow condition in the reservoir. More than 100 models were generated to create a basis for the study. The correlation trend detected between b-sigmoids and flow regime changes in Figure 9.51—the proposed method used for EUR estimation of Eagle Ford shale.

Masini et al. (2019) presented a study on using artificial intelligence for decline curve analysis. The automated DCA technique uses a cutting-edge clustering method to identify the data points in the production decline. Since decline curve analysis is a time-series sequence problem, XGBoost and recurrent neural network (RNN) is applicable to handling this regression. Anomalies data were removed with automatic DCA. A good agreement was found between observed and calculated data.

Boulis et al. (2012) used four analytical methods to estimate the ultimate recovery of shale gas wells located in Barnett, Haynesville, Marcellus, and Woodford. Application of DCA requires some modifications because b-parameters are often higher than 1. This is because adsorbed shale gas has a long transient flow period. The following authors modified the traditional Arps's decline curve method: Fetkovich

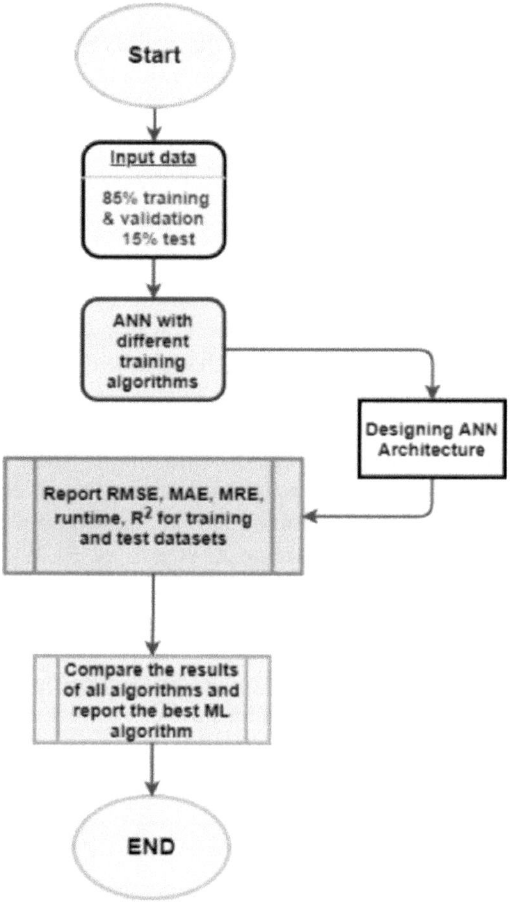

FIGURE 9.50 The workflow of the study.

Source: Rahmanifard et al., 2020

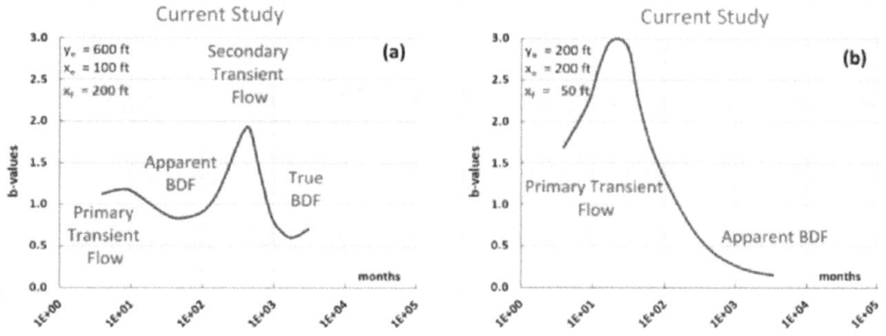

FIGURE 9.51 Correlation between b-sigmoids and flow regimes.

Source: Tugan & Weijermars, 2020

(1980), Doublet et al. (1994), and Agarwal (1997) (Agarwal et al., 1998). A typical example of a decline curve match in a shale gas well is shown in Figure 9.52. Model parameters for modified DCA models and conventional Arps's models are shown in Table 9.6.

The study showed that conventional hyperbolic decline overestimates the results for all the studied cases. However, it is worth noting that the hyperbolic model provides relatively good estimates when the wells are in boundary-dominated flow regime as compared with linear flow. Less than 5% error was obtained with modified decline curve analysis.

Strickland et al. (2011) investigated power-law exponential decline for shale gas wells in Haynesville and Barnett fields. Power-law exponential is also called stretched exponential. In Arps's hyperbolic model, the derivative of the decline rate was assumed constant. However, numerous shale gas wells showed that the b-factor varies for low permeable reservoirs. More specifically, the b-value decreases in the early time period, and it increases in the late time flow.

Kocoglu et al. (2020) presented a machine learning–based decline curve analysis for unconventional shale gas reservoirs. They incorporated a feed-forward neural network with power-law DCA to estimate 25 years' gas EUR potential of undrilled shale gas wells. The models were applied to 513 horizontal wells in the Marcellus shale reservoir. R^2 of 0.7754 for training data and 0.7655 for test data were obtained (Figure 9.53).

Odi et al. (2019) used a variable law model for decline curve analysis of Barnett shale wells. Due to the long-term transient flow of unconventional wells, it is challenging to apply traditional Arps decline curves. The proposed method was found to

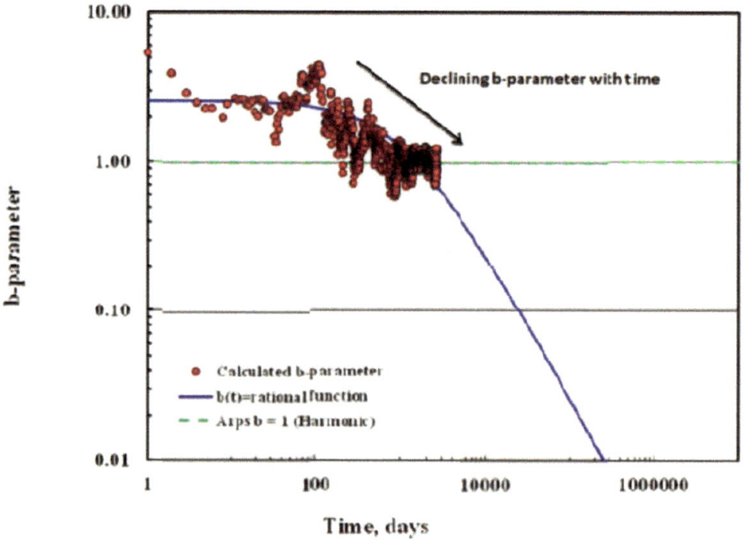

FIGURE 9.52 b-factor in a shale gas well.

Source: Boulis et al., 2012

TABLE 9.6
Comparison of Model Parameters for Conventional DCA and Modified DCA

	Marcellus Case Study		Haynesville Case Study		Barnett Case Study		Woodford Case Study	
Decline Curves	**Arps's Decline Curves**	**Modified Decline Curves**	**Arps's Decline Curves**	**Modified Decline Curves**	**Arps's Decline Curves**	**Modified Decline Curves**	**Arps Decline Curves**	**Modified Decline Curves**
Di	0.0008	0.0008	0.07	0.07	0.004	0.004	0.03	0.03
b	0.5	-	0.8	-	0.9	-	0.9	-
b0	-	1	-	1.4	-	0.5	-	0.5
bj	-	0.0002	-	0.00001	-	0.0009	-	0.0009
qi	2,800	2,800	14,000	14,000	3,000	3,000	4,400	4,400

Source: Boulis et al., 2012

conform to the long-term transient flow behavior of Barnett shale wells. The model validation was done by matching production rate, loss ratios, and the b-factors.

Artus et al. (2019) presented a study about flow regime–based decline analysis in unconventional reservoirs. They generalized the formulation to a succession of flow regimes to reproduce a succession of periods using power-law behavior. The base curve was incorporated with a rate-normalized pressure derivative on a log-log plot. In case of no available bottom-hole flowing pressure, linear flow and pseudo–steady state flow regimes were used for the base curve. However, the study focused on the cases in which bottom-hole flowing pressure was available. Rate normalized pressure was analyzed on a log-log plot to identify the flow regimes. For each flow regime, the corresponding physical parameters were calculated and used for the decline curve analysis.

Xi and Morgan (2019) combined decline curve analysis and geostatistics to forecast gas production in the Marcellus shale reservoir. The study aimed to estimate production forecasts at prospective undrilled sites based on existing wells in the same basin. To achieve this aim, Co-Kriging was employed to determine decline curve parameters. Co-Kriging is good at estimating linear unbiased prediction of parameters and estimating the uncertainty of the predictions. Using Duong's decline model, two years of production history matches of shale gas wells in Marcellus shale were constructed. Then, Co-Kriging was used to interpolate decline curve coefficients for new locations. After that, 25 years of gas EUR were mapped across the study area. The promising sites with high EUR were found mainly in the middle and northeast of the study area. The proposed method does not require costly data.

Chen et al. (2019) proposed a simulation-based DCA for volatile oil reservoirs in unconventional Permian shales. This method was applied to compensate for the limitations of the traditional DCA models. The proposed method used data analytics,

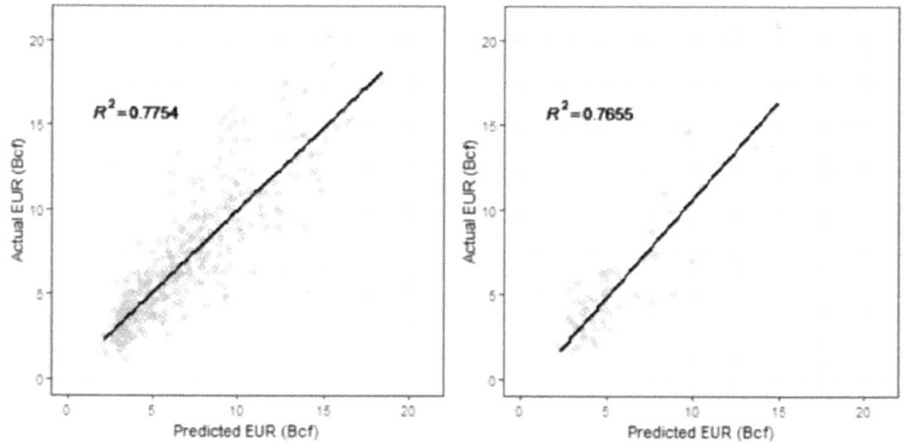

FIGURE 9.53 R square data for training and test data versus actual EUR.

Source: Kocoglu et al., 2020

including a wide range of data from the field such as well-log, PVT, geology, and production data. The reservoir simulation was performed for multiphase flow conditions. Then, the production forecasting was performed to create a database for DCA models. The model results were also verified with the material balance method. The created data was then used to formulate DCA models for the unconventional reservoir. The workflow of the proposed method is shown in Figure 9.54.

9.2.16 SEC—Revised Regulations and FASB

The requirements for oil and gas producing companies were issued in Financial Accounting Standards Board (FASB) statement number 19. Companies require the use of a form of the successful efforts method of accounting and disclosure of cost incurred, capitalized costs, and reserve quantities. The SEC released modernized and updated requirements for the reserve estimation of oil and gas after recent improvements in technology. The updated rules were effective for public companies in 2009. Oil- and gas-producing companies need to request to prepare disclosures pursuant to them. Also, financial statements found in forms 10-K and 20-F will be required for the fiscal year's ending report. The following rules were updated (Deloitte, 2020):

- The 12-month average price will be used to compare reserves instead of using the current daily price.

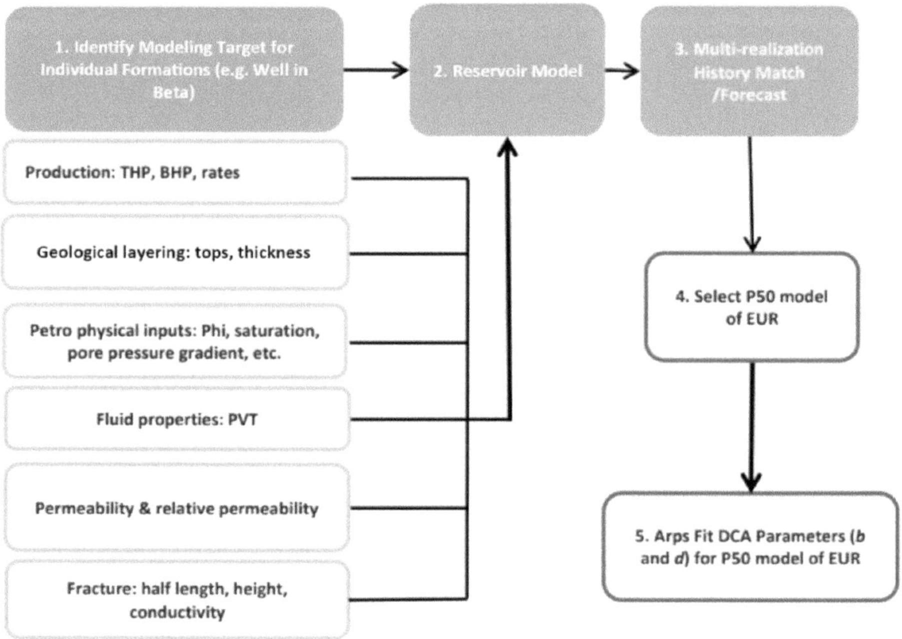

FIGURE 9.54 The workflow for dynamic modeling and DCA model.

Source: Chen et al., 2019

- To give more time to companies, the first-of-the-month time price will be used to prepare estimates.
- To extend the implementation of possible technology, reliable technologies are considered instead of the current specific tests.
- Unconventional resources such as bitumen, shale, and oil sands are included in oil and gas reserves.
- The definitions of proved, probable, and possible are supported by adding new rules.

9.2.17 Similarities to and Differences from PRMS

Fossil fuels are categorized based on technology, commodity properties, and production methods (conventional or unconventional). There is no strict definition of resources because of different assumptions, meanings, and derived terms. Speirs et al. (2015) compared different categorizations used by UNFC, PRMS, and Russian Federation Classification (RFC) in a classification scheme and generic resource hierarchy shown in Figure 9.55.

9.2.18 Additional Reporting Requirements (FASB)

The following supplementary information is required to be reported by publicly traded enterprises of oil and gas activities:

- The operation cost of oil and gas production
- Reserve quantities of oil and gas
- Cost of acquisition, exploration, and development activities
- Operation results
- Discounted future net cash flows related to reserves

9.2.19 The Canadian Oil and Gas Evaluation Handbook (COGEH) and the PRMS

The Canadian Oil and Gas Evaluation Handbook (COGEH) include the definition and estimation of reserves and resources. Disclosure standards for publicly traded oil and gas companies are set by Canadian Securities Administrators (CSA). Additionally, economic and financial analysis and benchmarking of reserves and resources are evaluated. Coalbed methane and bitumen reserves are delineated in detail. The handbook also discusses differences from PRMS.

According to COGEH, a project determines the assessment and classification of resources. The projects are defined based on the discovery and recovery activity of the oil and gas products. Project types are discovery, recovery pipelines, gathering, processing, and offshore projects. The projects are described as early, intermediate, and advanced stages. The times for proved undeveloped reserves and probable undeveloped reserves are three years and five years, respectively. Risk and uncertainty

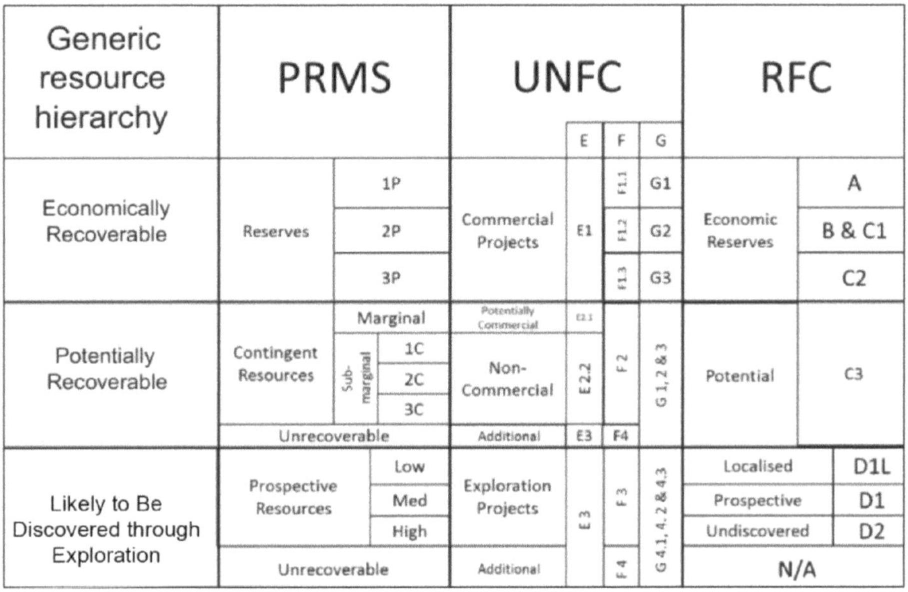

FIGURE 9.55 Classification scheme.

Source: Speirs et al., 2015

concepts are also used for resource classification. Low, best, and high (1P, 2P, and 3P) are described to account for uncertainty (Figure 9.56).

Some of the important differences between COGEH and PRMS are the following:

- Consumed in operation (CIO) is booked as reserves in PMRS. However, CIO is not booked as reserves in COGEH.
- The same conditions are applicable for proved and unproved reserves in PRMS, but COGEH applies different conditions for different reserves.
- PRMS suggests completing projects in five years, and it is based on the defined project. More than five years is also allowed in specific cases with documentation. COGEH provides specific guidance, which is three years for significant capital expenditures of proved reserves and five years for probable reserves.

9.2.20 UNFC—PRMS Provides the Hydrocarbon Specifications

United Nations Framework Classification (UNFC) provides specifications for resource classification. The bridging document includes estimated quantities to a specific class or sub-class, accounting for both the generics specifications and the source-specific requirements. The definitions of generic specifications need to be satisfied with high-level classification into viable, potentially viable, and non-viable projects (UNFC, 2009).

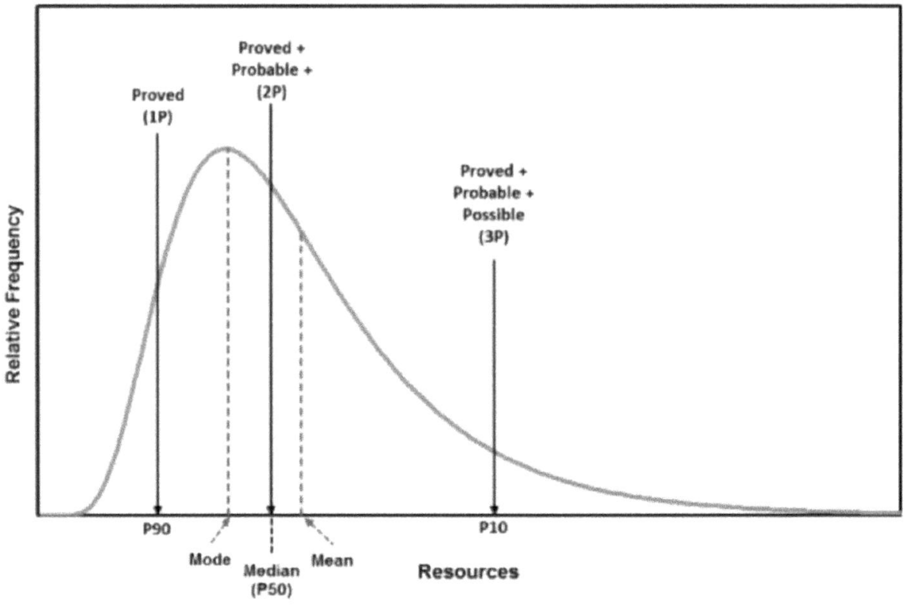

FIGURE 9.56 Probabilistic distribution for resources.

Source: GLJ Petroleum Consultants, 2019

Viable projects are approved and justified for development, and they are in production. All necessary approvals and contracts are committed. Potentially viable projects are limited to technical activities such as appraisal drilling and feasibility studies. A reasonable timeframe may be needed for developers with a high probability of achieving viability. Non-viable projects are usually in the very early stages of technical evaluation. Further data is required to create a basis (UNFC, 2009).

Ahlbrandt et al. (2004) studied the classification of oil and gas resources by harmonizing the United Nations Framework Classification (UNFC) for solid fuel and mineral resources, the World Petroleum Congress (WPC), the American Association of Petroleum Geologists (AAPG), and the Society of Petroleum Engineers (SPE). Three sets of criteria are used for categorization (Figure 9.57):

- Commercial viability
- Feasibility
- Geologic knowledge

Etherington and Ritter (2007) presented the 2007 SPE/AAPG/WPC/SPEE reserves and resources classification, definitions, and guidelines. The 2007 system was believed to incorporate best practices identified in the industry for reserve quantities. By considering previous guidelines studied by SPE/WPC/AAPG published in 1997, 2001, and 2004, the latest standards of reserve and

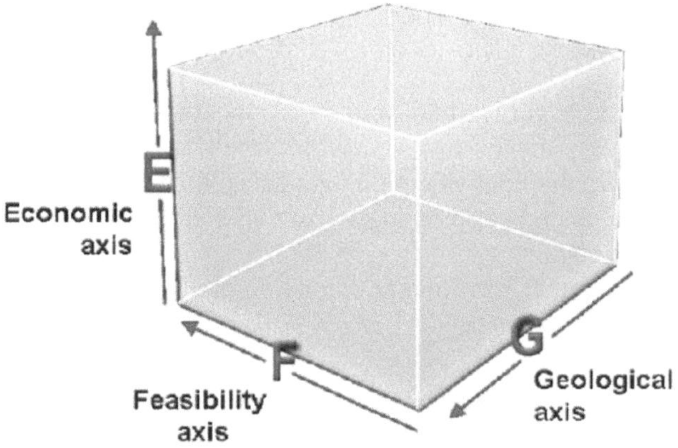

FIGURE 9.57 Principal element of the UNFC.

Source: Ahlbrandt et al., 2004

resource evaluation were delineated in (Etherington & Ritter, 2007). The concerns regarding the latest system were also discussed. One of the main deficiencies was clear guidelines for separating unproved reserves from contingent resources. Another important point was the applicability of guidelines to unconventional resources. The differences between deterministic and probabilistic approaches were unclear.

The production potential of unconventional resources is crucial for classifying them. This is obtained from tests or analogy between wells. Both methods require a detailed geological evaluation of the relevant reservoir sections. Elliott (2008) focused on the evaluation, classification, and reporting of unconventional resources. He grouped the evaluation of hydrocarbons into three steps: estimation, classification, and reporting. Due to the physical nature of unconventional reservoirs, the approaches used in conventional resources require some modifications to be used for unconventional resources. To illustrate, the term "discovery" shifts from drilling to testing. The size of unconventional resources is also questionable and needs to be addressed with more drilling data.

MacDonald et al. (2011) published a study on the classification of fossil energy and mineral reserves and resources by the United Nations Framework Classification in 2009. The UNFC (2009) aims to serve the following needs:

- Information to allocate capital appropriately by reducing costs, improving longsightedness through the application of lower risks for the financial communities.
- Capturing value efficiently within the established frameworks to serve host countries, shareholders, and stakeholders.
- Governments' help for the industry. Transferring the market prices to the wellhead with minimum loss.

- Formulating robust and longsighted policies for international energy and mineral studies.

The strengths of the UNFC (2009) were described in the study as following:

- A unified system of classification for all stakeholders.
- Harmonized energy terminology was found as the key for more reliable resource estimation.
- A common system for mineral extraction and petroleum extraction from unconventional resources.

Smith and Ray (2014) investigated a global classification standard for fossil energy and mineral reserves and resources. UNFC is a combined work that is conducted by petroleum and mining industries, regulatory agencies, and governments (Figure 9.58). The work is performed within the body of the United Nations for fossil fuel classification and reserve-resources assessment. The UNFC (2009) is a widely used reserves-resources classification system, which is fully compliant with SPE/WPC/AAPG/SPEEE in the world. UNFC started classification works in 1992 with classifications of solid minerals. The publication was in 1997, and a group of experts came together to extend the function of UNFC to oil and gas and uranium. Thus, UNFC-2004 was released. The SPE Oil and Gas Reserves Committee developed its "grand vision" to

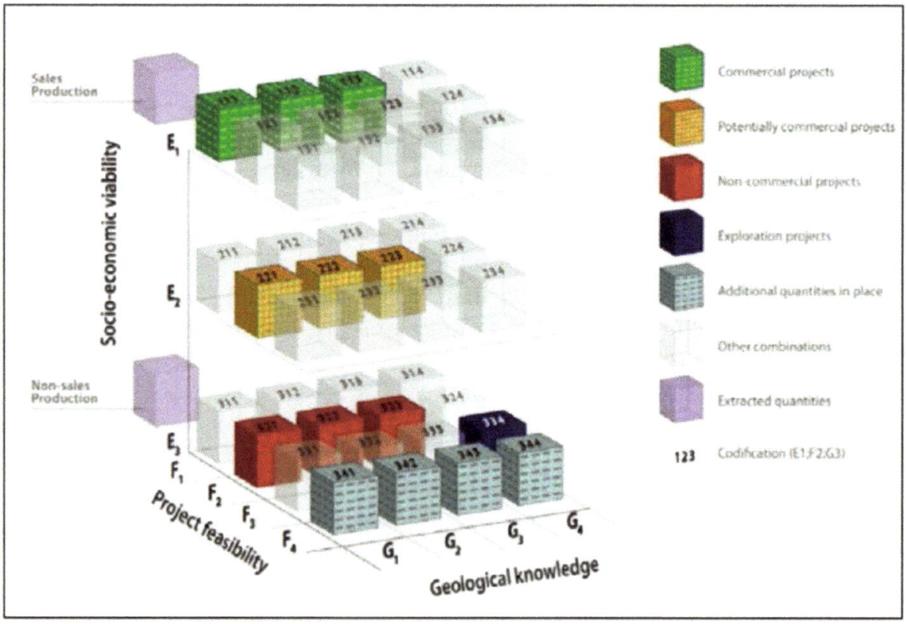

FIGURE 9.58 Examples of classes in UNFC-2009 categories.

Source: Smith & Ray, 2014

set universal reserve-resources definitions. The classification system was designed in a three-dimensional framework:

- Economic and social viability
- Project status and feasibility
- Geologic knowledge

9.2.21 Fuel and Flare Contract and Lease Term Royalty Non-Hydrocarbons Aggregation

Gas flare and venting have significant problems from the environmental and economic considerations. The combustion efficiency of the flaring system is directly related to the creation of potentially toxic air pollutants by emissions. The major products created as the result of flaring are toluene, ethylbenzene, xylenes, styrene, benzene, biphenyl, and fluorine. Most of the time, high-flaring oilfields are located far away from the public area. However, small communities are present in many nearby locations. Significant health risk arises for oilfield workers and local communities due to gas flaring or venting. To illustrate, Environmental Rights Actions (ERA) and the Climate Justice Programme (CJP) reported the impacts of toxic air pollutants emitted by gas flaring in the Niger Delta in 2005. The programs reported 49 premature deaths, 120,000 asthma attacks, and illnesses in 5,000 children associated with gas flaring from 17 onshore flow stations in Nigeria's Bayelsa State.

Guven et al. (2010) performed a study on gas flaring and venting. The impacts' extent and remedies for gas flaring were studied in detail. They found that the uncertainty in the magnitude of gas flaring and venting is extremely important to quantify the impacts of pricing and investment programs associated with gas flaring volumes and supplies. The reported volume of flared gas for 2004 and 2005 is shown in Table 9.7. As is clearly understood from Table 9.7, Nigeria, Russia, Iran, and Iraq are the highest gas flaring countries.

Most of the time, gas consumed in operations is not included in production reports. Flared gas may not be included in production and reserve estimation calculations. Gas consumed in operations should also be included in the gas reserve calculations. A footnote regarding consumed gas should be used to show the volume of gas used in reserve calculations.

9.2.22 What Do We Mean by Unconventional Resources?

Unconventional resources are the hydrocarbon resources that cannot be produced with conventional production techniques. Shale gas, shale oil, coalbed methane, and tight reservoirs are the main unconventional reservoirs. The common reservoir property of these formations is their extremely low permeability. The recovery factor of unconventional resources is very low because of low permeability, and a significant amount of gas remains in the reservoir as unrecovered. Hydraulic fracturing and horizontal well drilling are the latest technologies for developing unconventional reservoirs (Figure 9.59).

TABLE 9.7
Flared Gas Volume in 2004 and 2005 in Billions of Cubic Meters

Country	2004	2005
Nigeria	24.1	25.5
Russia	14.7	14.9
Iran	13.3	13.0
Iraq	8.6	7.2
Angola	6.8	6.4
Venezuela	5.4	5.4
Qatar	4.5	3.9
Algeria	4.3	3.5
USA	2.8	3.4
Indonesia	3.7	3.0
Kuwait	2.7	3.0
Kazakhstan	2.7	2.7
Equatorial Guinea	3.6	2.6
Azerbaijan	2.5	2.5
Libya	2.5	2.5
Mexico	1.5	2.5
Brazil	1.5	2.5
Congo	1.2	2.2
UK	1.6	1.6
Gabon	1.4	1.6
Total Top 20	109.3	109.8
Rest of the World	40.0	40.0
World Total	149.3	149.8

Source: Guven, 2010

The success of hydraulic fracturing depends on various parameters. However, sweet spot conditions are defined by IEA based on the experiences and several years of results in US major shale reservoirs. It was found that minerals with a high quartz and calcite content show more successful results than rocks with a majority of clay content. The thermal maturity of shale play was also found significant. Vitrinite indices from 0.7% to 1.0% refer to the oil generation window, and it is a wet gas generation window between 1.0% and 1.3%. Beyond 1.3% is the dry gas generation window; 2% of total organic content is usually considered as the minimum value for shale plays (Ashayeri & Ershaghi, 2015).

9.2.23 Approach to Assessing Reserves and Resources in Unconventional Reservoirs

Decline curve analysis and adsorption methods are the two important approaches for estimating unconventional reservoirs.

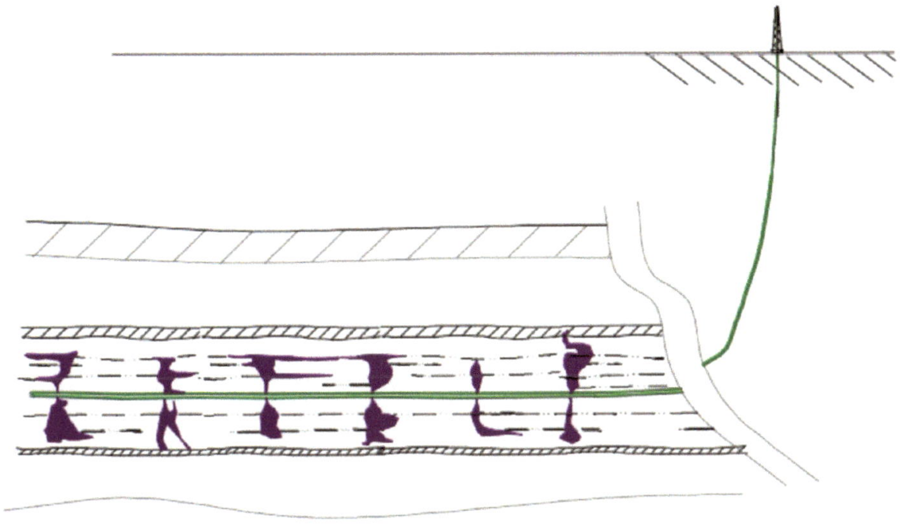

FIGURE 9.59 An illustration of multiple fractures in a horizontal well.

Source: Carter, 2009

Peacock (2018) outlined the issues considered during reserves assessment for multiple licenses in the same basin. Straddling the boundaries of petroleum accumulations is a widely known problem in petroleum fields. Resource unitization is proposed as the best solution for the assessment of straddling licenses. A typical unitization method uses hydrocarbon volumes in place with a single deterministic interpretation by following technical procedures. The applied method generally does not consider economics, and it is based on a specific time during the lifetime of the field. Considerations for unitized reserve estimations:

- The status of both sides
- Whether the wells are being drilled or in production
- Regulatory requirements based on the interested countries
- International boundaries

Unitization and resources and reserves assessment are distinct processes with different objectives. In reserves assessment, the amount of product or the economics of the project is assessed. However, unitization aims to maximize the production of a given project by optimizing related parameters. The main differences between reserve assessment and the unitization process are given in Table 9.8.

Canbaz et al. (2020) overviewed reserve evaluation in unconventional reservoirs. All the reserve evaluation techniques used for unconventional resources are delineated in detail in the study. The study comprehensively discussed the main shale resource basins in the world. The countries that hold the main shale gas reserves are listed in Table 9.9.

TABLE 9.8
The Main Differences between Reserves and Unitization

Reserves and Resources	Unitization and Redetermination
Range of estimates (low, best, high)	Single deterministic answer
Follows PRMS, SEC, COGEH, etc.	Follows technical procedures from UUOA
Based on economically recoverable volumes	Based on in-place volumes or similar
Should use all available data	Typically includes data in an agreed-on database
Affected by factors such as commerciality, development planning, prices, costs	Usually no consideration of commercial factors, development plans, etc.
May be different for each party	Must be same for all parties
Varies with time: e.g., annually	Determined only at fixed times: e.g., initial unitization plus one or more redeterminations

Source: Peacock, 2018

TABLE 9.9
Shale Gas Reserves of the Top 10 Countries

Rank	Country	Shale Gas (Billion Barrels)
1	China	75
2	Argentina	58
3	Algeria	32
4	US	27
5	Canada	26
6	Mexico	18
7	Australia	13
8	South Africa	13
9	Russia	9
10	Brazil	9
World Total		7,299

Source: Canbaz et al.

The general steps for reservoir evaluation methods are as follows:

- Performing core analysis and collecting related reservoir data
- Carrying out gas adsorption tests
- Determining fluid saturations and rock properties
- Conducting economic analyses using seismic data, mineral content, and PVT

Canbaz et al. (2020) gathered the evaluation methods for all unconventional reservoirs and resources. (See Table 9.10.)

TABLE 9.10
Unconventional Evaluation Methods

	Unconventional Oil and Gas Resource Evaluation Methods	
Resource Type	**China**	**Overseas**
Tight sandstone gas	Delphi method, geological analogy method, residual resource analysis method, accumulation-dispersion balance calculation method, formation-fluid abnormal pressure recovery method, accumulation condition analysis and prediction method, basin simulation method, sweet-spot scale sequence model method of deep-basin gas reservoir, volume method	FORSPAN method (USGS analogy method), resource network density method (ExxonMobil analogy method), stochastic simulation method, single-well reserve estimation method, statistical method (discovery process method and spatial resource distribution and prediction method)
Tight Oil	Graded resource abundance analogy method, EUR analogy method, small surface element method (volume method), basin simulation method	FORSPAN method, small surface element method
Shale gas	Resource abundance analogy method, EUR analogy method, volume method, adsorption method of unit rock gas	USGS analogy method, single-well reserves estimation method
Coalbed methane	Volume method	Volume method, analogy method
Gas hydrate	Volume method	Volume method
Oil shale	Volume method, pyrolysis simulation method	Volume method, spatial resource distribution and prediction method
Oil sands	Volume method	Volume method

Source: Canbaz et al., 2020

AbdelMawia and Hegazy (2015) studied important considerations for reserves estimation in unconventional resource plays. The reason for variations between different reserve estimations for the same asset was discussed in the study. The authors emphasized the lack of understanding of the flow and storage mechanisms in the unconventional resources. Figure 9.60 illustrates reserve estimations by four different experts for the same unconventional play. The following reasons for differences in reserve-resource estimations were studied:

- Reserves estimation using analogy may be inappropriate in case of lack of geological data.
- Different flow regimes in DCA may lead to different estimations.
- Aggregation of individual well results to project level.

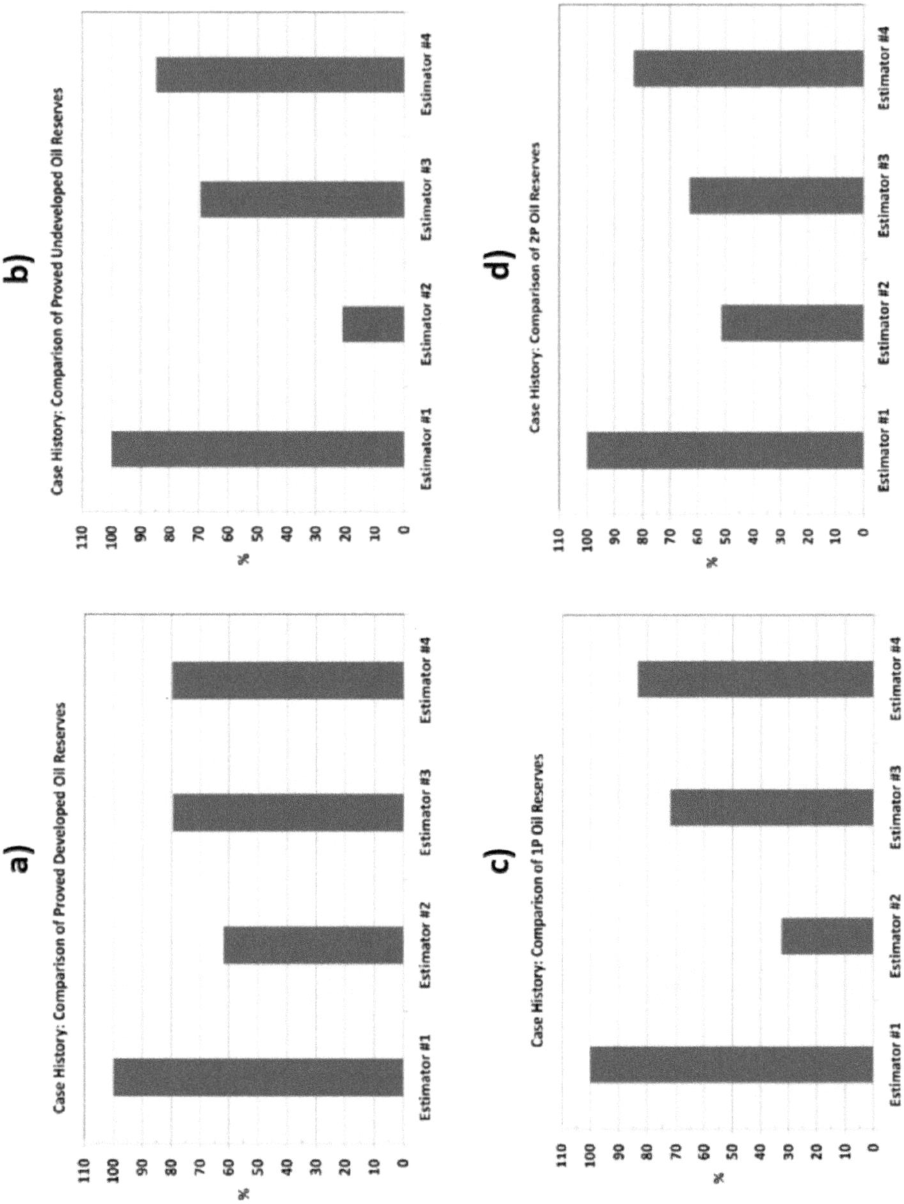

FIGURE 9.60 a) Proved developed, b) proved undeveloped, c) 1P, and d) 2P oil reserves: four different estimations of an unconventional resource.

Source: AbdelMawia & Hegazy, 2015

The study also focused on the reserve classification based on economic robustness (proved, probable, and possible).

Vassilellis and Elrafie (2016) studied geologic and dynamic indicators for reliable technology to estimate unconventional oil plays. They concluded the following:

- The performance estimation of resources is related to subsurface under-standing.
- Natural fracture systems determine the liquid-rich oil plays in subsurface areal variations.
- Subsurface indicators are static and dynamic indicators.
- Physical modeling and statistical multi-variable analysis can be used to establish a relation between subsurface indicators and multiple performance indicators.
- Subsurface and performance indicators can be used to construct composite dynamic maps.

Gong et al. (2013) conducted a study on the assessment of Eagle Ford shale oil and gas resources. Markov Chain Monte Carlo (MCMC) was employed to forecast reserve amount with probabilistic decline curves. P90 of 5.3 billion barrels of oil and P90 of 53.4 trillion cubic feet (TCF) total gas reserves were estimated. The study started with gathering data. Monthly allocated oil and gas production was analyzed for more than 800 vertical wells. Different decline models performed with probabilistic hind cast studies, and it was found that the Duong model and the Arps model had the best results among the available models. Thus, a combined decline mode was used for the boundary dominated flow regimes. The workflow of the study is shown in Figure 9.61.

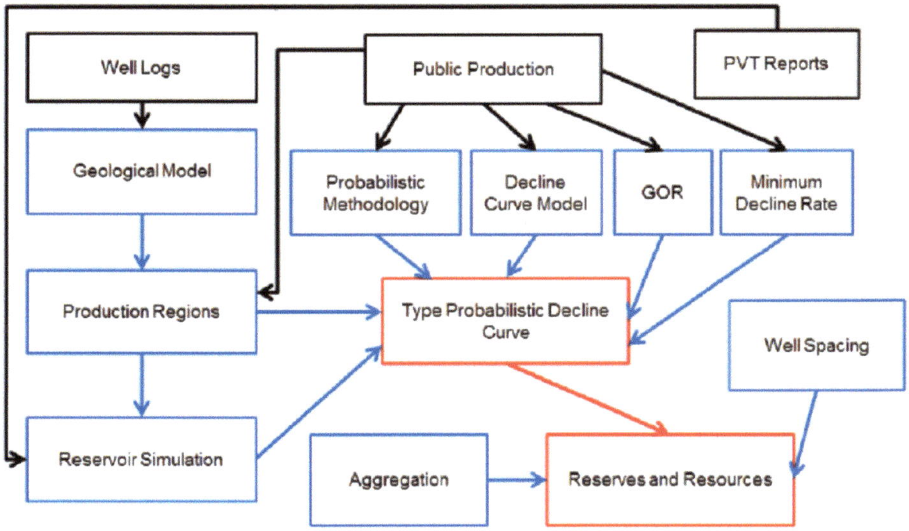

FIGURE 9.61 The workflow for assessment of Eagle Ford shale oil and gas resources.

Source: Gong et al., 2013

Moridis et al. (2019) proposed a methodology for resources classification with lower chances of commerciality to classes with a higher chance of commerciality and large uncertainty categories. They used Gaussian quadrature to create a relation between reserves categories and the PRMS matrix (Figure 9.62). Probabilistic decline curve analysis was performed with Monte Carlo simulation and the Arps hyperbolic model for 38 wells from a Permian Basin dataset.

The study started with prospective resources and moved upward through the PRMS matrix. According to the definitions of prospective resources, there is high uncertainty in prospective projects, and it decreases towards contingent resources and reserves. As illustrated in Figure 9.63, uncertainty decreases with increasing the amount of information.

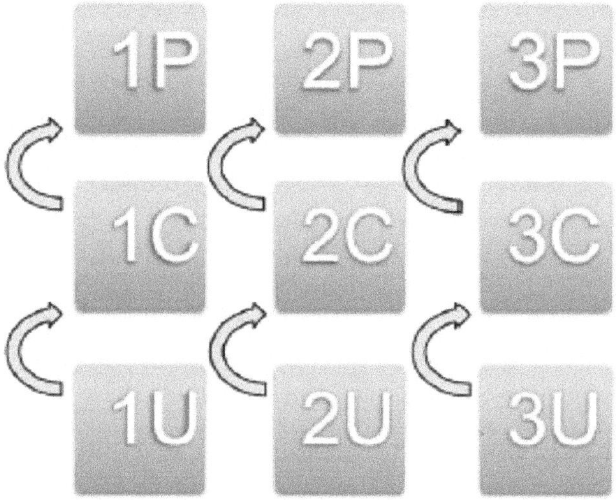

FIGURE 9.62 An illustration of the elements of the PRMS matrix.

Source: Moridis et al., 2019

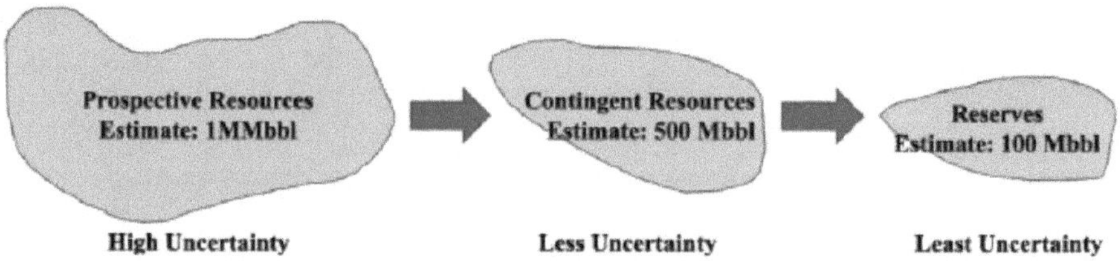

FIGURE 9.63 Initial estimates and uncertainties for prospective, contingent resources, and reserves.

Source: Moridis et al., 2019

Moridis et al. (2019) performed another study in 2019 estimating reserves and tracking the classification of reserves in unconventional reservoirs. They proposed systematic procedures for the classification of resources and reserve estimation methods in unconventional reservoirs. Well performance analysis, flow regimes, diagnostic plots, and multi-segment decline curve models were analyzed for the application of rate transient analysis and reservoir simulation. The movement from prospective resources to contingent resources was reported as nonlinear, which depends on various parameters such as technologies, oil prices, etc. The study concludes the following:

- The available data determines the estimation method to be used.
- Diagnostic plots and multi-segment DCA models are important steps for flow regime identification.
- Applying rate transient analyses and reservoir simulations are the two important methods for reserve estimation.

Each reserve estimation method requires certain information about the reservoir. Table 9.11 indicates the parameters needed for each estimation method.

Zaharova et al. (2018) proposed a single software solution for strategic decision-making based on resource analysis. The authors emphasized the need for market analysis software for the oil and gas industry in Russia and foreign markets. The software aggregates a large amount of information from different data sources. Online reserve-resource monitoring can significantly improve the quality of work for portfolio assets in the oil and gas industry. The drawbacks of the software are also discussed in the study.

9.2.24 What Constitutes a Discovery?

The presence of a hydrocarbon-bearing formation in a new field is proved well with the discovery. The discovery well is the first successfully drilled well completed with strong indications of hydrocarbon production. The discovered petroleum accumulation is determined through well testing, sampling, and logging. Thus, the existence of the recoverable amount of hydrocarbon reserves is proved with the evidence of producibility. Defining the production technology for the discovered reserves is crucial. If the discovered resource cannot be recovered with the current production technology, then the resource is defined as discovered unrecoverable with no contingent resources. Development of the technology may be sufficient for recovering these resources in the future.

9.2.25 Estimation of Contingent Resources and Reserves

Contingent resources are evaluated as of a given date for a known accumulation with current technology or future technology. However, contingent resources are not considered commercial currently. Economic status and certainty of contingent resources are used to classify them. NSAI (2019) published a report on the contingent natural gas resource of the Nigerian National Petroleum Corporation (NNPC) for the

TABLE 9.11
Data Required for Each Estimation Method

Estimation Method	Data Needed
Decline Curve Analysis (DCA)	Time
	Production rate (oil, gas)
Rate Transient Analysis (RTA)	Time
	Production rate (oil, gas)
	Pressure
	PVT properties
	Static reservoir properties
	Well length (heel to toe)
	Distance between perforations
	Number of fracture stages
Reservoir Simulation	Permeability
	Porosity
	Net-pay thickness
	Elevation
	Initial pressure
	Initial fluid saturations
	Formation volume factors
	Viscosity
	Solution gas-oil ratio
	Fluid densities
	Rock properties
	Relative permeabilities
	Capillary pressures
	Pore-volume compressibilities
	Well location
	Producing interval
	Production history of fluids present
	Bottom-hole pressure and productivity index
	Perforated lateral length
	Stage spacing

Source: Moridis et al., 2019

acquisition share of Essar Exploration and Production Limited (EEPL). The report is based on the following framework:

- The time period for minimum work
- Conversion of oil prospecting license to oil mining lease
- The commitment of the operator

NSAI (2019) presented risked gross (100%) and unrisked development contingent on oil resources to the Canadian Overseas Petroleum Limited interest in the Noa West Discovery. (See Table 9.12.)

TABLE 9.12
Risked and Unrisked Gross (100 %) Contingent Oil Resources

	Development Unclarified Contingent Light/Medium Oil Resources (Mbbl)				
	Unrisked		Risked		
Category	Gross (100%)	Company Gross(1)	Gross (100%)	Company Gross(1)	Pd (decimal)
Low Estimate (1C)	11,497.4	4,599.0	3,953.8	1,581.5	0.34
Best Estimate (2C)	16,072.9	6,429.2	5,527.3	2,210.9	0.34
High Estimate (3C)	20,653.3	8,261.3	7,102.4	2,841.0	0.34

Source: NSAI, 2019

9.2.26 How Far from a Discovery Can Contingent Resources Be?

Different institutions (SPE/AAPG/WPC) worked on thc classification system for reserve distinction. Commercial and sub-commercial hydrocarbon accumulations are critical for a reasonable reserve reporting based on a reasonable level of consistency. Contingent resources are not expected to be developed within a reasonable time period, which is around five years.

9.2.27 Estimation of Reserves Statistical Method Five-Year Rule

The five-year rule states that booked reserves can be classified as proved undeveloped reserves (PUDs) even if it takes more than five years. However, specific circumstances have to be justified and documented. The recent SEC's guidance limits the development of PUDs for more than five years (Sprague & Lee, 2024). It should be noted that the SEC has not clarified the exact starting point for the five-year period. It is either the time reserves are first booked as PUDs or the effective date of the final report, which is January 1, 2010. Companies have to provide evidence for adopting a development plan for a PUD location when it is booked. The development plan must include a statement that declares that the reserve location will be developed within five years.

9.2.28 How Do We Estimate and Incorporate Forecasts of Future Recovery?

Decline curve analysis (DCA) is a typical method for estimated ultimate recovery (EUR) of a shale gas well. Production data is sufficient to perform DCA. Empirical formulas have been developed for shale gas reservoirs since the 1940s. The most applicable empirical formulas are hyperbolic decline, power-law exponential decline, Doung's method, etc. Another important way to determine EUR is data-driven analytics.

The EUR of shale gas reservoirs depends on reservoir conditions and operating conditions. Numerical reservoir simulation suffers from flow mechanisms in shale formations. Thus, DCA is preferred to forecast EUR. The disadvantage of DCA is that the flow regime should be boundary dominated. However, shale gas wells usually do not have a long enough lifetime to reach boundary-dominated flow. The operating conditions also affect the DCA, which means that, in the case of drilling a new well or the shut-in of a well, the DCA of other wells is significantly affected.

9.2.29 Compliance with Requirements for Proved Reserves

To understand and improve the new rules of the SEC, the definitions and terms in PRMS have been revised. Proved oil and gas reserves are estimated by geoscientists and engineers with reasonable certainty considering economic production from a known reservoir. The following conditions should be met for a reserve to be defined (Gaffney, 2017):

- Fluid contacts and the identified area should be well defined.
- In case of absent data about fluid contacts, the proved reservoir is limited with the lowest known hydrocarbon penetrated during drilling.
- Additional oil production with improved recovery techniques is included in the reserve calculations only when there is a pilot test in the particular area of the reservoir.
- Current economic conditions, government regulations, technology, and operating conditions are taken into consideration for proved reserve evaluations. It is important to note that the economic producibility of reserves is calculated with a 12-month price. Some terms were added to the definition of the proved reserve.
- The lowest producible known hydrocarbon reserve and reliable technology are adopted to the proposed revision of proved reserves.

REFERENCES

AbdelMawia, A., & Hegazy, M. (2015). Important considerations for reserves estimation in unconventional resource plays. *Abu Dhabi international petroleum exhibition and conference* (p. D011S009R002). SPE-177444-MS. https://doi.org/10.2118/177444-MS

Agarwal, R. G., Gardner, D. C., Kleinstelber, S. W., & Fussell, D. D. (1998). Analyzing well production data using combined type-curve and decline curve analysis concepts. *SPE Reservoir Evaluation & Engineering*, *2*(5), 478–486.

Ahlbrandt, T. S., Blaise, J. R., & Kelter, D. (2004). Updated United Nations framework classification for reserves and resources of extractive industries. *SPE annual technical conference and exhibition* SPE-90839 https://doi.org/10.2118/90839-MS

Al-Rbeawi, S., & Owayed, J. (2019, March). Analytical and numerical analysis for temporal anomalous diffusion flow in unconventional fractured reservoirs. *SPE Middle East oil and gas show and conference* (p. D032S080R003). SPE.

Ampem-Darkoh, H. (2019). *Effects of fiscal systems on reserves recognition and investment metrics.* MS. Thesis. Department of Petroleum Engineering, African University of Science and Technology.

Artus, V., Houze, O., & Chen, C. C. (2019). Flow regime-based decline curve for unconventional reservoirs: Generalization to anomalous diffusion and power law behavior. *Unconventional resources technology conference*, URTeC 293. https://doi.org/10.15530/urtec-2019-293

Ashayeri, C., & Ershaghi, I. (2015). OPEC and unconventional resources. *SPE annual technical conference and exhibition*, SPE 174961-MS. https://doi.org/10.2118/174961-MS

Badia, M., Duro, M., Jorgensen, B. N., & Ormazabal, G. (2021). Disclosure regulation and competitive interactions: Evidence from the oil and gas industry. *The Accounting Review*, *96*(5), 1–29.

Blum, U., & Zhong, J. (2021). The loss of raw material criticality: Implications of the collapse of Saudi Arabian oil exports. *Intereconomics*, *56*, 362–370.

Boulis, A., Jayakumar, R., Lalehrokh, F., & Lawal, H. (2012). Improved methodologies for more accurate gas assessments. *SPE unconventional resources conference/gas technology symposium*, SPE 154981. https://doi.org/10.2118/154981-MS

Canbaz, C. H., Temizel, C., Ozesen, A., Gayde, S., & Akhasan, F. (2020). An overview of reserves and resources evaluation in unconventionals. *SPE annual caspian technical conference*, SPE 202577-MS. https://doi.org/10.2118/202577-MS

Carter, E. E. (2009). *Novel concepts for unconventional gas development of gas resources in gas shales, tight sands and coalbeds.* Carter Technologies RPSEA Subcontract 07122-7.

Chan, P., Etherington, J. R., & Aguilera, R. (2010). *A process to evaluate unconventional* resources. SPE Annual Technical Conference and Exhibition, SPE 134602. https://doi.org/10.2118/134602-MS

Chen, C., Liu, X., Girardi, A., McMullen, A., Gao, G., Bhattacharya, S., Richard, C., & Chowdhury, N. (2019, July). Using dynamic modeling to correct decline curve for liquid-rich shale reservoir. *SPE/AAPG/SEG unconventional resources technology conference* (p. D023S029R003). URTeC.

Cheng, Y., Lee, W. J., & McVay, D. A. (2008). Improving reserves estimates from decline curve analysis of tight and multi-layer gas wells. *SPE Reservoir Evaluation & Engineering*, *11*(05), 912–920.

Chorn, L., Yarusi, J., & Davis, S. D. R. (2013). Shale play development program using risk-reward mapping from geospatial analyses. *SPE annual technical conference and exhibition*, SPE 166130. https://doi.org/10.2118/166130-MS

Cipolla, C. L., Lolon, E. P., Erdle, J. C., & Rubin, B. (2010). Reservoir modeling in shale gas reservoirs. *SPE Reservoir Evaluation & Engineering*, *13*(04), 638–653.

Coll, C. (2019). Application of probabilistic and deterministic method for consistent reserves and resources estimation and reporting. *SPE Europec featured at EAGE conference and exhibition*, SPE-195465-MS. https://doi.org/10.2118/195465-MS

Cook, D. M. (2009). Technical evaluation of resource plays. *Offshore technology conference*, Houston, TX. OTC 20271.

Darabi, H., Ettehad, A., Javadpour, F., & Sepehrnoori, K. (2012). Gas flow in ultra-tight shale strata. *Journal of Fluid Mechanics*, *710*, 641–658.

Deloitte. (2020). *Oil and gas reporting modernization—A small entity compliance guide.* Retrieved November 10, 2020, from https://dart.deloitte.com/USDART/home/accounting/sec/sec-material-supplement/small-entity-compliance-guides/oil-gas-reporting-modernization-a-small

Deng, J., Zhu, W., & Ma, Q. (2014). A new seepage model for shale gas reservoir and productivity analysis of fractured well. *Fuel*, *124*, 232–240.

Diluzio, D. (2018). Global standard in reserves/resources classification and definitions. *Journal of Petroleum Technology*, *70*(09), 74–77.

Ding, D. Y., Wu, Y. S., Farah, N., Wang, C., & Bourbiaux, B. (2014). Numerical simulation of low permeability unconventional gas reservoirs. *SPE/EAGE European unconventional resources conference and exhibition* (Vol. 2014, No. 1, pp. 1–30). European Association of Geoscientists & Engineers. SPE 16711.

Doublet, L. E., Pandei, P. K., Mcoljum, T. J., & Blasingame, T. A. (1994). Decline curve analysis using type curves-analysis of oil well production data using material balance time. Application to field cases. *International petroleum conference and exhibition of Mexico*, SPE-28688, https://doi.org/10.2118/28688-MS

Duarte, J. C., Vinas, E. C., & Ciancaglini, M. (2014). Material balance analysis of naturally or artificially fractured shale gas reservoirs to maximize final recovery. *SPE Latin America and Caribbean petroleum engineering conference*, SPE 169480-MS. https://doi.org/10.2118/169480-MS

Elliott, D. C. (2008). The evaluation, classification, and reporting of unconventional resources. *SPE unconventional resources conference/gas technology symposium*, SPE 114160. https://dx.doi.org/10.2118/114160-MS

Etherington, J. R., & Ritter, J. E. (2007). The 2007 SPE/AAPG/WPC/SPEE reserves and resources classification, definitions, and guidelines: Defining the standard! *SPE hydrocarbon economics and evaluation symposium*, SPE 107693. https://doi.org/10.2118/107693-MS

Fathi, E., & Akkutlu, I. Y. (2013). Lattice Boltzmann method for simulation of shale gas transport in kerogen. *Society of Petroleum Engineers Journal, 18*(01), 27–37.

Fathi, E., Tinni, A., & Akkutlu, I. Y. (2012). Correction to Klinkenberg slip theory for gas flow in nano-capillaries. *International Journal of Coal Geology, 103*, 51–59.

Fatlawi, O. A., Mofazzal, M. H., Hicks, S., & Saeedi, A. (2016). Developed material balance approach for estimating gas initially in place and ultimate recovery for tight gas reservoirs. *Abu Dhabi international petroleum exhibition and conference*, SPE 183015-MS. https://doi.org/10.2118/183015-MS

Fetkovich, M.J. (1980). Decline Curve Analysis Using Type Curves., *J Pet Technology, 32*: 1065–1077. doi: https://doi.org/10.2118/4629-PA

Firanda, E. (2011, September). The development of material balance equations for coalbed methane reservoirs. *SPE Asia Pacific oil and gas conference and exhibition* (p. SPE-145382). SPE.

Friis, L. C., & Erlandson, M. (2015). Geology driven EUR prediction using deep learning. *SPE annual technical conference and exhibition*, SPE-174799-MS. https://doi.org/10.2118/174799-MS

Fulford, D. S., & Blasingame, T. A. (2013). Evaluation of time-rate performance of shale wells using the transient hyperbolic relation. *SPE Canada unconventional resources conference*, SPE 167242. https://doi.org/10.2118/167242-MS

Gaffney. (2017). *Proved reserves statement (SEC rules) certain properties in Asia*. Eni S.P.A.

Galvao, M., Hastenreiter, L., & Molina, J. (2012). Probabilistic reserves and resources estimation: A methodology for aggregating probabilistic petroleum reserves and resources. *SPE/EAGE European unconventional resources conference & exhibition-from potential to production*, SPE 152923. https://doi.org/10.2118/152923-MS

Gaurav, A. (2017, September). Horizontal shale well EUR determination integrating geology, machine learning, pattern recognition and multivariate statistics focused on the Permian basin. *SPE liquids-rich basins conference-North America* (p. D011S001R002). SPE-187494. https://doi.org/10.2118/187494-MS

Gherabati, S. A. (2018). Material-balance approach for determining drainage volume of multifracture unconventional oil wells. *Unconventional resources technology conference*, Houston, Texas, 23–25 July 2018 (pp. 1784–1798). Society of Exploration Geophysicists, American Association of Petroleum Geologists, Society of Petroleum Engineers. URTeC: 2901597.

GLJ Petroleum Consultants. (2019). *COGE handbook a new era*. CSUR Technical Luncheon.

Gong, X., Tian, Y., McVay, D. A., Ayers, W. B., & Lee, W. J. (2013). Assessment of Eagle Ford shale oil and gas resources. *SPE Canada unconventional resources conference*, SPE 167241. https://doi.org/10.2118/167241-MS

Guo, C., Xu, J., Wu, K., Wei, M., & Liu, S. (2015). Study on gas flow through nano pores of shale gas reservoirs. *Fuel*, *143*, 107–117.

Gupta, S., Fuehrer, F., & Jeyachandra, B. C. (2014). Production forecasting in unconventional resources using data mining and time series analysis. *SPE Canada unconventional resources conference*, SPE-171588-MS. https://doi.org/10.2118/171588-MS

Guven, B. B., Harriss, R., & Hertzmark, D. (2010). *Gas flaring and venting: Extent, impacts, and remedies*. James A. Baker III Institute for Public Policy Rice University.

Han, D., Kwon, S., Son, H., & Lee, J. (2019). Production forecasting for shale gas well in transient flow using machine learning and decline curve analysis. *Asia Pacific unconventional resources technology conference*, Brisbane, Australia, 18–19 November 2019 (pp. 1510–1527). URTeC-198198-MS.

Harrell, D. R. (2008). Achieving global acceptance of and compliance with a universal set of petroleum resources and reserves definitions: Are we there yet?, *J Pet Technol* 60 (2008): 34–39. doi: https://doi.org/10.2118/1008-0034-JPT.

He, L., Mei, H., & Hu, X. (2019). Advanced flowing material balance to determine original gas in place of shale gas considering adsorption hysteresis. *SPE Reservoir Evaluation & Engineering*, *22*(04), 1282–1292.

Indras, P. (2014). Applying Decline Curve Analysis in the Liquid-Rich Shales: Eagle Ford Study. Dissertation, Department of Petroleum Engineering, Texas A&M University. https://oaktrust.library.tamu.edu/items/75f58d31-9edb-461b-81d2-60c4fa0686dd

Javadpour, F. (2009). Nanopores and apparent permeability of gas flow in mudrocks (shales and siltstone). *Journal of Canadian Petroleum Technology*, *48*(08), 16–21.

Johnston, D. (2017). Changing fiscal landscape 2008–2017. *The Journal of World Energy Law & Business*, *10*(5), 415–443.

JPT Staff, Analyzing Well Production Data Using Combined Type-Curve-and Decline-Curve-Analysis Concepts. *J Pet Technology*, 50 (1998): 76–77. doi: https://doi.org/10.2118/1098-0076-JPT.

Kabir, S., Rasdi, F., & Igboalisi, B. (2011). Analyzing production data from tight oil wells. *Journal of Canadian Petroleum Technology*, *50*(05), 48–58.

Karacaer, C., & Onur, M. (2012, September). Analytical probabilistic reserve estimation by volumetric method and aggregation of resources. *SPE hydrocarbon economics and evaluation symposium* (p. SPE-162875). SPE.

Kocoglu, Y., Wigwe, M. E., Sheldon, G., & Watson, M. C. (2020, December). Machine learning based decline curve—Spatial method to estimate production potential of proposed wells in unconventional shale gas reservoirs. *Unconventional resources technology conference*, 20–22 July 2020 (pp. 499–517). URTEC.

Ledesma, D. (2009). The changing relationship between NOCS and IOCS in the LNG chain. Oxford Institute for Energy Studies. https://www.oxfordenergy.org/wpcms/wp-content/uploads/2009/07/NG32-TheChangingRelationshipBetweenNOCsandIOCsintheLNGChain-DavidLedesma-2009.pdf

Lee, J., & Sidle, R. (2010). Gas-reserves estimation in resource plays. *SPE Economics & Management*, *2*(02), 86–91.

Li, B., Billiter, T. C., & Tokar, T. (2020, December). Significant error reduction in machine-learning decline curve analysis for unconventional reservoirs. *Unconventional resources technology conference*, 20–22 July 2020 (pp. 82–101). URTEC.

Liang, Y., Liao, L., & Guo, Y. (2019, January). A big data study: Correlations between EUR and petrophysics/engineering/production parameters in shale formations by data regression and interpolation analysis. *SPE hydraulic fracturing technology conference and exhibition* (p. D022S003R003). SPE.

Liang, Y., & Zhao, P. (2019, September). A machine learning analysis based on big data for eagle ford shale formation. *SPE annual technical conference and exhibition* (p. D021S020R001). SPE.

MacDonald, D. G., Ross, J. G., Heiberg, S., Lynch-Bell, M. D., & Griffiths, C. (2011, May). The United Nations framework classification for fossil energy and mineral reserves and resources-2009. *Offshore technology conference* (p. OTC-21852). OTC.

Male, F., Aiken, C., & Duncan, I. J. (2018, September). Using data analytics to assess the impact of technology change on production forecasting. *SPE annual technical conference and exhibition* (p. D011S008R006). SPE.

Masini, E. B., Matinez, F. E., & Green, L. G. (2019). Fluid flow modeling in shale formations using the differential quadrature element method. *Fuel, 244*, 44–60.

McMichael, C. L., & Young, E. D. (2001). Reserve recognition under production-sharing and other nontraditional agreements. In *Guidelines for the evaluation of petroleum reserves and resources* (p. 107). https://www.spe.org/industry/docs/Guidelines-Evaluation-Reserves-Resources-2001.pdf

Moghaddam, R. N., & Jamiolahmady, M. (2017, June). Study of slip flow in shale and tight gas reservoir rocks using lattice Boltzmann method. *SPE Europec featured at EAGE conference and exhibition* (p. D041S009R002). SPE.

Molina, O., & Zeidouni, M. (2017, June). Analytical approach to determine the degree of interference between multi-fractured horizontal wells. *SPE Europec featured at EAGE conference and exhibition* (p. D031S007R003). SPE.

Morales, E., & Lee, W. J. (2015). Probabilistic reserves addition within a project. *SPE Economics & Management*, *7*(01), 16–21.

Moridis, N., Quist, M., Lee, W. J., Sim, W., & Blasingame, T. (2019, April). Defining and deriving the proper order of movements from prospective resources, to contingent resources, to reserves and back. *SPE western regional meeting* (p. D031S007R004). SPE.

Mousavi, S. M., Jabbari, H., Darab, M., Nourani, M., & Sadeghnejad, S. (2020, November). Optimal well placement using machine learning methods: Multiple reservoir scenarios. *SPE Norway subsurface conference* (p. D021S008R005). SPE.

Nascimento, J. H., Santos, A. A. S., & Schiozer, D. J. (2018, April). Dynamic uncertainties appraisal throughout a development project, applying PRMS indicators of resource and reserves categorization. *Offshore technology conference* (p. D041S051R002). OTC.

Negara, A., Salama, A., Sun, S., Elgassier, M., & Wu, Y. S. (2015, November). Numerical simulation of natural gas flow in anisotropic shale reservoirs. *Abu Dhabi international petroleum exhibition and conference* (p. D031S044R007). SPE.

NSAI. (2019). *Contingent natural gas resources of the NNPC.* https://nsps.org.ng/about/

Odi, U., Bacho, S., & Daal, J. (2019, July). Decline curve analysis in unconventional reservoirs using a variable power law model: A Barnett shale example. *SPE/AAPG/SEG unconventional resources technology conference* (p. D033S063R003). URTEC.

Ogilvie, J. S., & Keyser, W. L. (2004, October). Seismic considerations for classifying proved resources/reserves: West Africa example. *SEG international exposition and annual meeting* (p. SEG-2004). SEG.

Orozco, D., Fragoso, A., Selvan, K., Noble, G., & Aguilera, R. (2020). Eagle ford huff 'n'puff gas-injection pilot: Comparison of reservoir-simulation, material balance, and real performance of the pilot well. *SPE Reservoir Evaluation & Engineering*, *23*(01), 247–260.

Pan, Y., Zhou, P., Deng, L., & Lee, J. (2019, April). Production analysis and forecasting for unconventional reservoirs using Laplacian echo-state networks. *SPE western regional meeting* (p. D031S012R003). SPE.

Peacock, D. (2018, October). Reserves assessment in petroleum accumulations straddling boundaries: The relationship between reserves & resources and unitisation. *SPE Asia Pacific oil and gas conference and exhibition* (p. D011S006R001). SPE.

PRMS (Petroleum Resource Management System). (2007). SPE/WPC/AAPG/SPEE. https://www.spe.org/industry/docs/Petroleum-Resources-Management-System-2007.pdf

PRMS (Petroleum Resource Management System). (2018, June). SPE/WPC/SEG/AAPG/EAGE/SPEE/SPWLA, SPE website. https://www.spe.org/en/industry/petroleum-resources-management-system-2018/

Qiu, K., & Li, H. (2020, February). A new analytical model for production forecasting in unconventional reservoir considering the simultaneous matrix-fracture flow. *Asia Pacific unconventional resources technology conference*, Brisbane, Australia, 18–19 November 2019 (pp. 527–545). Unconventional Resources Technology Conference.

Rahmanifard, H., Alimohamadi, H., & Gates, I. (2020, July). Well performance prediction in Montney formation using machine learning approaches. *SPE/AAPG/SEG unconventional resources technology conference* (p. D023S042R002). URTEC.

Ross, J. G. (1997, March). The philosophy of reserve estimation. *SPE hydrocarbon economics and evaluation symposium* (p. SPE-37960). SPE.

Rudra, A. (2013). *Efficient techniques for decision support: Approximate query processing to parallel and value-based data mining.* Scholars' Press.

SEC. (2009). *Modernization of oil and gas reporting.* Securities and Exchange Commission. https://www.sec.gov/files/rules/final/2009/33-8995fr.pdf

Smith, S. T., & Ray, D. (2014, May). The United Nations framework classification for fossil energy and mineral reserves and resources–A global classification standard for energy resources. *SPE hydrocarbon economics and evaluation symposium* (p. D011S003R003). SPE.

Society of Petroleum Engineers. (2007). *Petroleum resources management system.* https://www.spe.org/industry/docs/Petroleum-Resources-Management-System-2007.pdf

Society of Petroleum Engineers. (2011). *Guidelines for application of the petroleum resources management system.* https://www.spe.org/industry/docs/PRMS_Guidelines_Nov2011.pdf

Speirs, J., McGlade, C., & Slade, R. (2015). Uncertainty in the availability of natural resources: Fossil fuels, critical metals and biomass. *Energy Policy*, *87*, 654–664.

Sprague, S. M., & Lee, W. J. (2024, September). PUD reserves converted in practice: Impact of the SEC's final rule. *SPE annual technical conference and exhibition* (p. D021S027R004). SPE.

Stalgorova, E., & Mattar, L. (2013). Analytical model for unconventional multifractured composite systems. *SPE Reservoir Evaluation & Engineering*, *16*(03), 246–256.

Stephenson, B., & Coflin, K. (2015, October). Guidelines for the handling of natural fractures and faults in hydraulically stimulated resource plays. *SPE Canada unconventional resources conference* (p. D011S003R001). SPE.

Strickland, R., Purvis, D., & Blasingame, T. (2011, June). Practical aspects of reserve determinations for shale gas. *SPE unconventional resources conference/gas technology symposium* (p. SPE-144357). SPE.

Tandon, S. (2019, October). Improved analysis of stimulated reservoir volumes in unconventional reservoirs using supervised learning techniques. *SPE eastern regional meeting* (p. D021S001R005). SPE.

Tugan, M. F., & Weijermars, R. (2020). Variation in b-sigmoids with flow regime transitions in support of a new 3-segment DCA method: Improved production forecasting for tight oil and gas wells. *Journal of Petroleum Science and Engineering*, *192*, 107243.

UNFC. (2009). *United Nations framework classification for fossil energy and mineral reserves and resources.* https://www.usgs.gov/publications/united-nations-framework-classification-fossil-energy-and-mineral-reserves-and

Vassilellis, G. D., & Elrafie, E. A. (2016, May). Geologic and dynamic indicators offering a reliable technology paradigm to reserve and resource estimation in unconventional oil plays. *SPE hydrocarbon economics and evaluation symposium* (p. D011S001R004). SPE.

Wagner, F. (2014). Renewables in future power systems: Implications of technological learning and uncertainty. Springer, https://doi.org/10.1002/ente.201500327

Xi, Z., & Morgan, E. (2019). Combining decline-curve analysis and geostatistics to forecast gas production in the Marcellus shale. *SPE Reservoir Evaluation & Engineering*, *22*(04), 1562–1574.

Xue, L., Liu, Y., Xiong, Y., Liu, Y., Cui, X., & Lei, G. (2021). A data-driven shale gas production forecasting method based on the multi-objective random forest regression. *Journal of Petroleum Science and Engineering, 196*, 107801.

Ye, Y., Yang, B., & Chongfuangprinya, P. (2020, October). GLOW: A novel co-simulation framework for integrated distribution network planning and data analysis using GridLAB-D. *2020 CIGRE Canada conference*, Toronto, Ontario.

Zaharova, O., Teplouhova, I., Islamuratov, M., Kisurina, A., Panfilova, E., Stepanova, V., & Ershov, A. (2018, October). Innovative strategy decision and portfolio assets analysis instrument–Permanent model of on-line reserves and resources monitoring. *SPE Russian petroleum technology conference* (p. D033S018R011). SPE.

Zhan, L., Guo, B., Yu, Y., Hu, J., & Wang, S. (2019). Source–reservoir chart for tight oil sweet spots evaluation and its application in Permian of Jimsar Sag, Junggar Basin. *Journal of Petroleum Exploration and Production Technology, 9*, 1723–1731.

Zhang, X. L., Xiao, L. Z., Guo, L., & Xie, Q. M. (2015). Investigation of shale gas microflow with the lattice Boltzmann method. *Petroleum Science, 12*, 96–103.

Zhao, K. & Du, P. (2020) A new production prediction model for multistage fractured horizontal well in tight oil reservoirs. *Advances in Geo-Energy Research, 4*(2): 152–161, doi: 10.26804/ager.2020.02.04.

Zhao, P., Dong, R., & Liang, Y. (2020, October). Regional to local machine-learning analysis for unconventional formation reserve estimation: Eagle Ford case study. *SPE annual technical conference and exhibition* (p. D021S006R004). SPE.

Zhou, H., & Lascaud, B. (2019, October). An integrated machine learning framework for optimizing unconventional resources development. *Unconventional resources technology conference*, Denver, Colorado, 22–24 July 2019 (pp. 3079–3087). Unconventional Resources Technology Conference (URTeC); Society of Exploration Geophysicists.

Zolotukhin, A. B. (2000, October). A novel approach to resources and reserves determination. *SPE annual technical conference and exhibition* (p. SPE-63199). SPE.

10 Surface Facilities, Metering, Measurements, and Testing in Unconventional Resources

Aliyu Mohammed Kaura and
Mohammed A Mohammed

10.1 INTRODUCTION AND CHALLENGES WITH SURFACE FACILITIES

Oil and gas surface facilities play a vital role in the energy industry. They are an essential part of the extraction, processing, and transportation of oil and gas from wells to refineries or other end users. These facilities consist of a complex network of pipelines, tanks, pumps, and other equipment. Pipeline works in particular are crucial for the efficient transportation of oil and gas from one point to another. With their sophisticated design and operation, oil and gas surface facilities help ensure a reliable and consistent supply of energy to meet the world's growing demand.

The successful extraction and processing of unconventional resources, such as shale gas and tight oil, relies heavily on the functionality of surface facilities. These surface facilities are specifically designed to cater to the unique characteristics and requirements of unconventional resources, which make them different from conventional oil and gas resources. Without these specialized surface facilities, extracting and processing unconventional resources efficiently and safely would be challenging.

Surface facilities required for the extraction and production of unconventional resources differ according to the type of resource being targeted. However, some of the commonly used components include well pads for drilling and completion activities, hydraulic fracturing equipment for fracturing the rock formations and creating pathways for the hydrocarbons to flow, water management systems for sourcing, treating, and disposing of water used in the drilling process, gas processing facilities for separating the hydrocarbon gases from other impurities and contaminants, and oil and water separation units for separating the oil and water produced from the well. These facilities are crucial for the efficient and safe production of unconventional resources (Figure 10.1).

DOI: 10.1201/9781003319955-10

Unconventional resource surface facilities involve a complex set of components and processes that are crucial for efficient and safe resource extraction. These may include drilling rigs, well pads, tanks, pipelines, compressors, separators, and other equipment necessary for the gathering, processing, and transportation of hydrocarbons. Additionally, a range of environmental, safety, and regulatory considerations must be taken into account throughout the design, construction, and operation of these facilities, including water management, air quality monitoring, waste disposal, and emergency response planning. Overall, the success of unconventional resource

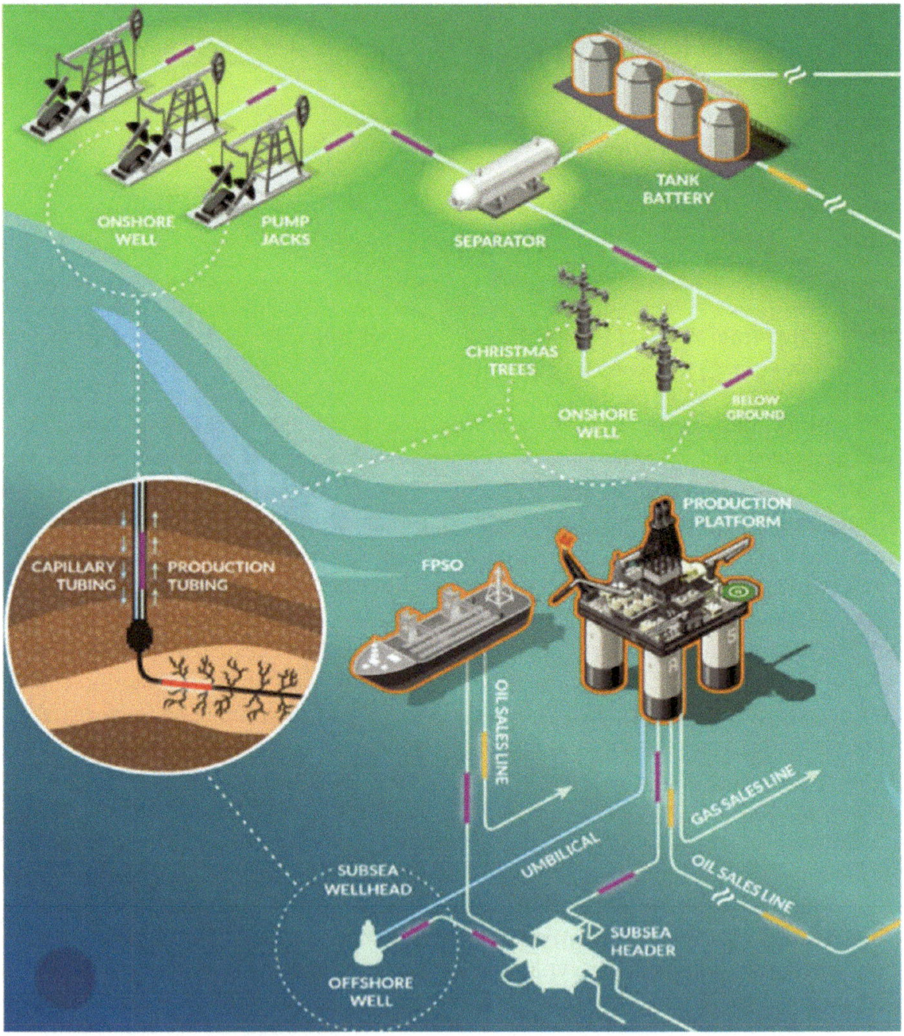

FIGURE 10.1 Surface facilities.

Source: Hydrocarbon Dynamics

development depends heavily on the proper management and integration of these various components and processes.

10.1.1 Well Pad Layout

One of the methods of extracting unconventional resources is through the use of horizontal drilling and hydraulic fracturing. To make the development of oil and gas fields with complex surface conditions possible, the oil industry relies heavily on the utilization of well pads, which serve a crucial role. These well pads allow for the efficient extraction of resources from the subsurface while minimizing the impact on the surface area (DeMong et al., 2013).

In order to facilitate the extraction of oil and gas, well pads are constructed on the surface to accommodate multiple wells in close proximity. The design of these well pads is crucial to ensure efficient operations while minimizing surface footprint, particularly in high-density developments. Well pads play a vital role in the oil industry, allowing for the development of oil and gas fields with complex surface conditions (Abramov, 2019).

Well pads are typically outfitted with several pieces of production equipment, including compressors, pumps, wellheads, separators, storage tanks, gathering lines, and hydraulic fracturing equipment. These components play a crucial role in effectively separating the produced fluids, such as oil, gas, and water, while maintaining the optimal pressure and flow rates required for efficient production. When designing a multi-well pad, one must consider various factors such as the type of reservoir, geological conditions, well spacing, and production rates to ensure optimal performance and maximum output (Abramov, 2019).

A well pad typically consists of several key components that work together to extract oil and gas from the ground. Some of the most common components include.

- Wellheads: These are the structures on top of each well that control the flow of oil or gas from the reservoir. Wellheads also provide a connection point for various types of equipment.
- Gathering lines: Pipelines that collect the oil, gas, and produced water from multiple wells on the pad and transport them to processing facilities.
- Separator units: These separate the produced fluids into oil, gas, and water. The separation process is vital for refining the extracted resources.
- Storage tanks: Tanks are used to temporarily store oil and water before transportation. They provide flexibility in managing varying production rates.
- Flare systems: In case of excess gas that cannot be processed or stored, flare systems are used to burn off the gas in a controlled manner, preventing the release of unburned methane into the atmosphere.
- Metering equipment: Instruments that measure the flow rates of oil, gas, and water. Accurate metering is crucial for production optimization and reporting.
- Control valves and manifolds: These regulate the flow of fluids between different components of the surface facilities, ensuring efficient and safe operation.

- Chemical injection systems: Injection of chemicals may be required to treat and manage issues such as corrosion, scale, or hydrate formation in the well and processing equipment.

Each of these components plays a critical role in the overall function of the well pad and must be carefully designed and maintained to ensure safe and efficient operation.

In the oil and gas industry, well pad development faces various challenges, including environmental concerns, regulatory compliance, and technical issues. Contamination of soil and water, habitat disruption, and air pollution are environmental challenges that need careful management during well pad construction and operation (Haïdar, 2011).

Also, well pads have technical challenges that involve drilling complexities, reservoir uncertainties, and well integrity issues. Efficient well pad design must address these factors to optimize production and minimize environmental impact. Additionally, addressing community concerns, such as noise and traffic, is essential for maintaining positive relations with local residents. Overall, balancing economic benefits with environmental and social considerations poses a continuous challenge in the oil and gas sector.

10.1.2 Hydraulic Fracturing

Hydraulic fracturing, or fracking, involves complex equipment and processes. Hydraulic fracturing, or fracking, is a complex process that involves the use of specialized equipment to extract oil or gas from deep underground rock formations. Although this process has revolutionized the energy industry, it comes with several challenges that require careful attention to ensure safe and efficient operations.

One of the most common challenges faced by fracking equipment is fluid leaks, which can occur due to faulty equipment or poor maintenance. These leaks not only result in significant economic losses but can also pose a threat to the environment and human health. In addition, equipment malfunctions such as pump failures, valve leaks, and blowouts can lead to costly downtime and safety hazards.

Wellbore integrity issues are another significant challenge in the fracking industry. As fracking involves injecting high-pressure fluids into the rock formations, it can create fractures that extend beyond the targeted reservoirs, leading to the contamination of nearby aquifers and other water sources. Moreover, maintaining proper waste disposal and managing the large volumes of water required for fracking present ongoing operational hurdles.

To address these challenges, regular maintenance of equipment and infrastructure is crucial. Advanced monitoring systems that can detect and prevent equipment failures can also minimize downtime and economic losses. Additionally, adherence to stringent safety regulations and best practices can help prevent accidents and ensure the safety of workers and the environment.

10.1.2.1 Surface Facilities Supporting Drilling and Completion

Drilling and completion processes in unconventional resources require surface facilities that face unique and complex challenges. These challenges include accurately characterizing the pore structure of the reservoir, understanding the fluid flow dynamics in the complex fracture-matrix network, designing effective hydraulic fracturing techniques, predicting fracture geometry, and analyzing reservoir intelligence. These challenges can be daunting, but with the right technology and expertise, they can be overcome to unlock the full potential of unconventional resources (Montagna et al., 2023).

10.1.2.2 Corrosion and Scale Challenges

Corrosion and scale challenges in surface facilities for unconventional resources, such as shale gas or oil sands, often arise due to the unique composition of extracted fluids. The presence of corrosive elements and minerals can lead to equipment degradation and reduced operational efficiency. Proper materials selection, corrosion inhibitors, and effective scale management are crucial for addressing these challenges in unconventional resource processing facilities. Regular monitoring and maintenance are essential to mitigate the impact of corrosion and scale on equipment integrity and overall production efficiency.

10.1.2.3 Challenges with Testing

Surface facilities testing for unconventional resources is a challenging task that requires addressing the unique properties and conditions associated with these resources. One of the key challenges in testing surface facilities for unconventional resources is the variability of fluid composition. The composition can vary widely from well to well and even within a single well, making it difficult to design and implement a standardized testing protocol.

Another challenge is the high temperatures and pressures typically encountered in unconventional resources. These conditions can put a strain on equipment and require the use of specialized materials and designs to prevent failure. In addition, complex reservoir conditions can make it challenging to accurately measure flow rates and other key parameters during testing.

Scale formation and deposition is another critical issue that must be addressed during testing of surface facilities for unconventional resources. Scale buildup can occur in the wellbore, casing, and surface equipment, which can reduce production rates and damage equipment. Effective scale management is essential to maintain optimal production rates and avoid costly equipment failures.

Finally, water management is a critical consideration when testing surface facilities for unconventional resources. Water is often used as part of the hydraulic fracturing process and must be carefully managed to avoid contamination of nearby water resources and to ensure the safe disposal of flowback water. Effective water management strategies are essential to ensure the long-term viability of unconventional resource development.

Overcoming these challenges requires a multidisciplinary approach, combining expertise in geology, engineering, chemistry, and environmental science to develop robust testing strategies tailored to the specific characteristics of unconventional resources.

10.1.2.4 Infrastructure Development

Establishing infrastructure for unconventional resources, such as shale gas or tar sands, can be a daunting task that demands a substantial upfront investment. The process involves building an entire supply chain from scratch, including pipelines, processing plants, and transportation systems. Designing and constructing this infrastructure can be a complex and time-consuming endeavor, requiring significant planning, engineering, and regulatory compliance. Moreover, the construction of such infrastructure may have significant environmental and social implications that must be carefully evaluated and addressed to ensure the safety and well-being of surrounding communities and ecosystems.

10.1.2.5 Water Management

Unconventional methods of extracting resources, such as oil and gas, often involve the use of large amounts of water, which can pose significant challenges in areas with limited water resources. Therefore, proper water management is crucial to minimize the environmental impact of these extraction techniques. This includes implementing water recycling and reuse systems, as well as utilizing alternative methods of extraction that require less water. By managing water resources effectively, we can ensure sustainable and responsible extraction practices that minimize harm to the environment.

10.1.2.6 Gas-Processing Facilities

Unconventional gas resources such as shale gas often contain a high proportion of natural gas. Surface facilities include gas processing units to remove impurities and separate valuable components, such as methane, ethane, propane, and butane. These components can be further processed and transported for various applications.

Gas-processing facilities for unconventional resources face challenges such as variable production rates, high initial water content, and the need for specialized technologies to handle complex hydrocarbon compositions. Additionally, managing environmental concerns, regulatory compliance, and optimizing operational efficiency pose ongoing difficulties in this context.

10.1.2.7 Oil and Water Separation Units

In unconventional resource facilities, oil and gas separation units face a range of challenges due to the complex and variable composition of the produced fluids. The separation of multiphase fluids, handling unconventional hydrocarbons like shale oil, and managing high levels of associated water pose significant technical and operational hurdles. In order to meet environmental standards, it is crucial to optimize the separation process while ensuring the economic viability of the facility. This requires a comprehensive understanding of the different components of the produced fluids, including their physical and chemical properties, and the ability to develop effective separation techniques that can accommodate the variability of the fluid composition. Furthermore, the development of innovative technologies and processes that can enhance the efficiency of the separation process and improve the quality of the separated products is an ongoing challenge in this context.

10.1.2.8 Technological Uncertainties

The extraction of unconventional resources frequently depends on the development of cutting-edge technologies. As these technologies are constantly evolving, there is a certain level of uncertainty surrounding their long-term effectiveness, which can have a significant impact on project planning and investment decisions. It is important to keep up with these technological advancements in order to make informed decisions and ensure the success of such projects.

10.1.2.9 Market Volatility

Unconventional resources such as shale gas and oil sands are particularly susceptible to fluctuations in global energy markets due to their relatively high production costs. These resources require significant capital investment and advanced technology to extract, which makes them more vulnerable to rapid changes in commodity prices. As a result, unexpected shifts in market conditions can have a significant impact on the economic viability of projects, making it important for companies to carefully manage their risks and adapt to changing market conditions. Addressing these challenges requires a multidisciplinary approach involving technology innovation, effective regulatory frameworks, and community engagement to ensure sustainable and responsible development.

10.1.2.10 Environmental Concerns

The exploration and exploitation of unconventional resources such as shale gas and oil sands can cause various adverse environmental impacts. These impacts include the disruption of wildlife habitats, contamination of water resources, and the release of greenhouse gases that contribute to climate change.

Unconventional resource extraction techniques such as hydraulic fracturing can potentially have environmental impacts, including water contamination (Haïdar, 2011) and induced seismicity (López-Comino et al., 2018). Therefore, surface facilities must incorporate effective environmental management practices to minimize these risks, such as proper wastewater treatment and monitoring systems. To mitigate environmental impact, surface facilities incorporate measures such as emission controls and noise reduction and measures to prevent soil and water contamination.

10.1.2.11 Community Opposition

Communities located near unconventional resource development sites may express resistance and opposition to these activities due to a variety of environmental, health, and social factors. These can include concerns regarding the impact of extraction processes on air and water quality, noise pollution, and the potential for hazardous chemicals to be released into the environment. Additionally, residents may be worried about the long-term effects of these activities on public health, including respiratory problems, increased cancer rates, and other negative health outcomes. Furthermore, local communities may be worried about the social and economic impacts of unconventional resource development, including changes in property values, increased traffic, and impacts on local infrastructure.

10.1.3 Metering and Measurement Applied to Unconventional Resources

Specialized metering equipment and techniques are often necessary in unconventional resource facilities due to their unique characteristics. For instance, shale gas extraction requires metering systems that can handle the high flow rates and variable compositions of the produced gases (Cahill, 2019). The purpose of these measurements is to ensure that the extraction and production processes are appropriately managed and controlled. Measuring the production of unconventional resources presents unique challenges compared to other industries. The mixture that comes out of a well is a constantly changing blend of oil, gas, water, and sediment (Cahill, 2019).

In the extraction of unconventional resources like shale gas and oil, accurate and precise measurements and metering play a crucial role. The significance of this process cannot be overstated as it allows for the optimization of resource extraction and ensures that the entire operation runs smoothly. A comprehensive operation and maintenance program must include a robust measurement and metering system that is capable of collecting and analyzing data in real time. This helps operators identify any issues early on and make necessary adjustments to improve efficiency and productivity. Moreover, a reliable measurement and metering system helps ensure safety and environmental compliance, which is of utmost importance in the extraction of unconventional resources (Sullivan et al., 2004).

Accurate quantification of production, optimization of resource recovery, operational safety, and environmental compliance all rely heavily on metering and measurement processes. However, the importance of precise metering extends far beyond these basic functions. It is also critical for monitoring and managing environmental impacts, such as methane emissions, that can arise from the extraction of unconventional resources. Precise metering is essential for ensuring the efficient and sustainable development of these resources, which is vital for both industry and the environment.

These processes include but are not limited to drilling, completion, stimulation, and production. The measurements and metering techniques are used to obtain data related to parameters such as flow rate, pressure, temperature, and fluid properties. This data is then analyzed to optimize the operations and improve the overall productivity of the resource. Thus, measurements and metering are integral components of the development and management of unconventional resources. Different metering devices are crucial to all unconventional facilities operations. In fact, there are several metering devices governing the flow of fluid, temperature, pressure, production, and composition.

10.1.4 Flow Rate Metering

The accurate measurement of flow rates is of utmost importance in the oil and gas industry. It is necessary for evaluating the productivity of a well, keeping track of fluid movement within the reservoir, and optimizing hydraulic fracturing operations. To facilitate this process, flow meters such as ultrasonic, turbine, or electromagnetic meters are installed at the wellheads and production facilities. These meters are capable of accurately measuring both the gas and liquid flow rates. This information is then used by engineers and operators to make informed decisions regarding the operation of the well and to ensure that it is functioning at its maximum potential.

10.1.5 Pressure and Temperature Metering

Temperature and pressure metering in unconventional resources facilities is crucial for ensuring safe and efficient operations in the extraction and production of resources like shale gas, tight oil, and coalbed methane. These metering processes involve the use of advanced instruments and technologies to monitor and measure temperature and pressure at various points in the production system. Monitoring systems are installed to measure downhole and surface pressures and temperatures in unconventional resource wells. These measurements are essential for evaluating reservoir performance, detecting operational issues, and optimizing artificial lift methods.

Function of temperature metering:

1. Downhole temperature monitoring: In the context of unconventional resources, downhole temperature monitoring is essential for assessing the thermal conditions of the reservoir. It provides valuable insights into the behavior of hydrocarbons at depth and aids in optimizing production strategies.
2. Surface equipment temperature monitoring: Monitoring the temperature of surface equipment, such as separators, compressors, and pipelines, is critical to prevent overheating, ensure equipment integrity, and maintain the efficiency of production processes.
3. Fluid characterization: Temperature measurement is also important for understanding the behavior of fluids under different thermal conditions. This is particularly relevant in unconventional resources, where accurate temperature data helps in predicting phase behavior and fluid properties.

Function of pressure metering:

- Reservoir pressure monitoring: Continuous measurement of reservoir pressure provides operators with essential information to optimize well productivity, plan hydraulic fracturing operations, and assess overall reservoir performance.
- Flowline and pipeline monitoring: Pressure sensors installed along flowlines and pipelines help in maintaining the required pressure levels for transporting fluids from the well to the processing facilities.
- Wellhead and surface equipment pressure: Monitoring the pressure at the wellhead and surface equipment ensures the safe and efficient operation of gathering, separation, and treatment facilities.

The integration of advanced sensors, gauges, and transmitters for temperature and pressure metering enables operators to make informed decisions, detect anomalies, and prevent potential safety hazards. Real-time data from these metering systems plays a crucial role in optimizing production, ensuring asset integrity, and adhering to regulatory standards.

Overall, temperature and pressure metering in unconventional resources facilities is fundamental for maximizing production efficiency, safeguarding personnel and equipment, and meeting environmental and safety requirements. Continuous monitoring and accurate measurement of temperature and pressure parameters contribute to the sustainable and responsible development of unconventional resources.

10.1.6 Composition Analysis

In the oil and gas industry, analyzing the composition of produced hydrocarbons is a critical task for ensuring product quality, tracking reservoir changes, and optimizing processing and treatment facilities. Gas chromatography, along with other analytical methods, is commonly used to measure the concentration of methane, ethane, propane, and other components present in the produced hydrocarbons. This detailed analysis provides valuable information on the hydrocarbon composition, including the presence of impurities, which is essential for making informed decisions to improve product quality and optimize production processes. The data obtained from these analytical methods is used by engineers and technicians to monitor and adjust processing and treatment facilities to ensure that the final product meets the desired specifications.

10.1.7 Production Metering

Measurement and metering systems play a crucial role in the oil and gas industry by helping accurately measure the production of hydrocarbons, including gas, oil, and water, from unconventional resource wells. These systems are designed to provide precise and reliable data that is essential for evaluating well performance, optimizing production strategies, and calculating royalties and taxes. By monitoring the flow rates, pressure, and composition of the fluids produced, these systems enable operators to make informed decisions about how to extract and process hydrocarbons in the most efficient and cost-effective way possible. Additionally, the data collected by these systems can be used to identify potential problems and to ensure compliance with environmental regulations and safety standards. Overall, measurement and metering systems are critical tools for maximizing the value of hydrocarbon resources while minimizing the risks and costs associated with their extraction and production.

10.1.8 Multiphase Metering

As the onshore oil and gas industry continues to evolve, there is a pressing need for highly precise, dependable, and economical methods of measuring the flow of these valuable resources. Conventional separation techniques have become increasingly cumbersome, especially when dealing with multiple phases in unconventional shale plays. The industry requires alternative solutions that can accurately and efficiently measure the flow of oil and gas in onshore operations. This is where innovative technologies like multiphase flow meters (MPFM) come in. These advanced devices can monitor the flow of oil and gas in real time, even in challenging conditions, such as high-pressure environments. With the ability to accurately measure the flow of oil, water, and gas simultaneously, MPFMs provide vital information for reservoir management, production optimization, and cost control. In addition, MPFMs can reduce the need for expensive well testing and increase safety by minimizing manual intervention. Ultimately, the adoption of cutting-edge measurement technologies like MPFMs will play a critical role in ensuring the profitability and sustainability of onshore oil and gas operations in the modern era (Ruden, 2017).

10.1.9 Environmental Monitoring

Measurement and metering systems play a crucial role in ensuring environmental sustainability and compliance. These systems are designed to meticulously monitor and quantify a wide range of environmental parameters, including air emissions, water usage, and wastewater discharge. The data collected through these systems is essential in supporting environmental compliance efforts, enabling regulatory bodies to evaluate the impact of industrial activities on the environment, and enforcing sustainable resource extraction practices that protect natural resources for future generations. Through these systems, companies can track their environmental footprints, identify areas where they can improve, and take proactive measures to minimize their impact on the environment.

10.1.10 Well Testing and Validation

In order to better understand the reservoir characteristics, evaluate the potential for well deliverability, and optimize production strategies in unconventional resources, it is common practice to conduct integrated measurement and testing programs. These programs utilize techniques such as drill stem testing (DST) and well interference testing to gain valuable insights into the subsurface reservoir. By analyzing the data obtained from these tests, engineers can make informed decisions about how best to develop and produce hydrocarbons from unconventional resources. These measurement and testing programs are critical to ensuring the economic viability of oil and gas projects in these types of reservoirs.

10.1.11 Data Integration and Automation

Advanced metering technologies, such as smart sensors and monitoring systems, provide a wealth of data that can be analyzed and leveraged for optimizing unconventional resource production and operations. By collecting and integrating real-time data from various sources, including production rates, flow rates, and reservoir pressure, operators can gain valuable insights into the performance of their assets and make informed decisions to enhance productivity and reduce costs.

In addition to real-time monitoring, data analytics tools can be used to identify patterns and trends in the data, providing operators with actionable intelligence for optimizing production and reducing downtime. This can include predictive maintenance models that use machine learning algorithms to forecast equipment failures before they occur or advanced reservoir simulation models that help optimize well placement and production rates.

Overall, measurements and metering play a critical role in ensuring the efficient and responsible extraction of unconventional resources. By leveraging advanced technologies and analytics, operators can gain a competitive edge in the market and achieve greater operational efficiency and sustainability.

10.1.12 Corrosion and Scale Detection

The safe and efficient extraction of unconventional resources is crucial for meeting the growing energy demands of the world. However, harsh operating conditions can lead to corrosion and scale formation, which can damage pipelines and equipment, posing a significant threat to safety and the environment. To ensure the safe and sustainable extraction of unconventional resources, specialized metering techniques such as corrosion and scale monitoring probes and sensors are utilized. These techniques help detect and measure the extent of corrosion and scale formation, enabling operators to take appropriate preventive measures and ensure compliance with regulatory requirements. Such metering techniques also promote environmental stewardship by minimizing the impact of resource extraction on the environment and ensuring the long-term viability of unconventional resources.

10.1.13 Latest Advances, Techniques, and Methods in Metering and Measurement

The oil and gas industry is constantly progressing and requires cutting-edge technologies to ensure that surface facilities are operating at their best. One crucial aspect of this is metering and measurement techniques, which help accurately monitor and control the flow of hydrocarbons. In recent years, these techniques have undergone significant advancements due to the development of new tools and technologies that enable more precise measurements and improved data analysis. Some of the latest techniques include ultrasonic flow meters, Coriolis flow meters, and multiphase flow meters, which can be used to measure the flow of oil, gas, and water in a single pipeline. Advancements in sensor technology have made it possible to measure a wider range of parameters, such as temperature, pressure, and fluid density, which can provide valuable insights into the performance of surface facilities.

In the future, metering and measurements in the oil and gas industry are expected to involve advancements in digitalization, the Internet of Things (IoT), and smart sensors. The integration of real-time data analytics and machine learning for accurate forecasting and optimization of production processes will likely be a key focus. Additionally, there may be a shift towards standardized protocols and interoperability to enhance data exchange between different measurement devices and systems. These advancements will likely reduce operational risks, improve efficiency, and ensure compliance with evolving regulatory standards.

10.2 TESTING

Surface facilities in unconventional resource operations rely heavily on testing to ensure safe and efficient operations. This encompasses a wide range of testing, including but not limited to integrity testing of wellheads, flowline and pipeline pressure testing, and environmental testing for air and water quality. The results of these tests inform decisions on equipment maintenance and replacement, production optimization, and compliance with regulatory requirements. Without proper testing, the safety and profitability of unconventional resource operations would be compromised.

10.2.1 Pipeline Testing

Once the installation of the pipeline is completed, it is crucial to test its safety and integrity. This testing process may include hydrostatic testing, whereby the pipeline is filled with water and pressurized to detect any potential leaks. Other tests are also conducted to ensure compliance with regulatory requirements and to verify the pipeline's overall integrity.

10.2.2 Flow Testing

Flow testing is a crucial process in the oil and gas industry involving the evaluation of the flow rates of various fluids, including oil, gas, and water. The main objective of flow testing is to understand the production capabilities of a well and to optimize the extraction processes to increase the yield and efficiency of the production. This process involves a wide range of techniques and tools that help in measuring the flow rates, pressure, temperature, and other critical parameters that affect the performance of the production system. Flow testing is a highly specialized field that requires expertise in fluid dynamics, instrumentation, and data analysis to ensure accurate and reliable results.

10.2.3 Water Quality Testing

Hydraulic fracturing requires large amounts of water, and ensuring that the water used meets quality standards is crucial. Water quality testing involves analyzing water samples to detect any contaminants or substances that may pose a risk to the environment, human health, or the fracking process itself. The testing process typically involves measuring pH, total dissolved solids, and other parameters to ensure that the water is suitable for use in hydraulic fracturing. In addition to testing the water before it is used, ongoing monitoring is also important to detect any potential environmental impact of the process. This may involve regular sampling of nearby water sources, such as rivers or groundwater wells, to ensure that they are not being contaminated by the fracking process. Overall, water quality testing is a critical aspect of hydraulic fracturing that helps ensure that the process is safe and environmentally responsible.

10.2.4 Pressure Testing

Pressure testing is a common practice used to ensure the structural integrity of pipelines, wellheads, and other oil and gas infrastructure. This process involves subjecting the equipment to a specified pressure level, which is usually higher than the operating pressure, to identify any potential leaks or weaknesses. Pressure testing is performed on newly installed equipment, as well as on existing equipment during maintenance or inspection activities. By pinpointing any issues in the equipment, pressure testing helps prevent costly and potentially hazardous failures, ensuring the safe and efficient operation of the equipment.

10.2.5 Chemical Analysis

In the field of oil and gas extraction, chemical analysis is a crucial process that involves assessing the composition of fluids produced during extraction. This process helps in optimizing extraction processes and ensuring compliance with environmental regulations. Through chemical analysis, it is possible to determine the precise composition of the fluids, including the concentration of various chemicals and minerals. This information is then used to optimize the extraction process, minimize waste, and comply with environmental regulations. Additionally, chemical analysis can help identify potential hazards associated with the fluids, which can be addressed to ensure the safety of workers and the environment.

10.2.6 Safety Systems Testing

One of the most crucial aspects of ensuring a safe working environment is regularly testing safety systems. This includes conducting frequent assessments of emergency shutdown procedures and fire detection/prevention systems to minimize potential risks. By constantly monitoring and evaluating these systems, companies can identify any potential issues or malfunctions and take corrective actions to prevent accidents or emergencies. Regular testing of safety systems not only helps ensure the safety of employees and the environment but also helps compliance with regulatory requirements and industry standards.

10.2.7 Electrical System Testing

Proper functioning of electrical systems is crucial for the smooth operation of surface facilities. Electrical system testing is an essential step towards ensuring the safety and reliability of the power supply. This includes a detailed inspection of the electrical wiring, grounding systems, protection devices, and overall system performance. The process involves various tests and measurements, such as insulation resistance, continuity, polarity, and voltage drop tests. The results of these tests help identify any potential issues or faults in the system, which can then be rectified to ensure optimal performance and safe operation of the electrical system.

10.2.8 Water Management Testing

Water management testing plays a crucial role in ensuring effective and sustainable unconventional resource development. Given the significant amount of water used in this type of resource development, testing water samples becomes necessary to assess its quality, evaluate treatment effectiveness, and ensure compliance with water disposal regulations. The process involves a range of analytical techniques, including chemical and biological analyses, to detect and quantify the presence of various contaminants such as heavy metals, organic compounds, and microorganisms that could potentially pose a threat to human health and the environment. Accurate water management testing helps in facilitating responsible water usage and mitigating the risks associated with unconventional resource development.

10.2.9 Corrosion Testing

The extraction of unconventional resources has become more prevalent, and it requires equipment and infrastructure that can withstand harsh fluids and operating conditions. Corrosion testing is a crucial process that evaluates the resistance of materials to corrosion, ensuring the integrity and longevity of surface facilities. This testing involves subjecting materials to different corrosive environments, such as acidic or saline solutions, to determine their susceptibility to corrosion. By examining the corrosion behavior of various materials, engineers can identify the best materials to use in constructing and maintaining surface facilities, reducing the likelihood of equipment failure and ensuring the safety of workers.

10.2.10 Emissions Monitoring

Continuous monitoring of air emissions from surface facilities is a crucial aspect of environmental management. It is imperative to assess the impact on air quality and adhere to environmental regulations to prevent any negative impact on human health and the environment. To achieve this, gas analyzers, particulate matter samplers, and other monitoring devices are deployed to measure emissions of greenhouse gases, volatile organic compounds (VOCs), and other pollutants. The collected data is analyzed to identify any potential issues and take necessary actions to mitigate the environmental impact. By ensuring continuous emissions monitoring, industries can minimize their carbon footprint and contribute to the overall sustainability efforts.

10.2.11 Geotechnical Assessments

Geotechnical assessments are a crucial part of any construction or infrastructure project that involves surface facilities. These assessments involve a comprehensive evaluation of the ground's stability, which includes detailed analysis of soil and rock properties, assessing slope stability, and designing appropriate foundations to mitigate risks associated with ground movement. Geotechnical engineers use advanced techniques to gather data on soil conditions, such as drilling boreholes and conducting laboratory tests. They also analyze the data to determine the ground's load-bearing capacity, which is essential for designing stable and safe foundations. The assessments also consider potential soil liquefaction, which can occur during earthquakes and can cause significant damage to structures. Overall, geotechnical assessments are vital in ensuring the safety and stability of surface facilities and protecting them from potential ground hazards.

10.2.12 Automation and Control System Testing

Surface facilities in the oil and gas industry increasingly rely on automation and control systems to manage their operations. In order to ensure the proper functioning of these systems, comprehensive testing is required. This testing involves assessing the functionality, integration, and responsiveness of the automated systems in controlling critical operations such as well control, monitoring, and emergency shutdown. By conducting thorough automation and control system testing, potential issues can be identified and addressed, ensuring the safe and efficient operation of surface facilities.

10.2.13 Well Integrity Testing

Ensuring the integrity and reliability of wellbore structures such as casing and cement barriers is a critical process in the oil and gas industry. This process, known as well integrity testing, plays a crucial role in ensuring the safe and effective containment of fluids within the well to prevent any leakage. The process involves various methods for evaluating the condition and effectiveness of the wellbore structures, including visual inspections, pressure testing, and acoustic and electromagnetic surveys. These methods provide valuable information on the condition and integrity of the well, allowing operators to make informed decisions about its safety and efficiency. Ultimately, well integrity testing is a vital component of the oil and gas industry, ensuring that wells are safe, reliable, and capable of delivering the energy resources that power our world.

10.2.14 Factors Impacting Testing

Several factors can affect metering and measurements in the oil and gas industry, including:

- Fluid properties: Changes in the composition, temperature, and pressure of the fluids can all have an impact on measurement accuracy.
- Flow conditions: The accuracy of metering can be impacted by irregular flow patterns, such as turbulence or velocity variations.
- Equipment calibration: Regular calibration is crucial to ensure accuracy in metering devices and measurement instruments.
- Pipeline integrity: Measurement errors may result from leaks or anomalies in pipelines.
- Environmental conditions: Weather conditions such as extreme temperatures or humidity may affect measurement accuracy.

- Maintenance and contamination: Measurement equipment contamination and fouling can be avoided with proper maintenance.
- Metering technology: Depending on the situation, accuracy can be affected by the metering technology employed, such as positive displacement, turbine, ultrasonic, or another type.
- Regulatory compliance: Adherence to industry standards and regulations is critical to ensuring accurate measurements and legal compliance.
- Human error: Operator training and adherence to proper procedures are crucial to minimize errors in measurement.
- Instrument drift: Instruments can become less accurate over time due to instrument drift. To prevent inaccuracies, it's important to regularly monitor and correct any drift.

10.2.15 Latest Advances, Techniques, and Methods in Testing

The oil and gas industry has been continuously evolving, and so have the techniques used for testing surface facilities. There are various methods and equipment available today that are more advanced and efficient than ever before. These include nondestructive testing methods such as ultrasonic testing, radiographic testing, and magnetic particle inspection. Additionally, advanced monitoring systems like acoustic monitoring and vibration analysis are being widely used to detect any potential abnormalities in the surface facilities. Overall, these latest techniques not only provide more accurate results but also ensure the safety and reliability of the facilities, thereby reducing the risk of any potential accidents or failures.

The future direction of surface facilities testing in the oil and gas industry is likely to focus on enhanced automation, integration of advanced sensors for real-time monitoring, and increased use of data analytics for predictive maintenance. Additionally, there may be a growing emphasis on environmentally sustainable practices and technologies to minimize the industry's ecological impact.

10.3 INORGANIC SCALE

The formation of scale, or scaling, occurs during several industrial operations, such as crystallization, distillation, evaporation, mixing, cooling, or heating of liquids (Hoang, 2015). Deposits formed under various conditions have different mechanical characteristics and are accordingly described by rather loose terminology: the term "scale" is used to describe a dense crystalline deposit well bonded to the surface while "soft scale," "powdery deposit," or "sludge" describes a porous and loosely adherent deposited layer (Griffith, 1982; Hoang, 2015). The solid layer may be either temporary or permanent, depending on whether it disappears at a later stage of the process or gradually increases when exposed for a long time to the same condition. Permanent scale formation poses significant industrial challenges, including diminished thermal efficiency in heat exchangers, induced under-deposit corrosion, heightened pressure drop and upstream pressure in piping, flow oscillations, cavitation, and flow blockage. Moreover, it impairs safe equipment operation, potentially contaminating fluids in contact with the deposits (Pritchard, 1988). The industrial sector bears a substantial cost for scaling, encompassing specialized equipment designs, energy loss, maintenance expenses, and lost production (Pritchard, 1988).

Inorganic scaling is a common problem in the oil and gas industry and can have detrimental effects on production rates and equipment integrity. It refers to the deposit or accumulation of inorganic minerals such as calcium carbonate, calcium sulfate, barium sulfate, and iron sulfide in oil and gas production systems. Scaling deposits form due to a combination of factors, including changes in temperature, pressure, and pH and the presence of incompatible fluids. When these conditions are met, the dissolved minerals can precipitate out of solution and adhere to the internal surfaces of pipelines, valves, pumps, and other production equipment.

Scale deposits can form due to three primary mechanisms. The first mechanism is the mixing of two incompatible brines, which results in the precipitation of minerals that can adhere to surfaces and form scales. The second mechanism is a change in conditions, such as temperature and pressure, that can cause minerals to dissolve and precipitate on surfaces. The third mechanism is brine evaporation, which can lead to the concentration of minerals and the formation of scales.

Surface scale formation generally involves two distinct processes: (1) deposition, characterized by heterogeneous nucleation and growth at surface asperities, and (2) adhesion, involving the adherence of pre-existing crystals, nucleated in the bulk solution, building up as a layer on the surface (BinMerdhah et al., 2010; Singh et al., 2018).

In the oil and gas industry, inorganic scales form when the formation water combines with various brines, including injection water (Hosny et al., 2009; Ohen et al., 2004). This mixing induces reactions between incompatible ions, altering the thermodynamic and equilibrium state of reservoir fluids. Subsequent supersaturation leads to the deposition of inorganic salts. Disturbances in equilibrium caused by changes in fluid composition through adverse mixing, as well as variations in temperature and pressure, can increase the degree of supersaturation, resulting in scaling. These scale deposits can obstruct the wellbore, diminishing equipment lifespan and impacting the integrity of components such as subsurface control valves (SSCVs), electrical submersible pumps (ESPs), and hydraulic actuators. The economic implications stem from reduced fluid flow, leading to lower oil production, and the substantial maintenance costs associated with replacing production lines (BinMerdhah et al., 2010; Singh et al., 2018).

According to Macadam and Parsons (2004), the nonproductive expenses associated with scale were approximated at \$0.8 billion in Great Britain, \$3 billion in Japan, and \$9 billion in the U. Understanding scale formation, recognizing its types, and controlling the scaling process in oil and gas fields are crucial. This knowledge can positively influence the development of new high-performance materials and the evolution of effective anti-scaling methods

Some of the common types of inorganic scales are calcium carbonate, calcium sulfate, calcium phosphate, barium sulfate, magnesium salts, and silica (Table 10.2) (Kamal et al., 2018). These scales are formed when the concentration of the dissolved salts exceeds their solubility limit due to factors such as water removal, concentration polarization, mixing of incompatible brines, and changes in temperature and pressure. The scales can either crystallize on the membrane surface or form in the bulk solution and then deposit on the membrane. The operating conditions, such as the flux, crossflow velocity, recovery, and pH, can influence the scaling process.

Preventing and mitigating inorganic scaling involves several approaches. One common method is to control the thermodynamic conditions under which scaling occurs. This can include adjusting the pH, temperature, and pressure to keep the minerals in solution. Chemical inhibitors are also used to prevent mineral deposition by interfering with crystal growth or modifying the surface properties to prevent adhesion.

Regular monitoring and analysis of the production system's chemistry and scaling tendencies are essential for effective scaling control. Various techniques such as water sampling, ion analysis, and predictive modeling can help identify scaling potential and guide mitigation strategies. Furthermore, periodic cleaning or mechanical removal of scale deposits may be required, depending on the severity of scaling and the specific equipment involved (Figure 10.2).

10.3.1 Scale Formation

In the oil and gas industry, inorganic scale formation is predominantly influenced by naturally occurring minerals and elements in production fluids. This process is initiated by the localized brine formation in the environment, exacerbated by the

FIGURE 10.2 Scale deposit in the pipeline.

Source: Scale | Energy Glossary, 2023

low solubility of specific inorganic salts. Scale deposits manifest through three key mechanisms, impacting both offshore and onshore environments. First, the mixing of two incompatible brines can cause the precipitation of salts and the formation of scale (Merdhah & Yassin, 2008). Secondly, changes in temperature and pressure can cause the solubility of salts to decrease, leading to precipitation and scale formation. Finally, brine evaporation can increase the concentration of salts, leading to precipitation and scale formation (BinMerdhah et al., 2010).

1. A decrease in pressure and/or increase in temperature of a brine goes to a reduction in the solubility of the salt (most commonly the scaled to precipitation of carbonate scales, such as CaCO3) (Heydrich et al., 2019).

$$Ca(HCO_{3)2} \leftrightarrow CO_2 + H_2O \tag{10.1}$$

2. The interaction between incompatible brines, often involving formation water abundant in cations like barium, calcium, and/or strontium, and sulfate-rich seawater leads to the precipitation of sulfate scales such as $BaSO_4$ (Heydrich et al., 2019).

$$Ba^{2+} (or\, Sr^{2+}\, or\, Ca^{2+} + SO_4^{2-} \leftrightarrow BaSO_4 \left(or\, SrSO_4 or\, CaSO_4\right) \tag{10.2}$$

Other fluid incompatibilities include sulfide scale, in which hydrogen sulfide gas mixes with iron-, zinc-, or lead-rich formation waters:

$$Zn^{2+} + H_2S \leftrightarrow ZnS + 2H^{2+} \tag{10.3}$$

3.Brine evaporation causes salt concentration to exceed the solubility limit, leading to salt precipitation. This scenario is evident in HP/HT gas wells, where the mixing of a dry gas stream with a low-rate brine stream results in dehydration and the frequent precipitation of NaCl.

However, formation of scale involves a number of stages, including precipitation, nucleation, an induction period, and growth (Hoang, 2015).

10.3.1.1 Precipitation

To begin, precipitation species must form. Within reservoir fluids, brine is commonly present and holds dissolved minerals such as calcium carbonate ($CaCO_3$), calcium sulfate ($CaSO_4$), barium sulfate ($BaSO_4$), and strontium sulfate ($SrSO_4$). When specific conditions arise, such as alterations in temperature, pressure, or pH, these salts surpass their solubility limit and begin to crystallize out of the solution, manifesting as solid scale deposits.

10.3.1.2 Nucleation

Nucleation is the initial stage of scale formation, in which small solid particles (nuclei) begin to form due to the supersaturation of minerals in the brine (Louhi-Kultanen & Ulrich, 2010). This process is influenced by factors like temperature, pressure, supersaturation levels, composition of the brine, flow dynamics, and the presence of nucleation sites (e.g., surface roughness, equipment imperfections). Two distinct nucleation mechanisms are recognized. Primary nucleation takes place in systems without crystalline matter, relying on the presence of tiny solid particles—either foreign bodies or spontaneously formed crystallites. The former is termed heterogeneous nucleation, and the latter is known as homogeneous nucleation. In contrast, secondary nucleation occurs when nuclei form near existing crystals in the vicinity (Hoang, 2015).

10.3.1.3 Induction Period

The induction period, as defined by Hoang (2015), is the period of time that elapses between the point supersaturation is achieved and when a precipitate is first detected (Hoang, 2015). It is influenced by factors like supersaturation degree, temperature, mixing intensity, and the presence of impurities or inhibitors. The significance of the induction time period is difficult to analyze although undoubtedly, it is related to the kinetics of nucleation.

10.3.1.4 Crystal Growth

During this stage, the minerals gradually deposit and grow into larger crystals as a result of various factors that influence their growth, including the residence time, temperature, and pressure conditions within the production system. The longer the minerals remain in the system, the more likely they are to grow and develop into larger crystals. Similarly, higher temperatures and pressures can also accelerate the growth process, resulting in larger and more complex crystals. Ultimately, the precise conditions during this stage play a critical role in determining the size, shape, and quality of the final product.

10.3.1.5 Deposition

Deposition is the final step, in which the larger crystals adhere to surfaces within the oil and gas production system, including wellbores, tubing, valves, and other equipment. This deposition can lead to the formation of scales that hinder the flow of hydrocarbons and reduce the efficiency of equipment.

The formation of scales in oil wells can vary significantly, both between different wells and within the same well at different depths. There are various types of oilfield scales, including sulfides (primarily iron sulfides), oxides (primarily iron oxides), sulfate, and carbonate scales, with some rare exceptions such as zinc sulfide (Berry et al., 2012). Table 10.2 provides a description of the most important and common oilfield scales. While the scaling tendency of certain scales, such as calcium sulfate and barite, is not affected by brine pH, others, like iron sulfides and carbonates, are pH dependent. It is worth noting that iron plays a crucial role in balancing the negative charge of hydroxyl group surfaces, and increasing the pH leads to a phenomenon called ion shielding, which increases the negative value as pH rises (Tan et al., 2013). In terms of solubility, most scales are either water or acid soluble. However, there are some scales that are not soluble in either water or acids. For example, sodium chloride is a typical

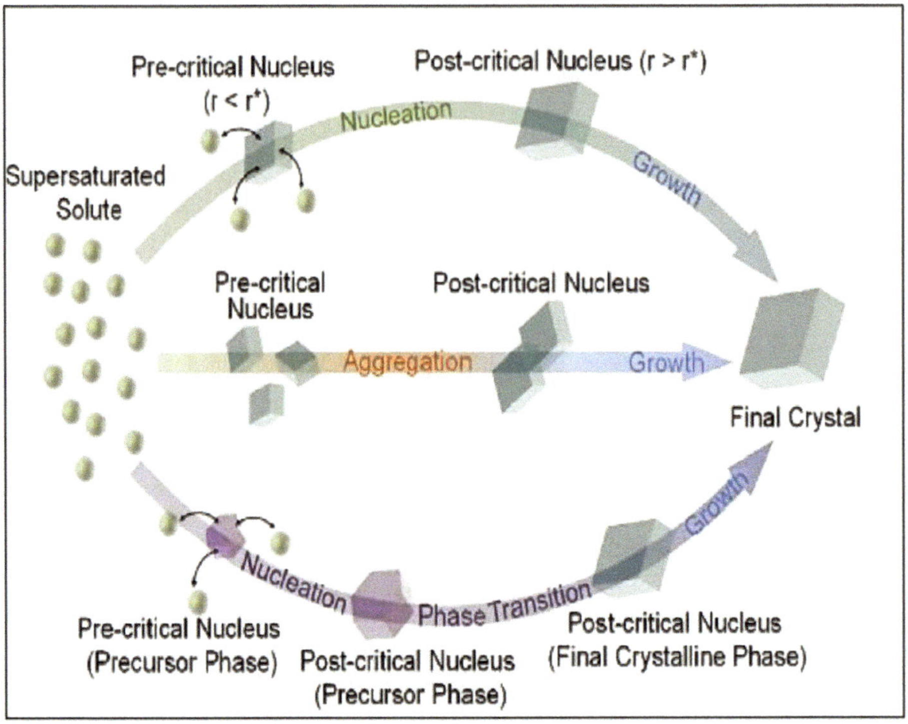

FIGURE 10.3 Schematic representation of inorganic scaling mechanism.

Source: Al-Gailani, 2021; Zhao et al., 2018

water-soluble scale while calcium carbonate, iron sulfide, and iron oxide are acid-soluble scales (Figure 10.3).

10.3.2 Conditions Influencing Formation of Inorganic Scaling

The formation of inorganic scale in the oil and gas industry is a common problem that arises due to the presence of various minerals and salts in reservoir fluids and production water. Inorganic scale refers to the solid deposits that can form on the surfaces of production equipment, pipelines, and wellbore sections. The conditions that contribute to the formation of inorganic scale include:

10.3.2.1 Temperature

Temperature has different impacts on different scale formations. It may either increase or decrease the rate of scale formation (Budz et al., 1985; Pritchard, 1988). The scaling tendency of calcium sulfate and calcium carbonate increases with rising temperatures. Strontium sulfate exhibits a comparable behavior in response to temperature variations. However, barium sulfate demonstrates a distinct trend—higher temperatures result in increased ionic strength of barium sulfate brine, consequently reducing its scaling tendency (Hajirezaie et al., 2017).

10.3.2.2 Pressure

Changes in pressure can cause changes in the solubility of minerals and salts, leading to the formation of scale (Heydrich et al., 2019). Pressure drops across chokes, valves, and other components can trigger the precipitation of scales (Budz et al., 1985; Hajirezaie et al., 2017). After prolonged oil production, wells may undergo substantial pressure drops, leading to the accumulation of significant $CaCO_3$ deposits as the equilibrium adjusts to counteract pressure changes (Frenier & Ziauddin, 2008). Additionally, the solubility of calcium, barium, and strontium sulfates rises with increased pressure, making scale formation more likely during pressure drops. This occurrence is particularly notable in perforations, chokes, and production tubing. The solubility of calcium carbonate is significantly influenced by the CO_2 content of water; an elevated CO_2 partial pressure diminishes calcium carbonate precipitation (Hajirezaie et al., 2017)

10.3.2.3 Supersaturation

The primary cause of inorganic scaling is supersaturation. When the solubility of a salt is exceeded, the salt precipitates and, as a result, different types of scales form in both porous media and wellbore tubing. (Hajirezaie et al., 2017). Increasing supersaturation accelerates the nucleation rate. If the supersaturation region is at the membrane–liquid interface, precipitation on the membrane surface is likely. Supersaturation condition is dependent on several other parameters, including pressure, temperature, concentration of ions, and pH (Hajirezaie et al., 2017).

10.3.2.4 pH of Water

Water pH levels play a significant role in the solubility of minerals and the formation of scales. The acidity or alkalinity of the water can either increase or decrease the solubility of minerals, which can lead to the precipitation of these minerals and the formation of scales. Extremely high or low pH levels can significantly promote the formation of scales, which can damage plumbing systems and other equipment that comes into contact with the water (Azizi et al., 2018). Therefore, it is essential to maintain proper pH levels to ensure that the water remains safe and free from mineral buildup.

10.3.2.5 Water Chemistry

The chemical composition of water plays a crucial role in determining the likelihood of scale formation (Al-Gailani, 2021). Dissolved minerals such as calcium, magnesium, and carbonate are the primary culprits behind scale buildup. The concentration of these minerals in water directly influences the potential for scale formation. Higher levels of these minerals increase the likelihood of scale buildup while lower levels reduce the risk. Therefore, it is essential to monitor the chemical composition of water to prevent scale formation and ensure the efficient operation of various water-based systems. These can cause well plugging and near wellbore formation damage from scale formation (Hamid et al., 2013).

10.3.2.6 Mixing of Fluids

When incompatible fluids come into contact, such as formation water, injection water, and produced fluids, they can undergo chemical reactions that promote scale formation (BinMerdhah et al., 2010). This is particularly true when incompatible water sources are used for injection or flooding operations (Azizi et al., 2018).

10.3.2.7 Residence or Contact Time

Prolonged contact time between the fluid and the surfaces can provide more opportunities for scale formation (Hoang, 2015). Slower flow rates, stagnation, and inadequate mixing can increase the residence time and enhance scale deposition.

10.3.2.8 Flow Rate and Turbulence

Flow characteristics such as velocity, turbulence, and flow patterns can affect the potential for scale formation. The flow rate of water passing through industrial equipment has a significant impact on the formation of scale. Water flow rate influences how fast minerals and other deposits accumulate in the equipment. Higher flow rates can help reduce the buildup of minerals and scale deposits while lower flow rates may lead to increased deposition, which can cause equipment to malfunction. Therefore, it's important to maintain a consistent flow rate to ensure that the industrial equipment performs optimally and lasts longer.

10.3.2.9 Surface Material and Roughness

When it comes to scale buildup in industrial equipment, it's important to consider the material and surface roughness of the equipment itself (Hamid et al., 2013; Heydrich et al., 2019). The type of material and the level of surface roughness can have a significant impact on the formation of scale. For instance, certain materials and smoother surfaces tend to be less prone to scale buildup, while others may be more susceptible. A polished surface exhibits fewer crystals at the conclusion of a crystallization experiment than a rough surface. This discrepancy arises due to heightened surface roughness, which elevates surface free energy, thereby fostering increased adhesiveness and facilitating the entrapment of crystals. Therefore, it's crucial to take into account the surface characteristics of industrial equipment when assessing the risk of scale formation and determining appropriate prevention measures.

10.3.3 Common Inorganic Scale in the Oil and Gas Industry

The issue of scale, one of the three major challenges in water-associated production alongside corrosion and gas hydrate (Kelland, 2009), poses a significant global concern. The economic impact of scale is substantial, estimated to exceed billions (Frenier & Ziauddin, 2008). Consequently, proactive measures are essential to prevent scale formation. This necessitates accurate prediction and modeling of scale, followed by the application of appropriate scale inhibitors. Table 10.1 outlines the most prevalent types of scale encountered in oilfields (BinMerdhah et al., 2010; Graham & Frigo, 2022; Kamal et al., 2018; Kelland, 2009; Li et al., 2017; Merdhah, 2007; Ohen et al., 2004); and their formation mechanism and reaction are discussed later.

TABLE 10.1
Different Types of Scale

Scale	Name	Chemical Formula
Sulfide	Pyrrhotite	Fe_7S_8
	Troilite	FeS
	Pyrite	FeS_2
	Marcasite	FeS_2
	Greigite	Fe_3S_4
	Galena	PbS
	Sphalerite	ZnS
	Mackinawite	Fe_9S_8
Carbonate	Calcite	$CaCO_3$
	Vaterite	$CaCO_3$
	Aragonite	$CaCO_3$
	Siderite	$FeCO_3$
	Dolomite	$CaMg(CO_3)_2$
	Aragonite	$CaCO_3$
Sulfates	Gypsum	$CaSO_4.2H_2O$
	Anhydrate	$CaSO_4$
	Barite	$BaSO_4$
	Hemihydrate	$CaSO_4.5H_2O$
	Celestite	$SrSO_4$
Oxide scales	Ferrous Hydroxide	$Fe(OH)_2$
	Ferrous Hydroxide	$Fe(OH)_3$
	Hematite	Fe_2O_3
	Magnetite	Fe_3O_4
	Akaganeite	α-FeOOH
	Goethite	β-FeOOH
	Lepidocrocite	γ-FeOOH
	Hibbingite	$Fe_2(OH)_3Cl$

Source: Kamal et al., 2018

10.3.3.1 Carbonate Scales

Carbonate scale in the oil and gas industry refers to the deposition of carbonate minerals on equipment surfaces. Calcium carbonate ($CaCO_3$) is a common scale formed in wells due to the reaction of bicarbonate ions with calcium ions. This can occur during the production of oil and gas, especially in reservoirs with high mineral content. In reservoir fluids, calcium (Ca^{2+}) reacts with bicarbonate ions ($2HCO_3^-$) present in water. The reaction results in the precipitation of calcium carbonate, often triggered by changes in temperature, pressure, or pH. This solid $CaCO_3$ can accumulates in wellbores and production equipment, causing scaling issues (Hajirezaie et al., 2017; Heydrich et al., 2019). The governing equation for calcium carbonate precipitation is as follows:

$$Ca^{2+} \leftrightarrow CO_3^{2-} + CaCO_3 \tag{10.4}$$

$$Ca^{2+} + 2HCO_3^- \leftrightarrow CaCO_3 + +H_2O \tag{10.5}$$

Calcite stands out as the most stable form of $CaCO_3$ scale under reservoir conditions, making it more prevalent than other calcium carbonate scales such as aragonite and vaterite (Hajirezaie et al., 2017). The formation of carbonate scale is intricately linked to factors like pH, temperature, calcium ion concentration, bicarbonate concentration, and ionic strength. The chemical equilibrium between ions and CO_2 in water plays a pivotal role in determining calcium carbonate scale deposition. This deposition involves a combination of calcium and bicarbonate ions (eq. 10.5) (Hajirezaie et al., 2017). The likelihood of $CaCO_3$ scale deposition increases with rising temperature and pH, and lower pressure (Hamid et al., 2016). Beyond calcium carbonate, iron carbonate scales have been documented in various types of downhole equipment (Amiri et al., 2013). Siderite (iron carbonate) scale primarily originates from shifts in pressure and temperature conditions, resulting in the loss of dissolved CO_2. This loss elevates the system's pH, diminishing the solubility of these minerals in the flowing fluid (Jordan et al., 2014).

10.3.3.2 Sulfate Scales

Sulfate scales form in situations in which incompatible brines mix, often occurring when seawater is utilized for maintaining pressure and extracting oil. The solubility of Group II metal ion (M2+) sulfates typically decreases down the group. Barium sulfate scales are particularly challenging to remove due to their high insolubility, resulting in extremely hard scale deposits. The reaction can be summarized as follows:

$$M^{2+} + SO_{2(aq)}^{2-} \leftrightarrow MSO_{4(aq)} \tag{10.6}$$

Sulfate scales, recognized as some of the toughest to handle, are prevalent in seawater injection applications (Hajirezaie et al., 2017). These scales exhibit minimal solubility in acids, making their removal challenging. Their deposition results from the combination of waters containing sulfate ions and those with barium, strontium, or calcium ions, commonly occurring when formation water and injected seawater mix (Frenier & Ziauddin, 2008). The relationship between these scales and the degree of supersaturation is crucial for predicting precipitated sulfate salts (Collins, 2002). Reservoirs undergoing seawater injection for enhanced oil recovery often face significant challenges with sulfate scale issues (Frenier & Ziauddin, 2008). Barium sulfate, strontium sulfate, and calcium sulfate are the predominant sulfate minerals encountered in the oil and gas industry (Li et al., 2017).

Barium sulfate ($BaSO_4$) occurs due to the mixing of formation water that is highly rich in barium cations and injection seawater enriched with sulphate anions (Heydrich et al., 2019). Barium ions (Ba^{2+}) present in the formation water combine with sulfate ions (SO_4^{2-}). This reaction forms solid barium sulfate ($BaSO_4$), which can precipitate out of the solution. Factors like temperature changes and pressure variations influence the precipitation of barium sulfate.

$$Ba^{2+} + SO_4^{2-} \rightarrow BaSO_4 \quad (10.7)$$

Strontium sulfate ($SrSO_4$): Similar to barium sulfate, strontium sulfate can precipitate and cause scaling problems in production systems. Similar to barium sulfate, strontium sulfate forms when strontium ions (Sr^{2+}) react with sulfate ions (SO_4^{2-}). The precipitation of strontium sulfate can occur in wells and production systems under specific chemical and physical conditions.

$$Sr^{2+} + SO_4^{2-} \rightarrow SrSO_4 \quad (10.8)$$

Calcium sulfate ($CaSO_4$) forms when sulfate ions react with calcium ions and can lead to scale deposition in oil and gas wells. Calcium ions (Ca^{2+}) combine with sulfate ions (SO_4^{2-}) to form solid calcium sulfate ($CaSO_4$). Changes in temperature, pressure, or fluid composition can trigger the precipitation of calcium sulfate, leading to scaling problems.

$$Ca^{2+} + SO_4^{2-} \rightarrow CaSO_3 \quad (10.9)$$

The most common sulfate scales found in oil and industries are calcium sulfate ($CaSO_4.xH_2O$), barium sulfate ($BaSO_4$), strontium sulfate ($SrSO_4$), and silicates (Li et al., 2017).

Although less common in the oilfield, other scale types include phosphate scale, silicate scale, and elemental sulfur. Calcium phosphate scale, prevalent in water treatment plants and cooling water systems, often arises from increased orthophosphate concentrations in cooling water due to extensive water usage. Notably, phosphate scales are not frequently encountered in oilfields.

Due to its relatively high solubility in water, calcium sulfate is chemically easier to remove than strontium and barium sulfate (Frenier & Ziauddin, 2008). Initially dispersed in production systems, scale crystals accumulate at nucleation centers, leading to the growth of stable, insoluble sulfate scales (Aberdeen et al., 2000). Rapid scale growth on uneven pipeline surfaces and production equipment such as pumps and valves can reach levels that block pipelines (Kelland, 2009).

The complexity of scale removal arises from deposits rarely being pure calcium carbonate, calcium sulfate, barium sulfate, or strontium sulfate. Instead, they typically consist of mixtures with inorganic components, corrosion products, organic scales, silica, and other impurities trapped in the inorganic lattice (Ohen et al., 2004).

10.3.3.3 Sulfide Scale

Sulfide scale refers to the formation of sulfide deposits on equipment surfaces, often pipelines and wellbores, due to the presence of hydrogen sulfide (H_2S) in the produced fluids. Iron sulfide (FeS), zinc sulfide (ZnS), and lead sulfide (PbS) are some common sulfide scales encountered in the oil and gas industry. These scales form as a result of reactions between the metal ions present in the produced fluids and hydrogen sulfide (H_2S) (Kamal et al., 2018).

The formation of common sulfide scales in the oil and gas industry involves reactions between metal ions in produced fluids and hydrogen sulfide (H_2S). Here are simplified representations of some reactions:

1. Iron sulfide (FeS):

$$Fe^{2+} + H_2S \rightarrow FeS + 2H^+ \tag{10.10}$$

2. Zinc sulfide (ZnS):

$$Zn^{2+} + H_2S \rightarrow ZnS + 2H^+ \tag{10.11}$$

3. Lead sulfide (PbS):

$$Pb^{2+} + H_2S \rightarrow PbS + 2H^+ \tag{10.12}$$

These reactions illustrate the precipitation of metal sulfides that can accumulate as scale on the surfaces of equipment and pipelines.

The predominant sulfide scale encountered is iron sulfide (FeS), a major scale in sour oil and gas wells. Its physical state ranges from viscous fluid to dry black powder, with characteristics influenced by pH, temperature, scale age, and pressure. Hydrogen sulfide may exist as a free gas in sour wells, originating from various sources such as degradation of drilling mud additives, reduction of sulfate ions, thermal degradation of organic sulfur compounds, acid treatment of sour wells, and sulfate-reducing bacteria (SRB) (Kamal et al., 2018; Marriott et al., 2016). Additionally, H_2S can be generated through the thermal degradation of sulfate-containing minerals.

10.3.4 Characterization of Inorganic Scale

The characterization of inorganic scaling in unconventional resource facilities involves understanding the chemical composition and physical properties of the scales that form. This includes identifying the type of scale (such as calcium carbonate, calcium sulfate, barium sulfate, or strontium sulfate), determining the crystal structure, examining the morphology, and assessing the extent of scaling within the facility's infrastructure.

Characterization is essential for developing effective mitigation and remediation strategies, as well as for designing anti-scaling treatments tailored to the specific properties of the scales in the oil and gas facilities. It allows for a better understanding of the scaling mechanisms and aids in the selection of appropriate scale inhibition or removal techniques.

Various analytical techniques are used for characterization, including scanning electron microscopy (SEM), X-ray diffraction (XRD), energy-dispersive X-ray spectroscopy (EDS), plasma-optical emission spectrometry (ICP-OES), X-ray fluorescence (XRF), and atomic force microscopy (AFM). These methods help in identifying the mineral composition, crystal size, shape, morphology, and distribution of the scales (Ali et al., 2022; Ling et al., 2022; Vazirian et al., 2016).

10.3.4.1 Identification of Scaling Compounds

Characterization of inorganic scale begins with identifying the types of minerals or salts causing scaling. Techniques such as X-ray diffraction (XRD), scanning electron microscopy (SEM), and energy-dispersive X-ray spectroscopy (EDS) can help determine the elemental composition and crystalline structure of the scaling compounds.

10.3.4.2 Morphological Analysis

Morphological characterization involves examining the physical characteristics of the scaling, including its morphology, texture, and adhesion to surfaces. Microscopic imaging techniques such as SEM and atomic force microscopy (AFM) can provide detailed insights into the surface topography and growth patterns of the scaling deposits.

10.3.4.3 Chemical Composition

Determining the chemical composition of scaling deposits is essential for understanding the specific ions and compounds involved in the scaling process. Elemental analysis techniques, such as inductively coupled plasma-optical emission spectrometry (ICP-OES) and X-ray fluorescence (XRF), can provide quantitative data on the elemental composition of the scaling (Jamaluddin & Massinai, 2018).

Characterization involves studying the factors influencing the nucleation and growth of scaling deposits. Understanding the thermodynamic and kinetic drivers of scaling formation, including factors such as supersaturation, pH, temperature, and pressure, is vital for developing effective mitigation strategies.

10.3.5 Impact of Scaling on Operations

Inorganic scaling can have a significant impact on the operation of oil and gas facilities, posing challenges and potential risks to production, equipment integrity, and overall operational efficiency (Hamid et al., 2013). Scales can form either downhole on wellbore, the surface of pipes, tubing, pumps, and surface equipment, leading to the blockage of production lines and equipment, as well as hindered fluid flow. These consequences can result in equipment failure, emergency shutdowns, heightened maintenance costs, and decreased production efficiency. Additionally, the failure of production equipment and instruments can create safety hazards (BinMerdhah et al., 2010; Yeboah et al., 1993). Some of the key impacts of scale on oil and gas facilities operation include:

10.3.5.1 Flow Restriction

Inorganic scales can accumulate on the internal surfaces of pipelines, wellbores, production tubing, and surface facilities, reducing the effective diameter and obstructing the flow of the hydrocarbons. This can result in decreased production rates, increased pressure differentials, and even complete blockages, leading to production losses and costly interventions.

10.3.5.2 Decreased Efficiency

Inorganic scale deposits can restrict the flow of hydrocarbons through production wells, pipelines, and processing equipment. This leads to a reduction in operational efficiency, as it requires higher pressures and pumping efforts to maintain the desired production rates. The presence of scale can also impede heat transfer in equipment like heat exchangers, resulting in decreased performance and increased energy consumption.

10.3.5.3 Decreased Heat Transfer Efficiency

Scale deposits on heat exchangers and other heat transfer surfaces reduce their efficiency, resulting in decreased heat transfer rates and increased energy consumption. This can lead to decreased system performance, increased operating costs, and potential equipment failures due to overheating.

10.3.5.4 Corrosion Promotion

Inorganic scale deposits create crevices and pockets where corrosive elements can accumulate, leading to localized corrosion of the underlying metal surfaces. This corrosion can weaken the structural integrity of the equipment and pipelines, potentially leading to leaks, breaches, and costly repairs.

10.3.5.5 Reduced Efficiency of Chemical Treatment

Inorganic scales such as calcium carbonate and calcium sulfate can accumulate on the surface of equipment and pipelines, forming a stubborn barrier that can impede the efficacy of chemical inhibitors or scale dissolvers. This, in turn, reduces the overall effectiveness of scale control and mitigation strategies, which require higher chemical dosages and more frequent interventions to maintain equipment performance and integrity. The presence of these scales can also lead to increased energy consumption, decreased heat transfer efficiency, and corrosion of the underlying surfaces. Thus, it is crucial to address the challenge of inorganic scales to ensure the smooth operation of industrial processes and prevent costly equipment failures.

10.3.5.6 Increase Operational Cost

The presence of inorganic scales, such as calcium carbonate, barium sulfate, and strontium sulfate, in oil and gas facilities can lead to a variety of operational challenges. Over time, these scales can accumulate on the internal surfaces of pipes, equipment, and other components, reducing flow rates, decreasing heat transfer efficiency, and potentially causing equipment failure. As a result, regular monitoring, cleaning, and maintenance activities are necessary to ensure the smooth operation of these facilities. However, these activities can incur significant costs, require additional manpower, and result in downtime during cleaning operations, which can further impact the overall productivity and profitability of the facility.

10.3.5.7 Equipment Damage and Failures

The accumulation of inorganic scales in equipment can lead to significant mechanical stress, which, in turn, can cause premature wear, fatigue, and failure of critical components such as valves, pumps, and compressors. This is due to the fact that the scales impede the flow of fluids and gases, leading to inefficient operation and increased pressure within the system. Over time, these conditions can result in unplanned production downtime, increased maintenance costs, and potential safety hazards. It is therefore crucial to regularly inspect and clean equipment to prevent the buildup of inorganic scales, ensuring the optimal performance and longevity of critical components.

10.3.5.8 Safety and Environmental Concerns

Scale deposits can also pose safety risks. Accumulation of scale in pipelines or equipment can create points of weakness, increasing the possibility of leaks or ruptures. This puts personnel and the surrounding environment at risk. Moreover, the chemicals used to mitigate and remove scale can have their own environmental impacts if not handled appropriately.

10.3.5.9 Compliance and Regulatory Considerations

Scaling issues may affect compliance with regulatory requirements and operational standards. Operators are obligated to manage scaling challenges in adherence to environmental, health, and safety regulations, with non-compliance potentially leading to regulatory penalties and reputational risks.

To effectively manage the impact of scale on oil and gas facilities operation, proactive measures such as scale monitoring, risk assessment, and mitigation strategies must be implemented. These measures help minimize the adverse effects of scale deposition and ensure the reliable and efficient operation of oil and gas production facilities.

10.3.6 Scale Mitigation and Removal

Scale mitigation and removal in the oil and gas industry are crucial for maintaining efficient operations. Various methods are employed, including chemical treatments, mechanical methods, and, occasionally, the use of advanced technologies. Chemical inhibitors are commonly injected into wells to prevent scale formation, while mechanical techniques such as pigging and descaling tools help remove existing scale deposits. It's essential to tailor the approach based on the specific characteristics of the reservoir and the types of scales encountered.

Scale removal techniques should be economical, fast, and non-damaging to the wellbore tubing and formation and must be able to prevent precipitation. The strength of the scale and texture play important roles in the selection of the appropriate removal technique.

10.3.6.1 Mechanical Method

Milling and high-pressure water jetting are common mechanical methods for treating scale deposition, as documented in studies (Brown, 1991). Easily accessible scale is milled out based on its location and nature, making these methods applicable primarily in the wellbore. Milling tools are effective for removing less compacted filter cakes or loosely formed depositions, after which high-pressure water jetting can further clean the system. This mechanical scale removal treatment proves efficient for less consolidated surface scales, offering an economical option (Brown, 1991).

Disadvantages of mechanical scale removal:

- If the milling process is not managed properly, unexpected damage will be caused due to long intervals between applications.
- Even though high-pressure water jetting can carry away most of the small particles, some hard bits or even large pieces cannot be lifted away, which could lead to further blockage due to gravity segregation (Eslinger et al., 2000).

In consideration of these disadvantages, further improvements in downhole tools and milling techniques have been made to optimize the removal process as well as reduce the risk caused by a failed treatment.

10.3.6.2 Chemical Methods

Chemical methods offer a cost-effective approach to removing various scales, applicable to both wellbore and formation descaling. Employing inorganic and organic chemicals, these methods require precise knowledge of scale composition and physical/chemical properties for optimal selection. The effectiveness of a scale-removal chemical is scale specific, and poor choices can accelerate recurrence. Chemical methods' efficiency relies on scale accessibility, with the surface area–to-volume ratio being a crucial parameter for effective and economical scale removal. Scales with large surface areas, high porosity, and hairlike projections exhibit high removal rates, facilitating enhanced chemical access to scale particles.

While both chemical and mechanical scale removal treatments can mitigate or eliminate mineral scale deposition, they are unable to control mineral precipitation or prevent it from occurring during subsequent production.

10.3.6.2.1 Acidizing

Acidizing is a method commonly used in the oil and gas industry to treat inorganic scale deposits. Hydrochloric acid (HCl) is the most commonly used acid for scale removal. It is effective in dissolving carbonate scales such as calcium carbonate ($CaCO_3$). Other acids, like hydrofluoric acid (HF), may be used for siliceous scales. The selected acid is injected into the wellbore or production system. This acid reacts with the scale deposits, causing dissolution.

In the case of carbonate scales, the acid reacts with the carbonate minerals to form water, carbon dioxide, and soluble salts. The reaction can be represented as follows for calcium carbonate:

$$CACO_3 + 2HCl \rightarrow CaCl_2 + CO_2 + H_2O \tag{10.13}$$

Acidizing is a carefully controlled process to avoid damage to the well and equipment. Parameters such as acid concentration, injection rate, and contact time are monitored and adjusted as needed. In addition to scale removal, acidizing is often used to enhance the permeability of the reservoir rock. This is known as matrix acidizing and involves dissolving minerals in the formation to improve fluid flow.

Sometimes, acidizing treatments include pre-flushing with a specific fluid to prepare the well for acid injection and post-flushing to clean up the well after the treatment.

Acidizing requires careful handling due to the corrosive nature of acids. Safety measures are implemented to protect workers and the environment. Acidizing is a well-established technique for treating scale issues, but its success depends on factors such as the type of scale, reservoir conditions, and the proper design and execution of the treatment.

10.3.6.2.2 Scale Inhibition

Inhibition is a process employed in the oil and gas industry to prevent the formation and accumulation of mineral scales in wells, pipelines, and production facilities. The inhibition method involves the use of chemicals known as inhibitors, which work by interfering with the crystallization process of the minerals, making it more difficult for them to adhere to surfaces and form scale. These inhibitors function by either modifying the crystal structure of the scales or forming a protective coating on surfaces to prevent scale adhesion.

The inhibition method is typically employed in two ways: continuous treatment and intermittent treatment. In continuous treatment, small amounts of the inhibitor are continuously dosed into the system to maintain an inhibitor concentration that is effective in preventing scale formation. This is commonly used in systems where scale formation is an ongoing issue. However, in intermittent treatment, inhibitors are added periodically in higher dosages to remove existing scale deposits. This method is used when the scale has already formed and needs to be removed. The inhibitors help in dissolving or loosening the scale, making it easier to remove through mechanical or chemical means.

Scale inhibitors are chemical compounds that interfere with the formation and growth of scale-forming minerals and can be an effective strategy for managing inorganic scaling. They are significantly effective and economical approaches to controlling scale deposition. These inhibitors are typically added to the produced fluids at various stages in the production process (Frenier & Ziauddin, 2008; Kamal et al., 2018; Li et al., 2017).

Scale inhibitors function by either modifying the crystal structure of the scales or forming a protective coating on surfaces to prevent scale adhesion. They can also act as dispersants, preventing agglomeration of scale particles and keeping them in a more dispersed and manageable state within the fluid. Commonly used inhibitors include phosphonates, polyphosphates, and organic acids. Before applying scale inhibitors, compatibility testing has to be carried out to ensure that the chosen inhibitor is suitable for the well conditions and does not interact adversely with other chemicals present in the system (Jordan et al., 2014). Also, the dosage and application depend on specific well conditions, and continuous monitoring is crucial for effective scale control.

- Polyphosphates inhibit scale formation by forming a protective coating on surfaces, preventing the deposition of scale-forming ions. They sequester metal ions and prevent them from reacting with other ions to form insoluble scales. Sodium hexametaphosphate and sodium tripolyphosphate are common types of polyphosphate inhibition chemicals.
- Phosphonate inhibition agents act as scale inhibitors by binding with metal ions and preventing them from forming scales; 1-hydroxyethylidene-1,1-diphosphonic acid (HEDP) and amino trimethylene phosphonic acid (ATMP) are common examples of phosphonate inhibition agents. Phosphonate inhibition agents are commonly used in treating calcium carbonate and sulfate scales in the oil and gas industry. They form stable complexes with metal ions, preventing scale precipitation.
- Silicate inhibition agents inhibit scale formation by reacting with calcium and magnesium ions to form insoluble precipitates. Silicate-based inhibitors work by forming a protective layer on metal surfaces, reducing the adhesion of scale-forming ions. Sodium metasilicate and sodium orthosilicate are examples of silicate inhibition agents. They are commonly used in the oil and gas industry as corrosion inhibitors.
- Zinc compounds act as corrosion inhibitors and can also help control scale formation, particularly in high-temperature environments. They are utilized to mitigate scaling issues that can occur in various production and processing facilities. These inhibitors, often applied in water injection systems, help prevent the deposition of scales like calcium carbonate and barium sulfate. Zinc ions interfere with the crystallization process, inhibiting scale formation and protecting equipment such as pipelines and valves from potential damage.
- Molybdate-based inhibitors are effective in preventing the formation of scales such as calcium carbonate and calcium sulfate, which can precipitate and cause issues in production equipment and pipelines. These inhibitors work by forming a protective film on metal surfaces, reducing the adhesion and precipitation of scale-forming ions. The use of molybdate inhibitors is part of the water-treatment process in the industry to control scaling and maintain the efficiency of production systems.
- Titanium-based inorganic scale inhibitors are used in the oil and gas industry for scale control. Titanium compounds such as titanium dioxide (TiO_2) are known for their ability to inhibit the precipitation of various scales, including calcium carbonate and calcium sulfate. These inhibitors are often employed in high-temperature applications. These inhibitors are particularly useful in water-treatment processes to prevent scaling in production equipment and pipelines.
- Titanium-based inhibitors function by interfering with the crystal growth of scale-forming minerals, thereby preventing them from adhering to surfaces. This helps maintain the integrity and efficiency of production systems by reducing the risk of scale buildup.
- Aluminum-based inorganic scale inhibitors work by interfering with the crystallization of scale-forming minerals, such as calcium carbonate or

sulfate. The aluminum ions help prevent the growth of scale deposits by modifying the crystal structure or forming complexes with the scale-forming ions. The use of aluminum-based inhibitors is advantageous due to their efficiency, cost-effectiveness, and environmental compatibility. However, it's essential to carefully monitor their concentration and potential side effects as excess aluminum can lead to other operational challenges or environmental concerns. Aluminum-based inorganic scale inhibitors are utilized for inhibiting scales such as calcium carbonate and silicate.
- Polymers play a crucial role in inorganic scale inhibition by forming a protective barrier on surfaces, preventing the deposition of mineral scales. Polymer inhibitors interact with ions in the water to hinder the formation and growth of scales like calcium carbonate or sulfate. Their adsorption on surfaces disrupts crystal nucleation and growth, thereby minimizing scale buildup in industrial processes.

Polyacrylic acid (PAA) is a polymer used for inorganic scale inhibition in the oil and gas industry. PAA works by adsorbing onto surfaces and interfering with the crystallization process of scales like calcium carbonate and calcium sulfate. In oilfield applications, injecting PAA into the reservoir or mixing it with produced water helps prevent scale deposition in pipelines, wellbores, and production facilities. This aids in maintaining efficient oil and gas production by minimizing the impact of scaling on equipment and infrastructure.

10.3.6.2.2.1 Mechanism of Scale Inhibition

Inorganic scale inhibition typically involves the use of specific chemicals, such as calcium carbonate or sulfate, to prevent the precipitation and deposition of inorganic salts, which can form scales in industrial processes (Figure 10.4). These inhibitors work through various mechanisms:

- Crystal modification: Inhibitors can alter the crystal structure of the scaling salts, making them less likely to adhere to surfaces and form scales.
- Dispersancy: Inhibitors may act as dispersants, preventing the agglomeration of scaling particles. This helps keep the particles in suspension, reducing their tendency to form deposits.
- Sequestration: Certain inhibitors sequester metal ions by forming stable complexes, preventing them from participating in scale formation reactions.
- pH Adjustment: Inhibitors may modify the pH of the system to create conditions unfavorable for scale formation.
- Threshold effect: Some inhibitors function by forming complexes with metal ions present in the scaling salts, raising the concentration required for precipitation. This higher threshold prevents scale formation.

The effectiveness of inorganic scale inhibitors depends on factors like dosage, water composition, and operating conditions in the specific industrial process.

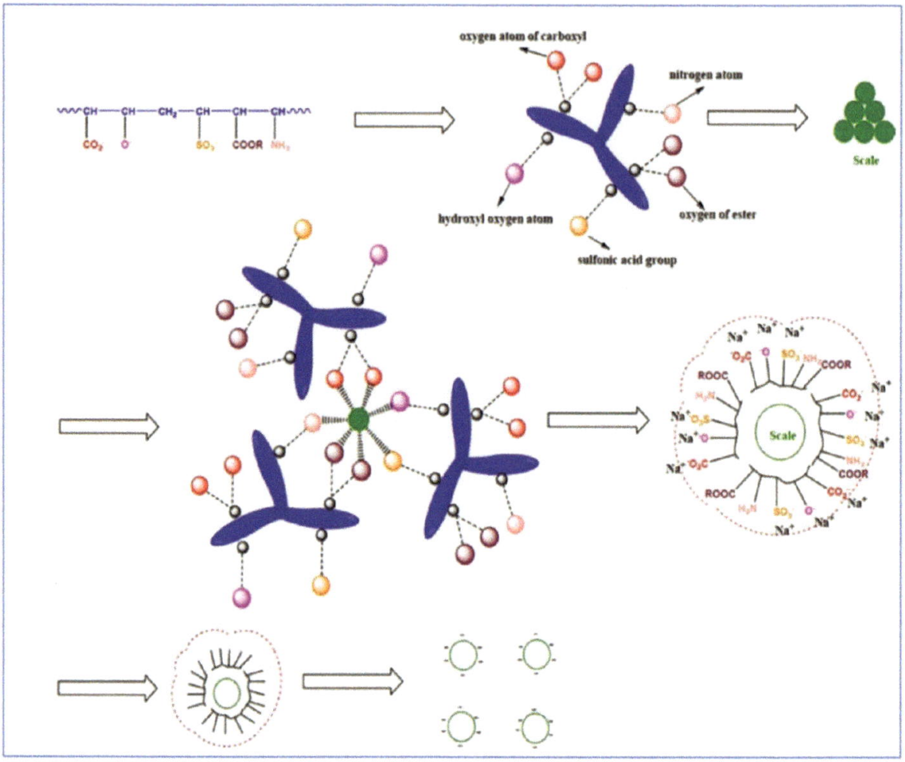

FIGURE 10.4 The scale inhibition mechanism of copolymer.

Source: Li et al., 2017

10.3.6.2.3 Chelation

Chelation is a chemical method similar to inhibition used in the oil and gas industry to treat inorganic scale. The method involves the use of chelating agents, such as ethylenediaminetetraacetic acid (EDTA) or hydroxy ethylenediaminetriacetic acid (HEDTA). These agents form complex molecules with metal ions found in scale deposits.

Chelating agents have multiple sites that can bond with metal ions. When introduced to the system, they surround and bind to metal ions within the scale, forming stable, water-soluble complexes. By forming these complexes, chelating agents prevent the metal ions from participating in the precipitation of scale compounds. This inhibits the formation of solid scale deposits.

Chelation not only inhibits scale formation but also helps disperse existing scale deposits. The chelating agents make the scale particles more soluble, facilitating their removal from surfaces and pipelines. They can be applied through continuous injection into the production system or through batch treatments, depending on the severity of scale deposition and system requirements.

The effectiveness of chelation depends on factors such as pH, temperature, and the concentration of the chelating agent. Optimization of these parameters is crucial for successful scale removal.

10.3.6.2.4 Sequestration

Sequestration is an effective method of preventing and mitigating the impact of inorganic scale in oil and gas production systems. It involves the use of chemicals known as sequestering agents or sequestrants. These agents form stable complexes with metal ions that are responsible for scale formation. Sequestrants have molecules with multiple sites capable of bonding with metal ions. When introduced into the system, they surround and bind to these metal ions, forming soluble and stable complexes, preventing the metal ions from participating in the precipitation of scale-forming minerals, thus inhibiting the formation of solid scale deposits.

Sequestrants not only prevent new scale formation but can also disperse existing scale deposits. The sequestering agents make the scale particles more soluble, facilitating their removal from surfaces and pipelines. Sequestration treatments can be applied continuously or intermittently, depending on the specific requirements of the well and the severity of scale issues.

Common sequestrants include various organic acids and polymeric compounds. The choice of sequestrant depends on factors such as the type of metal ions present in the scale and the conditions of the well.

10.3.6.2.5 Surfactants

Surfactants, or surface-active agents, are used in the oil and gas industry to treat inorganic scale. Surfactants can lower the surface tension of liquids and disrupt the bonding between scale particles. This dispersing action prevents the particles from aggregating and forming solid deposits. They can form emulsions with hydrophobic components of scale deposits, helping break down and disperse these components. This makes it easier to transport and remove the dispersed particles from the system.

Surfactants can act as wetting agents, improving the contact between the treatment fluid and the surfaces covered by scale. This enhances the effectiveness of other scale treatment chemicals. Surfactants are often introduced into wellbore or production systems along with other scale treatment chemicals.

The selection of surfactants depends on their compatibility with other treatment chemicals and the specific characteristics of the scale being treated.

While surfactants alone may not be the primary method for scale removal, they are often used in conjunction with other chemicals and methods to enhance the overall effectiveness of scale treatment in oil and gas production systems.

10.3.7 Latest Advances, Techniques, and Methods in Control of Inorganic Scale

The oil and gas industry often faces challenges posed by inorganic scale, which can lead to reduced production rates, equipment damage, and increased maintenance costs. In response, researchers and industry experts constantly strive to develop and implement the latest techniques for mitigating and controlling inorganic scale. These techniques encompass a range of approaches, including chemical inhibitors, mechanical removal, and electrochemical methods. Chemical inhibitors work by inhibiting the precipitation of scale-forming minerals while mechanical removal involves scraping or flushing the scale from equipment surfaces. Electrochemical methods rely on the application of an electric field to control the deposition of scale-forming minerals. Other techniques include the use of nanotechnology and biotechnology, which have shown promise in reducing scale formation and improving the overall efficiency of oil and gas operations. As the industry continues to evolve, the search for innovative solutions to address the challenges of inorganic scale will remain a top priority.

The future direction of inorganic scale control in the oil and gas industry is likely to involve innovative technologies such as smart inhibitors, nanotechnology, and advanced analytics. Smart inhibitors can autonomously adjust their performance based on changing conditions, providing more efficient scale control. Nanotechnology may enable the development of nanoscale materials for targeted scale inhibition. Additionally, advanced analytics and machine learning can enhance predictive modeling for better monitoring and proactive management of inorganic scale formation. Overall, a focus on precision, sustainability, and real-time data-driven solutions is expected to shape the future of inorganic scale control.

10.4 BACTERIAL AND BACTERIA-RELATED CORROSION

Corrosion is a natural process that leads to the gradual loss of metal due to the chemical reaction of fluids with metal surfaces. It is a destructive phenomenon that not only diminishes the aesthetic appeal of objects but also reduces their lifespan, (Ahmad, 2006) and it has always remained a key problem in oil and gas industries. Internal corrosion in lines carrying produced/separated fluids results from CO_2 / H_2S and in the lines carrying water for injection, fire mains, wash down, etc., from dissolved oxygen and microbiologically influenced corrosion (MIC) (Dawson et al., 2010). Over time, corrosion can weaken the structural integrity of the metal, leading to costly repairs or even failure of the affected components.

Oil and gas production aims to extract hydrocarbons from fields, reservoirs, and wells in a cost-effective and safe manner. This involves separating oil and gas from any produced water and transporting them through flow lines and gathering lines. The extracted hydrocarbons are then exported through pipelines to storage before being refined or transferred to a tanker or export pipeline. However, all these facilities can pose potential risks to health and safety, as well as environmental and commercial concerns. Even secondary facilities like water injection, firewater, and produced water disposal are subject to corrosion threats. While failure of these systems may not present immediate safety, health, or environmental risks, it could lead to a reduction or cessation of production, which could have commercial implications.

Bacteria-related corrosion, also known as microbiologically influenced corrosion (MIC), refers to the process by which microorganisms facilitate corrosion in industrial and natural environments (Stott, 2010). These microorganisms often form biofilms on metal surfaces, which can lead to accelerated corrosion rates and reduced material life. Understanding the mechanisms of bacteria-related corrosion requires a multidisciplinary approach as it involves the interaction of microbiology, chemistry, biochemistry, and metallurgy (Muthukumar et al., 2003). In particular, knowledge of microbial metabolism, electrochemistry, and material science is essential to effectively prevent and mitigate bacteria-related corrosion.

Bacteria-related corrosion it is not in itself a form of corrosion but rather a process that can influence and even initiate corrosion (Rawat et al., 2014). These microorganisms are considered tiny catalytic entities that have the capability of initiating and accelerating a wide range of biochemical reactions, which, in turn, accelerate most forms of corrosion, including uniform corrosion, pitting corrosion, crevice corrosion, galvanic corrosion, intergranular corrosion, dealloying, and stress corrosion cracking. They produce and secrete aggressive chemical agents such as acids or the consumption of substances that are important in corrosion reactions, such as oxygen or nitrite corrosion inhibitors. Alternatively, the physical presence of microorganisms may form a slime or poultice, leading to differential aeration corrosion or crevice corrosion. The process typically involves the formation of a biofilm on the surface of the metal, creating an electrochemical environment that accelerates corrosion (Khan et al., 2021). Over time, the biofilm can become thicker and more complex, leading to a more severe corrosion process that can compromise the structural integrity of the material.

Bacterial corrosion can occur in various environments, including soil and water, and even within industrial processes. It is a significant issue in many industries, including oil and gas, marine, and wastewater treatment, among others. This phenomenon usually occurs in water and soils with a pH range of 4 to 9 and at temperatures between 10° and 50°C (WebC+).

Unconventional resource facilities face a risk of microbiologically influenced corrosion (MIC), stemming from the presence of microorganisms and specific corrosion-promoting conditions. MIC manifests as pitting corrosion, deteriorating surfaces of pipes, tanks, or vessels. The formation of slime or tuberculation nodules can lead to blockages or reduced flow (Rawat et al., 2014). Systems with low flow or stagnant conditions are more prone to microbial growth. Microbiologically induced corrosion can impact and potentially fail various system types, resulting in premature metal component failure. Comprehensive understanding of the causes and effects requires insights into the chemical, metallurgical, and microbiological aspects of microbial corrosion.

10.4.1 Bacteria

True bacteria, unicellular microorganisms that fall under neither animal nor plant categories, belong to the distinct and primitive domain of prokaryotes (Stott, 2010). Typically a few micrometers in size, they exhibit diverse shapes such as spheres, rods, and spirals. While often found in pairs, clusters, or chains due to incomplete separation after cell division, these formations are not crucial to their existence. Bacteria are pervasive, thriving in nearly every conceivable habitat, including soil, underwater environments, and deep within the Earth's crust. Some are aerobic, requiring oxygen, while others are anaerobic, flourishing in its absence.

Bacterial cells, characterized by their diminutive size, notably lack a nucleus or other membrane-bound structures with specialized functions (organelles). Unlike higher organisms, bacteria exhibit a wide metabolic diversity, with various species deriving energy from a broad spectrum of chemical reactions (Khan et al., 2021). The majority are heterotrophs, relying on the oxidation of organic carbon fixed by other organisms rather than performing photosynthesis to fix their own carbon. However, a select group of specialized bacteria engages in inorganic reactions, involving the oxidation or reduction of nitrogen, sulfur, or iron. Despite being a minority in the biosphere, these chemolithotrophic bacteria, particularly in their activities, are of primary concern to corrosion engineers.

The activities of microorganisms are the driving force of biocorrosion. When bacteria grow on the walls of the pipes, they develop into colonies containing many types of bacteria, which help each other in life. Well-known types that cause corrosion include BRB and acid-producing bacteria, two common type in oil and gas. They are highly corrosive to pipe. SBR produce hydrogen sulfide, while APB generate, both (1) Sulfate-reducing bacteria (SRB), (2) sulfate-oxidizing bacteria (SOB), and (3) acid-producing bacteria (APB). Metal oxidation and bacterial reduction are two main causes of MIC, in which the SRB is considered to be the major cause of corrosion (Khan et al., 2021; Muthukumar et al., 2003; Rawat et al., 2014). Other types of bacteria are manganese-oxidizing bacteria and heterotrophic bacteria. Iron bacteria precipitate the iron present in solutions arising from corrosion and form a tubercle on top of a corrosion pit. The species of these bacteria are grouped by Rawat et al. (2014) and Saji et al. (2020) as follows:

- Sulfate-reducing bacteria (*Desulphovibrio*, *Desulphotomaculum*, and *Desulphomonas sp.*)
- Iron-reducing bacteria (*Gallionella, Crenothrix, Leptothrix, Gallionellea ferrugine* and *Ferrobacillus spp.*)
- Acid-producing bacteria (*Pseudomonas*, *Aerobacter*, and *Bacillus spp.*)
- Sulfur-oxidizing bacteria (*Acidithiobacillus Thiobacillus spp.*)
- Nitrate-reducing bacteria (*Pseudomonas aeruginosa*)
- Sulfate-reducing bacteria (SRB) (*Desulfovibrio*)

10.4.1.1 Sulfate-Reducing Bacteria (SRB)

In the oil and gas industry, SRB thrive in environments such as pipelines, storage tanks, and other equipment that handles water or fluid containing sulfate compounds. They produce hydrogen sulfide $H_2(H_2S)$ gas as they metabolize sulfate, which can lead to the formation of iron sulfide (FeS) compounds on metal surfaces. These compounds can cause pitting and accelerated corrosion, compromising the integrity of metal structures.

Sulfate-reducing bacteria (SRB) are most important group of microorganisms associated with corrosion (Figure 10.5) (Stott, 2010). These bacteria produce hydrogen sulfide (H_2S) gas as they metabolize sulfate, which can lead to the formation of iron sulfide (FeS) compounds on metal surfaces. These compounds can cause pitting and accelerated corrosion, compromising the integrity of metal structures. SRB are commonly found in anoxic environments such as deep-sea sediments, hydrothermal vents, oil reservoirs, and sulfur-rich environments like marshes and swamps. They can also be present in industrial settings such as pipelines and oil wells, where anaerobic conditions prevail.

SRB derive their energy from organic nutrients, grow in a pH range of 4.0 to 9.5 (Stott, 2010; WebCorr, 2014), and can withstand the pressures up to 500 atmospheres (Barton & Fauque, 2009; Schnieders, 2005). SRB reduce oxidized sulfur compounds such as sulfate, sulfite, and thiosulfate, as well as sulfur to H_2S. The oil and gas industry is badly affected by the sulfides generated by these SRB. Their presence can lead to the production of hydrogen sulfide, which is corrosive and can pose operational challenges. High concentrations of H_2S in reservoir fluid also cause reservoir souring (Stott, 2010), which happens in different parts of the world. It is therefore important to monitor and manage the presence of these bacteria to mitigate corrosion in infrastructure and equipment (Khan et al., 2021). SRB can be found everywhere, from more than 70 meters deep in clay to seawater (Hernandez et al., 1994). The reaction equation for SRB is (Khan et al., 2021):

$$SO_4^{2-} + CH_2COOH + 2H^{2+} \rightarrow HS^- + 2HCO_3^- + 3H^+ \quad (10.14)$$

$$HS^- + Me^{2+} \rightarrow MeS + H^+ \quad (10.15)$$

Me^{2+} can be, for example copper, zinc, etc.

Sulfate-reducing bacteria that produce hydrogen sulfide, which can cause sulfide stress cracking, include:

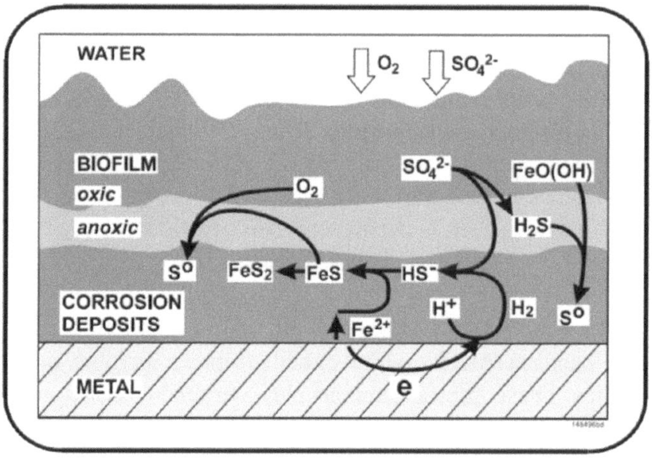

FIGURE 10.5 Sulfate-reducing bacteria corrosion of mild steel in the presence of oxygen is an acid corrosion.

Source: Lewandowski & Beyenal, 2008)

- *Acidithiobacillus* bacteria produce sulfuric acid; *Acidothiobacillus thiooxidans* frequently damages pipes.
- *Ferrobacillus ferrooxidans* directly oxidizes iron to iron oxides and iron hydroxides; the rusticles forming on the RMS *Titanic* wreck are caused by bacterial activity. Other bacteria produce various acids, both organic and mineral, or ammonia.

10.4.1.2 Sulfur-Oxidizing Bacteria (SOB)

Sulphate-oxidizing bacteria (SOB) are aerobic in nature, which means they need oxygen for their metabolism. SOB grow at low pH values and in environments like solids, soils, and water that contains sulfide minerals. SOB convert sulfur to sulfuric acid, a highly corrosive substance that rapidly corrodes metal surfaces on contact. The bacteria belong to an unusual group of microorganisms called chemolithotrophs (Stott, 2010). SOB play a crucial role in the biogeochemical cycling of sulfur and are found in various environments, including aquatic ecosystems, soil, and deep-sea hydrothermal vents. They utilize sulfate SO_4^{2-} as an electron acceptor in their energy metabolism. Sulfur-oxidizing bacteria (SOB) derive their energy by oxidizing inorganic sulfur compounds such as sulfides to sulfuric acid (eq. 10.16). These bacteria are capable of fixing carbon dioxide to build up their cell material, which makes them autotrophs. Essentially, SOB are self-sufficient organisms without the need for external sources of organic carbon (Stott, 2010). Both SOB and SRB can coexist in the system.

Sulfur-oxidizing bacteria (SOB) can contribute to corrosion by oxidizing elemental sulfur, producing sulfuric acid. The general chemical reaction involves the oxidation of sulfur to sulfuric acid:

$$S + 2O_2 + 2H_2O \rightarrow H_2SO_4 \tag{10.16}$$

This reaction occurs in the presence of sulfur-oxidizing bacteria, leading to the production of sulfuric acid, which can corrode metals and other materials. Corrosion caused by sulfur-oxidizing bacteria is a concern in various industries, especially those involving sulfur-rich environments.

Sulfate-oxidizing bacteria can be harnessed for bioremediation purposes, particularly in environments contaminated with sulfide compounds. By promoting the oxidation of sulfide to sulfate, these bacteria can contribute to the detoxification and remediation of sulfur-contaminated sites (Figure 10.6).

10.4.1.3 Metal-Reducing Bacteria (MRB)

Metal-reducing bacteria (MRB) exhibit either aerobic, anaerobic, or facultative aerobic characteristics. These microorganisms play a crucial role in converting insoluble ferric compounds into soluble ferrous ions. The insoluble nature of ferric ions contributes to corrosion control through chemical activity facilitated by ferric salts. In contrast, ferrous salts, being predominantly soluble, lead to the removal of protective layers, thereby directly promoting corrosion in the system. See Lloyd (2003) for the detailed reaction mechanism of microbial reduction of different metals,

FIGURE 10.6 Splash and crevice corrosion caused by barnacles attachment on bar screen in desalination.

Source: Saji et al., 2020)

10.4.1.4 Metal-Oxidizing Bacteria (MOB)

Metal-oxidizing bacteria, also known as iron oxidizing bacteria, have the ability to transform ferrous ions into ferric ions (Muthukumar et al., 2003). These bacteria thrive in aerobic environments, particularly in areas with low levels of oxygen. Furthermore, they can also oxidize manganous ions into manganic ions and deposit manganese dioxide as a result. However, the accumulation of these bacteria on metal surfaces can increase corrosion by promoting the deposition of cathodically reactive ferric and manganic oxides and consuming oxygen through their respiration (Beech et al., 2000). It is important to consider the impact of metal oxidizing bacteria on metal surfaces to prevent or mitigate the effects of corrosion.

When ferrous iron is in the presence of iron-oxidizing bacteria, it undergoes oxidation (Singh et al., 2018).

Some examples of metal oxidizing bacteria are *Acidithiobacillus*, *Leptospirillum*, *Gallionella*, *Thiobacillus ferrooxidans*, f*errooxidans*, *Sulfolobus acidocaldarius*, and *Hyphomicrobium* (Singh et al., 2018).

10.4.1.4.1 Iron-Oxidizing Bacteria (IOB)

IOB are bacteria that obtain their energy by oxidizing ferrous ions (Fe2+) present in the environment. These bacteria can form thick and adherent biofilms on metal surfaces, promoting corrosion by trapping corrosive agents and reducing the protective properties of passive films. IOB can be found in both aerobic and anaerobic environments, including water injection systems, pipelines, and storage tanks. Example of iron oxidizing bacteria are *Pseudomicrobium spp* and *Gallionella spp* (Muthukumar et al., 2003).

10.4.1.5 Iron-Reducing Bacteria

Iron-reducing bacteria play an important role in the corrosion of metal surfaces. These bacteria are capable of reducing ferric ions ($Fe3+$) to ferrous ions ($Fe2+$) (Muthukumar et al., 2003). Ferric ions are insoluble under normal conditions, but they can form soluble ferric salts at very low pH levels. These salts act as a protective layer on metal surfaces, preventing further corrosion due to chemical activity. However, when iron-reducing bacteria are present, they can reduce ferric salts to ferrous salts, which are mostly soluble in water. This results in the removal of the protective layer, exposing the metal surface to further corrosion. In other words, iron-reducing bacteria can directly promote corrosion by breaking down the protective layer and making the metal surface more vulnerable to chemical attacks.

One example of iron-reducing bacteria is *Pseudomonas spp.* This type of bacteria is commonly found in soil, water, and other environments rich in organic matter (Muthukumar et al., 2003). *Pseudomonas spp* can thrive in environments with low oxygen levels, and they can use ferric ions as an electron acceptor in their metabolic processes. As a result, they can play a significant role in the corrosion of metal structures such as pipelines, storage tanks, and bridges.

10.4.1.6 Acid-Producing Bacteria (APB)

Acid-producing bacteria (APB) can cause corrosion by producing organic and inorganic acids that lower the pH of the environment and dissolve the metal surface (Khan et al., 2021). Some examples of APB are *Pseudomonas*, *Aerobacter*, and *Bacillus spp* (Khan et al., 2021; Muthukumar et al., 2003). These bacteria are mostly aerobic and consume carbohydrates as their energy source. The acids they produce include carbonic acid, lactic acid, acetic acid, and others. The acidic conditions created by the bacteria can induce electrochemical reactions on metal surfaces, promoting localized corrosion processes such as pitting and crevice corrosion. APB can also degrade protective coatings, inhibitors, oils, and emulsions used to prevent corrosion. APB are commonly found in solids, soil, and water that contains sulfide minerals. They are also reported in some pipes and concrete carrying municipal water (Muthukumar et al., 2003).

10.4.2 Mechanism of Microbial Corrosion

The mechanism for formation of bacteria corrosion is a complex process involving the interaction between microorganisms and metal surfaces (Figure 10.7). The mechanisms of microbial corrosion of metals may be site specific and tend to vary depending on the microbial species involved, metal composition, and environmental conditions. Microorganisms may induce corrosion processes directly or indirectly.

The process starts with biofilm formation. Here, microorganisms, particularly bacteria, adhere to the metal surface and form a biofilm. This biofilm creates a localized microenvironment in which the bacteria can thrive and interact with the metal surface. The biofilm may be composed of a complex matrix of extracellular polymeric substances (EPS) produced by the bacteria, providing protection and stability for the microbial community.

Once the biofilm is established, the bacteria within the biofilm engage in metabolic activities. Some bacteria are capable of metabolizing substances present in the

TABLE 10.2
Some Anaerobic and Aerobic Bacteria Implicated in MIC

Bacteria	Temp. (°C)	pH	Characteristic	Function
Sulfate-reducing bacteria	20–40	5.0–9.0	Anaerobic	Reduce SO_2 4 to S and H_2S
Iron-reducing bacteria	10–50	4.0–8.0	Anaerobic	Reduce insoluble Fe III to soluble Fe II
Nitrate-reducing bacteria	5–40	5.5–9.5	Anaerobic	Reduce nitrate to ammonium and N_2
Methanogenic bacteria	20–40	5.4–7.2	Anaerobic	Produce methane using elemental iron as the sole source of hydrogen via cathodic depolarization
Acid-producing bacteria	20–40	4.5–8.0	Anaerobic	Produce fermentation metabolites, such as volatile fatty acids
Iron-oxidizing bacteria	20–40	6.0–8.0	Aerobic	Oxidize Fe II to Fe III
Sulfur-oxidizing bacteria	20–40	6.5–7.5	Aerobic	Release aggressive metabolites such as acetic, succinic, and sulfuric acid

Source: Lv & Du, 2018

environment, such as organic matter or metallic ions. As a result, metabolic byproducts such as organic acids, hydrogen sulfide, ammonia, or other corrosive substances may be secreted into the microenvironment.

The presence of bacteria and their metabolic activities can lead to localized changes in the electrochemical behavior of the metal surface. For instance, certain bacteria can facilitate the transfer of electrons between the metal and their metabolic processes, potentially promoting accelerated corrosion reactions. Additionally, the biofilm itself can create oxygen concentration cells, leading to differential aeration and promoting differential aeration corrosion.

Bacteria can also indirectly influence the corrosion process by altering the composition and pH of the surrounding medium, potentially creating conditions favorable for corrosion. The interactions between the bacteria, the metal surface, and the surrounding environment contribute to the initiation and progression of corrosion processes.

Various mechanisms have been proposed to explain the active participation of microbes in corrosion. Some microbes can produce metabolites that are acidic (for instance, sulfuric acid by sulfur-oxidizing bacteria) or facilitate local corrosion. Some of these mechanisms were documented by Ariwahjoedi et al. (2012, Kakooei et al. (2012), Khan et al. (2021), Knisz et al. (2023), Lewandowski and Beyenal (2008), Little et al. (2020), and Lv and Du (2018) and include the following: cathodic depolarization theory (Kuhr & Vlugt, 1964), anodic depolarization (Obuekwe et al., 1981), metal-binding exopolymers (Beech et al., 1996), volatile phosphorus compounds (Iverson, 1968), biocatalytic cathodic sulphate reduction (Gu et al., 2009), biomineralization (Little et al., 2000), and Romero three-stage mechanism (Fernandez de Romero, 2005).

10.4.3 Some Facilities Commonly Affected by Corrosion

Microbiologically influenced corrosion (MIC) can affect various surface facilities in the oil and gas industry, including:

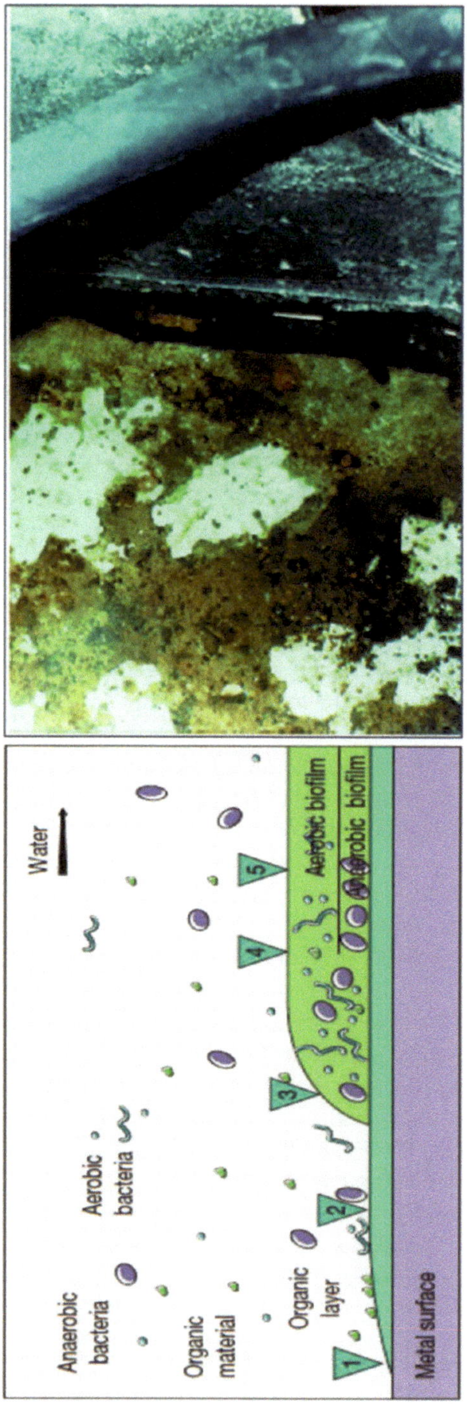

FIGURE 10.7 (a) Steps in biofilm formation, (b) Dense biofilm on a metallic surface.

Source: Stott, 2010

10.4.3.1 Pipelines

Both onshore and offshore pipelines are susceptible to bacteria-related corrosion, especially in environments where water is present. Microorganisms such as sulfate-reducing bacteria (SRB) and acid-producing bacteria can thrive in pipeline systems, leading to localized pitting, under-deposit corrosion, and general deterioration of metal surfaces.

10.4.3.2 Production Equipment

Bacteria-related corrosion can affect different types of surface production equipment, such as wellheads, separators, and valves. The presence of microorganisms such as hydrocarbon-related bacteria and iron-related bacteria in production fluids can lead to corrosion of equipment surfaces. This can result in reduced operational lifespan of the equipment and pose potential safety risks.

10.4.3.3 Downhole Equipment

Bacteria-induced corrosion can have detrimental effects on downhole equipment. Corrosive byproducts produced by certain bacteria can accelerate the degradation of metal components in downhole equipment, such as tubing and casing. SRB and other bacteria can produce corrosive byproducts such as hydrogen sulfide and organic acids, which promote corrosion of metal surfaces. Corrosion in downhole equipment can result in thinning of tubular components, integrity loss, and the potential for costly well interventions.

10.4.3.4 Heat Exchangers and Coolers

Bacteria-induced corrosion in heat exchangers and coolers in the oil and gas industry can lead to reduced heat transfer efficiency and equipment failure. Microbiologically influenced corrosion (MIC) can cause pitting, fouling, and the formation of biofilms on heat exchange surfaces. These effects result in decreased thermal performance, increased energy consumption, and higher maintenance costs.

The accumulation of microbial deposits can impede fluid flow, reduce heat transfer rates, and lead to localized corrosion. Regular monitoring, proper biocide treatment, and material selection resistant to microbial corrosion are crucial in preventing and mitigating these issues. Implementing effective maintenance practices helps ensure the optimal performance and longevity of heat exchangers and coolers in oil and gas facilities.

10.4.3.5 Flowlines and Surface Equipment

Bacteria-related corrosion can also affect surface infrastructure, including flowlines, surface piping, and production facilities. Microbial activity in water-containing systems can lead to under-deposit corrosion, general deterioration of metal surfaces, and the formation of biofilms that can obstruct flow and inhibit the effectiveness of corrosion protection coatings.

10.4.3.6 Separation Facilities

Oil/water separation facilities typically rely on heat to aid in the separation process. However, scale deposition can hinder heat transfer in these facilities, leading to severe under-deposit corrosion as temperatures become concentrated. In direct-fired heaters, the buildup of scale can cause a type of failure called creep rupture, as certain areas underneath the scale can reach extremely high temperatures. Additionally, if the separation equipment is open to the atmosphere, oxygen ingress can cause increased corrosion in the water phase.

10.4.3.7 Tanks/Water Storage

Oil and gas storage tanks, whether aboveground or underground, are vulnerable to corrosion caused by bacteria. Microorganisms can thrive in the water present at the bottom of these tanks, leading to the corrosion of the tank walls and floor. Furthermore, microbial activity can result in the formation of sludge and sediment, which further speeds up the corrosion process.

10.4.3.8 Water Injection and Disposal Systems

Water injection and disposal systems are crucial to oil and gas production operations. However, microbial growth in these systems can cause biofouling, leading to reduced system efficiency and microbiologically induced corrosion of infrastructure. Corrosion in injection and disposal wells, piping, and associated equipment can result in significant operational disruptions and maintenance costs.

10.4.3.9 Injection Flowlines and Wells

The corrosion mechanisms for producing well tubulars and flowlines are similar, mainly caused by H_2S, CO_2, and organic acids corrosion. Corrosion can occur underneath the deposits at the bottom of the lines, as well as under scale and under SRB colonies. Oxygen contamination can significantly increase corrosion, except for SRB, which are essentially anaerobic. Corrosion can still happen in areas with restricted oxygen entry, such as under scales and sludge. The aerobic bacteria slime can turn anaerobic and can then support SRB growth.

10.4.4 Characterization and Modeling of Bacteria-Related Corrosion

Effective characterization, monitoring, and inspection of microbial-induced corrosion (MIC) is vital for identifying the specific organisms causing corrosion and preventing, controlling, or mitigating these occurrences. By implementing robust systems for characterization, monitoring, and inspection, equipment failure and extensive maintenance can be avoided, particularly in industries such as oil and gas, in which downtime costs can be exorbitant. Identifying MIC can be challenging as it often results in failures similar to those of other forms of corrosion. However, with knowledge and experience in the field, it is now possible to determine which systems are more susceptible to MIC and analyze the pit morphology, appearance, and colors of corrosion products associated with MIC. For instance, black-colored and foul-smelling FeS corrosion products are indicative of sulfate-reducing bacteria while reddish-brown deposits are typical of iron-oxidizing bacteria in the system (Javaherdashti, 2008; Khan et al., 2021).

In the oil and gas industry, the characterization of bacteria-related corrosion (BRC) involves the use of various laboratory methods to understand the presence and behavior of corrosive bacteria in production environments. Some common laboratory methods applied for the characterization of bacteria corrosion include:

10.4.4.1 Bacteria Culture

For MIC scenarios, understanding the mechanism of biofilm formation and microbe population is of prime interest, and it can be done via bacteria culture. This method involves the isolation and cultivation of bacteria from samples collected from oil and gas production systems. Different types of media and culture conditions are used to promote the growth of corrosive bacteria, allowing for their identification and characterization.

While this method is effective in identifying bacterial species associated with biofilm formation, its reliability diminishes when certain species thrive while others do not, often due to constraints in replicating the natural culture environment. Little et al. reported an instance when sulfate-reducing bacteria (SRB) and oil-oxidizing bacteria coexisted, revealing inconsistencies between the observed and actual corrosion rates of mild steel in the culture, highlighting the limitations of this approach.

10.4.4.2 Molecular Biology Methods

Unlike culture, molecular biology methods are based on utilizing the genetic material of bacteria: i.e., DNA and RNA. Polymerase chain reaction (PCR) and quantitative PCR (qPCR) are used to detect specific genes associated with bacteria known to cause corrosion, such as sulfate-reducing bacteria (SRB) and acid-producing bacteria. This allows for the quantification of corrosive bacterial populations. Also, these methods indicate whether bacteria are influenced by or are resistant to biocides.

10.4.4.3 Molecular Biological Tools

In the molecular biological method, polymerase chain reaction (PCR) and quantitative PCR (qPCR) are used to detect specific genes associated with bacteria known to cause corrosion, such as sulfate-reducing bacteria (SRB) and acid-producing bacteria. This allows for the quantification of corrosive bacterial populations.

10.4.4.4 Microscopic Examination

Microscopic methods, employing staining and counting techniques, prove effective in quantifying bacteria in microbiologically influenced corrosion (MIC) issues. Notably, distinctive bacteria like filamentous iron bacteria and stalked iron oxidizers can be directly observed under a microscope (Javaherdashti, 2008). Nevertheless, microscopic examination falls short in identifying individual bacterial families that may share visual similarities, potentially overlooking crucial bacteria actively contributing to MIC problems.

Microscopic techniques play a crucial role in visualizing and analyzing the structure of biofilms formed by corrosive bacteria on metal surfaces. This enables a deeper understanding of attachment mechanisms and spatial distribution within biofilms. Advanced microscopic examinations, such as scanning electron microscopy (SEM), transmission electron microscopy (TEM), and environmental scanning and transmission electron microscopy (ESEM and ETEM), provide detailed insights into biofilm composition and development, surpassing the capabilities of standard microscopic examination.

10.4.4.5 DNA Sequencing and Analysis

Genomic and metagenomic sequencing techniques are used to analyze the genetic composition of microbial communities present in production environments. This provides valuable information about the diversity of corrosive bacteria and their potential for causing corrosion.

10.4.4.6 Chemical Analysis

Various chemical analysis methods, such as gas chromatography, liquid chromatography, and mass spectrometry, are used to detect and quantify metabolites produced by corrosive bacteria, including organic acids, sulfides, and other corrosive compounds.

10.4.4.7 Electrochemical Techniques

Electrochemical methods, including electrochemical impedance spectroscopy (EIS) and potentiodynamic polarization, are employed to assess the corrosion behavior of metal surfaces in the presence of corrosive bacteria. These techniques help in determining corrosion rates and mechanisms driven by microbial activity.

10.4.4.8 Material Testing and Evaluation

Laboratory-based material testing, including corrosion testing, exposure testing, and electrochemical testing, is conducted to assess the susceptibility of different materials to microbial corrosion. This helps in identifying corrosion-resistant materials for use in production systems.

10.4.4.9 Electrochemical Methods

There are several electrochemical techniques that can be used to investigate microbial-influenced corrosion (MIC) concerns. These techniques include weight loss testing (using coupons) and electrical and polarization resistance approaches, as well as potentiodynamic sweep methods. However, it has been noted by Gilbert and Herbert et al. that these methods are not particularly effective in identifying the specific type of organism responsible for the biodeterioration of a system, nor are they capable of accurately quantifying the number of organisms involved (Videla, 2002). This might be because these methods are designed to monitor uniform corrosion, which does not usually occur in MIC, in which the corrosion attack is mostly localized. Nonetheless, linear polarization coupled with other electrochemical methods can be used to obtain valuable data about biodeterioration in a system. Additionally, electrochemical impedance spectroscopy (EIS) is widely used for the analysis of sulfide films formed on steel during MIC. For more information on EIS, see the recent review paper by Kashkovskiy et al. (2019).

10.4.4.10 Environmental Simulation Testing

In order to better understand the effects of specific environmental factors on the activity and corrosion potential of bacteria, laboratory-scale simulation tests are carried out. These tests aim to replicate the conditions present in a production environment, such as temperature, pressure, and chemical composition. By doing this, researchers can observe how these factors impact the activity and corrosiveness of bacteria and develop strategies to mitigate any negative effects. Such research is particularly important in fields such as microbiology and industrial engineering, in which the behavior of bacteria can have a significant impact on product quality and safety.

10.4.4.11 Quick Check Tests

Quick check test methods are used to determine the presence of microbially influenced corrosion (MIC) and associated bacteria in a system. These methods provide a simple yes or no answer for (1) the occurrence of MIC and/or (2) the presence of microorganisms that can cause MIC. Some of the commonly used quick check test methods include hydrogenase tests, antibody tests, and fatty acid analyses.

By utilizing these laboratory methods for characterization, researchers and industry professionals can gain valuable insights into the nature of bacteria-related corrosion in oil and gas production systems.

10.4.5 Conditions Affecting Formation of Bacteria-Related Corrosion

Bacteria-related corrosion is ubiquitous in nature and grows at very rapid rates in soil, water, and air. It shows extreme tolerance to varying environmental conditions, such as acidic and alkaline pH and low and higher temperatures. Understanding these conditions can help prevent and reduce the impact of bacterial corrosion on infrastructure and facilities. Some of the key conditions that contribute to the formation of bacteria-related corrosion are:

10.4.5.1 Moisture and Nutrient Availability

Bacteria are incredibly versatile and can survive and grow in a wide range of environments. However, to sustain their life and growth, they require moisture and nutrients. In environments where water is present, such as soil, water systems, and process fluids, and organic materials or nutrients are available, bacteria can gather and create biofilms. These biofilms are colonies of bacteria that stick to each other and to surfaces, forming a slimy layer. As the biofilm grows, it can become more complex and can cause microbial corrosion.

10.4.5.2 Oxygen Levels

The presence or absence of oxygen plays a significant role in the types of bacteria that cause microbial-induced corrosion (MIC). Aerobic bacteria require oxygen to survive, and they play a crucial role in the initial stages of corrosion. These bacteria consume oxygen and generate acidic byproducts that can accelerate the corrosion process.

On the other hand, anaerobic bacteria thrive in low-oxygen or oxygen-deprived environments and contribute to the later stages of corrosion. These bacteria consume organic matter and produce hydrogen sulfide, which is a corrosive agent that can attack metal surfaces.

Both aerobic and anaerobic bacteria can contribute to the corrosion process, and their activities can interact in complex ways. For example, aerobic bacteria can create localized oxygen-depleted zones that promote the growth of anaerobic bacteria. The activities of both types of bacteria can ultimately lead to the degradation of metal structures.

10.4.5.3 pH Levels

Different types of bacteria have distinct pH preferences, and some are more corrosive and active in acidic conditions while others thrive in alkaline environments. For example, sulfate-reducing bacteria (SRB) are commonly found in acidic environments and are known to be highly corrosive. In contrast, nitrate-reducing bacteria (NRB) prefer alkaline conditions and tend to be less corrosive.

Fluctuations in pH levels can significantly affect the diversity and activity of bacteria involved in MIC. For instance, a sudden increase in pH can lead to the proliferation of alkaline-loving bacteria, which can accelerate corrosion rates. Conversely, a drop in pH can inhibit the growth of some bacteria, leading to a decline in corrosion activity.

Therefore, maintaining a stable pH level in the environment is crucial to mitigate the impact of MIC on infrastructure. This can be achieved through various measures, such as the use of corrosion inhibitors, biocides, or pH-adjusting chemicals.

10.4.5.4 Temperature

The temperature plays a vital role in bacterial activity and corrosion rates. The optimal temperature levels can significantly impact the growth of bacteria and their metabolic activities, leading to increased corrosion rates. Bacteria can thrive in a wide range of temperatures, and their metabolic rate can double with every 10°C increase in temperature. As a result, elevated temperatures can accelerate the corrosion of metal structures and equipment.

It is essential to maintain appropriate temperatures to minimize the potential for corrosion. The ideal temperature range for preventing bacterial growth and corrosion is between 20°C and 37°C (Rasol et al., 2015). Keeping the temperature within this range can help slow down bacterial activity and reduce the risk of corrosion. Any temperature outside this range could lead to bacterial growth, increased metabolic activity, and, ultimately, higher corrosion rates.

10.4.5.5 Material Susceptibility

The susceptibility of infrastructure and equipment to bacterial corrosion can be impacted by the type of materials used. Some metals and alloys are more susceptible to MIC while others show greater resistance. Additionally, the presence of protective coatings and inhibitors can play a role in determining the susceptibility of materials to bacterial corrosion.

To prevent bacterial corrosion, it is crucial for operators and engineers to have a good understanding of the underlying conditions. This enables them to take timely and proactive measures to control and mitigate the corrosion, as detailed in section 10.5.4.

10.4.6 Impact of Bacteria-Related Corrosion on Operations

Bacteria-induced corrosion is a serious problem that can affect various industries, including refining. This type of corrosion can cause issues in water-handling operations, oil and gas production, pipelining, refining, petrochemical synthesis, and other industrial sectors. Even commonly used metals and alloys, such as stainless steels, nickel, and aluminum-based alloys, as well as materials like concrete, asphalt, and polymers, can easily degrade due to microorganisms. Protective coatings, inhibitors, oils, and emulsions can also be susceptible to microbial degradation. As a result, bacterial corrosion can have a significant impact on the facilities and operations of these industries.

10.4.6.1 Infrastructure Integrity

Bacterial corrosion is a serious issue that can have significant consequences on the integrity of infrastructure components such as pipelines, well casings, and storage tanks. This type of corrosion is caused by the action of microorganisms on metal surfaces, leading to pitting, cracking, and other forms of damage. If left unchecked, it can compromise the structural integrity of these components, leading to leaks, ruptures, and overall degradation of the facility. This can pose a serious risk to operational safety and efficiency, as well as the environment and public health. Therefore, it is crucial for industries and organizations to take proactive measures to prevent and mitigate the effects of bacterial corrosion.

10.4.6.2 Production Interruption

Bacterial corrosion can cause significant damage to equipment and infrastructure, leading to unplanned downtime for production facilities. This, in turn, can result in costly repairs, maintenance, and production losses, ultimately affecting the bottom line of the production. Bacteria can eat away at metal surfaces, causing pitting and cracks that weaken the structure over time. The damage caused by bacterial corrosion can be difficult to detect and may require specialized testing to identify. It is important to implement preventive measures to minimize the risk of bacteria corrosion and ensure the longevity and reliability of their equipment and infrastructure.

10.4.6.3 Environmental and Safety Risks

Bacterial corrosion can cause serious leaks and failures in industrial equipment and infrastructure, which can lead to hazardous environmental and safety risks. These risks may include the release of contaminants into the environment, posing a threat to natural habitats and ecosystems, as well as endangering the health and safety of communities and workers living near the affected area. Therefore, it is crucial to take measures to prevent and control bacterial corrosion to ensure safe and sustainable operations in various industries.

10.4.6.4 Mitigation Costs

Unconventional resource facilities such as oil and gas pipelines are often vulnerable to bacterial corrosion. This type of corrosion can cause significant damage to the infrastructure and lead to costly repairs and downtime. To manage this issue, mitigation strategies such as biocide treatments, corrosion-resistant coatings, and monitoring programs can be implemented. These strategies aim to control the growth of bacteria and prevent them from causing corrosion. However, while these strategies are effective, they also come with additional operational and maintenance costs that need to be considered by the facility management.

10.4.6.5 Regulatory Compliance

Bacterial corrosion issues may need to be addressed to comply with environmental and safety regulations, requiring resources and efforts to implement control measures and demonstrate compliance.

To address the potential effects of bacterial corrosion on unconventional resource facilities, it is important for operators to proactively implement corrosion management and monitoring programs. This may involve regular inspection and maintenance, the use of corrosion-resistant materials, and the implementation of effective microbial control strategies to minimize the impact of bacterial corrosion on facility integrity and operations. Furthermore, collaboration with corrosion and microbiology experts can help develop tailored solutions to mitigate the risks posed by bacterial corrosion in unconventional resource facilities.

10.4.7 Mitigation and Control of Bacteria-Related Corrosion

There are several effective methods for preventing and managing microbial corrosion. These methods include the use of biocides, coatings, and cathodic protection systems, as well as the implementation of proper maintenance and inspection procedures. By taking these considerations into account and implementing appropriate measures, facilities can effectively mitigate the risks associated with microbial corrosion and ensure the durability and reliability of their infrastructure.

Corrosion control practices can differ based on the form and impact of corrosion and can depend on factors such as cost, simplicity, operation, and maintenance activities. There are three main corrosion control practices used in thermal desalination plants: environmental control, material selection, and corrosion monitoring. In the following sections, we will describe these practices in detail.

10.4.7.1 Biocide Treatment

Effective control of microorganisms is crucial in preventing the onset of microbial corrosion in water systems, pipelines, and equipment. Biocide treatment programs can be implemented to combat the activity of these corrosive microbes and reduce the risk of damage. Glutaraldehyde, isothiazolinones, and quaternary ammonium compounds, are commonly used biocides and are effective in controlling bacterial growth in water systems. They disrupt microbial activity, preventing the formation of biofilms and reducing the risk of MIC.

By choosing the right biocides and applying them correctly, it is possible to control the growth of microorganisms, minimize the risk of microbial corrosion, and maintain the integrity of water systems, pipelines, and equipment.

10.4.7.2 Monitoring and Control of Water Quality

Continuous monitoring of water quality is essential for preventing microbial corrosion in industrial systems. Microbial counts, chemical composition, temperature, and oxygen levels are among the key environmental factors that need to be closely monitored. Accurate tracking of these parameters can provide valuable insights into the conditions that promote microbial growth and corrosion. By implementing effective control measures, such as maintaining water quality within specified parameters, industries can significantly minimize the risk of microbial corrosion and ensure the longevity of their systems.

10.4.7.3 Cathodic Protection

One of the most widely used and effective methods of actively safeguarding a metal surface against corrosion is by polarizing it cathodically. Cathodic protection systems, such as impressed current and sacrificial anode systems, are specifically designed to help mitigate the impact of microbial corrosion on metallic infrastructure. These systems create a protective electrical field that suppresses corrosion activity, even in the presence of corrosive microbes. By lowering the potential of the surface to such an extent, oxygen reduction becomes the only or primary reaction that occurs on the cathodic surface. This results in an alkaline environment at the protected surface, and a calcareous layer may be formed, which provides an additional layer of protection and lowers the required protecting current even further.

The cathodic protection process can be carried out in two different ways: cathodic protection with sacrificial anodes and impressed cathodic current protection (ICCP). The principle for cathodic protection involves the formation of a galvanic cell, in which the anodic reaction dissolves, for example, aluminum anodes in a controlled way. As long as there are aluminum anodes left, the cathodic polarization is sustained on the metal surface to be protected. This process helps significantly reduce the rate of corrosion on the metal surface and prolongs the lifespan of the infrastructure.

10.4.7.4 Material Selection and Coatings

In order to reduce the likelihood of microbial corrosion on surfaces, it is important to carefully select materials that are corrosion resistant and to apply protective coatings and linings. Corrosion can cause significant damage to equipment and infrastructure over time, leading to costly repairs and replacements (Saji et al., 2020). By choosing the right materials and coatings, you can prevent or minimize this damage. It is also important to maintain coatings properly to ensure their effectiveness in the long term. Proper material selection and coating maintenance are key factors in preventing microbial corrosion and ensuring the longevity of equipment and infrastructure.

10.4.7.5 Pipeline Management

To reduce the risk of microbial corrosion in unconventional resources facilities, it's crucial to pay attention to proper pipeline design and management. This includes taking into account factors such as flow rates, water chemistry, and insulation.

Flow rates should be optimized to prevent stagnant water, which can encourage the growth of microbes that cause corrosion. It's also important to consider water chemistry as high levels of certain minerals and nutrients can promote microbial growth. Proper insulation can help maintain temperatures that are unfavorable to microbes

Implementing pigging and cleaning programs is another effective method of preventing microbial corrosion in pipelines. These programs can help remove microbial biofilms that can accumulate on the walls of pipes, which can lead to corrosion. Regular cleaning can also help prevent the buildup of sediment and other materials that can contribute to corrosion.

10.4.7.6 Microbiological Testing and Analysis

Regular testing and analysis of water samples, biofilms, and corroded surfaces are essential in managing bacterial induced corrosion. These tests can provide valuable insight into the presence and activity of corrosive microorganisms, including the type of microorganism, the extent of the corrosion, and the potential for future damage. Based on this information, targeted mitigation strategies can be developed to manage microbial corrosion and prevent further damage. These strategies may include changes in water chemistry, the use of biocides, and modifications to the design and materials of metal structures. Regular testing and analysis are crucial for maintaining the integrity of water systems and metal structures, preventing damage, and ensuring safe and efficient operations.

10.4.7.7 Regulatory Compliance

Compliance with regulatory standards and guidelines related to water quality, corrosion management, and environmental protection is essential for addressing microbial corrosion in unconventional resources facilities. This may include implementing best practices for microbial control and corrosion prevention to meet industry regulations.

10.4.7.8 Training and Awareness

Providing training and increasing awareness among facility personnel about the potential risks of microbial corrosion, as well as the importance of proper maintenance and mitigation measures, can help foster a proactive and vigilant workplace.

10.4.7.9 Chemical Inhibition

In the oil and gas industry, various bacteria corrosion inhibitors are used. The selection of the appropriate inhibitor depends on the type of system to be protected, whether it is a pumping oil well, a gas-lift well, a gas well, or a water flood system.

Chemical corrosion inhibitors are commonly used to protect metal surfaces from corrosion caused by bacteria. These inhibitors form a protective layer on the metal surface, hindering the corrosive effects of bacteria. Some of the commonly used corrosion inhibitors in the industry include biocides, oxygen scavengers, and film-forming corrosion inhibitors. Biocides are used to kill bacteria while oxygen scavengers remove oxygen from the system to prevent bacterial growth. Film-forming corrosion inhibitors create a protective film on the metal surface, which prevents bacteria from coming into contact with the metal.

It is important to select the right inhibitor for the specific system as each inhibitor has its own benefits and limitations. The choice of inhibitor depends on factors such as the type of bacteria present, the temperature and pressure of the system, and the composition of the fluid being transported. By using the appropriate corrosion inhibitor, the risk of MIC can be significantly reduced, ensuring safe and efficient operations in the oil and gas industry (Figure 10.8).

10.4.7.9.1 Phosphate-Based Inhibitors

Phosphate-based inhibitors have proven to be a reliable solution for controlling corrosion in cooling water systems. These inhibitors work by forming a protective layer on metal surfaces, which prevents corrosion from occurring. In addition, they also

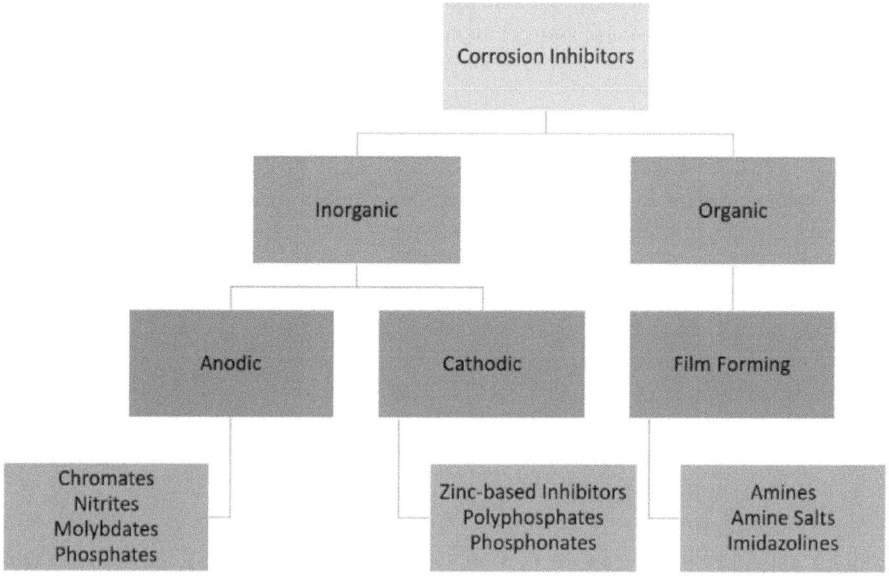

FIGURE 10.8 Types of corrosion inhibitors.

Source: Rasol et al., 2015

play a crucial role in limiting the growth of phosphate-reducing bacteria. These bacteria can thrive in environments where phosphates are abundant and can further contribute to corrosion by producing acidic byproducts. The use of phosphate-based inhibitors helps prevent this from happening by reducing the availability of phosphates, thereby limiting the growth of these harmful bacteria.

10.4.7.9.2 Nitrate-Based Inhibitors

Adding nitrate-based inhibitors to water systems can have a significant impact on the microbial population and the subsequent chemical reactions that occur within the system. Nitrate-based inhibitors create an environment that is conducive to the growth of nitrate-reducing bacteria. These bacteria utilize the nitrate in the water as a source of energy, and, in the process, they reduce the nitrate to nitrogen gas, which is released back into the atmosphere.

One of the significant benefits of the proliferation of nitrate-reducing bacteria is that they outcompete sulfate-reducing bacteria, which are known to produce corrosive sulfides. When sulfate-reducing bacteria are present in a water system, they convert sulfate to hydrogen sulfide, which can lead to corrosion of metal pipes and other infrastructure.

10.4.7.9.3 Molybdate Inhibitors

Molybdate-based inhibitors such as sodium molybdate are widely used to control the growth of sulfate-reducing bacteria. The inhibitors work by interfering with the microbial metabolism of sulfate-reducing bacteria, which is involved in the corrosion process. By inhibiting the growth of these bacteria, the inhibitors help prevent the formation of corrosive byproducts that can cause extensive damage over time.

10.4.7.9.4 Zinc-Based Inhibitors

Zinc-based inhibitors are highly effective in providing cathodic protection to metal surfaces against microbial-induced corrosion. This protection is achieved by using either zinc salts or zinc anodes, which undergo sacrificial corrosion to prevent the underlying metal surfaces from degradation. Zinc acts as a sacrificial anode and corrodes instead of the metal it is protecting, which is why it is referred to as cathodic protection. Zinc salts such as zinc oxide are often used as an anti-corrosion coating while zinc anodes are commonly used in marine environments to protect boats and other structures from corrosion.

10.4.7.9.5 Copper-Based Inhibitors

Copper-based inhibitors are known to possess strong antimicrobial properties owing to their ability to interfere with the metabolic processes of bacterial cells. They can effectively inhibit the growth and proliferation of a variety of bacterial strains, including those that are known to be resistant to conventional antibiotics. Copper ions or copper alloys are often used in industrial settings to control bacterial growth on surfaces and prevent microbial corrosion of oil and gas facilities. The use of copper-based inhibitors has been found to be a cost-effective and environmentally friendly alternative to traditional antimicrobial agents, making them an attractive option for a wide range of applications.

10.4.7.9.6 Polymeric Inhibitors

Polymeric inhibitors have been found to be effective in disrupting the formation of biofilms, which are complex communities of microorganisms that adhere to surfaces and are difficult to remove. These inhibitors, which can be in the form of various polymers and surfactants, work by interfering with the attachment of bacteria to metal surfaces, which, in turn, reduces the rate of corrosion. The use of polymeric inhibitors is a promising approach for preventing the formation of biofilms and protecting metal surfaces from corrosion.

10.4.8 Factors Affecting Inhibitor Selection

When selecting a corrosion inhibitor, there are several factors to consider. These include identifying the problem to be solved; identifying the corrosive species; and evaluating the pressure and temperature, velocity, production composition, water/oil ratio, salinity of water, and acidity of water and oil. It is best to opt for materials that exhibit high resistance to corrosion, especially in the presence of corrosive bacteria. Stainless steel and corrosion-resistant alloys are commonly used. Ensure that the selected materials are compatible with biocides and other chemicals used in corrosion treatment to reduce the risk of adverse reactions.

Select materials that can withstand the temperature and pressure conditions in the oil and gas production system without compromising their structural integrity. Consider the presence of hydrogen sulfide, a common corrosive component in oil and gas environments, and choose materials that resist H_2S corrosion. If dealing with sour service conditions (high H_2S content), use materials specifically designed for such environments, like duplex stainless steel.

10.4.9 Latest Advances, Techniques, and Methods in Control of Bacterial Corrosion

Corrosion caused by bacteria is a persistent problem for the oil and gas industry, leading to significant financial losses and damage. To tackle this issue, various techniques have been developed to mitigate and control bacteria growth in pipelines and storage tanks. These techniques include the use of biocides, corrosion inhibitors, and monitoring systems. Moreover, in recent years, non-chemical solutions such as ultrasonic technology and electrochemical methods have gained popularity. These methods offer a promising alternative by reducing bacteria corrosion while also having a minimal impact on the environment.

In the future, the oil and gas industry may explore advanced microbial monitoring technologies, develop biofilm-resistant materials, and create innovative biocides that have minimal environmental impact. Researchers could also focus on understanding microbial communities within pipelines and optimizing strategies to mitigate corrosion. By doing so, sustainable and effective solutions can be developed for the industry, ensuring a safer and more efficient operation.

REFERENCES

Aberdeen, M. C., Fletcher, P., Rosharon, A. J., King, T. G., Acock, A., Houston, W., Adrichem, W. Van, Zemlak, M., Barrett, S., Davies, I., Diakonov, J., Elphick, Cambridge, E., Burdylo, L., Huldal, I., Jasinski, R., & Jacobsen, S. (2000). *Fighting scale—Removal and prevention.* https://kh.aquaenergyexpo.com/wp-content/uploads/2023/09/Fighting-Scale-Removal-and-Prevention.pdf

Abramov, A. (2019). Optimization of well pad design and drilling—well clustering. *Petroleum Exploration and Development*, *46*(3), 614–620. https://doi.org/10.1016/S1876-3804(19)60041-8

Ahmad, Z. (2006). Principles of corrosion engineering and corrosion control. *Principles of Corrosion Engineering and Corrosion Control*, 1–656. https://doi.org/10.1016/B978-0-7506-5924-6.X5000-4

Al-Gailani, A. (2021). *Scaling of inorganic minerals from potable water in appliances and systems surface adhesion of lime scale in appliances and systems view project modelling of mineral scaling in geothermal power plant view project scaling of inorganic minerals from potable.* www.researchgate.net/publication/368563867

Ali, A., Chiang, Y. W., & Santos, R. M. (2022). X-ray diffraction techniques for mineral characterization: A review for engineers of the fundamentals, applications, and research directions. *Minerals*, *12*(2). https://doi.org/10.3390/min12020205

Amiri, M., Moghadasi, J., & Jamialahmadi, M. (2013). Prediction of iron carbonate scale formation in Iranian oilfields at different mixing ratio of injection water with formation water. *Energy Sources, Part A: Recovery, Utilization, and Environmental Effects*, *35*(13), 1256–1265.

Ariwahjoedi, B., Kakooei, S., & Ismail, C. (2012). Mechanisms of microbiologically influenced corrosion: A review. *World Applied Sciences Journal*, *17*(4), 524–531.

Azizi, J., Reza Shadizadeh, S., Khaksar Manshad, A., Jadidi, N., & Student, M. (2018). Effects of pH and temperature on oilfield scale formation. *Petroleum Engineering Iranian Journal of Oil & Gas Science and Technology*, *7*(3), 18–31. http://ijogst.put.ac.ir

Barton, L., & Fauque, G. D. (2009). Biochemistry, physiology and biotechnology of sulfate-reducing bacteria. *Advances in Applied Microbiology*, *68*, 41–98.

Beech, I. B., Bergel, A., Mollica, A., Flemming, H.-C., Scotto, V., & Sand, W. (2000, September). Simple methods for the investigation of the role of biofilms in corrosion. *Biocorrosion Network*, *BIOCORROSI*, 0–27. https://www.scribd.com/doc/175654787/Mic-Book

Beech, I. B., Cheung, C. W. S., Johnson, D. B., & Smith, J. R. (1996). Comparative studies of bacterial biofilms on steel surfaces using atomic force microscopy and environmental scanning electron microscopy. *Biofouling*, *10*(1–3), 65–77. https://doi.org/10.1080/08927019609386271

Berry, A., Dabrowski, T., & Lyons, K. (2012). The oil spill model OILTRANS and its application to the Celtic Sea. *Marine Pollution Bulletin*, *64*(11), 2489–2501.

BinMerdhah, A. B., Yassin, A. A. M., & Muherei, M. A. (2010). Laboratory and prediction of barium sulfate scaling at high-barium formation water. *Journal of Petroleum Science and Engineering*, *70*(1–2), 79–88. https://doi.org/10.1016/j.petrol.2009.10.001

Brown, J. A. (1991). *Modern manufacturing processes*. Industrial Press Inc.

Budz, J., Karpiński, P. H., & Nuruć, Z. (1985). Effect of temperature on crystallization and dissolution processes in a fluidized bed. *AIChE Journal*, *31*(2), 259–268. https://doi.org/10.1002/aic.690310213

Cahill, J. (2019). *Optimizing well performance in unconventional shale using cloud-enabled multiphase flow meters*. Online Resorces. www.emersonautomationexperts.com/2019/industry/oil-gas/optimizing-well-performance-unconventional-shale-cloud-enabled-multiphase-flow-meters/

Collins, I. R. (2002). A new model for mineral scale adhesion. *Society of petroleum engineers—international symposium on oilfield scale 2002*. https://doi.org/10.2118/74655-ms

Dawson, J. L., John, G., & Oliver, K. (2010). Management of corrosion in the oil and gas industry. *Shreir's Corrosion*, 3230–3269. https://doi.org/10.1016/B978-044452787-5.00168-2

DeMong, K., Longmuir, K., Affleck, B., Elgar, T., & Boulton, K. (2013). The evolution of high density pad design and work flow in shale hydrocarbon developments. *SPE Eastern Regional Meeting*, 25–39. https://doi.org/10.2118/165673-ms

Eslinger, D., Johnson, A. B., Mcdonough, A., & Oettli, M. (2000, April). A hybrid milling/jetting tool-The safe solution to scale milling. *SPE/ICoTA well intervention conference and exhibition* (p. SPE-60700). SPE.

Fernandez de Romero, M. (2005). The mechanism of SRB action in MIC based on sulfide corrosion and iron sulfide corrosion products. *NACE—international corrosion conference series*, April, 1–30.

Frenier, W. W., & Ziauddin, M. (2008). Formation, removal, and inhibition of inorganic scale in the oilfield environment. *Formation, Removal, and Inhibition of Inorganic Scale in the Oilfield Environment*. https://doi.org/10.2118/9781555631406

Graham, G. M., & Frigo, D. M. (2022). Inorganic mineral scale mitigation. *Flow Assurance*, 287–442. https://doi.org/10.1016/B978-0-12-822010-8.00010-6

Griffith, P. (1982). Fouling of heat transfer equipment. *Nuclear Technology*, *58*(1), 121–122. https://doi.org/10.13182/nt82-a32967

Gu, T., Zhao, K. & Nesic, S. (2009). A new mechanistic model for mic based on a biocatalytic cathodic sulfate reduction theory. *Corrosion 2009*, NACE-09390.

Haïdar, T. (2011). To frack or not to frack?: Meeting the environmental challenges of shale gas drilling. *Online Resorces.* www.oilandgasiq.com/drilling-and-development/articles/meeting-the-environmental-challenges-of-shale-gas

Hajirezaie, S., Wu, X., & Peters, C. A. (2017). Scale formation in porous media and its impact on reservoir performance during water flooding. *JNGSE, 39*, 188–202. https://doi.org/10.1016/J.JNGSE.2017.01.019

Hamid, S., De Jesús, O., Jacinto, C., Izetti, R., Pinto, H., Droguett, E., Edwards, C., Cassidy, J., Zhang, H., Dagenais, P., & Batocchio, M.A. (2016). A practical method of predicting calcium carbonate scale formation in well completions. *SPE Production & Operations, 31*(01), 1–11. https://doi.org/10.2118/168087-PA

Hamid, S., De Jesus, O., Jacinto, C. M., Izetti, R., Pinto, H., Droguett, E., Edwards, C., Cassidy, J., Zhang, H., Dagenais, P., & Batocchio, M. (2013). A practical method of predicting chemical scale formation in well completions. *Society of Petroleum Engineers—SPE Asia Pacific Oil and Gas Conference and Exhibition, APOGCE 2013: Maximising the Mature, Elevating the Young, 2*, 1481–1500.

Hernandez, G., Kucera, V., Thierry, D., Pedersen, A., & Hermansson, M. (1994). Corrosion inhibition of steel by bacteria. *Corrosion, 50*(8), 603–608. https://doi.org/10.5006/1.3293532

Heydrich, M., Hammami, A., Choudhary, S., Mockel, M., & Ratulowski, J. (2019). Impact of a novel coating on inorganic scale deposit growth and adhesion. *Proceedings of the Annual Offshore Technology Conference, 2019.* https://doi.org/10.4043/29218-ms

Hoang, T. A. (2015). Mechanisms of scale formation and inhibition. *Mineral Scales and Deposits: Scientific and Technological Approaches*, 47–83. https://doi.org/10.1016/B978-0-444-63228-9.00003-6

Hosny, R., Desouky, S. E. M., Ramzi, M., Abdel-Moghny, T., El-Dars, F. M. S., & Farag, A. B. (2009). Estimation of the scale deposits near wellbore via software in the presence of inhibitors. *Journal of Dispersion Science and Technology, 30*(2), 204–212. https://doi.org/10.1080/01932690802498658

Iverson, W. P. (1968). Corrosion of iron and formation of iron phosphide by Desulfovibrio desulfuricans. *Nature, 217*(5135), 1265–1267. https://doi.org/10.1038/2171265a0

Jamaluddin, D. A., & Massinai, M. A. (2018). X-ray fluorescence (XRF) to identify chemical analysis of minerals in Buton island, SE Sulawesi, Indonesia. *IOP Conference Series: Earth and Environmental Science, 118*(1). https://doi.org/10.1088/1755-1315/118/1/012070

Javaherdashti, R. (2008). *Microbiologically influenced corrosion: An engineering insight.* Springer.

Jordan, M. M., Williams, H., Linares-Samaniego, S., & Frigo, D. M. (2014, May). New insights on the impact of high temperature conditions (176 C) on carbonate and sulphate scale dissolver performance. *SPE international oilfield scale conference and exhibition* (p. D011S004R002). SPE.

Kakooei, S., Ismail, M. C., & Ariwahjoedi, B. (2012). Mechanisms of microbiologically influenced corrosion: A review. *World Applied Sciences Journal, 17*(4), 524–531.

Kamal, M. S., Hussein, I., Mahmoud, M., Sultan, A. S., & Saad, M. A. S. (2018). Oilfield scale formation and chemical removal: A review. *Journal of Petroleum Science and Engineering, 171*, 127–139. https://doi.org/10.1016/j.petrol.2018.07.037

Kashkovskiy, R., Strelnikova, K., & Fedotova, A. (2019). Application of electrochemical impedance spectroscopy to study hydrogen sulphide corrosion of steel and its inhibition: A review. *Corrosion Engineering Science and Technology, 54*(6), 493–515. https://doi.org/10.1080/1478422X.2019.1619979

Kelland, M. A. (2009). Production chemicals for the oil and gas industry. *Production Chemicals for the Oil and Gas Industry*, 1–439. https://doi.org/10.1201/9781420092974

Khan, M. A. A., Hussain, M., & Djavanroodi, F. (2021). Microbiologically influenced corrosion in oil and gas industries: A review. *International Journal of Corrosion and Scale Inhibition*, *10*(1), 80–106. https://doi.org/10.17675/2305-6894-2021-10-1-5

Knisz, J., Eckert, R., Gieg, L. M., Koerdt, A., Lee, J. S., Silva, E. R., Skovhus, T. L., An Stepec, B. A., & Wade, S. A. (2023). Microbiologically influenced corrosion—more than just microorganisms. *FEMS Microbiology Reviews*, *47*(5). https://doi.org/10.1093/femsre/fuad041

Kuhr, C. A. H. von W., & Vlugt, L. S. van der. (1964). The graphitization of cast iron as an electro-biochemical process in anaerobic soils. *Water*, *18*(16), 147–165.

Lewandowski, Z., & Beyenal, H. (2008). *Mechanisms of microbially influenced corrosion*. https://doi.org/10.1007/7142_2008_8

Li, J., Tang, M., Ye, Z., Chen, L., & Zhou, Y. (2017). Scale formation and control in oil and gas fields: A review. *Journal of Dispersion Science and Technology*, *38*(5), 661–670. https://doi.org/10.1080/01932691.2016.1185953

Ling, H., Montoya, J., Hung, L., & Aykol, M. (2022). Solving inorganic crystal structures from X-ray powder diffraction using a generative first-principles framework. *Computational Materials Science*, *214*. https://doi.org/10.1016/j.commatsci.2022.111687

Little, B. J., Hinks, J., & Blackwood, D. J. (2020). Microbially influenced corrosion: Towards an interdisciplinary perspective on mechanisms. *International Biodeterioration and Biodegradation*, *154*. https://doi.org/10.1016/j.ibiod.2020.105062

Little, B. J., Ray, R. I., & Pope, R. K. (2000). Relationship between corrosion and the biological sulfur cycle: A review. *Corrosion*, *56*(4), 433–443. https://doi.org/10.5006/1.3280548

Lloyd, J. R. (2003). Microbial reduction of metals and radionuclides. *FEMS Microbiology Reviews*, *27*(2–3), 411–425. https://doi.org/10.1016/S0168-6445(03)00044-5

López-Comino, J. A., Cesca, S., Jarosławski, J., Montcoudiol, N., Heimann, S., Dahm, T., Lasocki, S., Gunning, A., Capuano, P., & Ellsworth, W. L. (2018). Induced seismicity response of hydraulic fracturing: Results of a multidisciplinary monitoring at the Wysin site, Poland. *Scientific Reports*, *8*(1). https://doi.org/10.1038/s41598-018-26970-9

Louhi-Kultanen, M., & Ulrich, J. (2010). Industrial crystallization. *Chemical Engineering and Technology*, *33*(5), 718. https://doi.org/10.1002/ceat.201090019

Lv, M., & Du, M. (2018). A review: Microbiologically influenced corrosion and the effect of cathodic polarization on typical bacteria. *Reviews in Environmental Science and Biotechnology*, *17*(3), 431–446. https://doi.org/10.1007/s11157-018-9473-2

Macadam, J., & Parsons, S. A. (2004). Calcium carbonate scale control, effect of material and inhibitors. *Water Science and Technology*, *49*(2), 153–159. https://doi.org/10.2166/wst.2004.0112

Marriott, R. A., Pirzadeh, P., Marrugo-Hernandez, J. J., & Raval, S. (2016). Hydrogen sulfide formation in oil and gas. *Canadian Journal of Chemistry*, *94*(4), 406–413. https://doi.org/10.1139/cjc-2015-0425

Merdhah, A. B. (2007). The study of scale formation in oil reservoir during water injection at high-barium and high-salinity formation water. *Marine Science & Tchnology Seminar*, 1–189. http://scholar.google.com/scholar?hl=en&btnG=Search&q=intitle:The+study+of+scale+formation+in+oil+reservoir+during+water+injection+at+high-barium+and+high-salinity+formation+water#3

Merdhah, A., & Mohd. Yassin, A. (2008). Study of scale formation due to incompatible water. *Jurnal Teknologi*, *49*, 9–26.

Montagna, A. F., Cafaro, D. C., Grossmann, I. E., Ozen, O., Shao, Y., Zhang, T., Guo, Y., Wu, X. H., & Furman, K. C. (2023). Surface facility optimization for combined shale oil and gas development strategies. *Optimization and Engineering*, *24*(4), 2321–2355. https://doi.org/10.1007/s11081-022-09775-8

Muthukumar, N., Rajasekar, A., Ponmariappan, S., Mohanan, S., Maruthamuthu, S., Muralidharan, S., Subramanian, P., Palaniswamy, N., & Raghavan, M. (2003). Microbiologically influenced corrosion in petroleum product pipelines—A review. *Indian Journal of Experimental Biology*, *41*(9), 1012–1022.

Obuekwe, C. O., Westlake, D. W. S., Plambeck, J. A., & Cook, F. D. (1981). Corrosion of mild steel in cultures of ferric iron reducing bacterium isolated from crude oil—2. Mechanism of anodic depolarization. *Corrosion*, *37*(11), 632–637. https://doi.org/10.5006/1.3577551

Ohen, H. A., Williams, L. E., Lynn, J. D., & Ali, L. (2004). Assessment and diagnosis of inorganic-scaling potential using near-infrared technology for effective treatment. *SPE Production and Facilities*, *19*(4), 245–252. https://doi.org/10.2118/86474-PA

Pritchard, A. M. (1988). Fouling science and technology. *Fouling Science and Technology*. https://doi.org/10.1007/978-94-009-2813-8

Rasol, R. M., Noor, N. M., & Din, M. M. (2015). Effect of temperature in SRB growth for oil and gas pipeline. *Indian Journal of Science and Technology*, *8*(13). https://doi.org/10.17485/ijst/2015/v8i13/70635

Rawat, J., Sharma, N., & Khandelwal, A. (2014). Microbiological causes of corrosion. *Petroleum Technology Quarterly*, 1–7.

Ruden, L. A. (2017). *Multiphase metering advances in onshore fields*. Hart Energy. www.hartenergy.com/exclusives/multiphase-metering-advances-onshore-fields-176740

Saji, V. S., Meroufel, A. A., & Sorour, A. A. (2020). Corrosion and fouling control in desalination industry. *Corrosion and Fouling Control in Desalination Industry*, 1–406. https://doi.org/10.1007/978-3-030-34284-5

Scale | Energy Glossary. (2023). https://glossary.slb.com/en/terms/s/scale

Schnieders, J. H. (2005). Sulfate-reducing bacteria. *Water Well Journal*, *59*(11), 68–70.

Singh, V. K., Singh, A. L., Singh, R., & Kumar, A. (2018). Iron oxidizing bacteria: Insights on diversity, mechanism of iron oxidation and role in management of metal pollution. *Environmental Sustainability*, *1*(3), 221–231. https://doi.org/10.1007/s42398-018-0024-0

Stott, J. F. D. (2010). Corrosion in microbial environments. *Shreir's Corrosion*, 1169–1190. https://doi.org/10.1016/B978-044452787-5.00048-2

Sullivan, G. P., Pugh, R., Melendez, A. P., & Hunt, W. D. (2004). Metering for operations and maintenance. In *O&M best practices guide 2.0* (pp. 1–9). Pacific Northwest National Laboratory (PNNL), US Department of Energy.

Tan, H., Skinner, W., & Addai-Mensah, J. (2013). pH-mediated interfacial chemistry and particle interactions in aqueous chlorite dispersions. *Chemical Engineering Research and Design*, *91*(3), 448–456.

Vazirian, M. M., Charpentier, T. V. J., de Oliveira Penna, M., & Neville, A. (2016). Surface inorganic scale formation in oil and gas industry: As adhesion and deposition processes. *Journal of Petroleum Science and Engineering*, *137*, 22–32. https://doi.org/10.1016/j.petrol.2015.11.005

Videla, H. A. (2002). Prevention and control of biocorrosion. *International Biodeterioration and Biodegradation*, *49*(4), 259–270. https://doi.org/10.1016/S0964-8305(02)00053-7

WebCorr. (2014). Different types of corrosion—crevice corroson. *Online Resorces*, 1–3. www.corrosionclinic.com/types_of_corrosion/Corrosion Fatigue.htm

Yeboah, Y. D., Samuah, S. K., & Saeed, M. R. (1993). *Prediction of carbonate and sulfate scales in oilfields.* https://doi.org/10.2118/25586-ms

Zhao, J., Wang, M., Lababidi, H. M. S., Al-Adwani, H., & Gleason, K. K. (2018). A review of heterogeneous nucleation of calcium carbonate and control strategies for scale formation in multi-stage flash (MSF) desalination plants. *Desalination*, *442*, 75–88. https://doi.org/10.1016/j.desal.2018.05.008

11 Data Analytics, Machine Learning, and Artificial Intelligence in Unconventional Resources

Minou Rabiei, Kalyan Venugopal, Karthik Balaji, Cilia Abdelhamid, and Abdeldjalil Latrach

11.1 DATA SITUATION IN OIL AND GAS INDUSTRY

The data situation in the industry has all the characteristics of what we call big data with the famous 5V characteristics (Figure 11.1). For decades, companies have been collecting large volumes of structured and unstructured static and dynamic data from a variety of sources; in various formats like numeric, text, image, audio, video, or time series; and in different scales and different frequencies, including geology, petrophysics, and geophysics data from high-frequency drilling and production measurements to daily written operations reports. Companies have been using high-speed digital technologies such as seismic acquisitions devices and fluid front monitoring geophones and logging-while-drilling (LWD) and measurement-while-drilling (MWD) tools to transmit real-time data to the surface; optical fiber and advanced sensors are now being used in well tubulars to record different parameters like fluid pressure, temperature, and composition during production. There are also different types of hardware configuration data, maintenance data, and other event-related data that come from the field. There are also massive amounts of business data, such as internal financial and economic studies, and external competitors' news, such as mergers, acquisitions, leases, and major capital investments.

The wealth of data generated in the industry today can be a valuable source of important facts and information if they are effectively utilized using sophisticated data-driven analytics. These technics can support and/or substitute traditional analytical approaches by revealing the full potential of the data with the goal of improving performance and the increasing competitive edge for companies. By rapidly analyzing the incoming technical and business data and applying that information to complex models in real time, companies can generate strategic insights that help increase their operational performance while preventing problems.

DOI: 10.1201/9781003319955-11

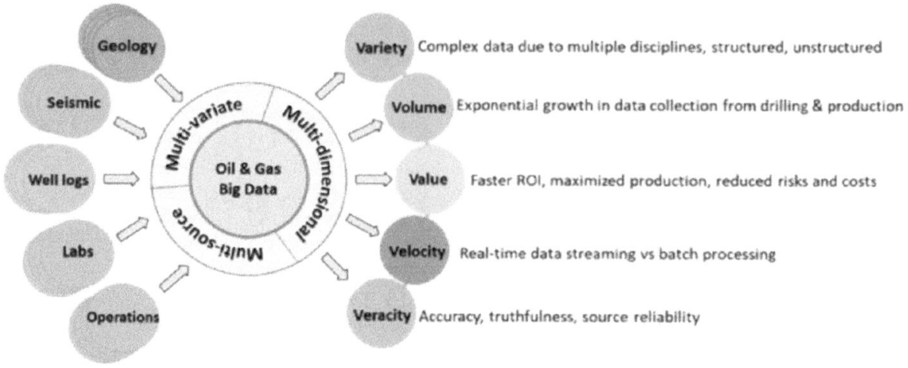

FIGURE 11.1 Characteristics of data in the oil and gas industry

11.1.1 What Is Data?

Data can be identified as a collection of data objects and their attributes. An attribute (also known as column, variable, field, dimension, or feature) is a property of an object. Examples in the upstream oil and gas industry could include rock lithology, wellhead temperature and pressure, etc. A collection of attributes describe an object (also known as row, record, sample, for instance). Table 11.1 displays an example of a dataset from the upstream oil and gas industry. The data has ten attributes (dimensionality of ten) and a sample of ten objects.

- FieldName is the name of the oil/gas field (qualitative/nominal attribute).
- DiscYear is the year in which the field was discovered (interval attribute).
- WaterDepthCat indicates the category (shallow, deep, ultra-deep) of the field based on the water depth (qualitative/ordinal attribute).
- The remaining attributes are quantitative (numerical, continuous).
- FieldGOR: Field gas-oil ratio in standard cubic feet (SCF)/stock tank barrels (STB)
- OrigOil: Initial oil in place in MMbbl
- CumProdOil: Cumulative oil produced from the field in MMbbl
- RemOil: Remaining oil available to be produced from the field in MMbbl
- OrigGas: Initial gas in place in bcf
- CumProdGas: Cumulative gas produced from the field in bcf
- RemOil: Remaining gas available to be produced from the field in bcf

11.1.2 Data Types and Examples

Structured data: Structured data that is tabular in nature is preferred in data analytics. The most common type of data model is a relational data model based on the mathematical concept of a relation that is represented as a table that contains rows

TABLE 11.1
Oil and Gas Reserves 2018, Gulf of Mexico

FieldName	Disc Year	Field GOR	Orig Oil	Orig Gas	CumProdOil	CumProdGas	RemOil	RemGas	WaterDepthCat
MC807	1989	1,332	1,741.9	2,320	1,426.6	1,828	315.3	492	DEEP
EI330	1971	4,086	469.3	1,917.2	459.2	1,896.5	10.1	20.7	SHALLOW
WD030	1949	1,653	596.7	988.5	592	975.5	4.7	13	SHALLOW
GI043	1956	4,359	402.7	1,744.5	380.8	1,656.1	21.9	88.4	SHALLOW
TS000	1958	79,938	46.2	3,692.3	45.5	3,647.1	0.7	45.2	SHALLOW
BM002	1949	1,052	548.6	577.5	545.5	575.1	3.1	2.4	SHALLOW
GC640	2002	670	555.6	372.5	379	248.2	176.6	124.3	DEEP
VR014	1956	65,255	47.9	3126.5	47.9	3,126.5	0	0	SHALLOW
GC743	1998	816	502.9	410.4	360.2	239.7	142.7	170.7	ULTRA-DEEP
MP041	1956	5,658	276.5	1,564.2	272.7	1,558.9	3.8	5.3	SHALLOW

Source: www.data.bsee.gov/Main/FieldReserves.aspx

(tuples) and columns (attributes). In a relation, each row (tuple) is uniquely identified by an attribute or a set of attributes known as relational keys. A great strength of the relational model is this simple logical structure backed by a sound theoretical foundation (Connolly, 2015). Table 11.1 is an example of structured data or a relation where the FieldName attribute is the relational key.

Unstructured data: Data that does not conform to a data model such as a relational data model is known as unstructured data. Chat conversations, emails, meeting notes, and text fields in daily rig reports are examples of unstructured data. Given the significant volume of unstructured data in the form of archived reports, the upstream oil and gas industry has been exploiting techniques such as machine learning–based optical character recognition (OCR) and text mining to extract both tabular data and text data for data mining purposes to extract valuable knowledge.

Sequential data: Data is termed sequential when the objects in the data are dependent on the other objects in the dataset. Time series is a common example of this, with each point reflecting an observation at a certain point in time. Other examples of sequential data are sensor data, genetic sequence data, meteorological data, video data, audio data, and text data. Well log data is an example of sequential data from the upstream oil and gas industry. Figure 11.2 shows examples of time series data from the oil and gas industry.

Geospatial data: Geospatial data is any type of data that contains a geographic component. This includes any data with a location component that is considered as point data (e.g., a location represented by latitude and longitude). Line data represents data that is inherently imbued with length (e.g., roads and rivers), and polygon data is used to represent an area using perimeters and boundaries (e.g., forests, cities, counties, states, and lakes) (4earthintelligence, 2021). Oil and gas wells and disposal wells are examples of geospatial point data. Oil and gas lease boundaries are examples of polygon geospatial data. Figure 11.3 shows an example of geospatial data from the upstream oil and gas industry.

11.2 KNOWLEDGE DISCOVERY THROUGH APPLICATION OF DATA-DRIVEN ANALYTICS

Data-driven analytics is a set of tools and techniques that provide the means for extracting knowledge and insights from data by integrating fragments of heterogeneous yet related information, recognizing the hidden patterns in it and quantifying the associated uncertainties. Before discussing how these techniques are applied in the oil and gas industry, it is important to understand the common terminology used in this field and the differences in meaning. Data analytics entails the use of quantitative data for effective decision-making. Data-driven models are often built using data alone and require an understanding of dependent and independent variables (Sankaran et al., 2020). Some of the advantages of data-driven models are:

- A precise understanding of the physical processes at play is not required.
- Insights into the processes are derived directly from the data by analyzing patterns.

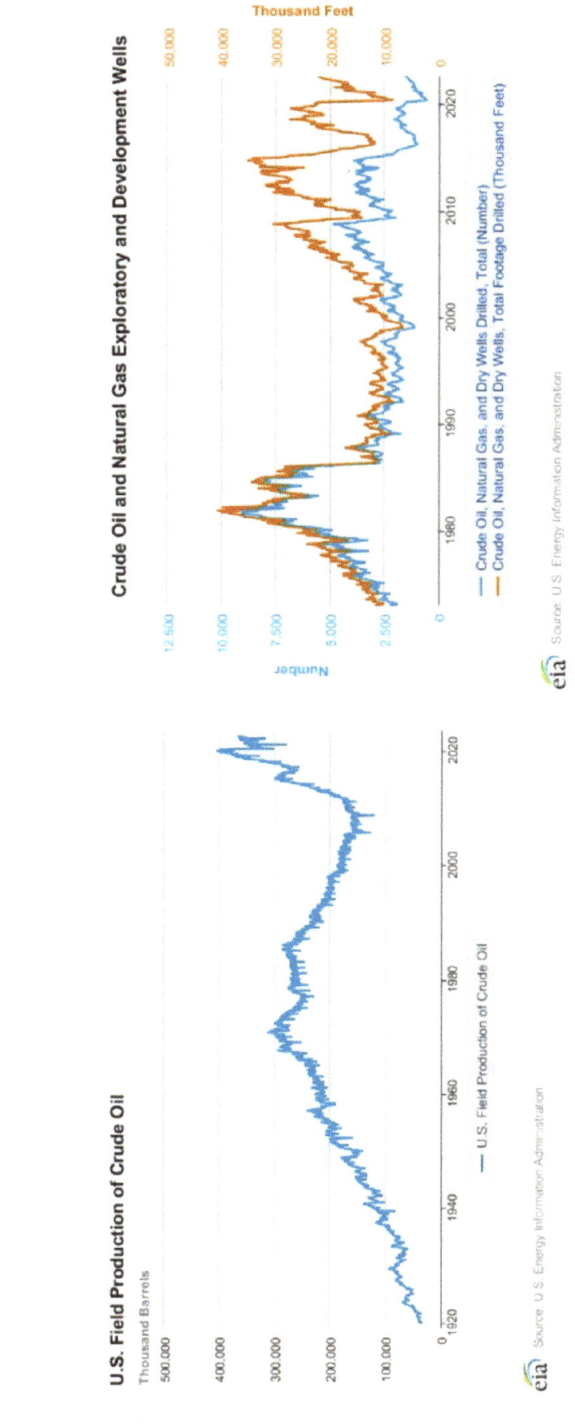

FIGURE 11.2 Time series data examples from the upstream oil and gas industry.

Source: US EIA

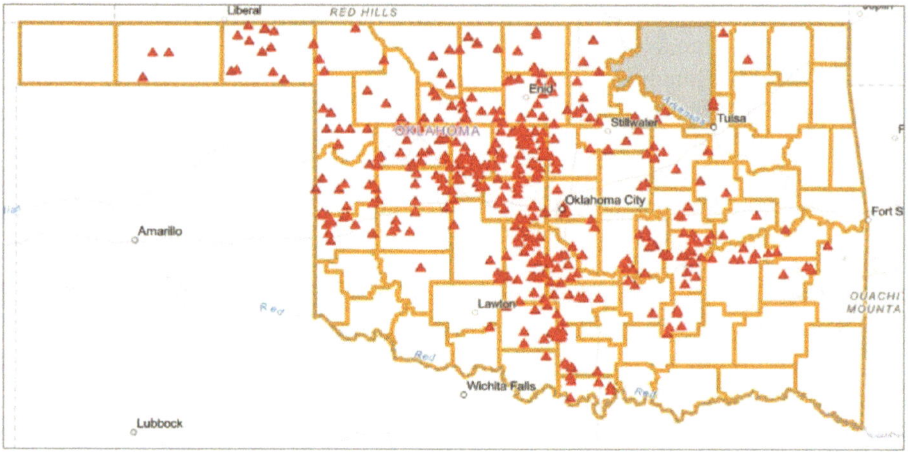

FIGURE 11.3 Oklahoma State Public Land Survey System (PLSS) boundaries (orange lines) and commercial underground injection control (UIC) wells (red solid triangle points).

Source: https://oklahoma.gov/occ/divisions/oil-gas/database-search-imaged-documents/occ-well-data-finder.html)

- Correlation is not identical to causality, but it can often lead to better understanding of causality.
- Recognizing recurring patterns can target an area for investigation and help understand the cause-and-effect process.
- Data-driven analytics are often limited to patterns seen in the historical training data and tend to perform poorly when extrapolated to new operating states. This limitation is usually addressed by retraining models over new patterns.
- The speed of data-driven models under favorable conditions is often very attractive, and they can add substantial business value if they offer a sound and timely decision.

Data mining is a key component in data-driven approaches, using advanced analytics techniques that go beyond counts, descriptive techniques, and simple rule-based methods. The term "data mining" encompasses a variety of data processing, manipulation, and analysis techniques as well as intelligent technologies applied to decision-making, such as artificial intelligence, machine learning, and deep learning. There seems to be a lot of confusion about the definition of these technologies, and while the meaning of each term is different, many people use these terms interchangeably.

Artificial intelligence is a broader field in computer science that describes computer applications that mimic the capabilities of the human mind and motor skills, such as learning, moving, communication, recognizing objects, understanding and responding to language, making decisions, and solving problems.

Machine learning is a subfield in AI that focuses on analyzing and learning from the data to make predictions without the need for explicit programming by humans. Humans can only analyze and understand the relationships between variables in a low dimensional dataset containing two or three variables. Machine learning uses algorithms that can change and improve over time by analyzing, processing, and learning from very large quantities of data with hundreds of variables and are able to make informed decisions based on the learned insights.

Deep learning is a young subfield of machine learning that uses a multi-layered structure of algorithms inspired by the structure and function of the brain called neural networks to solve tasks that traditional ML models are not able to solve, even when there is a lot of data. The network of neurons in this type of algorithm mimics the structure and functions of the human brain. The more complex (deeper) this network, the better the learning capabilities of the algorithm. Traditional ML techniques generally require a preprocessing step, called feature extraction, which provides an abstract presentation of the original raw data. An optimal feature extraction task requires detailed knowledge of the problem domain and is usually performed through several iterations of adapting and refining the given variables. Deep learning, on the other hand, does not require a separate preprocessing step as this task is already embedded in the learning process of neural networks.

The performance of most machine learning algorithms plateaus after a while no matter how many more data points are added to the dataset, but with deep learning algorithms, the performance of the model increases more and more as more data is added (Figure 11.4).

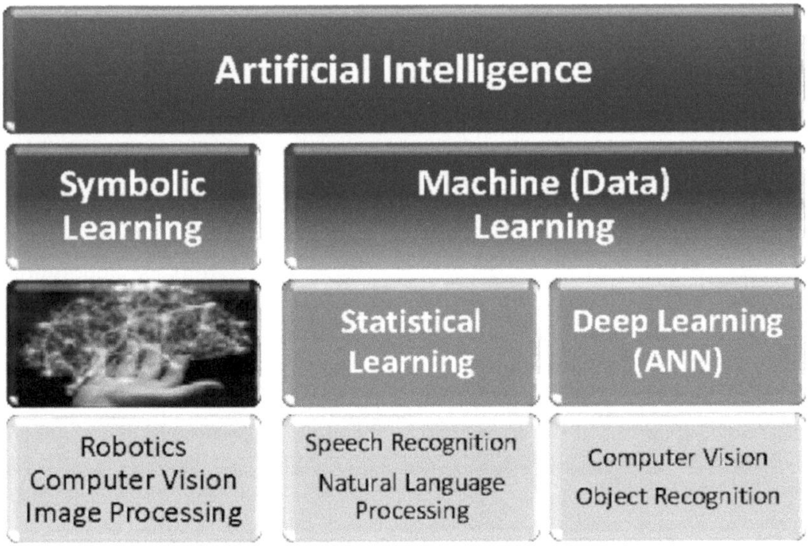

FIGURE 11.4 The relationship between AI, ML, and DL.

11.2.1 Cross-Industry Standard Process for Data Mining (CRISP-DM) Pipeline

Data mining stands at the confluence of the fields of statistics and machine learning without the constraints of lack of sufficient data or computing power (Shmueli, 2018). Figure 11.5 shows a visual representation of data mining.

The usual tasks of a data mining project are grouped together to form the data mining pipeline (Juodyte, 2017), which has inputs and a number of processing steps chained together in some way to produce an output. In the pipeline, each step logically follows the previous step, providing an outcome. The cross-industry standard process for data mining (CRISP-DM) pipeline shown in Figure 11.6 consists of six steps: business understanding, data understanding, data preparation, modeling, model evaluation, and model deployment.

An overview of the six CRISP-DM steps as explained by Juodyte (2017) is provided next:

Business understanding is the initial phase that focuses on understanding the problem domain, the relevant prior knowledge, and the requirements from a business perspective, then converting this information into a data mining problem definition, data mining goals, preliminary plan to achieve the objectives, and success criteria. Neglecting this step would likely result in a lot of wasted time answering the wrong questions.

Data understanding is the step in the data mining pipeline that starts with initial data collection and proceeds with activities that enable the project team to become

FIGURE 11.5 Data mining for data-driven decisions.

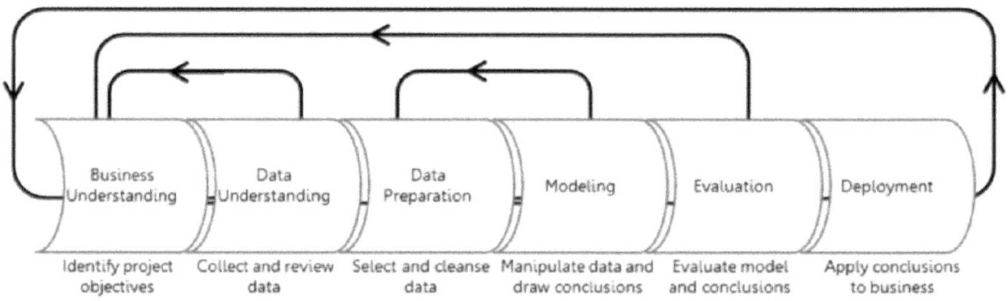

FIGURE 11.6 Cross-industry standard process for data mining (CRISP-DM).

Source: Juodyte, 2017

familiar with the data. If the data is available on a website, such data can be collected through web scraping. Data description using statistical techniques, exploratory data analysis (EDA), and data quality verification are key tasks in this step. EDA helps the project team gain a good understanding of the data, discover initial insights, identify potential data quality and inconsistency issues, and lay the foundation for the data cleaning and preparation activities. Exploratory data analysis can help identify patterns and trends hidden in the data (norm) and deviations from the patterns (exceptions). Using visual analytics to carry out EDA helps overcome the cognitive limitation faced when dealing with numbers and spreadsheets.

Data preparation is the phase that covers all activities needed to construct the final dataset that will serve as input to the modeling process from the initial raw data. The data mining pipeline (Figure 11.7) shows that the iterative nature of data mining and data preparation may be performed multiple times. Attribute selection and sample selection are key components of the data preparation task. Dimensionality reduction, addressing redundancy in attributes, data cleaning, and transformation are also key steps in the data preparation phase.

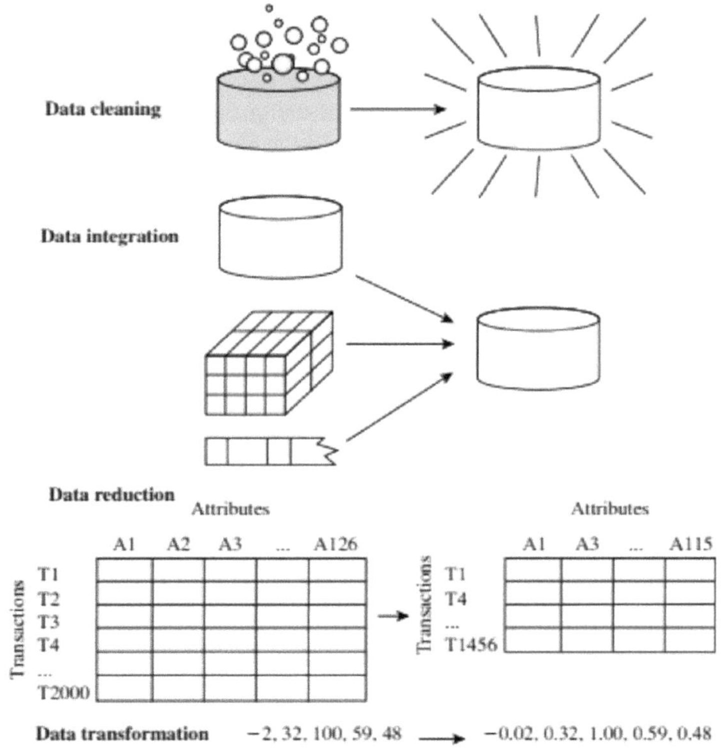

FIGURE 11.7 Data preprocessing steps.

Source: Han, 2012

According to Han (2012), there are several data preparation or preprocessing techniques. Data cleaning can be applied to remove noise and correct inconsistencies in data. Data transformations (e.g., normalization) may be applied; data are scaled to fall within a smaller range like 0.0 to 1.0. This can improve the accuracy and efficiency of mining algorithms involving distance measurements. Some ways to achieve data reduction are aggregation, elimination of redundant features, clustering, and dimensionality reduction. Principal component analysis (PCA) is a method of dimensionality reduction to reduce the number of random variables or attributes under consideration. Suppose that the data to be reduced consist of objects or rows described by n attributes or dimensions. PCA searches for k n-dimensional orthogonal vectors that can best be used to represent the data, where $k \leq n$. The original data are thus projected onto a much smaller space, resulting in dimensionality reduction. PCA often reveals relationships that were not previously suspected and thereby allows interpretations that would not ordinarily result. Attribute subset selection, which retains a subset of the initial attributes, is another method of dimensionality reduction in which irrelevant, weakly relevant, or redundant attributes or dimensions are detected and removed. This usually requires domain knowledge and a good understanding of the nature of the attributes. These techniques are not mutually exclusive, and they may work together. For example, data cleaning can involve transformations to correct wrong data, such as by transforming all entries for a date field to a common format. Figure 11.5 provides a summary of the various data preprocessing steps.

Data enhancement is the process of increasing the value of the initially available dataset by appending additional value-added knowledge. Enhancement links multiple datasets to pool information from multiple sources, with the intention of identifying actionable knowledge from the combined data. Data enhancement has the potential of increasing the quality of the final dataset and the predictive power of machine learning models trained using the enhanced final dataset (Loshin, 2012). Combining data from multiple sources to create the final dataset is known as data integration. Relational data can be combined using merge and join operations. Spatial data can be combined using spatial join operation, based on proximity of the features calculated using a distance measure. Distance between any two features is calculated as the shortest separation between them.

In data modeling phase, the prepared data is fed into a data-driven model to extract useful knowledge from the data. Discovering knowledge from data can be in the form of descriptive, diagnostic, predictive, and prescriptive analysis. Descriptive analysis is about identifying the underlying patterns or structures in the historical data and providing retrospective insight and understanding about the current status of the problem at hand. This insight will help in answering the "what," "where," and "when" questions about the matter under study: e.g., what issues happened during drilling operations, where they happened, and when. During diagnostic analyses, hypotheses are formulated based on the current or historical perceptions, theories, and behaviors. Diagnostic analysis helps in answering the "why" and "how" questions by identifying, analyzing, and validating the causal relationships between parameters under study. Once a deep understanding of the cause/effect relations within the parameters is cultivated, inferences will be derived from the identified

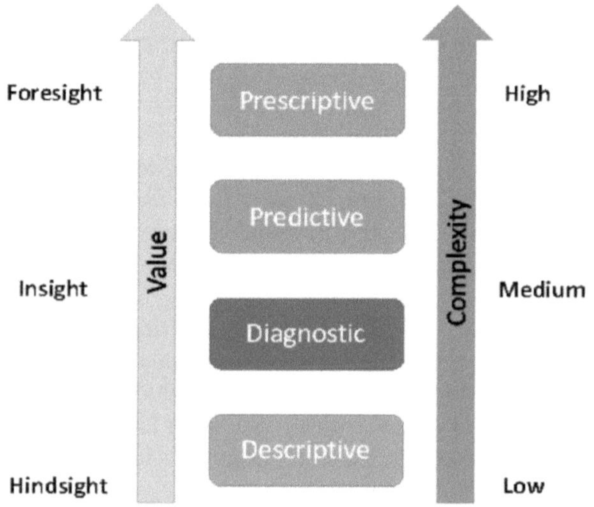

FIGURE 11.8 Data-driven model value versus complexity.

patterns. This means that the descriptive and diagnostic metrics identified in the previous steps will be used for more informed decision-making through predictive and prescriptive analyses. During predictive analysis, the established patterns will be used to answer the "what," "where," and "when" questions with regards to the potential future outcomes. The most advanced form of analysis is called prescriptive analysis, which will provide solutions on how the future status of a system can be enhanced. The knowledge gained through different types of analyses will yield recommendations on the best course of action to prevent a potential problem, deal with an issue, and/or enhance operation efficiency.

Therefore, the model is selected based on the purpose and the anticipated value identified during the business understanding phase and other factors, such as the size of the dataset available for training the model, speed of training, and interpretability of the results.

The learning process in data-driven models is generally done in three ways very similar to the strategies humans use for learning:

- Supervised learning: In this type of learning, a dataset consisting of pairs of input variables known as predictors or independent variable and an associated target/outcome variable known as dependent variable are used, from which the model can identify relationships between input and output variables. Once the model is trained, new inputs can be fed into the model, and the algorithm can predict an output. The learning process is controlled by what the user provides to the model, and therefore, it is very important to have balanced, reliable, and unbiased examples to the models to avoid bias. Supervised learning requires a lot of data to be able to learn the relationships, and obtaining sufficient reliably labeled data can be challenging and costly.

- Unsupervised learning: For unsupervised learning, only input data is provided to the model with no labeled example output to aim for. The algorithm conducts an iterative process, analyzing data without human intervention to identify patterns, similarities, and associations in order to group similar data points into clusters.
- Reinforcement learning: A different type of model training based on behavioral learning. The learning process is done through trial and error, and the algorithm receives positive or negative feedback, depending on how it approaches a predefined specific endpoint or goal.

The general tasks that ML models can do are classification, regression, and clustering. Classification comes under supervised learning and is the process of making categorical predictions using data. The model is required to assign the previously unseen observations to a class as accurately as possible, using the given input parameters. The regression task can help in predicting numerical values based on a given set of numerical and categorical data. The clustering task under unsupervised learning groups related items together into a number of groups such that data points are more similar to other data points in the same group and dissimilar to the data points in other groups.

In the following section a practical and real-world case study on how to source the data for a research project and steps taken in the preprocessing phase to prepare the data for the analysis is presented.

11.2.2 Case Study: A Data-Driven Proxy Model to Predict/Optimize the Performance of Planned Saltwater Disposal Wells in North Dakota

Data is the backbone of any data-driven modeling project. The case study in this section demonstrates the data understanding and data preparation steps of the data mining pipeline. The underground injection control (UIC) program, created by the US Environmental Protection Agency (EPA) under the Safe Drinking Water Act (SDWA) regulates disposal of produced water in injection wells, including permitting, construction, operation, and closure. Saltwater disposal (SWD) is a method of reinjecting produced water back into the subsurface for disposal purposes, and SWD wells are part of the UIC Class-II wells.

The objective of the research project was to explore the application of data mining techniques to gain insights into SWD well performance in the state of North Dakota (ND) and design a data-driven proxy model to predict/optimize the performance of planned SWD wells.

While significant time and effort were spent on data collection and preparation, the data understanding and data preparation steps were revisited many times during the project execution, during the exploratory data analysis activities, and also as part of model tuning and evaluation activities, as highlighted by the iterative nature of the CRISP-DM in Figure 11.2. Domain knowledge and expertise and the industry experience of the researcher and the advisor were critical in identifying the various data sources and selecting datasets and features that were relevant for the data driven modeling task.

TABLE 11.2
Datasets from Primary Data Source

Dataset	Type
1. Well Index	Flat File (Excel/CSV)
2. Log Tops	Flat File (Excel/CSV)
3. Monthly Disposal Data for SWD Wells	HTML
4. Well Scout Ticket	HTML

Primary data source: The oil and gas division of the North Dakota Industrial Commission (NDICOG) (www.dmr.nd.gov/dmr/oilgas) regulates the drilling and production of oil and gas in North Dakota and was the primary data source for the project. The following datasets were collected from NDICOG using the University of North Dakota (UND) premium subscription. Table 11.2 lists the various datasets used in this project and their types.

1. Well index file: This file contained a total of 39,455 objects at the time of download in October 2020. Each object represented a well drilled in the state of ND for various purposes, such as oil production, gas production, saltwater disposal, etc. The dataset contained a total of 27 attributes. Data size was reduced to 905 SWD wells for further analysis.
2. Log tops file: This file downloaded from the NDICOG website contained formation tops information for various formations penetrated by each well in the well index dataset. This dataset contained 31,201 samples and a total of 72 attributes. Each sample represented a well drilled in the state of ND that had log data. These two data were relational tables, and the primary key to combine them were the well API, a unique identifier of each well in the dataset. Each well also had a FileNumber attribute, which was another unique identifier.
3. Monthly disposal data for SWD wells: For each of the 905 SWD wells, the monthly disposal volumes and average injection pressures were available as time series data in HTML format. Web scraping was performed using the Python BeautifulSoup library to extract the monthly injection data for each of the wells for all the available periods. For the requirement of the project analysis, the time series data was aggregated so that there was one value of barrels of SWD disposed per well and one value of average injection pressure in pounds per square inch.

   After reducing the number of useful attributes (predictors) to six quantitative attributes (average monthly injection pressure, well's total depth, formation top depth of the disposal reservoir, well location (latitude and longitude), and well's surface elevation) and one qualitative attribute (well orientation) utilizing the researcher's domain knowledge, data cleaning, and data normalization, a base multi-linear regression model was trained to understand the relationship between the predictors and the target variable, the average monthly barrels of saltwater disposed. The goodness of fit value of 31.7% indicated that a relationship existed between the predictors and target and encouraged the researcher to explore data enhancement possibilities.

4. Well scout ticket data scraping: For each of the wells drilled in the state, NDICOG maintains well scout ticket data that contains additional essential information about the well from the time it is permitted through drilling, completion, and production. The file number for each well (primary key for data integration), casing size, casing depth, and top and bottom perforation depths were some of the additional features extracted from each SWD well's scout ticket through web scraping.

Secondary data sources: Leveraging his domain experience in the oil and gas industry, the researcher examined other secondary data sources listed in Table 11.3.

1. North Dakota Inyan Kara Isopach maps: From the various datasets extracted from the primary data source, one key subsurface feature available was the formation top depth of the Inyan Kara (K-IK) formation, the target disposal reservoir. This important secondary data source contained valuable thickness information of the K-IK formation across ND, where oil and gas wells, including disposal wells, were drilled as polygon (contour) spatial data in shape files format. Considering the SWD well point spatial data and the K-IK formation thickness polygon data and utilizing spatial join operation, the K-IK thickness feature was extracted for each of the SWD well available from the primary data source.
2. North Dakota roads: One of the critical factors in finalizing the location of an oil well, gas well, injection well, or disposal well in the state of ND or the entire US is the proximity of the well site to established roads in the state to minimize transportation-related hurdles and additional road construction costs. Hence, the researcher explored this secondary data source relating to the infrastructure of all the roads in the state of ND. The spatial join operation was used to identify the closest road to each of the SWD wells in the primary data, extract the distance of that road in meters from the well head location along with the road type, and integrate them as two additional features for the final dataset.
3. North Dakota oilfield waste disposal sites: Produced water (saltwater) needs to be treated prior to recycling or disposal to remove any residual

TABLE 11.3
Datasets from Secondary Data Sources

Dataset	Type	Source
1. North Dakota Inyan Kara Isopach Maps	Shapefile	North Dakota Geological Survey (Inyan Kara Maps)
2. North Dakota Roads	Shapefile	North Dakota GIS Hub
3. North Dakota Oilfield Waste Disposal Sites	Shapefile	North Dakota GIS Hub

hydrocarbons and other solids. This motivated the researcher to look for data relating to oilfield waste disposal sites that would treat produced water and prepare it for disposal. The spatial join operation was used to identify the closest disposal site to each of the SWD wells in the primary data, extract the distance of that site in meters, and integrate it as an additional feature for the final dataset. Primary data was enhanced with five additional features from the three secondary data sources, which improved the multi-linear regression model performance goodness of fit by 10%. This model performance was further enhanced by another 10% using the random forest ensemble-based algorithm. There is room to further improve the model performance through data enhancement by considering data relating to subsurface properties of the K-IK formation and other related information. This case study demonstrates the role domain knowledge plays in identifying relevant data sources for effective data enhancement and integration of various data types that were earlier introduced.

11.3 HOW IS DATA-DRIVEN ANALYTICS CHANGING THE OIL AND GAS INDUSTRY?

As was mentioned before, a hydrocarbon reservoir is a complicated, heterogeneous, multivariate, and multi-dimensional system of rocks and multiphase fluids with numerous dynamic and nonlinear interactions within. We also talked about the characteristics of data that is generated in exploration and production activities on a daily basis. The more high-quality data that is available, the higher the quality of the decisions will be since they will be made with less risk and uncertainty. Companies are increasingly seeking ways to find the most productive zones in a basin, increase performance, decrease costs, and reduce risks. Recent developments in computational intelligence, especially in the field of machine learning, have greatly expanded the capabilities in data analysis and modeling throughout a reservoir's life cycle. These models, coupled with cloud computing capabilities, can now use the real-time data generated from sensors, integrate it with legacy data to discover previously undetected patterns in the data, and construct predictive or descriptive models with good generalization abilities. They can align business intelligence with operational intelligence by identifying system components, their interactions, and the underlying causal relationships between them. These new intelligent capabilities can provide a comprehensive understanding of the reservoir heterogeneity, increasingly creating leverage and significant business value for companies by helping them gain more insights into the data for operation efficiency and improvements in risk analysis in core operations. Figure 11.8 displays some of the potential areas in which advanced data analytic techniques can improve the efficiency in unconventional operations.

In the following sections, some applications of data-driven technologies in various disciplines in the oil and gas industry are discussed, and practical examples from the literature will be provided.

11.4 OPTIMIZING EXPLORATION AND FIELD DEVELOPMENT

The process of oil and gas exploration is very expensive and exhaustive. Oil and gas companies are increasingly moving towards deeper and more geologically complex areas to find promising hydrocarbon deposits. Geophysical surveys are critical to understanding the subsurface environment and identifying suitable petroleum systems. Exploring such complex reservoirs requires a more in-depth understanding of indicative patterns or signatures in seismic attributes to identify hydrocarbon traps. The traditional approach in geophysical data interpretation has been a manual and tedious procedure of picking horizons on a geophysical section and then assessing the geological significance of the results to specify the depositional environment of sedimentary rock and locate a potential reservoir. Recent advancements in AI and ML technologies have enabled automation of this procedure, significantly accelerating the interpretation workflow, enabling the experts to focus on assessing the geological and geophysical significance of the selected horizons. Data-driven intelligent approaches have substantially increased the success rate of identifying the presence of hydrocarbon sources without the need for geologists and geophysicists to have a full understanding of the rock and fluid mechanisms or variations of reservoir properties in a field. Machine learning techniques have also been used in remote sensing and satellite imagery by identifying the random associations between different segments of remotely sensed images. By integrating data from various sources, including geological, seismographic, and magneto-graphic surveys, more robust spatio-temporal models of the subsurface are created, enabling more successful identification of potential hydrocarbon traps (Figure 11.9).

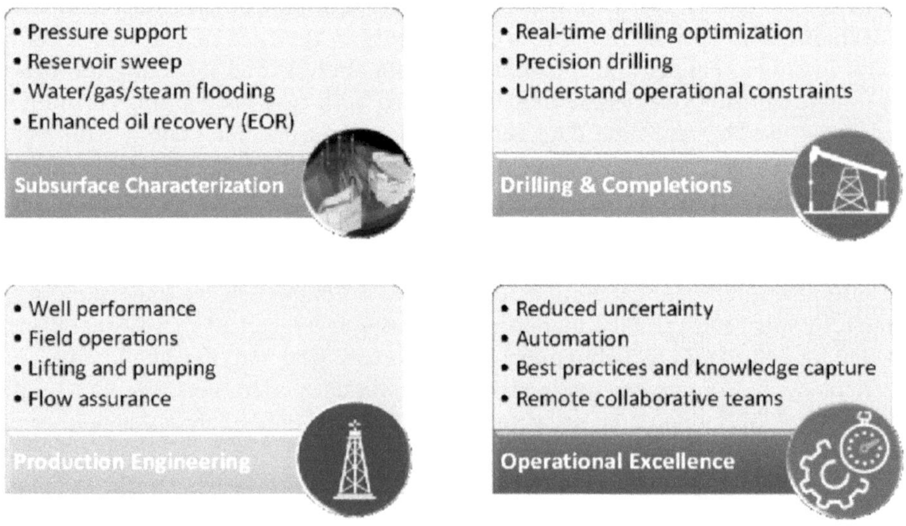

FIGURE 11.9 Improving operational efficiency in unconventional E and P.

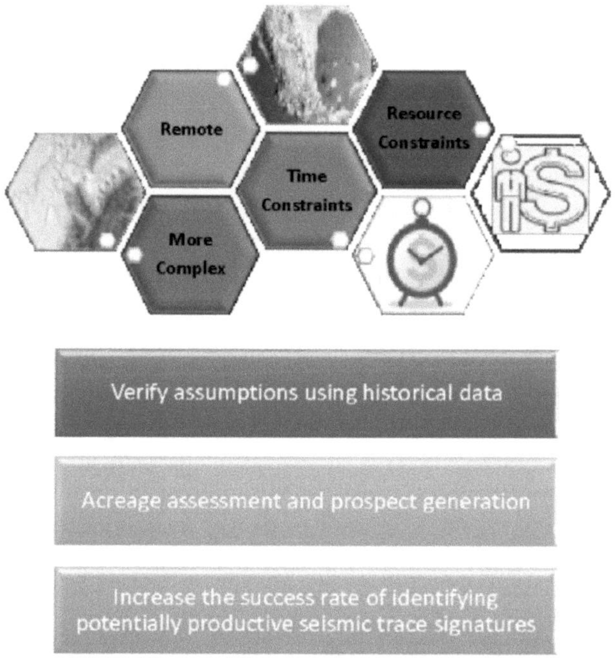

FIGURE 11.10 Example applications of data analytics in unconventional exploration activities.

11.5 AI AND ML IN SEISMIC DATA ANALYTICS

Seismic structural interpretation relies heavily on the domain expertise and experience of interpreters as well as visual cues of geologic structures, such as texture and geometry, to identify geologic structures of interest, such as faults or components of petroleum systems in general.

11.5.1 Seismic Data Acquisition

In general, cost and data quality are traded off when acquiring seismic data. Reducing the acquisition effort in the field and then addressing flaws in the recorded data through subsequent processes is one strategy to resolve these competing objectives. Compressive sensing has received a lot of attention in this field over the past several years (Herrmann, 2010; Mosher et al., 2014). This technique seeks to restore acceptable data quality from the collected data using effective sampling plans that may not always adhere to the Nyquist sampling theorem.

The practice of blended acquisition, also known as simultaneous source acquisition, is used extensively to improve data quality while preserving or even lowering project costs and duration (Abma et al., 2012; Beasley, 2008; Berkhout, 2008; Bouska, 2010; Nakayama et al., 2015). In most cases, these acquisition techniques are followed by a subsequent data recovery process like deblending and data reconstruction. Typically, these processes are presented as an inverse problem. A common strategy is to utilize seismic data's low-rank structure or sparse representation in some transform domain. It is commonly known that low frequencies are important for seismic data in a number of ways, including improved full-waveform inversion convergence, reduced side lobes of the wavelet, and deep source energy penetration (Kroode et al., 2013). However, low-frequency component emission is usually problematic in the field. For instance, it necessitates huge and specialized devices and additional shooting efforts (Dellinger et al., 2016). As a result, low-frequency extrapolation, a processing technique that creates missing low frequencies, is more useful than simply gathering them in the field. Numerous experiments have been done to extrapolate low frequencies: for example, by applying the recorded signal's envelope or convolving a broadband wavelet to an approximated reflectivity model (Wu et al., 2014; Zhang et al., 2017). Finding an underlying connection between the recorded and missing frequencies is still a very challenging undertaking.

Recent studies on synthetic data using ML to increase bandwidth have produced promising results, particularly for low-frequency extrapolation (Siahkoohi et al., 2019; Sun & Demanet, 2020). Considerable research has been conducted with the goal of improving current geophysics-based data recovery techniques by designing irregularities in collection parameters, such as spatial sampling and blending schemes (Mueller et al., 2015; Mosher et al., 2014). Nakayama & Blacquière (2021) investigated an ML method that sought to recover as much data as possible from recorded seismic data using a blended, sparsely sampled, and narrowband method. In terms of the spatial sampling of detectors and sources, Siahkoohi et al. (2019) used ML to solve a problem involving data reconstruction and showed that irregularly sampled data outperformed regularly sampled data. Baardman & Hegge (2020) introduced new deblending and blended quality control (QC) algorithms. They showed that the use of CNNs results in two hands-off algorithms without the need of a priori information. They also proved that the deblending scheme can be paired with data augmentation and transfer learning techniques for potentially increased performance.

ML plays an important role in data acquisition geometry design too. The goal of acquisition design is to identify the most cost-effective geometry that provides solutions to subsurface geology–related problems based on a number of objectives. Blacquière and Nakayama (2019) presented a related study, in which they used a genetic algorithm and convolutional neural network to model the optimum dataset from the parameters of an acquisition geometry, including the quantity and placement of sources and detectors, as well as the blending and dispersed source array (DSA) parameters. This design methodology led to better results than randomized acquisition geometries while satisfying the objectives of the seismic survey in terms of quality and economic factors.

11.5.1.1 Seismic Data Processing

Seismic data processing involves the compilation, organization, and conversion of wave signals into a visual map of the areas below the surface of the Earth. During seismic data processing, the raw data is transferred in a form that is optimal for extracting the signals. The seismic processing parameters selected when collecting seismic data and interpreting images determine how accurate an interpreted subsurface model will be. Seismic imaging is a technique used for subsurface structural interpretation and rock property inversion. Surface sources and overlapping receivers illuminate the subsurface structures from different angles and azimuths, which allows them to acquire a seismic survey. Reverse time migration is a shot-based migration algorithm that images each individual shot to illuminate their closeness under the sensors in the imaging aperture. Multiple single-shot images interfere during the stacking, creating a subsurface image, which results in a lower amplitude of incoherent swing noise. The resulting imaging is afflicted by random noise, coherent swing artifacts, residual multiples, and illumination irregularities (Zhang & Sun, 2009).

High-density acquisition, uniform illumination, and accurate velocity modeling are necessary to achieve the highest signal-to-noise ratio of the subsurface when stacking multiple single-shot images. The conditions mentioned here are hardly met in the deeper part of the image or in subsalt, where the signal is reduced due to very low illumination. In this situation, stacking does not suppress the noise, which leads to a poor primary signal with migration artifacts and coherent noise. Neural network can be applied directly on post-stack images, which makes it a powerful tool in different applications in post-migration seismic processing (Wang & Nealon, 2019). For the best output, some parametrization is required to adjust to different frequencies and grid sizes. The neural network improves poor reflections that are buried in strong noise, which makes the resulting image easier to interpret.

In these applications, training data are of high importance as the network learns the mapping from noisy and weak structures to cleaner events with more balanced amplitude. The trained network reduces migration artifacts such as conflicting dips and random and swing noise while preserving detailed geologic structures such as faults, diffractors, and salt flanks that may resemble the shape of the swing noise. As an example, deep neural network (DNN) has been used to improve weak primary reflections and reduce the noise, resulting in a favorable output for structural interpretation. Liu et al. (2020) proposed a deep learning approach to address absorption and dispersion issues, accelerating high-resolution imaging. In this work, selected features from the high-resolution results from conventional PSTM and de-absorption QPSTM were used to train the neural network, resulting in faster convergence and accelerated high-resolution imaging. In a similar approach, Chen et al. (2017) used a convolutional neural network (CNN) in the reconstruction of low-dose CT images by extracting certain features from the data to train the neural network, leading to an optimal output to nonlinear tasks.

Seismic data are usually noisy, which can lead to poor signal-to-noise ratio (SNR) and low resolution of seismic data, as well as untrusted seismic inversion results. Several signal-processing techniques, such as wavelet transform (Wu & Liu, 2008), f-x prediction filtering (Harris & White, 1997), and signal vector decomposition (Bekara & van der Baan, 2007), have been used to suppress random noise in seismic data. The traditional techniques commonly use the spatial coherence of seismic data to improve SNR at the expense of horizontal resolution, which attenuates tilt and bending events.

Principal component analysis (PCA), a statistical method commonly used for data reduction, can considerably improve the quality of the resultant 2D maps or

3D volumes by decreasing the noise and redundancy of the data, especially in noisy locations. PCA detects the principal or main elongation directions in a multi-attribute cross-plot and allows for an N-dimensional rotation to modify the input space (where the principal axes are orthogonal to one another). Dropping the dimensions with the lowest variability results in data reduction. Deep neural networks (DNN) have showed great success in image signal processing, particularly in 3D or higher-dimension seismic data denoising. Other examples of the use of ML techniques in noise suppression include convolutional neural networks (Jain & Seung, 2008), multi-layer perceptrons (Burger et al., 2012; Zheng et al., 2014), and a nonlinear reaction diffusion model (Chen & Pock, 2016). It is important to note that the deep learning techniques mentioned here can achieve remarkable results if the noise has Gaussian distribution; however, if the noise is non-Gaussian, the performance of these algorithms deteriorates.

The choice of training samples used for training neural networks for seismic denoising is critical. In the absence of clean data, the most advanced non-ML method available can be used to suppress the noise in a subset of seismic data to generate the desired samples. There are also some methods available for obtaining high-SNR training samples. For example, high-density exploration technology is used in some local areas, allowing for the acquisition of high-SNR seismic data. Alternatively, high-SNR data may be obtained from smaller regions with favorable surface conditions. In general, the underground structures in a large survey region are similar, which means that the structure of useful seismic imaging signals as well as the noise in that region are similar. Deep learning can use the data from a small high-SNR region to extract noise features and then apply these features to suppress noise in other regions. Results from different studies in seismic data denoising have shown that in some cases where conventional methods struggle to remove noise, a denoising convolutional neural network (3D-DnCnn) can successfully remove most of the noise while retaining the desired signals (Liu et al., 2018).

11.5.1.2 Seismic Data Analysis and Interpretation

Seismic attribute analysis is the process of extracting or deriving a quantity from seismic data that can help in identifying details that may be otherwise hidden in a conventional seismic image, resulting in a better geological or geophysical interpretation of the data. Seismic attribute calculations are used in conventional seismic interpretation techniques to determine geometries, lithologies, and reservoir properties. Seismic attributes are a response to the physical properties of rocks. As a result, whether used individually or in combination, the spatial distribution of and relationships between these responses, along with the geological context, help create a more reliable seismic interpretation.

The size of seismic data collected for oil exploration has dramatically increased, making structural interpretation more time consuming and labor intensive. Attribute extraction and structure identification are the two basic phases in structural interpretation. Many researchers have investigated the use of machine learning methods to develop automated or semi-automated procedures in order to interpret the geologic structures of subsurface.

The depth and speed of advancement in computer vision and image processing have significantly improved automated structural interpretations. Researchers in this field are now able to use machine learning approaches to interpret seismic data as images rather than signal traces. These machine learning models give repeatable and reliable result for seismic interpretation, eliminating the need for human interpreters to read massive amounts of data and comprehend the relationships between several types of data at once. Based on signal and image processing methods, attributes that capture the essential elements of seismic data are either explicitly defined or implicitly specified. Recently, algorithms that simulate the human visual system, like saliency detection (Shafiq et al., 2018) and color space analysis (Wang et al., 2014a), have been suggested to replicate the interpretation process by removing the features of seismic data that are most perceptually representative.

Nowadays, advanced seismic interpretations make use of complex machine learning algorithms that allow for faster and more accurate seismic data analysis than traditional single attribute studies performed by interpreters. Self-organizing maps (SOM) (Kohonen, 1989), geostatistics, and neural networks are among the methods used for enhancing the ability of interpreters to recognize patterns beyond what traditional techniques could provide (Strecker & Uden, 2002). SOM is a form of unsupervised neural network used for clustering, which has been used for seismic data classification using a 1D latent space (Coléou et al., 2012), computing spectral attributes in combination with PCA technique (Guo et al., 2009), and interpreting facies in various depositional environments (Marca & Bedle, 2022). Other studies have demonstrated the benefit of extending the SOM latent space to 2D and 3D by delineating elements in deep-water depositional environments with 2D and 3D color bars (de Matos et al., 2006; Roy & Marfurt, 2010; Roy, 2013).

The use of sequence stratigraphy in seismic interpretation has proven useful in basin analysis. It provides a framework for understanding stratigraphic evolution and is critical in predicting reservoir, seal, and source rock spatial distribution (Van Hoek et al., 2010). The goal of seismic stratigraphy interpretation is to show the subsurface geologic sequences using three-dimensional (3D) seismic surveys. Accurate stratigraphy delineation has numerous applications in robust reservoir exploration. Traditional procedures usually incorporate horizon tracking and geology modeling. In the horizon tracking step, a seismic interpreter selects a set of seismic reflectors, each of which represents one of the important geologic boundaries. The geologic modeling step provides a mathematical solution for stacking the selected seismic horizons in 3D space and reconstructing the sequence model by following the fundamental geologic rules. Because the underlying structures are complicated, the horizons in a 3D seismic dataset are frequently of low amplitude, a low signal-to-noise ratio, and/or a low degree of lateral continuity. The standard horizon tracker is very susceptible to two types of difficulties when done on such a low-quality horizon (Di et al., 2018). The recent success of machine learning in image processing, particularly convolutional neural networks (CNN), has brought new insights and tools for seismic interpreters. Enormous efforts have been made in efficiently utilizing deep CNNs for resolving typical problems in seismic pattern recognition and interpretation, such as saltbody delineation and fault detection (Di et al., 2020). CNN has been successfully used for estimating the relative geologic time (Geng et al., 2019) and investigating the feasibility of seismic horizon racking from few training images (Peters et al., 2019). CNNs have also been used in the delineation of subsurface saltbodies. The CNN constructs the mapping relationship between the seismic signals and the saltbodies, using the original seismic amplitude rather than manually selected seismic attributes. This method requires less interpretation time and avoids the use of less efficient seismic attribute to the greatest extent possible. More importantly, the CNN identifies, learns, and classifies saltbodies based on local seismic reflection patterns, allowing seismic sounds and processing artifacts from diverse patterns to be successfully identified and removed (Di et al., 2020).

However, there are two significant restrictions associated with these approaches. First, building a competent interpretation machine requires a large amount of training data, which is either not readily available in this domain or needs considerable effort from an expert seismic interpreter. Otherwise, overfitting might occur, which limits the CNN's performance to only the training data. Second, and more importantly, applying CNN typically requires the seismic sections used for training the algorithm to be fully labeled, which is not always possible in seismic data due to not only the intensive labor for full-sectional seismic annotation but also the interpretational uncertainties present in subsurface complexities (Di et al., 2020).

Generally, post-stack seismic data are of low quality due to seismic imaging algorithms and velocity models, which restricts the effectiveness of interpretation techniques. Researchers have suggested doing interpretation tasks on pre-stack seismic data, which are the input of time or depth migration and have less noise than migrated ones. According to experimental findings, fault detection using pre-stack seismic data offers a fresh and promising approach to structural interpretation and has a high computational efficiency (Wang et al., 2018). To identify and locate faults in pre-stacked seismic data, Zhang et al. (2014) introduced kernel-regularized least-squares fitting. The kernel ridge regression model and the Nyström approach were used by Lin et al., (2017) to extract faults from pre-stack seismic observations. Araya-Polo et al. (2017) suggested utilizing the Wasserstein loss function to train a deep neural network to learn the mapping relationship between pre-stack data and fault spaces (Wang et al., 2018). Xiong et al. (2018), used a convolutional neural network to develop a method to automatically detect and map fault zones using 3D seismic images.

Seismic investigation aims to map the inherent reservoir heterogeneities and develop a reservoir model with relevant attributes. The reflectivity of the underlying rocks affects the seismic response. Differences in the amplitude of the seismic response are typically interpreted as changes in the depositional environment, lithology, and fluid content. As an alternative, seismic facies analysis can be used to examine the nature of seismic data (Wrona et al., 2018). Seismic facies classification can be used for reservoir characterization since it produces images that contribute to the interpretation of geological settings and supplement data from windowed amplitude maps.

Facies classification is mainly used to extract geological information from seismic data, understand lateral and vertical distribution of seismic facies, and predict rock type distribution away from well control. Classification of lithofacies and their accurate representation in a 3D cellular geologic model is critical in understanding a field because permeability and fluid saturations for a given porosity and height above free water vary considerably among lithofacies (Dubois et al., 2003). The easiest way to learn about lithofacies is to obtain core samples of reservoir rock from wells, but this is rarely done because it is expensive. Since there are fewer cores available than there are wells in the field, there is a need for a method to determine lithofacies in wells without cores. Core lithofacies can be extrapolated from cored wells (training wells) to uncored wells by comparing the physical rock parameters determined by wireline logs (Dubois et al., 2007).

A typical machine learning workflow based on CRISP-DM is provided in a study by Abu-Mostafa (2012) for seismic facies classification. The main goal in this workflow is to "teach" a model how to recognize seismic facies in a seismic section, as depicted in the schematic shown in Figure 11.11, and the different steps are explained as follows:

Feature extraction: The goal of feature extraction is to learn as much as possible about the object under investigation. In order to achieve this, the so-called features or characteristics that characterize the object under examination are extracted. The features in this instance are statistical qualities of seismic attributes inside a moving window, and the object is a seismic section. Seismic attributes are appropriate features since they have been expressly created to highlight specific aspects of seismic reflection data (Chopra & Marfurt, 2007; Randen & Sønneland, 2005).

Training: The goal of training is to minimize the in-sample error or the discrepancy between the predicted and actual outcome. The model in this study is a multiclass classification model in which the output consists of four distinct classes of seismic facies. A model that is extremely complicated and overfitted as a result of the use of many features can create random mistakes or noise in the data. The model is regularized during training wherever it is feasible to prevent overfitting. The reduction of in-sample error typically occurs during training in machine learning.

Cross-validation: Typically, the available data is divided into two groups, one for training the model and one for testing the performance of the model. Ideally, one would want to maximize the number of instances in each group, first, to produce well-constrained models and second, to get accurate estimations of the model performance. However, the amount of available data in the real world is usually limited. Cross-validation is a technique commonly used for assessing the performance of ML models when there is limited data available. In this technique, data is randomly divided into *n* equal sections or folds. The model is then trained on *n-1* folds and assessed on the remaining *n*th group of data. This procedure is repeated *n* times, and the average performance of the *n*-trained models from the original data is considered to assess how well the model performs. The performance of each model is evaluated by generating a confusion matrix, which is a display of the actual classes against the predicted classes and calculating the model goodness metrics, such as average precision, recall, and the *f1* score criteria. Precision refers to a classifier's ability to predict classes correctly. Recall, also known as sensitivity, refers to a classifier's ability to locate all instances of a given class. The *f1* score is an equally weighted harmonic mean of precision and recall.

Over the past few decades, a variety of multivariate statistical approaches have been applied to address the problem of facies classification from wireline log curve measurements. Wolf and Pelissier-Combescure (1982) used PCA and cluster analysis in an automated process to determine facies while Busch et al. (1987) and Delfiner et al. (1987) applied discriminant function analysis to estimate facies. Baldwin et al. (1990) and Rogers et al. (1992) stepped away from the classical multivariate statistical methods and used non-parametric feed forward-back propagating neural networks for rock classification. Since then, neural networks have been utilized in an increasing number of facies classification problems (Kapur et al., 1998; Russell et al., 2002). Saggaf and Nebrija (2003) summarize the use of fuzzy mathematical techniques in geologic classification problems.

Several mathematical techniques have also been used for facies prediction from geophysical data such as well logs or seismic surveys. Traditionally, knowledgeable interpreters visually examined seismic waveforms, frequency, amplitude, phase, and geometric configurations to designate seismic facies on a variety of 2D lines. A seismic facies map would then be created by manually contouring the facies that had been displayed on the map. The advent of 3D seismic data and volumetric features has increased the quantitative and automated nature of such analysis (Zhao et al., 2015). The classification of facies using well log and seismic data inside the reservoir model is the goal of applying a classification approach. Conventional classification techniques classify facies in each location separately, which means the facies' value predicted at one place is unrelated to the facies' values estimated at a nearby location. Figure 11.11 depicts a general classification workflow for seismic facies interpretation.

Only the consistency of the observed data determines the continuity of the anticipated model. However, facies models must be spatially correlated to mimic the stratigraphy and the depositional processes (Lindberg & Grana, 2015). Similarly,

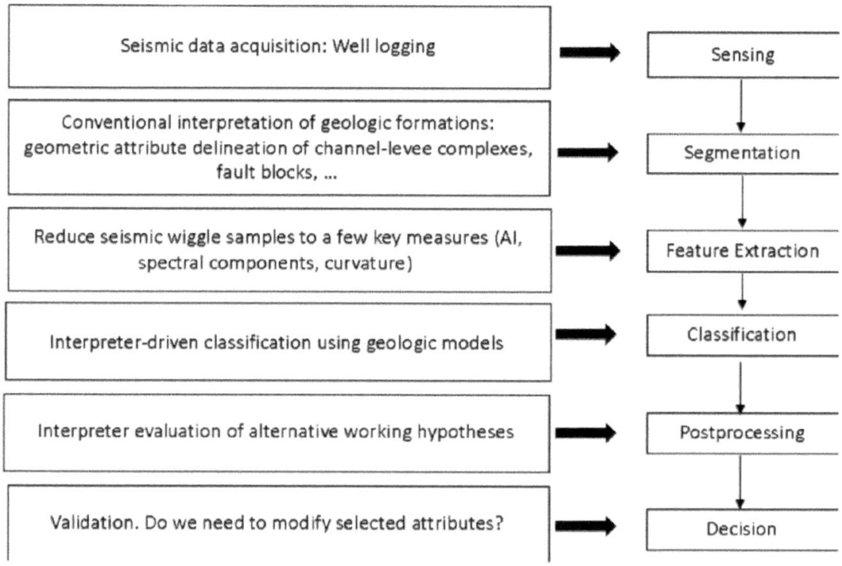

FIGURE 11.11 Classification workflow for seismic facies interpretation.

Source: Zhao et al., 2015

gravity governs the fluid effect inside facies; therefore, spatially correlated models must be imposed to maintain the geological ordering in facies sequences. Pattern recognition and clustering algorithms are the most often used techniques for facies classification (Hastie et al., 2009; Martinez & Martinez, 2001). Using these techniques, a collection of measurements is grouped into a number of clusters that correspond to the various facies. The idea is to categorize the measurements so there is a higher degree of similarity between measurements from the same facies than between measurements from different facies. The measurements typically include computed properties derived from the data as well as elastic and/or petrophysical parameters. Other examples of the use of data-driven approaches in facies classification are the Bayesian method (Doyen, 2007), statistical facies classifications based on maximum-likelihood standards; support vector machine algorithms (Grana et al., 2020); discriminant analysis techniques; and spatial statistical models, using Markov chain Monte Carlo (MCMC) approaches (Lindberg & Grana, 2015). Bayesian classification and neural networks have been used in a variety of depositional contexts, including the Marcellus gas shale reservoir and the marine turbiditic system. The effectiveness of these techniques in comparison to discriminant analysis has been demonstrated in several studies (Fan & Poh, 2007). Some of these techniques were initially created for the classification of well logs and have since been expanded to include inverted seismic data or similar seismic attributes (Mukerji et al., 2001; Avseth et al., 2010; Dubois et al., 2007).

Despite recent progress, there are still some issues remaining with facies classification, and research is still being conducted on the development of machine learning approaches, particularly those based on deep learning techniques for facies classification applications and geological modeling (Grana et al., 2020). Due to the non-uniqueness of the solution to the classification and inversion problem, assessing the uncertainty of the estimated model is particularly difficult. Furthermore, the absence of clear spatial statistical models makes it challenging to maintain the geological significance of the spatial distribution of the facies, such as the vertical transitions imitating the geological deposition. In comparison to standard pointwise classification, facies classification with underlying spatial statistics models incurs higher computational costs and necessitates making assumptions about the parameters of the statistical models, such as the variogram length or the prior transition probabilities (Grana et al., 2020). Therefore, a significant number of statistical realizations are necessary for the assessment of the uncertainty under these statistical assumptions. A training dataset, which might be a set of observations or a dataset predicted using physical models, is used by deep machine learning techniques to "learn" the inherent spatial interactions within the facies distribution.

11.6 IMPROVED OPERATIONAL EFFICIENCY AND REDUCED OPERATIONAL COSTS

Due to the nature and complexity of unconventional reserves and inherent rock and fluid properties, it is of paramount importance that a comprehensive analysis of influencing parameters and interpretation of reservoir behavior are performed in order to better understand the quality of the resource and get an accurate estimate of remaining reserves and forecast of future production from assets. With the expanding number of wells in a field and the increasing volumes of data collected every day, conventional deterministic approaches that are based on rigorous mathematical manipulation of physical phenomena are not very efficient in modeling unconventional reservoirs. Studying the complex system of actions, reactions, and interactions requires an approach that is capable of handling such complexities while capturing the dynamic nature of the system. Recent advances in AI and ML techniques have enabled consistent data integration, more efficient spatial analysis, robust simulation models, and better uncertainty quantification.

In the following sections, various applications of advanced data-driven techniques in drilling and production operations will be explored, and different use cases reported in the literature will be provided.

11.6.1 AI and ML in Drilling Operations

Besides hydraulic fracturing, advancement in directional and horizontal drilling techniques has been one of the main reasons behind the success of the recent shale boom. Currently, horizontal wells are drilled more efficiently and with a higher precision than ever before. Nevertheless, drilling is still the most critical, complex, costly, and sometimes notoriously unpredictable operation in the oil and gas industry, challenged with many safety issues, operational difficulties, and financial risks. Drilling costs represent nearly half of well expenditure, and the majority of drilling operation time is spent dealing with drilling problems and rig defects. This is due to the many factors that lead to a high uncertainty in the processes, ranging from poorly understood reservoirs to equipment failure and even human error. As a result, oil companies are continuously seeking ways to develop new drilling techniques and equipment and, ultimately, to completely autonomous drilling solutions.

For a long time, the drilling industry was ruled by empirical models and human experience. These empirical models are barely applicable across fields nowadays because drilling operations generate a lot of data, especially in modern rigs where sensors are ubiquitous, and digital storage is cheap. Conventional data processing techniques are slow, depend on human intervention, and do not have the power to process the myriad data generated today. In recent years, the field of AI and machine learning has made huge leaps and helped reach state-of-the-art results in various other fields. With the current trend in drilling automation, data-driven approaches have proved to be strong potential solutions for automating and optimizing directional drilling, such as optimal ROP, less tortuosity, and less human intervention.

These techniques have provided powerful tools to explore and gain insights from huge amounts of data where the traditional methods fail. Using holistic approaches based on data-driven analytics and the aggregation of historical and real-time data, companies can avoid a lot of costly issues in drilling operations. AI-based approaches can uncover relations between operational and non-operational data, identify key performance indicators, anticipate drilling problems, assess their causes, make real-time adjustments to influencing drilling parameters, and plan effective solutions to prevent or mitigate the problems. Any problem that involves an agent choosing among a set of possible actions, in order to maximize a certain reward, is a candidate problem to be addressed with ML-based approaches such as reinforcement learning. Directional drilling automation and optimization are problems that fit this description, since they involve drilling engineers taking certain actions to achieve the objectives of optimal ROP, minimal deviation from the planned trajectory, less tortuosity, etc. Historical and simulation data is usually used by expert drillers to improve the efficiency of directional drilling processes. In a study done by Pollock et al. (2018), the authors showed that a neural network–based model coupled with a reinforcement learning model (RL) was able to successfully replicate the decisions of expert drillers with an error margin of only 3%. Similar examples from the literature demonstrate the feasibility of using data-driven automated systems to match or even exceed the performance of expert human agents in similar situations.

Well placement optimization: The very first aspect of any drilling endeavor is determining the actual drilling location or well placement. Since the output of the drilling operation (i.e., amount of produced hydrocarbons) is a direct result of the location of the well, a lot of effort is put into optimizing well placement/arrangement. Despite its importance, well placement optimization is a difficult task due to the amount of planning and investigations that needs to be put into it and determining the optimum design from the large number of possible solutions (Wang et al., 2022b). There have been several studies on implementations of data-driven, machine learning-based approaches to this specific problem. In a study by Mousavi et al. (2020), xGBoost, an implementation of the gradient boosting technique, was compared to an experimental-based statistical method called central composite design (CCD). The comparison was done on three different scenarios of homogenous reservoir, heterogenous reservoir, and water flooding into a heterogenous reservoir, with the objective function being the net present value (NPV). xGBoost proved to be either comparable or superior and more reliable than the traditional method for optimizing well placement in each scenario.

Machine learning models, which are usually purely data driven, can also be used in conjunction with and extended by using physical knowledge about the problem at hand. The underlying physical aspect of the problem is used to impose constraints on the data-driven model to enhance its predictions and reliability. Theory-guided convolutional neural networks (TgCNN) are an example of such models. TgCNN can be used for subsurface flow modeling (Wang et al., 2022b), in which theoretical knowledge is used to guide and impose constraints on the training process of the data-driven model, achieving better accuracy even with a limited amount of data, which is always welcomed. The model was then combined with a genetic algorithm for well placement optimization, which achieved results comparable to simulation-based approaches.

Drilling optimization and rate of penetration prediction: Drilling companies always aim to finish the drilling successfully and as quickly as possible because it is a costly operation, and the financial impacts increase drastically with time. A successful and quick drilling operation is determined by a lot of factors, such as the choice of drill bit, drilling fluid, weight on bit, rotary speed, etc. Drilling performance can be measured by several factors, such as rate of penetration (ROP), which is a complex function of different variables. Drilling engineers are always looking for ways to predict and optimize ROP as a direct indicator of the speed of drilling.

The first step towards finding the optimum drilling parameters to maximize the rate of penetration is to build a model that can predict ROP for a given set of inputs (i.e., drilling parameters). This area of research is relatively rich, with a plethora of models and techniques that were employed to address this problem, including artificial neural networks, fuzzy logic models, ensemble modeling, Kalman filters, etc. Predictive models range from theoretical/physical to empirical and data-driven approaches. ROP modeling and prediction based on theoretical, analytical, and empirical approaches is a notoriously difficult task and may prove to be unreliable (Soares et al., 2016). This is mainly due to the complexity of the problem and its nonlinearity. As a result, researchers started investigating the use of alternative data-driven approaches to model ROP using various machine learning models. Machine learning models are well known for their ability to effectively model and fit highly nonlinear functions based on input data with minimal human intervention. In traditional methods, the driller needs to pick a model before application, which is not an easy task since various models perform very differently from one formation to another whereas machine learning models can learn almost any complex pattern without the need to select a specific one (Barbosa et al., 2019). Most studies on this topic concluded that machine learning–based models such as artificial neural networks and fuzzy logic models are superior to traditional analytical and empirical methods (Moradi et al., 2010; Arabjamaloei & Shadizadeh, 2011; Nybo & Azizi, 2013; Bataee et al., 2014; Wallace et al., 2015). These predictive models can also be used for optimizing the rate of penetration. However, it should be noted that this is not a simple optimization problem in which the goal is to achieve the highest possible rate of penetration. A high rate of penetration can lead to sever issues; therefore, many other variables should be taken into consideration, which ultimately yields a multi-objective optimization problem (Barbosa et al., 2019).

Drill bit selection is a complicated task that is guided by the to-be-drilled formation properties. The conventional approach is to use the data gathered from offset wells to aid the selection process (Bilgesu et al., 2001). A machine learning approach to drill bit selection is described in a work by Batruny et al. (2021), in which an ANN model was trained on a large dataset collected from offset wells to predict the rate of penetration (ROP) as an output. Different configurations of drill bit, weight on bit (WOB), rotary speed (RPM), bit hydraulics, and lithological properties were used as inputs to train the model. A combination of Monte Carlo and sensitivity analyses were then used to compare different bits under various conditions, and the optimal bit was eventually selected, resulting in a 32% improvement in ROP. There are not a lot of use cases for visual data analytics in the oilfield, but the work in Ekeregbe et al. (2021) describes an interesting application of this technology in evaluating the dullness of drilling bits. A convolutional neural networks–based model called You Only Look Once (YOLO) was used on a series of video inputs. The video input contained visuals of several drill bits on a drill site. The YOLO model was able to process the visual input and estimate the drilling bits' dullness from the frames of the video input. The advantage of this model was its ability grade multiple bits at once while avoiding the subjective evaluations humans are prone to making.

Drilling fluid engineering: Another very important aspect of the drilling operation is designing and managing the drilling fluid. The fluid (or mud) serves many critical functions, such as being the main tool for well control, cooling the drilling bit, transporting cuttings, etc. Therefore, assuring the quality of the mud and its ability to perform its functions is key for a successful and quick drilling operation.

Drilling fluids are specifically designed with a precise set of properties that will enable it to perform its assigned functions. Maintaining these properties is a very important task, but in practice, there are several factors that can lead to the deterioration of the quality of the mud, such as contamination by solids, contamination by reservoir fluids, or simply the deterioration of its quality over time. The diminishing of mud's quality will result in its inability to carry on its functions, eventually leading to a wide range of fluid-related issues such as loss of circulation, various well control events, wellbore instability, etc.

In order to maintain the drilling mud in an optimal state, it must be constantly monitored, its properties predicted, and its quality assessed.

Characterization of drilling fluid's physical and chemical properties is an essential part of drilling operations. Conventionally, this is done by mud engineers several times a day. Besides the sampling frequency and the repetitive nature of the task, a very important consideration is that the mud properties are determined at laboratory conditions, rather than downhole conditions as they should be. Prediction of drilling fluid rheology at downhole conditions is essential for avoiding fluid loss to the formation, which is a costly issue that can also hinder the drilling progress.

Several attempts have been made to employ machine learning and automated solutions to predict fluid properties or to estimate them at wellbore conditions; neural networks–based and fuzzy inference systems, combined with swarm and genetic optimization algorithms, proved to be powerful tools for predicting the fluid's rheological properties at downhole conditions with great accuracy (Ahmadi et al., 2018). This means that the repetitive task of manually analyzing the mud can be delegated to an automated or semi-automated system that can get the task done in the required downhole conditions, quicker and with greater accuracy.

Besides rheological properties, it is also important to monitor the solids content in the drilling fluid and distinguish the low- and high-gravity solids (LGS and HGS). The importance of this monitoring process is the need to keep these solids at optimum levels for effective hole cleaning and an overall efficient drilling process. To generate the required data for a data-driven approach, an experimental dataset that correlates the fluid's rheological properties with different solids content in that fluid can be created, as shown in the work by Gul et al. (2020). Several models, such as Ridge and Lasso regressions, K-nearest neighbors (KNN), and random forest (RF), were used on this dataset to predict the LGS and HGS content. These models were able to predict the solids content with a greater accuracy than more conventional procedures such as API retort measurements. It is important to note that these are relatively simple models. Contrary to other black boxes such as neural networks, the underlying structure and parameters of these white-box models can be studied, and researchers can get insights into how decisions are made.

With the advancements in the drilling world in the past few decades, it is now possible to drill deeper and more complex trajectories. But extended reach drilling (ERD) comes with its own set of problems and challenges. The maintenance of a precise mud weight becomes of greater importance as the mud window becomes narrower, and the risk of a kick or a loss increases. Several attempts have been made to precisely control the equivalent circulating density (ECD). However, it is not a straightforward task because ECD itself is a function of annular fluid pressure losses that need to be precisely predicted (Gul et al., 2019). Machine learning approaches have been used to address this problem using an appropriate dataset consisting of several parameters that can affect pressure loss, such as drill string configuration, motion eccentricity, rotation, and axial position. Ozbayoglu et al. (2018) showed that a neural network model was able to effectively predict the pressure loss with high prediction accuracy compared to both experimental and other models from the literature. In another study, random forest and XGBoost-based models were used to predict pressure losses in a drilling fluid with long-chain polymer additives. These additives significantly alter the fluid characteristics, and the current models used in the industry have been shown to generate inaccurate results (Gul et al., 2019). These ML models were able to accurately predict the frictional loss in this type of fluid and, in accordance with the experimental results, are proving to be a more reliable alternative to industry-adopted solutions.

Defect diagnostics and predictive maintenance: A typical drilling rig comprises varying and complex equipment. As is the case with any other industrial setting, maintaining the equipment in an optimal state is critical to ensure a successful operation and avoid non-productive time. This requires a dedicated team with an effective maintenance program that ensures all equipment is up and running at all times. Oftentimes, this is done through corrective or condition-based maintenance policies. While this is the easiest and most straightforward approach, it is almost always the least efficient. Corrective maintenance entails taking no actions until an equipment failure happens. It is mainly employed for minor equipment with a non-critical role for which sudden failure can be tolerated. Condition-based maintenance entails placing a hard threshold on a specific parameter (e.g., time, cycles, vibrations, pressure, etc.) beyond which a corrective action is taken to remedy the situation. It is employed for more critical equipment, for which sudden failures should be avoided at all costs. Nevertheless, condition-based maintenance is still lacking and not very effective for complex equipment, for which failure can happen as a combination of several factors that may not be easily detectable from direct sensor measurements.

Oil and gas companies have been capturing a lot of legacy data on equipment and operation failures. However, many of these legacy data lakes do not have the proper timestamps on the events that led to the occurrence of an issue. Any abnormal observation that can lead to a potential issue stems from complex interactions among multiple factors within the associated workflow. The more aware companies becomes of incidences and near-miss occurrences, the better insights they gain over their procedures, empowering proactive and ongoing risk management and safety efforts. As with many other industries, there have been several attempts to use machine learning for predictive maintenance in the oil and gas industry, which will be discussed in the following sections.

Predictive maintenance is a proactive approach to maintenance in which the goal is to assess the health state of certain equipment and even predict its remaining useful life (RUL). This approach enables maintenance teams to intervene at the right time: i.e., when the equipment shows signs of degradation but before any actual damage happens. Combining the legacy data lakes with real-time data streams collected from surface and downhole equipment, such as flow rates, temperature, pressure, vibration, etc., will provide the necessary situational awareness of particular issues and their root causes. This fused source of information can help not only in identifying a particular issue during production but also in anticipating failures and taking proactive action to avoid such undesirable failures (Figure 11.12).

Arguably the most important function of the drilling fluid is well control. It is of the greatest importance to keep the mud within a specific window in order to avoid fluid loss to the formation or fluid influx from it. Drilling fluid accounts for almost 40% of the total drilling cost (Ahmed et al., 2020). Lost circulation, therefore, is a serious event that can dramatically increase the cost of drilling. Detecting these events is not a straightforward task, and it can be challenging due to technical or human errors. In the continuous attempts to automate the drilling processes, automated and data-driven events detection approaches can provide powerful tools to minimize errors, especially from the human factor; cut unnecessary costs related to loss and influx; and minimize drilling time. Historic records are always a good place to start and develop a new data-driven model. Artificial neural networks (ANN), functional networks (FN), and fuzzy logic (FL) are some example applications of ML models used to correlate several drilling parameters (e.g., depth, hook height, hook load, flow pump) to the historic loss circulation records (Ahmed et al., 2020). All the aforementioned models accurately predicted zones of lost circulation, achieving correlation coefficients of 0.09 to 0.99. In other similar studies, random forest and SVM models successfully classified zones of influx and loss, achieving an accuracy of 90% (Shi et al., 2019).

By leveraging predictive models and sophisticated mobility solutions, companies can harness the available data to develop risk-sensing capabilities to monitor emerging risks and uncover potential problems before they fully hatch. So instead of relying on static maintenance schedules, companies can benefit from condition-based

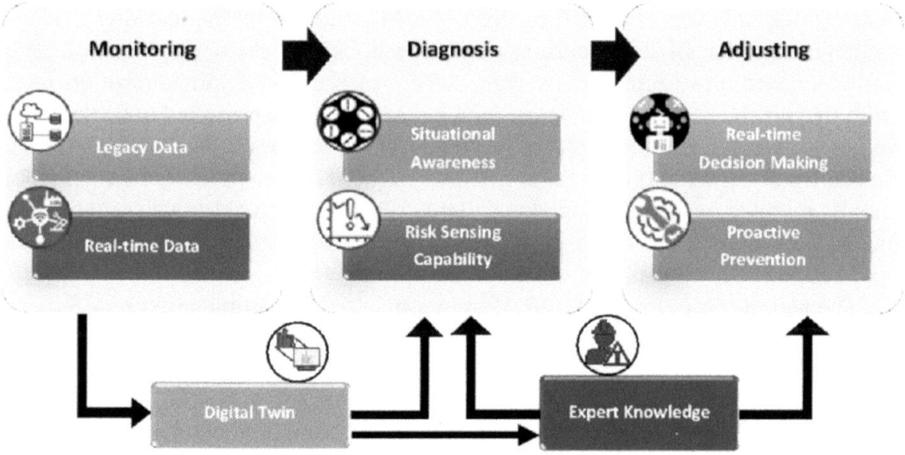

FIGURE 11.12 Defect diagnostics and predictive maintenance using AI and ML.

maintenance techniques. As an example, proper hole cleaning is critical to avoid many potential problems in drilling operations, such as pack-offs, stuck pipe incidents, and lost circulation events, and to ensure an efficient drilling process (Saini et al., 2020). Digital twins are a powerful tool that have been used to facilitate effective mud monitoring and cut transportation efficiency. A digital twin is a software representation or emulation of a real-world physical system based on mathematical and analytical modeling of its elements (Mayani et al., 2018). For a mud monitoring task, a digital twin acts as a digital representation of the different functionalities of the drilling fluid. This digital twin monitors downhole fluid properties in real time and aims to support real-time operations and suggest proper solutions to the potential problems. Building a digital twin requires developing predictive models that tackle different tasks of the drilling fluid and can continuously learn from new data and update the underlying models (Shirangi et al., 2020). Digital twins coupled with reinforcement learning create a system that can simulate different hole-cleaning scenarios in addition to predicting the impeding problems and proposing preventive actions (Saini et al., 2020). For this purpose, several variables such as equivalent circulating density and cuttings concentration are used to classify the state of the borehole. The reinforcement learning–based digital twin evaluates the borehole state and suggests possible remedial actions if necessary. These actions are based on several controllable surface parameters, such as flow rate, rotary speed, mud density, etc. The system is also smart enough to filter through unrealistic solutions (e.g., a solution that violates the actual mud window) and discard them (Saini et al., 2020).

Historic operational records are very useful to understand and model the equipment's behavior. These records can range from sensor data to written reports. Reported problems and failures can then be linked and correlated to some operational variables. For example, Lawal et al. (2021) demonstrates how historic drilling datasets were used to study the relationship between drilling and non-drilling events and mud motor failure. Mud motors are positive displacement motors (PDM) that transform the hydraulic energy from the drilling fluid into mechanical rotational energy, which turns the drilling bit. Failure of mud motors can cause huge losses and significantly increase drilling costs. In this study, machine learning models—namely, linear classifiers, decision trees, and boosted trees—were used to predict failures based on selected sensor data. After isolating the features, most correlated with the motor failure, a cumulative mud motor index was proposed to track the degradation of the mud motor, which, according to the authors, proved to be accurately indicative of degradation and failure. The strength of this method lies in its simplicity and its reliance on simple, already-available sensor data to calculate the cumulative mud motor index in real time, enabling the drillers to monitor and get instantaneous updates on the health state of the mud motor.

The authors in Belov et al. (2021) used a machine learning–based reduced order model (ROM) to replace the traditional and more complicated model of the power section in a mud motor. Despite the simplicity of power section, which encompasses a rotor and a stator, its modeling can be very problematic, requiring complicated and time-consuming multi-physics simulations (Belov et al., 2021). A reduced-order model is a simplified version of a more complex model that retains the general behavior of the original model while significantly reducing its computation time and practicality. The proposed ML-ROM model in this work takes several downhole drilling conditions as inputs and outputs the remaining useful life (RUL) of the motor. The simplicity of the alternative model enables RUL prediction in real time, unlike the more complicated and computationally expensive model.

Another hybrid approach to predictive maintenance is the augmentation of intelligent models with expert knowledge. Semantic data standards can be used to codify expert knowledge related to the knowledge derived from legacy and real-time data sources. A variation of this data-knowledge hybrid approach is the application of virtual collaborations, in which remote guidance from an experienced worker can be provided to a less experienced worker in real time. Operators can also utilize computer vision and augmented reality to get better views of their equipment, facilities, and operations, which is more cost effective than the prevailing processes of manual testing and inspection procedures. Such intelligent systems can also help in decentralizing the decision-making process so the person in charge of an operation does not need to be at a very high technical level to make a decision in critical situations. The insights and predictions provided by the system by integrating the historical data with real-time data enable the operator to make quick decisions in real-time scenarios.

11.6.2 Applications of AI and ML in Production and Field Operations

As was seen in the previous sections, the applications of data science in the fields of prospecting and drilling are plentiful. They have been used for proper structural identification of development prospects to set up resilient and smart wells. However, the purview of data science in unconventional development does not end there and is a crucial aspect in the continued development and operation of oilfield activity. This section focuses on various techniques related to data analytics in field development, production optimization, stimulation and workover, and oilfield maintenance activities. It also includes studies from professional conferences and peer-reviewed journals focusing on developing traditional long-standing analytics workflows, using new proprietary connected systems and digital technology, and adopting new applications of existing machine learning models in the oilfield.

Production forecasts and well operating conditions: Some of the earliest applications of machine learning (ML) and artificial intelligence (AI) in the oil and gas industry came from the usage of these technologies in the prediction of potential revenues generated by oil companies listed on the New York Stock Exchange (NYSE). The valuation of oil companies is based on oil reserve holdings and the performance of their assets in terms of the production of oil and gas. Traditionally, the prediction of oil and gas from reservoirs and wells depended on analysis techniques such as decline curve analysis (DCA). Even today, DCA is the cornerstone of production prediction (Carpenter, 2020). Hence, it is critical to examine the usage of AI to aid DCA applications in field applications or as a replacement for DCA for production predictions. Carpenter (2020) describes the work of Kianinejad et al. (2019), in which the authors present a case study using quantile regression based on AI technology. Quantile regression represents a methodology that calculates the conditional quantile of the response variable in data and helps in better correlation among the variables. The authors utilized this technique to better understand uncertainty in dealing with DCA predictions. This methodology was used by Kianinejad et al. (2019) to estimate DCA patterns in a giant mature sandstone reservoir field with a 70-year production history and more than 2,000 wells. The authors further used regular DCA analysis conducted by the production engineers on well sites to compare the results of their methodology. Their results had a very good correlation to the manual DCA procedure and significantly reduced processing time. This indicates the possibility of automating the process and significantly reducing processing costs related to manual intervention. In addition, the AI approach could classify wells into different groups according to their performance and characteristics to predict their ultimate recovery rates.

Unconventional reservoirs pose an even more significant challenge for real-time forecasting. The physics of fluid flow is a complex combination of processes due to the presence of micropores (< 2 nm), mesopores (2–50 nm), and fractures. Reservoir flow becomes dominated by complex physics that operates on different timeframes; more importantly, it results in poor recovery efficiencies. The traditional technologies utilized for such probabilistic modeling for a reservoir in traditional fields include history-matching techniques. History matching is building one or more sets of numerical models (representing a reservoir), which account for observed and measured data. However, artificial intelligence has recently emerged to fill the knowledge gap generated. Srinivasan et al. (2021) tried to address a workflow for forecasting production in unconventional reservoirs with reduced-order models of production patterns with transfer learning techniques. The fundamental problem of forecasting is predicting the long-term production of a site, given its short-term production. The computations involved in the prediction need to be accelerated to achieve the goal of near real-time forecasting. For all forecasting, past data is required with continuous updates to make short-term production estimations. These short-term production models are in terms utilized to create long-term production estimations. The short-term and long-term models must be efficient to be feasible and computationally viable. High-fidelity data were naturally ruled out of such models due to the computational overload and were replaced with simpler reduced-order models by Srinivasan et al. (2021). The machine learning inverse models (required for long-term production forecasts even in conventional techniques) are artificial neural networks used as transfer learning models. This type of ML model replaced data-driven ML models, which struggle with inconsistent or heterogeneous datasets. The transfer learning models introduce a novel approach based on ML techniques for real-time forecasting. The more significant picture problem solved is the extension of an ML technique across various unconventional reservoirs with fewer required variations to the underlying reduced-order models and ML architectures. This technique was initially applied to the Marcellus shale play.

A similar methodology was utilized by Thavarajah et al. (2022), in which the traditional DCA method was replaced by a flexible sequence-based deep learning (DL) framework for production forecasting. The authors utilized three techniques: an encoder-decoder architecture model and two traditional DL models, including DeepAR and temporal fusion transformer (TFT) models. The novel encoder-decoder model parses historical data to yield an encoded representation of production history and interprets this representation to build a production forecast. DeepAR is a supervised learning algorithm for time series forecasting that has an autoregressive design that utilizes previous observations as input for learning but does not allow for time-dependent features that are known only up to the present. DeepAR creates prediction intervals for future values using Monte Carlo samples. TFT is a recursive neural network that can predict time-dependent confidence intervals. Three input features for wells, including static features such as geologic characteristics, time-variant control variables like choke size, and production history. The model then uses these inputs to predict future values of long-term production for oil, gas, and brine, along with uncertainty quantification. The models were tested on multiple wells (208) in the Eagle Ford formations with good results. These techniques enable a single model utilization for step-wise prediction of various production facets, compensate for short production history with static and control variables, and help control production with the knowledge obtained by varying the control variables.

A similar transfer learning model to replace DCA analysis for predicting oil production at the Eagle Ford formation was utilized by Tadjer et al. (2021). The authors used the DeepAR methodology and compared it with the prophet time-series technique. The prophet time-series method is a Bayesian nonlinear univariate generative model developed for creating high-quality forecasting. This technique is often used to predict human behavior related to special events during specific times of the year. The methods are similar in that they are often used for pattern and sequence recognition. These two algorithms can be used to understand and predict the behavior of oil wells. The authors used these techniques to predict the distribution of future outcomes, specifically with time series data representing petroleum oil production, without having substantial knowledge regarding the extraction process or the physical relationship between the geological and dynamic parameters. These methods were tested in the Midland field with 22 naturally fractured wells, where the study showed that both models are capable of nonlinear short-term forecasting.

Various ML models for processing data to produce vertical flow performance (VPF) were utilized by Masini et al. (2020). The authors combined DCA with different big data and AI techniques to create a seamless prediction methodology for better results. The authors utilized clustering techniques for outlier identification, specifically density-based clustering (DBSACN) and ordering points to identify the clustering structure (OPTICS). Further, the long-short term memory algorithm (LSTM) and random forest techniques were utilized to model the prediction from the backward-looking regression. LSTM is a recurrent model that helps bridge the gaps in time lags. An enhanced neuro-fuzzy technique to optimize production and fluid flow analysis in wells in the Oredo oilfields in Nigeria was shown by Akanji et al. (2019). The neuro-fuzzy methodology uses fuzzy modeling to make discontinuous blocks of varying data formats to be fed into a five-layer feed-forward neural network designed in C++. The analysis uses various input parameters, including historical production data, reservoir rock and fluid properties, well geometry, architecture, completion profile, and surface data on overall well deliverability. The algorithm identified 368 patterns with a 0.86 error index.

A similar technique was utilized by Khan and Louis (2021), in which the authors used an artificial neural network (ANN) model in combination with a time series forecast. The model was trained with tubing head pressure and bottom-hole flowing pressure to predict future gas flow rates. The paper gives a preliminary view of a simple feed-forward network utilized for prediction. It provides further details on selecting various factors related to designing an ANN, including initialization, hidden layers, and activation functions. The overall analysis was applied in a tight gas reservoir and a shale gas reservoir. Mohmad et al. (2020) showcased a stepwise methodology to build a reservoir model to predict future flow rates through an integrated history matching and AI methodology. They integrated a simple ANN model into the history-matching formulation, which indicated multiple parameters, including oil rate, water rate, and wellhead pressure. The authors used a top-down history-matching technique that was implemented in "A" field in Malaysia with a 25-year history of production. The area has over 100 faults, is subdivided into various fault-based sections with over 200 producers, and is subject to water injection.

Another notable trend observed in the oil and gas literature is the utilization of a global optimization tool to tune the machine learning models. It is a setting created to optimize a model with other machine learning models, mostly with tools like genetic algorithms (GA), particle swarm optimization (PSO), and gradient descent (GD). Ahmadi et al. (2015) put forth an example of this analysis. The authors utilized a primary ANN mode in which the model's hyper-parameters were tuned to the optimal settings using a PSO tool. PSO is an evolutionary algorithm themed after a natural phenomenon to define the best parameters to achieve the optimal values for the objective statements. A feed-forward ANN tool within a PSO loop was installed to predict the performance of horizontal oil wells. The authors compared this model with another hybrid ML model that included a least square support vector machine (LSSVM) regressor whose hyperparameters were tuned using a genetic algorithm. It was found that there was a very good match between the modeling output and the actual data taken from the literature, and a very low average absolute error percentage was attained. It was determined that the LSSVM-GA model outperformed the PSO-ANN model.

Sand production during oil operations leads to several problems, including equipment maintenance and compressor issues (Ngwashi et al., 2021). A two-layered ANN model with a back-propagation algorithm was developed that used 11 geological and reservoir parameters associated with the onset of sanding, including depth, overburden, pore pressure, maximum and minimum horizontal stresses, well azimuth, well inclination, Poisson's ratio, Young's modulus, friction angle, and shale content. The methodology was utilized on Niger delta sandstone reservoir wells, and two different AI techniques, SVM and back-propagation ANN, were compared. ANN gave an accuracy, precision, recall, and F1 score of about 80% while the SVM performance was 94% for the four metrics. Similar studies are utilized for reducing the effect of oil and gas byproducts on overall operational procedures.

Another use for AI, commonly used in consort with flow prediction and forecasting, is estimating well performance in terms of well variable parameters. These can refer to both bottom-hole variable parameters like bottom-hole pressure and temperature and wellhead parameters such as choke size and wellhead pressures. These models generally work in consort with real-time parameters recording using sensor technology. This model is great for interpreting missing data, especially during sensor downtimes.

Sami and Ibrahim (2021) utilized AI to estimate flowing bottom-hole pressure, a key metric in evaluating oil and gas wells. The authors used a model to predict the multiphase flowing bottom-hole pressure using three different models, including random forest regressors, K-nearest neighbor classification, and ANN. The authors used a dataset of vertical wells from Middle Eastern fields consisting of nine production variables, including wellhead and bottom-hole parameters. The model identified ANN as the better model for estimating this parameter.

A similar methodology to estimate gas-liquid flow patterns using tree-based algorithms, including random forest, bagged-boost algorithm, and extreme gradient boosted models, was utilized by Mask et al. (2019). They created various dimensionless variables that were utilized to calibrate laboratory sample datasets into field-specific data parables to upscale. About 8,000 lab tests were conducted on 12 different experimental setups. The database for input variables included measurements of density, viscosity, temperature, pressure, velocity, surface tension, pipe diameter, angle of the pipe, and observed flow patterns. The predicted patterns included dispersed bubble, stratified smooth, stratified wavy, annular, intermittent, and bubble flow. The accuracy of the AI model was measured by up to 80% with the dimensionless variables.

Numerous examples in the literature utilize AI to predict flow, wellhead, and bottom-hole variable parameters. Jahanandish et al. (2011) presented an ANN model for predicting the bottom-hole flowing pressure and, consequently, the pressure drop in vertical multiphase flowing wells. A total of 413 field datasets collected from Iran fields were used to develop the ANN model, which had approximately 3.5% absolute average percent error and 0.9222 correlation coefficient. Li et al. (2014) used a combined BHP calculation procedure of multiphase correlations and trained ANN models. Compared to the multiphase correlations, the combined approach gave fairly accurate predictions, with an average absolute percent error of 23.0% and a standard deviation of 0.176. Ahmadi et al. (2016) used a least square support vector machine (LSSVM) method to develop a mechanistic approach to estimate or represent the BHP with high accuracy and low uncertainty. The proposed LSSVM model showed high integrity and performance based on the statistical criteria gained from the outcomes of the LSSVM approach. Moreover, a very low relative deviation of less than 6% was observed between the model estimations and the relevant actual BHP data. Ketmalee et al. (2019) developed a new approach to estimate gas production rates during downtimes when other indicators are unavailable. The authors used surface facility information, fluid properties, production conditions, and geological setup as input parameters to develop an ANN model with a 0.93 confidence level. Ketmalee et al. (2019) verified their approach with a full field test in 2018 and observed improved prediction accuracy in wells. The field test also revealed that it could predict phenomena in the well without human intervention and improve efficiency with production allocations. Reddicharla et al. (2021) went with a hybrid machine learning model. They developed a methodology to reduce well test rejections and generate a timely warning for operator intervention to recommence the well tests. This case study was applied in a mature field, producing for 40 years with a good volume of historical well test data. Parameters such as choke size, tubing head pressure and temperature, and flow line pressure and temperature were used as input parameters. Random forest and gradient methodology were applied to select the data to be fed into the ANN model. The results showed a reduction of at least 5% in well test rejection, resulting in an optimized resource allocation and improved back allocation process.

Artificial lift operations: One of the standard applications of data analytics and ML in oil and gas production is optimizing artificial lift operations. Artificial lift is an essential part of production post natural pressure decline in reservoirs. Various technics such as pattern recognition, parameter identification, and criteria selection are used to improve the operations' logistics and the lift operations' selection in the fields. One of the more advanced uses of computational innovation is edge computing, in which the deployment of ML models at the boundaries of the production network enables near-real-time optimization of networks. Saghir et al. (2020) showcased the deployment of an edge-computing framework using Industrial Internet of Things (IIOT) edge gateway and cloud platforms, where each tool performs the following tasks:

IIoT Edge Gateway:

- Capture dynagraph (a 3D-plotting utility) data from the remote telemetry unit (RTU) and convert the dynagraph shapes into images
- Run on-device ML models to predict the shape of the dynagraph card
- Compress and store dynagraph prediction data and push intermittently to the cloud platform

Cloud Platform:

- Store dynagraph data pushed by the edge gateway to the online cloud storage
- Display dynagraph data and the associated predictions on a user interface
- Run online clustering analysis model to check/verify if any new cluster of dynacards have been detected.
- Provide a model training interface to the operators to tag new dynagraph clusters.

When paired with augmented intelligence capabilities, these smart nodes allow subject matter experts to interact with ML models and help improve their accuracy over time. This, in turn, helps increase confidence in data-based results and enables operators to make informed decisions. Another such approach is put forward by Palmer and Turland (2016), in which the authors indicate how distributed systems, logic machines, and data analytics revolutionized production activity from shale formations. According to the authors, the concepts of big data, data mining, cloud computing, and predictive analytics enabled better insights into making informed decisions and helping resolve issues. According to the authors, the oil and gas industry took three significant steps to make big data an integral part of suck-rod pumps:

- Data consolidation: joining SCADA, well test, and well equipment databases
- Automated workflows: calculations using consolidated operational data to generate a "recommended optimization action" for each rod pump system
- Interactive data visualization: interface for decision-makers to view aggregated results, with the ability to drill down to specific well-level details

This improved system implementation provided increased unit runtimes, decreased unit cycling, and enhanced production/equipment surveillance. It increased staff productivity, thereby improving profitability by reducing costs related to artificial lift operations. In recent years, ML techniques have been commonly used on suck-rod operations. Peng et al. (2020) present an ML methodology to predict the real-time production rate from suck-rod data. In suck-rod pumps, a diagnostic device called a dynamometer is used to plot the measured load on the top rod in relation to the rod position as the pumping unit moves through a stroke cycle. The authors utilize two ANN models, a deep autoencoder to extract the feature representations from dynamometer cards, and another neural network to combine all related features to predict real-time production. The auto-encoder uses the features extracted from the dynamometer, such as pumping stroke and rate, pump, rod, wellbore, and reservoir parameters and production rates, enabling real-time production prediction models. Hand-crafted features can lose important information, whereas an autoencoder is designed to minimize information loss by learning high-level features that can be used to reconstruct the cards. The methodology was tested in an oilfield in China, providing an R-squared value of 0.92 and a mean average error of less than 0.5 in low-production wells.

The usage of analytics extends to other types of artificial lift techniques, too, including electronic submersible pumps (ESPs) and gas-lift pumps. Sarapulov and Khabibullin (2017) used data analytics in ESPs to identify cause-effect relationships (patterns) that precede an event, such as jamming, overheating, and unsuccessful start-up, to develop proactive measures to prevent these problems. They created an information model necessary to identify, formalize, and further analyze the chain of events preceding the failure of the ESP self-start, extracted from 200 million records, including emergency logs, from 1,649 wells. This information was presented in the form of generalized patterns of cause-effect chains, leading to failures of self-starting ESPs, which were further analyzed for the development of appropriate management solutions aimed at minimizing and/or preventing such shortcomings in the future. Kandziora (2019) takes this a step further by describing the role of AI and its importance in boosting the performance of ESPs in the field in cohort with other technological advances such as SCADA systems, fiber-optic networks, and real-time monitoring. Kandziora (2019) also provides an example of deploying such a system on a fleet of 30 ESP-installed wells with various input variables, including annulus pressure, oil production rate, motor current, motor speed, inlet oil pressure, inlet oil temperature and pump motor temperature for predicting ESP maintenance operations. Bermudez et al. (2021) showed the potential of AI in ESP utilization in the Gulf Cooperation Council's wells. The system was able to predict imminent and future failures, extend and monitor run life, and maximize the production of ESPs in real time. An analytical and an AI-based virtual flow meter (VFM) model were created, and model performances were compared. In the analytical VFM model, an optimization equation was solved based on the assumption of certain constants. At the same time, the AI model used historical data to make informed decisions on these variables using parameters from well completion, equipment specifications, and field sensor data. The system was implemented in two stages. In the first phase, an offline implementation with good predicting capabilities was performed. The second phase was a pilot project on 27 wells for six months. During the pilot implementation, a 20% to 30% increase in pump run life, a 5% increase in production, and long-term visibility into potential failures to support planning efforts were observed.

Gas lift pumps are another common artificial lift technique. Rozlan et al. (2022) showcase PETRONAS's digital intelligence artificial (DIAL) gas lift production optimization system. According to the authors, DIAL technology removes the need for well intervention by remotely adjusting gas lift valves based on the downhole monitoring of production parameters. The technology was installed in a 30-km offshore oilfield in East Malaysia with interbedded shale layers with a production history of over 50 years. A control system is integrated downhole automatically with various sensors that can be connected to a SCADA system. The assessed parameters used were technology design and functionality, scheduled maintenance and cost impact, surveillance performance, safety performance, production performance, and integration performance. Khan et al. (2019) utilized varying AI techniques: support vector machines, artificial neuro-fuzzy interference systems (ANFIS), and ANN to predict oil flow rates. Fourteen parameters were used as input data, including choke size, upstream pressure, upstream and downstream temperature, and API gravity. The ANN model was deemed the most fitting solution with a high accuracy rate. The model showcased how ANN solutions can be converted to traceable equations and optimized solutions rather than treating them as unknowns. Pennel et al. (2018) used AI to check for failure detection and model optimization in the artificial lift for both sucker rod and gas lift pumps. The model was trained using data from 800 wells. The ensemble model was designed in five stages:

- Evaluation of the data and machine state: From the IoT data available for rod pumps, thresholds were placed to ensure data quality and quantity. Specific criteria were specified related to each time period to ensure there was adequate information to make a prediction.
- Feature extraction: Generated features were compared to similar statistics during regular state operation over long periods of time patterns to enable comparison across wells.
- Base period diagnostic: This step focuses on identifying a period of high risk based on signal patterns.
- Event diagnostic: During this step, complete event information was gathered for continuous periods of high confidence, and the potential beginning, middle, and end of the event were marked.
- Event classification: Based on the scores of multiple models in the ensemble, a final prediction was made to resolve conflicting predictions.

Analytics in well completions: Completions are essential to field development and the production of existing and new oil and gas wells. Completions have become an art form dependent on varying factors and subject to geologic properties and fluid properties. If not done correctly, completion design can negatively affect production and cause constant well intervention. Researchers and engineers have brought AI and data analytics into completion designs and optimization. These AI and analytics techniques are used for various applications, such as learning the complex relationships between facies, suitability of completion options, and cluster productivity. Completions design is dependent on the reservoir formation and fluid characteristics. Traditionally, selecting zones for specific completion design or specialized completion architecture is an art and not data driven. However, the bridge between data and completion design is possible with AI and analytics. For example, Jacques et al. (2022) used a new type of logging system that used piecewise linear transformation (PLT) proxies combined with gradient boosting methodology to find zones ideal for hydraulic fracturing operations. Avoiding a black box–type learning process, the authors were able to find optimal conditions for completions by combining the results of the well test log with all available data to sort various facies based on fracturing ability. Similarly, Calad et al. (2022) used a hybrid model combining machine learning and reservoir physics to an entire field model. The physical model ensures that model predictions are always realistic and reliable. At the same time, the machine learning algorithm allows the utilization of different data types to make a prediction that cannot be directly integrated into the physical model. Calad et al. (2022) included data from 1,800 unconventional wells with varying features of input data like lateral length, well spacing, cluster spacing, frac fluid volume, pad/well location, injection profile, and so on. Different AI models were used, including SVR, Gaussian progress regression (GPR), and neural networks, which were able to locate regions of higher productivity and completion designs with high reliability and accuracy. The increased amount of data provided to these models will result in better performance; however, these models can also perform with lower levels of data using augmentation.

By leveraging basin-level knowledge captured by big data sculpting with uncertainty quantification, augmented AI can provide quick and science-based answers for completion optimization and assess the full potential of an asset in unconventional reservoirs. Ma et al. (2020) used augmented AI technology—basically, AI techniques boosted by imputation techniques—to fill in missing data to improve the predictability of ultimate recovery rates of wells. The prediction made by the system developed by Ma et al. (2020) has an 80% confidence interval. The input features are extracted from several unstructured data sources, including reports in which natural language processing (NLP) is used semantically (importance of words extracted). More than 100 features have been generated for the study and categorized into five groups: base, frac, detailed frac, well spacing, and geology features. Another approach is supplementing field data with hybrid proxy models to compensate for sparse datasets. Hydraulic fracturing designs in unconventional reservoirs require various considerations such as geology, fracture design, post-fracture completions, and production simulations. Success and failure data from previously fractured wells can be used to train models that can identify the best approaches to optimize operations in the future (Garjan & Ghaneezabadi, 2020). Hybrid numerical-machine learning approaches can help improve the overall production from hydraulically fractured wells by processing several input data while capturing the physics of hydraulic fracture propagation and flow simulation, as explained in Xue et al. (2019). This work used decision tree networks and gradient boosting to process various input data on geologic and near-well factors. They used proxy models to enable rapid history matching and populated additional data to create a rapid completion design estimation tool.

Eventually, the resulting completion designs from existing wells can also be used to showcase the performance of completion designs for fields. This is extremely useful in increasing frac hits and the ultimate recovery from stimulation activity. Nabors et al. (2020) used a data-driven cloud-computing methodology to create a toolbox to model, order, and rank key factors in completion designs, such as proppant and fluid intensity, perforation intervals, fracturing fluid types, and proppant loadings and types, for better well selection. The system was fueled by a deep learning neural network feeding into a gradient boost system to identify and rank the key parameters that would better boost production for a 12-month period post completion. The model was trained on the data from more than 12 fields and was validated using the Permian Basin wells. This type of analysis helps conveniently ascertain ground reality. Another example of utilizing analytics to determine the importance of completion parameters is Baki et al. (2021), in which AI models show the impact of completion parameters such as lateral length, frac type, number of stages, proppant, and fluid volume on production performance in unconventional reservoirs. The primary dataset was from the Eagle Ford formation, which included 7,282 horizontal wells from the basin between 2015 and 2020. The data had factors such as well locations and trajectory and fracture formation data. Three AI models were used: neural networks, SVM, and extreme gradient boosting. The extreme gradient boosting method showed the best results in predicting data. Wang et al. (2020) showed similar work in which they reasonably combined a physics-based model with varying AI models. The work done by Wang et al. (2020) shows the improved effect of geological parameters in improving the considerations of AI in the completions design of hydraulically fractured wells in the Permian Basin. Cross et al. (2021) investigated the effect of various factors on immediate production windows post hydraulic fracturing using shapely additive explanation (SHAP). The study was conducted over the Bakken, Eagle Ford, and Permian unconventional plays.

The potential of large amounts of existing data from various wells is considerable, as seen in works understanding the performance of hydraulically fractured wells. This technique can also be used in general well workover analysis too. Karnik et al. (2021) utilized a methodology to improve well workover estimations in various oil and gas companies to predict unusual events during operations. The authors focused on combining data from multiple unstructured data sources like logs, reports, dynacards, etc. The information was collected for over 2,000 producing wells over 50 years from 20 different file formats. The authors created a methodology to convert the data systematically and automatically into actionable information. An ANN model was developed to predict the workover class for each report with > 85% accuracy. Key performance indicators were recognized for the various events and production enhancement.

Analytics in stimulation operations: Stimulation is an essential part of production optimization in unconventional oil and gas fields, including activities like hydraulic fracturing and acid stimulation. Significant studies have been done on optimizing hydraulic fracturing and other stimulation techniques. The nature of unconventional reservoirs requires a complex relationship to be discovered between varying factors that affect the performance of stimulation operations, including geology, stimulation parameters, and completion designs. However, with the advances in AI and data analytics in stimulation technology, understanding the complex behavior of stimulation in unconventional reservoirs is increasing. This section will explore some common operations included as part of AI and analytics in stimulation that enhances production.

Machine learning can be used at all hydraulic fracturing stages from locating ideal spots for optimizing fracturing operations to post-fracturing operations and predictions. This ability to find complex patterns keeps aiding operations in unconventional reservoirs, especially well stimulation jobs like hydraulic fracturing. Azbukhanov et al. (2019) showcased the usage of ML in appropriate locations for performing hydraulic fractures in unconventional plays and fields. The authors developed an integrated approach to selecting candidate wells for fracturing using input such as starting oil and liquid production rates after one, three, and six months and an economic profitability indicator. Three separate ML models were used to model ideal candidates for hydraulic fracturing, including gradient boosting and random forest, which turned out to be the best models. These models had the least absolute error in their predictions for candidate well selections. The proposed approach was utilized in one of Rosneft's fields and has shown the potential to improve forecasting accuracy and economic efficiency. Such predictions' accuracy improves with increased production data and geological insights.

Wang et al. (2022a) utilized a new deep-learning methodology known as the artificial learning fracture model to process and predict the performance of hydraulic fractured wells in a quicker fashion. They use trained and simulated U-net and geomechanical models from reservoir simulators as training models for the deep learning model (three-dimensional models in two-dimensional slices). Further, the effect of well interference was also modeled using this approach. The model can estimate the flow in the Permian Basin to a decent degree. However, such models find trouble in transferring to different fields, especially since most modeling is learned from interpolation on simulated reservoir models. It is possible, however, to increase the efficacy of the AI model by combining field data with simulated models and geology. Such an example for selecting candidates for hydraulic fracturing is presented by Erofeev et al. (2021). The dataset included geology, transport and storage properties, depths, and oil/liquid rates before fracturing for target and neighboring wells was used. The study was used to predict the oil rates of wells six months into the future using gradient boosting. The data was gathered from eight oilfields with a comprehensive list of 700 and more well intervention incidents. The results were decent, with low mean average error rates. Such models can heavily inform future decision-making and help reduce the total number of intervention incidents in fractured well sites by choosing the well candidate with a lower probability of well intervention incidents.

Pumping rates in fracturing jobs are used for determining stages in hydraulic fracturing. This is essential for a successful fracturing job and ensuring maximum return on investment. Analytics has been used in many cases to identify better and inform systems of the initiation of various stages of pumping activity. Zhukova et al. (2022) utilized ML in data processing for real-time pumping operations during hydraulic fracturing and other treatment operations. The authors used the model to automate the classification of various stages of the treatment schedule performed by stage. Then the data labeling process is achieved with a signal processing application. The input data is related to pumping data to various stages of hydraulic fracturing and labels related to them. The authors compared the performance of two models, SVM and ANN, and the SVM outperformed the ANN in label classification using signal processing. The output labels from this exercise are used for labeling various stages of the fracturing operation, which is used for further data analysis and visualization for decision-making. This shows that ML can be used for preprocessing data workflow. Similar research on pumping data related to hydraulic fracturing was conducted by Ramirez and Iriarte (2019), using time-series analysis-based logistic regression to identify various hydraulic fracturing stages. This paper adds dimension to data-driven technology by adding time-series information to pumping stages. The model was trained with real-time field data, so the authors resorted to preprocessing steps such as loss-function implantation, smoothening techniques, and rate change analysis. The analysis provides an accuracy of 90% and shows a deviation from just 10 seconds from manually tagged classification periods. This type of analysis can further be utilized conversely to use pumping stages and fracture properties to determine wellhead parameters, as shown by Madusu and Rangarajan (2018). The authors used a similar time-series analysis to estimate surface pressure prediction during pumping operations for hydraulic fracturing. The authors utilized the LSTM algorithm, a recurring neural network on fluid and proppant rates on wells in the Permian Basin. The results showed the analysis could predict surface pressure at a 5% differential from recorded surface pressure. The significance of this study shows the forecasting and reaction of wells during hydraulic fracturing under varying conditions and compensates for sensor downtime.

The natural progression of such analysis is the eventual estimation of the production capability or the success of the frac jobs. Such studies can be formulated using data-driven analysis of existing wells along with additional supplementary data. Alarifi and Miskimins (2021) used a hybrid physics-based model with an ANN to predict the ultimate recovery from hydraulically fractured wells. The authors used data from 989 wells across four different unconventional reservoirs in the USA as training datasets. The ANN model was a feed-forward network with three hidden layers. It used data from drilling datasets, geology, and completion datasets to predict the recovery for a period of 3 months post-fracturing to up to 2 years post-fracturing. The physics-based models use DCA methodology to adapt variables specific to each well's production history. The absolute error value for prediction varies depending on the period for prediction: 3 months—0.85 to 0.95; up to 6 months—0.89 to 0.96; and up to 2 years—more significant than 0.94. Perrier et al. (2018) showed another model to showcase the effect of ML on the overall estimation of productivity from hydraulic fracturing. The authors used the data from over 600 wells and used ANN modeling to good effect to show the impact of varying geological and control variables on optimizing production from wells. Perrier et al. (2018) discuss the model's application to predict wells' performance and their classification. They discuss optimal models to involve geology as datasets in models and the effect of adding each feature to ML models. Further, they discuss the trouble of tuning hyperparameters specific to different fields. This shows the importance of improved data quality in model performance.

This type of analysis can be extended to other applications besides hydraulic fracturing. Liu et al. (2022) used a study to predict the productivity of acid fractures in a tight shale formation, tested in two fracturing operations in the Kuqa foreland area of the Tarim Basin in China. This model tests the performance of AI in analyzing the performance of acid treatments. Five AI models were initially used to test data modeling capabilities, including Logistic regression, SVM, AdaBoost, gradient boosting, and HistGradient Boosting Classifier. The Eigen data from eleven variables are considered for input, including logging porosity, logging Gamma, logging density, logging acoustic interpretation, logging deep resistivity, logging interpretation of gas saturation, fracture tendency, fracture dip, the angle between fracture and horizontal maximum principal stress, fracture density, and drilling fluid leakage. Logistic regression showed the highest accuracy in modeling the data with 0.67. Further Voting regressor was chosen as the method for optimizing parameter selection. Further, different models are chosen to optimize each parameter related to acid stimulation to select the best parameters while conducting operations, including optimizing proppant fracturing. Compared with the natural productivity, the productivity after stimulation was increased by 5.5 times and 21.5 times, respectively.

Besides these typical applications which are seen across several studies, each aspect of a stimulation job can be broken down to understand their performance and effect on the overall process. Simple analytics have been used to see the correlation between production and performance results with varying fracture parameters to see patterns for completion designs and fracturing optimization. Such a study was conducted by Bazan et al. (2021) to check the strategies which optimize cluster efficiency, fluid and mass placement, pressure management, and fracture half-lengths. This analysis can even be extended to the fluids used in stimulation treatments. Rastogi and Sharma (2019) used Machine learning models to see the impact of different chemicals used in hydraulic fracturing on the ultimate recovery from wells. Variable data such as production volumes, completion data, well design parameters, and petrophysical properties were used as input data for classification. Different models like SelectKBest (F Regression score), Tree-based regression, Mutual Info Regression, Recursive Feature Elimination (RFE), and Correlation-based Feature Selection (CFS) have been used by the authors. In this paper, Max Ingredient Mass (LBS) was utilized for the analysis, which ultimately indicated that different chemicals had different effects on productivity on other production windows in unconventional reservoirs.

Shut-in strategy is an essential aspect of hydraulic fracturing, which becomes extremely hard to understand in unconventional wells. A three-step process was utilized by Shahkarami et al. (2020) for developing shut-in strategies in multistage hydraulic fracturing operations. The steps involved analytics in assessing the effect of fracture interference, quantifying the long-term impact of shutting-in offset wells, and finally, using an ML model to optimize shut-in strategies in offset distance and communication. The dataset consists of time-series pressure data, daily production volumes, completion data, reservoir properties, and well trajectories for over 200 wells. This data was further analyzed to understand frac-hits, fracturing stage, and production history matching. The data was obtained from the Bakken and Three-Forks formations. Thus, it is seen that ML extends to providing strategies related to every aspect of the Hydraulic fracturing process.

The final push is the digitization and near-real-time utilization of these analytics and AI strategies in fieldwork. Many operators and service providers are working extensively on developing their operations in a smart digital fashion. Butler et al. (2022) showcased the usage of ML on fracturing activity in a live well site through cloud computing, edge computing, and sensors. This paper discusses a fully integrated and automated completion performed on the operator's four-well pad in the Bakken. The system reviews the effect of the automated system on completion performance, completion design, components of the system, and execution. Further, the

ML model was executed on live data to predict unusual events. The objective was two-fold: First was to test the automation of the frac fleet functionally; Second was to incorporate ML into completion designs in a real-time fracturing operation. The system has five main components: Digital workflows, Real-time data aggregation; Edge computing; ML and AI; AI-enabled equipment control. The controlled ML can help decide various completion parameters as fracturing steps. This paper represents a future of hydraulic fracturing operations where intelligent systems control fracturing operations in a "push-button" fashion, thus reducing potential errors and operating costs.

Enhanced Production Operations: Enhanced production technique has been used to improve production in different conventional and unconventional reservoirs. Reservoir simulation and economic feasibility operations have generally been used to monitor predictions from such operations for field development. However, ML, AI, and analytic workflows are being extensively used to account for heterogeneity and uncertainty in these operations. Please note that the paper in this section also covers a few applications in conventional fields.

Waterflood operation, a secondary-level recovery technique, is an operation necessary for optimization in unconventional reservoirs. Naturally, AI and analytics come into these fields as an aid to optimize and improve simulations and find better-operating parameters. Chen et al. (2019) proposed a method to predict the performance of water flooding activity using big data analytics and machine learning. The technique is used by superimposing the injector and producer model with existing well logging data and reservoir properties to form the least square principle and the particle swarm optimization algorithm. Synthetic data were used for training, providing an average error of 4.59%. The model was tested on the ZS36 offshore oilfield in the lower Liaongdang Bay in China. Different combinations of algorithms are used to improve the efficiency of these models. These include Miftakhov et al. (2020), where the authors utilized reinforcement learning, a neural network deep-learning system based on behavioral psychologists' theories. Deep Reinforcement Learning (RL) by pixel data is used to maximize the Net Present Value (NPV) of water flooding by changing the water injection rate. The optimization is tested on the 2D model, a vertical section of the SPE 10 model. It has been shown that RL can optimize water flooding in a 2D compressible reservoir with the 2-phase flow (oil-water) using pixel data. Using the methodology, the simulations showed a 15% increase in the method. Further analysis of this technique is first needed in 3-Dimensional modeling to check its feasibility, but the results indicate a possible means to optimize operations. An LSTM model was developed by Huang et al. (2021) to simulate the optimization of a carbonate reservoir. The model used 15 years of production data fed to the LSTM network to forecast oil production, gas-oil ratio (GOR), and water cut for a Water assisted gas injection operation. The method created a proxy model, which was 864% less intensive than a full-scale reservoir simulation and 261% less error rate than a reduced order model.

However, these models often lack field-data, and supplementing such reservoir models with wellhead information will lead to a more transferable model. Such an implementation was seen in the works of Timonv et al. (2021). They created a smaller dimension proxy model of wells for predicting oil production rates using ML models.

The process was tested in the oilfields of Western Siberia with 120 wells and underwater injection. Production data and properties from a computational reservoir model were input for the AI models, including random forest regressors, XGBoost, LSTM, and temporal convolutional networks. Other global optimization algorithms were used to solve differential equations, such as differential evolution, Limited memory Broyden Fletcher Goldfarb Shanno, Sequential Least Squares Programming, and Constrained Optimization BY Linear Approximations. The model tended to have better results with LSTM and TCN models in conjunction with the Limited memory Broyden Fletcher Goldfarb Shanno method.

Such application in secondary recovery operations can be extended to more complex issues, such as connectivity between wells or injector-producer combinations. Ujjwal et al. (2020) present a case study of advanced analytics applied to a mature waterflood field where available data was rapidly integrated into an integrated visualization dashboard providing an example of the integration of analytics with near-real-time data streams. Machine learning was used to identify injector producer connectivity and allocation factors. The data was obtained from a mature field with 15-year production history and contained over 100 producers. Varying data sources were used to create an analytical dashboard after data cleaning. The following attributes were identified as indications of producer injector connectivity: (a) Correlation in injection water salinity and produced water salinity, (b) Correlation in injection rate trends and produced liquid rate trends, (c) Clear jump in producer ESP intake pressure. Machine learning was used to cluster producers based on their produced water salinity trends, enabling shortlisting injectors potentially connected to them. Another machine learning algorithm was then used to estimate connectivity factors between producers and injectors based on their distance and correlation in injection-production trends. The paper indicates a 5% gain in oil rate production.

Carbon-dioxide (CO_2) injection into well sites and storage of CO_2 in reservoirs are done heavily in many areas of the world, especially in North America. It is a field of heavy scrutiny; thus, improving reservoirs' storage capability to reduce CO_2 leakage is essential. Katterbauer et al. (2021) utilized a 100-layer neural network (deep-learning system) in conjunction with genetic algorithms to optimize the cycle time for CO_2 injection in an enhanced oil recovery (EOR) process. The process is a process to improve the overall water-gas injection cycling. The deep learning algorithm had injection and production data as the only input criteria and several nonlinear optimization parameters tuned using genetic algorithms. Rich and lean-gas injection is also gaining traction worldwide to improve tertiary recovery factors from mature fields, especially in the Bakken and Permian basins. Thus, analyzing the composition of the injection stream to be integrated into complicated reservoir models is of the essence. Seikhanov et al. (2020) utilized a similar methodology to predict the gas injection (CO_2) production potential in injection wells. Various machine learning algorithms were tested to develop predictive models for the cyclic injection process using a dataset obtained from numerical simulation runs. Five operational parameters were used as inputs for the model: injection rate, injection time, soaking time, cycle rate, the composition of three injected gases (CO_2, N_2, CH_4), and output as the economic efficiency for each year derived by simulations. The model utilized were random forest regressor, neural networks, and linear regression, where random forest showed the best estimation of future prediction with an R-score of 0.93. These computational models are commonly used to reduce computation time and improve simulation efficiency.

One of the critical tools used in proper reservoir management is geological and fluid flow models of the reservoirs. These models informed the potential of the reservoirs, and new measures for continued maintenance of the reservoirs are based on these reservoir models. Hence, more efficient and better-performing models are essential, and AI is used to improve these models' performance. You et al. (2022) showed an example of such a hybrid methodology by using ML-assisted workflows to optimize a reservoir model. The model utilized here is the Farnsworth model from Texas, where they tried replacing multi-objective optimization with a Gaussian-support vector regression model to optimize proxy models of the reservoir simulator. Further, a multi-objective PSO generates the Pareto front of the CO_2 injection activity. The proposed work introduces a novel concept to increase the computational efficiency of the proposed workflow and guarantee the high accuracy of the obtained optimized results. It extends the utility of the model by helping to solve multiple control parameters. Thus, ML models in the study showcase their potential as stand-n for multi-objective solvers for complex numerical simulations.

High computational costs are the major drawback of high-fidelity reservoir models. However, techniques using AI have started to emerge to merge varying low-fidelity models to inform reservoir management decisions. He et al. (2021) also used ML models to enhance reservoir simulations and improve their useability and applicability. In their work, He et al. (2021) use ML models to generate surrogate models of high-fidelity reservoir simulation runs for faster implementation under varying conditions. The authors utilized different ANN models to simulate proxy models, and further overall implementation of these proxies into the production model was optimized for gas injection using a genetic algorithm. The work from this paper once again showcases the potential of ML to be used as a low-fidelity quick implantation solution for higher-fidelity numerical optimization models. These simulation models are also being utilized in thermal treatment applications, mainly to inform the decision of cycling steam treatments. A proxy reservoir model (low fidelity for quick deployment) was demonstrated by Izgec et al. (2022). They showed the implementation of the ANN model integrated with a full-physics model for a cyclic steam stimulation operation. The proxy model implemented using ML techniques lowers a full-scale reservoir numerical simulation model's overall computational model and provides good matches to the actual output from the numerical simulation. Another application of such ML models in improving physics-based numerical simulation is also seen in Arfie et al. (2022), where they use clustering algorithms to enhance the optimization of the production system on a thermal operation at the Schoonebeek Field.

Production String and Wellsite Maintenance: Downhole production string maintenance is vital to the continued operations of wellsites. It is essential to realize potential failure points and mitigate them if possible. Pattern recognition and understanding of time-dependent variables is a highly specialized skill within analytics in several industries. These practices are being increasingly implemented in the oilfield. Iriarte et al. (2021) published a paper describing methods to check abnormal pressure buildups and analyze the root cause of such events, especially during hydraulic fracturing. The authors implemented six different ML models using parameters related to the wellhead, bottom hole parameters, geology, and pumping rates. These models can recognize varying events based on ongoing pressure buildup and adjoining parameters, leading to less complex solutions to various problems. Automation played a crucial role in efficiently analyzing several hundred stages of data while considering multiple treatment channels and their interaction with the treating pressure. This study was employed in various wells in the Eagle Ford formation to analyze the reasons for abnormal pressure buildups. These layers identified significant changes in the slurry rate, proppant concentration, and friction reducer concentration.

This management is extended even for identifying well stimulation techniques in case of noticeable lower production. Often, however, the treatment provided to specific wells might not be ideal and can lead to well failures of further problems. Kellogg et al. (2018) gives an example of identifying proper treatment procedures. Kellogg et al. (2018) showed the use of ML in screening candidate wells for acid jobs for wellbore damage removal. The paper concentrates on a dataset from the Wilmington Oilfield in California, USA, where 500 producer wells data were used, which had a total of about 3900 acid jobs amongst them. The dataset was completed with information about Production/injection history, Reservoir properties, Acid type and volume, Delivery mechanism, Formation damage, Well completion design, and Surface facility properties. Different models were used for classifying wells to find ideal candidates for varying acid job treatments, including logistic regressor, SVM, random forests, and gradient boosting machines. Logistic regressors could produce decent results with a reasonable accuracy percentage (71% for producers and 77% for injectors). The model has recently been deployed in over 1000 wells in the Wilmington Field, which subject matter experts overlook.

Such pre-meditated intervention techniques can be seen even in the case of other issues like scale formation. Al-Hajiri and Alghamdi (2019) utilized the ML model to predict the probability of scale formation and use the uncertainty related to the scale level to form appropriate inhibition measures for wells. The model uses ionic composition, pH, sample collection and inspection dates, and scale formation events. This data is grouped on production conditions, followed by the calculation of chemical indices. The ML models (SVM, K-nearest neighbors, gradient boosting, and decision trees) are used to classify data to predict the probability of scaling along with uncertainty resulting from creating a confusion matrix. The uncertainty measurements inform the scale inhibition methods suggested by the program. The model even provides an expenditure report. An accuracy of 95% and a three-fold cost-saving potential are observed. A similar model for predicting barium sulfate and calcium carbonate scales in wells has been provided by Ugoyah et al. (2022). A decision tree-based ML model was used for classification, which provided great results for a dataset obtained from the Niger Delta and North Sea fields.

The well pads and processing units have a lot of parts, both static and dynamic. These parts must be constantly maintained to high standards to ensure sustained operations over long periods. Traditionally maintenance of this equipment has been a traditional inspection-based process of preventative and curative measures. However, with improved sensory systems, cloud computing, and data analytics, these maintenance and intervention workovers are taking on new forms. These measures ensure proper production facility maintenance with lower operational costs due to interventions. There are several examples of how AI, ML, and data analytics help maintain facilities, similar to those done in a larger industrial facility with the employment of digital twins, unmanned aerial systems, and SCADA systems. However, in this section, we will take a few examples of oilfield maintenance and well workovers being benefited from analytics and AI. Beduschi et al. (2021) showcased the usage of ML software in daily oil and gas field operations through a combination of modeling tools, including feature extraction and ML-enabled anomaly detection. The developed predictive tool has been implemented on operating sites and demonstrated the capability to generate accurate warnings and detect anomalies to be processed by the maintenance engineers. It has mainly shown efficacy in reducing flaring, reducing downtimes, and keeping maintenance costs in check by enabling predictive maintenance. The tool is individually implemented on different operating units, which uses Principal component analysis for feature extraction and data preprocessing and follows it up with kernel density estimation (KDE). The KDE tool enables the detection of deviation from normal operating conditions. If an anomaly is detected, the fault is isolated, and the specific sensor on which the variation has occurred evaluates the anomaly using a failure mode matrix, which showcases the most likely failure mode. The KDE is a non-parametric method to estimate the probability density function of a random variable. The model is trained on historical data and needs real-time sensor data for activation. This showcases the usage of ML in real time in helping promote preventative maintenance and reducing OPEX.

The usage of analytics also integrates well with other emerging technologies, such as unmanned aerial systems (UAS), digital twins, and augmented reality. Akbari (2021) describes the usage of various advanced technologies in offshore inspection and maintenance of oil and gas facilities. He suggests three significant advances:

- Application of digital twins: The implementation of a digital twin system is the orthogonal mapping of physical infrastructure into an equivalent digital version where every aspect of the facility can be supervised and checked through a computer. Further details are added to the system via edge computing and the power of sensors to have a real-life backup of a physical facility and remote access to the state of technology, reducing operational costs.
- Application of smart glasses: The usage of smart glasses enables virtual reality–based inspection and cross-checking of digital twins in a fashion that is otherwise fantasy-like. With its basic hands-free video streaming and communication functionality, the AR headset enabled a remote pre-commissioning, certification, and audit with vendors and independent certifiers during emergencies.
- Application programming interfaces (APIs): APIs are commonly used to sync various software and tools. The APIs tend to create a wrapper and convert the functionalities of one program into the context of another. An example of such an application would be GIS programs accessing CAD software to integrate infrastructure design. This enables easy access and communication between various applications. This allows easy integration of analytic workflows with digital twin infrastructure.

The usage of digital twins is slowly picking up. It will soon be poised to revolutionize the system's maintenance program by integrating techniques such as

SCADA, project management tools, and cloud computing. Another example of the integration of analytics with advanced technologies is shown by Shaji et al. (2022). They give a sample of computer vision applications in handling oil and gas facilities and how changes in images over periods of time help delineate various safety hazards in workspaces. The authors give an example of thermographic imaging of the workspace and how unmanned aerial systems (UAS) can be used for anomaly detection by comparing current images to ideal state images. The anomaly detection enables training convolutional neural networks that match hazard categories. Shaji et al. (2022) also explains the mechanisms for anomaly detection in processing units in an industrial facility where anomalies in operating conditions can be signaled out and reported as potential failures. The usage of ML can be extended to various other models for the maintenance and operation of oilfield activities. This includes the use of ML with other digital technology as well as traditional technology. This example extends to even the change in culture in many companies in developing the skills required for various positions. There has been a noted increase in the usage of automation and digital technology in all areas of oil and gas, as seen by the change in the required skill sets according to Clemens and Gruber (2020).

The production sites in the oil and gas industry are increasingly adopting digital technologies. The application of artificial intelligence and machine learning, which was purely academic till the early 2010s, is now becoming more and more applicable in real-world scenarios. As seen in several of the papers discussed, tangible benefits of implementing these technologies in the field are being seen:

- Reduced operational costs due to higher oversight
- Preparedness for unexpected behavior from wells
- Higher levels of health and safety standards
- Better preventative maintenance scheduled, leading to fewer downtimes
- Better digital infrastructure, leading to better remote control of units

These tangible benefits are being seen across the board with changes in trends in hiring strategies and business decisions. Access to data is influencing better decision-making across the board. With the improvement in technology and data sciences, there will be increased adoption of advanced analytics in production.

REFERENCES

4earthintelligence. (2021, May 10). *What is geospatial data*? 4EI. www.4earthintelligence.com/insights/what-is-geospatial-data/

Abma, R., Zhang, Q., Arogunmati, A., & Beaudoin, G. (2012). An overview of BP's marine independent simultaneous source field trials. *Society of exploration geophysicists international exposition and 82nd annual meeting 2012*, SEG 2012, 41–45. https://doi.org/10.1190/SEGAM2012-1404.1

Abu-Mostafa, Y. S. (2012). Machines that think for themselves. *Scientific American*, *307*. https://doi.org/10.1038/scientificamerican0712-78

Ahmadi, M. A., Galedarzadeh, M., & Shadizadeh, S. R. (2016). Low parameter model to monitor bottom hole pressure in vertical multiphase flow in oil production wells. *Petroleum*, *2*, 258–266. https://doi.org/10.1016/j.petlm.2015.08.001

Ahmadi, M. A., Shadizadeh, S. R., Shah, K., & Bahadori, A. (2018). An accurate model to predict drilling fluid density at wellbore conditions. *Egyptian Journal of Petroleum*, *27*(1), 1–10.

Ahmadi, M. A., Soleimani, R., Lee, M., Kashiwao, T., & Bahadori, A. (2015). Determination of oil well production performance using artificial neural network (ANN) linked to the particle swarm optimization (PSO) tool. *Petroleum*, *1*(2), 118–132. https://doi.org/10.1016/j.petlm.2015.06.004

Ahmed, A., Elkatatny, S., Ali, A., Abughaban, M., & Abdulraheem, A. (2020). Application of artificial intelligence techniques in predicting the lost circulation zones using drilling sensors. *Journal of Sensors*, *2020*.

Akanji, L. T., Dala, J., Bello, K., Olaekan, O., & Jadhawar, P. (2019). Application of artificial intelligence in well screening and production optimization in Oredo Oilfields, Niger Delta, Nigeria. *Paper presented at the SPE nigeria annual international conference and exhibition*, Lagos, Nigeria. https://doi.org/10.2118/198877-MS

Akbari, F. (2021). Intelligent digital twins and augmented reality in inspection and maintenance. *Paper presented at the Abu Dhabi international petroleum exhibition and conference*, Abu Dhabi. https://doi.org/10.2118/207659-MS

Alarifi, S. A., & Miskimins, J. (2021). A new approach to estimating ultimate recovery for multistage hydraulically fractured horizontal wells by utilizing completion parameters using machine learning. *SPE Production and Operations*, *36*(3), 468–483. https://doi.org/10.2118/204470-PA

Al-Hajiri, N. M., & Alghamdi, A. (2019). Scale prediction and inhibition design using machine learning techniques. *Paper presented at the SPE gas and oil technology showcase and conference*, Dubai. https://doi-org.ezproxy.library.und.edu/10.2118/198646-MS

Arabjamaloei, R., & Shadizadeh, S. (2011). Modeling and optimizing rate of penetration using intelligent systems in an Iranian southern oil field (ahwaz oil field). *Petroleum Science and Technology*, *29*, 1637–1648.

Araya-Polo, M., Dahlke, T., Frogner, C., Zhang, C., Poggio, T., & Hohl, D. (2017). Automated fault detection without seismic processing. *Leading Edge*, *36*(3), 208–214. https://doi.org/10.1190/TLE36030208.1

Arfie, M., Ghodke, N., & Groenbroek, K. (2022). Artificial intelligence for production optimization in schoonebeek thermal EOR field. *Paper presented at the SPE EuropEC—Europe energy conference featured at the 83rd EAGE annual conference and exhibition*, Madrid, Spain. https://doi-org.ezproxy.library.und.edu/10.2118/209670-MS

Avseth, P., Mukerji, T., & Mavko, G. (2010). *Quantitative seismic interpretation: Applying rock physics tools to reduce interpretation risk.* https://books.google.com/books?hl=fr&lr=&id=rKYhAwAAQBAJ&oi=fnd&pg=PR1&dq=quantitative+seismic+interpretation&ots=u4FI6BJtCo&sig=LM2pOxL9Qmh-CdoyepV2Np8hwY4

Azbukhanov, A. F., Kostrigan, I. V., Bondarenko, K. A., Semenova, M. N., Sereda, I. A., & Yulmukhametov, D. R. (2019). Selection of wells for hydraulic fracturing based on mathematical modeling using machine learning methods (Russian). *OIJ*, *11*, 38–42.

Baardman, R. H., & Hegge, R. F. (2020). Machine learning approaches for use in deblending. *Leading Edge*, *39*(3), 188–194. https://doi.org/10.1190/TLE39030188.1

Baki, S., Temizel, C., & Dursun, S. (2021). Well completion optimization in unconventional reservoirs using machine learning methods. *Paper presented at the SPE annual technical conference and exhibition*, Dubai. https://doi-org.ezproxy.library.und.edu/10.2118/206241-MS

Baldwin, J., Bateman, R., & Analyst, C. W.-T. L. (1990). Application of a neural network to the problem of mineral identification from well logs. *Onepetro.Org*. Retrieved September 28, 2022, from https://onepetro.org/petrophysics/article-abstract/170779/Application-Of-A-Neural-Network-To-The-Problem-Of

Barbosa, L. F. F., Nascimento, A., Mathias, M. H., & de Carvalho, J. A. (2019). Machine learning methods applied to drilling rate of penetration prediction and optimization—a review. *Journal of Petroleum Science and Engineering*, *183*, 106332.

Bataee, M., Irawan, S., & Kamyab, M. (2014). Artificial neural network model for prediction of drilling rate of penetration and optimization of parameters. *Journal of the Japan Petroleum Institute*, *57*(2), 65–70.

Batruny, P., Zubir, H., Slagel, P., Yahya, H., Zakaria, Z., & Ahmad, A. (2021). Drilling in the digital age: Machine learning assisted bit selection and optimization. https://doi.org/10.2523/IPTC-21299-MS

Bazan, L. W., Meyer, B. R., & Tran, V. P. (2021). Innovative analysis of treatment well data determines cluster efficiency, assesses fracture complexity, and provides fracture geometry in unconventional reservoirs. *Paper presented at the SPE/AAPG/SEG Asia pacific unconventional resources technology conference*, Virtual. https://doi-org.ezproxy.library.und.edu/10.15530/AP-URTEC-2021-208315

Beasley, C. J. (2008). A new look at marine simultaneous sources. *Leading Edge (Tulsa, OK)*, *27*(7), 914–917. https://doi.org/10.1190/1.2954033

Beduschi, F., Turconi, F., De Gregorio, B., Abbruzzesse, F., Tiozzo, A., Amabili, M., & Prospero, A. (2021). Optimizing rotating equipment maintenance through machine learning algorithm. *Paper presented at the Abu Dhabi international petroleum exhibition and conference*, Abu Dhabi. https://doi.org/10.2118/207657-MS

Bekara, M., & van der Baan, M. (2007). Local singular value decomposition for signal enhancement of seismic data. *Geophysics*, *72*(2). https://doi.org/10.1190/1.2435967

Belov, D., Ba, S., Liu, J. T., Kolyshkin, A., & Sergio Daniel Rocchio. (2021, October). Data-driven PHM solution for health monitoring of mud motor power sections while drilling. *Paper presented at the SPE Europec featured at 82nd EAGE conference and exhibition*, Amsterdam, The Netherlands. https://doi.org/10.2118/205219-MS

Berkhout, A. J. (2008). Changing the mindset in seismic data acquisition. *Leading Edge (Tulsa, OK)*, *27*(7), 924–938. https://doi.org/10.1190/1.2954035

Bermudez, F., Nahhas, N., Yazdani, H., Tan, M. L., & Shono, M. (2021). Unlocking the potential of electrical submersible pumps: The successful testing and deployment of a real-time artificially intelligent system, for failure prediction, run life extension, and production optimization. *Paper presented at the Abu Dhabi international petroleum exhibition and conference*, Abu Dhabi. https://doi.org/10.2118/207839-MS

Bilgesu, H. I., Al-Rashidi, A. F., Aminian, K., & Ameri, S. (2001, March). An unconventional approach for drill-bit selection. *Paper presented at the SPE middle east oil show*, Manama, Bahrain. https://doi.org/10.2118/68089-MS

Blacquière, G., & Nakayama, S. (2019). Optimum seismic acquisition geometry design with the help of artificial intelligence. *Onepetro.Org*, 117–121. https://doi.org/10.1190/segam2019-3202558.1

Bouska, J. (2010). Distance separated simultaneous sweeping, for fast, clean, vibroseis acquisition. *Geophysical Prospecting*, *58*(1), 123–153. https://doi.org/10.1111/J.1365-2478.2009.00843.X

Burger, H. C., Schuler, C. J., & Harmeling, S. (2012). Image denoising with multi-layer perceptrons, part 1: Comparison with existing algorithms and with bounds. *Journal of Machine Learning Research*. http://arxiv.org/abs/1211.1544

Busch, J. M., Fortney, W. G., & Berry, L. N. (1987). Determination of lithology from well logs by statistical analysis. *SPE Formation Evaluation*, *2*(4), 412–418. https://doi.org/10.2118/14301-PA

Butler, E., Pertuoso, D., Hua, G., & Stark, P. (2022). Automated hydraulic fracturing integrated with predictive machine learning. *Paper presented at the SPE hydraulic fracturing technology conference and exhibition*, The Woodlands, TX. https://doi-org.ezproxy.library.und.edu/10.2118/209165-MS

Calad, C., Raifee, J., Sarma, P., Zhao, Y., & Betancourt, D. (2022). Combining machine learning and reservoir simulation for robust optimization of completion design and well location of unconventional wells. *Paper presented at the SPE/AAPG/SEG unconventional resources technology conference*, Houston, TX. https://doi-org.ezproxy.library.und.edu/10.15530/urtec-2022-3723680

Carpenter, C. (2020). AI-based decline-curve analysis manages reservoir performance. *Journal of Petroleum Technology*, *72*(9), 58–59. https://doi.org/10.2118/0920-0058-JPT

Chen, C., Yang, M., Han, X., & Zhang, J. (2019). Water flooding performance prediction in layered reservoir using big data and artificial intelligence algorithms. *Paper presented at the Abu Dhabi international petroleum exhibition and conference*, Abu Dhabi. https://doi.org/10.2118/197585-MS

Chen, Y., & Pock, T. (2016). Trainable nonlinear reaction diffusion: A flexible framework for fast and effective image restoration. *Ieeexplore.Ieee.Org*. Retrieved September 29, 2022, from https://ieeexplore.ieee.org/abstract/document/7527621/?casa_token=eY-rVU9g62AAAAAA:4wAVzC2OHDkAiRTSN5KqUJI3ETQoMULuwbQpSZ1J4SwnEkUznkRHpv9byJK6kzag1uvTRBkvb5E

Chen, Y., Liu, J., Xie, L., Hu, Y., Shu, H., Luo, L., Zhang, L., Gui, Z., & Coatrieux, G. (2017). Discriminative prior—prior image constrained compressed sensing reconstruction for low-dose CT imaging. *Scientific Reports*, *7*(1), 1–17. https://doi.org/10.1038/s41598-017-13520-y

Chopra, S., & Marfurt, K. J. (2007). *Seismic attributes for prospect identification and reservoir characterization*. https://doi.org/10.1190/1.9781560801900

Clemens, T., & Gruber, M. V. (2020). Impact of digitalization on the way of working and skills development in hydrocarbon production forecasting and project decision analysis. *Paper presented at the SPE Europec*, Virtual. https://doi.org/10.2118/200540-MS

Coléou, T., Poupon, M., & Azbel, K. (2012). Unsupervised seismic facies classification: A review and comparison of techniques and implementation. *The Leading Edge*, *22*(10), 942–953. https://doi.org/10.1190/1.1623635

Connolly, T. M. (2015). *Database systems* (6th ed.). Pearson.

Cross, T., Niederhut, D., Cui, A., Sathaye, K., & Chaplin, J. (2021). Quantifying the diminishing impact of completions over time across the bakken, eagle ford, and wolfcamp using a multi-target machine learning model and SHAP values. *Paper presented at the SPE/AAPG/SEG unconventional resources technology conference*, Houston, TX. https://doi-org.ezproxy.library.und.edu/10.15530/urtec-2021-5633

de Matos, M. C., Osorio, P. L. M., & Johann, P. R. S. (2006). Unsupervised seismic facies analysis using wavelet transform and self-organizing maps. *Geophysics*, *72*(1). https://doi.org/10.1190/1.2392789

Delfiner, P., Peyret, O., & Serra, O. (1987). Automatic determination of lithology from well logs. *SPE Formation Evaluation*, *2*(3), 303–310. https://doi.org/10.2118/13290-PA

Dellinger, J., Ross, A., Meaux, D., Brenders, A., Gesoff, G., Etgen, J. T., Naranjo, J., Openshaw, G., & Harper, M. (2016). Wolfspar®, an "fwi-friendly" ultra-low-frequency marine seismic source. *SEG Technical Program Expanded Abstracts*, *35*, 4891–4895. https://doi.org/10.1190/SEGAM2016-13762702.1

Di, H., Gao, D., & Alregib, G. (2018). 3D structural-orientation vector guided autotracking for weak seismic reflections: A new tool for shale reservoir visualization and interpretation. *Interpretation*, *6*(4), SN47–SN56. https://doi.org/10.1190/INT-2018-0053.1

Di, H., Li, Z., Maniar, H., & Abubakar, A. (2020). Seismic stratigraphy interpretation by deep convolutional neural networks: A semisupervised workflow. *Geophysics*, *85*(4), WA77–WA86. https://doi.org/10.1190/GEO2019-0433.1

Doyen, P. M. (2007). *Seismic reservoir characterization: An earth modelling perspective*. https://doi.org/10.3997/9789073781771.

Dubois, M., Bohling, G., Byrnes, A., & Seals, S. (2003). *Extracting lithofacies from digital well logs using artificial intelligence, Panoma (council grove) field, Hugoton embayment, Southwest Kansas: Abstract*. http://archives.datapages.com/data/tgs/tgs-sp/data/051/051001/26_tgs-sp0510026.htm

Dubois, M. K., Bohling, G. C., & Chakrabarti, S. (2007). Comparison of four approaches to a rock facies classification problem. *Computers & Geosciences*, *33*(5), 599–617. https://doi.org/10.1016/J.CAGEO.2006.08.011

Ekeregbe, M. P., Khalaf, M. S., & Robello, S. (2021). Dull bit grading using video intelligence. *Paper presented at the SPE annual technical conference and exhibition*, Dubai, September. https://doi.org/10.2118/206254-MS

Erofeev, A., Orlov, D., Perets, D., & Koroteev, D. (2021). AI-based well ranking system for hydraulic fracturing events. *Paper presented at the SPE middle east oil and gas show and conference.* https://doi-org.ezproxy.library.und.edu/10.2118/204571-MS

Fan, L., & Poh, K. L. (2007). A comparative study of PCA, ICA and class-conditional ICA for naïve bayes classifier. *Lecture notes in computer science (Including subseries lecture notes in artificial intelligence and lecture notes in bioinformatics)*, 4507 LNCS, 16–22. https://doi.org/10.1007/978-3-540-73007-1_3/COVER

Garjan, Y. S., & Ghaneezabadi, M. (2020). Machine learning interpretability application to optimize well completion in montney. *Paper presented at the SPE Canada unconventional resources conference*, Virtual. https://doi-org.ezproxy.library.und.edu/10.2118/200019-MS

Geng, Z., Wu, X., Shi, Y., & Fomel, S. (2019). *Relative geologic time estimation using a deep convolutional neural network.* Onepetro.Org. https://doi.org/10.1190/segam2019-3214459.1

Grana, D., Azevedo, L., & Liu, M. (2020). A comparison of deep machine learning and Monte Carlo methods for facies classification from seismic data. *Geophysics*, *85*(4), WA41–WA52. https://doi.org/10.1190/GEO2019-0405.1

Gul, S., Johnson, M. D., Karimi Vajargah, A., Ma, Z., Hoxha, B. B., & Eric van Oort. (2019). A data driven approach to predict frictional pressure losses in polymer-based fluids. *Paper presented at the SPE/IADC international drilling conference and exhibition*, The Hague, The Netherlands, March. https://doi.org/10.2118/194132-MS

Gul, S., Vajargah, A., van Oort, E., Leulseged, A., & Cayeux, E. (2020). *Automated solids content determination in drilling and completions fluids.* https://doi.org/10.2118/201368-MS

Guo, H., Marfurt, K. J., & Liu, J. (2009). Principal component spectral analysis. *Geophysics*, *74*(4). https://doi.org/10.1190/1.3119264

Han, J. E. (2012). Data preprocessing. In J. E. Han (Ed.), *Data mining: Concepts and techniques* (pp. 83–120). Elsevier Incorporated.

Harris, P. E., & White, R. E. (1997). Improving the performance of f-x prediction filtering at low signal-to-noise ratios. *Geophysical Prospecting*, *45*(2), 269–302. https://doi.org/10.1046/J.1365-2478.1997.00347.X

Hastie, T., Tibshirani, R., & Friedman, J. (2009). *The elements of statistical learning: Data mining, inference, and prediction* (2nd ed.). Springer Series in Statistics.

He, X., Qiao, T., Santoso, R., Hoteit, H., Alsinan, M. M., & Kwak, H. T. (2021). Gas injection optimization under uncertainty in subsurface reservoirs: An integrated machine learning-assisted workflow. *Paper presented at the ARMA/DGS/SEG 2nd international geomechanics symposium*, Virtual. https://onepetro-org.ezproxy.library.und.edu/armaigs/proceedings/IGS21/All-IGS21/ARMA-IGS-21-009/473069?searchresult=1

Herrmann, F. J. (2010). Randomized sampling and sparsity: Getting more information from fewer samples. *Geophysics*, *75*(6). https://doi.org/10.1190/1.3506147

Huang, R., Wei, C., Li, B., Yang, J., Wu, S., Xu, X., Ou, Y., Xiong, L., Lou, Y., Li, Z., Deng, Y., & Zhang, C. (2021). Prediction and optimization of WAG flooding by using LSTM neural network model in middle east carbonate reservoir. *Paper presented at the Abu Dhabi international petroleum exhibition and conference*, Abu Dhabi. https://doi-org.ezproxy.library.und.edu/10.2118/207584-MS

Iriarte, J., McConnell, M., Hoda, S., Siegel, E., & Wolfbrandt, C. (2021). Employing a suite of machine learning algorithms in a holistic approach to trouble stage recognition and failure diagnostics. *Paper presented at the SPE hydraulic fracturing technology conference and exhibition*, Virtual. https://doi-org.ezproxy.library.und.edu/10.2118/204145-MS

Izgec, O., Li, Y., Javaheri, M., Tawfik, M., & Osako, I. (2022). A simulation augmented machine learning approach for cyclic steam stimulation development targeting lower carbon/higher return. *Paper presented at the SPE western regional meeting*, Bakersfield, CA. https://doi-org.ezproxy.library.und.edu/10.2118/209336-MS

Jacques, A., Daguet-Schott, C., Billat, A., & Brouard, B. (2022). Optimizing completions strategies using low-cost data learning. *Paper presented at the SPE/AAPG/SEG unconventional resources technology conference*, Houston, TX. https://doi-org.ezproxy.library.und.edu/10.15530/urtec-2022-3701794

Jahanandish, I., Salimifard, B., & Jalalifar, H. (2011). Predicting bottomhole pressure in vertical multiphase flowing wells using artificial neural networks. *Journal of Petroleum Science and Engineering*, *75*(4), 336–342. https://doi.org/10.1016/j.petrol.2010.11.019

Jain, V., & Seung, S. (2008). Natural image denoising with convolutional networks. *Proceedings.Neurips.Cc*. Retrieved September 29, 2022, from https://proceedings.neurips.cc/paper/2008/hash/c16a5320fa475530d9583c34fd356ef5-Abstract.html

Juodyte, M. (2017). *Overview: Data mining pipeline*. Technical University of Munich.

Kandziora, C. (2019). Applying artificial intelligence to optimize oil and gas production. *Paper presented at the offshore technology conference*, Houston, TX. https://doi.org/10.4043/29384-MS

Kapur, L., Lake, L., & Sepehrnoori, K. (1998). Facies prediction from core and log data using artificial neural network technology. *Onepetro.Org*. Retrieved September 28, 2022, from https://onepetro.org/SPWLAALS/proceedings-abstract/SPWLA-1998/All-SPWLA-1998/SPWLA-1998-PP/19730

Katterbauer, K., Marsala, A., & Al Qasim, A. (2021). A deep learning wag injection method for Co2 recovery optimization. *Paper presented at the SPE middle east oil and gas show and conference*. https://doi.org/10.2118/204711-MS

Karnik, S., Gupta, S., & Baihly, J. (2021). Machine intelligence for integrated workover operations. *Paper presented at the SPE/ICoTA well intervention conference and exhibition*, Virtual. https://doi.org/10.2118/204423-MS

Kellogg, R. P., Chessum, W., & Kwong, R. (2018). Machine learning application for wellbore damage removal in the Wilmington field. *Paper presented at the SPE western regional meeting*, Garden Grove, CA. https://doi-org.ezproxy.library.und.edu/10.2118/190037-MS

Ketmalee, T., Sirisawadwattana, J., Piyajunya, T., Ngamkamollert, K., Bandyopadhyay, P., & Whangkitjamorm, J. (2019). Use of artificial intelligence to measure gas flow rate, Bongkot asset. *Paper presented at the Abu Dhabi international petroleum exhibition and conference*, Abu Dhabi. https://doi.org/10.2118/197454-MS

Khan, H., & Louis, C. (2021). An artificial intelligence neural networks driven approach to forecast production in unconventional reservoirs—Comparative analysis with decline curve. *Paper presented at the international petroleum technology conference*, Virtual. https://doi.org/10.2523/IPTC-21350-MS

Khan, M. R., Alnuaim, S., Tariq, Z., & Abdulraheem, A. (2019). Machine learning application for oil rate prediction in artificial gas lift wells. *Paper presented at the SPE middle east oil and gas show and conference*, Manama, Bahrain. https://doi.org/10.2118/194713-MS

Kianinejad, A., Kansao, R., Maqui, A., Kadlag, R., Hetz, G., Ibrahima, F., Suicmez, V., Darabi, H., & Castineira, D. (2019). Artificial-intelligence-based, automated decline curve analysis for reservoir performance management: A giant sandstone reservoir case study. https://doi.org/10.2118/197142-MS

Kohonen, T. (1989). *Self-organizing feature maps* (pp. 119–157). https://doi.org/10.1007/978-3-642-88163-3_5

Kroode, T. F., Bergler, S., Corsten, C., de Maag, J. W., Strijbos, F., & Tijhof, H. (2013). Broadband seismic data: The importance of low frequencies. *Geophysics*, *78*(2), WA3–WA14. https://doi.org/10.1190/GEO2012-0294.1

Lawal, T., Ashok, P., van Oort, E., Zheng, D., & Isbell, M. (2021, March). Real-time prediction of mud motor failure using surface sensor data features and trends. *Paper presented at the SPE/IADC international drilling conference and exhibition*, Virtual. https://doi.org/10.2118/204099-MS

Li, X., Miskimins, J. L., & Hoffman, B. T. (2014). A combined bottom-hole pressure calculation procedure using multiphase correlations and artificial neural network models. *Presented at the SPE annual technical conference and exhibition*, Amsterdam, The Netherlands. https://doi.org/10.2118/170683-MS

Lin, Y., Wang, S., Thiagarajan, J., Guthrie, G., & Coblentz, D. (2017). Efficient data-driven geologic feature detection from pre-stack seismic measurements using randomized machine-learning algorithm. *Geophysical Journal International, 215*(3), 1900–1913.

Lindberg, D. V., & Grana, D. (2015). Petro-elastic log-facies classification using the expectation–maximization algorithm and hidden Markov models. *Mathematical Geosciences, 47*(6), 719–752. https://doi.org/10.1007/S11004-015-9604-Z

Liu, D., Wang, W., Chen, W., Wang, X., Zhou, Y., & Shi, Z. (2018). Random noise suppression in seismic data: What can deep learning do? *SEG Technical Program Expanded Abstracts, 2016–2020*. https://doi.org/10.1190/SEGAM2018-2998114.1

Liu, H., Cui, L., Liu, Z., Zhou, C., Yao, M., Ma, H., & Liu, Q. (2022). Using machine learning method to optimize well stimulation design in heterogeneous naturally fractured tight reservoirs. *Paper presented at the SPE Canadian energy technology conference*, Calgary, Alberta, Canada. https://doi-org.ezproxy.library.und.edu/10.2118/208971-MS

Liu, W., Cheng, Q., Liu, L., Wang, Y., & Zhang, J. (2020). Accelerating high-resolution seismic imaging by using deep learning. *Applied Sciences (Switzerland), 10*(7). https://doi.org/10.3390/APP10072502

Loshin, D. (2012). *Business intelligence: The savvy manager's guide*. Newnes.

Ma, Z., Davani, E., Ma, X., Lee, H., Arslan, I., Zhai, X., Darabi, H., & Castineira, D. (2020). Unlocking completion design optimization using an augmented AI approach. *Paper presented at the SPE Canada unconventional resources conference*, Virtual. https://doi-org.ezproxy.library.und.edu/10.2118/200000-MS

Madusu, S., & Rangarajan, K. P. (2018). Deep recurrent neural network DRNN model for real-time multistage pumping data. *Paper presented at the OTC arctic technology conference*, Houston, TX. https://doi-org.ezproxy.library.und.edu/10.4043/29145-MS

Marca, K. L., & Bedle, H. (2022). *User vs. machine-based seismic attribute selection for unsupervised machine learning techniques: Does human insight provide better results than statistically chosen*. Elsevier. Retrieved September 28, 2022, from www.sciencedirect.com/science/article/pii/B9780128222959000029

Martinez, W. L., & Martinez, A. R. (2001). *Computational statistics handbook with MATLAB*. https://doi.org/10.1201/9781420035636

Masini, S. R., Goswami, S., Kumar, A., Chennakrishnan, B., & Baghele, A. (2020). Artificial intelligence assisted production forecasting and well surveillance. *Paper presented at the offshore technology conference Asia*, Kuala Lumpur, Malaysia. https://doi.org/10.4043/30332-MS

Mask, G., Wu, X., & Ling, K. (2019). An improved model for gas-liquid flow pattern prediction based on machine learning. *Journal of Petroleum Science and Engineering, 183*, 106370. https://doi.org/10.1016/j.petrol.2019.106370

Mayani, M. G., Svendsen, M., & Oedegaard, S. I. (2018). Drilling digital twin success stories the last 10 years. *Paper presented at the SPE Norway one day seminar*, Bergen, Norway. https://doi.org/10.2118/191336-MS

Miftakhov, R., Al-Qasim, A., & Efremov, I. (2020). Deep reinforcement learning: Reservoir optimization from pixels. *Paper presented at the international petroleum technology conference*, Dhahran, Kingdom of Saudi Arabia. https://doi.org/10.2523/IPTC-20151-MS

Mohmad, N. L., Mandal, D., Amat, H., Sabzabadi, A., & Masoudi, R. (2020). History matching of production performance for highly faulted, multi layered, clastic oil reservoirs using artificial intelligence and data analytics: A novel approach. *Paper presented at the SPE Asia pacific oil and gas conference and exhibition*, Virtual. https://doi.org/10.2118/202460-MS

Moradi, H., Bahari, M., Bagher, M., Naghibi-Sistani, M., & Bahari, A. (2010). Drilling rate prediction using an innovative soft computing approach. *Scientific Research and Essays*, *5*, 1583–1588.

Mosher, C. C., Kaplan, T. S., & Janiszewski, D. F. (2014). Non-uniform optimal sampling for seismic survey design. *74th EAGE conference and exhibition incorporating EUROPEC 2012*. https://doi.org/10.3997/2214-4609.20148781

Mousavi, S. M., Jabbari, H., Darab, M., Nourani, M., & Sadeghnejad, S. (2020). Optimal well placement using machine learning methods: Multiple reservoir scenarios. *Day 2 Tue*, 3 November, SPE, November 2020.

Mueller, M. B., Halliday, D. F., van Manen, D. J., & Robertsson, J. O. A. (2015). The benefit of encoded source sequences for simultaneous source separation. *Geophysics*, *80*(5), V133–V143. https://doi.org/10.1190/GEO2015-0015.1

Mukerji, T., Avseth, P., Mavko, G., Takahashi, I., & González, E. F. (2001). Statistical rock physics: Combining rock physics, information theory, and geostatistics to reduce uncertainty in seismic reservoir characterization. *Leading Edge (Tulsa, OK)*, *20*(3), 313–319. https://doi.org/10.1190/1.1438938

Nabors, J., Voss, T., Bogdan, A., De Sario, C., Fu, D., & Khan, S. (2020). Basin-specific machine learning models for efficient completions optimization. *Paper presented at the SPE/AAPG/SEG unconventional resources technology conference*, Virtual. https://doi-org.ezproxy.library.und.edu/10.15530/urtec-2020-2770

Nakayama, S., & Blacquière, G. (2021). Machine-learning-based data recovery and its contribution to seismic acquisition: Simultaneous application of deblending, trace reconstruction, and low-frequency extrapolation. *Geophysics*, *86*(2), P13–P24. https://doi.org/10.1190/GEO2020-0303.1

Nakayama, S., Mercado, G., Benson, M. A., Belaid, K., & Garden, M. (2015). Field-wide implementation of time and distance separated source techniques on a 3D OBC survey offshore Abu Dhabi, UAE. *First Break*, *33*(4). https://doi.org/10.3997/1365-2397.2014026

Ngwashi, A. R., Ogbe, D. O., & Udebhulu, D. O. (2021). Evaluation of machine-learning tools for predicting sand production. *Paper presented at the SPE Nigeria annual international conference and exhibition*, Lagos, Nigeria. https://doi-org.ezproxy.library.und.edu/10.2118/207193-MS

Nybo, S. D., & Azizi, V. (2013). Real-time optimization of rate of penetration during drilling operation. *2013 10th IEEE international conference on control and automation (ICCA)*. IEEE, June.

Ozbayoglu, E. M., Erge, O., & Ozbayoglu, M. A. (2018). Predicting the pressure losses while the drillstring is buckled and rotating using artificial intelligence methods. *Journal of Natural Gas Science and Engineering*, *56*, 72–80.

Palmer, T., & Turland, M. (2016). Proactive rod pump optimization: Leveraging big data to accelerate and improve operations. *Proactive rod pump optimization: Leveraging big data to accelerate and improve operations*. https://doi.org/10.2118/181216-MS

Peng, Y., Zhang, J., Xiong, C., Shi, J., Zhao, R., Zhang, X., Xu, D., Li, Q., Deng, F., Chen, S., Liu, M., Chen, G., & Wang, C. (2020). Deep autoencoder-derived features applied in virtual flow metering for sucker-rod pumping wells. *Paper presented at the SPE/IATMI Asia pacific oil and gas conference and exhibition*, Bali, Indonesia. https://doi.org/10.2118/196288-MS

Pennel, M., Hsiung, J., & Putcha, V. B. (2018). Detecting failures and optimizing performance in artificial lift using machine learning models. *Paper presented at the SPE Western regional meeting*, Garden Grove, CA. https://doi.org/10.2118/190090-MS
Perrier, S., Delpeint, A., Shawuti, Z., & Shrestha, A. (2018). Machine-learning based analytics applied to stimulation performance in the Utica Shale: Case study and lessons learned. *Paper presented at the SPE/AAPG/SEG unconventional resources technology conference*, Houston, TX. https://doi-org.ezproxy.library.und.edu/10.15530/URTEC-2018-2902017
Peters, B., Granek, J., & Haber, E. (2019). Multiresolution neural networks for tracking seismic horizons from few training images. *Interpretation*, *7*(3), SE201–SE213. https://doi.org/10.1190/INT-2018-0225.1
Pollock, J., Stoecker-Sylvia, Z., Veedu, V., Panchal, N., & Elshahawi, H. (2018). Machine learning for improved directional drilling. https://doi.org/10.4043/28633-MS
Ramirez, A., & Iriarte, J. (2019). Event recognition on time series frac data using machine learning. *Paper presented at the SPE Western regional meeting*, San Jose, CA. https://doi-org.ezproxy.library.und.edu/10.2118/195317-MS
Randen, T., & Sønneland, L. (2005). Atlas of 3D seismic attributes. *Undefined*, *7*, 23–46. https://doi.org/10.1007/3-540-26493-0_2
Rastogi, A., & Sharma, A. (2019). Quantifying the impact of fracturing chemicals on production performance using machine learning. *Paper presented at the SPE liquids-rich basins conference—North America*, Odessa, TX. https://doi-org.ezproxy.library.und.edu/10.2118/197095-MS
Reddicharla, N., Rachapudi, S. R., Utama, I., Khan, F., Vanam, P. R., Al Nuimi, S. M., & Ali, M. A. S. (2021). A novel well test data analyzer and process optimizer using artificial intelligence and machine learning techniques. *Paper presented at the SPE annual technical conference and exhibition*, Dubai. https://doi.org/10.2118/206137-MS
Rogers, S. J., Fang, J. H., Karr, C. L., & Stanley, D. A. (1992). Determination of lithology from well logs using a neural network. *American Association of Petroleum Geologists Bulletin*, *76*(5), 731–739. https://doi.org/10.1306/BDFF88BC-1718-11D7-8645000102C1865D
Roy, A. (2013). *Latent space classification of seismic facies*. https://search.proquest.com/openview/a0470b8c2bea3972c626838eed06eca6/1?pq-origsite=gscholar&cbl=18750
Roy, A., & Marfurt, K. (2010). Applying self-organizing maps of multiple attributes, an example from the Red-Fork Formation, Anadarko Basin. *Onepetro.Org*. Retrieved October 9, 2022, from https://onepetro.org/SEGAM/proceedings-abstract/SEG10/All-SEG10/SEG-2010-1591/96363
Rozlan, M. R., Ghazali, M. H. M., Bakar, M. F., Ishak, M. F., Samuel, O. B., Misron, M. P., Ismail, M. S., Salleh, N. I., Rizwan, B., Amara, A. B., Faux, S., Azhari, A. H. A., & Muara, A. A. (2022). World'S first successful dual string installation of a digital intelligent artificial lift DIAL—Interventionless gas lift production optimization system, Offshore Malaysia. *Paper presented at the offshore technology conference Asia*, Virtual. https://doi.org/10.4043/31563-MS
Russell, S. D., Akbar, M., Vissapragada, B., & Walkden, G. M. (2002). Rock types and permeability prediction from dipmeter and image logs: Shuaiba reservoir (Aptian), Abu Dhabi. *AAPG Bulletin*, *86*(10), 1709–1732. https://doi.org/10.1306/61EEDD5A-173E-11D7-8645000102C1865D
Saggaf, M. M., & Nebrija, E. L. (2003). A fuzzy logic approach for the estimation of facies from wire-line logs. *American Association of Petroleum Geologists Bulletin*, *87*(7), 1223–1240. https://doi.org/10.1306/02260301019
Saghir, F., Gilabert, H., & Mancuso, B. M. (2020). Application of augmented intelligence and edge analytics in upstream production operations: An innovative approach for optimizing artificial lift systems performance. *Paper presented at the SPE annual technical conference and exhibition*, Virtual. https://doi.org/10.2118/201516-MS

Saini, G. S., Ashok, P., & Oort, E. V. (2020). Predictive action planning for hole cleaning optimization and stuck pipe prevention using digital twinning and reinforcement learning. *Paper presented at the IADC/SPE international drilling conference and exhibition*, Galveston, TX, March. https://doi.org/10.2118/199548-MS

Sami, N. A., & Ibrahim, D. S. (2021). Forecasting multiphase flowing bottom-hole pressure of vertical oil wells using three machine learning techniques. *Petroleum Research*, *6*(4), 417–422.

Sankaran, S., Matringe, S., Sidahmed, M., Saputelli, L., Wen, X.-H., Popa, A., & Dursun, S. (2020). *Data analytics in reservoir engineering.* Society of Petroleum Engineers.

Sarapulov, N. P., & Khabibullin, R. A. (2017). Application of big data tools for unstructured data analysis to improve ESP operation efficiency. *Presented at the SPE Russian petroleum technology conference*, Moscow, Russia. https://doi.org/10.2118/187738-MS

Seikhanov, A., Otynshish, D., Novopokrovskiy, G., & Artun, E. (2020). Comparison of machine learning algorithms for the development of a forecasting tool for cyclic gas injection in hydraulically-fractured wells. *Paper presented at the SPE annual caspian technical conference*, Virtual. https://doi-org.ezproxy.library.und.edu/10.2118/202507-MS

Shafiq, M. A., Alshawi, T., Long, Z., & AlRegib, G. (2018). The role of visual saliency in the automation of seismic interpretation. *Geophysical Prospecting*, *66*, 132–143. https://doi.org/10.1111/1365-2478.12570

Shaji, H. A., Bishnu, S. K., Mishra, T., Ramakrishna, M. V. S. S., Krishna, R. S. N., Thomas, K. R., & David, D. R. (2022). Artificial intelligence for automating and monitoring safety, efficiency and productivity in industrial facilities. *Paper presented at the offshore technology conference*, Houston, TX. https://doi-org.ezproxy.library.und.edu/10.4043/32018-MS

Shahkarami, A., Klenner, R., Stephenson, H., Siripatrachai, N., Kim, B., & Murrell, G. (2020). A data-driven workflow and case study for optimizing shut in strategies of adjacent wells during multi-stage hydraulic fracturing operations. *Paper presented at the SPE/AAPG/SEG unconventional resources technology conference*, Virtual. https://doi-org.ezproxy.library.und.edu/10.15530/urtec-2020-2753

Shi, X., Zhou, Y., Zhao, Q., Jiang, H., Zhao, L., & Yang, G. (2019). A new method to detect influx and loss during drilling based on machine learning. *Paper presented at the international petroleum technology conference*, Beijing, China, March. https://doi.org/10.2523/IPTC-19489-MS

Shirangi, M. G., Ettehadi, R., Aragall, R., Furlong, E., May, R., Dahl, T., Samnejad, M., & Thompson, C. (2020). Digital twins for drilling fluids: Advances and opportunities. *Paper presented at the IADC/SPE international drilling conference and exhibition*, Galveston, TX, March. https://doi.org/10.2118/199681-MS

Shmueli, G. E. (2018). *Data mining for business analytics: Concepts, techniques, and applications in R.* John Wiley & Sons.

Siahkoohi, A., Kumar, R., & Herrmann, F. J. (2019). Deep-learning based ocean bottom seismic wavefield recovery. *SEG technical program expanded abstracts*, 2232–2237. https://doi.org/10.1190/SEGAM2019-3216632.1

Soares, C., Daigle, H., & Gray, K. (2016). Evaluation of pdc bit rop models and the effect of rock strength on model coefficients. *Journal of Natural Gas Science and Engineering*, *34*, 1225–1236.

Srinivasan, S., O'Malley, D., & Mudunuru, M. K. (2021). A machine learning framework for rapid forecasting and history matching in unconventional reservoirs. *Scientific Reports*, *11*, 21730. https://doi.org/10.1038/s41598-021-01023-w

Strecker, U., & Uden, R. (2002). Data mining of 3D poststack seismic attribute volumes using Kohonen self-organizing maps. *The Leading Edge*, *21*(10), 1032–1037.

Sun, H., & Demanet, L. (2020). Extrapolated full-waveform inversion with deep learning. *Geophysics*, *85*(3), R275–R288. https://doi.org/10.1190/GEO2019-0195.1

Tadjer, A., Hong, A., & Bratvold, R. B. (2021). Machine learning based decline curve analysis for short-term oil production forecast. *Energy Exploration and Exploitation*, *39*(5), 1747–1769. https://doi.org/10.1177/01445987211011784

Thavarajah, R., Darabi, H., Ruta, N., Saleh, A., & Kianinejad, A. (2022). A deep learning framework for multi-horizon probabilistic production forecasting in unconventional reservoirs. *Paper presented at the SPE/AAPG/SEG unconventional resources technology conference*, Houston, TX. https://doi-org.ezproxy.library.und.edu/10.15530/urtec-2022-3723682

Timonv, A. V., Shabonas, A. R., & Schmidt, S. A. (2021). Field development optimization using machine learning methods to identify the optimal water flooding regime. *Paper presented at the SPE Russian petroleum technology conference*, Virtual. https://doi-org.ezproxy.library.und.edu/10.2118/206533-MS

Ugoyah, J. C., Ajienka, J. A., Wachikwu-Elechi, V. U. W., & Ikiensikimama, S. S. (2022). Prediction of scale precipitation by modelling its thermodynamic properties using machine learning engineering. *Paper presented at the SPE Nigeria annual international conference and exhibition*, Lagos, Nigeria. https://doi-org.ezproxy.library.und.edu/10.2118/212010-MS

Ujjwal, M., Tummala, P., Kumar, V., Singhal, M., & James, A. (2020). Rapid insights into waterflood management using advanced analytics and machine learning. *Paper presented at the SPE/IATMI Asia pacific oil and gas conference and exhibition*, Bali, Indonesia. https://doi-org.ezproxy.library.und.edu/10.2118/196460-MS

Van Hoek, T., Gesbert, S., & Pickens, J. (2010). Geometric attributes for seismic stratigraphic interpretation. *Leading Edge (Tulsa, OK)*, *29*(9), 1056–1065. https://doi.org/10.1190/1.3485766

Wallace, S. P., Hegde, C. M., & Gray, K. E. (2015). A system for real-time drilling performance optimization and automation based on statistical learning methods. *Day 1 Tue*, 15 September, SPE.

Wang, E., & Nealon, J. (2019). Applying machine learning to 3D seismic image denoising and enhancement. *Interpretation*, *7*(3), SE131–SE139. https://doi.org/10.1190/INT-2018-0224.1

Wang, G., Carr, T., Ju, Y., & Geosciences, C. L.-C. (2014a). *Identifying organic-rich Marcellus Shale lithofacies by support vector machine classifier in the Appalachian basin*. Elsevier. Retrieved September 28, 2022, from www.sciencedirect.com/science/article/pii/S0098300413003002?casa_token=Bja1zPF6144AAAAA:gOgnhifnL4iNa6wq0kWeOy4QlK-GVx1FCNKzoLMZaoOmOiZL9zzc1wkn6b8LpWSTYTzG0Nq6dA

Wang, J., Tan, Y., Liang, B., Min, X., Hu, C., Zhao, C., Wang, Y., Li, M., Chen, Y., Jiminez, G., & Khan, S. (2022a). Deep-learning-based prediction of post-fracturing permeability field for development strategy optimization in unconventional reservoirs. *Paper presented at the SPE/AAPG/SEG unconventional resources technology conference*, Houston, TX. https://doi-org.ezproxy.library.und.edu/10.15530/urtec-2022-3723603

Wang, L., Gale, J., & Sun, A. Y. (2020). Development of a machine-learning-based workflow for well completion optimization in Permian Basin. *Paper presented at the SPE/AAPG/SEG unconventional resources technology conference*, Virtual. https://doi-org.ezproxy.library.und.edu/10.15530/urtec-2020-3101

Wang, N., Chang, H., Zhang, D., Xue, L., & Chen, Y. (2022b). Efficient well placement optimization based on theory-guided convolutional neural network. *Journal of Petroleum Science and Engineering*, *208*, 109545.

Wang, Z., Di, H., Shafiq, M. A., Alaudah, Y., & Alregib, G. (2018). Successful leveraging of image processing and machine learning in seismic structural interpretation: A review. *Leading Edge*, *37*(6), 451–461. https://doi.org/10.1190/TLE37060451.1

Wolf, M., & Pelissier-Combescure, J. (1982). *Faciolog—Automatic electrofacies determination*. OnePetro.

Wrona, T., Pan, I., Gawthorpe, R. L., & Fossen, H. (2018). Seismic facies analysis using machine learning. *Geophysics*, *83*(5), O83–O95. https://doi.org/10.1190/GEO2017-0595.1/ASSET/IMAGES/LARGE/FIGURE7.JPEG

Wu, J., & Liu, C.-H. (2008). *Investigation of engine fault diagnosis using discrete wavelet transform and neural network*. Elsevier. Retrieved September 29, 2022, from www.sciencedirect.com/science/article/pii/S0957417407003454?casa_token=TR_zmiPDf6MAAAAA:sCBrFrx5p3EqDYHdawCo4-aoEFWJiCF5Y1AMf9Y5V4_1d6vvjARwCtyR2iX7n7qUkYHI2pqhAw

Wu, R. S., Luo, J., & Wu, B. (2014). Seismic envelope inversion and modulation signal model. *Geophysics*, *79*(3), WA13–WA24. https://doi.org/10.1190/GEO2013-0294.1

Xiong, W., Ji, X., Ma, Y., Wang, Y., Benhassan, N. M., Ali, M. N., & Luo, Y. (2018). Seismic fault detection with convolutional neural network. *Geophysics*, *83*(5), O97–O103. https://doi.org/10.1190/GEO2017-0666.1

Xue, H., Malpani, R., Agarwal, S., Bukovac, T., Mahesh, A. L., & Judd, T. (2019). Fast-track completion decision through ensemble-based machine learning. *Paper presented at the SPE reservoir characterisation and simulation conference and exhibition*, Abu Dhabi. https://doi-org.ezproxy.library.und.edu/10.2118/196702-MS

You, J., Ampomah, W., Tu, J., Morgan, A., Sun, Q., Wei, B., & Wang, D. (2022). Optimization of water-alternating-CO2 injection field operations using a machine-learning-assisted workflow. *SPE Reservoir Evaluation & Engineering*, *25*(2), 214–231. https://doi-org.ezproxy.library.und.edu/10.2118/203913-PA

Zhang, C., Frogner, C., Araya-Polo, M., & Hohl, D. (2014). Machine-learning based automated fault detection in seismic traces. *76th annual international conference and exhibition*, EAGE, Extended Abstracts, 807–811. http://doi.org.libproxy.uwyo.edu/10.3997/2214-4609.20141500.

Zhang, P., Han, L., Xu, Z., Zhang, F., & Wei, Y. (2017). Sparse blind deconvolution based low-frequency seismic data reconstruction for multiscale full waveform inversion. *Journal of Applied Geophysics*, *139*, 91–108. https://doi.org/10.1016/J.JAPPGEO.2017.02.021

Zhang, Y., & Sun, J. (2009). Practical issues in reverse time migration: True amplitude gathers, noise removal and harmonic source encoding. *First Break*, *27*(1). https://doi.org/10.3997/1365-2397.2009002

Zhao, T., Jayaram, V., Roy, A., & Marfurt, K. J. (2015). A comparison of classification techniques for seismic facies recognition. *Interpretation*, *3*(4), SAE29–SAE58. https://doi.org/10.1190/INT-2015-0044.1

Zheng, Y. U., Research, M., Zheng, Y., Capra, L., Wolfson, O., & Yang, H. (2014). Urban computing: Concepts, methodologies, and applications. *ACM Transactions on Intelligent Systems and Technology (TIST)*, *5*(38). https://doi.org/10.1145/2629592

Zhukova, K., Antonic, M., Solesa, M., & Camber, D. (2022). Data-driven model for measuring hydraulic fracturing efficiency by utilizing the real-time treatment data. *Paper presented at the international petroleum technology conference*, Riyadh, Saudi Arabia. https://doi-org.ezproxy.library.und.edu/10.2523/IPTC-22384-MS

12 Unconventional Shale Resources in Energy Transition and Digitalization

Amirsaman Rezaeyan, Yihuai Zhang, Tao Bai, Muhammad Arif, and Shams Kalam

12.1 INTRODUCTION

The global economy's dependency on fossil fuels faces critical challenges due to concerns about supply security and the impacts of climate change. The world's fossil fuel reserves are not unlimited, and the findings from the Intergovernmental Panel on Climate Change (IPCC) highlight the increasing risks of global warming, with fuel combustion–related greenhouse gas emissions being a significant contributor (Pörtner et al., 2023). The current industrial economy, heavily based on fossil fuels, is intrinsically linked to energy and climate change, with the transition to alternative energy sources being a pivotal focus. Despite rapid adoption of renewable energy, the global shift towards a low-carbon energy framework is still in its early stages. Consequently, there is an increasing interest in developing energy transition at global scales and identifying pathways towards sustainable, low-carbon production (Zheng et al., 2023).

Amidst this transition, unconventional shale resources, particularly shale gas and oil, have emerged as significant players. These resources are characterized by their unique geology and extraction methods, notably hydraulic fracturing and horizontal drilling. Shale reservoirs have become increasingly important, especially in regions like North America, as they offer a substantial energy source that can bridge the gap between declining conventional fossil fuels and the increasing renewable energy sector. The extraction and utilization of energy from unconventional shales present both challenges and opportunities. Environmental concerns such as water usage, potential groundwater contamination, and induced seismicity pose significant challenges, leading to regulatory and public scrutiny. Conversely, the economic benefits of shale energy, including job creation and energy security, are considerable. These resources can provide a more stable energy supply while being within carbon emission allowances during the transition to a low-carbon future (Weijermars et al., 2011).

DOI: 10.1201/9781003319955-12

The development and use of clean energy require significant technical and financial resources, which are often beyond the capacity of individual entities or businesses, necessitating support and decision-making from government agencies (Hainsch et al., 2022). There has been a considerable focus on evaluating clean energy policies, strategizing the use of clean energy technologies, and exploring pathways for energy transition. Alongside, research is also being directed towards understanding factors that could eliminate barriers to clean energy use and assessing the potential for clean energy application based on each country's resource endowment (Zheng et al., 2023). However, these studies often overlook the concept of embodied energy and its impact. Trade, as a crucial element, significantly influences the energy transition process. This underlines the need for a more comprehensive approach that draws from historical experiences and considers various influencing factors like political, economic, environmental, societal, and technological aspects in the energy transition roadmap (Zheng et al., 2023, Hainsch et al., 2022).

12.1.1 Unconventional Shale Resources

Unconventional shale resources are composed primarily (over 50% by weight or volume) of grains smaller than 62.5 μm. While the term "shale" is sometimes used more narrowly to describe visibly laminated and fissile varieties of these rocks, in this context, it is used to denote the entire class of fine-grained, layered sedimentary rocks and is often used synonymously with "mudrock." Shales form a major part of Earth's sedimentary rocks, which constitute about two-thirds of it and cover a quarter of the landmasses. Shales are predominant in both marine environments and terrestrial sequences and play a significant role in the interactions between fluids and rocks that determine the distribution and fate of elements within the upper crust and atmosphere. For instance, the chemical weathering of shale acts as a long-term global sink for carbon dioxide (CO_2). Consequently, the properties of shale are critical in influencing the dynamics between the atmosphere, hydrosphere, and sedimentary lithosphere in various scenarios. This makes shales indispensable for unconventional oil and gas extraction, as well as serving as top seals for conventional hydrocarbon traps in geological CO_2 storage and for the isolation of nuclear and other high-level wastes (Ilgen et al., 2017).

The properties of shales, both physical and chemical, are largely determined by the environment in which they were originally deposited, the alterations they undergo post deposition, and various engineering interventions. Shales are typically characterized by their stratified, low-permeability structure, primarily comprising fine-grained minerals. Notably, organic matter, such as kerogen, is a fundamental component of shale as a source rock (Ilgen et al., 2017). The porosity of shale and the sizes of its pores vary considerably, with nano-size pores constituting a significant portion of its overall porosity. Historically, the identification of shales has been heavily influenced by their clay mineral content, which plays a crucial role in determining their physical properties at different scales, from macroscopic field studies to microscopic laboratory analyses. Despite the prevalence of clay minerals, high-resolution petrographic studies have revealed that many shales are predominantly composed of non-clay minerals like quartz, feldspar, and calcite, which, along with organic matter, add to their intricate composition (Loucks et al., 2012). Post-depositional changes add to the complexity of shales' microfabric and texture, which involve a range of chemical and mechanical processes (Rezaeyan et al., 2023). These include compaction, cementation by materials such as carbonate and quartz, pressure-induced dissolution, and the alteration of grains: for instance, the transformation of feldspars into albite (Ilgen et al., 2017).

Conceptual models have been developed for the coupled thermal-hydraulic-mechanical-chemical-biological (THMCB) processes in unconventional shale resources. The compositional and textural complexity of shale leads to nonlinear and unpredictable physical and chemical processes, especially when subjected to both natural and anthropogenic perturbations. The interaction of chemical, mechanical, and transport processes, particularly evident in engineered systems, spans a vast range of scales, both temporally and spatially, from geological time scale to nanoseconds and from kilometers to nanometers, respectively (Rezaeyan et al., 2024). In shale, this complex interplay or process coupling can be exemplified by diagenesis, notably the transformation of mixed-layer clays containing smectite into versions richer in illite. This transformation, often accompanied by volume changes, releases significant amounts of dissolved silica, leading to quartz cementation in nearby reservoir sands and potentially in the shales themselves. These volume changes can induce fracturing in shales, altering the rock fabric and significantly affecting permeability (Ilgen et al., 2017). The distinct architecture of shale, influenced by diagenesis, controls fluid flow across scales and hosts diverse diffusion processes (Rezaeyan et al., 2022). The nano-channel and nanopore structures within shale, coupled with mixed wettability networks, dictate transport behavior. The biological activities in natural and engineered shale resources can lead to issues like reservoir plugging and equipment corrosion. Therefore, understanding and modeling these THCMB processes, including their volume changes and rates, are crucial, especially in applications like wellbore stability for oil and gas recovery, geological CO_2 and energy storage, and nuclear fuel repositories, where material removal or emplacement disrupts equilibrium. This complexity necessitates a multi-disciplinary approach for effective predictive modeling of shale behavior (Figure 12.1) (Ilgen et al., 2017; Barati et al., 2019).

Improved and enhanced oil and gas recovery from shale formations has significantly evolved due to advancements in both the understanding of and technology applied to these unconventional resources. Key to this advancement is hydraulic fracturing, a method that involves injecting high-pressure fluid to create fractures in the shale, thereby increasing permeability and facilitating the flow of hydrocarbons towards the wellbore. Horizontal drilling techniques enable longer lateral wells within the shale layers, maximizing contact with the hydrocarbon-bearing zones (Alfarge et al., 2020). Additionally, continuous and cyclic gas injection (e.g., CO_2 and lean gas) as well as nanomaterial and chemical flooding have been employed to enhance secondary oil and gas recovery. Recent innovations in enhanced recovery from shales include the utilization of CO_2-enhanced oil recovery (CO_2-EOR) techniques. In this method, CO_2 is injected into the shale formation, where it mixes

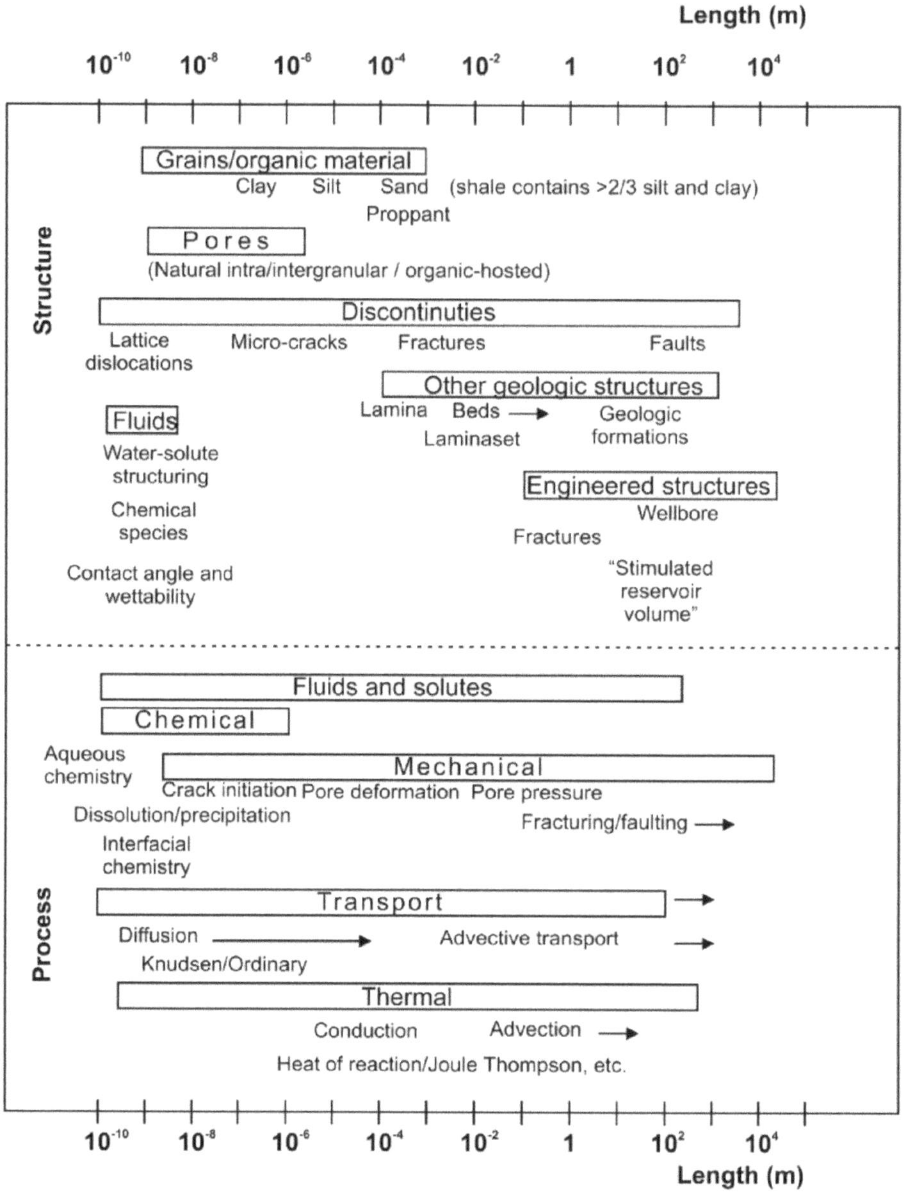

FIGURE 12.1 Structures, processes, and the relevant length scales for shale formations.

Source: Ilgen et al., 2017

with and mobilizes the trapped oil, thereby improving recovery rates (Abbasi et al., 2023). This method not only enhances oil recovery but also contributes to carbon capture and storage (CCS), aligning with broader environmental goals. The successful implementation of these enhanced recovery techniques in shale formations is underpinned by a deeper understanding of the rock's geomechanical properties, the chemistry of the fluids involved, and the interactions between the rock and the injected fluids. Continuous research and innovation are focused on optimizing these techniques to maximize recovery while minimizing environmental impacts, thereby ensuring a more sustainable approach to shale resource utilization in the context of global energy demands.

Unconventional shale resources have multifaceted roles in the global energy transition and achieving net-zero targets. On one hand, the exploitation of shale resources for oil and gas has led to a significant shift in the global energy landscape, contributing to energy security and economic growth. On the other hand, the environmental concerns associated with shale gas and oil, such as methane emissions and water usage, pose challenges to their alignment with net-zero objectives. As the world moves towards a more sustainable energy mix, integrating renewable energy sources with cleaner fossil fuel technologies, including natural gas from shale could be a transitional strategy. In this context, shales help balance immediate energy demands with long-term climate goals, acting as a bridge in the transition to a more sustainable, low-carbon future.

12.1.2 Energy Transition in Unconventional Shale Resources

As countries work to meet the Paris Agreement goals, transitioning to net-zero greenhouse gas (GHG) emissions by 2050, with its all technical-societal-economical-governmental aspects, has become imperative. This transition necessitates a significant reduction in emissions across all sectors and the adoption of negative emissions technologies, such as carbon capture and storage. Central to this transition is the electrification of end uses and decarbonization of the power sector. However, electricity alone cannot fulfill all energy needs (Cheng et al., 2023). Therefore, clean fuels like hydrogen (H_2), and natural gas (NG) are expected to play crucial roles in future net-zero economies (Shirizadeh et al., 2023). Additionally, natural gas can bridge fossil fuel to renewables since gas power plants can respond quickly to changes in load and renewable generation, unlike coal-fired and nuclear plants (Middleton et al., 2017). H_2 and NG can replace carbon intensive fossil fuels in various sectors, including power generation, heavy-duty transportation, and industries that are challenging to electrify (Cheng et al., 2023). Parallel to the push for net-zero, the depletion of conventional fossil fuel reserves has led to the exploration of unconventional shale resources such as shale gas. Shale gas is a future source of affordable energy, which is expected to improve energy security. Estimates suggest it could double current global gas reserves, adding nearly 7,299 trillion cubic meters to the total. Given that gas consumption is likely to increase by up to 50% by 2035, many countries are considering exploiting their shale gas deposits commercially on a large scale. This leads to low gas prices and (almost) self-sufficiency in gas supplies, contributing to the gross domestic product (GDP) and tax revenue. Despite offering new business opportunities, their production involves sophisticated techniques that are cost intensive and susceptible to market fluctuations (Cooper et al., 2018). Moreover, concerns regarding their environmental impact necessitate a careful evaluation of their sustainability, making the integration of shale development in gas supply chains a critical area for research and policy formulation. In the context of a net-zero future, the integration of clean fuels and the continued use of shale gas and oil present a complex challenge. While clean fuels are pivotal for reducing emissions and transitioning to renewable energy sources, the current dependence on fossil fuels, including shale resources, cannot be abruptly discontinued (Calderón et al., 2018). This necessitates a balanced approach that includes transitioning to lower-carbon versions of traditional energy carriers while gradually increasing the share of clean fuels in the energy mix. The current production of clean fuels primarily relies on unconventional shale gas resources, raising life cycle GHG concerns. For shale gas and oil, the economic and environmental assessments must consider the complex interactions between production technologies, market dynamics, and ecological implications (Calderón et al., 2018).

12.1.3 Adaptation and Transformation Strategies

Achieving net-zero CO_2 emissions by 2050 requires both technical adaptation and the transformation of unconventional shale resources. The increased availability of natural gas from shale, which emits approximately half the CO_2 of coal for the same heat energy, paves the road for the transition from coal- and oil-based to gas-based power generation. This transformation contributes to significant emissions reductions, such as US emissions returning to 1990 levels. Refracturing can transform low-productivity wells into high-performing ones. This approach carries significant potential for the rejuvenation of a vast number of shale gas wells across the United States. Refracturing notably lessens environmental impact through the use of non-aqueous working fluids such as supercritical CO_2 and natural gas. Given that current fracturing methods leave a large portion of shale gas resources untapped—with the average US shale play extracts just around 13% of the original gas in place—refracturing offers substantial prospects for boosting production in a cost-effective manner, avoiding the expenses of new wells, particularly when oil and gas prices rise, while maintaining a smaller environmental footprint. Total shale gas production from a well is mainly controlled by tail production (long-term production), rather than the initial production (short-term production). The tail production helps enhance longer-term profitability, increase energy security, and reduce CO_2 emissions (Middleton et al., 2017).

Depleted unconventional shale resources offer significant opportunities for CO_2 and hydrogen storage as well as geothermal heat production, contributing to the mitigation of greenhouse gas emissions. The transformation of depleted unconventional reservoirs into hubs for energy storage and low-carbon energy production relies heavily on their enormous capacity to accommodate host fluids, their ability to securely sequester gases, and the potential to reduce storage costs by leveraging existing infrastructure, such as hydraulic fractures and horizontal wells. Unconventional shale resources that are ideal for carbon sequestration are characterized by a high adsorption index, substantial kerogen and clay content, a significant solubility coefficient, a relevant inter-porosity flow coefficient, and a low mobility ratio. During sequestration, CO_2 exists in a multiscale hierarchical system, ranging from fractures to nanopores, and in various storage forms, including free gas, adsorbed gas, and dissolved gas (Figure 12.2). While hydraulic fractures networks enhance the mobility of

CO_2 from the wellbore to the reservoir, the ultra-low permeability and large specific surface area of shale reservoirs stabilize CO_2 for long-term and safe sequestration (Wang et al., 2023c). An unconventional reservoir candidate for hydrogen storage should possess sufficient stimulated reservoir volume post-hydraulic fracturing, have robust sealing capability to minimize the loss of stored gases under high working pressure, contain a reasonable amount of nanopores in the matrix rock to prevent the invasion of in situ hydrocarbons, and exhibit relatively higher permeability to guarantee effective storage capability (Wang et al., 2023b). Depleted shales resources and abandoned wells in unconventional shales present a viable option for geothermal heat production. Utilizing these existing infrastructures for geothermal electricity generation offers a cost-effective approach, reducing initial investment costs significantly. Repurposed wells not only lower the levelized cost of electricity generation but also minimize additional environmental impact as these wells, pipelines, and production systems are already established. The success of these geothermal endeavors relies heavily on the data collected during hydrocarbon production, which can be accessed from oil and gas databases. These provide insights into the potential for successful conversion to geothermal energy use with a broader trend towards developing renewable energy sources (Santos et al., 2022).

12.1.4 Science and Technology Advances

Advancements in science and technology, underpinned by interdisciplinary research, drive the global shift towards sustainable energy systems. Science and technology

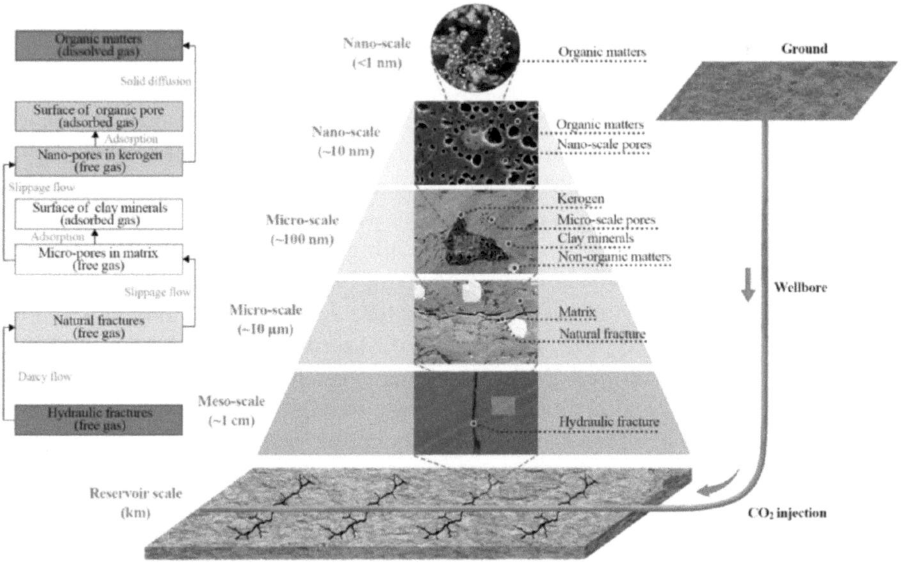

FIGURE 12.2 Multiscale CO_2 storage in shale reservoirs.

Source: Wang et al., 2023c

bridge the gap between theoretical, model-driven decarbonization pathways and the practical challenges encountered in real-world implementations. Innovations in geotechnical applications involving unconventional shale resources and repurposed depleted shale gas and oilfields are integral to the energy transition. These include advanced practices like CCS and innovative energy storage solutions. The electrification of energy storage and production systems, alongside the digital evolution of low-carbon technologies, further underscores this transition. By harnessing these cutting-edge techniques, the energy sector not only optimizes the use of existing resources but also facilitates a more sustainable and efficient future in energy production and management. Overall and most significantly, it establishes a new schema of energy asset management, which integrates the standard fair-value market analysis and corporate-owned real-option value estimated by digital optimal stopping techniques. This ongoing research, often a collaborative effort between academia, industry, and governments, leads to the creation of advanced solutions tailored to the specific energy challenges of different regions. For example, research efforts are supporting the US, China, Canada, the UK, the Netherlands, and Norway, to name a few, in the adoption and optimization of CCS technology. The integration of these scientific and technological advancements, backed by comprehensive research and coupled with environmental considerations, crafts realistic, effective, and inclusive strategies for the energy transition. The success of energy transition initiatives in unconventional shale resources depends on continued support from societies and governments and their willingness to adapt to and embrace these transformative changes (Wachsmuth et al., 2023).

12.1.4.1 Research

In-depth analysis of the academic literature on energy transition, employing bibliometric and systematic techniques (e.g., citation and co-citation analysis and dynamic co-citation analysis) emphasizes the role of research in the knowledge structure of energy transition and its evolution over time. Web of Science, Scopus, and international reports suggest the research and development (R&D) sector construct a reliable infrastructure that aims to enhance understanding of energy transition. This helps overcome public resistance and promote public acceptance of carbon pricing, carbon tax policies, and implementation of carbon pricing strategies (Nazari et al., 2024). Research efforts enable the development of policy frameworks tailored for energy transition, the evolution of infrastructure for renewable energy, dynamics of renewable supply, and projections for future energy demands. Energy transition research cannot be captured through prevalent sectors alone (e.g., energy engineering or environmental engineering). Energy transition research underscores the predominance of a comprehensive management and policy-based approach, highlighting its significance for decision-makers and stakeholders involved in the energy transition process. Recognizing the complex nature of energy transition, multidisciplinary research emphasizes the urgency to enhance renewable-based infrastructure and to engage consumers in sustainable consumption practices. This engagement is particularly critical in light of escalating environmental changes. The development of interdisciplinary research groups is recommended to enable the integration of diverse theoretical perspectives from various fields, thereby enriching future studies on this subject. A significant increase in publications related to energy transition is forecast, reflecting the field's growing importance and relevance. Energy transition plans are recognized as vital tools for navigating the rapidly changing environmental landscape, helping assess trade-offs and develop policies that align with broader sustainable development and climate objectives. Researchers, scientists, and authors devise strategies that utilize the technological, economic, and environmental knowledge available across society, aiming for cohesiveness and alignment of purpose (Harichandan et al., 2022). Geographical location plays a crucial role in the research landscape of energy transition (Figure 12.3). According to a country-wise analysis by Harichandan et al. (2022), which includes Scopus-indexed publications from 119 countries, the United States leads the field of energy transition research with 225 documents. It is closely followed by Germany with 201 documents and the United Kingdom with 193. European countries are significantly advanced in energy transition research compared to other continents. In Asia, China is prominent with 129 publications, far surpassing India, which has 35. Australia makes a notable contribution with 90 publications in this field. African nations such as South Africa with 21 publications, Nigeria with 12, and Kenya with 6 are increasingly engaging in research on energy transition. They recognize it as a key solution to the continent's energy poverty, focusing on renewable energy strategies (Harichandan et al., 2022).

Social sciences and humanities (SSH) are routinely excluded from discussions about energy strategies and transition. This can be rather problematic because citizens need awareness of the decarbonization of energy systems through the pursuit of more localized and collective forms of renewable energy production, reduced and more efficient energy usage, and voting for necessary regulatory changes. SSH suggests five key directions to foster energy transition. First, the significance of language and communication in shaping perceptions and practices in energy transition

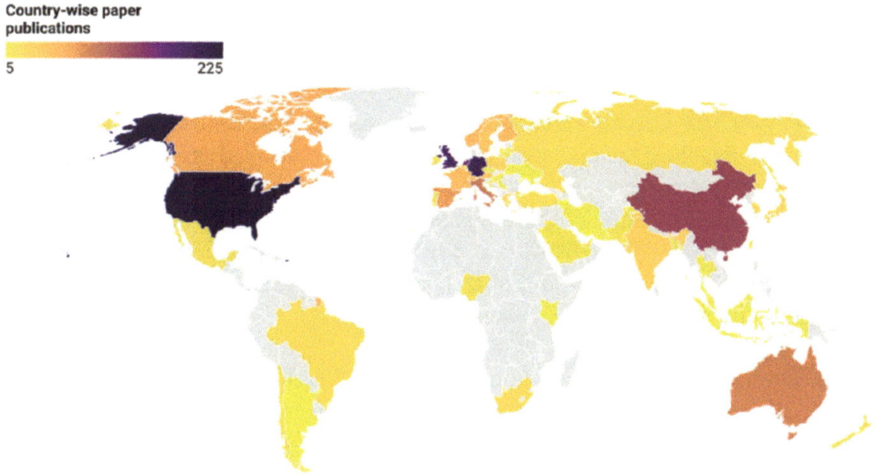

FIGURE 12.3 Global map of top 50 countries with publications on energy transition.

Source: Harichandan et al., 2022

studies, emphasizing the need for interdisciplinary research for effective collaborations. Second, with a focus on the distributional impacts and justice in energy transition, SSH advocates for research to ensure equitable and sustainable outcomes across different societal groups. Third, the importance of co-diffusion and co-adoption of technologies and lifestyles is highlighted, recognizing the energy transition as a complex social endeavor that requires a holistic approach. Fourth, SSH-informed methods and real-world applications (e.g., field experiments and living labs) gain insights more reflective of social realities. Lastly, SSH puts perspectives in addressing energy crises and shortages, understanding population reactions, and shaping political responses to energy security and renewable energy challenges (Brückmann et al., 2023).

The accelerated development of unconventional fields necessitates an R&D-focused approach, tailored to the unique subsurface conditions of shale resources. Emphasizing environmental responsibility, this strategy aims to stabilize reserves maturation, enhance well flow rates, reduce costs, and mitigate environmental impacts while addressing stakeholder concerns. Advanced geoscience and technology, underpinned by laboratory facilities, are crucial for conducting impactful research in hydraulic fracturing and petrophysical behavior. The framework for the unconventional development lifecycle includes iterative feedback loops to refine the learning cycle, optimizing net present value as subsurface uncertainties diminish over time through enhanced monitoring, analysis, and modeling (Weijermars et al., 2011). Further, true solutions to energy transition challenges require not only technical proficiency but also a comprehensive perspective and collective wisdom. International collaborations can help maximize the use of global expertise and resources. Collaborations ensure that solutions and lessons in the field of energy transition are shared on a global scale, rather than remaining confined to individual countries or regions. Collaboration and multidisciplinary approaches, integrated into social awareness, promote understanding and guide the evolving energy transition in unconventional oil and gas fields (Weijermars et al., 2011; Harichandan et al., 2022; Brückmann et al., 2023).

12.1.4.2 Digitalization

The significance of unconventional resources can be amplified by the potential of digitalization, enabling the control, monitoring, and optimization of resource exploitation and energy storage and production. Digital technologies can transform the management and utilization of these resources, enhancing efficiency, reducing environmental impact, and contributing to a more rapid and efficient transition to sustainable energy systems. The digitalization of unconventional resources involves the use of artificial intelligence, machine learning, robotics, automation, and cloud computing technologies. Digital twin models can be used in unconventional oil and gas development. They represent a convergence of technology, data, and analytics to create a dynamic, digital replica of real-world unconventional oil or gas fields. They help optimize the exploration and production process, enhance safety, reduce environmental impacts, and improve the overall efficiency of operations. The role of digitalization in the unconventional oil and gas sector is intertwined with energy conservation and emission reduction. On one hand, the expansion of the digital economy can lead to increased energy consumption, particularly in data-intensive operations like digital oilfield management. The increased use of electronic devices and data centers in the industry has raised concerns about their carbon footprint. On the other hand, the digitalization of unconventional industry has potential in driving efficiency, boosting productivity, and lowering production costs. For example, the deployment of advanced digital technologies such as artificial intelligence and automation in drilling and extraction processes can significantly enhance operational efficiency. Nevertheless, the synergy effect of digitalization and industrialization highlights the need for a balanced approach in the unconventional oil and gas sector. With the help of digital transformation and digital optimization throughout energy-efficient digital infrastructure and smart resource management, shale enterprises can balance the pollution emissions and energy consumption of their storage and production processes and the efficiency and productivity of their assets. Further, the accurate collection and dynamic management of data can also help them improve their sustainable performance (Ma & Lin, 2023; Wang et al., 2022a).

12.1.4.3 The Future of Energy Transition

As the world confronts the pressing challenges of climate change, a rapid transition to decarbonization is widely advocated. Yet the current energy infrastructure, still heavily reliant on fossil fuels, plays a pivotal role in this transformation. Fossil fuels accounted for approximately 80% of global energy consumption in 2019, according to the 2023 BP Energy Outlook (Figure 12.4). Although it is predicted that the share of fossil fuels will drop to 20% under net zero scenarios by 2050, new momentum in energy demand suggests the share of fossil fuels will hardly drop to further than 55% by 2050. This signifies both the continued importance of fossil fuels in daily economic functions and the necessary shift towards low-carbon technologies and infrastructures. This reliance on fossil energy, particularly during the transition phase, inevitably leads to an energy overhead in terms of CO_2 emissions. This factor is critical in determining the biophysical payback time, which is the duration required by alternative energy sources to offset the additional CO_2 emissions incurred during the construction and operation of green infrastructures and technologies (Manfroni et al., 2021).

Given the widespread carbon lock-in across primary and secondary production sectors, a swift and significant alteration of the current energy matrix to drastically cut CO_2 emissions appears improbable in the near future. Examining the current applications of fossil fuels, key industries—e.g., the unconventional oil and gas sector—predominantly rely on hydrocarbons. Additionally, the entrenched investment in existing technical capital, buildings, and infrastructures creates a substantial dependency on fossil fuels, which presents significant challenges to energy transition. Furthermore, rapid population growth in developing countries, coupled with a global pursuit of improved living standards, is poised to escalate the reliance on fossil energy and the resultant CO_2 emissions. Consequently, it is not unexpected that both the consumption of fossil energy and emissions are on the rise, despite global commitments such as the Paris Agreement and UNFCC and IPCC international meetings (Manfroni et al., 2021).

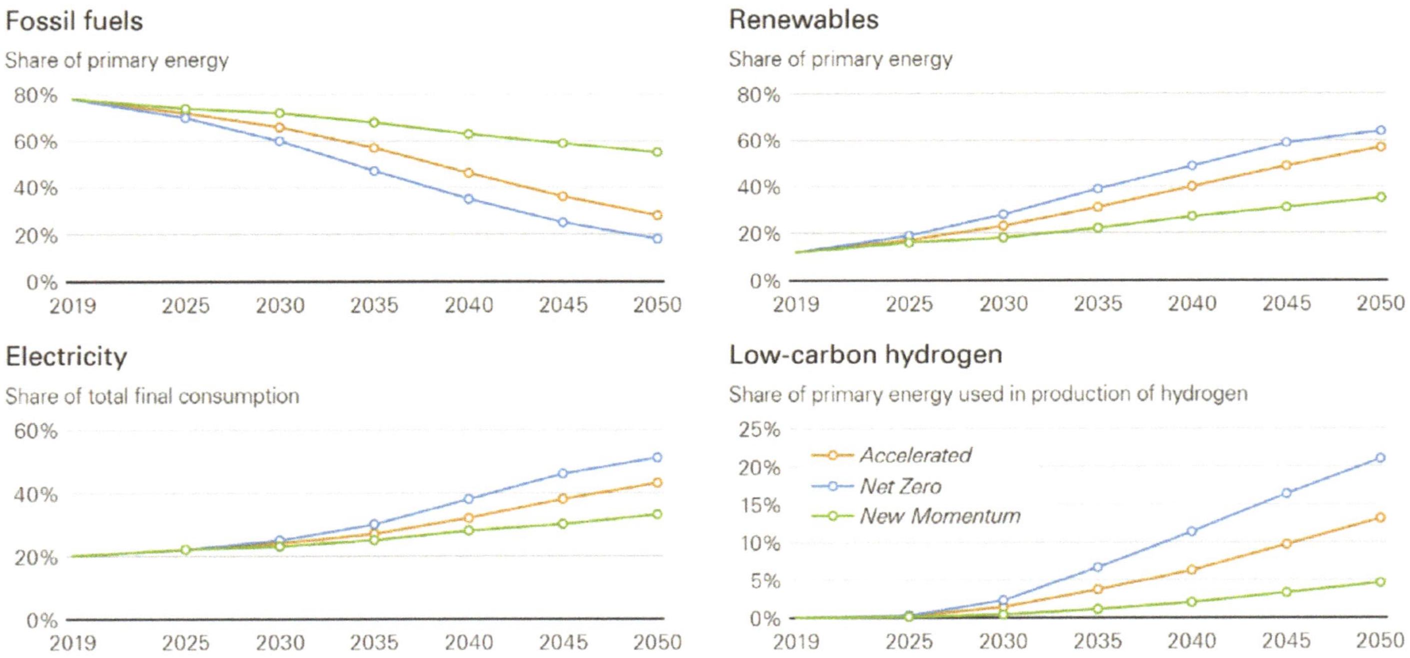

FIGURE 12.4 The future of global energy is dominated by four trends: declining role for hydrocarbons, rapid expansion in renewables, increasing electrification, and growing use of low-carbon hydrogen.

Source: 2023 BP Energy Outlook

At the global level, oil remains the leading primary energy source, comprising 33% of total primary sources as per the BP Energy Outlook, 2023. Instead of moving away from fossil fuels, the progressive aging (depletion) of conventional oilfields results in fossil transition or unconventionalization of oil supplies: e.g., unconventional oil shales. Transition to unconventional sources complicates the supply system, which leads to a decrease in energy return on investment and higher emissions per unit of net supply. This phase of unconventionalization is expected to be prominent between 2020 and 2060, a crucial period for climate change mitigation as identified by the IPCC. This trend poses significant challenges to the feasibility of a swift transition to renewable energy and the realization of climate change mitigation scenarios. Consequently, before committing to ambitious, drastic decarbonization targets for the global economy by 2050, it is crucial to assess the broader impacts of the oil sector's declining performance on low-carbon future aspirations. Combined methodological and systematic approaches (e.g., life-cycle assessment and integrated assessment modeling) are essential to develop effective climate change mitigation solutions and drive innovation and efficiency for a practical and clean energy transition (Manfroni et al., 2021).

12.2 SUSTAINABLE GEOENERGY PRODUCTION IN UNCONVENTIONAL SHALE RESOURCES

12.2.1 Technological Aspects

The sustainable production of geoenergy from unconventional shale resources depends significantly on technological innovations. Ongoing advancements in hydraulic fracturing, horizontal drilling, seismic imaging, well targeting, drilling fluids, and digital twins are crucial for increasing extraction efficiency, reducing costs, and minimizing environmental impact (Kang et al., 2020; Middleton et al., 2017; Hua & Gengsheng, 2012).

12.2.2 Horizontal Drilling

Horizontal drilling stands as a groundbreaking technology, radically transforming the extraction of unconventional shale resources (Yuan et al., 2013; Li et al., 2016). Traditional vertical drilling methods prove ineffectual when faced with shale gas and oil formations, which commonly extend horizontally (Brady et al., 1992). With horizontal drilling, a vertical descent is initially pursued until a particular depth is reached, then the drilling trajectory is gradually redirected to follow the oil- or gas-bearing stratum on a horizontal plane. This methodology opens up a vast area of the reservoir to the well, exponentially increasing its exposure to the hydrocarbon-rich regions. The considerable escalation in the well's exposure to the reservoir boosts the recovery rate of hydrocarbons, thus significantly amplifying the overall productivity and efficiency of the extraction process (Alexander et al., 2011). This innovative approach allows operators to tap into a larger portion of the formation, thereby yielding a greater volume of oil and gas than the traditional vertical drilling technique.

Horizontal drilling also introduces the possibility of drilling multiple horizontal wells from a single vertical well pad. This unique strategy reduces the need for multiple vertical wells, dramatically cutting down on surface disruptions. Less surface area is impacted, and the number of drilling sites required is minimized. This not only mitigates the ecological footprint of the operation but also cuts down on infrastructure costs and surface-level disturbances to local communities. Moreover, horizontal drilling offers benefits in reservoirs with complex geologies, where vertical wells may not provide sufficient access to the resource. By allowing operators to follow the contours of the resource layer, horizontal drilling ensures that even tricky formations can be effectively exploited.

The advent of horizontal drilling has revolutionized the extraction of unconventional shale resources. Its impact extends beyond merely enhancing resource recovery to include less surface disturbance, increased operational efficiency, and the potential to exploit complex reservoirs. This transformative technology continues to redefine the landscape of the hydrocarbon extraction industry.

12.2.2.1 Hydraulic Fracturing

Hydraulic fracturing, commonly known as fracking, has revolutionized the extraction of oil and gas from unconventional shale resources (Chen et al., 2021). The process involves the injection of a high-pressure fluid, typically a mixture of water, proppant (usually sand), and specific chemicals, into a wellbore drilled into a reservoir rock formation (Figure 12.5). This fluid causes the rock layer to crack or fracture, creating

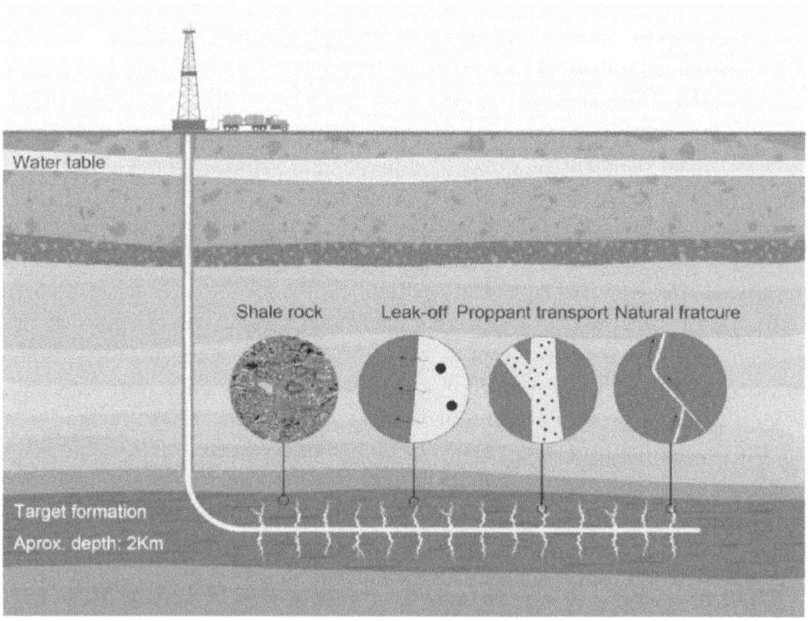

FIGURE 12.5 Horizontal well and hydraulic fracturing as a multiscale and multi-physics process.

Source: Chen et al., 2021

a network of fissures through which natural gas, oil, and other hydrocarbons can flow (Smith & Montgomery, 2015). The process has been refined over the years, with advancements in technology making it more efficient and effective. For instance, the use of 3D seismic imaging has improved the accuracy of identifying the best locations for drilling and fracking, reducing the risk of dry wells and minimizing environmental impact.

The proppant within the fluid serves a crucial role; once the fractures are created, it lodges within them, propping them open and preventing them from closing once the injection pressure is released. This allows hydrocarbons to continue flowing to the wellbore. The proppant used in the fracking fluid has also seen significant improvements. While sand is the most commonly used proppant, other materials like ceramic and resin-coated sand are being used to withstand higher pressures and temperatures. These advanced proppants can maintain the integrity of the fractures longer, increasing the well's productivity over time. Moreover, the chemicals used in the fracking fluid are being continually optimized to improve their effectiveness. These chemicals serve various purposes, such as reducing friction, preventing corrosion of the well casing, and inhibiting the growth of bacteria that could clog the fractures. Some companies are even developing more environmentally friendly chemicals to reduce the potential environmental impact of fracking (Chen et al., 2021).

The method of multistage fracking has undeniably revolutionized the industry (Carpenter, 2022). This transformative technique eschews the creation of a single elongated fracture. Instead, it partitions the wellbore into numerous segments, with each section undergoing its individual fracking process (Yao et al., 2018). This unique strategy provides a greater level of control over the fracturing procedure, making the overall process more efficient and targeted. By concentrating on one section at a time, operators can optimize the pressure and amount of fluid used for each stage, ensuring that the rock is fractured as thoroughly and effectively as possible (Zhao et al., 2022). This precise manipulation also mitigates the risks of unintentional damage to surrounding areas, significantly reducing the environmental footprint of the fracking operation. Moreover, multistage fracking aids in enhancing the overall yield of the reservoir. By breaking the operation down into distinct stages, operators can reach and fracture more of the reservoir rock, potentially exposing more hydrocarbon-bearing formations. This results in a higher probability of oil or gas recovery, making this process more productive and lucrative. Additionally, the multistage fracking process allows for real-time adjustments based on the feedback from each stage. As the wellbore's characteristics may vary along its length, this method enables operators to tailor the approach for each section to achieve the best possible outcome. This developing game-changing technique (see Figure 12.6 as an example of the advancing technique in multistage fracturing) promises a future in which reservoir utilization can be maximized while potential environmental impact can be minimized.

In recent years, there has been a burgeoning trend towards utilizing nanotechnology to enhance hydraulic fracturing stimulation (Yang et al., 2017). This is achieved by amplifying the viscosity of the fracturing fluid with the inclusion of nanoparticles within the reservoirs. Previous studies have shed light on the advantageous attributes of nanomaterials, characterized by their small size, large specific surface area,

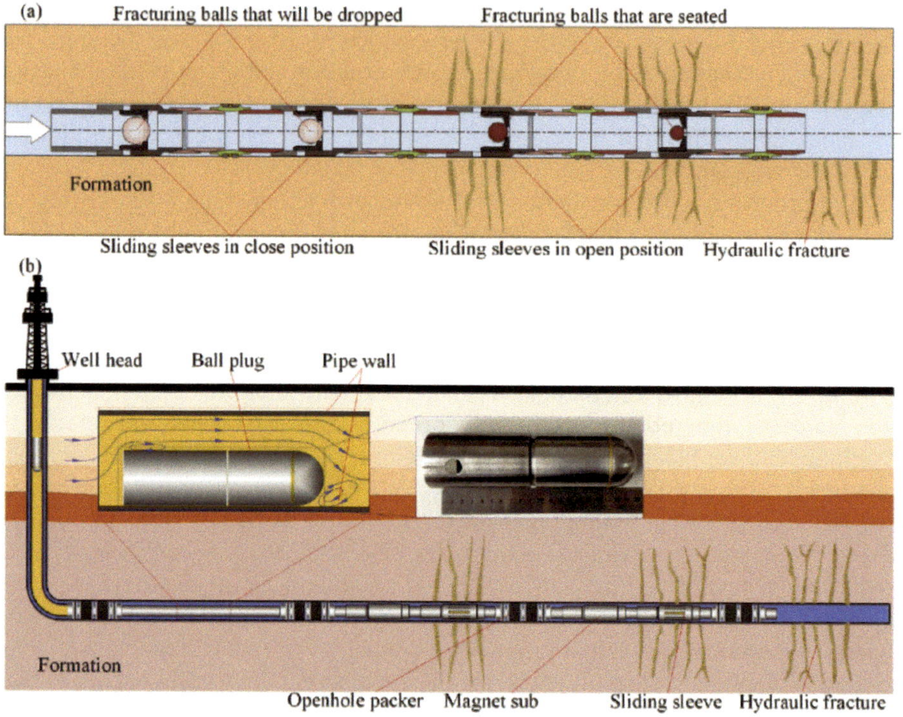

FIGURE 12.6 (a) Working principle of the traditional ball-drop fracturing system; (b) Downhole tool configuration of the proposed open-hole multistage fracturing system.

Source: Yao et al., 2018

magnetic properties, exceptional strength, and remarkable stability (Yekeen et al., 2019). These unique properties have proven useful in the development of various applications, such as nano-sensors, nano-proppants, gel breakers, and fluid loss control agents, all of which are critical for downhole operations.

A wide variety of nanoparticles can be employed for these purposes, ranging from common ones like aluminum oxide (Al_2O_3), copper oxide (CuO), and carbon nanotubes to more cost-effective options such as calcium carbonate ($CaCO_3$), silicon dioxide (SiO_2), and fly-ash nanoparticles. Figure 12.7 provides a comprehensive summary of the applications of nanoparticles for hydraulic fracturing in unconventional shale resources (Yekeen et al., 2019).

Fracking has been instrumental in unlocking previously inaccessible or economically unviable shale gas and oil reserves. By increasing the rock's permeability, fracking facilitates the migration of gas or oil to the wellbore, dramatically increasing the productivity of the well and the recoverable reserves. The future of fracking will depend on ongoing technological advancements and regulatory measures to address the environmental concerns associated with the process. As the industry continues to innovate and evolve, fracking is expected to remain a key player in the global energy landscape.

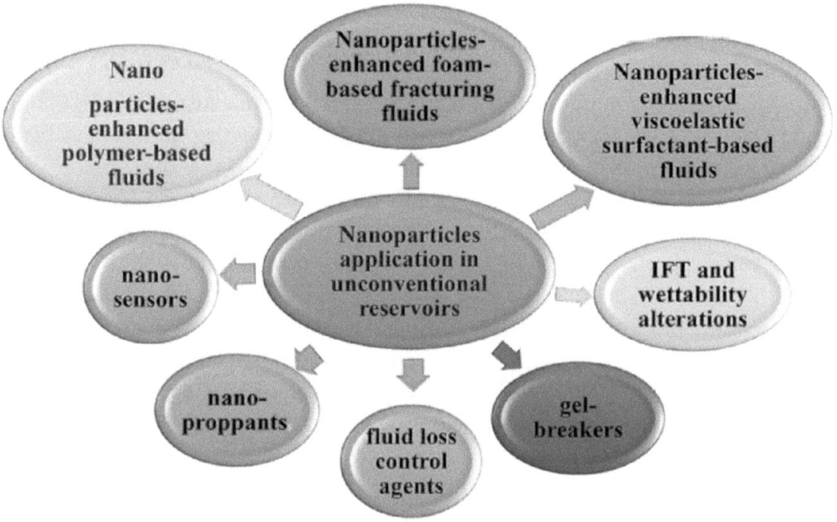

FIGURE 12.7 Summary of the applications of nanoparticles for hydraulic fracturing in unconventional shale resources.

Source: Yekeen et al., 2019

12.2.2.2 Other Technological Advancements and Digital Twins

The progression in fracking and drilling technologies has led to substantial enhancements in the efficacy and cost-efficiency of oil and gas extraction from unconventional shale resources. One of the technological leaps has been observed in seismic imaging, a tool that employs reflections of artificially induced shock waves to generate comprehensive images of subsurface geological structures. Modern iterations of this technology, such as 3D and 4D seismic imaging, provide granular, three-dimensional depictions of subsurface formations (Byerley et al., 2018; Maxwell, 2014; Nantanoi et al., 2022). This visual clarity allows for more accurate pinpointing of potential reservoirs and precise placement of wells, optimizing resource extraction and reducing unnecessary drilling.

Another area of significant advancement has been well targeting technology, including logging-while-drilling (LWD) systems (Meador, 2009; Liu & Liu, 2017). These systems have elevated drilling precision and control to new heights. Offering real-time data on various parameters such as pressure, temperature, and direction, these technologies enable operators to make necessary adjustments during the drilling process. This results in reduced drilling time, lower costs, and increased safety by providing crucial information to avoid potential drilling hazards.

Advancements also extend to the domain of drilling fluids or mud. Modern drilling fluids are tailored to perform several crucial functions, including lubricating and cooling the drill bit, transporting cuttings to the surface, preventing formation fluids from entering the wellbore, and maintaining wellbore stability (Hayatdavoudi, 1999). The continuous evolution and refinement of drilling fluids have led to the creation of more effective and environmentally friendly alternatives. These new-generation fluids mitigate environmental risks and contribute significantly to the overall sustainability of geoenergy production.

In addition, the development of digital technologies has also aided in the optimization of shale resource extraction (Xiao et al., 2021). The use of digital twins is part of a larger trend towards digitalization and automation in the shale resource industry (Wang et al., 2022b); the enablement of intricate simulation, functioning, and assessment is facilitated by the dynamic linkage of digital representations to their tangible counterparts. The exchange of information and data is facilitated through the interplay of physical components and their connections and services, as well as associated data and digital models (Figure 12.8). For instance, digital twins, a virtual replica of the physical drilling operation, can be utilized to simulate and predict various outcomes, leading to better decision-making and efficiency (Sircar et al., 2022). The advent of digital technologies has indeed facilitated the optimization of shale resource extraction.

Digital twins present a multifaceted application across the various stages of drilling operations (Said et al., 2022). These virtual representations afford operators an intricate visualization of subsurface reservoirs, as well as insights into their dynamic behaviors under an array of scenarios. Such capabilities contribute significantly to the enhancement of planning and the strategic design of drilling operations (Li et al., 2023). Utilizing an amalgamation of real-time data from current drilling sites, historical data from analogous operations, and sophisticated mathematical models, digital twin technology facilitates the prediction of optimal extraction strategies. Furthermore, it serves as a proactive measure to identify and address potential challenges and risks, effectively minimizing operational downtime.

The application of digital twins transcends the mere confines of drilling processes to encompass the production phase. Through simulative modeling of oil and gas flow within reservoirs and pipelines, this technology enables operators to refine production rates and curtail losses, culminating in a substantial augmentation of overall operational profitability.

Moreover, the fusion of digital twins with complementary digital technologies, such as artificial intelligence (AI) and the Internet of Things (IoT), amplifies their functionality (Panja et al., 2018; Syed et al., 2022). AI-driven algorithms, designed to extrapolate from historical data, can forecast future outcomes with heightened accuracy. IoT devices furnish real-time data from multifarious points within the operation, affording a comprehensive and integrative perspective on the entire process. Such synergistic integration renders the utilization of digital twins an indispensable asset in modern drilling and production operations. The development of technologies has greatly aided in the optimization of shale resource extraction. By simulating, predicting, and optimizing various outcomes, they lead to improved decision-making, increased efficiency, and cost savings. As the industry continues to adopt and integrate these technologies, we can expect further advancements in this area.

Overall, these advancements in technology contribute to the continuous improvement of drilling and hydraulic fracturing operations, offering higher efficiency, cost-effectiveness, and environmental sustainability. They represent a combination of evolving imaging techniques, precise well targeting technology, and advanced

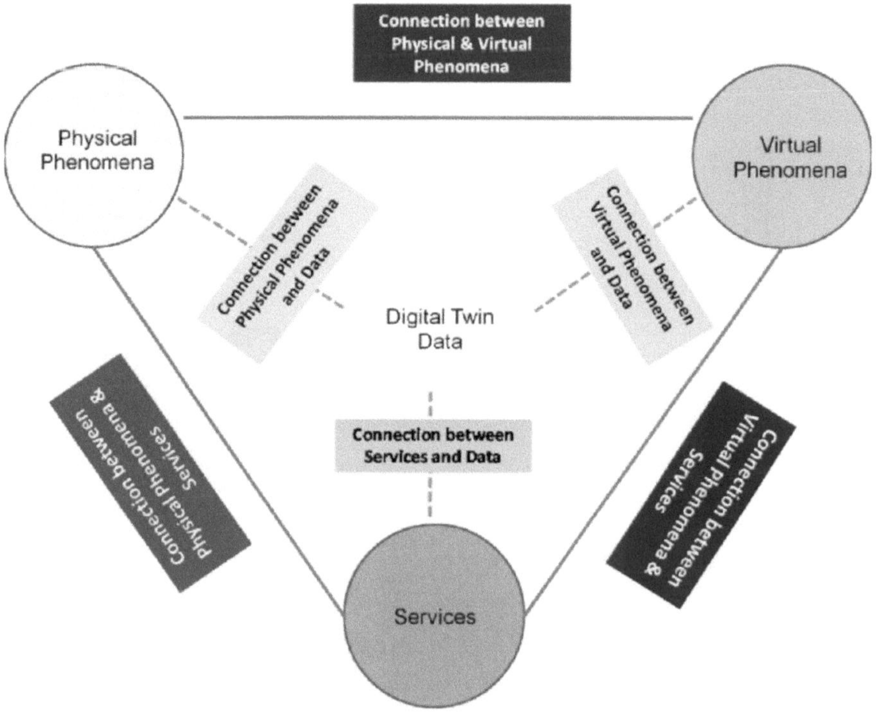

FIGURE 12.8 Digital twins (DT) representation.

Source: Sircar et al., 2022

drilling fluids, coupled with the power of digital technology and innovative extraction techniques. These technological leaps are driving the industry forward, paving the way for a more sustainable and profitable future in the extraction of unconventional shale resources.

12.3 ENVIRONMENTAL IMPLICATIONS

It's crucial to acknowledge and thoroughly consider the substantial environmental implications that come with this pursuit. See Figure 12.9 for the environmental concerns for hydraulic fracturing in unconventional shale production (Howarth et al., 2011a). This part will explore in depth the three major environmental impacts that accompany unconventional shale resource production: namely, water contamination, greenhouse gas emissions, and induced seismicity.

12.3.1 Water Contamination

Hydraulic fracturing, while an effective method of resource extraction, has raised valid apprehensions regarding the potential contamination of groundwater and

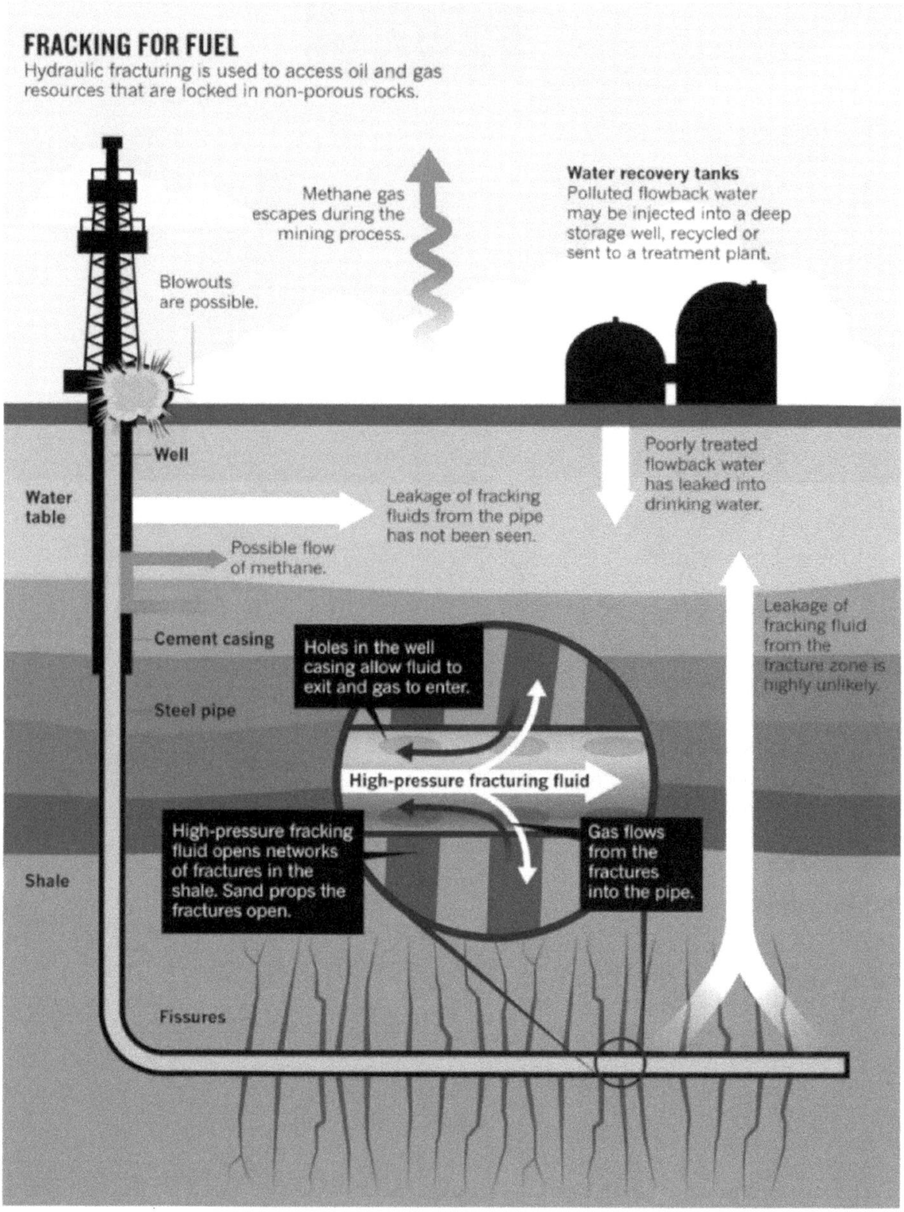

FIGURE 12.9 The environmental concerns for the hydraulic fracturing in unconventional shale production.

Source: Howarth et al., 2011a

surface water (Vengosh et al., 2013). This concern largely revolves around two primary risks. First, fracking fluids, which contain a mixture of potentially hazardous chemicals, may migrate from the drilling well into the groundwater supply. Second, methane, a key constituent of natural gas, might escape from improperly constructed wells, leading to contamination of both groundwater and the atmosphere (Vengosh et al., 2014; Vidic et al., 2013; Meakin et al., 2013).

Addressing and mitigating these risks is not only a focus of ongoing research but also a pivotal area for regulatory attention (Olmstead, 2015; Souther et al., 2014; Pietersen et al., 2021). A crucial part of the solution lies in the enhancement of well construction practices. By implementing multiple layers of protective casing and cement, the risk of leaks can be substantially curtailed. These protective barriers function as physical obstacles that prevent fracking fluids and natural gas from seeping into groundwater and the surrounding environment.

Another potential mitigation strategy involves treating and recycling the water produced during hydraulic fracturing operations (Estrada & Bhamidimarri, 2016). Figure 12.10 shows the issues for wastewater in unconventional shale production by hydraulic fracturing. This step can considerably decrease the volume of waste fluid that needs disposal, thereby lowering the potential for surface water contamination. This environmentally friendly approach not only reuses water resources but also lessens the pressure on water disposal systems. Moreover, the development and implementation of advanced monitoring systems are critical for early leak detection and ensuring well integrity. These systems can provide real-time data on well operations, enabling swift interventions should any signs of leakage or structural compromise arise. The potential for developing non-toxic or less toxic alternatives to the chemical additives used in fracking fluids is another area of research. These alternatives can significantly reduce the potential harm should a leakage into the groundwater occur.

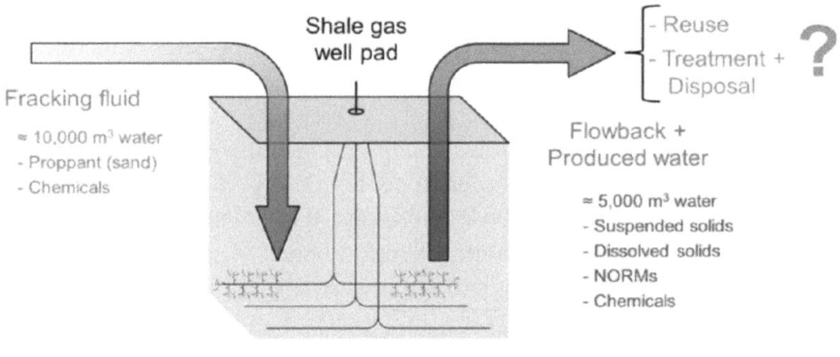

FIGURE 12.10 The issues for wastewater in unconventional shale production by hydraulic fracturing.

Source: Estrada & Bhamidimarri, 2016

While hydraulic fracturing presents tangible risks to groundwater and surface water, the concerted efforts in research and regulatory focus aim to develop and implement solutions that can substantially mitigate these risks. Advancements in well construction techniques, wastewater management, monitoring systems, and non-toxic fracking fluid alternatives promise a more sustainable future for hydraulic fracturing operations.

12.3.2 Greenhouse Gas Emissions

The extraction of unconventional shale gas and oil comes with a considerable environmental price tag, particularly in the form of substantial greenhouse gas emissions (Wang et al., 2011; Howarth et al., 2011b; Howarth, 2015). Among these emissions, methane takes precedence as a potent greenhouse gas. It carries a global warming potential exceeding that of carbon dioxide by more than 25 times over a century-long period. Methane emissions are primarily attributable to deliberate venting, equipment leaks, and the flowback process following hydraulic fracturing.

Compared to conventional gas extraction, shale gas extraction has the potential to generate even higher levels of emissions (Howarth et al., 2011b). This discrepancy stems from the greater well depths associated with shale gas extraction and the larger quantities of flowback water that must be managed. Yet it is crucial to note that, when burned, natural gas emits less carbon dioxide than coal, positing it as a potentially less polluting energy source—provided methane leaks can be sufficiently curtailed.

There is a multifaceted approach to addressing these emissions. One crucial effort involves enhancing the effectiveness of leak detection and repair. By spotting and addressing leaks promptly, operators can drastically cut down on the amount of methane that escapes into the atmosphere. Capturing gas during the flowback period represents another significant stride towards mitigating these emissions. This process can repurpose otherwise wasted methane, reducing the overall emission footprint.

On a broader level, the implementation of more rigorous regulatory standards plays a pivotal role in reducing these emissions. Stricter policies can compel companies to adopt more sustainable practices and implement the necessary precautions to minimize methane leaks.

Beyond these efforts, considerable research is currently ongoing to better quantify these emissions and to pinpoint the most economically viable mitigation strategies. With a deeper understanding of these emissions, the industry can adopt more effective solutions to minimize their environmental impact.

While unconventional shale gas extraction is still considered a significant contributor to greenhouse gas emissions, especially methane, targeted efforts in improving technology, operations, and regulations, coupled with ongoing research, can pave the way for significant reductions in these emissions. These collective efforts hold promise for a future in which these resources can be extracted in a more environmentally conscious and sustainable manner.

12.3.3 Induced Seismicity

Fracking is associated with a rise in seismic activity, a phenomenon known as "induced seismicity." Figure 12.11 illustrates the induced seismicity occurrence in

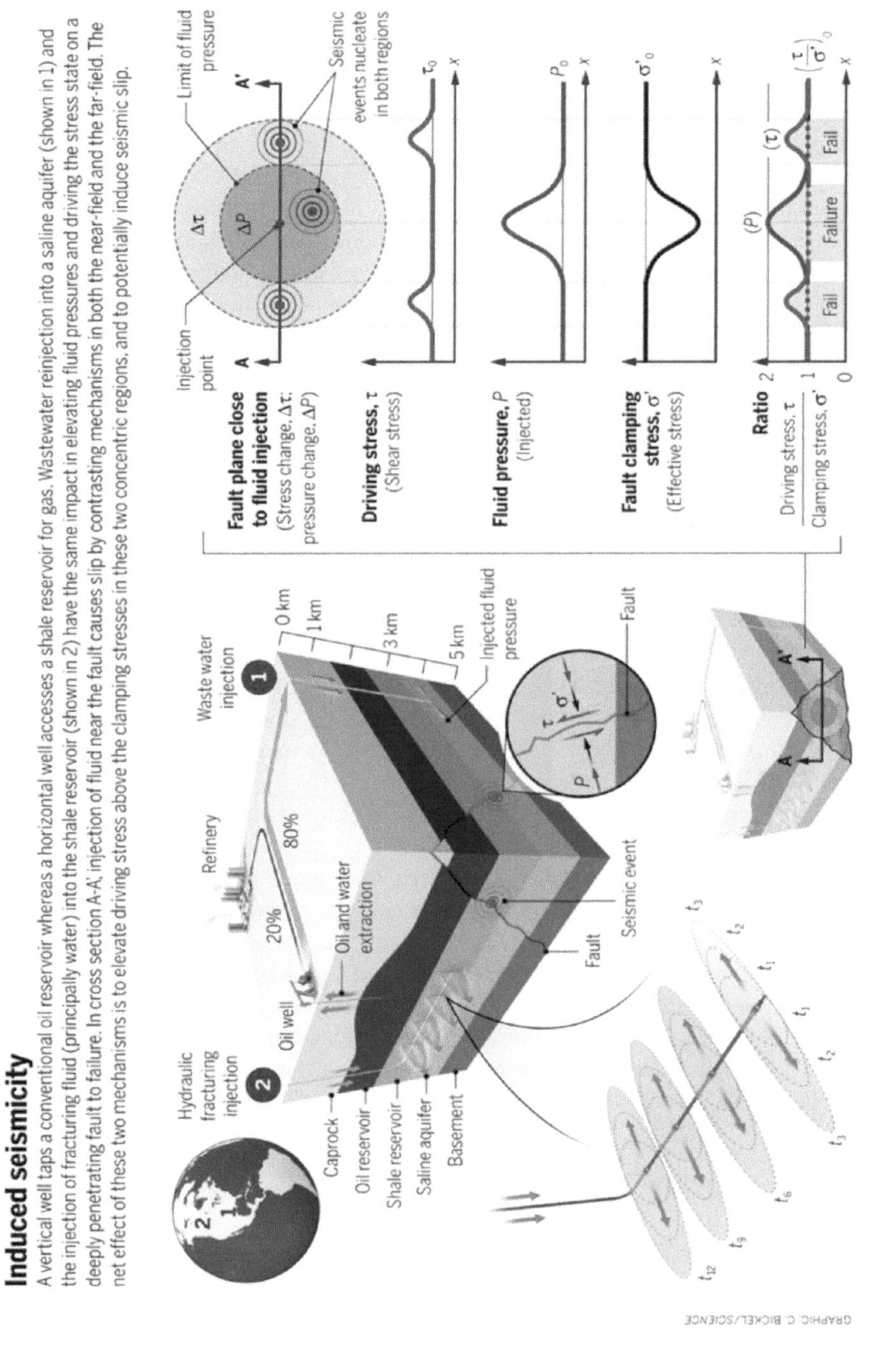

FIGURE 12.11 Induced seismicity in shale gas production.

Source: Elsworth et al. 2016; graphic by C. Bickel/Science

shale gas production. While most of the earthquakes instigated by this process are of such minimal magnitude that they are imperceptible at the surface, there exists a legitimate risk that fracking might trigger larger seismic events. These larger earthquakes have the potential to cause notable damage and stir public concern, creating a significant challenge for the continued use of this technology (Li et al., 2019; Rutqvist et al., 2013; Eaton & Schultz, 2018; Warpinski et al., 2012; Elsworth et al., 2016).

Comprehending the underlying mechanisms of induced seismicity is crucial, and this has become a dynamic area of research. One aspect that researchers are closely investigating is the role of pre-existing geological faults. When these faults are subjected to the high-pressure injection of fracking fluids, they may slip, leading to seismic events. Understanding this interaction between fracking and pre-existing faults can help predict where and when seismic events might occur, thus allowing operators to mitigate the risks.

Another important consideration is the changes in subsurface pressures that are brought about by fracking. The injection of fluids increases pore pressure in the rock, which can induce seismicity by reducing the friction on existing faults. Monitoring these pressure changes and their impact on the subsurface formations is therefore crucial in managing induced seismicity (Yu et al., 2021).

In terms of mitigation strategies, there are several options. Careful site selection is paramount; choosing locations with minimal pre-existing faults can drastically reduce the risk of inducing seismic events. Moreover, monitoring seismic activity during fracking operations allows for real-time adjustments to be made if unusual activity is detected. Technologies such as microseismic monitoring and passive seismic monitoring can help detect even small seismic events that might indicate a larger event is imminent.

Adaptive management strategies in which fracking operations are modified or halted based on seismic monitoring data are also effective in reducing the risk of induced seismicity (Hammond & O'Grady, 2017). Some jurisdictions have implemented "traffic light" systems (Baisch, 2013), in which operations are halted or modified if seismic activity exceeds a certain level.

Induced seismicity is a significant concern that needs careful attention in the production of unconventional shale resources; we need proactive management and continuous research to mitigate its risks effectively. As the industry moves forward, ongoing research, technological advancements, and regulatory improvements will play pivotal roles in reducing these risks and ensuring the sustainability and public acceptance of shale gas and oil extraction. The path forward will be one that balances the benefits of unconventional shale resources against their environmental impacts, striving to achieve a sustainable and responsible future for this important industry.

12.3.4 Economic Factors

The exploitation of unconventional shale resources has dramatically reshaped the energy landscape, with significant economic implications. The economic factors associated with sustainable geoenergy production in unconventional shale resources encompass aspects of energy security, job creation, and economic growth, as well as economic challenges. Each of these dimensions carries weight in assessing the overall economic impact of shale gas and oil extraction.

12.3.5 Energy Security

The unconventional shale revolution has indeed had a significant impact on energy security, especially in countries like the US that boast vast shale resources. The historical reliance of the US on foreign oil and gas imports has witnessed a dramatic shift due to the surge in domestic shale gas and oil production (Kobek et al., 2015; Andreas, 2015; Westphal et al., 2014; Aguilera & Radetzki, 2014).

Extraction of shale resources provides a stable, secure energy supply. This not only safeguards a country against the potential disruptions and price volatility often associated with foreign energy imports but also fosters self-reliance in energy production. This sense of energy independence can have transformative implications for a nation's economic stability and growth. By strengthening local industries, it can create jobs and stimulate economic growth, with the benefits felt across different sectors of the economy.

Moreover, this energy independence can positively influence geopolitical relations. Dependence on energy-rich nations often results in a power imbalance, which can potentially lead to a host of political and economic challenges. Reducing this reliance can lead to an improved geopolitical position for energy-independent nations, freeing them from potential pressures and manipulations.

The reduction in energy imports also has a significant impact on a nation's balance of trade. By producing energy domestically, countries can reduce their trade deficits, thereby strengthening their currency and providing a boost to the overall economy.

In addition to the direct economic and geopolitical benefits, energy independence also allows for greater control over environmental standards and practices. This could lead to better enforcement of regulations, improved environmental stewardship, and more sustainable practices in the energy sector.

Lastly, the unconventional shale revolution is also influencing the global energy market dynamics. It is not only diversifying the global energy supply but also affecting global energy prices. The increased competition can lead to more favorable energy prices for consumers worldwide.

The unconventional shale revolution has brought about far-reaching changes in energy security, particularly in countries that possess abundant shale resources. The resulting energy independence has numerous economic, geopolitical, and environmental benefits (Goldthau & Hoxtell, 2012). It's a revolution that's reshaping not only the domestic energy landscape but also global energy market dynamics, with effects likely to be felt for many years to come. As countries continue to navigate their unique energy challenges, the ongoing exploration of shale resources will remain a crucial part of the broader conversation on energy security and sustainability.

12.3.6 Job Creation and Economic Growth

The economic benefits of the shale boom are undeniable. Shale gas and oil extraction have been significant job creators, especially in regions with significant shale resources (Bilgili et al., 2020). These jobs span a range of sectors, from direct employment in extraction operations to indirect jobs in supply chains and induced jobs from the spending of industry employees and suppliers.

The shale industry has also made significant contributions to economic growth, such as Texas (Hartley et al., 2015).The lower energy costs resulting from increased shale gas and oil production can provide a competitive advantage to energy-intensive industries, further driving economic growth (Solarin & Bello, 2020).

12.3.7 Economic Challenges

While the advantages are many, the extraction of shale gas and oil also faces several economic hurdles. The shale industry is marked by pronounced price volatility, a phenomenon deeply ingrained in the structure of the oil and gas industry. This volatility can be attributed to several factors such as the cyclical nature of the oil and gas industry, technological innovations, changes in regulations, and the fluctuating dynamics of the global market (Salygin et al., 2019; Wang et al., 2014).

The cyclical pattern is especially pronounced in this industry, which is susceptible to "boom-and-bust" cycles (Mayfield et al., 2019). These cycles are periods of swift expansion and high prices, often followed by a downturn when prices plummet. These fluctuations can have a profound impact, especially on regions that rely heavily on the industry for economic sustenance. Communities can experience rapid growth during boom periods, with increases in jobs, income, and population. But when the bust occurs, these same communities can face job losses, decreased income, population outflows, and strained public services.

Furthermore, the profitability of shale gas and oil extraction is far from uniform and depends on various factors, such as underlying geology, extraction technology, and market conditions. The geology of a region determines the size and quality of the reserves, which, in turn, affect the cost and yield of extraction (Zendehboudi & Bahadori, 2016). Advances in extraction technologies can lower costs and increase yields, but they require significant capital investments. Market conditions, including the prices of oil and gas and the availability of infrastructure for transport and processing, also have a direct impact on profitability. These considerations can affect the economic feasibility of individual wells and have substantial implications for the stability and resilience of the industry as a whole.

In addition, the high upfront capital costs and long payback periods associated with shale exploration and production can pose significant financial risks. Companies in the sector often operate with high levels of debt, making them vulnerable to shifts in investor sentiment and credit availability.

Lastly, the environmental costs associated with shale gas and oil extraction can't be ignored. The potential costs of water contamination, air pollution, greenhouse gas emissions, and induced seismicity, among others, need to be factored into economic calculations. Proper management and mitigation of these environmental impacts can require additional expenditures, potentially affecting the industry's profitability.

Thus, while the extraction of unconventional shale resources brings enormous economic benefits—improved energy security, creation of jobs, and a boost to the GDP—these gains are weighed against economic challenges, including price volatility, boom-and-bust cycles, and variability in profitability. Striking the right balance calls for continued research and policy development. This will be crucial in harnessing the full economic potential of shale resources, managing these challenges, and ensuring that the industry grows in a sustainable and resilient manner. As we move forward, it will be essential to adopt an integrated approach that factors in not only the economic but also the social and environmental aspects of shale gas and oil extraction.

12.4 SUSTAINABILITY AND FUTURE PROSPECTS

The role of unconventional shale resources in the broader energy mix and the path towards a sustainable future is a subject of extensive debate. The sustainability and future prospects of shale resources are interlinked with several factors, such as energy transition, technological developments, and evolving policies.

12.4.1 Transition to Cleaner Energy

Shale gas/oil has often been touted as a "bridge fuel" in the transition to a cleaner, low-carbon energy future (Howarth, 2014; Schrag, 2012). When burned, natural gas emits about half as much carbon dioxide as coal, making it a more environmentally friendly choice for electricity generation.

Moreover, the flexibility of gas-fired power plants, which can be ramped up and down more quickly than many other types of plants, makes them an excellent partner for renewable sources. They can provide reliable power when wind or solar generation is low.

However, to fully realize the potential of natural gas in the transition to a low-carbon future, addressing the challenge of methane leaks during the extraction and distribution of shale gas will be crucial (Howarth et al., 2012; Howarth, 2014). Additionally, the potential application of carbon capture and storage (CCS) technologies could further reduce emissions from gas and oil extraction, thus enhancing their sustainability.

12.4.2 Future Developments

The sustainability of shale gas and oil extraction can be significantly improved through technological and policy advancements. Technological progress can lead to more efficient extraction techniques, minimizing the environmental footprint of operations. For instance, advances in drilling technologies, including horizontal drilling and hydraulic fracturing, have already greatly improved the efficiency of shale gas extraction.

On the policy front, the development of stricter environmental regulations can minimize the environmental impacts of shale gas and oil extraction. This might include more robust regulations on well construction and wastewater management or stricter standards for methane emissions.

Moreover, policy incentives for cleaner energy sources, including renewable energy technologies and CCS, can help in balancing the energy mix. These policies could range from direct subsidies to market mechanisms like carbon pricing.

In conclusion, the sustainability and future prospects of unconventional shale resources hinge on the successful integration of technological advancements, policy decisions, and the commitment to transition to cleaner energy sources. These elements combined can contribute to the broader aim of achieving a sustainable energy future.

12.5 CONCLUSION

The extraction of unconventional shale resources has indisputably reshaped the global energy landscape. Technological advancements, particularly hydraulic fracturing and horizontal drilling, have unlocked significant reserves of natural gas and oil, making extraction economically viable.

The environmental implications of shale gas and oil extraction, however, necessitate careful attention and management. Concerns regarding water contamination, greenhouse gas emissions, and induced seismicity underscore the need for rigorous regulation and improved industrial practices. Despite these challenges, the prospect of shale gas and oil as a "bridge fuel" towards a low-carbon future provides a key incentive for ongoing research and investment in this domain.

Economic considerations present a complex picture. While the shale boom has significantly bolstered energy security and contributed to economic growth, it also presents challenges in the form of price volatility and susceptibility to boom-and-bust cycles. Yet, with careful management and policy intervention, these challenges can be mitigated.

The sustainability of unconventional shale resources largely hinges on balancing environmental, economic, and energy security considerations. As the world continues to grapple with climate change, the role of natural gas in reducing carbon emissions can be pivotal, provided the issue of methane leakage is adequately addressed. Further, technologies like carbon capture and storage may play an increasingly vital role in reducing emissions from the sector.

Technological innovation is expected to continue to shape the industry, potentially leading to more efficient extraction techniques, reducing the environmental impact, and improving the economic viability of wells. Moreover, policy advancements such as stricter environmental regulations and incentives for cleaner energy can significantly contribute to making shale gas and oil extraction more sustainable.

Overall, the unconventional shale resource industry stands at a crossroads, balancing significant benefits against pressing challenges. The path forward must be navigated carefully, with an emphasis on sustainability and responsibility as these resources can play a significant role in the global energy matrix. The pursuit of sustainable geoenergy production in unconventional shale resources demands ongoing research, technological advancement, and policy development to meet the dual goals of energy security and environmental protection.

12.6 EMISSION REDUCTION IN UNCONVENTIONAL SHALE RESOURCES

12.6.1 Drilling Optimization

As the global community persistently confronts the complexities of climate change and ecological preservation, various industries are placing greater emphasis on mitigating greenhouse gas emissions and lessening their environmental impact (Porter et al., 2020). Among these industries, the energy sector, notably the oil and gas domain, assumes a paramount role in contributing to worldwide emissions owing to its heavy reliance on fossil fuels for energy generation.

12.6.1.1 Emissions during Drilling of Shales

Drilling and completion phases in unconventional shale reservoirs are considered risky in terms of emissions due to various factors that can contribute to releasing greenhouse gases and other pollutants (Bilgen & Sarıkaya, 2016). The reasons drilling in unconventional reservoirs poses emission risks include:

1. Methane leakage: One critical concern is the potential for methane release from the drilling mud. The drilling mud can come into contact with methane-bearing formations during the drilling process, and, if not adequately managed, methane can dissolve into the mud and be released into the atmosphere when the mud returns to the surface (Ren et al., 2019).
2. Volatile organic compound (VOC) emissions: Some drilling mud formulations contain volatile organic compounds (VOCs), which can evaporate during the drilling process (Rana, 2008).
3. Formation characteristics: Unconventional shale formations may have higher levels of organic matter and natural gas content. Drilling through these formations can release trapped hydrocarbons, including methane, which contribute to emissions (Yong et al., 2019).
4. Shale instability: Shale formations can be sensitive to drilling fluids, which can cause swelling and dispersion, leading to formation damage and the potential for methane and other hydrocarbon emissions (Saboori et al., 2019).
5. Mud additives and chemicals: Various additives and chemicals are incorporated into the drilling mud to optimize its properties and performance. These additives may contain compounds that can be released into the atmosphere, leading to emissions of harmful substances (Al-Shargabi et al., 2022).
6. Mud disposal and handling: After drilling, the used drilling mud needs to be properly managed and disposed of. If not handled correctly, it can lead to environmental contamination, including the potential for leaching harmful substances into the soil and groundwater (Ayele et al., 2014).

To address these emission risks effectively, both the oil and gas industry and regulatory authorities must concentrate their efforts on enhancing drilling techniques, integrating advanced emission detection and control technologies, optimizing wellbore integrity, and implementing policies that encourage the adoption of best practices for emission reduction. Furthermore, research and innovation play crucial roles in the development and implementation of cleaner technologies and drilling fluids, thus minimizing the environmental repercussions of drilling activities in unconventional reservoirs.

12.6.1.2 Challenges Associated with Conventional Drilling Fluids in Shales

Geologically, shale formation is composed mainly of fine clay-size particles. Like oil-bearing shale, this finely layered structure can lead to various challenges to wellbore stability during drilling. When drilling in overbalanced conditions, the pressure of the drilling fluid gradually infiltrates the shale formation. However, due to low permeability, only a small amount of filtrate enters the wellbore. As a result, pore fluid pressure near the wellbore increases, leading to less stability in the wellbore. This issue is especially critical in horizontal drilling, which is widely utilized to extract hydrocarbon resources from shale reservoirs. Typically, oil-based mud (OBM) is the preferred drilling fluid system for such wells. Technically, OBM is favored for its ability to maintain wellbore stability during drilling in inclined or horizontal sections.

Additionally, it offers excellent lubricity for fluid and filter cake performance during horizontal drilling and casing operations. However, the primary concern with using OBM lies in its potential environmental impact due to the base oil (e.g., diesel) and brine components, which can lead to disposal challenges. OBM contains oil-based additives that can lead to environmental damage if not appropriately managed, and its interactions with shale formations may cause wellbore instability. Disposal of oil-based cuttings is more complex and expensive, and complying with environmental regulations can complicate drilling operations (Young & Friedheim, 2013). As a result, the industry has been exploring an alternative drilling fluid system: a WBM that offers performance benefits comparable to OBM but with a more favorable environmental profile. However, when deploying WBM, achieving wellbore stability involves a different approach to shale inhibition than conventional situations. While traditional shale inhibitors (e.g., amines, silicates, or salts) are commonly used due to shale's high chemical reactivity, many shale gas or shale oil plays exhibit lower reactivity, with the primary concern being the shale's fissile nature. This suggests that the traditional additives are no longer enough to effectively block the micropores and cracks; thus, there is need for new additives (Young & Friedheim, 2013).

12.6.1.3 Environmentally Friendly Drilling Fluids

The primary focus revolves around the properties of drilling fluids and their associated additives, intending to achieve environmentally friendly outcomes and minimize emissions. By carefully selecting and optimizing these components, we aspire to develop drilling fluids that are effective in their intended purpose and environmentally sustainable. Reducing emissions and mitigating the environmental impact of drilling operations are crucial in promoting a more sustainable approach to energy exploration. Through rigorous research and innovation, we seek to create drilling fluids that align with eco-conscious practices while maintaining their efficiency and performance. Table 12.1 summarizes research conducted on eco-friendly chemicals for use in drilling operations in shale formations while the subsequent sections describe the critical features of these additives.

12.6.1.3.1 TiO2-bentonite Nanocomposite (TNBT)

Researchers conducted a study to address drilling challenges in water-sensitive formations, particularly shale rich in smectite. A custom water-based mud containing

TiO_2/bentonite nanocomposite was tested, and its effectiveness in terms of lubricity, wellbore stability, inhibition of shale and clay swelling, and stabilization of clay and shale were evaluated (Aftab et al., 2020a). The optimal concentration of TNBT was 0.8 g, offering improved lubricity due to the TiO_2 nanoparticles' high surface area–to-volume ratio. The nanocomposite mud effectively inhibited shale and clay swelling, and bentonite's expansive properties contributed to its viscosity, reducing water infiltration and further suppressing swelling. The findings showcased an innovative approach to addressing drilling challenges using nanocomposite technology in mud formulation.

12.6.1.3.2 Tween 80-ZnO Nanoparticle Composite

The objectives of incorporating a composite of Tween 80-ZnO nanoparticles are twofold. The primary goal is to study the immediate toxicity of drilling fluids, followed by enhancing the rheology, lubricating qualities, and clay inhibition performance of water-based drilling fluids (WBDF) (Aftab et al., 2020b). In acute toxicity tests, the waste from WBDF showed a remarkable 100% survival rate for white-leg shrimp (the marine species tested). Introducing 0.7 g of T80ZnO led to significant enhancements in the drilling fluid's characteristics (Aftab et al., 2020b). The addition of this nanomaterial resulted in a 12% increase in plastic viscosity (PV) due to its negative surface charge, which fostered repulsion forces and stability, thereby enhancing viscosity. The yield point (YP) witnessed an impressive 71% improvement, while both 10-minute and 10-second gel strengths (GS) increased significantly by 32% and 54%, respectively. The inclusion of metal oxide nanosolids contributed to improved heat transfer properties and desirable gelling and yield strength in the drilling fluids. Furthermore, by incorporating 0.6 g of T80ZnO nanoparticles into the water-based drilling fluid, the researchers achieved optimized heated clay swelling inhibition (Aftab et al., 2020b). This resulted in a 9% enhancement in clay inhibition compared to conventional WBDF and a 17% improvement when compared to using fresh water. The success in inhibiting clay swelling can be attributed to two crucial factors: the effective plugging of pores in the clay and the bridging of interparticle pores between the existing drilling additives and the introduced nanomaterial.

12.6.1.3.3 Nano-Silica SiO2

The nanoparticles have shown promising potential for several industrial and scientific applications, including their applications for subsurface systems. The addition of nano-silica to drilling fluids has demonstrated enhanced performance for drilling shale oil and gas reservoirs. When nano-silica is incorporated, it improves fluid rheological properties, stability, and lubricity, leading to more efficient drilling. It also helps reduce fluid loss and filter cake formation, which, in turn, improves wellbore stability and reduces damage to the surrounding formation. The nanoscale size of silica allows it to penetrate and strengthen the formation matrix, providing even better borehole stability and preventing formation collapse (Young & Friedheim, 2013). Figure 12.12 shows the performance of nano-silica particles in controlling plugging during drilling in Atoka shale. The nanoparticles provide a perfect seal, preventing methane from being emitted due to micro-cracks and pores.

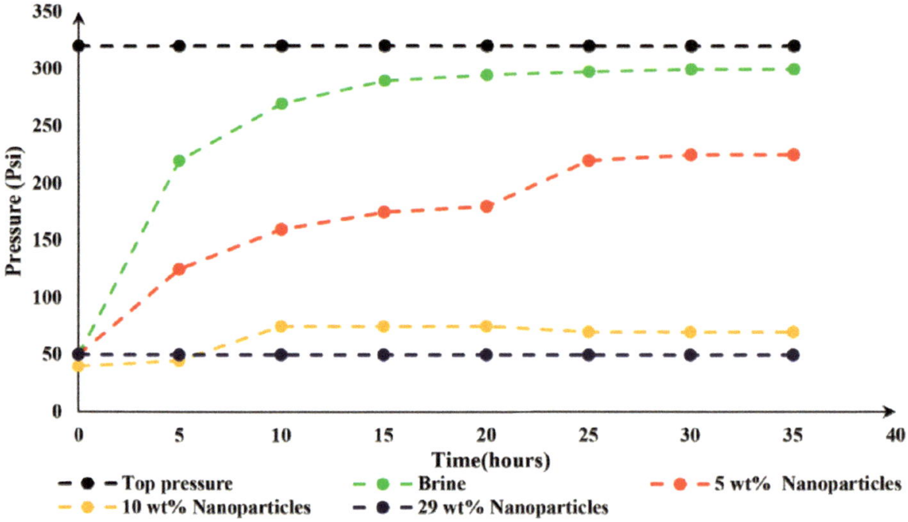

FIGURE 12.12 Plugging effect with different nano-silica concentrations on Atoka shale.

Source: Young & Friedheim, 2013

12.6.1.3.4 Graphene Oxides (GO) and Glycerol Deep Eutectic Solvent (Gly-DES)

Improving the lubricating ability of water-based drilling fluids is vital for maximizing the reach and efficiency of horizontal wells in unconventional oil and gas reservoirs, particularly shale formations. The main hurdle is the significant friction encountered, especially in long horizontal wells. Using graphene oxides (GO) and glycerol deep eutectic solvent (Gly-DES) as environmentally friendly lubricants overcomes the friction issues as the results showed that both GO and Gly-DES have a more pronounced lubricating effect, even at lower concentrations, than conventional lubricants (Ma et al., 2021). At a concentration of 0.25 wt %, GO reduced the adhesion coefficient of the filter cake by 93.33% and the friction coefficient of the drilling fluid by 48.47%.

12.6.1.3.5 Multi-Walled Carbon Nanotube (MWCNT)

To tackle high-temperature conditions, introducing MWCNT into the drilling fluid resulted in notable improvements in its performance (Ismail et al., 2017). In Saraline drilling fluid, the 10-sec gel strength increased by 33%, and the filtrate loss volume dropped to 10% after aging at 250°F. Moreover, both Saraline and Sarapar drilling fluids exhibited enhanced plastic viscosity by 6% and 27%, respectively, at 350°F after MWCNT addition. Additionally, for Sarapar drilling fluid, the filtrate loss volume decreased by 19% following 16 hours of aging at 250°F. Overall, MWCNT had a positive impact on the drilling fluids' rheological properties in elevated temperatures.

TABLE 12.1
Summary of Studies on Environmentally Friendly Chemicals for Drilling Applications in Shale Rocks

Reference	Shale Formation	Drilling Fluid Additive Used	Mud Base	Performance Factor Improved	Key Observations
(Aftab et al., 2020a)	Marang and Endau shale samples, Malaysia	TiO_2-bentonite nanocomposite	Water-based mud	Shale swelling was reduced	The reduction in swelling was linked to higher mud viscosity and improved nanopore plugging qualities, leading to decreased water infiltration.
(Aftab et al., 2020b)	Heated clay plugs	Tween 80-ZnO nanoparticle composite	Water-based mud	The optimal clay inhibition in the drilling fluid was achieved by introducing 0.6 g of T80ZnO nanoparticles.	The acute toxicity tests showed a 100% survival rate of white-leg shrimp in the waste of the water-based drilling fluid (WBDF).
(Young & Friedheim, 2013)	Marcellus shale Mancos shale Texas gas shale, US	Nano-silica SiO_2	Fresh water–based mud	Effective plugging into the micropores and cracks	The key to the success of this technology lies in its distinctive shale inhibition mechanism, which employs specially designed nano-size particles to seal the pores and micro-cracks that exist or are induced in unconventional shale formations.
(Ma et al., 2021)	N/A	Graphene Oxides (GO) and glycerol deep eutectic solvent (Gly-DES)	Water-based mud	Increase in lubricity efficiency	GO and gly-DES exhibited the ability to decrease the coefficient of friction due to their remarkable capacity to create a durable protective film on the surfaces of both the metal and the filter cake on contact. This protective film effectively reduces friction between the surfaces.
(Ismail et al., 2017)	N/A	Multi-walled carbon nanotubes (MWCNT)	Synthetic-based drilling fluid prepared using Sarapar and Saraline fluids	Filtration loss was reduced	Adding MWCNT brings about a considerable improvement in the rheological properties of Saraline and Sarapar drilling fluids. Particularly at higher temperatures, the inclusion of MWCNT results in significant enhancement of plastic viscosity, yield point, and gel strength.
(Abdo et al., 2014)	N/A	ZnO-clay nanocomposite	Water-based mud	Enhancement of stability in high-temperature conditions	The drilling fluids maintained their rheological properties consistently, even when exposed to elevated temperatures.

12.6.1.3.6 ZnO-Clay Nanocomposite

Zinc oxide (ZnO) nanoparticles have been the subject of extensive investigation for a considerable period. However, recent advancements in nanotechnology and material synthesis sparked a renewed interest in these structures. The incorporation of ZnO-clay nanocomposite into drilling fluids results in a notable steadiness in viscosity and yield point as temperature and pressure increase. In contrast, the absence of nanocomposites under identical conditions leads to a significant decline in both viscosity and yield point. This illustrates the impressive capability of ZnO-clay nanocomposites to confer stable colloidal properties to clays when combined, effectively hindering clay flocculation in water and ensuring substantial stability for the drilling fluids. It is important to highlight that by adjusting the size and composition of ZnO-clay nanocomposites, a flexible water-based drilling fluid can be tailored to meet diverse requirements, encompassing viscosity, yield point, gel strength, density, and more (Abdo et al., 2014).

12.7 EMISSION REDUCTION FOR HYDRAULIC FRACTURING IN SHALE RESERVOIRS

12.7.1 Conventional Fracturing Fluids

Conventional fracturing fluid refers to the widely used traditional types of fracturing fluid in oil and gas well fracturing operations, mainly water-based fluid. Water-based fracturing fluid is composed of 99.5% water and sand (as proppant), while the remaining 0.5% comprises chemical additives. The additives in fracturing fluid mainly include crosslinkers, breakers, fluid-loss additives, stabilizers, bactericides, surfactants, and clay stabilizers (Annevelink et al., 2016). Such fluids have the advantages of high viscosity, strong sand-carrying capacity, low friction, and low filtration loss. These fluids are utilized in most fracturing treatments today because water is generally an easily accessible and cost-effective resource, making water-based fracturing fluid relatively economical. They are relatively simple to transport and handle, requiring less complex equipment. However, there are several concerns related to this technique, such as a huge amount of water usage, clay swelling, formation damage, difficulty in flowback, and so on (Godec et al., 2014; Kalam et al., 2021). In some regions, the treatment and recycling of water in fracturing fluid may increase operational costs. As a result of these difficulties, scientists have been actively exploring alternative methods for fracturing that do not involve the use of water. These include gas fracturing, cryogenic fluids fracturing (liquid CO_2 and liquid N_2), supercritical CO_2 fracturing, foam fracturing, etc. The following section will discuss green fracturing fluids.

12.7.2 Green Fracturing Fluids

Due to the great demand for water resources, environmental impact, and the poor performance of water-based fracturing fluids, the emergence and development of anhydrous fracturing technology have been promoted (Nianyin et al., 2021). Thus, green fracturing fluid refers to a more environmentally friendly and sustainable type of fracturing fluid used in oil and gas well fracturing operations (Faber et al., 2021). It primarily emphasizes reducing adverse environmental impacts, minimizing water usage, and promoting sustainable resource utilization. The components of green fracturing fluids are designed to be easily broken down and degraded, reducing residual and environmental pollution in underground settings. Additives and chemicals in these fracturing fluids are designed to have minimal impact on the environment and ecosystems (Krishnan et al., 2022).

12.7.3 Foam Fracturing

This foam-based waterless fracture technique employs a fluid composed primarily of two phases: dispersed and continuous. The dispersed phase constitutes the fluid internal portion, ranging from 55% to 95%, and may consist of air, nitrogen, or carbon dioxide (Reidenbach et al., 1986). Meanwhile, the continuous phase comprises liquid substances like oil, acid, alcohol, or gel. CO_2-based foam fracturing is an attractive technique for fracking shale formations as it has the added benefit of utilizing CO_2 to reduce greenhouse gas emissions. Khan et al. evaluated CO_2 foam fracturing in various shale formations with different brittleness indexes (Khan et al., 2022). CO_2 foam fracturing yielded a low fracture initiation pressure in the highly brittle Mancos shale. In addition to reducing greenhouse gas emissions, CO_2 foam has several benefits, such as no usage of water, no formation damage, rapid post-cleanup operation, etc. Despite numerous merits, however, there are several shortcomings of CO_2 foam as well. The CO_2 foam exhibits a notably low apparent viscosity, which impacts the foam quality. The use of nanoparticles and polymers can aid in addressing this challenge. A study showed that hydrolyzed polyacrylamide (HPAM) and nano-silica were effective in achieving a foam quality of 90% with a supercritical CO_2 foam displaying an apparent viscosity of 100 cp (Li et al., 2015). The synergy of CO_2 foam with certain chemicals can help in optimizing the CO_2 fracturing process in shale formations.

12.7.4 Cryogenic Fracturing

Cryogenic fluids are substances that exist at significantly low temperatures. The most common types of cryogenic fluids are liquid nitrogen and liquid carbon dioxide. Liquid nitrogen fracturing is a promising technique due to its eco-friendly nature, ample availability, ultra-low temperature, and ability to generate multiple fractures. One of the most interesting inherent properties is its remarkably low temperature of –346°F, which induces significant thermal stress, leading to the formation of secondary fractures at right angles to the primary fractures. Such findings have been presented in past research (Cai et al., 2015; Tran et al., 2012; Zhang et al., 2019). In contrast, liquid CO_2 fracturing also presents an attractive opportunity to reduce greenhouse gas emissions. The temperature of liquid carbon dioxide is –109°F. Injecting liquid CO_2 into shale formations leads to the creation of fractures on rock surfaces through thermal shock, like the process of liquid nitrogen injection. Liquid

CO_2 also aids in reducing crude oil viscosity. Although cryogenic fracturing is a successful technique for fracturing shale reservoirs, the operational procedures require a special design to tackle the ultra-low temperatures.

12.7.5 Supercritical CO_2 Fracturing

Supercritical CO_2 fluids are different from both gases and liquids and have many unique physicochemical properties. Supercritical fluids possess a higher density than that of a liquid state, and this density rises as pressure increases. They exhibit characteristics of both gases and liquids, encompassing attributes of both states. For CO_2, the supercritical state occurs when the pressure and temperature reach or exceed 1070.38 psi and 87.98 °F, respectively. Supercritical CO_2 is gaining significant interest in the context of waterless fracturing in shale formations due to its enhanced stimulation outcomes, reduced pumping requirements, and fewer limitations related to blending. Since supercritical CO_2 fluid does not contain water, the problem of clay swelling is fundamentally avoided. In addition, the characteristics of low viscosity and large diffusion coefficient of supercritical CO_2 fluid can make the reservoir produce many microfracture networks and improve the conductivity of the reservoir. Moreover, within shale gas reservoirs, CO_2 will displace the adsorbed methane due to its higher adsorption capacity, thereby enhancing the recovery of unconventional natural gas effectively (Zhang et al., 2023a). CO_2 will also react with the formation and dissolve minerals, which can significantly increase the size of the original pores, thereby enhancing the permeability and porosity of the formation (Jiang et al., 2016). Therefore, supercritical CO_2 fracturing is expected to become an environmentally friendly, efficient, and safe new fracturing method.

Besides numerous advantages, there also exist challenges linked with supercritical CO_2 fracturing (Moridis, 2018; Shi et al., 2023): (1) the cost of CO_2 fracturing fluids is significantly higher than freshwater for hydraulic fracturing; (2) CO_2 capturing and transportation increases the fracturing's indirect expenses; (3) this technology is still in development, lacking a fully refined and successfully established system; (4) at reservoir conditions, CO_2 has significantly lower viscosity than, leading to inadequate proppant-carrying performance, (5) the decreased viscosity of CO_2 also results in notably enhanced frictional resistance within small-diameter casings; (6) throughout the CO_2-based fracturing procedure, the potential for gas leakage exists, leading to both wastage of CO_2 and potential environmental concerns.

12.7.6 Pulse Fracturing

Pulse fracturing is a relatively recent method that employs dynamic loading thermochemical reactions or high-energy gases to induce fracturing. This method is also known as high-energy gas fracturing, propellant fracturing, or tailored-pulse fracturing. It generates a load roughly ten times higher than the in situ stress level, yet it usually remains below the rock's plastic or compaction limit (Safari et al., 2015). A pulse-peak-pressure load is administered within the span of milliseconds to initiate and propagate multiple fractures radially outward from the borehole. Plasma pulse fracturing is a waterless fracturing technique in which electrical discharge is used to frack the reservoir. It is implemented through a two-step procedure. In the initial stage, pulsed plasma generates shockwaves directed towards the reservoir rock, inducing the formation of fractures that enhance the internal resonance of the reservoir rock fluid. Secondly, high-temperature plasma beams are generated to effectively dissolve components like asphaltenes and waxes present in the oil (Agarwal & Kudapa, 2023). This fracturing process is quick, does not need chemicals, and uses electrical energy for fracturing, thereby reducing formation damage. Nevertheless, this technology lacks comprehensive understanding, necessitating extensive research to explore various aspects. Table 12.2 presents the key advantages and disadvantages of waterless fracturing techniques.

12.7.7 Carbon Dioxide–Assisted Enhanced Recovery in Unconventional Shale Resources

CO_2 has been utilized in flooding conventional oil and gas reservoirs for decades to enhance hydrocarbon recovery and reduce greenhouse gas emissions. Such a technique can also be used to recover shale oil and gas resources (SOGR). The CO_2 flooding in SOGR presents unique complexities due to the complex nature of the shale microstructure. Therefore, the mechanism of conventional CO_2 flooding will be different from that of SOGR CO_2 flooding. This section focuses on the utilization of CO_2 in SOGR with different modes.

12.7.7.1 CO_2-EOR in Shale Oil Resources

The challenging conditions of shale reservoirs, characterized by ultra-low permeability, lead to significant issues such as inefficient sweep efficiency and premature breakthrough of injected fluids. Experimental studies showed that the traditional oil recovery technique—namely, water flooding—is not feasible for shale formations (Du & Nojabaei, 2019). There are several methods to improve the oil recovery from shales. Numerous studies have highlighted that miscible gas EOR stands out as the most effective approach for shales. Among the miscible gases, such as CO_2, N_2, CH_4, and C_2H_4, scholars have directed their attention toward CO_2 due to its superior performance (Lashgari et al., 2019; Sambo et al., 2023; Sheng, 2015) and the associated utilization in the context of carbon capture storage and utilization (CCUS). Under appropriate pressure and temperature conditions, there is a higher miscibility between oil and gas, resulting in reduced viscosity and lower IFT. Additionally, it serves as an effective approach to addressing future climate change concerns by averting the release of CO_2 emissions into the atmosphere (Kalam et al., 2020; Wang et al., 2023a).

The following main strategies utilize CO_2 in shale oil resources:

- Continuous CO_2 injection
- Water-alternating CO_2 injection
- CO_2 huff-n-puff

TABLE 12.2
Summary of Studies on Waterless Fracturing

Waterless Fracturing Technique	Key Findings: Merits	Key Findings: Drawbacks	Reference
Foam Fracturing	Higher proppant carrying capacity Minimal formation damage Rapid post-cleanup operation	Foam quality might need improvement by adding certain chemicals, which will increase the cost of the overall process.	(Khan et al., 2022; Li et al., 2015)
Cryogenic Fracturing	Environmentally friendly Formation of multiple fractures due to extremely low temperature Liquid nitrogen temperature: −346°F Liquid CO_2 temperature: −109°F Added benefit of liquid CO_2: oil viscosity reduction	Requirement of special operational procedures and facilities Possibility of CO_2 leakage to the atmosphere	(Cai et al., 2015; Tran et al., 2012; Z. Zhang et al., 2019; Kalam et al., 2021)
Supercritical CO_2 Fracturing	Environmentally friendly Improved shale gas recovery and CO_2 storage due to competitive adsorption mechanism Relatively lower pumping-pressure requirement	Elevated frictional pressure due to the need for reaching supercritical state conditions. Limited capacity to transport sand due to its low viscosity. CO_2 capturing and transportation increase the fracturing cost. Potential for gas leakage.	(Zhang et al., 2023a; Moridis, 2018; Shi et al., 2023; Kalam et al., 2021)
Pulse Fracturing	Creation of multiple fractures in a very short time	Restricted scope of usage. Relatively new technology needs investigation to understand fully.	(Safari et al., 2015; Agarwal & Kudapa, 2023)

12.7.7.1.1 Continuous CO_2 Injection

In continuous gas flooding, gas is injected consistently into the injection well while oil is produced simultaneously from the production well. The development of miscibility between gas and oil is achieved through a multi-contact process involving mechanisms such as vaporization, condensation, or a combination of both, as shown in Figure 12.13. While this method has been extensively investigated for conventional reservoirs (Bayat et al., 2016; Zhang et al., 2013), there are limited studies on continuous gas flooding in shale reservoirs (Du & Nojabaei, 2019). This scarcity is primarily attributed to the occurrence of severe fingering effects that can take place in the fractures and matrix system. Furthermore, a significant challenge hindering the application of continuous gas flooding in ultra-low permeability shale reservoirs is the issue of low injectivity. A large amount of gas is required, with high operating and maintenance costs in continuous flooding.

12.7.7.1.2 CO_2 Huff-n-Puff

The CO_2 huff-n-puff method involves a single well serving as both the injector and producer during the process. CO_2 continuous injection and water-alternating-CO_2 injection methods have been widely used in conventional reservoirs. However, in shale oil resources, it is very difficult to use such methods due to ultra-low permeability, leading to a huge pressure drop in the matrix system. Consequently, pushing the oil from the injector to the producer would be challenging. To overcome this challenge, the CO_2 huff-n-puff method is the most suitable technique to exploit shale oil resources. Both experimental and simulation studies also show that the CO_2 huff -n-puff method results in higher oil recovery than CO_2 continuous flooding (Wan et al., 2014; Yu et al., 2017). This method consists of three stages: (1) huff, (2) soak, and (3) puff. Figure 12.14 presents a schematic showing the stages of the huff-n-puff process (Hawthorne et al., 2013).

FIGURE 12.13 Schematic showing the mechanism of the CO_2 miscibility process.

Source: Lashgari et al., 2019

First stage—huff: Steps 1 and 2 constitute the injection stage. In the initial stages of CO_2 injection (step 1), the CO_2 flows rapidly, primarily through fractures rather than the rock matrix itself (Figure 12.14). As the injection progresses, the CO_2 gradually permeates the rock matrix due to the pressure gradient created by the injection (step 2). However, this initial permeation of CO_2 into the rock matrix may have certain effects on oil production. On the one hand, it could potentially reduce oil production by carrying oil that was initially near the surface deeper into the rock matrix. On the other hand, the oil swelling caused by the presence of CO_2 could lead to an increase in oil production during the pressurization process.

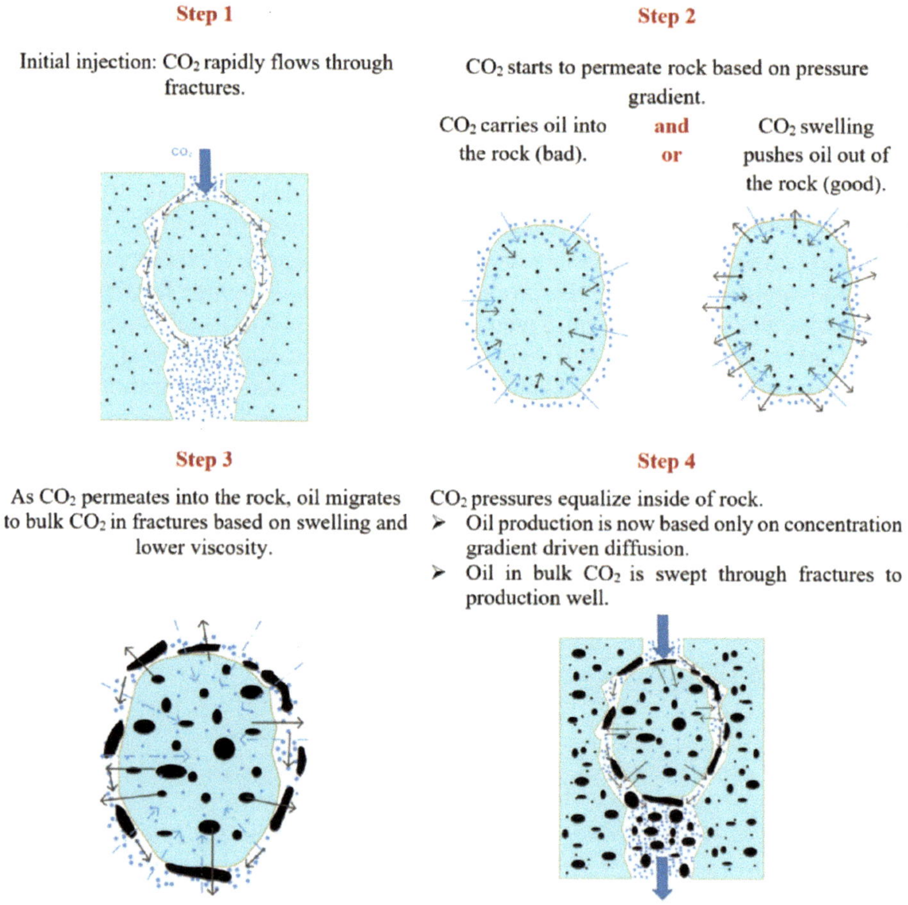

FIGURE 12.14 The conceptual steps of a CO_2 huff -n'-puff process. Steps 1 and 2 constitute the huff stage, step 3 is the soaking stage, and step 4 is the puff stage.

Source: Hawthorne et al., 2013

Second stage—soak: Step 3 shows the soaking stage, which is the well shut-in period (Figure 12.14). During this period, the injected CO_2 dissolves in the oil, causing the oil volume to expand and its viscosity to decrease.

Third stage—puff: Step 4 shows the oil production stage. In this stage, the CO_2 pressure starts to equalize across the rock matrix (Figure 12.14). During this stage, the oil swelling, reduced oil viscosity, and potential formation of a CO_2/oil miscible phase persist, further improving the mobilization of the oil. As pressure equilibrium approaches, the dominant process for oil recovery may shift to concentration-driven diffusion of hydrocarbons in CO_2 moving from the rock interior to the bulk CO_2 in the fractures.

Several factors can affect the performance of the CO_2 huff -n'-puff method, including but not limited to asphaltene deposition, matrix permeability, injection rate, slug size, and soaking time. Asphaltene deposition may impair permeability. Therefore, chemical inhibitors must be used to reduce the deposition of asphaltene. On the other hand, higher matrix permeability, increased injection rates, and optimal slug size will help in improving the efficiency of a CO_2 huff-n-puff process. Table 12.3 summarizes CO_2-EOR methods in shale oil resources, including mechanisms and shortcomings.

12.7.7.2 CO_2-EGR in Shale Gas Resources

Shale reservoirs are renowned for their significant organic content, comprising approximately 2% to 10% of the overall weight of organic-rich shales, while TOCs as high as 23% are also reported (Arif et al., 2017). The thermal maturity primarily governs this organic content variation. Large quantities of gas are contained within shale reservoirs, where it exists as adsorbed gas on the finely dispersed organic matter's surface (Pu et al., 2018). Furthermore, in these formations, "free" methane gas, which is not adsorbed, also exists in fracture porosity, intergranular microporosity, and micropores within the kerogen (Godec et al., 2014). Therefore, shale gas can be in the form of free gas within natural fractures, adsorbed gas in the matrix, and dissolved gas in kerogen, as shown in Figure 12.15 (Zhang et al., 2018).

Note that the minerals, organic matter, and microfractures formulate the microstructure of shales, and it is of utmost significance to precisely characterize this microstructure for gas content and transport evaluation (Arif et al., 2021a).

The discovery of shale gas has drawn increased focus as a promising target to increase gas production. Horizontal drilling in conjunction with the hydraulic fracturing technique has been extensively applied to exploit shale gas resources. Nevertheless, a sharp decline after some years of production, as observed in Figure 12.16, highlights the importance of implementing enhanced gas recovery (EGR) techniques (Baihly et al., 2015). In this tertiary recovery technique, CO_2 is injected into the reservoir as a displacing fluid, which helps release adsorbed methane gas and improves gas production.

There are several components of methane gas recovery mechanisms for the CO_2 EGR technique. The most famous is the adsorption of CO_2 leading to the desorption of methane gas. Such competitive adsorption behavior is illustrated in Figure 12.17 (Godec et al., 2014).

TABLE 12.3
Summary of CO_2-EOR Methods in Shale Oil Resources

Method	Key Mechanisms Involved	Limitations	Suitability for Shale Oil Resources	Reference
Continuous CO_2 Injection	• Vaporization • Condensation • Combination of vaporization and condensation	• Fingering effects • Requirement for a large volume of gas • High capital, operational, and maintenance costs	No	(Du & Nojabaei, 2019; Jia et al., 2019)
Water-Alternating-CO_2 Injection	• Three-phase WAG hysteresis • Mobility control • Oil vaporization and swelling	• Problems at distant wells • Early gas breakthrough in fractured shale formations	No	(Belazreg et al., 2019; Khoshsima et al., 2023)
CO_2 Huff-n-puff	• Miscibility • Viscosity reduction • Oil swelling • Molecular diffusion • Depressurization • Vaporization of components • Adsorption and/or absorption of CO_2 in kerogen • Wettability alteration • IFT reduction	• Asphaltene precipitation • Formation damage • Corrosion • High cost	Yes	(Hasanzadeh et al., 2023; Reza Azin & Amin Izadpanahi, 2022; Sambo et al., 2023)

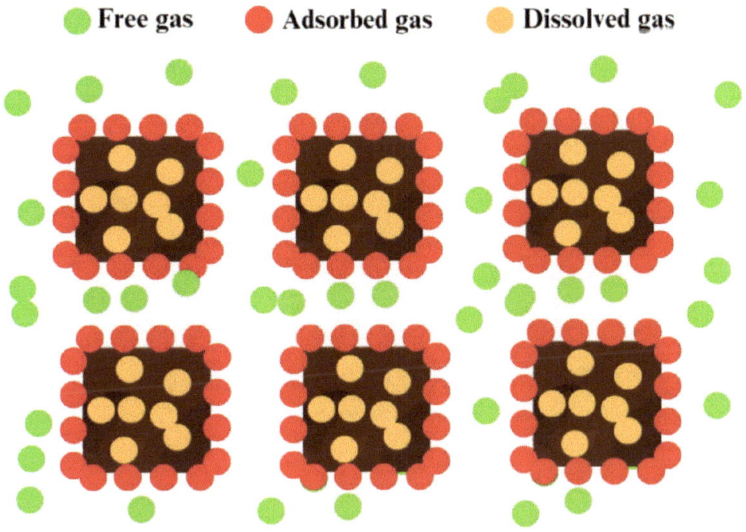

FIGURE 12.15 A typical gas storage system in shale gas resources, including free gas, adsorbed gas, and dissolved gas.

Source: Zhang et al., 2018

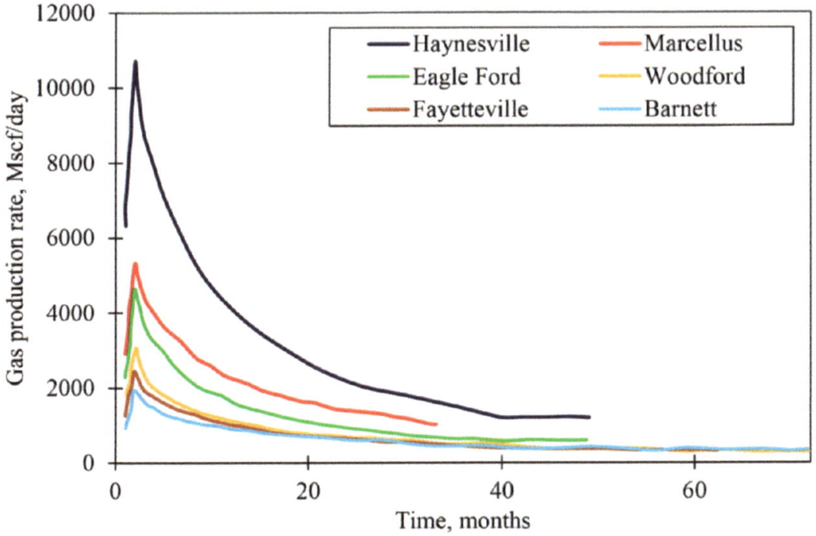

FIGURE 12.16 A comparison of several shale gas basins showing a decline in the gas production rate.

Source: Baihly et al., 2015

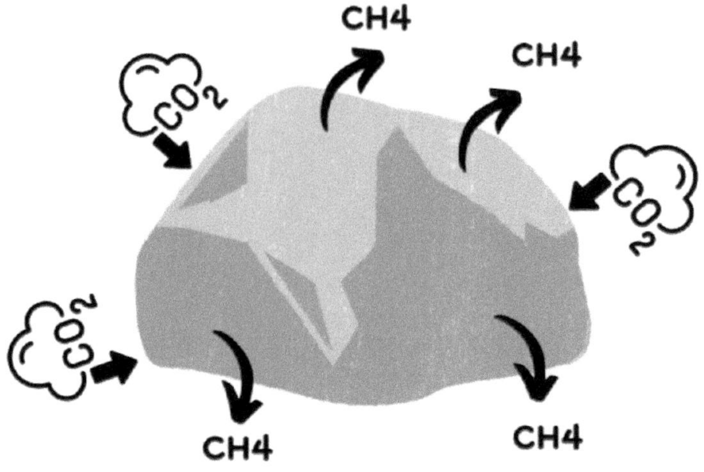

FIGURE 12.17 Adsorption-desorption mechanism in a CO_2 EGR process in shales.

Source: Godec et al., 2014

Kim et al. numerically investigated the CO_2 injection in shale gas reservoirs considering several factors, such as multi-component adsorption, dissolution, molecular diffusion, and geomechanics (Kim et al., 2017). They also compared flooding, huff-n-puff, and no-injection schemes and found a higher recovery percentage of methane due to geomechanical effects in the case of flooding, as shown in Figure 12.18. Mohagheghian and co-authors numerically explored the CO_2 EGR process considering several mechanisms: competitive adsorption of numerous components, viscous flow, gas slippage, Knudsen diffusion, pore size variation, and real gas effect (Mohagheghian et al., 2019). The major recovery mechanisms contributing to CO_2 EGR are summarized in Table 12.4.

12.7.8 Transformation of Unconventional Shale Resources for Energy Transition

As we embark on the epoch of energy transition, unconventional shale resources emerge as a pivotal element in bridging the gap between traditional energy consumption and sustainable alternatives.

Unconventional shale resources primarily comprise oil and natural gas trapped in the fine-grained sedimentary rocks often found in complex formations, which are unresponsive to conventional production techniques. Unlike their conventional counterparts, the extraction of these resources necessitates advanced technology, rendering it both challenging and essential. The abundance and untapped potential of shale resources offer a beacon of promise, ensuring energy security and contributing significantly to the world's energy mix. They represent a crucial source of energy that continues to fill the growing demand, especially in regions grappling with a scarcity

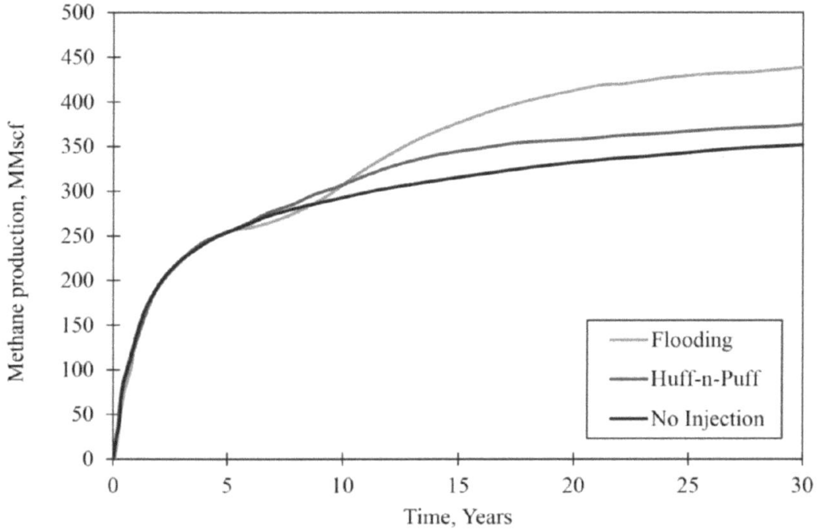

FIGURE 12.18 Production of methane gas for the Barnett shale models under different conditions (flooding, huff-n-puff, and no injection).

Source: Kim et al., 2017

of conventional resources (Davies et al., 2014; Castro-Alvarez et al., 2018; Vengosh et al., 2014).

As the global community strives to decarbonize and shift toward renewable energy, shale resources present an intriguing transition pathway (Cotton et al., 2014; Cuppen et al., 2019; Lau et al., 2021). They are not only a bridge between conventional fuels and cleaner alternatives but also an enabler of technological advancements that are pushing the boundaries of sustainable energy practices. By unlocking unconventional reservoirs and adopting responsible production methods, we are opening new horizons to realize energy transition goals effectively. This section will focus on three pivotal domains:

Geothermal energy system: By tapping into the thermal energy stored in the shale formations, innovative geothermal energy systems can be deployed, offering an efficient and renewable source of heat and power.

Geological energy (hydrogen) storage: The unique geological characteristics of shale make it a viable candidate for large-scale hydrogen storage. The utilization of shale as a storage formation or a cape rock for storing hydrogen may provide a new frontier in energy management, complementing the growing emphasis on hydrogen as a clean energy carrier.

Geological CO_2 storage: Another transformative application lies in the storage of carbon dioxide in shale formations. This method of geological CO_2 storage offers a practical solution to mitigating greenhouse gas emissions, thereby playing a substantial role in global climate change mitigation efforts.

TABLE 12.4
Shale Gas Recovery Mechanism of a CO_2 EGR Process

Mechanism	Reasoning	Reference
Viscous flow of the free gas	• Collisions between the molecules.	(Mohagheghian et al., 2019)
Klinkenberg effect	• Gas slippage.	(Mohagheghian et al., 2019)
Molecular diffusion/Knudsen diffusion	• Molecular collisions with the pore walls. • Knudsen diffusion is a modified molecular diffusion that accounts for the impact of porosity and tortuosity on the diffusion process. • Knudsen diffusion becomes more significant when the pore size is comparable to the molecules' mean free path.	(Kim et al., 2017; Mohagheghian et al., 2019)
Multi-components competitive adsorption	• Kerogen exhibits a preference for adsorbing CO_2 over hydrocarbon gas. • Inorganic pores will also exhibit a preference for CO_2 adsorption on their surface. • Consequently, adsorption of CO_2 leads to the desorption of methane gas.	(Kadoura et al., 2016; Mohagheghian et al., 2019; Sun et al., 2021)
Pore size variation	• Pore enlargement occurs due to the compressibility of shale pores.	(Guo et al., 2015; Mohagheghian et al., 2019)
Real gas effect	• Consideration of the non-ideal behavior of gases.	(Mohagheghian et al., 2019)
Dissolution	• Dissolution of gas components in the formation water according to the Henry's law.	(Kim et al., 2017)
Geomechanical effects	• Stress-dependent properties. • The permeability of natural fractures undergoes an increase during CO_2 injection until the injection is halted, after which it gradually decreases again.	(Kim et al., 2017)

The transformation of unconventional shale resources transcends traditional energy paradigms, illuminating a path toward a more sustainable and resilient energy future. Through an exploration of these multifaceted applications, this section will deepen our understanding of how shale resources can be harnessed in a manner that aligns with the global vision of clean and accessible energy for all.

12.7.8.1 Geothermal Energy System

Geothermal energy systems harness the Earth's innate heat, a heat that finds its origins in the radioactive decay of particles, convection currents, and, to a lesser extent, the primordial heat from the Earth's formation. Traditionally, these systems exploit natural heat from underground reservoirs to generate electricity or provide direct heating (Shahsavar et al., 2023); see Figure 12.19 for different types of geothermal energy applications. The main benefits of geothermal energy include (Huenges & Ledru, 2011; Barbier, 2002; Dickson & Fanelli, 2013):

- Sustainability: As long as the Earth exists, it will emanate heat, making geothermal energy a limitless renewable resource.
- Low emissions: Geothermal plants emit significantly fewer greenhouse gases than fossil fuel power plants, positioning it as a cleaner energy option.
- Reliability: Unlike solar or wind energy, geothermal energy is consistent and not dependent on external weather conditions.

12.7.8.1.1 Potential for Extracting Geothermal Energy from Unconventional Shale Resources

Shale is a sedimentary rock primarily composed of compacted clay and silt particles. Its unique thermal properties make it stand out in the geothermal context. Understanding the composition and genesis of shale provides insight into its peculiar attributes. Formed over millions of years, these rocks have undergone significant compression, which has expelled much of the original pore water and led to the

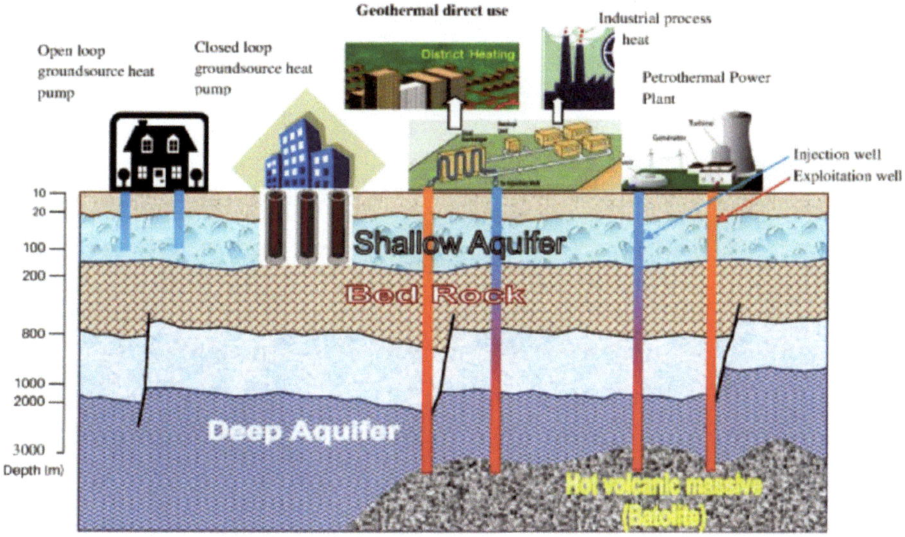

FIGURE 12.19 Schematic illustration of different types of geothermal energy applications.

Source: Shahsavar et al., 2023

compact arrangement of its mineral constituents. This tight packing gives shale its dense molecular structure. When combined with the clay minerals that dominate its composition, it results in exceptionally low permeability; see Figure 12.20 for thin-section images of a oil shale sample (Saif et al., 2017). Such a characteristic implies that fluids, including water and gases, find it difficult to move freely within the rock matrix (Zendehboudi & Bahadori, 2016). As a consequence, shale acts as an excellent insulator, almost akin to a thermal blanket. Heat that is generated or stored in or beneath the shale layer finds it challenging to migrate or dissipate through the rock. This is in stark contrast to more permeable rock formations, in which heat can readily transfer with fluid movement (Cutright, 2011; Mao & Zhang, 2018; Li et al., 2017). Additionally, the fine-grained nature of shale, coupled with its organic content, further augments its heat retention capacity. These organic materials, remnants of ancient microorganisms and plant matter, can break down and generate heat on reaching certain depths and temperatures. This endogenic heat adds to the geothermal gradient, thereby enhancing the overall thermal profile of the shale rock (Mulyanto et al., 2018; Nurhandoko, 2013; Ross & Bustin, 2009).

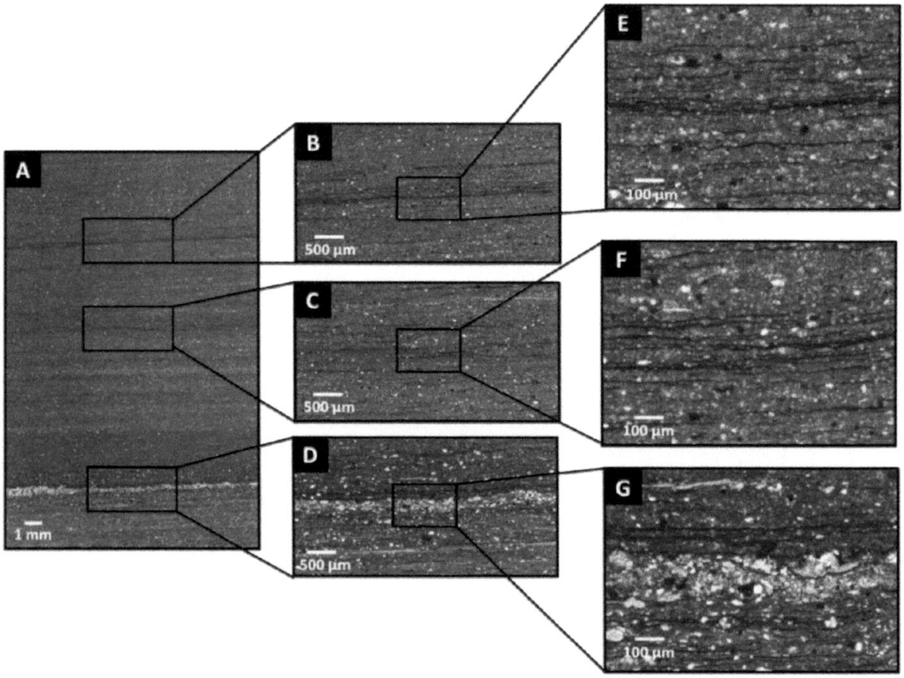

FIGURE 12.20 Example of thin-section images of oil shale from the organic-rich Green River formation at low (A), medium (B–D) and high (E–G) magnification. The images taken in plane polarized light (PPL) show a series of laminated structures of organic material in a distribution of fine-grained inorganic minerals.

Source: Saif et al., 2017

The depths at which shale formations commonly reside often coincide with regions of elevated temperature. Typically, shale formations are nestled at considerable depths, often ranging from a few hundred meters to several kilometers beneath the surface. This geothermal gradient, coupled with shale's heat-preserving characteristics, makes them substantial reservoirs of geothermal energy. Such formations, when accessed, can offer a steady and reliable source of heat energy without the intermittence commonly associated with other renewable energy sources (Yuan et al., 2023). Figure 12.21 shows the temperature distribution at a depth of 3,500 meters in the 48 states of the US. It highlights areas with high temperatures, which include states like Texas, Oklahoma, Louisiana, and North Dakota, where oil and gas activities are active (Augustine & Falkenstern, 2014; Wang et al., 2018).

Another aspect that boosts the potential of shale for geothermal energy is its vast spread and thickness in certain regions. Regions with significant shale deposits, such as the Marcellus or Barnett shales in the United States, can potentially offer expansive zones for heat extraction. When tapped efficiently using advanced drilling and heat exchange techniques, these vast shale beds can provide significant amounts of geothermal energy over extended periods (Thoram & Ehlig-Economides, 2011; Richards & Blackwell, 2012). Much of the infrastructure and knowledge base developed for shale gas extraction can be repurposed or adapted for geothermal energy extraction. This includes drilling equipment, expertise in handling shale formations,

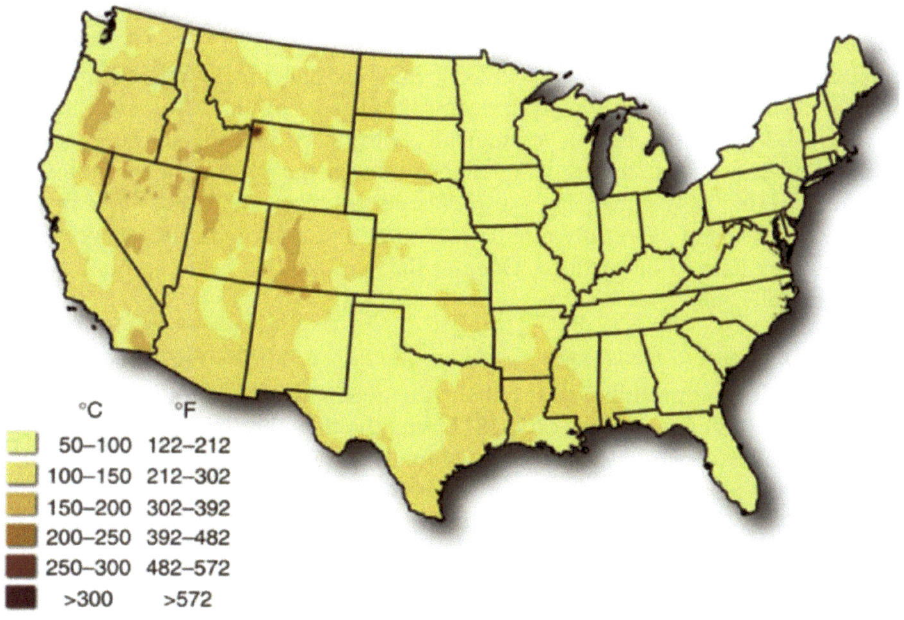

FIGURE 12.21 Map displaying the temperature of the 48 states of the United States at a depth of 3,500 meters.

Source: Augustine & Falkenstern, 2014

and even some elements of the logistical framework. Such a synergy could lead to cost savings and accelerate the deployment of geothermal projects in shale-rich regions.

In addition to the environmental benefits of a clean and renewable energy source, tapping into the geothermal potential of shale formations can present economic advantages. As the technology matures and economies of scale come into play, the cost per unit of energy extracted from shale-based geothermal sources could become competitive, even against conventional fossil fuels. Moreover, as shale gas extraction faces criticism for environmental concerns, pivoting to geothermal applications can extend the utility and life of shale formations, aligning them more closely with global sustainability goals.

12.7.8.1.2 Current Practices and Technologies Help Harnessing Geothermal Energy from Shale Resources

Harnessing geothermal energy from unconventional shale resources often requires technologies similar to those employed in the current shale gas/oil production, including:

Horizontal drilling: One of the primary techniques used in shale gas extraction, horizontal drilling allows access to larger portions of the shale formation, thereby enhancing the extraction of geothermal heat (Cui et al., 2017; Zhang et al., 2014; Feng et al., 2015).

Hydraulic fracturing: This is a common method of inducing fractures in shale rock to increase its hydrocarbon production (Pandey et al., 2014). In the context of geothermal energy, fracturing improves rock permeability and aids in the flow of heat.

Enhanced geothermal systems (EGS): EGS are specifically engineered to harness heat from rocks with low permeability (Olasolo et al., 2016), such as shale. In the EGS process, a fluid is strategically introduced into the subsurface. Under controlled conditions, this fluid injection induces the reopening of pre-existing fractures, effectively enhancing the rock's permeability. With this increased permeability, the fluid can easily circulate through the fractured rock, absorbing and transporting heat to the surface. Once at the surface, this heat can be utilized to generate electricity.

12.7.8.1.3 Current Progress of Geothermal Energy in Shale Resources

Due to the low permeability of the shale, the potential use of geothermal resources relies on hydraulic fracturing or EGS. Currently, there are many deep geothermal projects under construction/proposed around the world, such as the FORGE in Utah, United States; Habanero Pilot Plant in South Australia's Cooper Basin; United Downs Deep Geothermal Power Project in Cornwall, UK; and the Basel Project in Switzerland. Although primarily focused on granite rock formations, their experiences could also be applied to shale. One of the most famous shale reservoirs, the Eagle Ford shale in Texas, could have great potential for geothermal energy extraction (Richards & Blackwell, 2012). The Eagle Ford shale, with a stretch about 400 miles long and 50 miles wide, is predominantly recognized for its prolific oil and gas reserves. One of the primary advantages the Eagle Ford shale boasts is the existing infrastructure and technical expertise resulting from decades of hydrocarbon exploration and production. This pre-established framework can be leveraged, with necessary modifications, to facilitate geothermal energy extraction, potentially accelerating deployment and reducing initial costs. Indeed, utilizing traditional oilfield infrastructure for geothermal energy extraction has been common practice in recent years (Xin et al., 2012; Lund & Boyd, 2016; Li & Sun, 2015; Wang et al., 2018). See Table 12.5; these experiences can be applied to future geothermal energy extraction from unconventional shale reservoirs.

Moreover, geothermal energy extraction from shale reservoir can be combined with in situ shale gas production; recently, hybrid geothermal energy and hydrocarbon resources production by repurposing horizontal wells in shale gas reservoirs in Horn River Basin, British Columbia, Canada, has been proposed. The depth of the reservoir in the Horn River Basin (HRB) varies between 1,900 and 3,100 meters, with temperatures in the shale formations ranging from 80° to 160°C. This suggests that the shale gas reservoir could also function as a viable geothermal energy source (Walsh, 2013). Figure 12.22 illustrates the bottom-hole temperatures (BHTs), along with the corresponding temperature gradients. The average thermal gradient in the HRB is 46.4°C/km, which is higher than the global average of 25°C/km by a margin of 20°C. BHTs in the basin vary from 12°C to 178°C but most commonly fall within the 100°C to 130°C range at depths between 2,000 and 2,500 meters.

This strategy of using the heat in the HRB shale gas reservoir reduces the major well drilling cost of geothermal wellbores and has the advantage of co-producing hydrocarbon and geothermal resources by sharing one facility, providing a practical solution for converting high enthalpy hydrocarbon reservoirs to geothermal energy producers in supporting the low carbon energy transition (Yuan et al., 2023).

TABLE 12.5
Examples of Worldwide Oilfield Geothermal Utilization Projects

Oilfield Location	Geothermal Resource	Type of Utilizations	Reference
Albania	Abandoned wells	Greenhouse heating	(Lund & Boyd, 2016)
Styria, Austria	Abandoned wells	Spa resorts	(Lund & Boyd, 2016)
Algyő, Hungary	Produced water	Enhanced oil recovery	(Lund & Boyd, 2016)
Shengli, China	Abandoned wells	Spacing heating	(Junrong et al., 2015)
Daqing, China	Produced water	Spacing heating	(Wang et al., 2016)
Huabei, China	Abandoned wells	Heat-tracing oil Gathering	(Wang et al., 2016)
Wyoming, US	Produced water	Power generation	(Nordquist & Johnson, 2012)
Huabei, China	Produced water	Power generation	(Xin et al., 2012)

Source: Wang et al., 2018

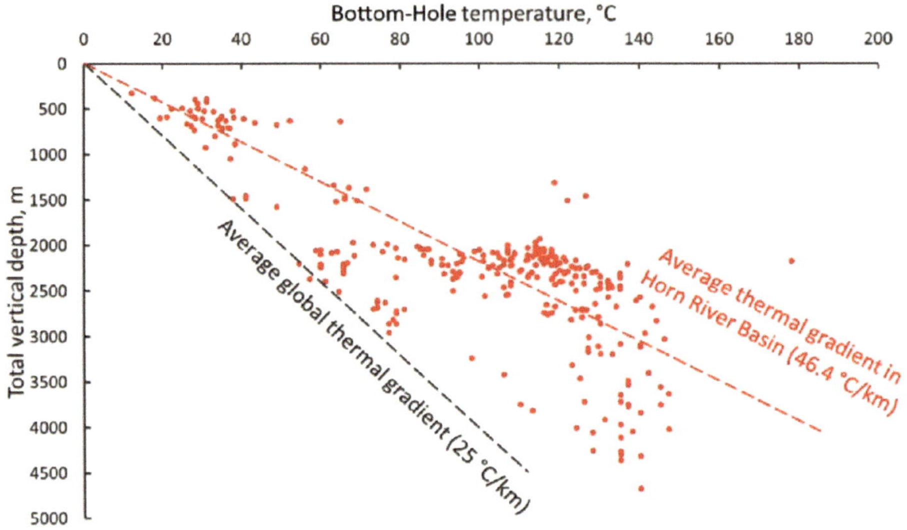

FIGURE 12.22 The geothermal gradient of the Horn River Basin measured from the wells.

Source: Yuan et al., 2023

12.7.8.1.4 Challenges and Future Prospects for Geothermal Energy Extraction from Shale

However promising, there are challenges to harnessing geothermal energy from shale:

Environmental concerns: Using hydraulic fracturing, as with current shale gas/oil production, has raised environmental concerns, particularly related to potential groundwater contamination and induced seismic activity (Ellafi et al., 2020; Spellman, 2012).

High initial costs: The infrastructure and technology required for geothermal extraction from shale demand substantial initial investments, and the return from the geothermal energy may lower than the cost.

Looking ahead, the prospects for geothermal energy extraction from shale are promising. As technological advancements refine extraction techniques and the energy sector continues its shift towards sustainable options, unconventional shale resources stand poised to play a pivotal role in the geothermal landscape.

In conclusion, the transformation of unconventional shale resources to facilitate geothermal energy systems symbolizes significant strides towards achieving a sustainable energy matrix. While challenges persist, the potential benefits and ongoing technological evolution render this avenue one filled with hope and promise for the future.

12.7.8.2 Geological Energy (H_2) Storage

As the global community intensifies its efforts to transition from fossil fuels to cleaner, renewable energy sources, the issue of energy storage has taken center stage. One of the inherent challenges with renewable energy—be it solar, wind, or tidal—is its intermittency. The sun doesn't always shine, and the wind doesn't always blow.

Consequently, there's an escalating need for efficient, large-scale energy storage solutions, and this is where geological energy storage comes into play (Ibrahim et al., 2008; Amrouche et al., 2016; Olabi, 2017; Barton & Infield, 2004).

Geological energy storage involves using underground formations to store energy, leveraging the Earth's natural geology to our advantage. These subterranean reservoirs might include depleted oil and gas fields, deep saline aquifers, or even caverns. They offer immense volume at depths where pressures can facilitate the long-term storage of energy with a minimal surface footprint (Carneiro et al., 2019; Bauer et al., 2013; Crotogino et al., 2018).

Among the various energies that can be stored geologically, hydrogen stands out as a prime candidate. As the most abundant element in the universe, hydrogen, when used as a fuel, produces water as its only byproduct, making it an attractive, eco-friendly solution. There are compelling reasons to focus on hydrogen for geological storage (Aftab et al., 2022; Tarkowski, 2019; Hassanpouryouzband et al., 2022; Zhang et al., 2023b; Heinemann et al., 2018; Jahanbakhsh et al., 2024):

Density and energy potential: Although hydrogen has a low energy density by volume compared to natural gas, its energy density by weight is remarkably high. This means that when stored under pressure, significant amounts of energy can be kept in confined spaces.

Flexibility and integration: Hydrogen can be both an energy carrier and a storage medium. It can be produced when there's an excess of renewable energy (via processes like electrolysis) and then stored underground. Later, it can be either used directly as a fuel or converted back to electricity.

Infrastructure reuse: Many regions around the world have mature oil and gas industries with infrastructure that's nearing the end of its operational life. These depleted reservoirs can be repurposed for hydrogen storage, capitalizing on existing knowledge and reducing initial setup costs.

Safety and containment: While hydrogen's small molecule size and mobility present challenges, geological formations, especially when combined with engineered barriers, can provide safe and secure long-term containment.

Economic and Environmental Benefits: As the demand for green hydrogen (produced using renewable energy) grows, large-scale storage solutions will become vital. Geological storage offers a cost-effective way to buffer the supply and demand while supporting the decarbonization of various sectors, from transportation to industry.

12.7.8.2.1 The Possibility of Hydrogen Storage in Shale Resources

The potential for hydrogen storage in shale reservoirs has garnered significant attention and research in recent years (Bechara et al., 2022; Wei et al., 2023). For example, Singh (2022) demonstrated the viability of using depleted horizontal shale gas wells as hydrogen storage repositories through numerical modeling. The study specifically focused on a hydraulically fractured well with representative properties in the Haynesville shale formation, as illustrated in Figure 12.23. From a technical standpoint, depleted shale gas wells offer several advantages over traditional underground hydrogen storage methods, including superior containment, elimination of the need for cushion gas, absence of water in the storage space, and a relatively high purity of stored hydrogen.

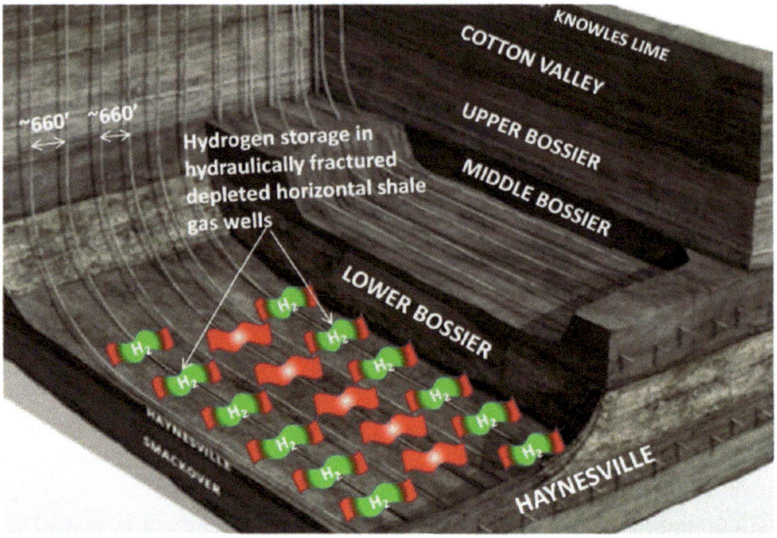

FIGURE 12.23 The geology of Haynesville shale and the storage of hydrogen in depleted shale gas wells that have been hydraulically fractured. The red and green patches indicate hydraulic fractures and injected hydrogen, respectively.

Source: Singh, 2022

While the potential for hydrogen storage in shale formations is promising and offers several advantages, numerous challenges remain to be addressed. The complex geometry and uncertain volume of fractures, coupled with the unique flow characteristics of hydrogen in a shale matrix, make it difficult to accurately estimate storage capacity using conventional volumetric methods. This is because hydrogen not only flows through artificially created fractures but may also be adsorbed onto the matrix surface through diffusive mechanisms (Wei et al., 2023). Additionally, the recovery rates for such adsorbed hydrogen are not yet well understood. Therefore, further research and pilot projects are essential for gaining a deeper understanding of the dynamics of hydrogen storage in shale formations, as well as for developing technologies and methodologies to overcome the associated challenges.

12.7.8.2.2 Shale as Caprock in Hydrogen Geological Storage

The utility of shale as a caprock—an impermeable geological formation that prevents the upward migration of stored substances—proves to be particularly reliable for hydrogen storage in porous rock, including deep saline aquifers and depleted oil and gas fields (Mahdi et al., 2021). Several characteristics contribute to shale's effectiveness as a caprock:

Low permeability: One of the defining traits of shale is its extremely low permeability, typically in the range of nano-Darcies. This low permeability functions as a natural barrier, effectively preventing the escape of gases, including hydrogen. In the context of hydrogen storage, this ensures minimal leakage and thus maximizes storage efficiency over the long term.

Geological abundance: Shale is one of the most prevalent sedimentary rock types on Earth. Its ubiquity provides a natural advantage; many existing and potential underground storage facilities already have overlying shale layers. These layers can serve as robust natural seals without the need for artificial barriers, thereby reducing both risk and cost in hydrogen storage projects.

Historical precedence: The role of shale as a natural caprock is not new; shale has acted as a seal for hydrocarbons such as oil and natural gas over geological time scales. This long history of natural containment lends credence to its reliability, offering a proven track record that can help mitigate perceived risks associated with hydrogen storage.

12.7.8.2.3 Challenges with Shale in Hydrogen Storage

While shale formations offer several advantages as caprocks for underground hydrogen storage, it is crucial to recognize the unique challenges that also exist (Tarkowski & Uliasz-Misiak, 2022; Raza et al., 2022). These challenges primarily arise from the distinctive properties of hydrogen and the complex geochemistry of shale.

One of the most pressing concerns is hydrogen's small molecular size. Although shale formations have low permeability, the minimal size of hydrogen molecules could potentially allow for diffusion through microfractures or nanopores in the shale (Liu et al., 2022). This is especially a concern when significant pressure differentials are applied, increasing the likelihood of leakage over time.

Another concern relates to how hydrogen interacts with the minerals and organic content with shale formations (Zeng et al., 2023). The behavior of hydrogen under geological storage conditions, especially with regard to its physical, geochemical, and microbial interactions, remains insufficiently understood (Bensing et al., 2022; Hassannayebi et al., 2019; Hassanpouryouzband et al., 2021). As depicted in Figure 12.24, a range of interactions could occur during the storage period that might compromise the sealing capabilities of the shale. However, existing laboratory studies have not provided direct evidence of geochemical reactions involving hydrogen with shale caprocks (Shi et al., 2020). This underscores the need for further research in this area.

The mechanical integrity of shale can also be compromised by the repeated cycles of hydrogen injection and withdrawal. This cyclical process could induce microfractures or even alter the shale's inherent permeability over time, leading to potential leaks (Liu et al., 2020). Long-term storage thus poses a risk of compromising the structural integrity of the shale formation, necessitating comprehensive integrity tests and ongoing monitoring.

Overview, shale, with its inherent low permeability and geological abundance, presents a promising storage place or caprock in hydrogen geological storage. While challenges exist, they are not insurmountable. As more research elucidates the specifics of hydrogen-shale interactions and technology advances, it's likely that shale will play a central role in the burgeoning hydrogen storage landscape, ensuring that this clean energy carrier remains safely ensconced beneath our feet.

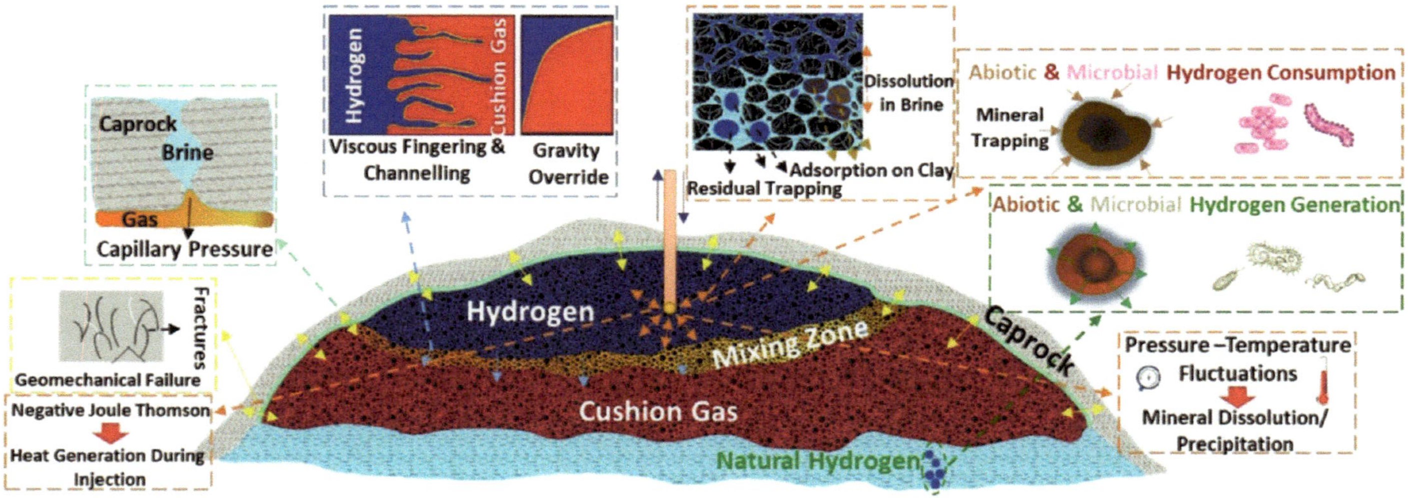

FIGURE 12.24 Graphical illustration of different physical, geochemical, and microbial reactions in hydrogen geological storage.

Source: Hassanpouryouzband et al., 2021

12.7.8.3 Geological CO_2 Storage

As concerns about climate change continue to escalate, various strategies aimed at mitigating greenhouse gas emissions are increasingly being adopted. One effective approach is the capture and storage of carbon dioxide (CO_2), a significant contributor to the greenhouse effect. This method, often referred to as carbon capture and storage (CCS), involves injecting CO_2 deep into underground geological formations. It is increasingly viewed as a promising solution for reducing human-made CO_2 emissions and combating global warming (Celia, 2017; Holloway, 2007; Bickle, 2009; Baines & Worden, 2004). Numerous research studies and projects have been conducted over the past few decades to better understand and implement this technology. There are four primary mechanisms by which carbon is trapped in geological formations (Bradshaw et al., 2007; Sun et al., 2020; Han et al., 2010): (1) Structural trapping is the primary mechanism in the initial stages of injection. CO_2 is trapped beneath impermeable rock layers, known as caprocks or seals, due to buoyancy; (2) Residual trapping: as CO_2 moves through the rock's pore spaces, small amounts are left behind, trapped by capillary forces; (3) In dissolution trapping, the injected CO_2 will dissolve in the formation water, resulting in a denser, downward-moving brine that ensures further secure storage; (4) Mineralization: in the long term, CO_2 can react with minerals in the host rock, leading to the precipitation of stable carbonate minerals, ensuring permanent storage. Potential storage sites include deep saline aquifers, depleted oil and gas reservoirs, and unminable coalbeds (Lebedev et al., 2017; Zhang et al., 2016; Jing et al., 2022). Traditional methods of CO_2 geological storage often employ porous rocks such as sandstone and carbonate rock with using shale as the caprock. However, this section will primarily focus on unconventional shale resources as potential storage formations. See Figure 12.25 for the concept of injected supercritical (sc) CO_2 in shale formation (Lyu et al., 2021).

12.7.8.3.1 Benefits of Using Shale Formation for CO_2 Storage

Abundance and utilization of existing infrastructure distribution: As mentioned in previous sections, one of the key advantages of using shale formations for CO_2 storage is their global abundance. Shale rock formations are geographically widespread, thereby offering vast and widely distributed potential storage spaces for CO_2. This geographic ubiquity can facilitate regional strategies for carbon capture and storage, making it more practical and cost effective for different countries or jurisdictions to invest in this technology. An abundant supply of these formations ensures that capacity will not be a limiting factor in the large-scale adoption of carbon capture and storage methods.

Shale formations have been a focal point in the exploration and exploitation of hydrocarbons, particularly in the development of unconventional oil and gas reserves. As a result, a significant amount of infrastructure—ranging from drilling rigs and pipelines to data collection and monitoring systems—is already in place (Bielicki et al., 2018; Tao & Clarens, 2013; Khosrokhavar et al., 2014). Utilizing this existing infrastructure for CO_2 storage can substantially lower the initial capital expenditures required for launching carbon capture and storage projects.

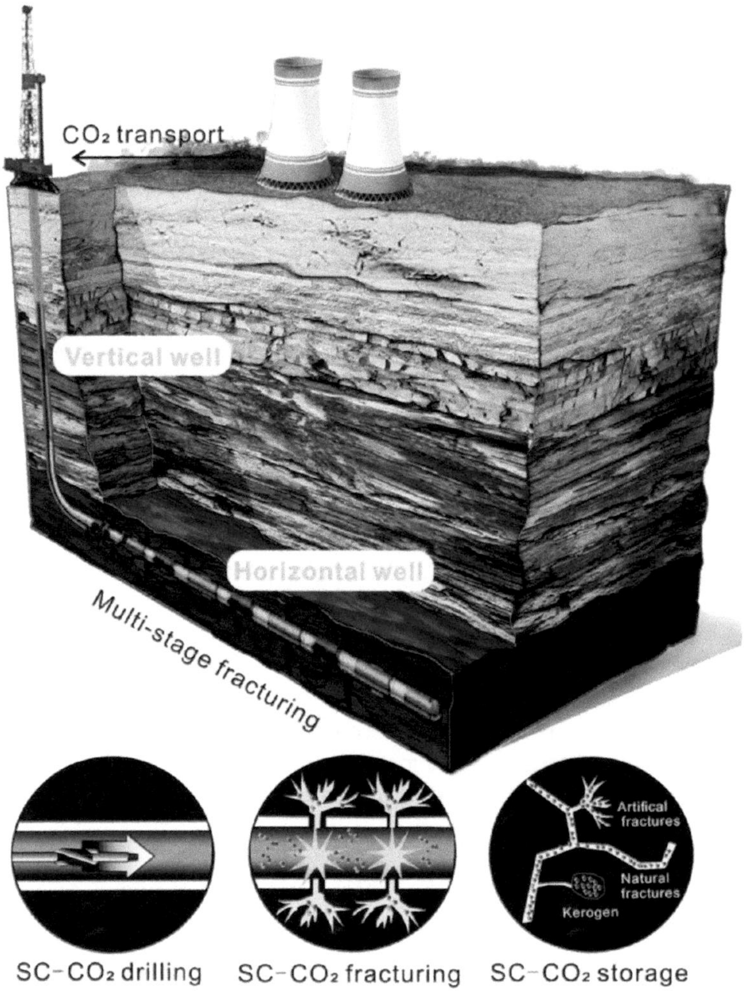

FIGURE 12.25 The role of injected sc CO_2 in shale formation.

Source: Lyu et al., 2021

The years of hydrocarbon extraction activities in shale formations have generated a robust body of knowledge. This includes detailed geological mapping, reservoir characterization, and an understanding of fluid dynamics within these formations; see Figure 12.26 for the gas flow mechanism in a shale formation. Such information can significantly de-risk and accelerate the implementation of CO_2 storage projects (Song et al., 2011). Engineers and geologists can leverage this data to optimize injection techniques, monitor storage stability, and ensure the long-term viability of CO_2 sequestration in these formations (Arif et al., 2021b; Al-Yaseri et al., 2021).

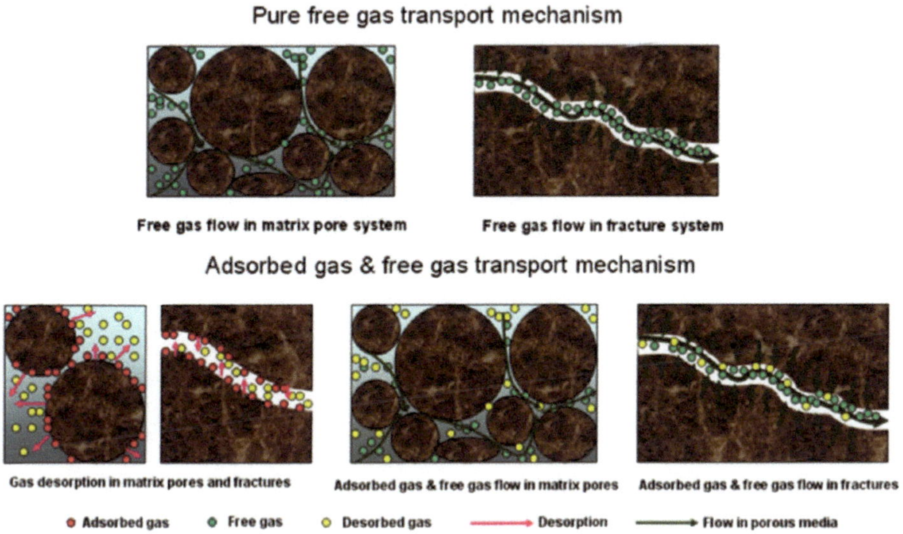

FIGURE 12.26 Gas flow mechanism in a shale formation.

Source: Song et al., 2011

Moreover, the existing infrastructure is often situated near industrial sources of CO_2 emissions, such as power plants and factories, further streamlining the capture and transportation processes. This proximity can lead to synergies that make the entire process more efficient, reducing both the financial and environmental costs.

Natural sealant: Another benefits of using shale formations for CO_2 storage lies in their inherent impermeability. Unlike other rock formations, the microstructure of shale acts as a natural barrier, effectively preventing the migration of gases (Zhou et al., 2016; Sakhaee-Pour & Bryant, 2012). This property makes shale an ideal caprock in traditional CO_2 storage methods, but it also means that shale formations themselves can serve as secure repositories for CO_2. Given that one of the primary concerns in CO_2 storage is the potential for leaks, the natural sealant properties of shale offer an added layer of security. The risk of CO_2 escaping into the atmosphere or infiltrating underground water sources is substantially reduced when it's stored in shale formations. This natural sealing capability can enhance the long-term viability and environmental safety of carbon capture and storage projects.

Combined with enhanced gas recovery: One of the most compelling benefits of utilizing shale resources for CO_2 storage is the potential for enhanced gas recovery (EGR) (Iddphonce et al., 2020; Mohagheghian et al., 2019). By injecting CO_2 into depleted shale gas reservoirs, not only is the greenhouse gas securely stored, but the natural gas (CH_4) is also extracted, thereby improving the economic viability of both the storage project and the gas field (Godec et al., 2014) (Figure 12.27).

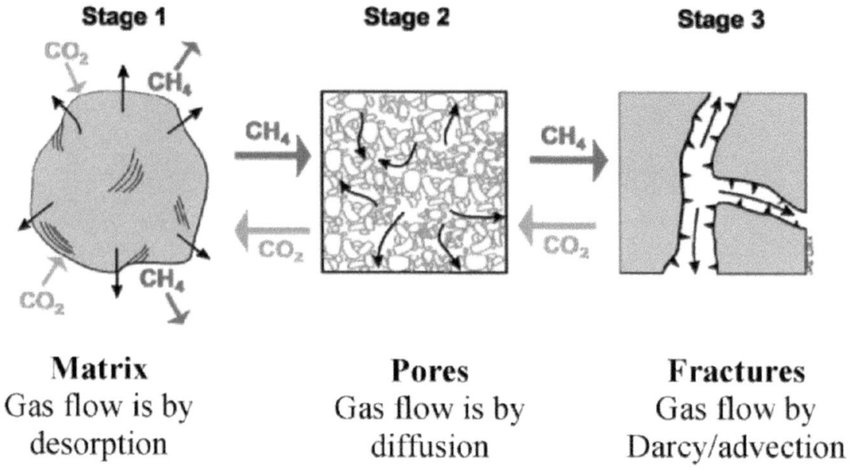

FIGURE 12.27 Schematic diagram showing flow dynamics of CO_2 and CH_4 in shale formation following CO_2 injection.

Source: Godec et al., 2014

The simultaneous execution of CO_2 storage and EGR enables a co-optimized use of subsurface resources. This two-pronged approach addresses both environmental concerns and energy security. Stored CO_2 serves the dual purpose of enhancing gas recovery while being sequestered securely underground, helping transition to a more sustainable energy landscape (Godec et al., 2013; Wu et al., 2019).

At present, the viability of CO_2 projects is dependent on balancing costs and benefits. Unfortunately, private enterprises have been unable to realize significant profits from traditional carbon storage initiatives, which has led to a decrease in investment. However, the emergence of EGR coupled with carbon storage holds promise for reducing operational and capital costs, resulting in a more cost-effective approach when compared to traditional CO_2 storage operations (Schaef et al., 2014). Globally, there are potentially recoverable shale gas resources estimated at 188 Tcm, which could enable the storage of up to 740 Gt of CO_2. Among these, up to 71 Tcm (2,505 Tcf) may be economically viable for production through EGR and could enable the economic storage of 281 Gt of CO_2 (Godec et al., 2014) (Table 12.6).

Multiple trapping mechanisms: Shale formations offer structural, residual, dissolution, and mineralization trapping mechanisms, as is common in many geological storage solutions (Romanov et al., 2015; Dai et al., 2023; Lu et al., 2023; Al-Yaseri et al., 2021; Pearce et al., 2022). However, shale adds another dimension to this: adsorption trapping, a feature not as readily available or effective in other porous rock types (Merey & Sinayuc, 2016; Arif et al., 2017).

The most unique and potentially game-changing trapping mechanism in shale is adsorption. Unlike in other trapping mechanisms in which CO_2 is in a free phase or dissolved in formation fluids, adsorption involves the adherence of CO_2 molecules directly to the surface of the shale rock. It has been proved that the injected CO_2 can

TABLE 12.6
Estimate of Recoverable Resources and CO_2 Storage Potential in Global Shale Gas Basins by Region

Region	Country	Risked Gas in Place (Tcm)	Risked Technically Recoverable Potential (Tcm)	Risked Technically Achievable CO_2 Storage Potential (Gt)	Risked Economically Recoverable Potential (Tcm)	Risked Economically Achievable CO_2 Storage Potential (Gt)
North America	United States	93	24	134	9	51
	Canada	42	11	43	4	16
	Mexico	67	19	72	7	27
	Subtotal	202	55	249	21	95
South America	Northern South America	3	1	3	0	1
	Southern South America	126	34	119	13	45
	Subtotal	129	35	122	13	46
Europe	Poland	22	5	19	2	7
	Eastern Europe	8	2	7	1	3
	Western Europe	43	11	47	4	18
	Subtotal	73	18	72	7	27
Africa	Central North Africa	53	14	55	5	21
	Morocco	8	2	6	1	2
	South Africa	52	14	52	5	20
	Subtotal	112	30	113	11	43
Asia	China	145	36	132	14	50
	India/Pakistan	14	3	11	1	4
	Turkey	2	0	2	0	1
	Subtotal	160	40	144	15	55
Oceania	Australia	39	11	39	4	15
Grand Total		**717**	**188**	**740**	**71**	**281**

Source: Godec et al., 2014

induce the shale swelling effect and then close the microfractures inside the shale matrix at reservoir conditions. Figure 12.28 shows the shale fractures closed after the supercritical CO_2 flooding by in situ micro-CT scanning at reservoir conditions (Yu et al., 2023). Such unique adsorption trapping has two critical advantages: (1) Enhanced security: The ability of adsorption to close microfractures enhances the seal integrity of the shale formation and locks the adsorbed CO_2 inside the shale matrix. This adds an extra layer of security against potential leaks, making CO_2 storage more reliable and reducing environmental risks. Indeed, the experimental data also showed that the shale permeability dropped significantly during CO_2 injection (Figure 12.29); (2) Maximizing storage capacity: The direct adherence of CO_2 to the shale rock surface can significantly increase the amount of CO_2 that can be securely stored in a given volume. This makes better use of available storage spaces and could be economically beneficial.

12.7.8.3.2 Challenges for CO_2 Storage in Shale Formations

Although CO_2 storage in shale formations is promising, challenges remain. First, shale formations are highly diverse, varying greatly in terms of porosity, permeability, and mineral content (Semnani & Borja, 2017; Chen et al., 2015). This variability poses a challenge to accurately characterizing the geological formation, which is essential for safe and efficient CO_2 storage. Additionally, while utilizing existing infrastructure and the potential for enhanced gas recovery can help, the upfront costs of setting up CO_2 capture and storage can still be significant, particularly for newer, unexplored shale formations. Furthermore, monitoring the movement and status of stored CO_2 in microfractures and pore spaces within shale can be complex (Louk et al., 2017), requiring rigorous monitoring protocols to ensure secure storage, which can add to operational complexities and costs. Finally, public perception of hydraulic fracturing in shale formations has been mixed, with concerns about groundwater contamination and induced seismic and environmental impact (Rivard et al., 2014; Israel et al., 2015), which may impact the acceptability of CO_2 storage in shale formations.

In summary, the utilization of unconventional shale resources for CO_2 storage presents several benefits. It can effectively contribute to the management of global CO_2 emissions while also potentially improving methane recovery from these formations through increased pressure from CO_2 injection. With diligent site selection, reliable monitoring, and continuous technological advancements, shale formations have the potential to play a significant role in the future of carbon capture and storage.

12.7.8.4 Conclusion

The quest for sustainable energy solutions and climate change mitigation is a global priority that requires the intelligent utilization of all available resources. This chapter has aimed to shed light on the transformative potential of unconventional shale resources in driving the energy transition. From serving as repositories for renewable energy in the form of geothermal systems and hydrogen storage to acting as secure

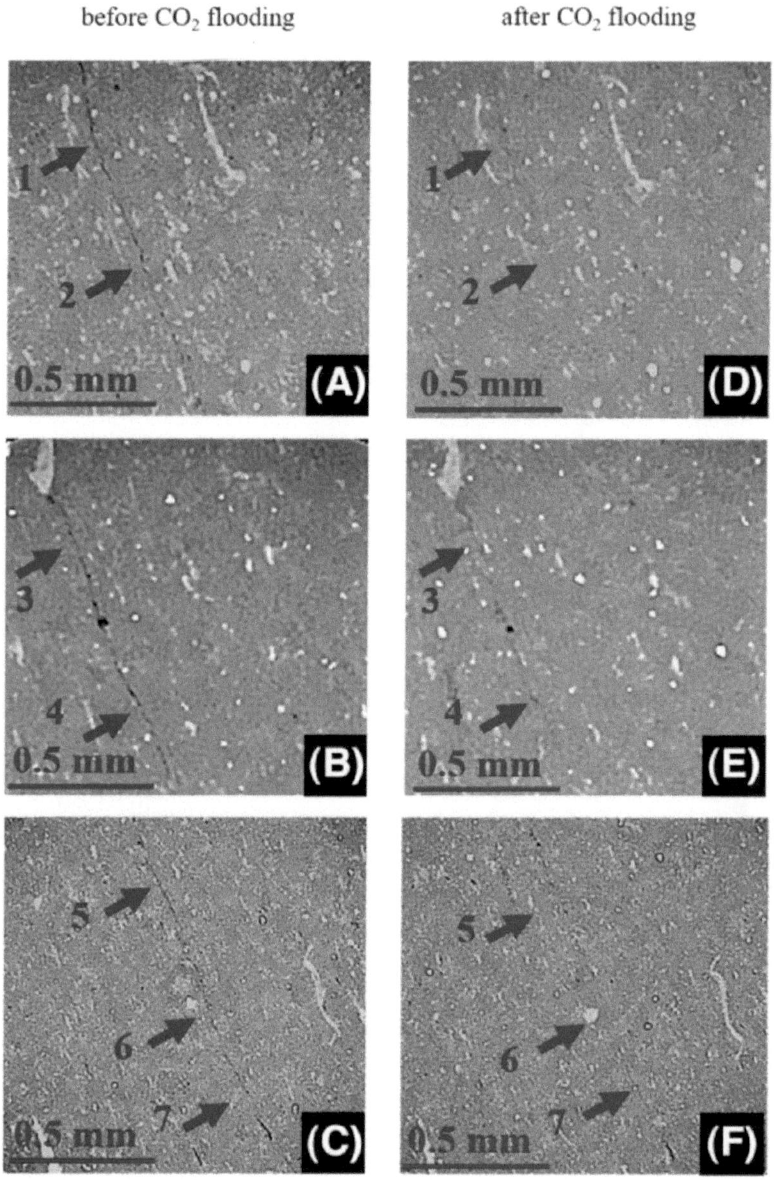

FIGURE 12.28 The shale matrix before (A–C) and after (D–F) supercritical CO_2 injection. The fracture is dark black while high-density minerals are shown as white, and the shale matrix (clay minerals) appears as gray. The blue arrows and numbers indicate fractures that were no longer detectable after CO_2 flooding.

Source: Yu et al., 2023

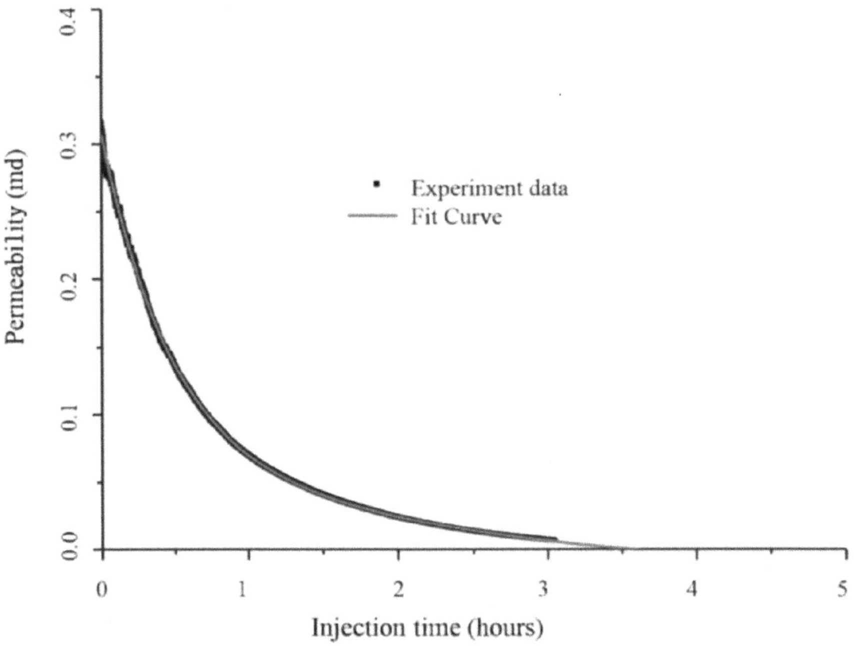

FIGURE 12.29 The measurement of dynamic permeability in relation to the injection time of CO_2.

Source: Yu et al., 2023

vaults for carbon capture and storage, shale formations have emerged as versatile and invaluable assets in the modern energy landscape.

We explored how shale formations, often considered merely hydrocarbon reservoirs, can be repurposed to tap geothermal energy. Their unique geological characteristics make them suitable for heat extraction, thereby contributing to the diversification of energy resources and the reduction of greenhouse gas emissions.

The chapter also delved into the prospects of using shale formations for hydrogen storage. As hydrogen emerges as a key player in a decarbonized future, the geological attributes of shale—such as porosity and impermeability—offer promising solutions for large-scale, secure hydrogen storage.

Finally, we examined the multifaceted advantages of storing CO_2 in shale formations, from leveraging existing infrastructure to the benefit of multiple trapping mechanisms, including the unique adsorption trapping. The natural sealant properties and the potential for enhanced gas recovery further solidify shale as an effective option for CO_2 storage, thereby aiding in global carbon-management efforts.

While significant strides have been made in understanding and utilizing shale for these applications, challenges such as regulatory hurdles, public acceptance, and economic viability remain. Nonetheless, the future prospects are encouraging, underpinned by ongoing technological advancements and a growing emphasis on collaborative research and policy reform.

In summary, unconventional shale resources hold transformative potential for multiple applications in the energy transition. As we navigate the complexities of global energy demands and climate change imperatives, the versatile and abundant shale formations stand as compelling contributors to a more sustainable, resilient, and decarbonized energy future.

12.7.9 Digitalization in Unconventional Shale Resources

12.7.9.1 Digitalization in Field Development

The concept of digitalization or digital transformation has gained more attention from the gas and oil industry all over the world as it is believed to be a tool that could help with reducing the financial burden and increasing operation efficiency, which is an important factor for oil companies to survive encountering a suddenly sharp decline in the price of oil. However, there is no clear definition of digitalization in the literature, and its explanation varies for different industries and communities. In the petroleum industry, digitalization refers to utilizing technologies like artificial intelligence and machine learning, robotics and automation, and cloud computing to add more value to business operations by increasing the productivity of the workforce and improving the decision-making process (Gurianov et al., 2019). Most current digital applications focus on the development of offshore resources and unconventional resources. Specific examples include robotics and automation, drones and unmanned vehicles, digital twins, big data analytics, machine learning and artificial intelligence, and cloud and cognitive computing.

Robotic technology provides an opportunity to reduce physical operation in the field and thus the requirements for the working crew. This is particularly important for offshore field development, some advanced robotics (Barrett, 1987), and automated offshore drilling systems (FUJITA & Sato, 1990) have been developed to perform exploration and investigation work and conduct field tests. The robotic drilling system (RDS) consists of an electric roughneck and pipe handler, a dynamic robot control system, and a drill floor robot, which aims to construct a fully electric and robotic drill floor with minimum human intervention. The field feasibility study indicates that it can achieve an estimated time saving of 20 to 60 days per rig yearly (Cappuccio et al., 2019). On the other hand, automated correlations and algorithms have been integrated with multiple disciplines to construct a complex management system and automate the data acquisition and analysis process, which has been extensively applied in the upstream sector to evaluate reservoir performance and optimize production (Bakdi et al., 2020).

Drones and unmanned vehicles have been widely used in oil and gas fields to monitor and inspect the state of field infrastructure such as offshore rigs, pipelines, and storage tanks (Puripanda et al., 2020). This technology has also been used during the phase of exploration and appraisal development. Due to inaccessible high elevations, for instance, Heo et al. (2018) utilized the images obtained by a drone to investigate the surface discontinuity of a rock. One thing that should be noted is that both drones and unmanned vehicles are still controlled by individuals through the central operation system. Thus, the current development trend aims to get rid of human intervention and utilize artificial intelligence to fully automate the anomaly detection process.

The concept of digital twins in the petroleum industry refers to constructing a digital model for any real physical process or object, which then can be used as a reference to test different operations until achieving the expected performance and also as a real-time data hub to support continuous offsite monitoring (Zborowski, 2018). This technology has been adopted for optimizing the drilling process and electrical submersible pump (ESP) design. For example, it was used by Saini et al. (2018) to predict the influence of the rate of penetration and time of non-drilling operations on the drilling performance with real-time data and physics-based models. It also allowed simulators to consider the state of ESP under different operating conditions, including failure and transient flow, to optimize its mechanical structure (Lastra & Aramco, 2019).

Big data analytics refers to inferring invisible patterns or trends from a large volume of data obtained from different sources, which then can be used to formulate analytical models for making predictions and helping with future planning (Perrons & Jensen, 2014). Intuitively, this technology is mostly suitable for processing seismic data as terabytes of data can be recorded in one hour, and the preprocessing and denoising process is also time consuming (Haroon et al., 2018). It has also been extensively tested in the fields of reservoir modeling, reservoir management, and drilling optimizations (AlBar et al., 2018). Specifically, a project named "shale analytics" tried to use big data analytics for over 3,000 wells in Marcellus, Utica, Niobrara, and Eagle Ford, and some general trends like the number of fracturing clusters needed, the optimum stage length, and well spacing can be inferred.

Machine learning and artificial intelligence are the oldest digitalization techniques used in the petroleum industry for exploration and production development. Both have become very popular recently due to the massive advancement of computational power and enormous amounts of data with improved resolution. Different types of machine-learning models, including supervised learning, unsupervised learning, and reinforcement learning, have been developed for geoscience, petroleum exploration, and production (Tahmasebi et al., 2020). On the other hand, the development of cloud and cognitive computing enables the reservoir simulation process to be completed with guaranteed performance and safety and to process continuously generated data in a timely, efficient manner.

All these digitalization technologies will or currently play a significant role in the development of unconventional resources. Particularly, hydraulic fracturing is a necessary procedure to maximize the productivity of these resources. Although its development cost is usually high, it also opens the door for digital transformation to reduce the risk and loss and increase operational efficiency.

12.7.9.2 Smart Well

The smart well technology emerged in 1997 and was first deployed at the Saga Petroleum's Snorre oilfield in the Norwegian sector of the North Sea. It can provide real-time pressure and temperature measurements from downhole gauges permanently installed close to the sand face and the flexibility to adjust flow profiles and optimize reservoir recovery through the remotely controlled high-fidelity flow-control valves (Gao et al., 2007). Since then, hundreds of oil and gas fields have benefited from a great improvement in reservoir management strategies supported by this novel technology.

Figure 12.30 shows the main components of a generic smart well installed during completion, which include downhole flow-control devices through interval control valves (ICV) and internal control devices (ICD), downhole sensors for sending signals to the transmitters, electronic transmitters for sending signals to the controllers, and packers to isolate sections. The most important components are the downhole valves, which can be classified into passive, autonomous passive, and reactive-actionable valves according to their designed functionalities. Typical passive valves are ICDs, which can produce an equalized pressure drop across the lateral length to get rid of early water or gas breakthroughs. Autonomous ICDs (AICDs), one example of autonomous passive valves, also control the fluid flow through the internal discs based on the centrifugation principle or differential density (Greci et al., 2014). The reactive-actionable valves, mainly represented by ICVs, work to control the downhole flow rate by turning the valve on and off or tuning the choke size position.

The downhole equipment is connected to the tubing through the cable and sends data to the control panel at the surface. When the smart well is under operation, a system called supervisory control and data acquisition (SCADA) is applied to collect pressure and temperature data in real time. Based on the data received, the production engineers then identify which segment is flowing and decide which valve should be tuned for optimizing reservoir production, which can be realized by sending back a signal to the control panel to automatically change the choke position. Such an

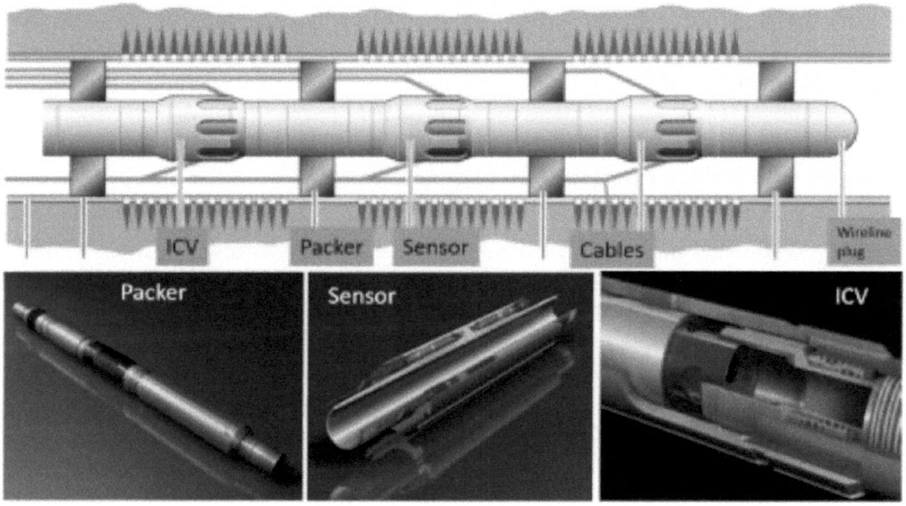

FIGURE 12.30 Main components of a generic smart well.

Source: Carvajal et al., 2018

intelligent system is capable of monitoring downhole operating conditions including flow; pressure; temperature; phase composition and water pH; describing the spatial distribution of reservoir characteristics from the well, including resistivity and acoustic impedance; and controlling the inflow and outflow rates of segregated well segments (Glandt, 2005). Specific field applications include controlling the inflow of gas and water fluids, controlling commingling wells, auto-gas lift injection, EOR/IOR optimization, and monitoring production in real time (Carvajal et al., 2018).

With the movement of oil and gas exploration into unconventional reservoirs like shale gas, horizontal wells have been extensively drilled to maximize productivity, and accordingly, many advanced horizontal well completion tools and instruments have been developed to automate the downhole operation. For instance, gravel packing has a good sand control effect and a long validity period for gas wells in deep water, but wellbore shrinkage and packer installation in long-reach horizontal wells may result in the problem of low packing rate and easy sand plugging. The diverting screen technology can continuously fill through the diverting tube and the wash-down valve even in sand plugging by direct circulation after removing contaminated mud by acid cleaning in the horizontal section (Saebi et al., 2010).

On the other hand, as the staged fracturing technique becomes mature for shale gas production through the horizontal well, various completion technologies also have been tested in the fields, including open hole packer plus sliding sleeve staged fracturing technology, which opens one sliding sleeve by dropping one ball and then combines open hole packer and sliding sleeve staged fracturing (Shaw, 2011); pump down drillable bridge plug staged fracturing technology, which suits for hydraulic fracturing with limitless stages and for casing completion (Smith & Starr, 2008); switchable staged fracturing sliding sleeve staged fracturing technology, which consists of full-bore switchable sliding sleeve staged fracturing technology and the ball injection repeated switchable sliding sleeve staged fracturing technology (Chunpeng et al., 2023); and coiled tubing hydrojet perforation annulus sand fracturing technology, which utilizes the hydraulic jet to complete the sand fracturing through the casing-coiled tubing annulus (Rosales et al., 2010).

12.7.9.3 Remote Sensing and Seismic Monitoring

Remote sensing is the technique of indirectly acquiring information about objects or phenomena near the surface of the Earth and in the atmosphere by observing the radiation reflected or emitted, typically through aircraft and satellites. It has proven to be an effective tool for monitoring environmental qualities over different ecosystems and providing detailed information like land use, infrastructure, and population density to guide urban planning. Remote sensing data also are highly useful and widely used in the oil and gas industry, ranging from upstream applications like identifying potential geological structures to downstream applications like mapping and optimizing pipelines. Due to the financial burden and safety risks brought about by the field crews, this technique has been extensively applied during the exploration stage to map the hydrocarbon system under the Earth's surface, particularly in remote areas with complex topography and insufficient infrastructure (Lord, 2017).

Most importantly, a wealth of satellite data is both publicly and commercially available, including Landsat data (Loveland & Dwyer, 2012) from the United States

Geological Survey (USGS), Sentinel data (European Space Agency, 2023) from the European Space Agency (ESA), Prisma data (Candela et al., 2016) from the Italian Space Agency (ASI), and WorldView-3 and WorldView-4 from Maxar. Moreover, with the help of cloud computing (Yan et al., 2018), the online processing platforms allow large volumes of satellite data to be effectively visualized with high spatial resolution and extremely short processing times. Typical platforms include Google Earth Engine, Amazon Web Services, NASA Earth Exchange, and Microsoft Azure.

During the exploration phase for unconventional shale resources, the remote sensing data usually is combined with the regional geological survey to target favorable areas for hydrocarbon exploration and production. These areas typically are characterized by complicated geological settings and ground surface conditions and limited petroleum geological data due to a low degree of exploration. For example, when pinpointing the prospecting regions of the Lower Cambrian shale gas resources in South China, the remote sensing data from Sentinel-2 and Landsat8 satellites were applied to estimate the normalized difference vegetation index (NDVI) over the study and to avoid the ecological reserves (Xu et al., 2021). On the other hand, the remote sensing data has been utilized to identify fractures and their direction and density for fractured oilfields at the northern Qinghai-Tibet plateau in China (Wang et al., 2020). Fractures of different scales were characterized by applying satellite data with different types and resolutions, including the ETM data from the United States Landsat (with a resolution of 15 and 30 meters) and the SPOT data from the French satellite (with a resolution of 10 meters). It was then combined with the interpretation of outcrops, downhole cores, and well-logging data, which represented a systematic approach to characterizing subsurface fractural reservoirs. The technique of remote sensing has also been applied to estimate methane emissions during the production stage of unconventional shale resources.

The wild applications of horizontal drilling and hydraulic fracturing techniques to exploit unconventional shale resources also provide an opportunity to reduce greenhouse gas emissions because the combustion of oil and gas produces less CO_2 into the atmosphere compared to coal with high carbon content (U.S. Energy Information Administration, 2022). However, the fugitive methane release during well completion and production process (Howarth et al., 2011) put a question mark against this benefit of climate effect mitigation as methane is the second most important anthropogenic greenhouse gas. Samples obtained at the onshore natural gas site can provide direct measurements of methane emissions from a single source, but it is found that the reported values are considerably underestimated based on measurements of tall tower flask samples and aircraft profiles (Caulton et al., 2014; Miller et al., 2013). In contrast, remote sensing provides an approach to estimating the methane emissions over a broader area and evaluating regional differences resulting from different regulations and practices.

The methane in the atmosphere can be measured through the satellite by observing the sunlight reflected from Earth. Figure 12.31 shows the mechanism of methane detection and measurement through ground-based and satellite-based remote sensing. When the sunlight travels through the Earth's atmosphere, it will change as methane can absorb some specific wavelengths within the spectrum of sunlight. The resulting spectra changes from the top to the bottom of the atmosphere are expressed

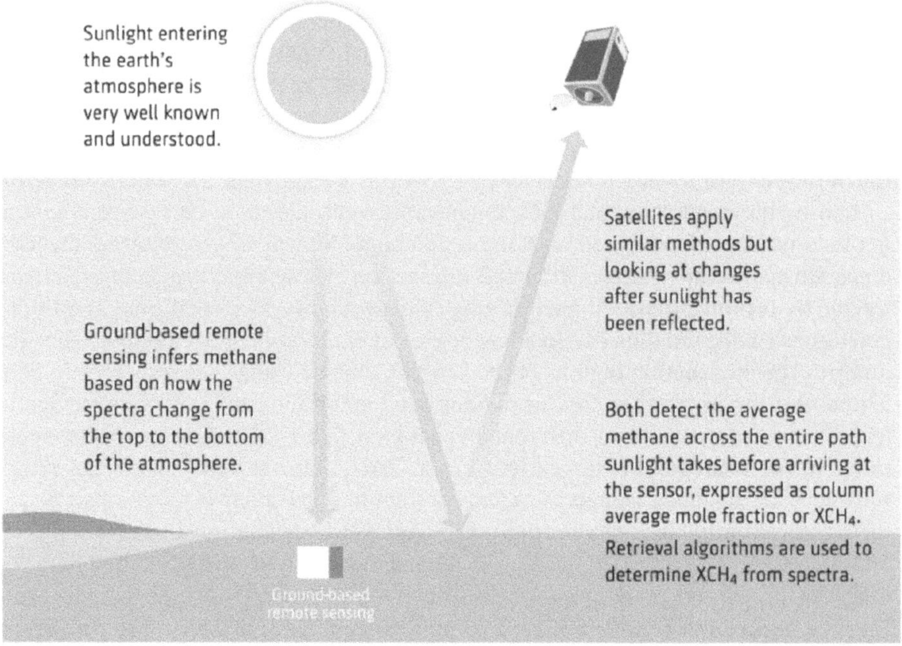

FIGURE 12.31 Methane detection through remote sensing.

Source: National Institute of Water and Atmospheric Research, 2023

as column average mole fraction or XCH_4 to represent the average methane across the entire travel path. Satellite-based remote sensing uses a spectrometer attached to the satellite and measures the spectra changes for the sunlight reflected from Earth. Compared to ground-based remote sensing, it provides more flexibility for calibrating and validating global methane emissions. Some principal instruments for this mission include the scanning imaging absorption spectrometer for atmospheric chartography (SCIAMACHY), Japanese greenhouse gases observing satellite (GOSAT), and MethaneSAT (Jacob et al., 2016).

The Bakken and Eagle Ford formations are two of the fastest-growing regions with production from unconventional shale resources in the United States. Schneising et al. (2014) retrieved satellite observations from SCIAMACHY and developed a mass-balance approach to detect positive methane anomalies associated with oil and gas industries in these regions. They found that the methane emissions have increased by 990 ± 650 kt CH_4 yr^{-1} and 530 ± 330 kt CH_4 yr^{-1} between the periods 2006 through 2008 and 2009 through 2011, which correspond to leakages of 10.1% ± 7.3% and 9.1% ± 6.2% in terms of energy content. It indicates that the current inventories likely underestimate the methane emissions from these two formations. Other researchers (Varon et al., 2021) also successfully utilized data from the Sentinel-2 multispectral instrument (MSI) to quantify anomalously large methane point sources. Compared to the previous regional-scale study, this project mostly focused on evaluating its local influence in terms of a higher spatial resolution (with a fine pixel resolution of 20 meters) and a higher temporal resolution (with a rapid revisit rate of two to five days), which enables these satellite-observing systems to function as early alert systems for identifying industrial facilities with anomalously large emissions. Most recently, the first single-blind controlled methane release testing of satellite-based methane emissions was completed by five independent teams (Sherwin et al., 2023), which was a necessary step for acceptance and use by policymakers and stakeholders.

While the technical advancement of hydraulic fracturing facilitates the development of unconventional shale resources all over the world, the induced seismicity also brings a lot of concerns to the public. The major mechanism of hydraulic fracturing is injecting high-pressure fluid into the subsurface to create or extend the fracture network and thus increase reservoir permeability and productivity. Fractures are usually developed near the wellbore and can extend tens or hundreds of meters away from the wellbore if there are some pre-existing natural fractures/faults. This spatial characteristic is also manifested by the field seismicity monitoring data (Keranen & Weingarten, 2018). Figure 12.32 shows a schematic diagram of induced

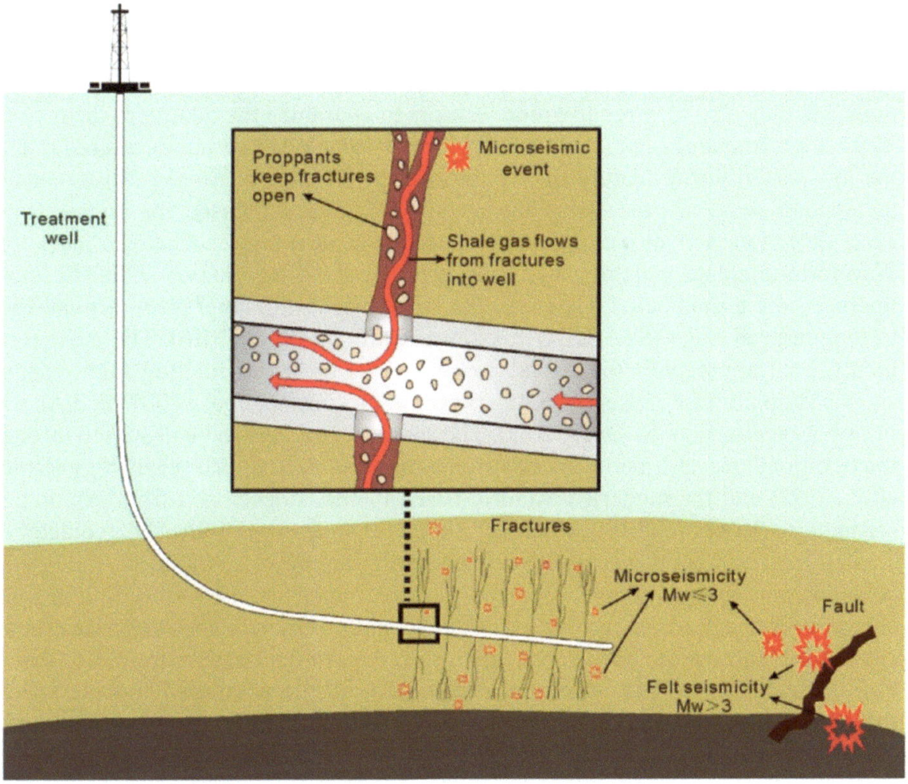

FIGURE 12.32 Schematic diagram of shale hydraulic fracturing and induced seismicity.

Source: Li et al., 2019

seismicity from hydraulic fracturing stimulation. Many anthropologic activities like wastewater injection and CO_2 sequestration also can induce seismicity and share the same mechanism of fracturing rocks by injecting high-pressure fluid. Most induced seismicity are microseismic events with a moment magnitude (Mw) less than 3, but higher-magnitude (Mw > 3) seismic events can be expected and are usually felt at the surface if pre-existing faults are reactivated. The occurrence of injection-induced felt seismicity depends on both geological conditions and field operational parameters, including the existence of large-scale faults, a fluid pathway connecting injection zone and faults, and sufficient pressure increments by injection.

Monitoring the induced seismicity currently is an essential step to characterize unconventional shale resources and to manage and mitigate the potential related risks. In the field, microseismic monitoring has been widely applied as a standard geophysical tool to monitor the fracture geometry to evaluate the stimulated reservoir volume (SRV) and optimize the field operations, which usually consist of three steps: acquisition, processing, and interpretation. During the acquisition phase, the monitoring network/array is optimized to enhance the detectability of weak seismic signals and the reliability of resolved seismic source parameters. Besides traditionally used geophones, distributed acoustic sensing (DAS) has also proven to be sensitive enough to record microseismic signals, even under harsh downhole conditions (Karrenbach et al., 2017). Surface installations of seismometers are preferred due to the operational challenges and the high expense of monitoring for downhole installations, but they also may require more sensors to guarantee the quality of the records as the high-frequency components of recorded seismic signal are decreased due to the low-pass filtering feature of the subsurface (Duncan & Eisner, 2010). During the second stage of processing, the recorded data is processed for microseismic event detection, arrival picking, source location, and source mechanism inversion. Waveform-template matching has been applied on a field dataset with 70% more microseismic events being detected, compared to the traditional short-term average to long-term average (STA/LTA) functions (Caffagni et al., 2016). The most commonly used approaches for arrival picking focus on differentiating microseismic signals from ambient noise, including polarization (Fischer et al., 2007) and Akaike information criterion (St-Onge, 2011). The source locations can be identified through the traditional travel-time inversion originating from earthquake seismology (Ishida et al., 1912) and the modern waveform-based methods developed from exploration seismology (Li et al., 2018), which lay a foundation for describing the geometry of created fractures based on the interpreted spatial-temporal distribution. Fault plane solutions and seismic moment tensors are two important methods to express the source mechanism for interpreting the state of natural fracture and subsurface stress, calculating parameters for discrete fracture networks, and estimating SRV (Eisner et al., 2010). For the last stage of interpretation, all the obtained parameters from the previous stage are integrated to estimate the growth of fractures and guide the subsequent fracture-stimulation operations.

Undoubtedly, utilizing hydraulic fracturing for unconventional shale resource production should be done under the premise that public safety should be guaranteed. In practice, the most straightforward parameter for the public to understand the influence of induced seismicity is its magnitude. It is also one of the most important references for operators and policy-makers to develop appropriate procedures to mitigate potential risks. Currently, the most widely used risk management tool for tight reservoir hydraulic fracturing is the induced seismicity traffic light protocol (IS-TLP) (Kao et al., 2018). The earliest version of this tool was developed in the United Kingdom in 2013 (Green et al., 2012), where the transition between light levels only depended on the magnitude, and yellow light and red light corresponded to the resulting local magnitude greater than 0 and 0.5, respectively. In the United States, the transition rules are specific to each state's regulations and site.

REFERENCES

Abbasi, M., Abdollahi, F., Abhishek, R., Abqory, M. H., Aditiawati, P., Afinanisa, Q., & Al-Fahdi, E. (2023). *Challenges and recent advances in sustainable oil and gas recovery and transportation*. Elsevier Inc.

Abdo, J., Zaier, R., Hassan, E., Al-Sharji, H., & Al-Shabibi, A. (2014). ZnO-clay nanocomposites for enhance drilling at HTHP conditions. *Surface and Interface Analysis*, *46*(10–11). https://doi.org/10.1002/sia.5454

Aftab, A., Ali, M., Arif, M., Panhwar, S., Saady, N. M. C., Al-Khdheeawi, E. A., Mahmoud, O., Ismail, A. R., Keshavarz, A., & Iglauer, S. (2020a). Influence of tailor-made TiO2/API bentonite nanocomposite on drilling mud performance: Towards enhanced drilling operations. *Applied Clay Science*, *199*. https://doi.org/10.1016/j.clay.2020.105862

Aftab, A., Ali, M., Sahito, M. F., Mohanty, U. S., Jha, N. K., Akhondzadeh, H., Azhar, M. R., Ismail, A. R., Keshavarz, A., & Iglauer, S. (2020b). Environmental friendliness and high performance of multifunctional tween 80/ZnO-nanoparticles-added water-based drilling fluid: An experimental approach. *ACS Sustainable Chemistry and Engineering*, *8*(30). https://doi.org/10.1021/acssuschemeng.0c02661

Aftab, A., Hassanpouryouzband, A., Xie, Q., Machuca, L. L., & Sarmadivaleh, M. (2022). Toward a fundamental understanding of geological hydrogen storage. *Industrial & Engineering Chemistry Research*, *61*, 3233–3253.

Agarwal, M., & Kudapa, V. K. (2023). Plasma based fracking in unconventional shale—A review. *Materials Today: Proceedings*, *72*, 2791–2795. https://doi.org/10.1016/j.matpr.2022.06.420

Aguilera, R. F., & Radetzki, M. (2014). The shale revolution: Global gas and oil markets under transformation. *Mineral Economics*, *26*, 75–84.

AlBar, A. H., Alotaibi, B. M., Asfoor, H. M., & Nefai, M. S. (2018). A journey towards building real-time big data analytics environment for drilling operations: Challenges and lessons learned. *Society of petroleum engineers—SPE Kingdom of Saudi Arabia annual technical symposium and exhibition 2018*, SATS, 23–26. https://doi.org/10.2118/192285-MS

Alexander, T., Baihly, J., Boyer, C., Clark, B., Waters, G., Jochen, V., Le Calvez, J., Lewis, R., Miller, C. K., & Thaeler, J. (2011). Shale gas revolution. *Oilfield Review*, *23*, 40–55.

Alfarge, D., Wei, M., & Bai, B. (2020). *Fundamentals of enhanced oil recovery methods for unconventional oil reservoirs*. Elsevier.

Al-Shargabi, M., Davoodi, S., Wood, D. A., Al-Musai, A., Rukavishnikov, V. S., & Minaev, K. M. (2022). Nanoparticle applications as beneficial oil and gas drilling fluid additives: A review. *Journal of Molecular Liquids*, *352*. https://doi.org/10.1016/j.molliq.2022.118725

Al-Yaseri, A., Abdulelah, H., Yekeen, N., Ali, M., Negash, B. M., & Zhang, Y. (2021). Assessment of CO2/shale interfacial tension. *Colloids and Surfaces A: Physicochemical and Engineering Aspects*, *627*, 127118.

Amrouche, S. O., Rekioua, D., Rekioua, T., & Bacha, S. (2016). Overview of energy storage in renewable energy systems. *International Journal of Hydrogen Energy*, *41*, 20914–20927.

Andreas, J.-J. (2015). The shale revolution in the US and its impact on energy markets, energy security, and the US energy transition. *International report of the konrad-adenauer-stiftung*. Berlin: Konrad-Adenauer-Stiftung.

Annevelink, M. P. J. A., Meesters, J. A. J., & Hendriks, A. J. (2016). Environmental contamination due to shale gas development. *Science of the Total Environment*, *550*, 431–438. https://doi.org/10.1016/j.scitotenv.2016.01.131

Arif, M., Mahmoud, M., Zhang, Y., & Iglauer, S. (2021a). X-ray tomography imaging of shale microstructures: A review in the context of multiscale correlative imaging. *International Journal of Coal Geology*, *233*, 103641. https://doi.org/10.1016/j.coal.2020.103641

Arif, M., Zhang, Y., & Iglauer, S. (2021b). Shale wettability: Data sets, challenges, and outlook. *Energy & Fuels*, *35*, 2965–2980.

Augustine, C., & Falkenstern, D. (2014). An estimate of the near-term electricity-generation potential of coproduced water from active oil and gas wells. *SPE Journal*, *19*, 530–541.

Ayele, Y. Z., Barabadi, A., & Barabady, J. (2014). Drilling waste handling and management in the High North. *IEEE international conference on industrial engineering and engineering management*. https://doi.org/10.1109/IEEM.2013.6962496

Azin, R., & Izadpanahi, A. (2022). *Fundamentals and practical aspects of gas injection* (R. Azin & A. Izadpanahi, Eds.). Springer International Publishing. https://doi.org/10.1007/978-3-030-77200-0

Baihly, J., Malpani, R., Altman, R. M., Lindsay, G., & Clayton, R. (2015). Shale gas production decline trend comparison over time and basins revisited. *Proceedings of the 3rd unconventional resources technology conference*. https://doi.org/10.15530/urtec-2015-2172464

Baines, S. J., & Worden, R. H. (2004). Geological storage of carbon dioxide. *Geological Society, London, Special Publications*, *233*, 1–6.

Baisch, S. (2013). Seismic hazard associated with shale-gas-fracking? The Bowland shale case study. *75th EAGE conference & exhibition-workshops*. European Association of Geoscientists & Engineers, cp-349-00031.

Bakdi, S., Kannan, N., Masini, S. R., & Chennakrishnan, B. (2020). Automated well correlation using machine learning and facial recognition techniques. *Society of petroleum engineers—Abu Dhabi international petroleum exhibition and conference 2020*, ADIP. https://doi.org/10.2118/203301-MS

Barati, R., Cai, J., Cui, G., Dehghanpour, H., Elsworth, D., Fan, Y., Feng, R., Gan, Q., Ge, X., Habibi, A., He, S., Hu, X., Hu, Q., Javaheri, A., Jia, B., Jin, B., Li, J., & Liang, Y. (2019). *Petrophysical characterization and fluids transport in unconventional reservoirs*. Elsevier Inc.

Barbier, E. (2002). Geothermal energy technology and current status: An overview. *Renewable and Sustainable Energy Reviews*, *6*, 3–65.

Barrett, R. W. (1987). Advanced robotics under water: The DTI initiative. In *Advances in underwater technology, ocean sclence and offshore Eng~neer~ng* (Vol. 14). OnePetro.

Barton, J. P., & Infield, D. G. (2004). Energy storage and its use with intermittent renewable energy. *IEEE Transactions on Energy Conversion*, *19*, 441–448.

Bauer, S., Beyer, C., Dethlefsen, F., Dietrich, P., Duttmann, R., Ebert, M., Feeser, V., Görke, U., Köber, R., & Kolditz, O. (2013). Impacts of the use of the geological subsurface for energy storage: An investigation concept. *Environmental Earth Sciences*, *70*, 3935–3943.

Bayat, M., Lashkarbolooki, M., Hezave, A. Z., & Ayatollahi, S. (2016). Investigation of gas injection flooding performance as enhanced oil recovery method. *Journal of Natural Gas Science and Engineering*, *29*, 37–45. https://doi.org/10.1016/j.jngse.2015.12.047

Bechara, E., Gamadi, T., Hussain, A., & Emadibaladehi, H. (2022). Effect of hydrogen exposure on shale reservoir properties and evaluation of hydrogen storage possibility in depleted unconventional formations. *Unconventional resources technology conference*, 20–22 June. Unconventional Resources Technology Conference (URTeC), 1734–1748.

Belazreg, L., Mahmood, S. M., & Aulia, A. (2019). Novel approach for predicting water alternating gas injection recovery factor. *Journal of Petroleum Exploration and Production Technology*, *9*(4), 2893–2910. https://doi.org/10.1007/s13202-019-0673-2

Bensing, J. P., Misch, D., Skerbisch, L., & Sachsenhofer, R. F. (2022). Hydrogen-induced calcite dissolution in Amaltheenton Formation claystones: Implications for underground hydrogen storage caprock integrity. *International Journal of Hydrogen Energy*, *47*, 30621–30626.

Bickle, M. J. (2009). Geological carbon storage. *Nature Geoscience*, *2*, 815–818.

Bielicki, J. M., Langenfeld, J. K., Tao, Z., Middleton, R. S., Menefee, A. H., & Clarens, A. F. (2018). The geospatial and economic viability of CO2 storage in hydrocarbon depleted fractured shale formations. *International Journal of Greenhouse Gas Control*, *75*, 8–23.

Bilgen, S., & Sarıkaya, İ. (2016). New horizon in energy: Shale gas. *Journal of Natural Gas Science and Engineering*, *35*. https://doi.org/10.1016/j.jngse.2016.09.014

Bilgili, F., Koçak, E., & Bulut, Ü. (2020). The shale gas production and economic growth in local economies across the US. *Environmental Science and Pollution Research*, *27*, 12001–12016.

Bradshaw, J., Bachu, S., Bonijoly, D., Burruss, R., Holloway, S., Christensen, N. P., & Mathiassen, O. M. (2007). CO2 storage capacity estimation: Issues and development of standards. *International Journal of Greenhouse Gas Control*, *1*, 62–68.

Brady, B., Elbel, J., Mack, M., Morales, H., Nolte, K., & Poe, B. (1992). Cracking rock: Progress in fracture treatment design. *Oilfield Review (Netherlands)*, *4*.

Brückmann, G., Berger, S., Caviola, H., Hahnel, U. J. J., Piana, V., Sahakian, M., & Stadelmann-Steffen, I., with the Swiss Social, S., & Humanities Energy Research, G. (2023). Towards more impactful energy research: The salient role of social sciences and humanities. *PLoS Climate*, *2*, e0000132.

Byerley, G., Monk, D., Aaron, P., & Yates, M. (2018). Time-lapse seismic monitoring of individual hydraulic frac stages using a downhole DAS array. *The Leading Edge*, *37*, 802–810.

Caffagni, E., Eaton, D. W., Jones, J. P., & van der Baan, M. (2016). Detection and analysis of microseismic events using a Matched Filtering Algorithm (MFA). *Geophysical Journal International*, *206*(1), 644–658. https://doi.org/10.1093/GJI/GGW168

Cai, C., Li, G., Huang, Z., & Chi, H. (2015). A waterless fracturing treatment: Liquid nitrogen fracturing and its application prospect. *"Proceedings" of "OilGasScientificResearchProjects" Institute, SOCAR*, *3*, 35–40. https://doi.org/10.5510/OGP20150300250

Calderón, A. J., Guerra, O. J., Papageorgiou, L. G., & Reklaitis, G. V. (2018). Disclosing water-energy-economics nexus in shale gas development. *Applied Energy*, *225*, 710–731.

Candela, L., Formaro, R., Guarini, R., Loizzo, R., Longo, F., & Varacalli, G. (2016). The PRISMA mission. *International geoscience and remote sensing symposium (IGARSS)*, November, 253–256. https://doi.org/10.1109/IGARSS.2016.7729057

Cappuccio, P., Burrafato, S., Maliardi, A., Maccarini, G. R., Taccori, D., Dalla Costa, R., Raunholt, L., & Larsen, Ø. (2019). Full robotic drill floor as advanced rig automation. *Society of petroleum engineers—Abu Dhabi international petroleum exhibition and conference 2019*. ADIP. https://doi.org/10.2118/197854-MS

Carneiro, J. F., Matos, C. R., & Van Gessel, S. (2019). Opportunities for large-scale energy storage in geological formations in mainland Portugal. *Renewable and Sustainable Energy Reviews*, *99*, 201–211.

Carpenter, C. (2022). Study summarizes 20 years of horizontal multistage completions. *Journal of Petroleum Technology*, *74*, 60–62.

Carvajal, G., Maucec, M., & Cullick, S. (2018). Smart wells and techniques for reservoir monitoring. *Intelligent Digital Oil and Gas Fields*, 249–289. https://doi.org/10.1016/B978-0-12-804642-5.00007-4

Castro-Alvarez, F., Marsters, P., De León Barido, D. P., & Kammen, D. M. (2018). Sustainability lessons from shale development in the United States for Mexico and other emerging unconventional oil and gas developers. *Renewable and Sustainable Energy Reviews*, *82*, 1320–1332.

Caulton, D. R., Shepson, P. B., Santoro, R. L., Sparks, J. P., Howarth, R. W., Ingraffea, A. R., Cambaliza, M. O. L., Sweeney, C., Karion, A., Davis, K. J., Stirm, B. H., Montzka, S. A., & Miller, B. R. (2014). Toward a better understanding and quantification of methane emissions from shale gas development. *Proceedings of the National Academy of Sciences of the United States of America*, *111*(17), 6237–6242. https://doi.org/10.1073/PNAS.1316546111/SUPPL_FILE/PNAS.201316546SI.PDF

Celia, M. A. (2017). Geological storage of captured carbon dioxide as a large-scale carbon mitigation option. *Water Resources Research*, *53*, 3527–3533.

Chen, B., Barboza, B. R., Sun, Y., Bai, J., Thomas, H. R., Dutko, M., Cottrell, M., & Li, C. (2021). A review of hydraulic fracturing simulation. *Archives of Computational Methods in Engineering*, 1–58.

Chen, L., Lu, Y., Jiang, S., Li, J., Guo, T., & Luo, C. (2015). Heterogeneity of the Lower Silurian Longmaxi marine shale in the southeast Sichuan Basin of China. *Marine and Petroleum Geology*, *65*, 232–246.

Cheng, F., Luo, H., Jenkins, J. D., & Larson, E. D. (2023). The value of low- and negative-carbon fuels in the transition to net-zero emission economies: Lifecycle greenhouse gas emissions and cost assessments across multiple fuel types. *Applied Energy*, *331*, 120388.

Chunpeng, W., Yucai, W., Weixiang, C., Liangyu, R., Xueqin, H., Shuzhe, S., Zhen, N., Bohong, W., & Yanna, Z. (2023). An integrated well completion technology for low permeability reservoirs in the middle east: Fracturing, injection control and production control. *Proceedings of the SPE/IADC middle east drilling technology conference and exhibition*, May. https://doi.org/10.2118/214553-MS

Cooper, J., Stamford, L., & Azapagic, A. (2018). Economic viability of UK shale gas and potential impacts on the energy market up to 2030. *Applied Energy*, *215*, 577–590.

Cotton, M., Rattle, I., & Van Alstine, J. (2014). Shale gas policy in the United Kingdom: An argumentative discourse analysis. *Energy Policy*, *73*, 427–438.

Crotogino, F., Schneider, G.-S., & Evans, D. J. (2018). Renewable energy storage in geological formations. *Proceedings of the Institution of Mechanical Engineers, Part A: Journal of Power and Energy*, *232*, 100–114.

Cui, G., Ren, S., Zhang, L., Ezekiel, J., Enechukwu, C., Wang, Y., & Zhang, R. (2017). Geothermal exploitation from hot dry rocks via recycling heat transmission fluid in a horizontal well. *Energy*, *128*, 366–377.

Cuppen, E., Pesch, U., Remmerswaal, S., & Taanman, M. (2019). Normative diversity, conflict and transition: Shale gas in the Netherlands. *Technological Forecasting and Social Change*, *145*, 165–175.

Cutright, B. (2011). *The transformation of tight shale gas reservoirs to geothermal energy products*. Southern Methodist University Geothermal Laboratory (Prime); BEG.

Dai, X., Wei, C., Wang, M., Ma, R., Song, Y., Zhang, J., Wang, X., Shi, X., & Vandeginste, V. (2023). Interaction mechanism of supercritical CO2 with shales and a new quantitative storage capacity evaluation method. *Energy*, *264*, 126424.

Davies, R. J., Almond, S., Ward, R. S., Jackson, R. B., Adams, C., Worrall, F., Herringshaw, L. G., Gluyas, J. G., & Whitehead, M. A. (2014). Oil and gas wells and their integrity: Implications for shale and unconventional resource exploitation. *Marine and Petroleum Geology*, *56*, 239–254.

Dickson, M. H., & Fanelli, M. (2013). *Geothermal energy: Utilization and technology*. Routledge.
Du, F., & Nojabaei, B. (2019). A review of gas injection in shale reservoirs: Enhanced oil/gas recovery approaches and greenhouse gas control. *Energies*, *12*(12), 2355. https://doi.org/10.3390/en12122355
Duncan, P. M., & Eisner, L. (2010). Reservoir characterization using surface microseismic monitoring. *Geophysics*, *75*(5). https://doi.org/10.1190/1.3467760
Eaton, D. W., & Schultz, R. (2018). Increased likelihood of induced seismicity in highly overpressured shale formations. *Geophysical Journal International*, *214*, 751–757.
Eisner, L., Williams-Stroud, S., Hill, A., Duncan, P., & Thornton, M. (2010). Beyond the dots in the box Microseismicity-constrained fracture models for reservoir simulation. *The Leading Edge*, *29*(3), 326–333. https://doi.org/10.1190/1.3353730
Ellafi, A., Jabbari, H., Tomomewo, O. S., Mann, M. D., Geri, M. B., & Tang, C. (2020). Future of hydraulic fracturing application in terms of water management and environmental issues: A critical review. *SPE Canada unconventional resources conference*. SPE, D053S011R001. https://doi.org/10.2118/199993-MS
Elsworth, D., Spiers, C. J., & Niemeijer, A. R. (2016). Understanding induced seismicity. *Science*, *354*, 1380–1381.
Estrada, J. M., & Bhamidimarri, R. (2016). A review of the issues and treatment options for wastewater from shale gas extraction by hydraulic fracturing. *Fuel*, *182*, 292–303.
European Space Agency. (2023). Sentinel Technical Guide.
Faber, A.-H., Brunner, A. M., Dingemans, M. M. L., Baken, K. A., Kools, S. A. E., Schot, P. P., de Voogt, P., & van Wezel, A. P. (2021). Comparing conventional and green fracturing fluids by chemical characterisation and effect-based screening. *Science of the Total Environment*, *794*, 148727. https://doi.org/10.1016/j.scitotenv.2021.148727
Feng, Y., Tyagi, M., & White, C. D. (2015). A downhole heat exchanger for horizontal wells in low-enthalpy geopressured geothermal brine reservoirs. *Geothermics*, *53*, 368–378.
Fischer, T., Bouskova, A., Eisner, L., & Le Calvez, J. (2007). Automated P- and S-wave picking of micro earthquakes re-corded by a vertical array. *69th European Association of Geoscientists and Engineers Conference and Exhibition 2007: Securing the Future. Incorporating SPE EUROPEC*, *3*, 1802–1806. https://doi.org/10.3997/2214-4609.201401880/CITE/REFWORKS
Fujita, H., & Sato, M. (1990). *Development of automated offshore drilling system*. British Maritime Technology.
Gao, C., Rajeswaran, T., of Technology, C. U., Nakagawa, E., & Petroleum Resources, C. (2007). A literature review on smart-well technology. https://doi.org/10.2118/106011-MS
Glandt, C. A. (2005). Reservoir management employing smart wells: A review. *SPE Drilling & Completion*, *20*(4), 281–288. https://doi.org/10.2118/81107-PA
Godec, M., Koperna, G., Petrusak, R., & Oudinot, A. (2013). Potential for enhanced gas recovery and CO2 storage in the Marcellus Shale in the Eastern United States. *International Journal of Coal Geology*, *118*, 95–104.
Godec, M., Koperna, G., Petrusak, R., & Oudinot, A. (2014). Enhanced gas recovery and CO2 storage in gas shales: A summary review of its status and potential. *Energy Procedia*, *63*, 5849–5857. https://doi.org/10.1016/j.egypro.2014.11.618
Goldthau, A., & Hoxtell, W. (2012). The impact of shale gas on European energy security. *Global Public Policy Institute (GPPi) NR*, *14*.
Greci, S., Least, B., & Tayloe, G. (2014). Testing results: Erosion testing confirms the reliability of the fluidic diode type autonomous inflow control device. *Society of Petroleum Engineers—30th Abu Dhabi International Petroleum Exhibition and Conference, ADIPEC 2014: Challenges and Opportunities for the Next 30 Years*, *6*, 4045–4064. https://doi.org/10.2118/172077-MS

Green, C. A., Styles, P., & Baptie, B. J. (2012). *Preese Hall shale gas fracturing review and recommendations for induced seismic mitigation.* www.gov.uk/government/publications/preese-hall-shale-gas-fracturing-review-and-recommendations-for-induced-seismic-mitigation

Guo, C., Wei, M., & Liu, H. (2015). Modeling of gas production from shale reservoirs considering multiple transport mechanisms. *PLoS ONE, 10*(12), e0143649. https://doi.org/10.1371/journal.pone.0143649

Gurianov, A., Katashov, A., Ovchinnikov, K., Saprykina, K., Novikov, I., Malyavko, E., & Kiselev, V. (2019). Big data in field development projects. *Society of petroleum engineers—SPE Russian petroleum technology conference 2019*, RPTC 2019, 22–24. https://doi.org/10.2118/196862-MS

Hainsch, K., Löffler, K., Burandt, T., Auer, H., Crespo Del Granado, P., Pisciella, P., & Zwickl-Bernhard, S. (2022). Energy transition scenarios: What policies, societal attitudes, and technology developments will realize the EU Green Deal? *Energy, 239*, 122067.

Hammond, G. P., & O'Grady, Á. (2017). Indicative energy technology assessment of UK shale gas extraction. *Applied Energy, 185*, 1907–1918.

Han, W. S., Mcpherson, B. J., Lichtner, P. C., & Wang, F. P. (2010). Evaluation of trapping mechanisms in geologic CO2 sequestration: Case study of SACROC northern platform, a 35-year CO2 injection site. *American Journal of Science, 310*, 282–324.

Harichandan, S., Kar, S. K., Bansal, R., Mishra, S. K., Balathanigaimani, M. S., & Dash, M. (2022). Energy transition research: A bibliometric mapping of current findings and direction for future research. *Cleaner Production Letters, 3*, 100026.

Haroon, S., Alyamkin, S., & Shenoy, R. (2018). Big data-driven advanced analytics: Application of convolutional and deep neural networks for GPU based seismic interpretations. *Society of petroleum engineers—Abu Dhabi international petroleum exhibition and conference 2018*, ADIPEC. https://doi.org/10.2118/193259-MS

Hartley, P. R., Medlock III, K. B., Temzelides, T., & Zhang, X. (2015). Local employment impact from competing energy sources: Shale gas versus wind generation in Texas. *Energy Economics, 49*, 610–619.

Hasanzadeh, M., Izadpanahi, A., & Naghizadeh, A. (2023). Carbon dioxide huff-n-puff. In *Gas injection methods* (pp. 171–198). Elsevier. https://doi.org/10.1016/B978-0-12-822302-4.00007-7

Hassannayebi, N., Azizmohammadi, S., De Lucia, M., & Ott, H. (2019). Underground hydrogen storage: Application of geochemical modelling in a case study in the Molasse Basin, Upper Austria. *Environmental Earth Sciences, 78*, 1–14.

Hassanpouryouzband, A., Adie, K., Cowen, T., Thaysen, E. M., Heinemann, N., Butler, I. B., Wilkinson, M., & Edlmann, K. (2022). Geological hydrogen storage: Geochemical reactivity of hydrogen with sandstone reservoirs. *ACS Energy Letters, 7*, 2203–2210.

Hassanpouryouzband, A., Joonaki, E., Edlmann, K., & Haszeldine, R. S. (2021). Offshore geological storage of hydrogen: Is this our best option to achieve net-zero? *ACS Energy Letters, 6*, 2181–2186.

Hawthorne, S. B., Gorecki, C. D., Sorensen, J. A., Steadman, E. N., Harju, J. A., & Melzer, S. (2013, November 5). Hydrocarbon mobilization mechanisms from upper, middle, and lower Bakken reservoir rocks exposed to CO2. *All Days.* https://doi.org/10.2118/167200-MS

Hayatdavoudi, A. (1999). Changing chemophysical properties of formation and drilling fluid enhances penetration rate and bit life. *SPE international conference on oilfield chemistry.* SPE, SPE-50729-MS. https://doi.org/10.2118/50729-MS

Heinemann, N., Booth, M., Haszeldine, R. S., Wilkinson, M., Scafidi, J., & Edlmann, K. (2018). Hydrogen storage in porous geological formations–onshore play opportunities in the midland valley (Scotland, UK). *International Journal of Hydrogen Energy, 43*, 20861–20874.

Heo, I., Park, K., Lee, J., Lee, J., & Chang, B. (2018). *Cases of investigation on rock slope and analysis on displacement utilizing a drone*. OnePetro.

Holloway, S. (2007). Carbon dioxide capture and geological storage. *Philosophical Transactions of the Royal Society A: Mathematical, Physical and Engineering Sciences, 365*, 1095–1107.

Howarth, R. W. (2014). A bridge to nowhere: Methane emissions and the greenhouse gas footprint of natural gas. *Energy Science & Engineering, 2*, 47–60.

Howarth, R. W. (2015). Methane emissions and climatic warming risk from hydraulic fracturing and shale gas development: Implications for policy. *Energy and Emission Control Technologies*, 45–54.

Howarth, R. W., Ingraffea, A., & Engelder, T. (2011a). Should fracking stop? *Nature, 477*, 271–275.

Howarth, R. W., Santoro, R., & Ingraffea, A. (2011b). Methane and the greenhouse-gas footprint of natural gas from shale formations: A letter. *Climatic Change, 106*, 679–690.

Howarth, R. W., Santoro, R., & Ingraffea, A. (2012). Venting and leaking of methane from shale gas development: Response to Cathles et al. *Climatic Change, 113*, 537–549.

Hua, D. D. Z. C. Y., & Gengsheng, W. Y. L. X. C. (2012). Progress and prospects of shale gas exploration and development in China. *Acta Petrolei Sinica, 33*, 107.

Huenges, E., & Ledru, P. (2011). *Geothermal energy systems: Exploration, development, and utilization*. John Wiley & Sons.

Ibrahim, H., Ilinca, A., & Perron, J. (2008). Energy storage systems—Characteristics and comparisons. *Renewable and Sustainable Energy Reviews, 12*, 1221–1250.

Iddphonce, R., Wang, J., & Zhao, L. (2020). Review of CO2 injection techniques for enhanced shale gas recovery: Prospect and challenges. *Journal of Natural Gas Science and Engineering, 77*, 103240.

Ilgen, A. G., Heath, J. E., Akkutlu, I. Y., Bryndzia, L. T., Cole, D. R., Kharaka, Y. K., Kneafsey, T. J., Milliken, K. L., Pyrak-Nolte, L. J., & Suarez-Rivera, R. (2017). Shales at all scales: Exploring coupled processes in mudrocks. *Earth-Science Reviews, 166*, 132–152.

Ishida, T., Chen, Y., Bennour, Z., Yamashita, H., Inui, S., Nagaya, Y., Naoi, M., Chen, Q., Nakayama, Y., & Nagano, Y. (1912). Probability method for the determination of earthquake epicentres from the arrival time only. *Bull Stint Louis University, 8*(11), 60. https://doi.org/10.1002/2016JB013365

Ismail, A. R., Sulaiman, W. R. W., Jaafar, M. Z., Aftab, A., Razi, A. A., & Ibupoto, Z. H. (2017). The application of MWCNT to enhance the rheological behavior of drilling fluids at high temperature. *Malaysian Journal of Fundamental and Applied Sciences, 12*(3). https://doi.org/10.11113/mjfas.v12n3.423

Israel, A. L., Wong-Parodi, G., Webler, T., & Stern, P. C. (2015). Eliciting public concerns about an emerging energy technology: The case of unconventional shale gas development in the United States. *Energy Research & Social Science, 8*, 139–150.

Jacob, D. J., Turner, A. J., Maasakkers, J. D., Sheng, J., Sun, K., Liu, X., Chance, K., Aben, I., McKeever, J., & Frankenberg, C. (2016). Satellite observations of atmospheric methane and their value for quantifying methane emissions. *Atmospheric Chemistry and Physics, 16*(22), 14371–14396. https://doi.org/10.5194/ACP-16-14371-2016

Jahanbakhsh, A., Louis Potapov-Crighton, A., Mosallanezhad, A., Tohidi Kaloorazi, N., & Maroto-Valer, M. M. (2024). Underground hydrogen storage: A UK perspective. *Renewable and Sustainable Energy Reviews, 189*, 114001.

Jia, B., Tsau, J.-S., & Barati, R. (2019). A review of the current progress of CO2 injection EOR and carbon storage in shale oil reservoirs. *Fuel, 236*, 404–427. https://doi.org/10.1016/j.fuel.2018.08.103

Jiang, Y., Luo, Y., Lu, Y., Qin, C., & Liu, H. (2016). Effects of supercritical CO2 treatment time, pressure, and temperature on microstructure of shale. *Energy, 97*, 173–181. https://doi.org/10.1016/j.energy.2015.12.124

Jing, Y., Rabbani, A., Armstrong, R. T., Wang, J., Zhang, Y., & Mostaghimi, P. (2022). An image-based coal network model for simulating hydro-mechanical gas flow in coal: An application to carbon dioxide geo-sequestration. *Journal of Cleaner Production*, *379*, 134647.

Junrong, L., Rongqiang, L., & Zhixue, S. (2015). Exploitation and utilization technology of geothermal resources in oil fields. *Proceedings world geothermal congress 2015.* https://pangea.stanford.edu/ERE/db/WGC/papers/WGC/2015/28005.pdf

Kadoura, A., Narayanan Nair, A. K., & Sun, S. (2016). Molecular dynamics simulations of carbon dioxide, methane, and their mixture in montmorillonite clay hydrates. *The Journal of Physical Chemistry C*, *120*(23), 12517–12529. https://doi.org/10.1021/acs.jpcc.6b02748

Kalam, S., Afagwu, C., Al Jaberi, J., Siddig, O. M., Tariq, Z., Mahmoud, M., & Abdulraheem, A. (2021). A review on non-aqueous fracturing techniques in unconventional reservoirs. *Journal of Natural Gas Science and Engineering*, *95*, 104223. https://doi.org/10.1016/j.jngse.2021.104223

Kalam, S., Olayiwola, T., Al-Rubaii, M. M., Amaechi, B. I., Jamal, M. S., & Awotunde, A. A. (2020). Carbon dioxide sequestration in underground formations: Review of experimental, modeling, and field studies. *Journal of Petroleum Exploration and Production Technology*. https://doi.org/10.1007/s13202-020-01028-7

Kang, Z., Zhao, Y., & Yang, D. (2020). Review of oil shale in-situ conversion technology. *Applied Energy*, *269*, 115121.

Kao, H., Visser, R., Smith, B., & Venables, S. (2018). Performance assessment of the induced seismicity traffic light protocol for northeastern British Columbia and western Alberta. *The Leading Edge*, *37*(2), 117–126. https://doi.org/10.1190/TLE37020117.1

Karrenbach, M., Ridge, A., Cole, S., Boone, K., Kahn, D., Rich, J., Silver, K., & Langton, D. (2017). DAS microseismic monitoring and integration with strain measurements in hydraulic fracture profiling. *SPE/AAPG/SEG unconventional resources technology conference 2017.* https://doi.org/10.15530/URTEC-2017-2670716

Keranen, K. M., & Weingarten, M. (2018). Induced seismicity. *Annual Review of Earth and Planetary Sciences*, *46*, 149–174. https://doi.org/10.1146/ANNUREV-EARTH-082517-010054

Khan, J. A., Padmanabhan, E., Haq, I. U., Yekeen, N., Mumtaz, M., & Prajapati, S. (2022). Experimental assessment of fracture initiation and wettability in low and high brittle shales by CO2 foam fracturing fluids. *Energy & Fuels*, *36*(15), 8288–8300. https://doi.org/10.1021/acs.energyfuels.2c00310

Khoshsima, A., Sedighi, M., & Mohammadi, M. (2023). Enhanced oil recovery by water alternating gas injection. In *Gas injection methods* (pp. 295–316). Elsevier. https://doi.org/10.1016/B978-0-12-822302-4.00001-6

Khosrokhavar, R., Griffiths, S., & Wolf, K.-H. (2014). Shale gas formations and their potential for carbon storage: Opportunities and outlook. *Environmental Processes*, *1*, 595–611.

Kim, T. H., Cho, J., & Lee, K. S. (2017). Evaluation of CO2 injection in shale gas reservoirs with multi-component transport and geomechanical effects. *Applied Energy*, *190*, 1195–1206. https://doi.org/10.1016/j.apenergy.2017.01.047

Kobek, M. L. P., Ugarte, A., & Aguilar, G. C. (2015). Shale gas in the United States: Transforming energy security in the twenty-first century. *Norteamérica*, *10*, 7–38.

Krishnan, A., Mohd Aji, A. Q., Maulianda, B., Mohshim, D. F., & Barati, R. (2022). Study of D-limonene as novel green hydraulic fracturing surfactant in shale gas reservoir. *Journal of Natural Gas Science and Engineering*, *103*, 104588. https://doi.org/10.1016/j.jngse.2022.104588

Lashgari, H. R., Sun, A., Zhang, T., Pope, G. A., & Lake, L. W. (2019). Evaluation of carbon dioxide storage and miscible gas EOR in shale oil reservoirs. *Fuel*, *241*, 1223–1235. https://doi.org/10.1016/j.fuel.2018.11.076

Lastra, R., & Aramco, S. (2019). Electrical submersible pump digital twin, the missing link for successful condition monitoring and failure prediction. *Society of petroleum engineers—Abu Dhabi international petroleum exhibition and conference 2019*, ADIP. https://doi.org/10.2118/197156-MS

Lau, H. C., Ramakrishna, S., Zhang, K., & Radhamani, A. V. (2021). The role of carbon capture and storage in the energy transition. *Energy & Fuels*, *35*, 7364–7386.
Lebedev, M., Zhang, Y., Sarmadivaleh, M., Barifcani, A., Al-Khdheeawi, E., & Iglauer, S. (2017). Carbon geosequestration in limestone: Pore-scale dissolution and geomechanical weakening. *International Journal of Greenhouse Gas Control*, *66*, 106–119.
Li, C.-R., Rao, S., Hu, S.-B., Wang, J.-Y., Wei, Z.-H., Huang, Q.-Q., Hu, D.-F., & Guo, X.-S. (2017). Present-day geothermal field of the Jiaoshiba shale gas area in southeast of the Sichuan basin, SW China. *Chinese Journal of Geophysics*, *60*, 617–627.
Li, K., & Sun, W. (2015). Modified method for estimating geothermal resources in oil and gas reservoirs. *Mathematical Geosciences*, *47*, 105–117.
Li, L., Becker, D., Chen, H., Wang, X., & Gajewski, D. (2018). A systematic analysis of correlation-based seismic location methods. *Geophysical Journal International*, *212*(1), 659–678. https://doi.org/10.1093/GJI/GGX436
Li, L., Tan, J., Wood, D. A., Zhao, Z., Becker, D., Lyu, Q., Shu, B., & Chen, H. (2019). A review of the current status of induced seismicity monitoring for hydraulic fracturing in unconventional tight oil and gas reservoirs. *Fuel*, *242*, 195–210. https://doi.org/10.1016/J.FUEL.2019.01.026
Li, Q., Wang, J., & Yin, H. (2023). Intelligent stuck pipe type recognition using digital twins and knowledge graph model. *Applied Sciences*, *13*, 3098.
Li, X., Gao, D., Zhou, Y., & Zhang, H. (2016). Study on open-hole extended-reach limit model analysis for horizontal drilling in shales. *Journal of Natural Gas Science and Engineering*, *34*, 520–533.
Li, Y., DiCarlo, D., Li, X., Zang, J., & Li, Z. (2015). An experimental study on application of nanoparticles in unconventional gas reservoir CO 2 fracturing. *Journal of Petroleum Science and Engineering*, *133*, 238–244. https://doi.org/10.1016/j.petrol.2015.05.023
Liu, H., & Liu, H. (2017). Logging-while-drilling (LWD). In *Principles and applications of well logging* (pp. 211–236). Springer.
Liu, J., Wang, S., Javadpour, F., Feng, Q., & Cha, L. (2022). Hydrogen diffusion in clay slit: Implications for the geological storage. *Energy & Fuels*, *36*, 7651–7660.
Liu, Y., Ma, T., Wu, H., & Chen, P. (2020). Investigation on mechanical behaviors of shale cap rock for geological energy storage by linking macroscopic to mesoscopic failures. *Journal of Energy Storage*, *29*, 101326.
Lord, B. (2017). Remote sensing techniques for onshore oil and gas exploration. *The Leading Edge*, *36*(1), 24–32. https://doi.org/10.1190/TLE36010024.1
Loucks, R. G., Reed, R. M., Ruppel, S. C., & Hammes, U. (2012). Spectrum of pore types and networks in mudrocks and a descriptive classification for matrix-related mudrock pores. *AAPG Bulletin*, *96*, 1071–1098.
Louk, K., Ripepi, N., Luxbacher, K., Gilliland, E., Tang, X., Keles, C., Schlosser, C., Diminick, E., Keim, S., & Amante, J. (2017). Monitoring CO2 storage and enhanced gas recovery in unconventional shale reservoirs: Results from the Morgan County, Tennessee injection test. *Journal of Natural Gas Science and Engineering*, *45*, 11–25.
Loveland, T. R., & Dwyer, J. L. (2012). Landsat: Building a strong future. *Remote Sensing of Environment*, *122*, 22–29. https://doi.org/10.1016/J.RSE.2011.09.022
Lu, Y., Liu, Y., Tang, J., Jia, Y., Tian, R., Zhou, J., Chen, X., Xu, Z., & Cheng, Q. (2023). Water wettability alternation of CO2-water-shale system due to nanoparticles: Implications for CO2 geo-storage. *International Journal of Greenhouse Gas Control*, *124*, 103836.
Lund, J. W., & Boyd, T. L. (2016). Direct utilization of geothermal energy 2015 worldwide review. *Geothermics*, *60*, 66–93.
Lyu, Q., Tan, J., Li, L., Ju, Y., Busch, A., Wood, D. A., Ranjith, P. G., Middleton, R., Shu, B., & Hu, C. (2021). The role of supercritical carbon dioxide for recovery of shale gas and sequestration in gas shale reservoirs. *Energy & Environmental Science*, *14*, 4203–4227.

Ma, J., Xu, J., Pang, S., Zhou, W., Xia, B., & An, Y. (2021). Novel environmentally friendly lubricants for drilling fluids applied in shale formation. *Energy and Fuels*, *35*(9). https://doi.org/10.1021/acs.energyfuels.1c00495

Ma, R., & Lin, B. (2023). Digitalization and energy-saving and emission reduction in Chinese cities: Synergy between industrialization and digitalization. *Applied Energy*, *345*, 121308.

Mahdi, D. S., Al-Khdheeawi, E. A., Yuan, Y., Zhang, Y., & Iglauer, S. (2021). Hydrogen underground storage efficiency in a heterogeneous sandstone reservoir. *Advances in Geo-Energy Research*, *5*(4), 437–443.

Manfroni, M., Bukkens, S. G. F., & Giampietro, M. (2021). The declining performance of the oil sector: Implications for global climate change mitigation. *Applied Energy*, *298*, 117210.

Mao, L., & Zhang, Z. (2018). Transient temperature prediction model of horizontal wells during drilling shale gas and geothermal energy. *Journal of Petroleum Science and Engineering*, *169*, 610–622.

Maxwell, S. (2014). *Microseismic imaging of hydraulic fracturing: Improved engineering of unconventional shale reservoirs*. Society of Exploration Geophysicists.

Mayfield, E. N., Cohon, J. L., Muller, N. Z., Azevedo, I. M., & Robinson, A. L. (2019). Cumulative environmental and employment impacts of the shale gas boom. *Nature Sustainability*, *2*, 1122–1131.

Meador, R. A. (2009). Logging-while-drilling a story of dreams, accomplishments, and bright futures. *SPWLA annual logging symposium*. SPWLA, SPWLA-2009-43458.

Meakin, P., Huang, H., Malthe-Sørenssen, A., & Thøgersen, K. (2013). Shale gas: Opportunities and challenges. *Environmental Geosciences*, *20*, 151–164.

Merey, S., & Sinayuc, C. (2016). Analysis of carbon dioxide sequestration in shale gas reservoirs by using experimental adsorption data and adsorption models. *Journal of Natural Gas Science and Engineering*, *36*, 1087–1105.

Middleton, R. S., Gupta, R., Hyman, J. D., & Viswanathan, H. S. (2017). The shale gas revolution: Barriers, sustainability, and emerging opportunities. *Applied Energy*, *199*, 88–95.

Miller, S. M., Wofsy, S. C., Michalak, A. M., Kort, E. A., Andrews, A. E., Biraud, S. C., Dlugokencky, E. J., Eluszkiewicz, J., Fischer, M. L., Janssens-Maenhout, G., Miller, B. R., Miller, J. B., Montzka, S. A., Nehrkorn, T., & Sweeney, C. (2013). Anthropogenic emissions of methane in the United States. *Proceedings of the National Academy of Sciences of the United States of America*, *110*(50), 20018–20022. https://doi.org/10.1073/PNAS.1314392110/SUPPL_FILE/PNAS.201314392SI.PDF

Mohagheghian, E., Hassanzadeh, H., & Chen, Z. (2019). CO2 sequestration coupled with enhanced gas recovery in shale gas reservoirs. *Journal of CO2 Utilization*, *34*, 646–655.

Moridis, G. (2018). Literature review and analysis of waterless fracturing methods. https://doi.org/10.2172/1468605

Nantanoi, S., Rodríguez-Pradilla, G., & Verdon, J. (2022). 3D seismic interpretation and fault slip potential analysis from hydraulic fracturing in the Bowland Shale, UK. *Petroleum Geoscience*, *28*, petgeo2021–057.

National Institute of Water and Atmospheric Research. (2023, May 10). *MethaneSAT: Detecting and quantifying methane emissions by satellite*. https://niwa.co.nz/atmosphere/methanesat

Nazari, M., Asgary, A., Nezami, I. M., & Ghayourisales, S. (2024). From resistance to resilience: A comprehensive bibliometric analysis of carbon pricing public acceptance. *Energy Research & Social Science*, *107*, 103340.

Nianyin, L., Jiajie, Y., Chao, W., Suiwang, Z., Xiangke, L., Jia, K., Yuan, W., & Yinhong, D. (2021). Fracturing technology with carbon dioxide: A review. *Journal of Petroleum Science and Engineering*, *205*, 108793. https://doi.org/10.1016/j.petrol.2021.108793

Nordquist, J., & Johnson, L. (2012). Production of power from the co-produced water of oil wells, 3.5 years of operation. *Geothermal resources council transactions, geothermal resources council 2012 annual meeting* (pp. 207–210). https://publications.mygeoenergynow.org/grc/1030230.pdf

Nurhandoko, B. E. B. (2013). *Shale gas reservoir characterization strategy*. HAGI-IAGI, Indonesia.

Olabi, A. (2017). *Renewable energy and energy storage systems*. Elsevier.

Olasolo, P., Juárez, M., Morales, M., & Liarte, I. (2016). Enhanced geothermal systems (EGS): A review. *Renewable and Sustainable Energy Reviews*, *56*, 133–144.

Olmstead, S. (2015). *Wastewater and shale formation development: Risks, mitigation, and regulation*. CRC Press.

Outlook, B. E. (2023). *BP energy outlook* (July ed.). BP Energy.

Pandey, S., Chaudhuri, A., Kelkar, S., Sandeep, V., & Rajaram, H. (2014). Investigation of permeability alteration of fractured limestone reservoir due to geothermal heat extraction using three-dimensional thermo-hydro-chemical (THC) model. *Geothermics*, *51*, 46–62.

Panja, P., Velasco, R., Pathak, M., & Deo, M. (2018). Application of artificial intelligence to forecast hydrocarbon production from shales. *Petroleum*, *4*, 75–89.

Pearce, J., Raza, S., Baublys, K., Hayes, P., Firouzi, M., & Rudolph, V. (2022). Unconventional CO2 storage: CO2 mineral trapping predicted in characterized shales, sandstones, and coal seam interburden. *SPE Journal*, *27*, 3218–3239.

Perrons, R. K., & Jensen, J. W. (2014). Data as an asset: What the upstream oil & gas industry can learn about big data from companies like social media. *Proceedings—SPE Annual Technical Conference and Exhibition*, *1*, 714–721. https://doi.org/10.2118/170633-MS

Pietersen, K., Chevallier, L., Levine, A., Maceba, T., Gaffoor, Z., & Kanyerere, T. (2021). Prospective policy safeguards to mitigate hydrogeological risk pathways in advance of shale gas development in the Karoo basin, South Africa. *Groundwater for Sustainable Development*, *12*, 100499.

Porter, L., Rickards, L., Verlie, B., Moloney, S., Lay, B., Latham, B., Anguelovski, I., Pellow, D., Porter, L., Bosomworth, K., Rickards, L., Verlie, B., Moloney, S., Lay, B., Latham, B., Bcnuej, I. A., Pellow, D. N., Naarm/Birrarung-ga, Porter, L., . . . Lay, B. (2020). Climate justice in a climate changed world. *Planning Theory and Practice*, *21*(2). https://doi.org/10.1080/14649357.2020.1748959

Pörtner, H. O., Roberts, D. C., Adams, H., Adler, C., Aldunce, P., Ali, E., & Fischlin, A. (2023). Intergovernmental panel on climate change. In *Climate change 2022—Impacts, adaptation and vulnerability: Working group II contribution to the sixth assessment report of the intergovernmental panel on climate change*. Cambridge: Cambridge University Press.

Pu, H., Wang, Y., & Li, Y. (2018). How CO2-storage mechanisms are different in organic shale: Characterization and simulation studies. *SPE Journal*, *23*(3), 661–671. https://doi.org/10.2118/180080-PA

Puripanda, N. K., Nooraldeen, Y., Derbas, Z., Alnowakhtha, A., & Abdulsalam, A. (2020). Best practice of utilizing drones for surveying and mapping in the Bahrain oil field. *Society of petroleum engineers—Abu Dhabi international petroleum exhibition and conference 2020*. ADIP. https://doi.org/10.2118/203077-MS

Rana, S. (2008). Facts and data on environmental risks-oil and gas drilling operations. *SPE Asia Pacific oil and gas conference and exhibition 2008—"Gas now: Delivering on expectations"*, 1. https://doi.org/10.2118/114993-ms

Raza, A., Arif, M., Glatz, G., Mahmoud, M., Al Kobaisi, M., Alafnan, S., & Iglauer, S. (2022). A holistic overview of underground hydrogen storage: Influencing factors, current understanding, and outlook. *Fuel*, *330*, 125636.

Reidenbach, V. G., Harris, P. C., Lee, Y. N., & Lord, D. L. (1986). Rheological study of foam fracturing fluids using nitrogen and carbon dioxide. *SPE Production Engineering*, *1*(1), 31–41. https://doi.org/10.2118/12026-PA

Ren, X., Hall, D. L., Vinciguerra, T., Benish, S. E., Stratton, P. R., Ahn, D., Hansford, J. R., Cohen, M. D., Sahu, S., He, H., Grimes, C., Fuentes, J. D., Shepson, P. B., Salawitch, R. J., Ehrman, S. H., & Dickerson, R. R. (2019). Methane emissions from the Marcellus shale in Southwestern Pennsylvania and Northern West Virginia based on airborne measurements. *Journal of Geophysical Research: Atmospheres, 124*(3). https://doi.org/10.1029/2018JD029690

Rezaeyan, A., Kampman, N., Pipich, V., Barnsley, L. C., Rother, G., Magill, C., Ma, J., & Busch, A. (2023). Evolution of pore structure in organic-lean and organic-rich mudrocks. *Energy & Fuels, 37*, 16446–16460. https://doi.org/10.1021/acs.energyfuels.3c02180

Rezaeyan, A., Kampman, N., Pipich, V., Barnsley, L. C., Rother, G., Magill, C., Ma, J., & Busch, A. (2024). Compaction and clay content control mudrock porosity. *Energy, 289*, 129966. https://doi.org/10.1016/j.energy.2023.129966

Rezaeyan, A., Pipich, V., Ma, J., Leu, L., Seemann, T., Rother, G., Barnsley, L. C., & Busch, A. (2022). Predicting fluid flow regime, permeability, and diffusivity in mudrocks from multiscale pore characterisation. *Transport in Porous Media, 141*, 201–229. https://doi.org/10.1007/s11242-021-01717-9

Richards, M. C., & Blackwell, D. D. (2012). *Developing geothermal energy in Texas: Mapping the temperatures and resources* (pp. 351–363). Gulf Coast Association of Geological Societies Transactions.

Rivard, C., Lavoie, D., Lefebvre, R., Séjourné, S., Lamontagne, C., & Duchesne, M. (2014). An overview of Canadian shale gas production and environmental concerns. *International Journal of Coal Geology, 126*, 64–76.

Romanov, V., Soong, Y., Carney, C., Rush, G. E., Nielsen, B., & O'Connor, W. (2015). Mineralization of carbon dioxide: A literature review. *ChemBioEng Reviews, 2*, 231–256.

Rosales, S. J., Hernandez, R. A., Lynch, C. Y., & Bustamante, E. (2010). Effectiveness of water jet fracturing on a horizontal open hole completion: Case study. *Proceedings—SPE Annual Technical Conference and Exhibition, 4*, 3421–3428. https://doi.org/10.2118/134855-MS

Ross, D. J., & Bustin, R. M. (2009). The importance of shale composition and pore structure upon gas storage potential of shale gas reservoirs. *Marine and Petroleum Geology, 26*, 916–927.

Rutqvist, J., Rinaldi, A. P., Cappa, F., & Moridis, G. J. (2013). Modeling of fault reactivation and induced seismicity during hydraulic fracturing of shale-gas reservoirs. *Journal of Petroleum Science and Engineering, 107*, 31–44.

Saboori, R., Sabbaghi, S., & Kalantariasl, A. (2019). Improvement of rheological, filtration and thermal conductivity of bentonite drilling fluid using copper oxide/polyacrylamide nanocomposite. *Powder Technology, 353*. https://doi.org/10.1016/j.powtec.2019.05.038

Saebi, S., Tyutikov, S., Machado, A., Curteis, C., Rivero, M. E., Khaled, M. Z., & Habib, M. S. (2010). Design, execution, and results of the longest multizone cased-hole gravel-pack completion in Malaysia. *Society of Petroleum Engineers—SPE Asia Pacific Oil and Gas Conference and Exhibition 2010, APOGCE 2010, 1*, 77–86. https://doi.org/10.2118/130785-MS

Safari, R., Gandikota, R., Mutlu, O., Ji, M., Glanville, J., & Abass, H. (2015). Pulse fracturing in shale reservoirs: Geomechanical aspects, ductile/brittle transition, and field implications. *SPE Journal, 20*(6), 1287–1304. https://doi.org/10.2118/168759-PA

Said, M. M., Pilgrim, R., Rideout, G., & Butt, S. (2022). Theoretical development of a digital-twin based automation system for oil well drilling rigs. *SPE Canadian energy technology conference*. SPE, D021S025R003. https://doi.org/10.2118/208902-MS

Saif, T., Lin, Q., Bijeljic, B., & Blunt, M. J. (2017). Microstructural imaging and characterization of oil shale before and after pyrolysis. *Fuel, 197*, 562–574.

Saini, G., Ashok, P., van Oort, E., & Isbell, M. R. (2018). Accelerating well construction using a digital twin demonstrated on unconventional well data in North America. *SPE/AAPG/SEG unconventional resources technology conference 2018*, URTC. https://doi.org/10.15530/URTEC-2018-2902186

Sakhaee-Pour, A., & Bryant, S. L. (2012). Gas permeability of shale. *SPE Reservoir Evaluation & Engineering, 15*, 401–409.

Salygin, V., Guliev, I., Chernysheva, N., Sokolova, E., Toropova, N., & Egorova, L. (2019). Global shale revolution: Successes, challenges, and prospects. *Sustainability*, *11*, 1627.

Sambo, C., Liu, N., Shaibu, R., Ahmed, A. A., & Hashish, R. G. (2023). A technical review of CO2 for enhanced oil recovery in unconventional oil reservoirs. *Geoenergy Science and Engineering*, *221*, 111185. https://doi.org/10.1016/j.petrol.2022.111185

Santos, L., Dahi Taleghani, A., & Elsworth, D. (2022). Repurposing abandoned wells for geothermal energy: Current status and future prospects. *Renewable Energy*, *194*, 1288–1302.

Sapto Mulyanto, B., Dewanto, O., & Rizky, S. (2018). *Determining layer oil shale as new alternative energy sources using core analysis and well log method*. http://repository.lppm.unila.ac.id/43582/

Schaef, H. T., Davidson, C. L., Owen, A. T., Miller, Q. R., Loring, J. S., Thompson, C. J., Bacon, D. H., Glezakou, V. A., & Mcgrail, B. P. (2014). CO2 utilization and storage in shale gas reservoirs: Experimental results and economic impacts. *Energy Procedia*, *63*, 7844–7851.

Schneising, O., Burrows, J. P., Dickerson, R. R., Buchwitz, M., Reuter, M., & Bovensmann, H. (2014). Remote sensing of fugitive methane emissions from oil and gas production in North American tight geologic formations. *Earth's Future*, *2*(10), 548–558. https://doi.org/10.1002/2014EF000265

Schrag, D. P. (2012). Is shale gas good for climate change? *Daedalus*, *141*, 72–80.

Semnani, S. J., & Borja, R. I. (2017). Quantifying the heterogeneity of shale through statistical combination of imaging across scales. *Acta Geotechnica*, *12*, 1193–1205.

Shahsavar, A., Ghadamian, H., Bahri, A., Amirian, H., Shakouri, M., Bahramara, S., & Adl, M. (2023). Renewable energy for enhanced building energy flexibility. In *Building energy flexibility and demand management*. Elsevier.

Shaw, J. (2011). Benefits and application of a surface-controlled sliding sleeve for fracturing operations. *Proceedings—SPE Annual Technical Conference and Exhibition*, *6*, 4727–4734. https://doi.org/10.2118/147546-MS

Sheng, J. J. (2015). Enhanced oil recovery in shale reservoirs by gas injection. *Journal of Natural Gas Science and Engineering*, *22*, 252–259. https://doi.org/10.1016/j.jngse.2014.12.002

Sherwin, E. D., Rutherford, J. S., Chen, Y., Aminfard, S., Kort, E. A., Jackson, R. B., & Brandt, A. R. (2023). Single-blind validation of space-based point-source detection and quantification of onshore methane emissions. *Scientific Reports*, *13*(1), 1–10. https://doi.org/10.1038/s41598-023-30761-2

Shi, W., Zhu, L., Guo, M., Huang, Z., Wang, G., Lin, L., He, L., Liao, Y., He, H., & Gong, J. (2023). Assessment of CO2 fracturing in China's shale oil reservoir: Fracturing effectiveness and carbon storage potential. *Resources, Conservation and Recycling*, *197*, 107101. https://doi.org/10.1016/j.resconrec.2023.107101

Shi, Z., Jessen, K., & Tsotsis, T. T. (2020). Impacts of the subsurface storage of natural gas and hydrogen mixtures. *International Journal of Hydrogen Energy*, *45*, 8757–8773.

Shirizadeh, B., Villavicencio, M., Douguet, S., Trüby, J., Bou Issa, C., Seck, G. S., D'Herbemont, V., Hache, E., Malbec, L.-M., Sabathier, J., Venugopal, M., Lagrange, F., Saunier, S., Straus, J., & Reigstad, G. A. (2023). The impact of methane leakage on the role of natural gas in the European energy transition. *Nature Communications*, *14*, 5756.

Singh, H. (2022). Hydrogen storage in inactive horizontal shale gas wells: Techno-economic analysis for Haynesville shale. *Applied Energy*, *313*, 118862.

Sircar, A., Nair, A., Bist, N., & Yadav, K. (2022). Digital twin in hydrocarbon industry. *Petroleum Research*, *8*(2), 270–278.

Smith, D., & Starr, P. (2008). Method to pump bridge/frac plugs at reduced fluid rate. *Proceedings—SPE International Symposium on Formation Damage Control*, *1*, 341–348. https://doi.org/10.2118/112377-MS

Smith, M. B., & Montgomery, C. (2015). *Hydraulic fracturing*. CRC Press.

Solarin, S. A., & Bello, M. O. (2020). The impact of shale gas development on the US economy: Evidence from a quantile autoregressive distributed lag model. *Energy, 205*, 118004.

Song, B., Ehlig-Economides, C., & Economides, M. J. (2011). Design of multiple transverse fracture horizontal wells in shale gas reservoirs. *SPE Hydraulic fracturing technology conference*. OnePetro.

Souther, S., Tingley, M. W., Popescu, V. D., Hayman, D. T., Ryan, M. E., Graves, T. A., Hartl, B., & Terrell, K. (2014). Biotic impacts of energy development from shale: Research priorities and knowledge gaps. *Frontiers in Ecology and the Environment*, *12*, 330–338.

Spellman, F. R. (2012). *Environmental impacts of hydraulic fracturing*. CRC Press.

St-Onge, A. (2011). Akaike information criterion applied to detecting first arrival times on microseismic data. *SEG Technical Program Expanded Abstracts*, *30*(1), 1658–1662. https://doi.org/10.1190/1.3627522

Sun, J., Chen, C., Hu, W., Cui, J., Jiang, L., Liu, Y., Zhao, Y., Li, W., & Song, Y. (2021). Asymmetric competitive adsorption of CO2/CH4 binary mixture in shale matrix with heterogeneous surfaces. *Chemical Engineering Journal*, *422*, 130025. https://doi.org/10.1016/j.cej.2021.130025

Sun, Q., Ampomah, W., Kutsienyo, E. J., Appold, M., Adu-Gyamfi, B., Dai, Z., & Soltanian, M. R. (2020). Assessment of CO2 trapping mechanisms in partially depleted oil-bearing sands. *Fuel*, *278*, 118356.

Syed, F. I., Alshamsi, A., Dahaghi, A. K., & Neghabhan, S. (2022). Application of ML & AI to model petrophysical and geomechanical properties of shale reservoirs–A systematic literature review. *Petroleum*, *8*, 158–166.

Tahmasebi, P., Kamrava, S., Bai, T., & Sahimi, M. (2020). Machine learning in geo- and environmental sciences: From small to large scale. *Advances in Water Resources*, *142*, 103619. https://doi.org/10.1016/j.advwatres.2020.103619

Tao, Z., & Clarens, A. (2013). Estimating the carbon sequestration capacity of shale formations using methane production rates. *Environmental Science & Technology*, *47*, 11318–11325.

Tarkowski, R. (2019). Underground hydrogen storage: Characteristics and prospects. *Renewable and Sustainable Energy Reviews*, *105*, 86–94.

Tarkowski, R., & Uliasz-Misiak, B. (2022). Towards underground hydrogen storage: A review of barriers. *Renewable and Sustainable Energy Reviews*, *162*, 112451.

Thoram, S., & Ehlig-Economides, C. (2011). Extending shale gas well life with low grade geothermal power. *GRC Transactions*, *35*, 553–560.

Tran, D., Settari, A., & Nghiem, L. (2012). *Initiation and propagation of secondary cracks in thermo-poroelastic media*. American Rock Mechanics Association.

U.S. Energy Information Administration. (2022). *U.S. energy-related carbon dioxide emissions, 2021*. https://www.eia.gov/environment/emissions/carbon/

Varon, D. J., Jervis, D., McKeever, J., Spence, I., Gains, D., & Jacob, D. J. (2021). High-frequency monitoring of anomalous methane point sources with multispectral Sentinel-2 satellite observations. *Atmospheric Measurement Techniques*, *14*(4), 2771–2785. https://doi.org/10.5194/AMT-14-2771-2021

Vengosh, A., Jackson, R. B., Warner, N., Darrah, T. H., & Kondash, A. (2014). A critical review of the risks to water resources from unconventional shale gas development and hydraulic fracturing in the United States. *Environmental Science & Technology*, *48*, 8334–8348.

Vengosh, A., Warner, N., Jackson, R., & Darrah, T. (2013). The effects of shale gas exploration and hydraulic fracturing on the quality of water resources in the United States. *Procedia Earth and Planetary Science*, *7*, 863–866.

Vidic, R. D., Brantley, S. L., Vandenbossche, J. M., Yoxtheimer, D., & Abad, J. D. (2013). Impact of shale gas development on regional water quality. *Science*, *340*, 1235009.

Wachsmuth, J., Warnke, P., Gambhir, A., Giarola, S., Koasidis, K., Mittal, S., Nikas, A., Vaillancourt, K., & Doukas, H. (2023). Co-creating socio-technical scenarios for net-zero emission pathways: Comparison of five national case studies. *Renewable and Sustainable Energy Transition*, *4*, 100064.

Walsh, W. (2013). Geothermal resource assessment of the Clarke lake gas field, fort nelson, British Columbia. *Bulletin of Canadian Petroleum Geology*, *61*, 241–251.

Wan, T., Meng, X., Sheng, J. J., & Watson, M. (2014, April 12). Compositional modeling of EOR process in stimulated shale oil reservoirs by cyclic gas injection. *All Days*. https://doi.org/10.2118/169069-MS

Wang, G. H., Ma, X. S., Zhang, Y. S., Wang, D. A., Lv, J. T., Zhang, M., & Wang, X. R. (2020). Reservoir fractures predicted with remote sensing data in Yingxiongling of West Qaidam Basin. *Springer Series in Geomechanics and Geoengineering*, 2108–2119. https://doi.org/10.1007/978-981-15-2485-1_192/FIGURES/10

Wang, J., Ma, X., Zhang, J., & Zhao, X. (2022a). Impacts of digital technology on energy sustainability: China case study. *Applied Energy*, *323*, 119329.

Wang, J., Ryan, D., & Anthony, E. J. (2011). Reducing the greenhouse gas footprint of shale gas. *Energy Policy*, *39*, 8196–8199.

Wang, K., Yuan, B., Ji, G., & Wu, X. (2018). A comprehensive review of geothermal energy extraction and utilization in oilfields. *Journal of Petroleum Science and Engineering*, *168*, 465–477.

Wang, L., Zhang, Y., Zou, R., Zou, R., Huang, L., Liu, Y., Meng, Z., Wang, Z., & Lei, H. (2023a). A systematic review of CO2 injection for enhanced oil recovery and carbon storage in shale reservoirs. *International Journal of Hydrogen Energy*. https://doi.org/10.1016/j.ijhydene.2023.06.099

Wang, Q., Chen, X., Jha, A. N., & Rogers, H. (2014). Natural gas from shale formation–the evolution, evidences and challenges of shale gas revolution in United States. *Renewable and Sustainable Energy Reviews*, *30*, 1–28.

Wang, S., Chen, F., Wu, Y.-S., & Nasrabadi, H. (2023b). The potential of hydrogen storage in depleted unconventional gas reservoirs: A multiscale modeling study. *International Journal of Hydrogen Energy*, *48*, 16007–16019.

Wang, S., Yan, J., Li, F., Hu, J., & Li, K. (2016). Exploitation and utilization of oilfield geothermal resources in China. *Energies*, *9*, 798.

Wang, X., Zhang, G., Tang, W., Wang, D., Wang, K., Liu, J., & Du, D. (2022b). A review of commercial development of continental shale oil in China. *Energy Geoscience*, *3*, 282–289.

Wang, Y., Dai, Z., Chen, L., Shen, X., Chen, F., & Soltanian, M. R. (2023c). An integrated multi-scale model for CO2 transport and storage in shale reservoirs. *Applied Energy*, *331*, 120444.

Warpinski, N. R., Du, J., & Zimmer, U. (2012). Measurements of hydraulic-fracture-induced seismicity in gas shales. *SPE Production & Operations*, *27*, 240–252.

Wei, M., Wei, S., Duan, Y., & Wang, H. (2023). Blasingame decline theory for hydrogen storage capacity estimation in shale gas reservoirs. *International Journal of Hydrogen Energy*, *48*, 13189–13201.

Weijermars, R., Drijkoningen, G., Heimovaara, T. J., Rudolph, E. S. J., Weltje, G. J., & Wolf, K. H. A. A. (2011). Unconventional gas research initiative for clean energy transition in Europe. *Journal of Natural Gas Science and Engineering*, *3*, 402–412.

Westphal, K., Overhaus, M., & Steinberg, G. (2014). The US shale revolution and the Arab Gulf States: The economic and political impact of changing energy markets. https://www.swp-berlin.org/publications/products/research_papers/2014_RP11_wep_ovs_sbg.pdf

Wu, M., Ding, M., YaO, J., Li, C., Li, X., & Zhu, J. (2019). Development of a multi-continuum quadruple porosity model to estimate CO2 storage capacity and CO2 enhanced shale gas recovery. *Journal of Petroleum Science and Engineering*, *178*, 964–974.

Xiao, H., He, L., Li, J., Zou, C., & Shao, C. (2021). Permeability prediction for porous sandstone using digital twin modeling technology and Lattice Boltzmann method. *International Journal of Rock Mechanics and Mining Sciences*, *142*, 104695.

Xin, S., Liang, H., Hu, B., & Li, K. (2012). A 400 kW geothermal power generator using co-produced fluids from Huabei oilfield. *Geothermal Resources Council Transactions*, *36*, 219–223.

Xu, X., Shi, W., Zhai, G., Xia, X., Zhang, X., Liu, S., Chen, X., Zhang, C., & Meng, F. (2021). A novel approach of evaluating favorable areas for shale gas exploration based on regional geological survey and remote sensing data. *Journal of Natural Gas Science and Engineering*, *88*, 103813. https://doi.org/10.1016/J.JNGSE.2021.103813

Yan, J., Ma, Y., Wang, L., Choo, K. K. R., & Jie, W. (2018). A cloud-based remote sensing data production system. *Future Generation Computer Systems*, *86*, 1154–1166. https://doi.org/10.1016/J.FUTURE.2017.02.044

Yang, X., Shang, Z., Liu, H., Cai, J., & Jiang, G. (2017). Environmental-friendly salt water mud with nano-SiO2 in horizontal drilling for shale gas. *Journal of Petroleum Science and Engineering*, *156*, 408–418.

Yao, B., Ding, Q., Hou, Y., Liu, S., & Zhang, S. (2018). A new openhole multistage hydraulic fracturing system and the ball plug motion in a horizontal pipe. *Journal of Natural Gas Science and Engineering*, *50*, 11–21.

Yekeen, N., Padmanabhan, E., Idris, A. K., & Chauhan, P. S. (2019). Nanoparticles applications for hydraulic fracturing of unconventional reservoirs: A comprehensive review of recent advances and prospects. *Journal of Petroleum Science and Engineering*, *178*, 41–73.

Yong, L. I., Yang, J., Pan, Z., Meng, S., Wang, K., & Niu, X. (2019). Unconventional natural gas accumulations in stacked deposits: A discussion of upper Paleozoic coal-bearing strata in the East Margin of the Ordos Basin, China. *Acta Geologica Sinica (English Edition)*, *93*(1). https://doi.org/10.1111/1755-6724.13767

Young, S., & Friedheim, J. (2013). Environmentally friendly drilling fluids for unconventional shale. *Offshore mediterranean conference and exhibition 2013*, OMC 2013.

Yu, H., Harrington, R. M., Kao, H., Liu, Y., & Wang, B. (2021). Fluid-injection-induced earthquakes characterized by hybrid-frequency waveforms manifest the transition from aseismic to seismic slip. *Nature Communications*, *12*, 6862.

Yu, H., Zhang, Y., Lebedev, M., Meng, K., Chen, S., Verrall, M., Johnson, L., & Iglauer, S. (2023). Swelling-induced self-sealing mechanism in fractured cap rock: Implications for carbon geosequestration. *AAPG Bulletin*, *107*, 1091–1104.

Yu, Y., Li, L., & Sheng, J. J. (2017). A comparative experimental study of gas injection in shale plugs by flooding and huff-n-puff processes. *Journal of Natural Gas Science and Engineering*, *38*, 195–202. https://doi.org/10.1016/j.jngse.2016.12.040

Yuan, J.-L., Deng, J.-G., Tan, Q., Yu, B.-H., & Jin, X.-C. (2013). Borehole stability analysis of horizontal drilling in shale gas reservoirs. *Rock Mechanics and Rock Engineering*, *46*, 1157–1164.

Yuan, W., Chen, Z., Kong, B., & Zhao, G. (2023). Hybrid geothermal energy and hydrocarbon resources production by repurposing horizontal wells in shale gas reservoirs in Horn River Basin, British Columbia, Canada. *Geoenergy Science and Engineering*, *227*, 211913.

Zborowski, M. (2018). Finding meaning, application for the much-discussed "digital twin". *Journal of Petroleum Technology*, *70*(6), 26–32. https://doi.org/10.2118/0618-0026-JPT

Zendehboudi, S., & Bahadori, A. (2016). *Shale oil and gas handbook: Theory, technologies, and challenges*. Gulf Professional Publishing.

Zeng, L., Vialle, S., Ennis-King, J., Esteban, L., Sarmadivaleh, M., Sarout, J., Dautriat, J., Giwelli, A., & Xie, Q. (2023). Role of geochemical reactions on caprock integrity during underground hydrogen storage. *Journal of Energy Storage*, *65*, 107414.

Zhang, B., Zhang, C. P., Ma, Z. Y., Zhou, J. P., Liu, X. F., Zhang, D. C., & Ranjith, P. G. (2023a). Simulation study of micro-proppant carrying capacity of supercritical CO2 (Sc-CO2) in secondary fractures of shale gas reservoirs. *Geoenergy Science and Engineering*, *224*, 211636. https://doi.org/10.1016/j.geoen.2023.211636

Zhang, L., Shan, B., Zhao, Y., Du, J., Chen, J., & Tao, X. (2018). Gas transport model in organic shale nanopores considering Langmuir slip conditions and diffusion: Pore confinement, real gas, and geomechanical effects. *Energies*, *11*(1), 223. https://doi.org/10.3390/en11010223

Zhang, P., Brodie, J., Daae, V., Erbas, D., & Duncan, E. (2013, September 3). BP North sea miscible gas injection projects review. *All Days*. https://doi.org/10.2118/166597-MS

Zhang, Y., Bijeljic, B., Gao, Y., Goodarzi, S., Foroughi, S., & Blunt, M. J. (2023b). Pore-scale observations of hydrogen trapping and migration in porous rock: Demonstrating the effect of Ostwald ripening. *Geophysical Research Letters*, *50*, e2022GL102383.

Zhang, Y., Lebedev, M., Sarmadivaleh, M., Barifcani, A., & Iglauer, S. (2016). Swelling-induced changes in coal microstructure due to supercritical CO2 injection. *Geophysical Research Letters*, *43*, 9077–9083.

Zhang, Y.-J., Li, Z.-W., Guo, L.-L., Gao, P., Jin, X.-P., & Xu, T.-F. (2014). Electricity generation from enhanced geothermal systems by oilfield produced water circulating through reservoir stimulated by staged fracturing technology for horizontal wells: A case study in Xujiaweizi area in Daqing Oilfield, China. *Energy*, *78*, 788–805.

Zhang, Z., Mao, J., Yang, X., Zhao, J., & Smith, G. S. (2019). Advances in waterless fracturing technologies for unconventional reservoirs. *Energy Sources, Part A: Recovery, Utilization and Environmental Effects*, *41*(2), 237–251. https://doi.org/10.1080/15567036.2018.1514430

Zhao, J., Ren, L., Jiang, T., Hu, D., Wu, L., Wu, J., Yin, C., Li, Y., Hu, Y., & Lin, R. (2022). Ten years of gas shale fracturing in China: Review and prospect. *Natural Gas Industry B*, *9*, 158–175.

Zheng, S., Zhou, X., Tan, Z., Liu, C., Hu, H., Yuan, H., Peng, S., & Cai, X. (2023). Assessment of the global energy transition: Based on trade embodied energy analysis. *Energy*, *273*, 127274.

Zhou, J., Liu, G., Jiang, Y., Xian, X., Liu, Q., Zhang, D., & Tan, J. (2016). Supercritical carbon dioxide fracturing in shale and the coupled effects on the permeability of fractured shale: An experimental study. *Journal of Natural Gas Science and Engineering*, *36*, 369–377.

Index

D

E

H

I

J

K

L

M

N

O

P

Q

R